QUANTUM MECHANICS

SERIES IN PHYSICS

General Editors:

J. DE BOER, Professor of Physics, University of Amsterdam
H. BRINKMAN, Professor of Physics, University of Groningen
H. B. G. CASIMIR, Director of the Philips Laboratories, Eindhoven

Monographs:

B. BAK, Elementary Introduction to Molecular Spectra
H. C. BRINKMAN, Application of Spinor Invariants in Atomic Physics
W. ELENBAAS, The High Pressure Mercury Vapour Discharge
S. R. DE GROOT, Thermodynamics of Irreversible Processes
E. A. GUGGENHEIM, Thermodynamics
E. A. GUGGENHEIM, Boltzmann's Distribution Law
E. A. GUGGENHEIM, and J. E. PRUE, Physicochemical Calculations
H. A. KRAMERS, Quantum Mechanics
H. A. KRAMERS, The Foundations of Quantum Theory
I. PRIGOGINE, The Molecular Theory of Solutions
E. G. RICHARDSON, Relaxation Spectrometry
L. ROSENFELD, Nuclear Forces
L. ROSENFELD, Theory of Electrons
J. L. SYNGE, Relativity: The Special Theory
J. L. SYNGE, The Relativistic Gas
H. UMEZAWA, Quantum Field Theory

Edited Volumes:

J. BOUMAN (editor), Selected Topics in X-Ray Crystallography
J. M. BURGERS and H. C. VAN DE HULST (editors), Gas Dynamics of Cosmic Clouds. A Symposium
C. J. GORTER (editor), Progress in Low Temperature Physics. Volume I-II
Z. KOPAL (editor), Astronomical Optics and Related Subjects
K. SIEGBAHN (editor), Beta- and Gamma-Ray Spectroscopy
J. G. WILSON (editor), Progress in Cosmic Ray Physics. Volume I-II-III

QUANTUM MECHANICS

BY

H. A. KRAMERS
LATE PROFESSOR OF
THEORETICAL PHYSICS UNIVERSITY, LEIDEN

1957

NORTH-HOLLAND PUBLISHING COMPANY - AMSTERDAM
INTERSCIENCE PUBLISHERS INC. - NEW YORK

Original title of part one: Die Grundlagen der Quantentheorie.
Original title of part two: Quantentheorie des Elektrons und der Strahlung.

Translated by D. ter Haar

Part one of this book is published separately
under the title of
THE FOUNDATIONS OF QUANTUM THEORY

SOLE AGENCY FOR U.S.A.
INTERSCIENCE PUBLISHERS, INC., NEW YORK

PRINTED IN THE NETHERLANDS
BY BOEKDRUKKERIJ VOORHEEN GEBROEDERS HOITSEMA, GRONINGEN

PREFACE

With this first volume of the Hand- und Jahrbuch der chemischen Physik the editor wishes to give an exposition of the theoretical basis of our present knowledge of the structure and properties of atomic systems. The centre of such an exposition had clearly to be a discussion of modern quantum theory and its applications to the Rutherford model of the atom. We have restricted ourselves to those parts of quantum theory which can be considered to form a well-founded, connected whole. It seemed to me that the best way to do this would be to present a textbook describing the unified, physical points of view of this theory. The present representation is thus neither historical nor axiomatic. Starting from experimental evidence and theoretical considerations about the wave nature of free particles we develop modern quantum theory more or less heuristically, using as far as possible the approach suggested by Bohr. We have considered only the most important concepts and methods, and references to the literature are mostly to basic papers.

The apparent lack of mathematical morals which is contritely pointed out repeatedly in the text is not exclusively due to the incompetence of the author. Physical morals, even (or rather especially) in their purest form, that is, unencumbered by pedagogic afterthoughts, do not live happily together with their mathematical relations in the restricted mansion of the human mind—and neither in the restricted volume of a monograph. We have not used the theory of groups explicitly. All the same I am well aware of the fact that the mathematical arguments—especially in the second part where the proofs have not always been given in all detail—may sometimes put a strain on the reader.

In the first part it is shown how the classical mechanics of point particles can be generalised into a consistent quantum mechanics. We draw special attention to the properties and use of the Schrödinger equation and to transformation theory. The last chapter of this part deals with the methods of perturbation theory.

In the second part we treat those extensions of quantum mechanics which are necessary in order that it can be applied to problems of atomic and molecular structure: electron spin, Pauli principle, and radiation theory.

We pay special attention to the mathematical description of the spin, the spinors, and we used for this purpose van der Waerden's representation of the Dirac theory.

In the chapter on the Pauli principle the emphasis is on general rules referring to spin multiplets derived by the simplest possible means and on a discussion of the most important examples of the very rich store of applications.

In the chapter on radiation theory we first of all derive by means of the so-called semi-classical method the general formulae for absorption and emission. After that we discuss in detail the quantisation of the radiation field and emphasise especially that Dirac's radiation theory cannot be considered by itself to be a quantisation of the classical theory of electrons as it is able to describe—in contradistinction to this theory—only the "secular" interaction between radiation and particles.

I would like to express my gratitude to the editors and publishers for their assistance and suggestions and also to Dr. G. P. Ittmann and Dr. F. Bloch for their help with the preparation of the first part.

Leiden, August 1937. H. A. Kramers.

TRANSLATOR'S PREFACE

When I was asked whether I would be willing to prepare a translation of Kramers' monograph, and thus complete an English edition of all his published works, I agreed for several reasons, even if it meant the hazardous task of translating from one foreign language into another. The main reason was that I felt that this book still represents the best available exposition of quantum theory and that the English speaking world was the poorer for not having it readily available. Also, in this book, as much as in some of his papers, Kramers showed some delightfully elegant methods which might otherwise be lost to the physics world in general. I may refer to such gems as the discussion at the end of § 66, the discussion of the improper eigenstates in § 23, or the symbolic method discussed in Chapter 7.

The translation is a literal one, except where the literature has been brought up to date and the few places where the text was rather cryptic.

I am indebted to Dr. F. J. Belinfante, Dr. D. Bijl and Dr. N. G. van Kampen for drawing my attention to misprints and cryptic passages. I am especially indebted to Dr. A. W. Ross for reading through the whole of the translation and removing the worst barbarisms.

Oxford, Summer 1956. D. ter Haar.

CONTENTS

PART TWO:

QUANTUM THEORY OF THE ELECTRON AND OF RADIATION

GLOSSARY OF SYMBOLS

This glossary contains a list of the more important symbols occurring in the text. The page number indicates where they are introduced.

a ,	component of null vector	257
a_1 ,	radius of first Bohr orbit	184
$\mathbf{a}_k$, $\mathbf{a}_k$,	Jordan-Wigner matrices, Jordan-Klein matrices 328,	332
A ,	amplitude function	19
$A_{l\to k}$,	Einstein transition probability	411
$\boldsymbol{A}$,	spin vector	228
$\mathfrak{A}$,	vector potential.	84
b ,	component of null vector	257
$B_{l\to k}$,	Einstein transition probability	400
$\boldsymbol{B}$,	spin vector.	229
c ,	component of null vector	257
c ,	velocity of light	6
c_k ,	expansion coefficients	65
C ,	constant given by equation (1.30)	53
E ,	energy.	7
E_{nr} , E_{r} ,	non-relativistic and relativistic energy	13
$\mathfrak{E}$,	electric field	7
F ,	observable	128
F_{op} ,	observable	128
F_{e} ,	eigenvalue of an observable	128
$F(t)_{\text{op}}$,	time-dependent operator.	159
$\boldsymbol{F}$,	force.	7
$\mathfrak{F}$,	given by expression (6.119)	271
g ,	degree of degeneracy	67
h ,	Planck's constant.	2
$\hbar$,	Dirac's constant	20
H , H_{op},	Hamiltonian	42
H_s ,	spin Hamiltonian	233
$H_{kk'}$,	energy matrix element	155
$\mathscr{H}$,	Hamiltonian function	34
$\mathfrak{H}$,	magnetic field	7
J ,	probability density	27
$\boldsymbol{J}$,	total angular momentum	236

l	, angular momentum quantum number	173
m	, rest mass	7
m	, magnetic quantum number	169
$\boldsymbol{M}$	, angular momentum	168
n	, principal quantum number	185
p_k	, generalised momentum	42
p'_k	, defined by equation (4.9)	125
p_s	, spin canonical momentum	233
$\boldsymbol{p}$	, momentum vector	87
P_l^m	, spherical harmonics	172
$\boldsymbol{P}_{\text{op}}$	, electrical polarisation	401
$\boldsymbol{P}$	, momentum vector	7
$\mathscr{P}_l^m$	, normalised spherical harmonics	176
q_k	, generalised coordinate	42
q_s	, spin canonical coordinate	233
R_{nl}	, radial part of hydrogen atom eigenfunctions	186
$\boldsymbol{S}$	, probability current density	32
$\boldsymbol{S}$	, spin vector	229
t	, time coordinate	6
$T_m^{(\mu)}$	, transformation matrix	133
u	, spinor component	258
U	, potential energy	42
$U_m^{(\mu)}$	, unitary matrix	134
v	, velocity	6
v	, group velocity	17
v	, spinor component	258
$\boldsymbol{v}$	, velocity vector	7
w	, phase velocity	10
$W(E)$	, energy distribution density	71
W_1	, ionisation energy of hydrogen-like atom	184
$\boldsymbol{x}$	, position vector	20
$Y_l(\boldsymbol{\omega})$	, Laplacian spherical harmonic	171
Z	, nuclear charge number	182

α , ratio of magnetic moment to angular momentum 228
α , fine structure constant 303
$\boldsymbol{\alpha}$, Dirac operator 286
$\boldsymbol{\alpha}_\lambda$, $\boldsymbol{\alpha}_\lambda^*$, annihilation and creation operators. 424
$\boldsymbol{\beta}_\lambda$, $\boldsymbol{\beta}_\lambda^*$, annihilation and creation operators. 424
γ_k , expansion coefficient 210
δ_k , Weierstrass symbol 103
$\delta(x)$, Dirac's delta function. 105
ε , given by equation (6.156) 282
ζ , spin function. 335
η , unit spinor 242
λ , wave length 9
μ_B , Bohr magneton 243
$\boldsymbol{\mu}$, magnetic moment. 228
ν , frequency 9
$\nu_{ll'}$, Bohr frequency 161
$\boldsymbol{\pi}$, kinetic momentum 268
ξ , unit spinor 242
ϱ , density function 119
ϱ , Jacobian. 124
ϱ_{op} , charge density operator 398
$(\varrho \boldsymbol{v})_{op}$, current density operator. 398
$\boldsymbol{\rho}$, given by equations (6.153) 282
$\boldsymbol{\sigma}$, wave vector 9
$\boldsymbol{\sigma}$, given by equations (6.134) 276
$\boldsymbol{\sigma}_\lambda$, wave vector 396
Φ , scalar potential. 84
ψ , wave function 13
ψ_+ , ψ_- , wave function components. 279
ψ_1 , ψ_2 , ψ_3 , ψ_4 , wave function components 281
$\chi_+^\dagger$, $\chi_-^\dagger$, wave function components. 281
$\boldsymbol{\omega}$, direction 170
Ω , linear operator 100
$\Omega_{kk'}$, matrix representation of operator. 151

$\wedge$	indicates a vector product	7
$*$	indicates complex conjugates	15
$d\mathbf{x}$	indicates a volume element in coordinate space	20
$d\boldsymbol{\sigma}$	indicates a volume element in wave number space	20
—	indicates an expectation value	111
$\sim$	indicates an adjoint operator	103
$\doteqdot$	indicates approximate equality	14

INTRODUCTION

The concepts and theories of classical physics enable us to classify and describe in more or less detail an extensive field of physical phenomena. Classical theory was, however, unable to explain many phenomena and classes of phenomena which often showed clearly a consistent pattern and of which some had been known for a long time. We may point as an example to typical chemical phenomena. The existence of the chemical elements with their individual properties and with the ability to form compounds again with particular properties could not be interpreted by or based on classical physics. It is true that classical physics seemed to present a consistent picture—at least until Rutherford in 1911 proposed the planetary model of the atom based on his experimental results—and in very many cases classical ideas and theories could be used together with the idea that atoms and molecules existed to give a successful interpretation of observable data. As examples we may mention the kinetic theory of gases, the interpretation of thermodynamics by Boltzmann and Gibbs in terms of statistical mechanics, or the theory of the elastic and optical properties of crystals. The laws and connexions derived were, however, always of a *general* nature and the *particular* values of the constants for various substances which occurred in those laws had to be taken from experimental data. These circumstances pointed to the incompleteness of our knowledge of the material universe in the framework of classical physics.

Apart from such phenomena which showed up only an incompleteness of classical theories, there became known a large number of phenomena—especially since 1890—which seemed to be in direct or indirect contradiction to the concepts of classical physics. By this we mean that these phenomena could not only not be understood, but also seemed to be inconsistent with an attempt to extend the classical concepts in a consistent manner. Theoretical considerations of these phenomena led to the birth and further development of *quantum theory*. It started in 1900 with Planck's investigations of the empirical laws of black body radiation and was indicated by the introduction of the *universal quantum of action*, Planck's constant h, into physical theories. The most important steps in its further development were Einstein's considerations (1905) of the photoelectric effect and of the specific heats of solids and Bohr's theory (1913) of the hydrogen spectrum. While black body radiation, photo-effect and specific heats referred to general proporties of radiation and matter and their interaction, Bohr's theory is the first important step in the explanation of the specific properties of the chemical atom. This theory was put forward about two years after Rutherford's discovery of the atomic nucleus and it is important to remind ourselves that in the light of Rutherford's model of the atom the stability of chemical atoms with their specific properties is a phenomenon which is as difficult to understand and as contradictory as, for instance, the photo-effect or black body radiation. Only the incorporation of the universal quantum of action enabled a further development of atomic theory.

The possibility of treating theoretically phenomena of which it is stated that they are inconsistent with classical concepts seems at first sight to lead to a paradox. It must be remembered that the phenomena themselves, that is, the collection of experimental data obtained from suitable pieces of apparatus, do not show a paradox and that the use of classical concepts was essential in describing them. The difficulties occurred, as we mentioned a moment ago, only when one tried to apply the classical concepts too consistently. We are indebted to the recent detailed critical considerations about the limitations of the consistent application of these concepts in physics for removing the apparent contradictions to which the use of the quantum of action in describing physical phenomena often seemed to lead. The limits of the applicability of the classical particle concept are simply expressed by the *Heisenberg relations* *, and the peculiar nature of modern quantum theory is clearly illustrated by Bohr's concept of *complementarity* which implies that physical laws referring to events in space-time are *complementary* to those referring to energy-momentum relationships (or, to those referring to a causal chain of phenomena), that is, they exclude each other in as far as we are referring to an exact determinability by means of

* W. Heisenberg, *Z. Phys.*, **43**, 172, 1927; *The physical principles of quantum theory*, University of Chicago Press, 1930.

measurements *. The nucleus of these new ideas is a critique of the concept of observation and the earlier difficulties were mainly due to the fact that by extrapolating the classical concepts one tried to obtain a description of the universe in which one could talk of "objective", real events in space-time. However, the discussion of the empirical physical laws has shown us in fact that such an extrapolation is not permitted and that every measurement is connected with an interaction between the measuring apparatus and the observed system which cannot be described objectively and which is thus, as it were, irrational.

It has been found to be possible to build up a mathematical apparatus which to a large extent is suited to a quantitative formulation of the known physical phenomena. In this respect modern quantum theory can be considered to be a logical and relatively closed structure which shows an inner consistency as far as the great majority of empirical phenomena is concerned. The situation discussed a moment ago implies, however, that in order to obtain a physical interpretation of the equations we must use concepts which are always adapted to the experiment considered in such a way that in the case of phenomena occurring under the same external circumstance (if such a state of affairs can at all be defined physically) the possible descriptions, according to whether we use one way of observing the system or another, may be mutually exclusive. As an example we may mention the simultaneous applicability of the concept of "particles" and of "waves" in the discussions of light rays or matter beams. Furthermore the peculiar character of quantum theory entails that its predictions about the quantitative result of future measurements are always *probability statements*, just as if we are dealing with an objective world in which only statistical laws operate or at any rate can be observed to operate.

We want to discuss here briefly the often mentioned question in how far, knowing the results of quantum theory, we may still speak of determinism in nature. One can take from the above discussion, which tried to give the point of view put forward by Bohr, the following idea: there is absolutely no place for determinism in the usual classical sense as far as purely physical phenomena are concerned. As Heisenberg emphasised in his first paper on the uncertainty relations this is not due to the fact that the laws of nature contain an acausal aspect but to the impossibility of evaluating causal connexions quantitatively in a world where the data referring to such connexions can never all possess at the same time sharply defined values because of the occurrence of the elementary quantum of action.

* N. Bohr, *Naturwiss.*, **16**, 245, 1928; **17**, 483, 1929; **18**, 73, 1930; *Atomic theory and the description of nature*, Cambridge University Press, 1934.

Contrary to the above idea some people defend the following one: "... the conviction that all phenomena are causally connected is the basis of all exact research and complete determinism is hardly anything but a precise, consistent formulation of this conviction. Any resignation as far as this is concerned is desperation. If the quantum theorist refers to the lack of precision which as a matter of principle is connected with all observations, one sees only too plainly that the incompleteness of our knowledge is promoted to a virtue in a scientifically unjustifiable manner." In the same way many people objected to the atomistic theory of matter on the basis of the following considerations: "... even if one accepts the atomic picture, atoms themselves could never be observable; scientifically this means that atoms do not exist." All the same since that time atomic theory has won through especially because of newly discovered phenomena made possible by the development of experimental techniques.

We may remark the following as far as those two contradicting approaches are concerned. The statement that phenomena are determined has a meaning only if they can be defined and that can only be done by using the analysis of experiences and observations. If determinism has a physical meaning and is not just a metaphysical concept we must go back to the concepts of classical physics, especially those of space and time. Quantum physics leads now to the following: "... the classical space-time concept can be retained (at least to a large extent) without leading to contradictions. The same is true for the concepts of energy, momentum, charge, and so on. However, this does not mean that phenomena can be incorporated in the classical framework of 'phenomena being determined in space-time,' but it means that the simples description of those phenomena (which involves *all* means of definition which we can use) leads to leaving a certain amount of 'freedom', or, to incorporate a 'certain free choice of nature.' " In this 'technical' sense there is no determinism. As Heisenberg and Bohr have emphasised a similar situation occurs in Einstein's theory of relativity. In this theory it is seen that one can retain both the classical concept of time and the classical concept of space, but that technically one must expect that simultaneity is not defined or cannot be defined.

The future development of physics will lead probably to much which is not yet known. New phenomena will be discovered and solutions will be found for problems which at the moment defy our attempts to solve them. Our ideas about the 'free choice of nature' will be deepened. The opinion that the future theory will again approach closely the classical idea of phenomena determined in space-time, can be compared with the opinion that electromagnetic phenomena can be explained by a mechanical model or that it will be possible to determine absolute simultaneity.

We have not discussed in this introduction any technical details to which we hope to return in the body of the text. We shall also have ample opportunity to discuss the problem of the limitations of the applicability of modern quantum theory.

CHAPTER I

QUANTUM THEORY OF FREE PARTICLES

§ 1. Masspoints in classical physics

In classical physics the concept of a *mass point* is introduced as an idealisation of experience. In general one describes events in both space and time. A masspoint is then first of all characterised by the fact that at any time it is localised in one point in space. This point is determined by its three spatial co-ordinates x, y, z, and in principle one can measure the values of these co-ordinates; such a measurement presupposes the physical definition of an absolute system of co-ordinates. The existence of rigid bodies makes this definition possible in a very simple manner. It should also be possible to measure the time at which the position is measured; this is possible if there exist stationary clocks with absolute time scales. One further assumes that the position of a mass point changes continuously with time and that continuously changing velocities, accelerations,.. can in principle be measured.

Apart from the *kinematic aspect* of localisability in space and time, a mass point also possesses a *dynamic aspect.* One might choose for this its inertial mass. The introduction of mass and the assumption that it is constant made it is possible in Newtonian physics to describe causal connexions, although one must at the same time introduce the concept of force and Newton's third law — action equals reaction — a law based on experience. This law can be considered to be the principle of *conservation of momentum,* after one has introduced momentum as the product of mass and velocity. The principle of *conservation of energy,* the universal importance of which in physics became clear only in the nineteenth century, can be obtained in Newtonian physics only as a consequence of the special character of the forces occurring in nature.

The development of electrodynamics and of the theory of electrons has contributed greatly to the clarification of the principles of the conservation of energy and momentum. It was shown that one could retain these principles by a suitable generalisation of the definitions of energy and momentum. Momentum *densities* and energy *densities* enter into these theories since they are field theories, but the theory of relativity especially has shown how one can retain the concept of a mass point which, on the one hand, can be definitely localised in space and time and which, on the other hand, possesses a well defined energy and momentum. Typical for the theory of relativity was the proof of the close relation between energy and momentum, since these quantities transform as the components of a four-vector. Einstein's theorem of the equivalence of energy and mass led to a more or less final answer to the problem of the dynamic aspect of a mass point. One could consider the energy-momentum vector as describing the dynamic aspect of a mass point and the theory was constructed in such a way that one had to take this vector simply proportional to the four-vector of the Minkowsky velocity — without disagreement with experimental facts; the factor of proportionality was then the constant rest mass of the mass point. In accordance with the whole structure of classical physics one could measure in principle the energy and momentum of a mass point by suitable experimental arrangements in which the interaction with other mass points was observed and interpreted using the conservation laws. It is of interest to notice that the unchanging rigid bodies which we introduced to determine our co-ordinate system in space are the less suited for the measurement of momentum the larger their mass is, that is the better they are suited for their original purpose. However, we should not forget that once its mass is known the energy and momentum of a mass point can be determined, as we mentioned before, by a determination of its space and time co-ordinates, such as, for instance, is involved in the measurement of a velocity.

For future use we give here the principal facts from relativistic mechanics *. The movement of a particle in four-dimensional space-time (co-ordinates x, y, z, t) corresponds to a one dimensional curve, called the world line of the particle. If c is the velocity of light and v the particle velocity, we define the line element $d\tau$ (eigentime element) along the world line by the equation $c^2d\tau^2 = c^2dt^2-dx^2-dy^2-dz^2 = c^2dt^2[1-(v^2/c^2)]$. It follows that $d\tau$ is invariant against a Lorentz transformation, or, the value of $d\tau$ is independent of the special choice of the Galilean system of reference. Hence it follows that the four components of the so called "Minkowski velocity",

$$u_x = dx/d\tau = \beta dx/dt, \quad u_y = dy/d\tau = \beta dy/dt, \quad u_z = dz/d\tau = \beta dz/dt, \quad u_t = dt/d\tau = \beta,$$
$$\beta = [1-(v^2/c^2)]^{-\frac{1}{2}},$$

* See, for instance, C. Møller, *The Theory of Relativity*, Oxford University Press, 1953.

transform during a Lorentz transformation according to the same linear laws as the co-ordinates x, y, z, t.

Consider now a collision between two particles which both before and after collision may be regarded as free. If one assumes the existence of a relativistically generalised law of conservation of momentum it follows from a simple analysis that the only quantities which can correspond to the classical momentum components $m dx/dt, \ldots$, are given by the analogous expressions $m dx/d\tau = m\beta dx/dt, \ldots$, where m is the rest mass of the particle. Only this definition has as a consequence that conservation of momentum in one system of reference leads to conservation of momentum in all systems of reference. It follows also that the sum of the expressions $m dt/d\tau$ for both particles is the same before and after their collision. Writing $m dt/d\tau = m[1-(v^2/c^2)]^{-\frac{1}{2}} = m + \frac{1}{2}mv^2/c^2 + \ldots$, we see that we are dealing with the conservation of energy and that $mc^2 dt/d\tau$, apart from a constant term mc^2 (the "rest energy" of the particle), to a first approximation is equal to the classical kinetic energy $\frac{1}{2}mv^2$.

Denoting the momentum vector by $\boldsymbol{P}$, with components P_x, P_y, P_z, and the (relativistic) energy by E we have

$$\boldsymbol{P} = m\boldsymbol{v}[1-(v^2/c^2)]^{-\frac{1}{2}}, \quad E = mc^2[1-(v^2/c^2)]^{-\frac{1}{2}}.$$

In the case of a free particle we have the following relation between the momentum and energy ("kinetic momentum" and "kinetic energy")

$$E^2/c^2 - P^2 = mc^2.$$

If the mass point is accelerated, the (Newtonian) force $\boldsymbol{F}$ acting on it is defined by the equation $d\boldsymbol{P}/dt = \boldsymbol{F}$.

From this equation and the two preceding ones it follows that $dE/dt = (\boldsymbol{F} \cdot \boldsymbol{v})$, that is, the scalar product of force and path is equal to the work done, completely analogous to the situation in classical mechanics.

The force $\boldsymbol{F}$ on a charged mass point (charge e) in an electromagnetic field $\mathfrak{E}$, $\mathfrak{H}$ is given by the equation $\boldsymbol{F} = e[\mathfrak{E} + ([\boldsymbol{v} \wedge \mathfrak{H}]/c)]$, where the term $e[\boldsymbol{v} \wedge \mathfrak{H}]/c$ is the so-called Lorentz force and where radiative reaction terms are neglected. For quantum theoretical applications it is important to put the equations of motion in the Hamiltonian canonical form. This will be done in § 25.

The concept of mass points has played a very important part in the interpretation of experiments which showed the atomistic structure of matter. First of all we mention the kinetic theory of gases, where one is dealing mainly with neutral atoms or molecules and where one can obtain far-reaching conclusions without having to know much about the interactions between the particles. If we consider, however, the experiments which proved the existence of electrons and nuclei and which investigated their properties, we see that we are dealing with entities for which the possesion of a third aspect apart from the kinematic and dynamic aspects of mass points, namely, *electrical charge* is essential. This aspect is necessary and to a large extent also sufficient to describe the interaction of elementary particles with other particles or systems using the dynamical conservation laws, in a space-time description. From a theoretical point of view the introduction of electrodynamical conceptions immediately raises the question of whether it is possible to justify the simultaneous use of Maxwell's field theory and of the concept of electrical elementary particles. Classical physics never found a satisfactory answer to this problem. However, considerations of the electromagnetic mass, showing how small the linear dimensions of elementary particles must be, justified to a large extent the usual procedure of using practically always the classical laws of interaction even when particles approach each other at distances small compared to atomic dimensions.

§ 2. The de Broglie quantum postulate for free mass particles

The phenomena, which originally led to the quantum theory, were not concerned with free elementary particles. However, shortly after de Broglie had introduced in 1924 his quantum theoretical hypothesis about the *wave character of material particles*, experimental physicists succeeded in performing experiments with free particles which could most easily be interpreted as showing diffraction and interference of waves. They thus afforded the simplest possible confirmation of the de Broglie hypothesis and not only qualitatively but also quantitatively. The first results of this kind were obtained by Davisson and Germer [1] in 1926, when they studied reflexion of electrons by a nickel single crystal. Soon afterwards G. P. Thomson [2] found diffraction rings produced by electrons passing through thin foils. Since then many physicists have worked with great success in this field [3]. Stern and coworkers have for instance, shown the wave nature of hydrogen and helium molecular beams in experiments where these beams were reflected by crystals and artificial gratings and the wave nature of neutron beams has been successfully applied in neutron diffraction experiments.

If one wants to build up quantum theory systematically, it is to be recommended to start with a discussion of these experiments concerning free particles and their theoretical interpretation. Afterwards one can then proceed to the treatment of bound particles

[1] C. J. Davisson and L. H. Germer, *Phys. Rev.*, **30**, 705, 1927.

[2] G. P. Thomson, *Proc. Roy. Soc.*, **A 117**, 600, 1928.

[3] For a general review of many aspects of this field see, for instance, Z. G. Pinsker, *Electron diffraction*, Butterworth, London, 1953.

and of the interaction between particles and radiation. We have to emphasise straightaway that a "free particle" is an idealisation from experience. Strictly speaking one can never find out anything about a "free" particle, since one only obtains information when the particles are subject to interaction with apparatus and thus lose their "free" status. All the same, one can reach conclusions about the properties of free particles from such experiments, as long as in the theoretical interpretation one can treat the apparatus up to a point as an external system, and can forget its influence on the particle.

Earlier experiments with free electrons, nuclei, and so on, seemed to indicate that one could treat these particles as the mass points in classical physics. Wilson chamber photographs show not only their localisability, but also the applicability of the conservation laws — compare, for instance, Blackett's photographs of nuclear collisions [1]. The latter experiments on diffraction and interference of material particles, which we mentioned a moment ago, contradict the idea of classical mass points. However, they can be interpreted theoretically as follows — in accordance with de Broglie's hypothesis [2]: *The presence of a free material particle is connected with the presence of a measurable wave field in space and time. This connexion is such that to a particle of momentum* $\boldsymbol{P}$ *(components* P_x, P_y, P_z,*) and energy* E *there corresponds a plane monochromatic wave, travelling in the direction of the momentum vector. The wave vector* $\boldsymbol{\sigma}$ *and frequence* ν *are given by the equations* $\boldsymbol{\sigma} = \boldsymbol{P}/h$ *and* $\nu = E/h$ *where* h *is the universal quantum of action.*

The definitions of wave vector and frequency are such that the quantities used to describe the wave phenomenon depend on the space and time co-ordinates through a factor of form

$$\cos\left[2\pi(\sigma_x x+\sigma_y y+\sigma_z z-\nu t)+b\right]. \tag{1}$$

The vector character of $\boldsymbol{\sigma}$ follows immediately from the condition that $\sigma_x x+\sigma_y y+\sigma_z z$ must be invariant against rotations in ordinary three-dimensional space. The quantities σ_x, σ_y, and σ_z are simply the number of waves per unit length in the x-, y-, or z-direction; they are thus completely analogous to the frequency ν which is equal to the number of waves per unit time. Introducing the wave length λ, the direction cosines α, β, γ of the direction of propagation

[1] P. M. S. Blackett, *Proc. Roy. Soc.*, **A 103**, 78, 1923; **A 107**, 360, 1924.
[2] L. de Broglie, *Thesis*, Paris, 1924.

and the period τ, expression (1) can be written as follows

$$\cos\left[2\pi\left(\frac{\alpha x+\beta y+\beta z}{\lambda}-\frac{t}{\tau}\right)+b\right]. \tag{2}$$

Comparing expressions (1) and (2) shows immediately that the following relations hold

$$\sigma_x=\alpha/\lambda,\ \ \sigma_y=\beta/\lambda,\ \ \sigma_z=\gamma/\lambda,\ \ \sigma=(\sigma_x^2+\sigma_y^2+\sigma_z^2)^{1/2}=\lambda^{-1},\ \ \nu=\tau^{-1}, \tag{3}$$

where σ, the absolute value of the wave vector, is the "wave number", that is, the number of waves per unit length in the direction of propagation. The wave number is equal to the reciprocal of the wave length, just as the frequency is the reciprocal of the period.

The velocity with which planes of equal phase travel, the so-called phase velocity, is given by the equation

$$w=\lambda\nu=\nu/\sigma. \tag{4}$$

We shall call the de Broglie hypothesis the *quantum postulate for free particles*. It is expressed mathematically by the equations

$$\boldsymbol{P}=h\boldsymbol{\sigma},\ \ E=h\nu. \tag{5}$$

Using equations (3) and (5) we get for the wave length of matter waves the de Broglie relation

$$\lambda=h/P, \tag{6}$$

where $P=(P_x^2+P_y^2+P_z^2)^{1/2}$ is the absolute value of the momentum. If we use for the relation between momentum and velocity (v) the non-relativistic equation $P=mv$, equation (6) leads to

$$\lambda=h/mv. \tag{7}$$

If the velocity is not small compared to the velocity of light c, one has $P=mv[1-(v^2/c^2)]^{-1/2}$, and instead of equation (7) we have

$$\lambda=\frac{h}{mv}\left[1-\frac{v^2}{c^2}\right]^{1/2}. \tag{8}$$

The frequency and phase velocity satisfy in the non-relativistic case, where $E=\frac{1}{2}mv^2$, the equations

$$\nu=E/h=mv^2/2h,\ \ w=\nu\lambda=\tfrac{1}{2}v. \tag{9}$$

According to the theory of relativity a mass point at rest possesses an energy mc^2, and in general we have $E=mc^2[1-(v^2/c^2)]^{-1/2}$. In this way we get for the frequency and phase velocity the relations

$$\nu=E/h=mc^2/h[1-(v^2/c^2)]^{1/2},\ \ w=\nu\lambda=c^2/v. \tag{10}$$

We thus see that according to the theory of relativity the phase velocity is always larger than the velocity of light. We shall return in later sections to the difference between equations (9) and (10).

We must more closely analyse the statement that the experiments on material beams can be interpreted by the presence of a wave field. We are interpreting here the experimental results by analogy with the interpretation of optical interference experiments. The spatially periodic structure of the crystal or the grating on which the electron or molecular beam is impinging can be checked experimentally. If we now assume that the impinging particle corresponds to a wave phenomenon which will lead to coherent secondary spherical waves through interaction with the crystal or grating, and if we assume that we may superpose waves — as in optics — then we can expect in general the same manifestations of selective reflexion of the waves as we know them from optics. These reflexions have been observed in the sense that the particles are leaving the crystal or grating predominantly in those directions in which the intensity of the theoretical wave phenomenon is greatest for a suitable choice of the wave length. This has been observed by blackening of photographic plates, charging of Faraday cages, or increase of pressure in closed vessels. As in optics, one measures intensity maxima, but one does not really probe a wave field, that is, one establishes phase relations and measures in that way a wave length, but one does not even approximately measure phases themselves. There exists thus no immediate basis for a detailed physical interpretation of the quantity or quantities used to describe the wave field mathematically.

To confirm quantitatively de Broglie's relation (6) we need to know the momentum of the particles, the wave lengths of which are measured. This is obtained, since one knows, on the one hand, from the geometry of the apparatus the direction from which the particles approach the crystal or grating and thus the direction of their momentum. On the other hand, we know their energies and can thus from the usual formulae calculate the absolute value of the momentum. For electrons the energy follows directly from the accelerating voltage and for molecules from the temperature of the source.

The fact that in order to determine the momentum of the impinging particle we use the classical concept of a particle is by itself sufficient to make the result of the interference experiment self-contradictory. The same applies to the interpretation of the observations used to determine the interference pattern. Let us consider the case where we are dealing with electron beams and where the electrons are counted in a Geiger counter with a small opening. In this manner the electrons are counted one by one in a spatially localised position and this experiment shows immediately that electrons are discrete elementary particles.

From the above analysis which was completely based on experience we see also that the wave character of the electrons does not show up because there is a great number of electrons present in the beam at each moment. Of course, the interference pattern only shows up after many electrons have been counted, but the result is independent of the speed with which they are counted.

The necessity to ascribe to a particle the aspect of an extension in space and time as in a wave field cannot be avoided because of the interference experiments. The necessity of ascribing to the same particle a precise localisability during the interpretation of these experiments, using classical concepts, now introduces a contradiction, since in classical physics one demands that the description of one "object" should be unique. The contradiction, however, enters when one demands a complete objective description of the phenomenon; we have just seen from an analysis of the experiments that the use of various concepts to describe the properties of particles is adjusted to the nature of the apparatus. As long as one does not demand a complete description of the interaction between apparatus and particle, there are no logical contradictions.

In the following we shall discuss how one can obtain a consistent interpretation of experiments involving free particles. We start from the idea that the dynamic properties of mass particles can only be defined in connexion with the identification of a de Broglie wave field. This assumption entails that we ascribe to these properties a fundamental well defined role in our description of nature. Furthermore, we assume that they are immediately connected, as in classical physics, with the results of certain measurements in space and time. In contradistinction to the classical case we are now mainly concerned with the measurements of interference patterns and not of velocities. We shall see that nevertheless we can give a natural interpretation of the classical relations between momentum and energy, and velocity. These relations now appear as laws of an approximate nature. The limitations of these laws are not shown up in macroscopic experience due to the smallness of the universal quantum of action (compare § 6).

§ 3. Superposition of de Broglie waves

A plane travelling monochromatic wave can be described mathematically by a wave function ψ (or by a number of such functions, but for the time being we only use one function). As we mentioned before such wave functions depend on the space co-ordinates and the time as follows

$$\psi = B \cos [2\pi(\sigma_x x + \sigma_y y + \sigma_z z - \nu t) + b], \tag{11}$$

where B is the amplitude, b a phase constant, σ_x, σ_y, σ_z the wave numbers, and ν the frequency. According to the definition of a plane wave, equation (11) is valid for all values of x, y, z, t between $-\infty$ and $+\infty$. The function ψ given by equation (11) does not enable us to distinguish between different points in space and time; the intensity of the wave motion is everywhere the same. However, we can still choose the constants B, b, σ_x, σ_y, σ_z, and ν. Adding two or several ψ-functions of the form (11) we can obtain new wave functions. Such a procedure corresponds to applying the superposition principle to de Broglie waves; we remind ourselves that this principle played an essential part in the interpretation of the interference experiments, and we must therefore demand as a matter of principle that the superposition of plane monochromatic waves has a meaning in physics. Using such a superposition we are able to construct mathematically wave systems with different intensities at different space-time points, and once again such a construction is essential for the interpretation of the interference experiments.

In the classical theory of mass particles we have a definite relation between momentum and energy of a free particle; this relation contains the mass m. In Newtonian mechanics the energy of a particle at rest is assumed to vanish, like its momentum, and the two are related as follows

$$E_{\text{n.r.}} = P^2/2m = (P_x^2 + P_y^2 + P_z^2)/2m, \tag{12}$$

where the subscript "n.r." indicates that we are dealing with a non-relativistic energy definition. In relativity theory we have, however, (compare § 1)

$$(E_{\text{r}}^2/c^2) - (P_x^2 + P_y^2 + P_z^2) = m^2c^2, \tag{13}$$

where c is the velocity of light. This relation expresses the invariance of the absolute value of the energy-momentum vector under a Lorentz transformation. Equation (13) goes over into equation (12) for $P \ll mc$, if we write

$$E_{\mathrm{r}} = mc^2 + E_{\mathrm{n.r.}}\,, \tag{14}$$

where mc^2 is the rest energy, and if we neglect a term $E^2_{\mathrm{n.r.}}/c^2$ on the left hand side of equation (13). Both in the relativistic and in the non-relativistic case the energy depends on the absolute value of the momentum in such a way that the speed v of the particle is given by the equation

$$v = dE/dp \tag{15}$$

Equation (15) follows immediately from the fact that in both cases the change in the energy of the particle dE during a time interval dt is equal to the scalar product of the distance travelled $\boldsymbol{ds}$ and the force acting on the particle $\boldsymbol{F}$, or, $dE = (\boldsymbol{F} \cdot \boldsymbol{ds})$, while at the same time the change in momentum $\boldsymbol{dP}$ is given by the equation $\boldsymbol{dP} = \boldsymbol{F}dt$. We thus have $dE = (\boldsymbol{v} \cdot \boldsymbol{dP}) = vdP$, since $\boldsymbol{v} // \boldsymbol{P}$.

The fact that relation (13) or (14) leads to a relation between the frequency and the wave numbers of the de Broglie waves, that is, to a "dispersion law" for these waves, is essential for the development of quantum mechanics. Combining equations (5) and (12) or (13) we have

non relativistic: $$\nu_{\mathrm{n.r.}} = h(\sigma_x^2+\sigma_y^2+\sigma_z^2)/2m, \tag{16}$$

relativistic: $$\nu_{\mathrm{r}}^2/c^2 - (\sigma_x^2+\sigma_y^2+\sigma_z^2) = m^2c^2/h^2. \tag{17}$$

$$\nu_{\mathrm{r}} \doteq \nu_{\mathrm{n.r.}} + mc^2/h. \tag{18}$$

Both in the non-relativistic and in the relativistic case the mass m and the quantum of action enter only in the combination m/h into the relation between frequency and wave-numbers.

The above mentioned experiments on interference of matter waves do not prove the correctness of these relations, since one only measures in those cases an interference pattern in space, but not the frequency or phase velocity of the waves so that one cannot establish that the frequency of these waves is really the relativistic frequency ν_{r} and not $\nu_{\mathrm{n.r.}}$. However, if one presupposes the relativistic invariance of the matter waves, it is necessary that the expression $\sigma_x x + \sigma_y y + \sigma_z z - \nu t$ is invariant under Lorentz transformations and from the relation $\boldsymbol{P} = h\boldsymbol{\sigma}$ follows then necessarily $E_{\mathrm{r}} = h\nu_{\mathrm{r}}$ so that one must conclude that, indeed, ν_r must be considered to be the wave frequency. In the following we shall see, however, that even without having explicitly recourse to the postulate of relativistic invariance one is led to accept equation (16) or (17) in order that one can interpret sensibly the classical connexion between energy-momentum and velocity in quantum theory; we shall see then that the phase velocity has no physical significance.

The most general wave field obtained by a superposition of waves of the form (11) can be described by the equation

$$\psi = \iiint_{-\infty}^{+\infty} B(\sigma_x, \sigma_y, \sigma_z) \cos\{2\pi[\sigma_x x+\sigma_y y+\sigma_z z-\nu(\sigma_x, \sigma_y, \sigma_z)t] \\ +b(\sigma_x, \sigma_y, \sigma_z)\}\, d\sigma_x d\sigma_y d\sigma_z\,, \tag{19}$$

where B and b are arbitrary functions of σ_x, σ_y, σ_z, and where ν is a function of σ_x, σ_y, σ_z given by equations (16) or (17). If we use the non-relativistic formula (16) ν is a unique function of σ_x, σ_y, σ_z. However, if we use equation (17) ν will be real for real values of σ_x, σ_y, σ_z, and $|\nu_r| > mc^2/h$; however, the sign of ν_r is undetermined. This lack of uniqueness corresponds to the fact that the sign of the energy E_r in equation (13) is undetermined. In classical physics we can only accept, in accordance with non-relativistic mechanics, positive values of the energy of a free particle. Consequently we shall for the time being assume that we can develop the theory assuming ν_r always to be positive.

From the theory of Fourier integrals it follows that one can always determine B and b in such a way that at a given time, for instance, $t = 0$, ψ is equal to an arbitrarily given function of x, y, z. For the sake of simplicity we write equation (19) in the form

$$\psi = \mathrm{Re} \iiint A \exp\left[2\pi i(\sigma_x x+\sigma_y y+\sigma_z z)-\nu t\right] d\sigma_x d\sigma_y d\sigma_z\,, \tag{20}$$

where Re indicates the real part, while A is the complex function, $B \exp ib$. At $t = 0$ we have

$$\psi_0 = \mathrm{Re} \iiint A \exp\left[2\pi i(\sigma_x x+\sigma_y y+\sigma_z z)\right] d\sigma_x d\sigma_y d\sigma_z\,. \tag{21}$$

If ψ_0 is given, A must satisfy the equation (compare eq. (28) below)

$$\tfrac{1}{2}[A(\sigma_x, \sigma_y, \sigma_z)+A^*(-\sigma_x, -\sigma_y, -\sigma_z)] \\ = \iiint_{-\infty}^{+\infty} \psi_0 \exp\left[-2\pi i(\sigma_x x+\sigma_y y+\sigma_z z)\right] dx\,dy\,dz, \tag{22}$$

where an asterisk (*) indicates here, and henceforth, the conjugate complex, or,

$$A = \iiint \psi_0 \exp\left[-2\pi i(\sigma_x x+\sigma_y y+\sigma_z z)\right] dx\,dy\,dz+C, \tag{23}$$

where C satisfies the relation $C(\sigma_x, \sigma_y, \sigma_z) = -C^*(-\sigma_x, -\sigma_y, -\sigma_z)$, but is otherwise an arbitrary function of σ_x, σ_y, σ_z. That the amplitudes and phases are not uniquely determined by ψ_0 is due to the fact that we tried to describe the wave field by only one real function. It is therefore necessary, as we shall see below, to employ complex wave functions in the quantitative development of the theory.

The fact that one can choose at a given time the value of the wave function obtained by superposition of plane waves indicates how one should connect the concept of matter waves with the classical idea of localisation in space and time of a particle. The physical interpretation of the wave field when describing the interference experiments gave the following qualitative picture. The probability to find particles at a given point in space (corresponding, for instance, to a wee hole in a solid body) is large or small according to whether the wave function assumes in that point large or small values. Accordingly one can assign to a particle which, according to experiments, is localised at a given time in the neighbourhood of a given point in space a wave field of such a nature that the wave function (19) at that time is appreciably different from zero only in the neighbourhood of the given point in space. One usually expresses this as follows. A localised particle can be described by a *wave packet.* The more precise the localisation which has to be described the smaller one must choose the packet. The form of the packet, that is, the form of the wave function at $t = 0$ can still be chosen in many different ways, and the functions $A(\sigma_x, \sigma_y, \sigma_z)$ in (20) can thus still be quite different. One can now try to construct a packet from monochromatic plane waves in such a way that to a first approximation only such wave numbers and frequencies occur which differ very little from each other. In other words, we want to choose $A(\sigma_x, \sigma_y, \sigma_z)$ in such a way that it is appreciably different from zero only in the neighbourhood of σ_x^0, σ_y^0, σ_z^0 corresponding to a wave vector $\boldsymbol{\sigma}^0$. Such a packet is clearly very suitable for the description of a mass point which at $t = 0$ is both approximately localised in space and also possesses an approximately well defined momentum $(h\boldsymbol{\sigma}^0)$; that is, such a packet illustrates a physical situation which to a certain approximation corresponds to the kinematic and dynamic specification of a mass point in classical physics when use is made of the quantum postulate for free particles. When we talk in quantum theory about a wave packet, we have practically always such a *special wave packet* in mind.

§ 4. Properties of special wave packets

We shall now consider in more detail the mathematical properties of such a special wave packet. First of all we investigate how such a packet develops in time and we shall use semi-qualitative arguments based on the treatment of wave phenomena in classical physics. The kinematic picture of our packet is that of a wave group since it consists of a number of travelling successive plane "wave crests" and "wave troughs," the amplitudes of which at any time are appreciably different from zero only inside a finite region of space. The group as a whole travels in the direction perpendicular to the planes of constant phase with a velocity which is called the *group velocity* and which in general is different from the *phase velocity* with which the separate crests and troughs travel. The value of the group velocity is determined by the dispersion law of the waves, which gives the relation between the frequency ν and the absolute value σ of the wave vector, and is given by the equation

$$v = d\nu/d\sigma. \tag{24}$$

This derivative depends on σ and one must choose for σ a value characterising the wave numbers of the constituent waves. Since these wave numbers extend over an interval in the neighbourhood of $\boldsymbol{\sigma}^0$, of the order of magnitude $\Delta\sigma$, v is only defined up to an amount of the order of $(d^2\nu/d\sigma^2)\Delta\sigma$.

Equation (24) can easily be derived when one bears in mind that the wave group will be at that point in space where the constituent monochromatic plane waves have the same phase; at other points in space they will annihilate each other through interference. In order that a point will at all times be at the position of the wave group, it must move with such a velocity that two arbitrarily chosen monochromatic wave components of the group continually show the same phase difference at that point. If the components of the velocity of the point are v_x, v_y, v_z the phase of a plane wave (11) will change per unit time at that point by the amount $2\pi(\sigma_x v_x+\sigma_y v_y+\sigma_z v_z-\nu)$. This change will be the same for two plane waves with approximately the same σ_x, σ_y, σ_z, and ν-values, if we have

$$v_x d\sigma_x+v_y d\sigma_y+v_z d\sigma_z-d\nu = 0. \tag{25}$$

Since ν depends only on the absolute value σ of the wave vector, we see first of all that $\boldsymbol{v}$ will be perpendicular to all vectors $\boldsymbol{d\sigma}$ which are perpendicular to $\boldsymbol{\sigma}$ * or, that $\boldsymbol{v}//\boldsymbol{\sigma}$. We then get from equation (25) $vd\sigma - d\nu = 0$, and equation (24) follows.

Multiplying numerator and denominator in equation (24) by h and using equations (5) we can write equation (24) in the form $v = dE/dP$. Comparing this with equation (15) we see that *the group velocity of the matter waves is equal to the velocity of the mass point in classical theory.* Thus for the first time we note that quantum theory satisfies the *correspondence principle.* Originally introduced into the theory of spectra by Bohr in 1918, this principle states that quantum theory comprises necessarily all those characteristics of classical theory which do not clash with its essentially non-classical basis. In the present case we must in accordance with de Broglie's postulate (5) exclude the classical precise connexion between velocity and energy-momentum, which in the construction of classical mechanics and in the classical definition of energy and momentum plays such an important part. However, we see that now, in accordance with the correspondence principle, this connexion emerges as a law of nature with approximate or rather asymptotic validity.

Before we discuss in more detail the properties of wave packets and their physical significance we wish to amplify and round off the mathematical aspects of the problem. First of all we come back to the remark in section 3 that the wave field is not uniquely determined by giving the value of ψ at one moment, as long as we are dealing with *one* real wave function; it is thus senseless to ask for the wave field at a later time, as long as we have not given at $t = 0$ the value of a second function connected with ψ, for instance $\partial\psi/\partial t$ (compare the analogous situation in the classical theory of vibrations). It is, however, mathematically simpler to introduce in addition to the wave function given by equation (19) a second wave function

$$\psi' = \iiint B \sin\left[2\pi(\sigma_x x + \sigma_y y + \sigma_z z - \nu t) + b\right] d\sigma_x d\sigma_y d\sigma_z. \qquad (26)$$

This function ψ' differs from ψ only in that the cosine is changed into a sine, or in that the phase constant b, the absolute value of which is immaterial, has been decreased by $\frac{1}{2}\pi$. It is convenient

* This can be seen as follows. If $\boldsymbol{\sigma}' = \boldsymbol{\sigma} + \boldsymbol{d\sigma}$ and $\boldsymbol{d\sigma} \perp \boldsymbol{\sigma}$, then $\sigma' = \sigma$ up to quantities of the second order in $d\sigma$ and hence $\nu' = \nu$, or $d\nu = 0$ and thus from eq. (25) $\boldsymbol{v} \perp \boldsymbol{d\sigma}$.

to combine the two functions given by equations (19) and (26) into one complex wave function with real part ψ and imaginary part $i\psi'$. In the following we shall denote this complex wave function by ψ, where now

$$\psi = \iiint A \exp\,[2\pi i(\sigma_x x+\sigma_y y+\sigma_z z-\nu t)]\,d\sigma_x\,d\sigma_y\,d\sigma_z\,, \tag{27}$$

and A is again the complex amplitude function. One could take $-2\pi i$ instead of $2\pi i$ in the exponent (compare the discussion in § 3), but our choice is the usual one. If we now give the value ψ_0 of ψ at $t = 0$, A will be uniquely determined as follows *

$$A = \iiint \psi_0 \exp\,[-2\pi i(\sigma_x x+\sigma_y y+\sigma_z z)]\,dx\,dy\,dz. \tag{28}$$

We assume always that the integrals on the right hand side of equations (27) and (28) converge.

We can now study the properties of the special wave packets, or wave groups, using equations (27) and (28). Such a wave group is characterised by the fact that $|\psi_0|$ is appreciably different from zero only in the neighbourhood of x_0, y_0, z_0, while the same holds for $|A|$ only in the neighbourhood of σ_x^0, σ_y^0, σ_z^0. This can obviously be arranged by choosing for ψ_0 the following expression

$$\psi_0 = C(x, y, z) \exp\,[2\pi i(\sigma_x^0 x+\sigma_y^0 y+\sigma_z^0 z)]. \tag{29}$$

In equation (29) C is a "smooth" function of the space co-ordinates and its absolute value is different from zero practically only in a region U near x_0, y_0 z_0, while its phase changes only very little in U.

We now get from equations (28) and (29)

$$A=\iiint C \exp\{-2\pi i[(\sigma_x-\sigma_x^0)x+(\sigma_y-\sigma_y^0)y+(\sigma_z-\sigma_z^0)z]\}dx\,dy\,dz. \tag{30}$$

If $\sigma_x-\sigma_x^0$ becomes so large that $(\sigma_x-\sigma_x^0)x$ can vary by a few units within U, A will obviously be negligible compared to its value for $\boldsymbol{\sigma} = \boldsymbol{\sigma}^0$. Thus, if we denote by Δx the approximate extension of U in the x-direction and by $\Delta\sigma_x$ the range of σ_x-values for which A is appreciably different from zero, we have approximately

$$\Delta x \,.\, \Delta\sigma_x \approx 1, \tag{31a}$$

* Equation (28) follows from the theory of Fourier integrals. In the same theory one also proves $\iiint |\psi|^2\,dx\,dy\,dz = \iiint |A|^2\,d\sigma_x\,d\sigma_y\,d\sigma_z$.

and similarly

$$\Delta y \,.\, \Delta\sigma_y \approx 1, \quad \Delta z \,.\, \Delta\sigma_z \approx 1. \tag{31b}$$

Equations (31) are valid for packets for which we have tried to restrict as much as possible the extension of the constituent waves in σ-space for a given extension in ordinary space. For an arbitrary choice of packets we shall in general find the products $\Delta x \,.\, \Delta\sigma_x, \ldots$ large compared to one. We have thus the general relation

$$\Delta x \,.\, \Delta\sigma_x \gtrapprox 1. \tag{32}$$

This relation can be derived elementarily and practically without any mathematics as follows. The superposition of two (infinitely long) one-dimensional sine waves of equal amplitude with wave numbers which are only slightly different, say by $\Delta\sigma = \sigma_1 - \sigma_2$, gives rise to beats. The component with the larger wave number contains one more wave than the other component over the distance between two subsequent beat minima, or, $\sigma_1\Delta x - \sigma_2\Delta x = \Delta\sigma \,.\, \Delta x = 1$. If one now superposes a third sine wave whose wave number does not differ from σ_1 and σ_2 more than by an amount of the order of $\Delta\sigma$ and whose phase is the same as the phase of the first two waves at a point M in the middle of the two beat minima considered a moment ago, the resultant wave train will have an appreciable amplitude over a region on both sides of M of the order of Δx. One has to look much further away than Δx from M to find another point where the amplitude is again appreciable. The superpositions of a great many sine waves, all with wave numbers in the same region $\Delta\sigma$ and all with practically the same amplitude and with practically the same phase at M will lead to a wave train with an amplitude which is practically only in the immediate region of M with an extension $\Delta x = \Delta\sigma^{-1}$ appreciably larger than anywhere else. This comes about since everywhere else the phases of the component waves are practically always uniformly distributed from 0 to 2π, and the waves cancel one another.

If we define $\Delta x, \Delta\sigma_x, \ldots$ properly, we can determine exactly the minimum value of the product $\Delta x \,.\, \Delta\sigma_x$. The following definition is the most convenient one. First of all we multiply ψ_0 by a constant chosen such that *

$$\int |\psi_0|^2 \, d\mathbf{x} = 1. \tag{33}$$

We then define

$$x_0 = \bar{x} = \int x \, |\psi_0|^2 \, d\mathbf{x}, \tag{34}$$

$$(\Delta x)^2 = \overline{(x - x_0)^2} = \int (x - x_0)^2 \, |\psi_0|^2 \, d\mathbf{x}. \tag{35}$$

The quantity x_0 measures the position of the wave packet and Δx its extension.

* For the sake of simplicity we use here and henceforth the following abbreviated notation: $\int$ for $\iiint_{-\infty}^{+\infty}$, $d\mathbf{x}$ for $dx\,dy\,dz$, and $d\boldsymbol{\sigma}$ for $d\sigma_x\,d\sigma_y\,d\sigma_z$.

From equations (33) and (28) it follows that

$$\int |A|^2 \, d\boldsymbol{\sigma} = 1. \tag{36}$$

We introduce now the following definitions

$$\sigma_x^0 = \overline{\sigma_x} = \int \sigma_x \, |A|^2 \, d\boldsymbol{\sigma}, \tag{37}$$

$$(\Delta\sigma_x)^2 = \overline{(\sigma_x - \sigma_x^0)^2} = \int (\sigma_x - \sigma_x^0)^2 \, |A|^2 \, d\boldsymbol{\sigma}. \tag{38}$$

The quantity σ_x^0 measures the average wave number represented in the wave packet and $\Delta\sigma_x$ its extension in σ-space. The equations for the y- and z-components are analogous. It is possible to prove * that the minimum value which $\Delta x \,.\, \Delta\sigma_x$ can attain is given by the equation

$$(\Delta x \,.\, \Delta\sigma_x)_{\min} = 1/4\pi. \tag{39}$$

This minimum is attained for all three co-ordinates if ψ_0 and A are given by the expressions

$$\left.\begin{aligned} \psi_0 &= (8\tau_1\tau_2\tau_3)^{1/4} \exp\{-\pi\tau_1(x-x_0)^2 - \pi\tau_2(y-y_0)^2 - \pi\tau_3(z-z_0)^2 \\ &\qquad + 2\pi i(\sigma_x^0 x + \sigma_y^0 y + \sigma_z^0 z)\}, \\ A &= (\tfrac{1}{8}\tau_1\tau_2\tau_3)^{-1/4} \exp\{-\pi(\sigma_x-\sigma_x^0)^2/\tau_1 - \pi(\sigma_y-\sigma_y^0)^2/\tau_2 \\ &\quad -\pi(\sigma_z-\sigma_z^0)^2/\tau_3 - 2\pi i[x_0(\sigma_x-\sigma_x^0) + y_0(\sigma_y-\sigma_y^0) + z_0(\sigma_z-\sigma_z^0)]\}. \end{aligned}\right\} \tag{40}$$

The τ_i are arbitrary constants. If equation (39) is to be valid for any arbitrary direction in space, the three τ_i must be equal. One can thus construct "most favourable" wave packets for any value of σ_x^0 and x_0. These wave packets correspond to Gaussian distributions, both for $|\psi_0|^2$ and for $|A|^2$.

§ 5. The Heisenberg relations

As a result of the quantum postulate for free particles we can no longer connect a wave packet with the presence of a mass point with a sharply defined momentum and a sharply defined energy. In the packet all possible wave vectors $\boldsymbol{\sigma}$, and hence all possible momentum values $h\boldsymbol{\sigma}$, are represented. In special wave packets or wave groups, however, preponderantly those momentum values are found which are in the neighbourhood of $\boldsymbol{P}^0 = h\boldsymbol{\sigma}^0$, in such a way that the differences $\boldsymbol{\Delta P}$ between $\boldsymbol{P}$ and the preponderant momentum

* See, for instance, W. Heisenberg, *Physical principles of quantum theory*, University of Chicago Press, 1930; R. W. Ditchburn, *Proc. Irish Acad.*, **39**, 58, 1930.

value $\boldsymbol{P}^0$ will be of the order of magnitude $\hbar\boldsymbol{\Delta\sigma}$. Multiplying equation (39) by $\hbar$ gives us the following relation between the margins within which the position of a mass point and at the same time its momentum are defined, *

$$\Delta P_x \,.\, \Delta x \leqq \tfrac{1}{2}\hbar,\ \Delta P_y \,.\, \Delta y \leqq \tfrac{1}{2}\hbar,\ \Delta P_z \,.\, \Delta z \leqq \tfrac{1}{2}\hbar. \qquad (41)$$

Equations (41) express the so-called *Heisenberg relations* (see also the Introduction). They give an indication in how far one can ascribe values simultaneously to the spatial co-ordinates of a mass point and to the components of its momentum. In classical physics one could in principle determine these values accurately, but in quantum theory they can only be defined within a certain margin, in such a way that one finds a reciprocal relation between the lack of definition of the position and that of the momentum. According to the Heisenberg relations the product of the unsharpness of the x-coordinate and the unsharpness of the x-component of the momentum can, as far as orders of magnitude are concerned, not be smaller than the quantum of action.

In order that these considerations have a physical meaning, one must demand that in describing an observation performed on a free particle one can never ascribe to a particle at a given time values of its position and momentum with greater precision than would correspond to the Heisenberg relations. Consider, for instance, the experiment where one measures the interference pattern produced by electrons reflected from a lattice. In that case one must demand that both the orbits of the electrons which leave the cathode and the moment when they leave the cathode will be defined with sufficient imprecision so that one can still use the picture of a wave field extended both in the direction of propagation and perpendicular to it, as this picture is needed to produce the interference pattern. One sees immediately that in a normal experiment nothing contradicts the assumption of such an imprecision. If one, for instance, had put a very narrow slit near the cathode, one could assume that the orbit along which the electron leaves the cathode will be very accurately determined; however, the path of the electron from the slit to the lattice can again be assumed to be undetermined, corresponding to a diffraction of the matter waves

* Since in many formulae of quantum theory Planck's constant occurs in the combination $h/2\pi$, Dirac has introduced the symbol $\hbar : \hbar = h/2\pi = 1{\cdot}05.10^{-27}$ g cm^2 sec^{-1} $= 1.05.10^{-34}$ kg m^2 s^{-1}. We shall nearly always use Dirac's constant.

by the slit. If one had made an arrangement to open the slit only during a short period, so that a wave train containing only very few waves could pass the slit, the interference pattern will then not be found according to the theory. In the same way one must assume that an electron entering a Geiger counter through a small opening will, indeed, have a lack of precision of the component of its momentum parallel to the plane of the opening as soon as it has entered the counter. This lack of precision will be connected with the dimensions of the hole in accordance with the Heisenberg relations. The direction along which the electron approaches the opening may have been very accurately determined, but this precision is lost as soon as the opening is passed, corresponding once again to a diffraction of matter waves. If, moreover, one is able to determine accurately the moment when the electron is captured, the absolute value of the momentum of the captured electrons will be defined less accurately, the more accurately one has determined the moment of capture, independent of whether or not the absolute value of the momentum was well defined before the capture.

We now proceed with the discussion of the mathematical properties of wave packets. We saw, how the future behaviour of a packet is completely determined, if the wave function ψ_0 at $t = 0$ is given, once complex wave functions have been introduced. If we are dealing with a special wave packet, we see first of all that this packet will move as a whole with the group velocity $v = d\nu/d\sigma$. Using equations (27) and (29) we can give a more rigorous derivation of this result. We wish to compare the shape of the function ψ in equation (27) at a time t with its shape at $t = 0$ under the assumption that the values of $\boldsymbol{\sigma}$ which must be considered lie all in the neighbourhood of $\boldsymbol{\sigma}^0$.

We write $\boldsymbol{\sigma} = \boldsymbol{\sigma}^0 + \boldsymbol{\Delta\sigma}$, $\sigma = \sigma^0 + \Delta\sigma$. If the components of $\boldsymbol{\Delta\sigma}$ are small compared to the components of $\boldsymbol{\sigma}^0$, we have approximately * $\Delta\sigma = (\boldsymbol{\alpha} \cdot \boldsymbol{\Delta\sigma})$, where $\boldsymbol{\alpha} = \boldsymbol{\sigma}/\sigma$.

Using the dispersion law $\nu = \nu(\sigma)$ we have $\nu \doteq \nu^0 + (d\nu/d\sigma)_0 \Delta\sigma = \nu^0 + v d\sigma$.

We get now from equation (27)

$$\begin{aligned}\psi \doteq \exp[2\pi i\{(\boldsymbol{\sigma}^0 \cdot \boldsymbol{x})\nu^0 t]\int A(\boldsymbol{\sigma}^0 + \boldsymbol{\Delta\sigma}) \exp\{2\pi i(\boldsymbol{\Delta\sigma} \cdot \boldsymbol{x} - \boldsymbol{\alpha} v t)\}\, d\boldsymbol{\Delta\sigma} \\ = \exp[2\pi i\{(\boldsymbol{\sigma}^0 \cdot \boldsymbol{x}) - \nu^0 t\}]\, C(\boldsymbol{x} - \boldsymbol{\alpha} v t).\end{aligned} \tag{42}$$

* This result follows from the equations $\sigma^2 - 2\sigma\Delta\sigma \doteq \sigma^{0^2} = (\boldsymbol{\sigma}^0 \cdot \boldsymbol{\sigma}^0) \doteq (\boldsymbol{\sigma} \cdot \boldsymbol{\sigma}) - 2(\boldsymbol{\sigma} \cdot \boldsymbol{\Delta\sigma})$.

This equation will be valid for sufficiently small values of t. The function C with its argument $\boldsymbol{x}-\boldsymbol{\alpha}vt$ must clearly be identical with the function C of equation (29). This follows immediately by putting $t = 0$. Equation (42) expresses the fact that the wave packet, the position of which is determined by the values of $\boldsymbol{x}$ for which $|\psi|$, that is, $|C|$ differs appreciably from zero moves with a velocity $v = d\nu/d\sigma$ in the direction of the wave vector. The exponential multiplying factor shows that the planes of constant phase move across the wave packet with the phase velocity ν^0/σ^0.

One can also talk about the extension in time of a wave packet in the following sense. At a point of space through which the packet passes the wave function will possess appreciable values during a finite time interval Δt. Due to the dispersion law Δt will not be independent of the extension of the wave packet in space, but we have the relation

$$(d\nu/d\sigma)\Delta t = v\Delta t = \Delta l, \tag{43}$$

where Δl is the extension of the wave packet in the direction of propagation. We have neglected here the case where v is so small that the approximate equation (42) is no longer valid during the whole of Δt. Defining by $\Delta\nu$ the margin in ν which due to the dispersion law is related to the variation $\Delta\sigma$ of the wave number σ, we see that $\Delta\nu$ is a measure of the lack of precision of the definition of the frequency of the wave packet. Clearly

$$\Delta\nu \approx (d\nu/d\sigma)\Delta\sigma, \tag{44}$$

and combining equations (43) and (44) we get

$$\Delta\nu\,.\,\Delta t \approx \Delta\sigma\,.\,\Delta l \gtrapprox 1. \tag{45}$$

We could also have derived equation (45) directly without using the dispersion law by considering ψ as a function of t.

Multiplying equation (45) by h and denoting by ΔE the uncertainty with which the energy of a localised mass point is defined we get

$$\Delta E\,.\,\Delta t \gtrapprox h, \tag{46}$$

where we have used the fundamental relation $E = h\nu$.

This relation joins those of equation (41) as the fourth Heisenberg relation. Up to a point it is not really independent of the first three, as the relationship between energy and momentum of a free particle implies a relationship between the unsharpness in the energy and that in the momentum components. The physical meaning of equation (46) is that it should be impossible to conclude from an experiment performed on a free particle that this particle passed a given plane within a time interval $(t, t+\Delta t)$ (or, that *only* during a period Δt was it present in a given region of space) and at the same time possessed an energy between E and $E+\Delta E$, unless the product of Δt and ΔE was at least of the order of h. As an example we may refer to the case discussed earlier, where an electron passed a slit which was opened only during a very short period Δt. We remarked at the time that the margin within which the absolute value of the momentum was determined would be the larger the shorter Δt. This has, however, as a consequence that the unsharpness ΔE in the definition of the energy would also be larger and one can easily check that equation (46) is satisfied. In any case it seems extremely unlikely that actually such simple thought experiments can be performed.

In sections 16 and 28 we shall return to the general physical implications of the Heisenberg relations for more complicated systems.

§ 6. The approximate validity of Newton's first law

We have just seen in the previous section that a wave packet as a whole will move with a group velocity $v = d\nu/d\sigma$. The derivation was, however, only approximate. This follows immediately from the fact that the value of σ for which $d\nu/d\sigma$ has to be evaluated is only defined within a certain margin since the wave numbers of the monochromatic waves which together form the packet vary within a finite interval. In fact, while the wave packet moves along it will continuously change its shape and extension in space. The way in which this change takes place depends on the choice of ψ at $t = 0$. If one waits long enough, the packet will certainly be increasingly smeared out over space since the phase relations between the constituent waves which determined the spatial boundaries of the original wave packet, will no longer apply because of the dispersion. If one starts at $t = 0$ with a packet of the "most favourable" form, given by equation (40), it will start to disperse immediately. In that case one can follow the behaviour of ψ exactly as long as one uses the non-relativistic dispersion formula, and one can write down the equations which give at any t the shape of the packet.

We consider for the sake of simplicity the one- dimensional case. Let ψ at $t = 0$ be given by $\psi_0 = (2\tau)^{1/4} \exp[-\pi\tau x^2 + 2\pi i\sigma_0 x]$. The function $A(\sigma)$ is then given by [see eq. (40)] $A = (2/\tau)^{1/4} \exp[-\pi(\sigma-\sigma_0)^2/\tau]$. Using the relation $\nu = h\sigma^2/2m$ we have at time t

$$\psi(t) = \int_{-\infty}^{+\infty} A \exp[2\pi i(\sigma x - \nu t)]\, d\sigma = \int_{-\infty}^{+\infty} A \exp[2\pi i\{\sigma x - (h\sigma^2 t/2m)\}]\, d\sigma$$

$$= (2\tau)^{1/4}[1+(ih\tau t/m)]^{-1/2} \exp\{[-\pi\tau x^2 + 2\pi i(\sigma_0 x - \nu_0 t)]/[1+(ih\tau t/m)]\},$$

$$|\psi(t)|^2 = \{2\tau/[1+(h\tau t/m)^2]\}^{1/2} \exp\{-2\pi\tau(x - v_0 t)^2/[1+(h\tau t/m)^2]\}, \quad v_0 = h\sigma_0/m.$$

The "centre of gravity" of the packet moves with the velocity v_0 and its "width" increases proportional to $[1+(h\tau t/m)^2]^{1/2}$.

These considerations show in how far one can still attach in quantum theory a meaning to the concept of the motion of a free mass point. In classical physics one expected that in principle the position of a mass point at different times could be measured to any degree of accuracy, and that in that way one could define the orbit of a mass point. In the case of a free particle this orbit would correspond to a constant velocity vector, in accordance with Newton's first law. By accurately determining the position of the particle at two different times one could determine this velocity accurately, and hence one could predict the position of the particle at a later time with an accuracy which was in principle unlimited. The fact that Newton's first law already contains a dynamic part — even though the actual value of the mass plays no role in its formulation — entails that the situation is different in quantum theory. In fact, one must demand that the properties of wave packets should illustrate the natural limits for the application of Newton's first law. In particular the dispersion of the wave packet means that it will be impossible to predict accurately from measurements at $t = t_0$ the position of a mass point at a later time $t = t_1$. The accuracy of such a prediction will be the smaller, the longer the time interval (t_0, t_1).

That one can in principle predict the position of a free particle at a later time is a simple example of the causal description in space-time of classical physics. The preceding considerations show that in quantum theory one must renounce the possibility of such a description, since otherwise one would arrive at results in disagreement with basic experimental evidence. In this connexion we may remark that the statement is not that one can not measure the velocity of a free particle, or that the momentum of a free particle varies in time. In fact one can define a purely kinematic velocity, if in an experiment one has measured the position of a free particle at two different times, by taking the distance between the two positions divided by the time interval. On the other hand, the fact that to a free particle of well defined momentum there corresponds a wave field with a well defined wave vector, implies that by definition the momentum of such a particle is considered as a quantity constant in time. We have here an example of the *complementary* relation which exists according to quantum theory between the relations in space-time and the conservation laws of momentum and energy. If the position is measured accurately one loses all knowledge about the momentum and if the momentum is measured one must forego the knowledge of the position in space-time. A classical law, such as Newton's first law, where both kinematic and dynamic attributes play a role, can only be valid up to a point; its limitation is governed by the universal quantum of action.

§ 7. The quantitative formulation of probability laws

The situation we have just described entails, as we stressed earlier, that in quantum theory in general one cannot predict the result of an experiment with certainty. One can instead speak only of the *probability* that such or such a result is to be expected. We have already mentioned the way to interpret interference experiments using the de Broglie hypothesis. One should expect that the probability of finding a particle is appreciably different from zero only in those regions of space where the wave movement is intensive. Thus the concept of probability plays an essential role in the formulation of observable laws right from the beginning. We shall show now how one can give in the case of free particles a quantitative formulation of these laws.

We start from the idea that in the case of an experiment involving only one free particle, while the experimental apparatus can be considered to be external systems which are not measurably affected by the particle, the physical situation can be described by a de Broglie wave-field such as the one described by equation (27), that is, a superposition of plane waves. We are now faced with the problem of finding an expression for the intensity $J(\boldsymbol{x}, t)$ of the field which can play the role of a probability function. This function must be such that $J d\boldsymbol{x}$ is the probability of finding the particle within the volume element $d\boldsymbol{x}$ at time t. One must demand that J is always positive and that the integral of J over the whole of the space available to the particle is equal to unity, that is, independent of t. The simplest possibility is given by the expression

$$J = |\psi^2| = \psi^*\psi, \tag{47}$$

that is, the probability function is simply the absolute square of the wave function. Using this definition one can, indeed, construct the theory. To justify this choice we consider expression (27) and remind ourselves that $A(\boldsymbol{\sigma})$ will not change its value in a situation where during a time interval t the particle definitely does not interact with any apparatus. Provided the integral $\int |A|^2 d\boldsymbol{\sigma}$ converges, it will stay constant during this interval. However, as long as the integrals converge, one has at any time t

$$\int |A|^2 d\boldsymbol{\sigma} = |\psi|^2 d\boldsymbol{x}. \tag{48}$$

The integral on the right hand side of equation (48) remains thus also constant during the time interval t. The only additional requirement that $|\psi|^2$ can play the role of a probability function is that the wave function is *normalised*,

$$\int |\psi|^2 d\boldsymbol{x} = 1. \tag{49}$$

If the particle interacts with an apparatus, for instance, when it is reflected by a crystal lattice, A is clearly changing. We must thus expect that a detailed theory of the interaction will be such that the integrals in equation (48) do not change their value. The general theory to be discussed later on, indeed, satisfies this requirement quite clearly (see §§ 14 and 15).

In an actual application of $|\psi|^2$ as *probability density* one is dealing with experiments where it is clear from the experimental set-up what shape the wave function initially possessed so that one wants to draw conclusions about the probability of finding a given value of the position of the particle at a later time. We shall restrict the discussion for the time being to the case where the particle will not have interacted with other systems during the time interval under consideration, since only for that case can we put our considerations on a quantitative basis.

The initial form of the wave function can be considered as known, for instance, in the following idealised case. A fixed wall is bombarded by particles with a well defined energy travelling in a well defined direction starting from a sufficiently distant source. In the wall there are one or more holes the shape of which is arbitrary and variable. For the sake of simplicity we shall assume that the holes are only opened during a finite time interval (t_0, t_1). One notices that in discussing such an experimental set-up one assumes the existence of ideal rigid bodies and clocks. If such an assumption is made, it is possible, as we emphasised earlier, to define an absolute co-ordinate system in space-time. For the moment we shall not enter into a discussion of such details as the exact shape of the matter waves which have passed the holes — just as in the case of Fresnel's treatment of diffraction one does not need to worry about the detailed properties of the light waves at the surface of the diffracting body. In this case we can assume that the wave function is known at t_1 * and thus also at a later time, provided always that at least one particle has passed the holes **. A certain uncertainty enters since one can always multiply the wave function with a phase factor $e^{i\alpha}$. This does, however, not make any difference as far as the physical situation is concerned, as the factor drops out when we construct the expression $\psi^*\psi$. If we now want actually to apply the probability function $\psi^*\psi$ to predict the results of a measurement, we consider a measurement which enables us to decide whether or not at t_2 the particle is present inside a given region of space. Restricting ourselves for the time being to the use of rigid bodies, we imagine that at t_2 this region of space is suddenly enclosed in a vessel with fixed walls so that afterwards one can investigate the presence or absence of a particle, for instance, by a charge determination. Knowledge of the wave function at t_2 enables us to calculate the probability $\int |\psi|^2 d\boldsymbol{x}$ where the integration extends over the region under consideration, that the particle has actually been captured. If, for instance, there was only one hole of constant dimensions which was only opened during a short time interval (t_0, t_1) one would immediately conclude that very soon after t_1 the particle would certainly be in the im-

* We do not want to discuss how one can determine ψ at t_1. The situation is analogous to the case where one determines optical diffraction when one must take dispersion into account. The easiest way is to use the differential equation for the wave function which we shall discuss later on. Compare also footnote on page 56.

** The question also arises as to the probability that 0,1, or several particles have passed through. This question is naturally related to the probability distribution of particles in space. In order not to complicate the discussion we shall not consider this problem here.

mediate neighbourhood of the hole, just as in classical theory. If one waits longer, the result of our measurements of the position will depart the more from the classically expected result the more the wave packet which passed through changed its shape. If, for instance, the dimensions of the hole were small compared to the wave length, one will have a large probability that the particle will be deflected over a large angle. If the hole were large compared to the wave length, but only opened for a very short time so that only a few waves could pass through, one would expect a larger probability that the measurement of the position of the particle, in the manner we just described, would lead to a velocity completely different from the classically expected velocity. In short, we have been able to improve on our earlier qualitative considerations of the movement of free particles by introducing a quantitative probability law which determines the result of future experiments. It is, of course, implied in the whole situation that one can only test this law by repeating the measurement of the position many times under the same conditions. This corresponds to the fact that the confirmation of de Broglie's hypothesis was only possible through experiments involving many particles.

The peculiar deviations from the classical theory, which according to quantum theory are characteristic of the behaviour of free particles, can be studied by an analysis of several cases of experiments of the type just considered. For instance, one can illustrate the deflexion of particles by crystals or artificial lattices by considering an experiment where a parallel beam of free particles with fixed energies are bombarding a fixed, thin wall with a great number of parallel equidistant slits. One can then observe at some distance from the wall the deflexion pattern well-known from optics. In interpreting such an experiment it has no meaning to ask through which slit one or another particle has passed.

One should not restrict the possibility of defining quantitative probability laws to the prediction of the results of a measurement where one investigates the presence of a particle in a given region in space. It is true that we have considered only that kind of experiment because we were concerned with applying the probability function $\psi^*\psi$. These measurements are very academic and completely different from experiments whch are actually used to prove the presence of separate particles. One catches particles on photographic plates or in counters or similar devices. This kind of measurement can be schematised, if we want to consider only such experiments where ideal rigid bodies are introduced, by imagining the measuring apparatus to be a fixed, thin wall with a small hole. The actual measurement now consists of determining whether or not within a given time interval during which the hole was opened a particle has passed into a vessel behind the wall. We demand that the theory enables us to calculate from the wave function the probability of finding a positive result. The prescription for the calculation must, of course, be in accordance with the prescription for those calculations involving measurement of position.

One can satisfy this last requirement if one is able to define a real vector field $\boldsymbol{S}$ which is completely determined by the properties of the scalar field ψ at the same point and which satisfies an equation of continuity of the form

$$\operatorname{div} \boldsymbol{S} + \partial J/\partial t = 0, \tag{50}$$

so that for any closed surface we have

$$\oint (\boldsymbol{S} \cdot \boldsymbol{df}) = -(d/dt) \int J d\boldsymbol{x} \tag{51}$$

We can then interpret $\boldsymbol{S}$ in such a way that $(\boldsymbol{S} \cdot \boldsymbol{df})$ is the probability current density through a surface element $\boldsymbol{df}$, that is, the probability per unit time that the particle has passed $\boldsymbol{df}$ in the direction of the positive normal minus the probability per unit time that it has passed $\boldsymbol{df}$ in the opposite direction. If one wants to apply this definition to the capture experiment discussed a moment ago one proceeds as follows. To begin with, at t_1, the vessel is closed and the wave function inside the vessel is zero, that is, the particle is definitely not inside. Then, at t_2, the hole is opened and the matter waves pass through, until at t_3 the hole is closed. The wave function itself is known in the opening, since it was known at t_1 and since we may neglect any reflexions by the apparatus itself. The integral (51) is now extended over the vessel and will give a contribution practically only from the opening. From equation (51) and from our interpretation of $\psi^*\psi$ it follows now that the probability that our schematised capturing experiment leads to a positive result between t_2 and t_3 is given by the integral $\int_{t_2}^{t_3} dt \int (\boldsymbol{S} \cdot \boldsymbol{df})$, where the last integral extends over the opening. If the interval (t_2, t_3) is so large that part of the matter waves can leave the vessel again after having been reflected by the walls, one can, of course, no longer use the $\boldsymbol{S}$ following from the original wave field and the capturing probability will be smaller.

Let us now investigate whether one can define a vector with the above properties. We shall proceed heuristically by considering to begin with the case where the wave function is in a given region of space and within a finite time interval is to a first approximation given by the expression

$$\psi = A \exp[2\pi i\{(\boldsymbol{\sigma} \cdot \boldsymbol{x}) - \nu t)\}]. \tag{52}$$

In the interior of a wave packet of the "most favourable" shape [see eq. (40)] this case is realised for small values of τ. According to our earlier considerations one can assume that one is dealing with a particle the velocity of which is well defined and is equal to $(d\nu/d\sigma)(\boldsymbol{\sigma}/\sigma)$. The probability that the particle within a time interval dt crosses a surface element $dydz$ perpendicular to the x-axis can now be put equal to the probability that the particle is within a prism with base $dydz$ and height $(d\nu/d\sigma)(\sigma_x/\sigma)dt$, apart from a factor $dydzdt$; this probability is thus equal to

$$\begin{aligned} S_x &= \psi^*\psi \,.\, 2(d\nu/d\sigma^2)\sigma_x \\ &= (-i/\pi)(d\nu/d\sigma^2)\psi^*(\partial\psi/\partial x) = (i/\pi)(d\nu/d\sigma^2)(\partial\psi^*/\partial x)\psi \\ &= (2\pi i)^{-1}(d\nu/d\sigma^2)[\psi^*(\partial\psi/\partial x) - \psi(\partial\psi^*/\partial x)]. \end{aligned} \tag{53}$$

Analogous expression can be written down for S_y and S_z. The question is now whether and how one can generalise this result so that it can be applied to any arbitrary ψ in such way that $\boldsymbol{S}$ depends only on ψ and its space derivatives. If we use the non-relativistic dispersion law (16) we can easily answer this question. In this case $d\nu/d\sigma^2$ is a constant,

$$d\nu/d\sigma^2 = h/2m. \tag{54}$$

We can now write down for the probability current density the following expression

$$\boldsymbol{S} = (i\hbar/2m)[\psi \operatorname{grad} \psi^* - \psi^* \operatorname{grad} \psi]. \tag{55}$$

One can easily check by using equation (27) that this expression satisfies the equation of continuity (50). We have used the last member of equation (53) in order that $\boldsymbol{S}$ be always real.

It is interesting to note that it is the simple form of the dispersion law which enabled us to find so easily an expression for $\boldsymbol{S}$ satisfying our requirements. This would not have been possible for any arbitrary dispersion law. One can indeed show (compare § 44) that in general it is impossible to generalise expression (53).

In constructing the theory we need consider only the dispersion law which occurs in nature. We encounter here, however, the peculiar circumstance that the non-relativistic dispersion law is the only one which gives us a simple definition of the probability current density. The relativistic dispersion law (17), which can be written in the form

$$\nu = [m^2c^2/h^2 + c^2\sigma^2]^{1/2}, \tag{56}$$

and which we should really have used as the basis of our theory, would have led to great difficilties.

This difficulty is real. It is the first indication in our approach that the foundations which we have been using are insufficient to extend the theory so as to incorporate the requirements of the theory of relativity. These difficulties are surmounted to a large extent in Dirac's theory of the spinning electron *. In this theory one uses several wave functions instead of just one and the theory can in a natural way explain the existence of the spin of an electron.

§ 8. The Schrödinger wave equation

Up to now we have considered the wave field as a superposition of plane travelling waves. The simple form of the dispersion law (16) or (17) enables us, however, to treat the wave function as a solution of a simple *linear partial differential equation.* This equation will express in its pregnant mathematical formalism the motion of matter waves. We obtain this differential equation by observing that the function

$$\psi = A \exp\{2\pi i[(\boldsymbol{\sigma} \cdot \boldsymbol{x}) - \nu t]\}, \tag{57}$$

which corresponds to a plane travelling wave, satisfies the following relations

$$\partial\psi/\partial x = 2\pi i\sigma_x\psi, \quad \partial^2\psi/\partial x^2 = (2\pi i)^2\sigma_x^2\psi, \quad \partial\psi/\partial t = -2\pi i\nu\psi, \ldots$$

If, therefore, the dispersion law is of the form $F(\boldsymbol{\sigma}, \nu) = 0$ with F a polynomial in σ_x, σ_y, σ_z, and ν, ψ will satisfy the differential equation **

$$F([2\pi i]^{-1}\nabla, \ -[2\pi i]^{-1}(\partial/\partial t))\,\psi = 0. \tag{58}$$

* P. A. M. Dirac, *Proc. Roy. Soc.*, **A117**, 610, 1928; **118**, 351, 1928.

** We use the symbol ∇ for the gradient operator with components $\partial/\partial x$, $\partial/\partial y$, $\partial/\partial z$.

In this equation F is a *differential operator* which is obtained from the polynomical F by substituting $(2\pi i)^{-1}\nabla$ and $-(2\pi i)^{-1}(\partial/\partial t)$ for $\boldsymbol{\sigma}$ and ν. As equation (58) is a linear differential equation, any linear combination of plane waves which obey the same dispersion law will satisfy this equation. This means that equation (58) is the *differential equation for the matter waves of a free particle.*

As the dispersion law follows from the classical relation between momentum and energy by using the de Broglie relations (5) ($\boldsymbol{P} = h\boldsymbol{\sigma}$, $E = h\nu$) one can also obtain equation (58) as follows. Start from the classical relation between momentum and energy,

$$\mathscr{H}(\boldsymbol{P}, E) = 0, \tag{59}$$

where $\mathscr{H}$ is a polynomial. Now make the following substitutions

$$\boldsymbol{P} \rightarrow -i\hbar\nabla,\ E = i\hbar(\partial/\partial t). \tag{60}$$

We then get for the differential equation for matter waves

$$\mathscr{H}(-i\hbar\nabla,\ i\hbar(\partial/\partial t))\psi = 0. \tag{61}$$

Using the non-relativistic formulae (12) or (16) we have the following differential equation

$$(-i\hbar)^2(2m)^{-1}\nabla^2\psi - i\hbar(\partial\psi/\partial t) = 0, \text{ or, } \nabla^2\psi + (2mi/\hbar)(\partial\psi/\partial t) = 0. \tag{62}$$

The relativistic formulae (13) or (17) would have led to

$$\begin{gathered}(-i\hbar)^2[\nabla^2\psi - c^{-2}(\partial^2\psi/\partial t^2)] + m^2c^2\psi = 0, \text{ or,}\\ \nabla^2\psi - c^{-2}(\partial^2\psi/\partial t^2) - (mc/\hbar^2)\psi = 0.\end{gathered} \tag{63}$$

Equation (62) is called the non-relativistic Schrödinger wave equation and equation (63) the relativistic Schrödinger wave equation *.

Even if one is dealing with problems where the non-relativistic approach provides a sufficiently satisfactory approximation, one can easily take into account the fact that the energy of a free particle at rest is not equal to zero, but to mc^2. In that case one uses equations (14) or (18) and obtains instead of equation (62)

$$\nabla^2\psi + (2mi/\hbar)(\partial\psi/\partial t) - 2(mc/\hbar)^2\psi = 0. \tag{64}$$

As far as physical applications are concerned equation (64) leads to the same results as equation (62). This follows from the fact that any solution of equation (64) corresponds to a solution of equation

* E. Schrödinger, *Ann. Phys.*, **81**, 112, 133, 1926.

(62) multiplied by a factor exp $(imc^2t/\hbar)$. This factor drops out whenever one constructs real quantities such as those of equations (47) and (55).

Let us now consider in some more detail the non-relativistic wave equation. We shall for the sake of simplicity use equation (62). This equation contains only the first time derivative of ψ. This means that, in accordance with our earlier considerations, ψ at any time can be determined from the values of ψ at one given time. We now draw attention to the fact that all the solutions of equation (62) will be complex functions * because $\partial\psi/\partial t$ is multiplied by a purely imaginary factor. The fact that in the non-relativistic case one is led to the simple wave equation (62) as soon as one is willing to introduce complex functions is a justification of this introduction. Earlier we had done this mainly for the sake of simplicity, although even then the fact that the absolute value of the phase could not have physical significance played an important role. The meaninglessness of the phase is clear from equations (47) and (55) which give us predictions about results of physical measurements. It is also clear that any solution ψ of equation (62) can be multiplied by an arbitrary phase factor $\exp(i\alpha)$.

Using the Schrödinger equation (62) we can easily prove the equation of continuity (50), using for J and $\boldsymbol{S}$ the expressions (47) and (55). We multiply equation (62) by ψ^*,

$$\psi^*\boldsymbol{\nabla}^2\psi + (2mi/\hbar)\,\psi^*(\partial\psi/\partial t) = 0, \tag{65}$$

and consider the imaginary part of the left hand side of this equation. This can be done by writing the complex conjugated equation,

$$\psi\boldsymbol{\nabla}^2\psi^* - (2mi/\hbar)\,\psi(\partial\psi^*/\partial t) = 0, \tag{66}$$

and subtracting equation (66) from (65). The result is

$$\psi^*\boldsymbol{\nabla}^2\psi - \psi\boldsymbol{\nabla}^2\psi^* = \boldsymbol{\nabla}\,.\,(\psi^*\boldsymbol{\nabla}\psi - \psi\boldsymbol{\nabla}\psi^*) = -(2mi/\hbar)\,(\partial\psi^*\psi/\partial t),$$

and multiplication by $\hbar/2mi$ gives us the equation of continuity in the form

$$(\boldsymbol{\nabla}\,.\,[(\hbar/2mi)(\psi^*\boldsymbol{\nabla}\psi - \psi\boldsymbol{\nabla}\psi^*)]) + \partial\psi^*\psi/\partial t = 0. \tag{67}$$

Using equation (62) we can discuss the propagation of wave groups in a slightly different way. However, we shall discuss this

* We define a complex function as a function of a real variable which in general takes on complex values.

point in § 27 for wave equations of a more general nature and we refer the reader to that discussion.

Our interpretation of $\psi^*\psi$ as probability density requires that we can consider only such solutions of equation (62) for which the integral $\int \psi^* \psi d\boldsymbol{x}$ taken over the whole of the available space, is finite and equal to unity. Such solutions are called *normalised wave functions*. Even so it is often of advantage to use ψ-functions which do not satisfy this requirement. Such a situation occurs when we are considering experiments or processes which extend over a long time interval and involve many particles, while the physical situation remains the same during that interval. An example of such a case are the Davisson-Germer experiments where over an arbitrarily long period an arbitrarily large number of particle will be registered. If in such a case the interaction between the particles can be neglected, it is possible to use unnormalised wave functions ψ. In that case $\psi^*\psi$ is to be interpreted as the average number of particles per unit volume and $\boldsymbol{S}$, defined by equation (55), as the average number of particles crossing a unit area perpendicular to $\boldsymbol{S}$ per unit time. This procedure which seems justified on physical grounds can be justified completely when one uses the theory of interacting particles which will be discussed later on. We may also mention at this point the following often used application. One interprets a monochromatic de Broglie wave (57) as a physical situation where particles of energy $h\nu$ which are uniformly distributed over space with an average density $\psi^*\psi$ are flying in the direction of the wave vector with a well defined velocity $d\nu/d\sigma$.

Let us now consider the relativistic wave equation (63). This equation is real in contradistinction to equations (62) or (64) so that both the real and the imaginary part of a complex solution would themselves be solutions. However, the wave equation contains the second time derivative of ψ so that the value of ψ at one time does not suffice to determine t at a later time. Previously we had decided by discussing equation (27) that the value of ψ at a given time alone should be sufficient. This discrepancy originates from the assumption that ν given by the dispersion law was a unique function of $\boldsymbol{\sigma}$. This assumption leads in the relativistic case to the conclusion that only positive values of ν can be considered. Equation (63) can, however, be solved by plane waves of the form (57) with either positive or negative ν. The general solution of equation (63) when written in the form of a Fourier integral will therefore not be of the form (27) but will be the sum of two similar expressions with positive ν-values in the first and negative ν-values in the second expression. It is true that (27) (with positive ν-values only) is still a solution of equation (63), but the particular properties of this solution which distinguish it from the general solution are not necessarily found in the properties of ψ at a given time.

One might infer that one should give up the idea of a differential equation of the form (63) in the relativistic case. However, there are several reasons to retain it for the time being. To begin with we refer to the discussion at the end of the previous section. Secondly, we shall see later on in the discussion of the quantum theory for bound particles that to use only *real* values of σ_x, σ_y, σ_z, and ν satisfying the dispersion law is too restrictive. It is all right in the case of free particles, but the possibility of describing particles in fields of force is connected with lifting this restriction *.

For instance, one may have to consider imaginary values of σ in describing matter waves in a field free region of space. If one now uses the dispersion law (17), one is led to ν-values which are smaller than mc^2/h, and one can no longer exclude certain values of ν. Values of ν smaller than $-mc^2/h$ correspond again to real wave numbers, and it is impossible to exclude the possibility of travelling matter waves corresponding to negative ν-values. These considerations indicate that in a relativistic description of matter waves one should, indeed, retain equation (63) as a heuristic basis.

Of a theory on such a basis one should demand that it be possible to construct from the wave function describing the matter waves a real, positive probability density J and a real probability current density $\boldsymbol{S}$. They should satisfy the equation of continuity (50) and $\boldsymbol{S}$ and J should transform covariantly, that is, as $\boldsymbol{x}$ and t, under a Lorentz transformation. At the same time one must expect that in the limiting case where the energy of the particle differs only slightly from mc^2 the expressions for $\boldsymbol{S}$ and J will approach the nonrelativistic expressions (47) and (55). This programme can not be realised when one introduces only one scalar complex wave function, satisfying equation (63). One can construct the following

* A clear indication of this is the fact that the wave function is not zero in those regions of space where according to classical theory the kinetic energy would be zero (see § 18).

four-vector *

$$\begin{aligned} \boldsymbol{S} &= (\hbar/2mi)[\psi^*\nabla\psi-\psi\nabla\psi^*], \\ J &= -(\hbar/2mic^2)[\psi^*(\partial\psi/\partial t)-\psi(\partial\psi^*/\partial t)]. \end{aligned} \tag{68}$$

Multiplying equation (63) by ψ^* and considering as before the imaginary part of this equation, one easily checks that $\boldsymbol{S}$ and J satisfy the equation of continuity. Secondly, the expression for $\boldsymbol{S}$ is identical with (55) while the expression for J goes over into expression (47) if one assumes that the frequencies occurring in the Fourier expansion of ψ are all nearly equal to mc^2/h, so that ψ contains t practically only in the form of a factor exp $[-imc^2t/\hbar]$. All the same, equations (68) are unsatisfactory since J according to its physical meaning should always be positive, while expression (68) for J can take on both positive and negative values. This follows immediately from the fact that one can prescribe both ψ and $\partial\psi/\partial t$ at a given time.

Dirac has shown, however, that one can satisfy the physical requirements most simply by describing the matter waves of a particle through *four* complex wave functions. These functions satisfy four simultaneous first order differential equations so that the values of these functions at one time are sufficient to determine their values at any other time. They transform according to certain linear transformations under a Lorentz transformation and they lead automatically to the existence of the electron spin. We shall discuss the Dirac theory in chapter 6.

§ 9. THE QUANTUM THEORY OF FREE PARTICLES AND THE LAWS OF CONSERVATION OF MOMENTUM AND ENERGY

In this section we want to discuss briefly the relation between the description of the properties of free particles through wave fields and the conservation of momentum-energy. In classical physics the introduction of energy and momentum of a particle was intimately connected with the existence of their laws of conservation. The fundamental role played by energy and momentum in the formulation of the laws of quantum theory for free particles is connected with the fact that also in quantum theory the conservation laws are considered to be a principal feature of all physical

* These expressions were proposed independently by O. Klein (*Z. Phys.*, **41**, 414, 1927),E. Schrödinger (*Ann. Phys.*, **82**, 267, 1927), and W. Gordon (*Z. Phys.*, **40**, 121, 1927).

phenomena. Up to now we have only considered such experiments where either only one particle or a number of noninteracting particles were involved. The essential role of the conservation laws could therefore not yet be displayed. In the discussion of the propagation of wave packets, however, we could show the restricted validity of Newton's first law by considering the complementary relation between momentum and position.

In the next chapter we shall discuss interacting particles and we shall then give a discussion of the range of validity of the conservation laws. However, it is important to consider the connexion of the description of the properties of a free particle, as given in this chapter, with the conservation laws, since some essential characteristics of quantum theory will emerge. We consider an experiment where particles with a well defined energy emerging from a source at a great distance bombard a wall with one or several holes. Originally each particle has a well defined momentum. After it has passed through the holes this is no longer the case. The presence of all possible wave vectors in the Fourier components of the wave packet which has passed the wall corresponds to a decrease in the possibility of assigning to it a well defined momentum. If, for instance, there are parallel equidistant slits in the wall, its action as a grating will entail that one has a choice between a number of different directions of propagation corresponding to different possible momentum vectors of the particle. One can now catch the particle by a suitable choice of apparatus and it will be clear that the particle has moved in only one of the possible directions. One might think that there would be no conservation of momentum during such an experiment. This conclusion is, however, false. The deflexion of the particle is caused by the interaction of the grating with the particle and one is not led to any contradictions if one assumes that the reaction of the particle on the wall consisted in a transfer of momentum which exactly compensated the observed change of momentum. The change of momentum of the grating can in principle not be observed since we considered the grating to be an idealised, rigid body at rest, that is, a body which can give us the geometry of the experiment with infinite precision and which therefore has an infinite mass. Purely classically speaking this means that finite changes of momentum can not be observed (compare the remark in §7). After the experiment is over one can thus with impunity assume that a well defined transfer of momentum between particle and grating has taken place. According to Bohr's complementary principle one cannot describe this interaction any more precisely, that is, it has no meaning to ask through which slit the particle has passed or when (and where) the transfer of momentum took place. This is an example of the ''irrational'' aspects of the interaction between apparatus and object which is typical of quantum theory.

CHAPTER II

NON-RELATIVISTIC QUANTUM THEORY OF BOUND PARTICLES

§ 10. Bound particles in classical physics

The simplest case of a bound particle in classical physics is that of a single mass point under the influence of external forces. If the force is $\boldsymbol{F}$, the equations of motion are

$$m\ddot{\mathbf{x}} = \dot{\boldsymbol{P}} = \boldsymbol{F}, \tag{1}$$

as long as we are treating the problem non-relativistically.

We now put the question whether, and if so, how this problem can be treated in quantum theory. Following up our treatment of the theory of free particles we shall assume that the situation can be described at all times by a complex wave function and that once the force is known one can calculate uniquely the change of the wave function with time. In view of the correspondence principle it is to be expected that the application of our formalism to a situation described in terms of a special wave packet (wave group) will give the following result as a first approximation. The wave packet will move through space in such a way that its acceleration is that given by Newton's law (1). If $\boldsymbol{F} = 0$, we get in classical physics Galilei's law of inertia. We saw in § 6 that this law was only asymptotically valid in quantum theory corresponding to the lack of definition of the velocity of the wave packet (which is gradually smearing out). Analogously Newton's law, force = mass times acceleration, will in quantum theory have only asymptotic validity, corresponding to a lack of definition of the acceleration.

In the case where the external forces can be derived from a potential function, depending on the space co-ordinates, and perhaps also on the time, and *only* in this case *, it is possible to satisfy the above requirements simply and uniquely, namely, by writing down the Schrödinger equation which describes the varia-

* This fact shows how fundamental a role is played by the conservation of energy in quantum theory.

tion of the wave function in time. If $U(\boldsymbol{x}, t)$ be the potential function, we have

$$\boldsymbol{F} = -\nabla U, \tag{2}$$

and the classical equations of motion can be put in canonical form

$$\dot{p}_k = -\partial H/\partial q_k, \quad \dot{q}_k = \partial H/\partial p_k, \qquad k = 1, 2, 3, \tag{3}$$

with

$$\left.\begin{aligned} &q_1 = x,\ q_2 = y,\ q_3 = z;\ p_1 = P_x,\ p_2 = P_y,\ p_3 = P_z;\\ &H(p_k, q_k) = T+U,\ T = (P_x^2+P_y^2+P_z^2)/2m. \end{aligned}\right\} \tag{4}$$

§ 11. THE SCHRÖDINGER EQUATION AND ITS CONNEXION WITH THE HAMILTON EQUATION

From the Hamilton function $H(p_k, q_k, t) \equiv H(\boldsymbol{P}, \boldsymbol{x}, t)$ we can form a so-called *energy operator* by the prescription (compare eq. (1.60))

$$p_k \rightarrow -i\hbar(\partial/\partial q_k), \quad k = 1, 2, 3. \tag{5}$$

This energy operator is then of the form

$$H_{\text{op}} \equiv H(-i\hbar\partial/\partial q_k, q_k, t) \equiv H(-i\hbar\nabla, \boldsymbol{x}, t). \tag{6}$$

If we denote once again by ψ the (complex) wave function, we get for the Schrödinger equation of our problem the following homogeneous linear differential equation

$$H_{\text{op}}\psi = i\hbar\partial\psi/\partial t, \tag{7}$$

which equation can be written as

$$-(\hbar^2/2m)\nabla^2\psi+U\psi = i\hbar\partial\psi/\partial t. \tag{8}$$

We first of all notice that this equation reduces to the Schrödinger equation for a free particle (1.62) by taking $U = 0$. Next we see that it is also a natural generalisation of the case of a free particle when U is constant in a given region of space and in time, albeit different from zero. In this case if is easy to generalise the considerations of § 8. Since the particle behaves in this region as if it were free, one concludes that the wave function can again be constructed by the superposition of monochromatic plane waves. However, the dispersion law for these laws is no longer equation (1.16), since the total energy of the particle is now equal to the kinetic energy plus the constant amount U. Instead of equations

(1.12) and (1.16) we now have

$$E_{\text{n.r.}} = (\boldsymbol{P}^2/2m)+U, \tag{9}$$

$$\nu_{\text{n.r.}} = (h\boldsymbol{\sigma}^2/2m)+(U/h). \tag{10}$$

The prescription (1.60) and (1.61) for obtaining the differential equation for the wave function from the relation between momentum and energy leads to equation (8) when applied to equation (9). At the same time we notice that the generalisation from constant to varying U is obtained most simply by accepting equation (8), which is valid in the former case, to be valid also in the latter case.

The prescription (5) and (7) by which we obtain the Schrödinger equation from the Hamiltonian is practically the same as the one which was used in the case, of a free particle. In both cases we consider a polynomial in the components of the momentum to be an operator by considering each momentum component to be a demand for partial differentiation with respect to the canonically conjugate space co-ordinate (plus multiplication by $-i\hbar$). There is still the difference that in the previous chapter the energy could be interpreted explicitly as $i\hbar\partial/\partial t$, that is, the time was treated in the same way as the space co-ordinates with the negative of the energy as its canonical conjugate. However, one can get rid of this difference by a small purely formal modification of the canonical equations (3) and (4). One introduces a fourth pair of canonically conjugate variables q_0, p_0 where q_0 is the time, and one writes the equations of motion in the form

$$\begin{gathered} dq_k/d\tau = \partial\mathscr{H}/\partial p_k, \quad dp_k/d\tau = -\partial\mathscr{H}/\partial q_k \quad (k=0,1,2,3); \\ \mathscr{H} = H+p_0. \end{gathered} \tag{11}$$

If t is contained explicitly in H, it has to be replaced by q_0. From $dq_0/d\tau = \partial\mathscr{H}/\partial p_0 = 1$, it follows that the new parameter τ is equal to q_0, that is, the time coordinate apart from an unimportant additive constant. The equations (11) for $k = 1, 2, 3$ are thus identical with equation (3). The meaning of p_0, which varies in time according to the equation $\dot{p}_0 = dp_0/d\tau = -\partial\mathscr{H}/\partial q_0 = -\partial H/\partial t$, follows most easily from the following consideration. According to equations (11) $\mathscr{H}$ will not depend on τ, and thus not on the time. This means that the absolute value of $\mathscr{H}$ is immaterial to the motion of the particle and can always be chosen to be equal to zero. This last choice fixes the undetermined additive constant in p_0 in such a way that always $p_0 = -H$, or, the momentum p_0

which is conjugate to the time q_0 is always equal to the negative value of the energy function.

One can express this as follows. The equations of motion of the system are written down in the canonical form using the Hamilton equation (instead of using a Hamiltonian),

$$\mathscr{H}(q_k, p_k) \equiv H + p_0 = 0. \qquad (k = 0, 1, 2, 3) \tag{12}$$

The negative of the energy p_0 and the time q_0 appear here as completely equivalent to the other pairs of canonical variables.

Especially with regard to the relativistic treatment of problems in mechanics we want to draw attention to the following. If $\mathscr{H}'(\neq \mathscr{H}) = 0$ is a consequence of $\mathscr{H} = 0$, which means that everywhere in $p_0, p_1, \ldots, q_2, q_3$-space where $\mathscr{H} = 0$ one has also $\mathscr{H}' = 0$ (special cases are $\mathscr{H}' = \mathscr{H}^2$ or $\mathscr{H}' = \mathscr{H}\mathscr{G}$ with arbitrary $\mathscr{G}$), equations (11) will be equivalent to the equations

$$dq_k/d\tau' = \partial\mathscr{H}'/\partial p_k, \quad dp_k/d\tau' = -\partial\mathscr{H}'/\partial q_k, \qquad k = 0, 1, 2, 3. \tag{11'}$$

The parameter τ' is now no longer the same as τ, that is, as t, and its meaning depends on the choice of $\mathscr{H}'$.

The proof of the above statement follows from the fact that one can obtain a solution of equations (11) by solving the following problem from the calculus of variations. Determine on the hypersurface $\mathscr{H} = 0$ in $p_0 \ldots q_3$-space those curves for which the integral $\int \Sigma p_k dq_k$ (end points with fixed q_k-values) is an extremum. These curves are clearly also described by equations (11′).

It is clear that one can also obtain the Schrödinger equation (7) by using the following prescription. The left hand side of the Hamilton equation (12) is made into an operator $\mathscr{H}_{\text{op}}$ by the procedure

$$p_k \to -i\hbar\partial/\partial q_k, \qquad k = 0, 1, 2, 3, \tag{13}$$

which for the special case of $k = 0$ reduces to $p_0 \to -i\hbar\partial/\partial q_0$, or, $E \to i\hbar\partial/\partial t$. One now asks the wave function ψ to satisfy the equation

$$\mathscr{H}_{\text{op}}\psi = 0. \tag{14}$$

One sees immediately that this derivation of the Schrödinger equation is formally the same as the derivation in the previous chapter. We then used the relation (1.59) between momentum and energy, but this relation is just the Hamilton equation for a free particle.

§ 12. The motion of wave groups under the influence of external forces

We want to show in this section that asymptotically the motion of wave groups which satisfy equation (8) obey Newton's law for the motion of a mass point. As in § 4 we describe the wave group at $t = 0$ by the following wave function

$$\psi_0 = C(\boldsymbol{x}) \exp [2\pi i(\boldsymbol{\sigma}^0 \cdot \boldsymbol{x})], \tag{15}$$

where C is a smooth function which takes on appreciable values only in a region u in the immediate neighbourhood of $\boldsymbol{x}_0$. We now recollect that if U were constant and equal to U_0 inside u, the particle would behave as if it were free so that to a first approximation equation (1.42) would be valid. This means that after a short time interval Δt the wave group has moved a distance $\boldsymbol{v}\Delta t$, where $\boldsymbol{v}$ is the classical velocity of a particle with momentum $h\boldsymbol{\sigma}^0$. Its wave function would then be

$$\psi'_{\Delta t} = C(\boldsymbol{x}-\boldsymbol{v}\Delta t) \exp [2\pi i\{(\boldsymbol{\sigma}^0 \cdot \boldsymbol{x})-\nu^0\Delta t\}], \tag{16}$$

where $h\nu^0-U_0$ is equal to the classical kinetic energy. In the case where U varies inside u we write

$$U-U_0 = ([\boldsymbol{\nabla} U]_0 \cdot [\boldsymbol{x}-\boldsymbol{x}_0]), \tag{17}$$

where $U_0 = U(\boldsymbol{x}_0)$, and we assume that we do not need to take any higher order terms into account.

Writing down equation (8) once for ψ', that is, with $U = U_0 =$ constant, and once for ψ, using the fact that at $t=0$ $\psi = \psi' = \psi_0$, we get up to terms of the first order in Δt, $\psi_{\Delta t} = \psi'_{\Delta t} - (i/\hbar)(U-U_0)\psi_0\Delta t$, which to the same approximation can be written as $\psi_{\Delta t} = \psi'_{\Delta t} \exp [-(i/\hbar)(U-U_0)\Delta t]$, which means that the difference between $\psi_{\Delta t}$ and $\psi'_{\Delta t}$ can be described as a change of phase. Both the phase of $\psi'_{\Delta t}$ [see equation (16)] and $U-U_0$ are linear in $\boldsymbol{x}$, so that the same is true for the phase of $\psi_{\Delta t}$ and the terms in $\boldsymbol{x}$ in this phase are of the form $2\pi i(\boldsymbol{x} \cdot \{\boldsymbol{\sigma}^0-(\Delta t/h)[\boldsymbol{\nabla} U]_0\})$.

From the quantum theoretical interpretation of the phase this means that the momentum of the wave packet (which is only approximately defined) has changed by an amount $-(\boldsymbol{\nabla} U)_0\Delta t$, that is, by exactly the same amount which in classical physics corresponds to the action of a potential on a particle at $\boldsymbol{x}_0$.

One must take this change of the group velocity into account during a subsequent time interval $(\Delta t, 2\Delta t)$. The wave group will thus show in its motion — as far as this motion is defined — the classical acceleration.

If U does not depend on t, ν^0 in equation (16) will be equal to the clasical energy, which is a constant, divided by h. The variation of the wave function with t is thus still mainly described by a factor

$\exp(-2\pi i\nu^0 t)$, at any rate at those points where the wave function is appreciably different from zero.

§ 13. The physical meaning of the wave function

We come now to the important question of the physical meaning of the wave function ψ. The situation is as follows. From a number of physical measurements and connected theoretical considerations one has reached the conclusion that in some cases one is dealing with a physical system which classically would be described as a mass point in a field of force (for instance, a hydrogen atom considered as an electron in the electrical field of a proton). Classically the situation can be described at any moment by giving the values of six parameters, either the three position coordinates and the three momentum coordinates, or six suitably chosen functions of these variables. The causal character of Newton's law enabled one in principle to determine the values of these parameters at any time from their values at one given time. From the considerations of the case of a free particle it follows immediately that in quantum theory one should not expect to be able to define precisely six such values at a given time. All the same one must take into account that in interpreting an experiment which is used to investigate the properties of the system one can characterise the interaction between apparatus and system only by using classical concepts. In particular one uses the terms position, or momentum, or velocity, or angular momentum of a particle in a field of force, since we have no other attributes at our disposal.

In general a finite time interval will be involved in the interaction between system and apparatus and even classically in order to describe the interaction one would need to introduce explicitly the potential fields in the system and the forces of interaction. Using the process of physical abstraction one can, however, consider idealised experiments which are particularly suited to investigate and define the physical state of a system at a precisely determined moment. We remind ourselves of the experiment where suddenly a region of space is surrounded by fixed walls and then tested for the presence of a particle (see § 7). Another example would be the sudden lifting of the field of force to enable us to measure the velocity or momentum of the particle. If one feels the introduction of such "instantaneous experiments" to be justified also in quantum theory — if not, one should reconsider what is meant by the statement "the system consists of a particle in a field of force" — one is led to the following statement. *The physical situation in a mechanical system is characterised by a wave function ψ in such a way that this function enables us to enunciate probability laws for the distribution of any mechanical quantity over its possible values.* In arriving at this statement both the interpretation of the wave function of a free particle and the expectation of the occurrence of probability laws have played a role. A mechanical quantity or *observable* is any quantity which has a meaning in the corresponding classical problem; such a quantity will in general be a function of the position and momentum coordinates.

It has turned out to be possible to develop the theory — at any rate as long as we restrict the discussion to the non-relativistic case — in such a way that the wave function which satisfies the Schrödinger equation describes completely the physical situation at any moment. This is to be understood in the sense that it can give an answer in the form of a probability statement to all questions about the situation which would have made sense classically. The general treatment of this problem will follow in chapter 4 where we shall also discuss (§ 43) the time dependence of mechanical quantities. At the moment we shall discuss the probability laws only for position and momentum.

§ 14. Probability density and probability current density

First of all we want a probability function $J(\mathbf{x})$ which can be interpreted as a probability density. This means that, as in the case of a free particle, $J d\mathbf{x}$ should be the probability that the particle at time t is within the volume element of space $d\mathbf{x}$. The conditions which must be satisfied by J can again be met by the definition

$$J = \psi^* \psi = |\psi|^2. \tag{18}$$

To begin with, J is always positive. Secondly we must demand that the normalisation condition

$$\int J d\mathbf{x} = \int \psi^* \psi \, d\mathbf{x} = 1, \tag{19}$$

where the integral extends over the whole of space, is satisfied. Using the Schrödinger equation (8) one can, however, easily prove that, provided the integral $\int J d\mathbf{x}$ is convergent, it is independent of time so that equation (19) will always be satisfied as soon as it is satisfied at one time.

The proof is analogous to the one in § 8. One considers the imaginary part of the equation obtained by multiplying the Schrödinger equation by ψ^*,

$$-(\hbar^2/2m)\psi^*\nabla^2\psi+U\psi^*\psi = i\hbar\psi^*\partial\psi/\partial t.$$

Substracting from this equation its conjugate complex and multiplying the result by $i/\hbar$ we have

$$(\hbar/2mi)(\psi^*\nabla^2\psi-\psi\nabla^2\psi^*) = (\nabla\cdot[\hbar/2mi][\psi^*\nabla\psi-\psi\nabla\psi^*]) = -\partial\psi\psi^*/\partial t,$$

or,

$$(\nabla\cdot \boldsymbol{S})+\partial J/\partial t = 0, \tag{20}$$

where $\boldsymbol{S}$ is defined by the equation

$$\boldsymbol{S} = (\hbar/2mi)(\psi^*\nabla\psi-\psi\nabla\psi^*), \tag{21}$$

which is the same equation as expression (1.55).

Integrating equation (20) over the whole of space it follows that $\int J d\boldsymbol{x}$ is independent of time, provided the integral converges and provided that at the same time $\boldsymbol{S}$ vanishes sufficiently rapidly at infinity [compare eq. (1.51)].

The existence of the equation of continuity (20) leads us immediately to the physical interpretation of $\boldsymbol{S}$ as the probability current density, just as in the case of a free particle. This means that $(\boldsymbol{S}\cdot d\boldsymbol{f})dt$ is the probability that the particle within the time interval dt passes the surface element $d\boldsymbol{f}$ in the direction of the positive normal minus the probability that it passes in the opposite direction.

An idealised experiment to test the passing of a particle through $d\boldsymbol{f}$ is strictly speaking not one of the "instantaneous experiments" discussed in the previous section since in principle such an experiment will stretch over a finite, even if small, period. The fact that nevertheless the instantaneous value of ψ enters into expression (21) is clearly connected with the fact that classically the velocity of the particle depends only on the instantaneous momentum and not on the force. We refer to § 44 for a more general discussion of the probability current density.

One often interprets the probability density J as follows. The particle is no longer localised at a well defined point of space, but it is, so to say, extended over the whole of space according to the density function J. If it possesses a charge e, one says that this charge is smeared out over space in a definite way, and one speaks of a continuous *charge cloud* with a *charge density* eJ and a *current density* $e\boldsymbol{S}$.

Although this picture often fits naturally into the mathematical description of the action of the particle on other particles or on external systems (compare § 52), it is tempting to introduce too easily a much too simplistic interpretation of quantum mechanics. This interpretation uses models which can be described in classical terms, but does not do justice to the basic characteristics of quantum theory. Especially we want to warn against the idea that one must abandon the corpuscular nature of elementary particles and that according to quantum theory one must instead consider a theory with continuous charge and current densities.

§ 15. The momentum probability distribution

We want now to consider the function which will give us the probability that the momentum of the particle at a given moment lies in the interval $(\boldsymbol{P}, \boldsymbol{P}+d\boldsymbol{P})$. Heuristically one has a very simple solution to this question, a solution which can be retained in the further development of the theory. Once again we expand the wave function $\psi_0(\boldsymbol{x})$ at $t = 0$ in a Fourier integral *,

$$\psi_0 = \int A(\boldsymbol{\sigma}) \exp\left[2\pi i(\boldsymbol{\sigma}\cdot\boldsymbol{x})\right] d\boldsymbol{\sigma}, \tag{22}$$

where the integration extends on the whole of wavenumber space. We then have the equations [compare eqq. (1.28) and (1.48)]

$$A = \int \psi_0 \exp\left[-2\pi i(\boldsymbol{\sigma}\cdot\boldsymbol{x})\right] d\boldsymbol{x}, \tag{23}$$

$$\int \psi^*\psi d\boldsymbol{x} = \int A^* A d\boldsymbol{\sigma} = 1. \tag{24}$$

If at $t = 0$ the field of force is suddenly made to vanish, one is dealing with a free particle and one can use interference experiments — which must be repeated several times under the same conditions — to measure the wave numbers. From the existence of a wave vector with the well defined value $\boldsymbol{\sigma}$ one concludes the occurrence of a corresponding momentum $\boldsymbol{P} = h\boldsymbol{\sigma}$. According to equation (22) one will find a spread of values of $\boldsymbol{\sigma}$. Their relative weight, and thus also the momentum probability distribution, is determined by the amplitude function $A(\boldsymbol{\sigma})$. If we take the analogous classical problem of the energy distribution of a wave motion into account, it follows from equation (24) that the only possible assumption is that

$$A^* A d\boldsymbol{\sigma} = (|A|^2/h^3) d\boldsymbol{P}, \tag{25}$$

* We write $d\boldsymbol{\sigma}$ for a volume element $d\sigma_x d\sigma_y d\sigma_z$ in wave number space and similarly in equation (25) $d\boldsymbol{P}$ for $dP_x dP_y dP_z$.

is the probability that the momentum lies within the range $(\boldsymbol{P}, \boldsymbol{P}+d\boldsymbol{P})$. The function A follows immediately from ψ through equation (23). Other possibilities must be excluded, since A is completely arbitrary, apart from the normalisation condition (24). Moreover, a closer analysis of the interference experiments mentioned a moment ago leads uniquely to equation (25), if we use the laws for free particles. Mathematically this analysis will be the same as that by which, for instance in optics, one clarifies the physical significance of the distribution of the energy of light over the different wave lengths.

One can in principle consider an experiment which derives information about the instantaneous momentum of a particle by letting it collide with another particle (or possibly a light quantum). If it is possible to make the period during which the collision takes place sufficiently small, and if one can determine sufficiently accurately the momentum of the other particle before and after the collision, one can obtain information about the original momentum of the first particle by using the conservation laws in interpreting the experiment, which must be repeated several times under the same conditions. We do not want to discuss the analysis of such experiments any further. These experiments are less academic than the above mentioned idealised experiment where the field of force was suddenly made to vanish. However, they are less direct and in principle not any different, at any rate not as long as we consider them from the point of view of the present day non-relativistic quantum theory which we are discussing here.

In the case of a free particle A and thus the momentum distribution function is constant in time. In the case of a particle in a field of force A will in general vary with time. This corresponds in classical physics to the fact that the momentum is constant in the first case, but can change in the second case.

§ 16. The uncertainty relations; the uncertainty in energy

As far as the simultaneous definability of position and momentum of a particle is concerned, the situation is clearly not different from the case of a free particle. If, for instance, $|\psi_0|^2$ differs appreciably from zero only in a small interval Δx, that is, if the x-coordinate of the particle is uncertain with an uncertainty Δx, then the uncertainty ΔP_x in the x-component of its momentum must be at least of the order $h/\Delta x$. In short, as far as position and momentum coordinates are concerned the Heisenberg relations (1.41) which we discussed in § 5 are still valid.

The situation is slightly different as far as the fourth uncertainty relation (1.46) is concerned. This relation expressed the reciprocity of the definability of energy and time,

$$\Delta E \,.\, \Delta t \gtrapprox h. \tag{26}$$

In the case of a free particle where the energy depended only on the momentum this relation could be derived simply as a consequence of the first three relations and was, for instance, the uncertainty with which we could define the moment that the particle passed a given surface (§ 5). As soon as there is a field of force, equation (26) can be interpreted in so simple a manner only if we interpret ΔE as the uncertainty ΔT of the kinetic energy T and if we at the same time allow only values of Δt which are so small that within Δt the force does not appreciably influence T. All the same one can attribute a general validity to equation (26) in close connexion with the general validity of the law of conservation of energy. Generalising the quantum mechanical treatment of free mass points and taking into account the considerations of § 12 we introduce the following basic postulate of quantum theory. *The definability of the energy of a system is always connected according to* $E = h\nu$ *with the definability of the frequency of a wave phenomenon.* From this postulate it follows formally (compare § 33) that if for some problem a physical system "occurs" or "is considered" only during a finite period of time Δt, one can define the frequency of any wave phenomenon connected with this system only within a margin $(\Delta t)^{-1}$ *. The energy can thus be defined only with a margin $\Delta E = h\Delta\nu = h/\Delta t$. The physical significance of this relation is that *if in interpreting a physical process one assigns an energy E to a system at time t, the product of the margins within which E and t can be given must be at least of the order h.* **

§ 17. Energy eigenvalues and eigenfunctions

Consider now the energy of the system consisting of a mass point

* This corresponds to the mathematical theorem that the Fourier decomposition of $C(t) \exp(-2\pi i\nu t)$, where $C(t)$ is a smooth function which is appreciably different from zero only in the interval $t_0 < t < t_0+\Delta t$, will contain frequencies only in a neighbourhood of ν of the order $\Delta\nu = (\Delta t)^{-1}$ (compare the discussion in § 33).

** Especially it follows that the moment when a system changes its energy through interaction with another system will be the less well defined the more sharply the change in energy is defined.

in a field of force and apply the quantum postulate $E = h\nu$. A physical situation corresponding to a state of well defined energy must according to this postulate have a wave function which is periodic in time with a frequency $\nu = E/h$. We write thus

$$\psi(\boldsymbol{x}, t) = \varphi(\boldsymbol{x}) \exp[-iEt/\hbar). \tag{27}$$

The time dependent exponential does correspond not only to the way in which time occurred in the wave function of a free particle [compare eqq. (1.27) and (1.57)], but also to the approximate description in § 12 of the motion of a wave packet in a field of force [compare eq. (16)]. Introducing expression (27) into the Schrödinger equation (8) we obtain for the time independent function φ the differential equation

$$[(-i\hbar)^2/2m]\boldsymbol{\nabla}^2\varphi + U\varphi = E\varphi, \text{ or, } \boldsymbol{\nabla}^2\varphi + (2m/\hbar^2)[E-U]\varphi = 0. \tag{28}$$

This differential equation has clearly a meaning only when U does not depend on time, that is, when the field of force is constant in time. It is called the *time independent Schrödinger equation* which was derived by Schrödinger* in his first paper on wave mechanics. The time dependent equation (8) followed later **.

In general, for arbitrary U, it is impossible to give a solution of equation (28) in closed form. It presents us usually with a complicated mathematical problem. However, one can obtain a general insight into the nature of the solutions for a certain class of U-functions and draw attention to peculiarities which are of the greatest importance for quantum theory. Especially it appears that *in a given case practically always only a small fraction of all mathematically possible solutions can have a physical meaning. These physically significant solutions correspond to definite values or intervals of E. The corresponding E-values are called "energy eigenvalues" and the corresponding solutions the "Schrödinger eigenfunctions."*

We can immediately find a criterion for the physical significance of a solution of equation (28). One must expect that the wave function $\psi = \varphi \exp(-iEt/\hbar)$ will satisfy the normalisation condition,

$$\int |\psi|^2 d\boldsymbol{x} = \int |\varphi|^2 d\boldsymbol{x} = 1, \tag{29}$$

since only in that case can ψ be interpreted to characterise the

* E. Schrödinger, *Ann. Phys.*, **79**, 361, 1926.
** E. Schrödinger, *Ann. Phys.*, **81**, 109, 1926.

physical situation in a field of force. If one finds for a given E-value a solution for which $\int |\varphi|^2 d\boldsymbol{x}$ is finite, albeit different from 1, one can obtain a normalised solution by multiplying φ by a suitable factor.

The situation is the following for all cases of a one-particle problem which occur in practice. The eigenvalues E for which equation (28) possesses a normalisable solution form a finite or infinite discrete set, that is, if E_0 is one of those eigenvalues, then one can find a finite interval $E_0-\delta < E < E_0+\delta$ (δ positive) such that within that interval E_0 is the only eigenvalue corresponding to a normalisable solution.

We shall illustrate this by discussing a simple case. First of all, we restrict ourselves to the one-dimensional equation which is mathematically analogous to equation (28) but which is much simpler,

$$\varphi''+C(E-U)\varphi = 0, \quad \varphi'' = d^2\varphi/dx^2, \quad C = 2m/\hbar^2, \tag{30}$$

where U is a given "potential function" of the one variable x, while φ also depends only on x, and C is a positive constant. We need consider only real φ as both the real and the imaginary part of a complex solution φ will be solutions of equation (30). We look for solutions for which $\int_{-\infty}^{+\infty} \varphi^2 dx$ converges.

Let now U be a smooth function with just one minimum at $x = 0$ for which $U(0) = 0$ and let $U \to +\infty$ in some way for both $x \to +\infty$ and $x \to -\infty$. The potential of a one-dimensional harmonic oscillator, $U = \alpha x^2$, is a concrete example of such a function. From the equation $\varphi''/\varphi = -C(E-U)$ one sees that for any x for which $\varphi(x) \neq 0$, the curve representing $\varphi(x)$ will be convex or concave towards the x-axis according to whether $C(E-U)$ is negative or positive. If $C(E-U)$ were equal to $-a^2$ (a a real constant), equation (30) would have two solutions $\exp(\pm ax)$. If, on the other hand, $C(E-U)$ were equal to $+a^2$, the two solutions would be $\cos ax$ and $\sin ax$. We now introduce the following nomenclature: in a region where $E-U < 0$, φ behaves "exponentially", while in a region where $E-U > 0$, φ behaves "oscillatorily." For a positive value of E, $E-U$ will have two zeros, x_1 and x_2 ($x_1 < 0 < x_2$, see fig. 1) which we call the classical turning points, since in the corresponding classical problem a particle with energy E would go up and down between x_1 and x_2. In region I, $-\infty < x < x_1$, and in region III, $x_2 < x < \infty$, φ behaves exponentially, and in region II, $x_1 < x < x_2$, oscillatorily. At x_1 and x_2 the φ-curve has points of inflexion. Let now at a point x_0 in region I the values of φ and φ' be given. The solution of equation (30) is then completely determined. For fixed $\varphi(x_0)$ the behaviour of the φ-curve to the left of x_0 depends clearly on the value of $\varphi'(x_0)$. If $\varphi'(x_0)$

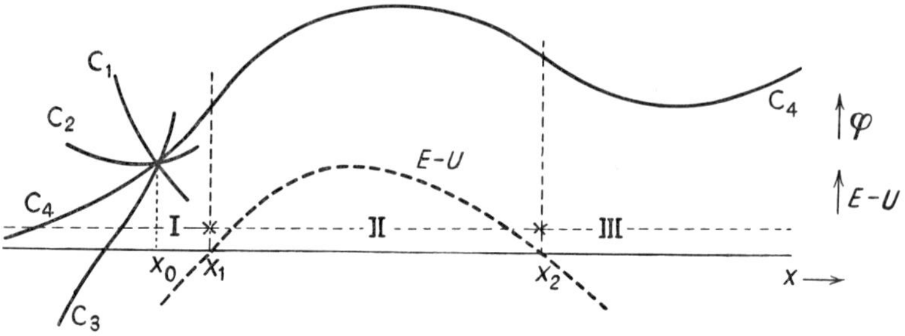

Fig. 1.

is negative or positive, but not very large, $\varphi(x)$ will go to $+\infty$ for $x \to -\infty$, without ever becoming negative (curves C_1 and C_2 in fig. 1). If, however, $\varphi'(x_0)$ is positive and large, the φ-curve will cut the x-axis and go to $-\infty$ for $x \to -\infty$ (curve C_3 in fig. 1). Obviously there is exactly one value of $\varphi'(x_0)$ for which $\varphi(x)$ becomes asymptotically equal to zero for $x \to -\infty$ (curve C_4) and this will be the only solution which can correspond to a normalisable φ.

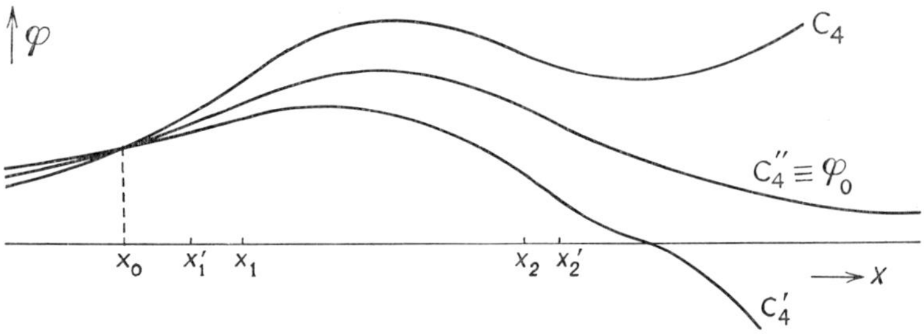

Fig. 2.

Every other solution which goes to zero for $x \to -\infty$ will differ from this solution only by a constant factor. This factor is determined by the choice of $\varphi(x_0)$. The fact that one can obtain another solution by multiplying a given solution by a constant follows from the fact that equation (30) is a linear homogeneous one. Let us now see how C_4 behaves for $x > x_1$. At x_1 it has a point of inflexion and its behaviour is oscillatory in region II. We have chosen a value of E so small that the curve does not cut the x-axis in region II. Its behaviour in region III is again exponential, and in general such that $\varphi \to +\infty$ or $-\infty$ for $x \to +\infty$. This means that in general the solution is not normalisable. In figure 1 C_4 corresponds to an E-value so small that in region III φ stays everywhere positive and goes to $+\infty$. If we now choose E slightly larger (fig. 2, curve C_4'), the turning points will be further apart (x_1' and x_2'), the curvature of the φ-curve will be larger in region II, and the value of φ' in x_2' will now be smaller than its value in x_2 was in the previous case. The curve C_4' in figure 2 corresponds to the case where the φ-curve cuts the x-axis in region III and goes to $-\infty$ for $x \to +\infty$. Obviously, there exists an E-value, E'', between E and E' such that the corresponding C_4''-curve just goes to zero for $x \to \pm\infty$. This particular value of E clearly corresponds to a normalisable solution, since any solution which goes to zero for $x \to \pm\infty$ should do so exponentially, when $U \to \infty$ for $x \to \pm\infty$.

We have in this way shown qualitatively the existence of a discrete eigenvalue and its eigenfunction. One sees easily that there is no eigenvalue smaller than the one just found. However, there are larger eigenvalues and an infinite number of them. One sees this easily by constructing for increasing values of E solutions of equation (30) which are analogous to C_4. The next normalisable solution will be one for which the φ-curve is curved so strongly in region II that

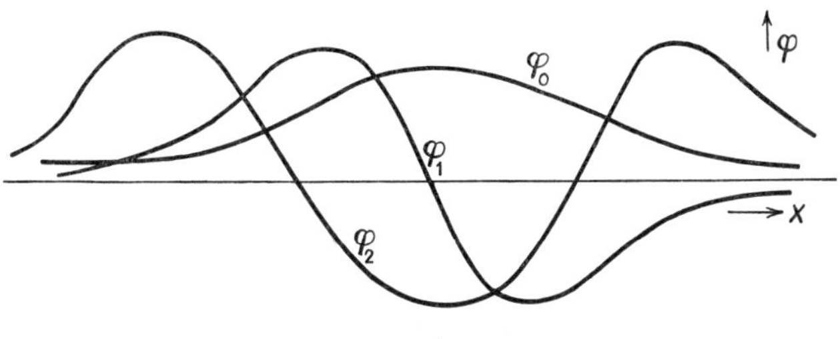

Fig. 3.

it cuts the x-axis, is still negative at the second turning point, stays negative in region III and finally approaches the x-axis asymptotically. This eigenfunction has one zero. Similarly there will be eigenvalues corresponding to solutions with 2, 3, . . ., n, . . . zeros (fig. 3). We can use the number of zeros as index, in such a way that the eigenvalue E_n corresponds to an eigenfunction φ_n

with exactly n zeros between the turning points. We have clearly that $E_0 < E_1 < E_2 < \ldots$ and $E_n \rightarrow \infty$ for $n \rightarrow \infty$.

In our example, U had only one minimum. However, one sees easily by using the same qualitative considerations that for any U which goes to ∞ for $x \rightarrow +\infty$ and for $x \rightarrow -\infty$, there exists a set of discrete eigenvalues E_n which increase with increasing value of n and go to ∞ for $n \rightarrow \infty$. The index n again gives the number of zeros of the corresponding φ.

A particularly simple case is the one of the square box (fig. 4), where $U = \infty$ for $x \leqq x_1$ and $x \geqq x_2$ and $U = 0$ for $x_1 < x < x_2$. We can consider this to be the limiting case of a function U' which

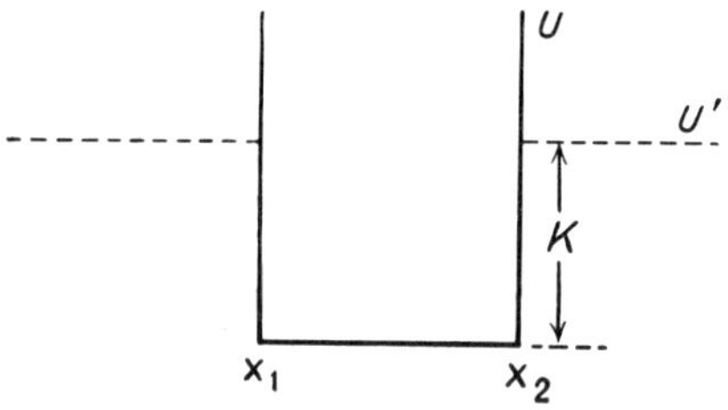

Fig. 4.

is zero for $x_1 < x < x_2$, but which is equal to a constant K for $x \leqq x_1$ or $x \geqq x_2$, as K goes to ∞. We can now solve equation (30) exactly in all these regions I, II and III. At x_1 and x_2, φ and φ' will be continuous, but φ'' will be discontinuous. In region I we have $\varphi = A \exp\{x[C(K-E)]^{1/2}\}$ from which follows $\varphi'(x_1)/\varphi(x_1) = [C(K-E)]^{1/2}$. In order that φ stays finite in region II we must ask that for $K \rightarrow \infty$, $\varphi(x_1) \rightarrow 0$. The classical one-dimensional problem which corresponds to this U is the presence of two idealised fixed walls at x_1 and x_2 between which the particle is reflected. Formally the corresponding quantum mechanical case is the mathematical problem of finding a solution of the wave equation of a free particle which satisfies the boundary conditions $\varphi(x_1) = \varphi(x_2) = 0$ *. This problem is identical with the classical problem of stationary vibrations of a homogeneous string which is fixed at x_1 and x_2. If we put $x_1 = 0$, $x_2 = a$, the solution is

$$\varphi_n = A \sin(\pi n x/a), \quad n = 1, 2, \ldots \tag{31}$$

* Similarly one sees easily that the problem of a free particle moving among rigid bodies entails the solution of the Schrödinger equation under the condition that the wave function is zero at the surface of these bodies.

The wave number has the value $\sigma = n/2a$ and the energy eigenvalues are given by the equation

$$E = h^2\sigma^2/2m = h^2n^2/8ma^2, \tag{32}$$

where now the index n is chosen so as to be one larger than the number of zeros of φ_n between 0 and a.

The second example we shall consider is that of the so-called "one-dimensional harmonic oscillator." The potential function is simply $U = \alpha x^2$ and equation (30) has the form $\varphi'' + C(E - \alpha x^2)\varphi = 0$. One can show * that this equation possesses the following solutions

$$\begin{aligned} \varphi_n &= \exp(\tfrac{1}{2}\xi^2)D^n \exp(-\xi^2) = H_n(\xi^2)\exp(-\tfrac{1}{2}\xi^2), \quad D = d/d\xi, \\ \xi &= (C\alpha)^{1/4}x, \quad E_n = (n+\tfrac{1}{2})(4\alpha/C)^{1/2}, \qquad n = 0, 1, 2, \ldots \end{aligned} \tag{33}$$

It is easily checked that these functions φ_n possess the properties of eigenfunctions. They go to zero for $|x| \to \infty$ and have n zeros. The function $\exp(-\xi^2)$ has no zeros, $D\exp(-\xi^2)$ one, $D^2\exp(-\xi^2)$ two, and so on. If one writes down explicitly the n-th derivative of $\exp(-\xi^2)$ one obtains the second form of φ_n, where the $H_n(\xi)$ are polynomials of the n-th degree with only odd or only even powers of ξ. They are the so-called Hermitean polynomials. The eigenvalues E_n are equidistant at distances $(4\alpha/C)^{1/2}$.

According to classical mechanics a particle of mass m in a potential αx^2 will show a harmonic oscillation with a frequency ω independent of its amplitude and equal to $(\alpha/2\pi^2 m)^{1/2}$. Substituting for C its value $2m/\hbar^2$ one finds

$$\xi = 2\pi x(m\omega/h)^{1/2}, \quad E_n = (n+\tfrac{1}{2})h\omega. \tag{34}$$

Using considerations similar to those used up to now we can also discuss the occurrence of eigenvalues and eigenfunctions for certain classes of three-dimensional problems. Let us consider the case where the potential energy U can be split up into three functions which each depend only on one coordinate, $U = U_1(x) + U_2(y) + U_3(z)$. Equation (28) can now be solved by using "separation of coordinates." We try a solution of the form $\varphi = \varphi^{(1)}(x)\varphi^{(2)}(y)\varphi^{(3)}(z)$, and find that each factor satisfies an ordinary differential

* One has $D^{n+2}\exp(-\xi^2) = D^{n+1}[-2\xi\exp(-\xi^2)] = -2\xi D^{n+1}\exp(-\xi^2) - (2n+2)D^n\exp(-\xi^2) \ldots$ (A). Put $D^n\exp(-\xi^2) = \chi$. We then have from equation (A) $\exp(\tfrac{1}{2}\xi^2)[D^2\chi + 2\xi D\chi + (\xi^2+1)\chi] + \exp(\tfrac{1}{2}\xi^2)(2n+1-\xi^2)\chi = 0$, or, $D^2[\exp(\tfrac{1}{2}\xi^2)\chi] + (2n+1-\xi^2)[\exp(\tfrac{1}{2}\xi^2)\chi] = 0$, or, $(C\alpha)^{-1/2}\varphi'' + [2n+1-(C\alpha)^{1/2}x^2]\varphi = 0$.

equation *,

$$\left.\begin{aligned} &\varphi^{(1)''}+(2m/\hbar^2)(E^{(1)}-U_1)\varphi^{(1)} = 0,\\ &\varphi^{(2)''}+(2m/\hbar^2)(E^{(2)}-U_2)\varphi^{(2)} = 0,\\ &\varphi^{(3)''}+(2m/\hbar^2)(E^{(3)}-U_3)\varphi^{(3)} = 0,\\ &E = E^{(1)}+E^{(2)}+E^{(3)}. \end{aligned}\right\} \tag{35}$$

If each of the three U-functions goes to infinity for increasing absolute values of its argument, but otherwise is well-behaved, each of the equations (35) will possess an infinite number of discrete eigenvalues and corresponding eigenfunctions. Let them be denoted by $E_k^{(1)}, E_l^{(2)}, E_m^{(3)}$ and $\varphi_k^{(1)}, \varphi_l^{(2)}, \varphi_m^{(3)}$. We then get the following normalisable solution, that is, eigenfunction, of equation (28)

$$\varphi_{klm} = \varphi_k^{(1)}(x)\varphi_l^{(2)}(y)\varphi_m^{(3)}(z), \tag{36}$$

corresponding to the eigenvalue

$$E_{klm} = E_k^{(1)}+E_l^{(2)}+E_m^{(3)}. \tag{37}$$

The numbers k, l, m can independently take on all integer values from 0 to ∞.

A simple special case is given by the problem of a free particle inside a rectangular box with rigid walls. One has to find solutions of the Schrödinger equation for a free particle which vanish at the walls. Let the edge lengths be a, b, and c. Using equations (31), (32), (36), and (37) one finds that the (unnormalised) eigenfunctions and eigenvalues are

$$\varphi = \sin\frac{\pi kx}{a}\sin\frac{\pi ly}{b}\sin\frac{\pi mz}{c}, \tag{38}$$

$$E = \frac{h^2}{8m}\left(\frac{k^2}{a^2}+\frac{l^2}{b^2}+\frac{m^2}{c^2}\right). \tag{39}$$

§ 18. Stationary states

A physical situation which is characterised by a solution of the Schrödinger equation of the form $\psi = \varphi \exp(-iEt/\hbar)$ with normalisable ψ and which thus in accordance with the quantum postulate $E = h\nu$ corresponds to a well defined energy of the

* Substitution into equation (28) and division by φ leads to the equation

$$[(\varphi^{(1)''}/\varphi^{(1)})-(2m/\hbar^2)U_1]+[\,.\,.\,.\,]+[\,.\,.\,.\,]+(2mE/\hbar^2) = 0,$$

and this equation can be satisfied only when each of the three expressions in square brackets is equal to a constant.

system under consideration is called a *stationary state* of the system. This nomenclature was introduced by Bohr in 1913 when he introduced his famous quantum postulates to interpret line spectra. The first of his postulates demanded the introduction of a number of discrete states, that is, states the energies of which differ by finite amounts, and assumed that the energy of the system in each of these stationary states possessed a well defined value and that each change of energy of the system corresponded to a process whereby the system made a complete transition from one stationary state to another. The second Bohr postulate referred to the case where such a transition was connected with the emission or absorption of electromagnetic radiation. It stated that the emitted or absorbed radiation would be monochromatic and that its frequency multiplied by Planck's quantum of action should be equal to the amount by which the energy changed during the process. Quantum mechanics allows us a more quantitative interpretation and description of Bohr's stationary states. We have just seen what this description is in the case of a mass point in a potential field.

Before the development of quantum mechanics one tried in close connexion with Bohr's original investigations to illustrate the stationary states qualitatively as well as quantitatively by identifying them with certain orbits from classical mechanics. By introducing the so-called *quantisation rules* one seemed able to select from the continuum of classically possible orbits a discrete set of them which could be considered to correspond to stationary states. This method was especially successful for the hydrogen atom. Apart from the fact that there were many points whose physical basis was unclear, it became gradually clear that this method was unsuited to deal quantitatively with complicated atomic systems. Also, it was unable to account for the finer details of the hydrogen spectrum.

In his first investigation of the wave theory of matter in 1924 de Broglie * emphasised the close analogy between the quantisation rules of the Bohr theory and the conditions for interference of wave phenomena. In particular he showed the equivalence of the statement in the old quantum theory that "among all classically possible circular orbits of a particle in a central field of force those orbits

* L. de Broglie, *Thesis*, Paris, 1924.

correspond to stationary states for which the angular momentum is equal to an integer times $\hbar$ "with the statement that in a stationary state the length of the orbit, that is, the circumference of the circle, will be an integral multiple of the wave length corresponding to the classical momentum *.

We owe the wave equation (28) to Schrödinger. This equation enables us to characterise quantitatively the wave phenomenon which corresponds to a stationary state. It shows that the above considerations of de Broglie are correct in principle, but give only an indication of the real situation. Schrödinger was able to show in 1926 in his first paper ** that, indeed, in the case of the hydrogen atom equation (28) possesses discrete eigenvalues E. These eigenvalues are exactly the same as the ones following from the Bohr theory. However, the illustration of the corresponding stationary states by classical orbits seems not to do justice to the physics of the problem and the quantitative success of the quantisation rules seems to be mathematically fortuitous.

We shall later on discuss the application of the Schrödinger equation to the hydrogen atom and to other special one-particle systems and the connexion between the quantum mechanical determination of the stationary states and the application in the old quantum theory of the quantisation rules on classical orbits. At this moment we are concerned only with an indication of how de Broglie's postulate of the wave nature of material particles, quantitatively formulated as a law for wave propagation through the Schrödinger equation, leads in a natural manner to the concept of stationary states which played such an extremely important role in the development of early atomic physics.

At this point we wish to draw attention to two consequences of the wave mechanical interpretation of stationary states which illustrate in more detail the typical difference with the old quantum theory. We start from our previous example (figs. 2 and 3) where we had drawn qualitatively the behaviour of the wave functions $\varphi_0, \varphi_1, \varphi_2, \ldots$ First of all we note that these functions attain values different from zero everywhere, that is, also in the regions outside the classical turning points ($x < x_1$, $x > x_2$), where $E-U$ is negative. From the physical interpretation of the wave function it follows that this means that there exists a finite probability of finding the particle in places which it could never reach in the classical picture when the system possessed the energy corresponding to the stationary state considered, since the kinetic energy would be negative. One should note at the same

* Angular momentum = momentum $\times$ radius = $n\hbar$ is equivalent to: circumference = 2π $\times$ radius = n $\times$ wave length = nh/momentum.

** E. Schrödinger, *Ann. Phys.*, **79**, 361, 1926; compare also § 46.

time, however, that this probability is not very large as φ decreases exponentially, and it decreases with increasing number of zeros. At first it looks as if one meets here with a violation of the principle of conservation of energy. One must, however, take into account that in interpreting an experiment which purports to be able to show the presence of a particle at such a place one must give up the demand of an accurate determination of the momentum, and thus the kinetic energy of the particle. This must be done in such a way that one assumes that the action of the experimental apparatus is accompanied by an uncontrollable exchange of energy with the system.

A second peculiarity which has no analogue in the classical picture is the occurrence of zeros of the wave functions φ_1, φ_2, . . . At such places the particle can never be found according to quantum mechanics while according to classical physics every point between the turning points would be reached.

To complete our present historical interlude we may mention that Heisenberg * in 1925 independent of de Broglie and starting from completely different premises showed a way to modern quantum mechanics. Heisenberg started from the Bohr postulates and Bohr's correspondence principle and consciously avoided introducing any assumptions about the describability of stationary states in terms of classical orbits. He arrived at a method of treating atomic problems which soon afterwards was developed into a definite discipline — especially through a paper by Born and Jordan **. It looked originally as if this discipline — which is often called matrix mechanics as it uses from the start the mathematical methods of matrix algebra — was completely different from Schrödinger's wave mechanical treatment of atomic problems. However, Schrödinger himself *** soon showed the mathematical equivalence of the two methods. Both methods play their role in present day quantum mechanics. We shall return to a discussion of matrix mechanics in § 43.

§ 19. The superposition principle in quantum mechanics

An arbitrarily chosen solution of the Schrödinger equation (8) obtained, for instance, by choosing arbitrarily the shape of ψ at $t = 0$ will clearly in general not correspond to a stationary state of the form (27). Let us consider such a general solution in more detail. First of all we note the following general property of the solutions of equation (8). If ψ_1 and ψ_2 are two different solutions

* W. Heisenberg, *Z. Phys.*, **33**, 879, 1925.

** M. Born and P. Jordan, *Z. Phys.*, **34**, 858, 1925.

*** E. Schrödinger, *Ann. Phys.*, **79**, 734, 1926; see also C. Eckart, Phys. Rev., **28**, 711, 1926.

of equation (8), then any linear combination, $\psi_3 = a_1\psi_1 + a_2\psi_2$, with a_1 and a_2 arbitrary complex numbers, will also be a solution. This is clearly a consequence of the fact that equation (8) is a linear homogeneous differential equation. If ψ_1, ψ_2, and ψ_3 are all three normalised, we can express this fact as follows. *The physical situation characterised by ψ_3 is obtained by a combination or a "superposition" of the physical situations ψ_1 and ψ_2* *. Clearly we can combine not only two, but a large number, possibly even an infinite number of situations to form a new physical situation.

The principle of superposition was the basis of the more elementary application of de Broglie's theory. However, it is part of the very basis of the whole edifice of modern quantum mechanics and it has no analogy, as regards its physical meaning, in classical physics.

As far as mathematics is concerned the present principle is obviously equivalent to the superposition principle which is applied to wave phenomena in material media or in vacuo. However, the wave functions occurring in classical physics are not connected with any lack of definability of physical quantities.

Let us in particular consider the physical situation obtained by the superposition of two stationary states. We denote the two normalised eigenfunctions by φ_1 and φ_2 and their eigenvalues by E_1 and E_2 and we consider the case where $E_1 \neq E_2$. The wave function obtained by superposition is

$$\psi_3 = c_1\varphi_1 \exp\left[-iE_1 t/\hbar\right] + c_2\varphi_2 \exp\left[-iE_2 t/\hbar\right]. \tag{40}$$

In order that ψ_3 be normalised one cannot choose c_1 and c_2 arbitrarily, but they must satisfy a certain relation. It follows easily that

$$1 = \int |\psi_3|^2 d\boldsymbol{x} = |c_1|^2 \int |\varphi_1|^2 d\boldsymbol{x} + |c_2|^2 \int |\varphi_2|^2 d\boldsymbol{x} + 2\mathrm{Re}\left\{c_1 c_2^* \exp\left[-i(E_1 - E_2)t/\hbar\right] \int \varphi_1\varphi_2^* d\boldsymbol{x}\right\}, \tag{41}$$

where Re indicates the real part. From equation (19) it follows that equation (41) is satisfied for all values of t if it is valid for one value. Since we have assumed $E_1 \neq E_2$ the last integral on the right hand side of equation (41) must vanish, or,

$$\int \varphi_1\varphi_2^* \, d\boldsymbol{x} = 0. \tag{42}$$

* In his book "*The principles of quantum mechanics*" (Oxford University Press, 1930) Dirac starts from the principle of the superposition of states. To emphasise the difference between classical physics and quantum mechanics we shall often call "physical situation" what is called "state" by Dirac.

In mathematical language one says that φ_1 and φ_2 are orthogonal to each other *.

Since we have assumed both φ_1 and φ_2 to be normalised it follows from equation (41) that

$$|c_1|^2+|c_2|^2 = 1. \tag{43}$$

The physical situation described by ψ_3 is clearly no stationary state. From the point of view of the quantum postulate $E = h\nu$, both E_1 and E_2 are "represented". In analogy with our earlier discussion of physical quantities, such as position and momentum of a particle, we can now ask what will be the probabilities that either E_1 or E_2 is the energy of the system. The sum of these two probabilities must obviously be equal to 1. Equation (43) suggests $|c_1|^2$ and $|c_2|^2$ as the obvious choice for these probabilities. This choice has been proved to be correct. In chapter 4 we shall meet it again in a more general context.

Physically the occurrence of two energy values means that if we perform an experiment which involves explicitly the use of the law of conservation of energy, we are led either to a value of E_1 or to a value E_2 for the energy of the system. As an example of such an experiment we may mention the collision of the system with a free particle whose energy before and after the collision is measured. In accordance with the fourth Heisenberg relation the duration of the collision must be sufficiently long, or the exact moment at which it takes place sufficiently inaccurately known, that one can distinquish with sufficient certainty between E_1 and E_2. A more detailed analysis of such an experiment can follow only after we have discussed in detail the quantum mechanical treatment of several interacting particles and at the same time have discussed the validity of the energy conservation law.

* This orthogonality follows immediately from equation (28). We have

$$\nabla^2\varphi_1+(2m/\hbar^2)(E_1-U)\varphi_1 = 0, \quad \text{(A)}; \qquad \nabla^2\varphi_2+(2m/\hbar^2)(E_2-U)\varphi_2 = 0, \quad \text{(B)}.$$

Taking the conjugate complex of (B), multiplying it by φ_1 and subtracting it from (A) multiplied by φ_2^* we have

$$\varphi_2^*\nabla^2\varphi_1-\varphi_1\nabla^2\varphi^* = (\nabla\cdot[\varphi_2^*\nabla\varphi_1-\varphi_1\nabla\varphi_2^*]) = (2m/\hbar^2)(E_2-E_1)\varphi_2^*\varphi_1.$$

Integrating this last equation over a finite volume of space we get

$$\oint (\boldsymbol{df}\cdot[\varphi_2^*\nabla\varphi_1-\varphi_1\nabla\varphi_2^*]) = (2m/\hbar^2)(E_2-E_1)\int \varphi_2^*\varphi_1 d\boldsymbol{x},$$

where $\oint \boldsymbol{df}$ indicates a surface integral. If we now let the volume expand to infinity, we may assume that φ_1 and φ_2 vanish sufficiently strongly at infinity that the left hand side goes to zero and equation (42) follows.

In the early period of the development of quantum theory before 1925 the concept of stationary states often led to difficulties. Usually it was tacitly assumed that the energy of an atomic system in a given physical situation was always well defined — by analogy with the classical case. The result was that one had to assume that interaction processes during which the energy of the system is changed could not be described classically, even though the correspondence principle enables us to show the analogy of certain aspects of these processes and classical interactions. The present idea that the energy of a system need not always be well defined gives us a way out of this difficulty. It leads to a consistent quantum mechanical treatment of atomic phenomena, or rather, it forms an indispensable part of this treatment.

§ 20. THE REPRESENTATION OF AN ARBITRARY PHYSICAL SITUATION AS THE SUPERPOSITION OF STATIONARY STATES

Our previous considerations lead to the following questions. *Is it possible to consider any arbitrary given physical situation of a one-particle system always to be a superposition of stationary states?* This question clearly has a meaning only if the potential energy does not depend on time, since otherwise there are no stationary states, as was mentioned before. We shall assume that this condition is fulfilled.

Let us first of all assume that the answer to this question is for a given system in the affirmative. Let us denote the discrete eigenvalues of the system by $E_1, E_2, \ldots, E_k, \ldots$ According to our basic assumption we can now write the wave function corresponding to an arbitrarily chosen physical situation of the system in the form

$$\psi = \Sigma_k c_k \varphi_k \exp\left(-iE_k t/\hbar\right). \tag{44}$$

In this equation the φ_k are normalised eigenfunctions corresponding to eigenvalues E_k. We shall assume that ψ satisfies the normalisation condition (19). We saw in § 17 that there may be an infinite number of eigenvalues and we shall therefore expect that in general the right hand side of equation (44) will be an infinite series. Let us now in particular assume that we choose ψ at $t = 0$ equal to an arbitrary normalised function ψ_0. From equation (44) we then have for $t = 0$

$$\psi_0 = \Sigma_k c_k \varphi_k. \tag{45}$$

This equation entails that *any arbitrary function can be expanded in terms of eigenfunctions of the time-independent Schrödinger equation.* The c_k which in general will be complex numbers are the expansion coefficients.

Multiplying both sides of equation (45) by their conjugate complex and integrating over the whole of space, one obtains the following equation, provided one may exchange the order of integration and summation,

$$\int |\psi_0|^2 d\boldsymbol{x} = \Sigma_k |c_k|^2 \int |\varphi_k|^2 d\boldsymbol{x} + \Sigma\Sigma_{k \neq l} c_k c_l^* \int \varphi_k \varphi_l^* d\boldsymbol{x}.$$

Since ψ_0 and all the φ_k are normalised and because of the orthogonality relation (42) we get from this equation

$$\Sigma_k |c_k|^2 = 1. \tag{46}$$

This equation can be interpreted analogously to equation (43) as follows. In the given physical situation all energy values E_k are represented and $|c_k|^2$ is the probability that we can assign the energy E_k to the system.

In order that this interpretation has a meaning we must clearly demand that the c_k are uniquely determined, that is, that expansion (45) is possible in essentially only one way. The only freedom which we can allow is to multiply each φ_k by an arbitrary phase factor $\exp(i\alpha_k)$ and the corresponding c_k by $\exp(-i\alpha_k)$. This will, however, change neither the normalisation of the φ_k nor the absolute value of the c_k.

The coefficients c_k follow easily from ψ_0. To see this we shall assume for the moment that to each eigenvalue E_k there corresponds, apart from a phase factor, only one normalised eigenfunction. This is, for instance, clearly the case for the previously discussed example of a particle in a one-dimensional field of force. Multiply now equation (45) by φ_l^* and integrate over the whole of space. The result is

$$\int \varphi_l^* \psi_0 d\boldsymbol{x} = \Sigma_k c_k \int \varphi_l^* \varphi_k d\boldsymbol{x}. \tag{47}$$

The integral with $k = l$ on the right hand side of this equation will be equal to one and the other integrals will vanish because of equation (42) so that we have finally

$$c_l = \int \varphi_l^* \psi_0 d\boldsymbol{x}. \tag{48}$$

§ 21. Degenerate stationary states; degree of degeneracy

In contrast to the assumption just made it can happen that the time-independent Schrödinger equation (28) possesses for a given eigenvalue E_k a number of normalisable solutions which differ not just by a constant factor. We can illustrate this by considering the case mentioned at the end of § 17, where $U(x)$ could be written as a sum, $U_1(x)+U_2(y)+U_3(z)$. In this case we could write down eigenfunctions of the form $\varphi_k^{(1)}(x)\varphi_l^{(2)}(y)\varphi_m^{(3)}(z)$ corresponding to an eigenvalue $E_k^{(1)}+E_l^{(2)}+E_m^{(3)}$. It may well happen that this last sum takes on the same value for two different combinations k, l, m, and k', l', m'. In that case we have two essentially different eigenfunctions $\varphi_k^{(1)}\varphi_l^{(2)}\varphi_m^{(3)}$ and $\varphi_{k'}^{(1)}\varphi_{l'}^{(2)}\varphi_{m'}^{(3)}$ which correspond to the same eigenvalue. In such a case one calls the eigenvalue — and also the corresponding stationary state — *degenerate*. If there corresponds only one eigenfunction to an eigenvalue, this eigenvalue is called *nondegenerate*.

Let now φ_1 and φ_2 be two essentially different eigenfunctions corresponding to the same eigenvalue E_k. Since they satisfy the same linear homogeneous differential equation (28), we see that any linear combination of φ_1 and φ_2,

$$\varphi_k = c_1\varphi_1+c_2\varphi_2, \tag{49}$$

will also satisfy the same equation. In this way one can construct an infinite number of eigenfunctions φ. However, whenever one chooses any three functions from this set, φ_{k1}, φ_{k2}, φ_{k3}, one can always find three numbers c_{k1}, c_{k2}, c_{k3}, which are not all three equal to zero, such that $c_{k1}\varphi_{k1}+c_{k2}\varphi_{k2}+c_{k3}\varphi_{k3}=0$. There are now two possibilities. Either there are no functions other than those given by equation (49) which satisfy equation (28) for the given value of E, or there are other functions. In the first case one calls the eigenvalue — and the corresponding stationary state — *two-fold degenerate*. In the second case we proceed as follows. Let φ_3 be an eigenfunction which does not satisfy equation (49). Any linear combination of φ_1, φ_2, and φ_3,

$$\varphi_k = c_1\varphi_1+c_2\varphi_2+c_3\varphi_3, \tag{50}$$

will again be an eigenfunction of equation (28) and now one can for any four functions of the form (50) find a linear equation of the form $c_{k1}\varphi_{k1}+c_{k2}\varphi_{k2}+c_{k3}\varphi_{k3}+c_{k4}\varphi_{k4} = 0$. There are again two possibilities. Either all eigenfunctions corresponding to the eigenvalue under consideration are of the form (50) or not. In the first case the eigenvalue is called *threefold degenerate*, in the second case the degree of degeneracy will be higher. In general an eigenvalue is said to be g-fold degenerate, if there always exists a linear relation

$$\Sigma_{s=1}^{g+1} c_{ks}\varphi_{ks} = 0 \tag{51}$$

for any arbitrarily chosen $g+1$ eigenfunctions, while in general such a relation can not be found for a smaller number of eigenfunctions. Clearly one has $g = 1$ for a non-degenerate eigenvalue.

In the example we discussed a moment ago the degree of degeneracy will be equal to g if there are g combinations k, l, m corresponding to the same total of $E_k^{(1)}+E_l^{(2)}+E_m^{(3)}$. If in particular $U^{(1)}$, $U^{(2)}$, and $U^{(3)}$ were identical, we have $E_k^{(1)} = E_k^{(2)} = E_k^{(3)}$, and any eigenvalue for which $k \neq l \neq m \neq k$ will be at least sixfold degenerate corresponding to the six permutations of k, l, m, since obviously $E_k^{(1)}+E_l^{(2)}+E_m^{(3)} = E_l^{(1)}+E_k^{(2)}+E_m^{(3)} = \ldots$

The degree of degeneracy is never infinite for the case of stationary states of a one-particle system.

Let us now return to the question of the coefficients c_k in the expansion (45) of an arbitrary function in terms of eigenfunctions. According to our basic assumption this expansion should be possible. However, we do not know how to choose our φ_k if E_k is degenerate. Let E_k be g_k-fold degenerate. This means that its eigenfunction can be written in the form

$$\varphi_k = \Sigma_s c_{ks}\varphi_{ks}, \tag{52}$$

where the φ_{ks} are g_k linearly independent, but otherwise arbitrary, eigenfunctions, and where s runs from 1 to g_k. It is possible to choose these φ_{ks} always in such a way that they are normalised and orthogonal to one another, that is, φ_{ks} is orthogonal to $\varphi_{ks'}$ if $s \neq s'$. This can be done in an infinity of ways. If a given set φ_{ks} does not satisfy these requirements, one can always by a simple standard method construct a set of g_k functions φ'_{ks} which do possess these properties and which are linear in the φ_{ks}. This last set can then be used instead of the φ_{ks} to represent the φ_k.

This method proceeds as follows. We first of all construct a set of mutually orthogonal eigenfunctions φ''_{ks} which are not necessarily normalised, as follows

$$\varphi''_{k1} = \varphi_{k1}, \ (\alpha); \quad \varphi''_{k2} = \varphi_{k2} - a\varphi''_{k1}, \ (\beta); \quad \varphi''_{k3} = \varphi_{k3} - b\varphi''_{k1} - c\varphi''_{k2}, \ (\gamma); \ \ldots$$

In equation (β) one must determine a in such a way that φ''_{k2} is orthogonal to φ'_{k1}, that is, $\int \varphi''^{*}_{k1}\varphi_{k2}\,d\boldsymbol{x} - a\int \varphi''^{*}_{k1}\varphi''_{k1}\,d\boldsymbol{x} = 0$. In equation ($\gamma$) one determines b and

c in such a way that φ''_{k1} and φ''_{k2} are orthogonal to φ''_{k3}, that is, both $\int \varphi''^*_{k1} \varphi_{k3} d\mathbf{x} - b \int \varphi''^*_{k1} \varphi''_{k1} d\mathbf{x}$ and $\int \varphi''^*_{k2} \varphi_{k3} d\mathbf{x} - c \int \varphi''^*_{k2} \varphi''_{k2} d\mathbf{x}$ must be equal to zero. One can clearly go on like this. By multiplying the φ''_{ks} by suitable factors one gets a set of mutually orthogonal and normalised functions φ'_{ks}.

In order that the functions φ_k given by equation (52) are themselves normalised, the coefficients c_{ks} must satisfy the condition $\Sigma_s |c_{ks}|^2 = 1$. Instead of the expansion (45) in terms of unknown eigenfunctions we can now write down an expansion in terms of known eigenfunctions by using equation (52),

$$\psi_0 = \Sigma_k c_k \Sigma_s c_{ks} \varphi_{ks}. \tag{53}$$

We can obtain the coefficients $c_k c_{ks}$ in equation (53) immediately by proceeding similarly to the derivation of equation (48), namely by multiplying equation (53) by φ^*_{ks} and integrating over the whole of space. The result is

$$c_k c_{ks} = \int \varphi^*_{ks} \psi_0 \, d\mathbf{x}. \tag{54}$$

The probability with which we can assign the value E_k to the energy of the system is still $|c_k|^2$ since $\Sigma_s |c_k c_{ks}|^2 = |c_k|^2 \Sigma_s |c_{ks}|^2 = |c_k|^2$.

One sees that, in accordance with our physical requirements, the coefficients given by equations (48) or (54) are uniquely determined, apart from an unimportant phase factor. If this were not the case, that is, if there were a second expansion $\psi_0 = \Sigma c'_l \varphi_l$, one finds that $\Sigma_l (c'_l - c_l) \varphi_l = 0$, or, that there exists a linear relation between the φ_k. This, however, is impossible because the φ_k are mutually orthogonal. *

§ 22. Unnormalisable eigenfunctions of free particles

The foregoing discussion is valid provided it is always possible to expand any arbitrary function according to equation (45) in terms of eigenfunctions. If this is, indeed, possible we say that the eigenfunctions form a complete or closed orthonormal set. In certain simple cases one may assume this to be the case, for instance, when the potential energy goes to infinity for $|x| \to \infty$. It is, however, easy to find examples where the expansion is clearly impossible. Consider the case of a free particle. In this case there are no stationary states at all in the sense of our previous definition.

* Multiplication by φ^*_l and integration over the whole of space leads immediately to $c'_l - c_l = 0$ for all l.

That is, the time independent Schrödinger equation (28), which now has the simple form

$$\nabla^2\varphi + (2m/\hbar^2)E\varphi = 0, \tag{55}$$

possesses no solutions for which φ is normalisable. A simple solution of this equation for a positive E is

$$\varphi = \exp\,[2\pi i(\boldsymbol{\sigma}\cdot\boldsymbol{x})], \tag{56}$$

$$(\boldsymbol{\sigma}\cdot\boldsymbol{\sigma}) = 2mE/h^2, \tag{57}$$

which corresponds to a plane de Broglie wave of a particle with energy E, provided $\boldsymbol{\sigma}$ is a real vector, as we may assume to be the case,

$$\psi = \exp\{2\pi i[(\boldsymbol{\sigma}\cdot\boldsymbol{x}) - (Et/h)]\}. \tag{58}$$

A more general solution of equation (55) is obtained by superposition of solutions of the form (56),

$$\varphi = \int a(\boldsymbol{\sigma})\exp\,[2\pi i(\boldsymbol{\sigma}\cdot\boldsymbol{x})]\,d\boldsymbol{\omega}, \tag{59}$$

where $d\boldsymbol{\omega}$ is an element of solid angle in $\boldsymbol{\sigma}$-space and where the $\boldsymbol{\sigma}$ satisfy equation (57). Unless a is identically equal to zero, $\int|\varphi|^2\,d\boldsymbol{x}$ will diverge. Special solutions such as (56) which can be obtained as limiting cases of equation (59) have the same property.

We can prove the divergence of $\int|\varphi|^2\,d\boldsymbol{x}$ as follows. Let $A(\boldsymbol{\sigma})$ be a function which is zero everywhere except within a spherical shell of thickness Δ and radius $\sigma = [2mE/h^2]^{1/2}$. We can now write equation (59) in the form $\varphi = \lim_{\Delta\to 0}\int A\,\exp[2\pi i(\boldsymbol{\sigma}\cdot\boldsymbol{x})]d\boldsymbol{\sigma}$. In order that ψ remains finite and non-vanishing in the limit we must have $A = a/\Delta$, where a is a finite function of $\boldsymbol{\sigma}$. We then get $\int|\varphi|^2\,d\boldsymbol{x} = \lim_{\Delta\to 0}\int|A|^2\,d\boldsymbol{\sigma} = \lim_{\Delta\to 0}\,(2mE/h^2\Delta)\int|a|^2\,d\boldsymbol{\omega} \to \infty$.

It is nevertheless possible to give a physical meaning to solutions such as (56) or (59). We can, for instance, imagine a situation to exist where at $t = 0$ we have a normalised wave function which inside a sphere of volume V is equal to $V^{-1/2}\exp[2\pi i(\boldsymbol{\sigma}\cdot\boldsymbol{x})]$, and outwith V equal to zero. If V is very large and is thought to increase and go to infinity, we are clearly considering the case of a wave packet the momentum — and thus also the energy — of which is increasingly well defined. Its time-dependence within the sphere will then for increasing periods of time be given by equation (58) apart from a multiplying factor $V^{-1/2}$ on the right hand side. We shall in such a case call (58) an improper stationary state and (56) an improper eigenfunction of the time-independent Schrödinger equation corresponding to an improper eigenvalue E. In the same sense, expression (59) will also be an improper eigenfunction of the Schrödinger equation (55).

An arbitrary function can always be written as a superposition of improper eigenfunctions of the form (59). This is immediately clear if we remember that any arbitrary function could always be written as the superposition of waves of the form (56) [see equations (1.27) and (1.28)]. In this case we do not have a sum, but an integral,

$$\psi_0 = \int A(\boldsymbol{\sigma}) \exp\,[2\pi i(\boldsymbol{\sigma}\cdot\boldsymbol{x})]\,d\boldsymbol{\sigma}, \tag{60}$$

with A satisfying equation (1.28). The integration in equation (60) is done in such a way that we first of all integrate for fixed values of $\sigma^2 = (\boldsymbol{\sigma}\cdot\boldsymbol{\sigma})$ over all possible angles and then over σ, or E, from 0 to ∞,

$$\psi_0 = \int_0^\infty \{\sigma^2(d\sigma/dE)\int d\boldsymbol{\omega} A \exp\,[2\pi i(\boldsymbol{\sigma}\cdot\boldsymbol{x})]\}dE = \int_0^\infty \varphi(\boldsymbol{x}, E)\,dE. \tag{61}$$

Using equation (57) we have

$$d\sigma/dE = m/h^2\sigma = (m/2h^2)^{1/2}E^{-1/2}. \tag{62}$$

The expression within braces in the second member of equation (61) is a function of $\boldsymbol{x}$ and of a continuous parameter E. We have denoted it by $\varphi(\boldsymbol{x}, E)$ and it is an improper eigenfunction of the form (59). The wave function at any time is now given by the equation

$$\psi = \int \varphi(\boldsymbol{x}, E) \exp\,(-iEt/\hbar)\,dE. \tag{63}$$

This equation is equivalent to equation (1.27) and expresses the fact that for a free particle *an arbitrary situation can always be represented as a superposition of improper stationary states.* It is the counterpart of equation (44) which referred to the analogous superposition of proper stationary states.

We may at this point pass a few general remarks. First of all we draw attention to the fact that the improper eigenfunctions are only a small fraction of all possible solutions of equation (55). They could all be built up out of functions of the form (56). However, this last expression was considered only in the case of real $\boldsymbol{\sigma}$, as we emphasised. Expression (56) is still a solution of equation (55) even if we consider complex vectors $\boldsymbol{\sigma}$, as long as equation (57) is satisfied. However, those solutions of equation (55) are not used in the expansion (63). The reason for this is clearly that as soon as $\boldsymbol{\sigma}$ is complex one can find directions in space such that $\varphi \to \infty$, if $\boldsymbol{x}$ moves into such a direction. The improper eigenfunctions are just those solutions of equation (55) which remain everywhere finite.

Our second remark concerns the eigenvalues E. According to its physical meaning one always assumes E to be real. However, the condition of real $\boldsymbol{\sigma}$ means that E will always be positive. We see thus that the set of improper eigenvalues is a continuum, but it is only *part* of the continuum of all real numbers.

Thirdly, we see that the improper eigenvalues are degenerate, since to each positive E there correspond infinitely many, linearly independent eigenfunctions corresponding to the infinity of points on the sphere (57). Only $E = 0$ is non-degenerate; this eigenvalue corresponds to the (improper) eigenfunction $\varphi =$ constant.

Fourthly, let us consider the energy probability distribution in a given physical situation. When we analysed in § 20 a situation by expanding arbitrary functions in terms of normalised proper eigenfunctions we could derive equation (46) and connect it with the energy probability distribution. We can proceed now in an analogous manner. There is now, however, a particular mathematical complication which will not become very clear in the present case of a free particle in three dimensions because of the infinite degeneracy, and we shall return to this point later on. Using equation (24) we can easily find the energy distribution function. As in the transition from equation (60) to (61) we write

$$1 = \int |A|^2 d\boldsymbol{\sigma} = \int dE\{\sigma^2(d\sigma/dE)\int |A|^2 d\boldsymbol{\omega}\} = \int W(E)\,dE. \quad (64)$$

The expression in braces which we have denoted by $W(E)$ depends only on that part of the wave function which according to equation (61) corresponds to the eigenvalue $E = \hbar^2\sigma^2/2m$. We can thus interpret $W(E)dE$ as the probability that the energy of the particle lies between E and $E+dE$.

Lastly we draw attention to the following interpretation of the non-normalisable wave functions $\psi = \varphi \exp(-iEt/\hbar)$ with φ given by equation (59). We can refer this function to a steady physical situation where we have an infinite number of free particles with energy E and absolute value of momentum $\hbar\sigma$. The probability of finding one of these particles in a volume element $d\boldsymbol{x}$ is $|\psi|^2 d\boldsymbol{x}$ and the probability that the direction of the momentum of a particle lies within the solid angle $d\boldsymbol{\omega}$ is proportional to $|a|^2 d\boldsymbol{\omega}$. This interpretation which is a generalisation of the case discussed in § 8 has a physical meaning only when the average density of the particles is so small that one can neglect their mutual interactions.

§ 23. Improper stationary states in an external field of force

We shall now investigate the existence of improper eigenvalues and eigenfunctions in the case of systems in an external field of force. We shall again consider the one-dimensional Schrödinger equation (30) but this time consider functions U which are essentially different from the type of function discussed in § 17.

Let U be a smooth monotonic function of x which goes to zero for $x \rightarrow \infty$ and to ∞ for $x \rightarrow -\infty$ (see fig. 5). In the corresponding

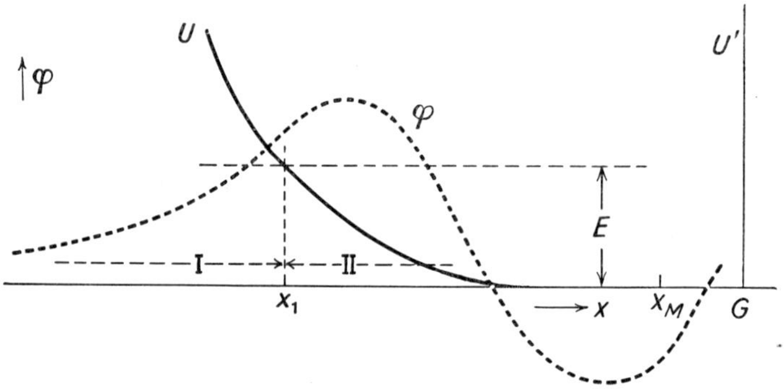

Fig. 5.

classical case we would be dealing with a particle which could occur only in region II where $E-U$ is positive (we assume E to be positive). The particle would emerge from infinity with a finite velocity, turn back at x_1 where $E-U = 0$, and disappear again to infinity.

Remembering the procedure of § 17 we now consider for a given value of E a real solution $\varphi(x, E)$ of equation (30) constructed in such a way that $\varphi(x, E)$ which behaves "exponentially" in region I, goes to zero for $x \rightarrow -\infty$. In region II which extends to infinity, φ has an oscillatory behaviour. For the sake of simplicity we shall assume that U is exactly equal to zero for $x > x_M$. In that case the solution of the Schrödinger equation for large values of x will be

$$\varphi = A \sin [x(CE)^{1/2} + \alpha], \tag{65}$$

where α depends on the value of E and the exact form of U, while the value of A is unimportant since equation (30) is linear and homogeneous in φ. This solution of equation (65) distinquishes itself among all possible solutions for the given value of E that it remains everywhere finite. However, it is not normalisable as follows immediately from equation (65) and clearly there exist no proper stationary states.

The solutions just considered can be considered, by analogy with the case discussed in § 22, to be the improper eigenfunctions of the system. There is one and only one such eigenfunction for every positive value of E. We have thus a continuum of improper non-degenerate energy eigenvalues.

If we assume that it is justified to draw physical conclusions from the one-dimensional case, we can consider the improper stationary states which correspond to these eigenfunctions to be limiting cases of physical situations. Let the situation at $t = 0$ be characterised by a wave function ψ_0 which is identical with one of the improper eigenfunctions $\varphi(x, E)$ for $-\infty < x < G$, where G is some large positive value of x. Let ψ_0, however, for $x > G$ converge so rapidly to zero that ψ_0 is normalisable. In any interval $-\infty < x < G_0$, where G_0 is fixed and smaller than G, the wave function will clearly be given approximately by $\psi_0 \exp(-iEt/\hbar)$, as long as t is not too large, and the larger G the better will be the approximation and the longer will be the time interval for which it is valid. Another way to consider $\varphi(x, E)$ as a limiting case is to consider the field of force to be the limiting case of a field the potential of which, U', is equal to U for $-\infty < x < G$, while $U' = \infty$ for $x \geqq G$ (see fig. 5). This potential corresponds to the introduction of a rigid elastic wall at a distance G from the origin.

The potential U' belongs to the class of one-dimensional fields of force discussed in § 17. It gives rise to an infinite number of discrete stationary states. The corresponding eigenfunctions can be obtained for $-\infty < x < G$ by choosing E such that the corresponding function $\varphi(x, E)$, which was defined a moment ago and which is everywhere finite, vanishes at $x = G$, which is the boundary condition at an infinitely high potential wall. Provided G is larger than x_M, the eigenvalues will be the solutions of

$$G[CE_n]^{1/2}+\alpha(E_n) = n\pi, \quad n = \text{integer}. \tag{66}$$

When G increases, the distance between two consecutive E_n will clearly decrease; in the limit $G \to \infty$ one obtains the improper eigenfunctions.

This second limiting process enables us to discuss in a simple manner the question whether an arbitrary function can be expanded in terms of improper eigenfunctions. In the case of the potential U' any arbitrary normalisable function which is zero for can be expanded in terms of the proper eigenfunctions $\varphi(x, E_n)$,

$$\psi_0(x) = \Sigma_n c_n \varphi(x, E_n). \tag{67}$$

If we now want to know what happens to this infinite series when G goes to ∞, it is an obvious assumption that we must expect an integral expansion of $\psi_0(x)$,

$$\psi_0(x) = \int c(E)\varphi(x, E)dE. \tag{68}$$

This limiting process which is a natural generalisation of the well known formal transition from Fourier series to Fourier integrals will be considered in some more detail.

Let us first of all assume that the $\varphi(x, E_n)$ which occur in equation (67) are chosen to be equal to the improper eigenfunctions $\varphi(x, E)$ in the interval $-\infty < x < G$. This means that they are in general not normalised. The functions $\varphi(x, E)$ have been chosen once and for all, that is, the value of the multiplying constant which is involved [A in equation (65)] has been fixed, for instance by putting A always equal to unity. The normalised eigenfunction corresponding to an energy E_n can be written in the form $F_G^{-1/2}\varphi(x, E_n)$ where F is a function of E given by the equation

$$\int_{-\infty}^{G} [\varphi(x, E)]^2 dx = F_G(E). \tag{69}$$

If G goes to infinity, F also goes to infinity and this proportionately. We introduce now a function $c_G(E)$ by the equation

$$c_G(E) = \int_{-\infty}^{G} \varphi(x, E)\,\psi_0(x)\,dx, \tag{70}$$

and we shall assume that for $G \to \infty$, $c_G(E)$ converges uniformly to a function $c_\infty(E)$. We can now write the expansion (67) in the form

$$\psi_0(x) = \Sigma_n[c_G(E_n)/F_G(E_n)]\,\varphi(x, E_n). \tag{71}$$

If G is very large, E_n and E_{n+1} will be very close to each other. We denote $E_{n+1} - E_n$ by ΔE and assume that $F_G(E)\Delta E$ converges to a function, $K(E)$, of E if $G \to \infty$. In that case the expansion (71) goes over into an integral

$$\psi_0(x) = \int_0^\infty [c_\infty(E)/K(E)]\varphi(x, E)dE = \int_0^\infty c(E)\varphi(x, E)dE, \quad c(E) = c_\infty(E)/K(E). \tag{72}$$

The existence of the function K can be seen as follows. From equations (69) and (65) it follows that for large values of G, $F_G(E) \doteq \frac{1}{2}A^2G$. Differentiation of equation (66) with respect to n leads to the following approximate expression for ΔE, as long as G is large: $\frac{1}{2}(C/E)^{1/2}\Delta E \,.\, G \doteq \pi$. Combining these two results we have that for $G \to \infty$,

$$\lim_{G\to\infty} F_G(E)\Delta E = K(E) = \pi A^2(E/C)^{1/2}. \tag{73}$$

This function depends on the normalising factor A which we can still choose freely. We make now the following choice

$$A = (C/\pi^2E)^{1/4}, \tag{74}$$

so that $K = 1$ and we obtain [compare equations (70) and (72)]

$$\psi_0(x) = \int_0^\infty c(E)\varphi(x, E)dE, \quad c(E) = \int_{-\infty}^{+\infty}\varphi^*(x, E)\psi_0(x)dx. \tag{75}$$

Although up to now φ was supposed to be real, we have written in the second of equations (75) φ^* in order that equations (75) would remain valid, if we multiply $\varphi(x, E)$ by an arbitrary phase factor $\exp(i\alpha)$. This way of choosing A according to equation (74) so that $K(E) = 1$ is called *normalisation of the improper eigenfunctions.*

If we normalise the $\varphi(x, E)$ in this manner so that they are completely determined apart from an unimportant complex phase factor, equations (75) will be completely analogous to the equations (45) and (48) valid in the case of discrete eigenvalues.

The validity of equations (75) is intimately connected with the so-called "improper orthogonality" of the improper eigenfunctions. Substituting from the first equation into the second and changing the order of integration we get

$$c(E) = \int_0^\infty dE'c(E')\int_{-\infty}^{+\infty} dx\,\varphi^*(x, E)\varphi(x, E').$$

From this last equation it follows that the integral

$$f(E, E') = \int_{-\infty}^{+\infty} dx\,\varphi^*(x, E)\varphi(x, E'), \tag{76}$$

which is a function of E and E' has the following property. It can formally be considered to be the limiting case for $\varepsilon \to 0$ of a function f' which is zero for $E' < E - \varepsilon$,

$E+\varepsilon < E'$, which is positive in the interval $E-\varepsilon < E' < E+\varepsilon$, and for which $\int_{E-\varepsilon}^{E+\varepsilon} f'(E, E')dE' = 1$, independent of ε (Dirac's δ-function; see § 29). By using the fact that $\varphi(x, E)$ and $\varphi(x, E')$ satisfy the Schrödinger equation (30) we get for the integral (76) taken between x_1 and x_2 [compare the proof of eq. (42)] the expression $[\{\varphi^*(x_2, E)\varphi'(x_2, E') - \varphi'^*(x_2, E)\varphi(x_2, E')\} - \{(\varphi^*(x_1, E)\varphi'(x_1, E') - \varphi'^*(x_1, E)\varphi(x_1, E'\}] / C(E'-E)$. When we let x_1 go to $-\infty$ the second expression within braces will vanish, but the first expression will not vanish for $x_2 \rightarrow \infty$, but will remain finite, although behaving as an oscillating function. If at $x = x_2 = G$ there were a rigid wall, as we assumed in our derivation of equations (75), $\varphi(x, E)$ and $\varphi(x, E')$ would both be zero for $x = x_2$ and we would be back to the case of strictly orthogonal functions.

As far as the mathematics is concerned, the discussion can be simplified by the introduction of the so-called eigendifferentials for a discussion of which we refer to § 36.

If the physical situation of our one-particle system is characterised by a wave function $\psi_0(x)$ at $t = 0$, the wave function at any time t will clearly be given by the expression

$$\psi(x, t) = \int_0^\infty c(E)\varphi(x, E) \exp(-iEt/\hbar)dE. \tag{77}$$

The integrand is a wave function of an improper stationary state of energy E. Equation (77) expresses the fact that the situation considered can be described as a superposition of such stationary states.

Let us now consider the probability $W(E)dE$ that the energy of the system lies between E and $E+dE$. Using equations (75) we write

$$\int_{-\infty}^{+\infty} |\psi|^2 dx = \int_{-\infty}^{+\infty} |\psi_0|^2 dx = \int_0^\infty dE\, c(E) \int_{-\infty}^{+\infty} dx\, \psi_0^* \varphi(x, E)$$
$$= \int_0^\infty dE\, c(E)\, c^*(E). \tag{78}$$

If we assume as usual that ψ is normalised, we see that we have the following relation

$$\int_0^\infty |c|^2 dE = 1, \tag{79}$$

which is analogous to equation (46). In accordance with the interpretation of equations (43), (46) and (64) we now identify $W(E)$ with $|c(E)|^2$.

The example which we have just discussed would not be changed qualitatively, if the potential function were not monotonic, but possessed a finite number of maxima and minima while still always being positive (see fig. 6).

We shall now discuss briefly a few other types of potential U which might occur in the one-dimensional Schrödinger equation (30). First of all, let U be a smooth monotically decreasing function which for $x \to -\infty$ converges to U_0 and for $x \to +\infty$ converges to zero (fig. 7). For values of E between zero and U_0 we can construct

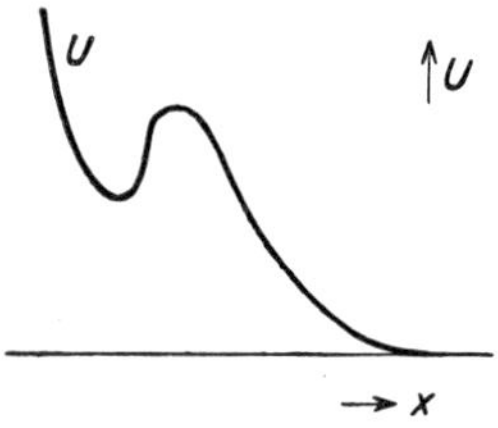

Fig. 6.

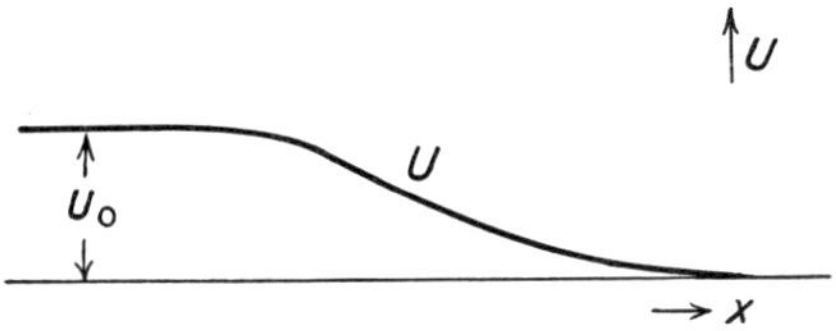

Fig. 7.

in only *one* way a real solution of equation (30) which remains finite (and goes to zero) for $x \to -\infty$. For such energy values we obtain improper eigenfunctions of the same type as the ones discussed earlier in this section. For $E > U_0$, however, *every* solution of equation (30) has an oscillatory character for all values of x. For a given value of E one can thus construct two linearly independent solutions which are everywhere finite and which are thus improper eigenfunctions. This case corresponds classically to a particle moving from $-\infty$ to $+\infty$ or from $+\infty$ to $-\infty$ and each energy value is two-fold degenerate, while for $0 \leqq E \leqq U_0$ (classically: reflexion of a particle) each E-value was nondegenerate. In the limit where $U_0 = 0$, that is, $U = 0$ everywhere, we are dealing with a one-dimensional free particle and one can take $\cos 2\pi\sigma x$ or $\sin 2\pi\sigma x$ as the eigenfunctions with $E = h^2\sigma^2/2m$.

An arbitrary normalisable function $\psi_0(x)$ can be expanded in terms of the eigenfunctions $\varphi(x, E)$,

$$\psi_0(x) = \int c(E)\varphi(x, E)\,dE. \tag{80}$$

For $E > U_0$ the $\varphi(x, E)$ are, however, not uniquely determined. They are of the form

$$\varphi(x, E) = c_1(E)\varphi_1(x, E) + c_2(E)\varphi_2(x, E), \tag{81}$$

where $\varphi_1(x, E)$ and $\varphi_2(x, E)$ are two linearly independent eigenfunctions, constructed in some way or other, while $c_1(E)$ and $c_2(E)$ in general will be complex numbers so that $\varphi(x, E)$ in general will no longer be a real function. By a suitable choice of $\varphi_1(x, E)$

and $\varphi_2(x, E)$ one can always arrange it so that the following equations, which are analogous to the second equation (75), are valid

$$c(E) = \int \varphi^*(x, E)\, \psi_0(x)\, dx,\ \ 0 \leqq E < U_0; \qquad (82)$$
$$c(E)\, c_s(E) = \int \varphi_s^*(x, E)\, \psi_0(x)\, dx,\ \ U_0 < E,\ \ s = 1, 2.$$

If equations (82) were not valid, but, for instance, the following relation held,

$$\int \varphi_s^* \psi_0\, dx = c(E)\, \Sigma_r K_{sr}(E) c_s(E),\ \ s,\ r = 1, 2, \qquad (83)$$

where the determinant K_{sr}, provided the φ_s are linearly independent, is not zero, then one can always change the φ_s in such a way that they will be valid. Take thereto $\varphi_s = \Sigma_r L_{sr}(E) \varphi_r'$ with the L's satisfying the relation $\Sigma_t L_{st}{}^* L_{rt} = K_{sr}$. One can always do this, if K_{sr} is a Hermitean matrix with positive eigenvalues (compare the general discussion of eigenvalue and eigenfunction problems in § 37). The Hermitean character of K_{sr} follows from the equation $\int |\psi_0|^2 dx = \int (\Sigma_{sr} c_s{}^* K_{sr} c_r) |c|^2 dE$, which is valid for all functions ψ_0 and which follows from equations (80), (81) and (83) by a calculation similar to the one which led to equation (78).

If equations (82) hold, we say that all eigenfunctions are orthonormal. Especially $\varphi_1(x, E)$ and $\varphi_2(x, E)$ are orthogonal to each other in the improper sense. If the c_s are chosen such that always $|c_1|^2 + |c_2|^2 = 1$, expression (81) itself is a normalised eigenfunction and $|c(E)|^2$ is the probability density function for the energy, where $c(E)$ is the coefficient occurring in equation (80).

We could have considered this last problem as the limiting case of a system with discrete eigenvalues, by taking U to be infinite for $x < -G$ and $x > G$ and then letting $G \to \infty$. A great disadvantage of this procedure would have been that for all values of G the eigenvalues would have been non-degenerate while two-fold degeneracy occurs in the limiting case.

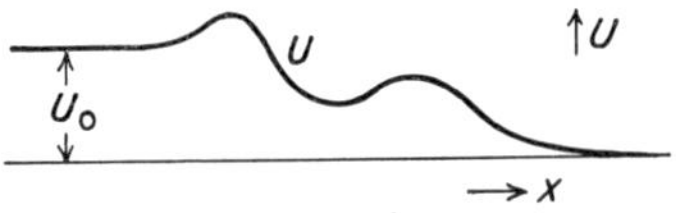

Fig. 8.

Nothing would be changed qualitatively in the last example, if U instead of decreasing monotonically from U_0 to zero would have possessed a finite number of maxima and minima without, however, ever becoming negative (see fig. 8).

A new case arises when U goes to ∞ for $x \to -\infty$ and to 0 for $x \to \infty$ while it possesses a minimum $-U_0$ for some value of x

(fig. 9). We now must consider separately the cases $E < 0$ and $E > 0$. For $E < -U_0$ each solution of equation (30) is everywhere "exponential" and will become infinite either for $x \to +\infty$ or for $x \to -\infty$. There are thus certainly no stationary states for $E < -U_0$.

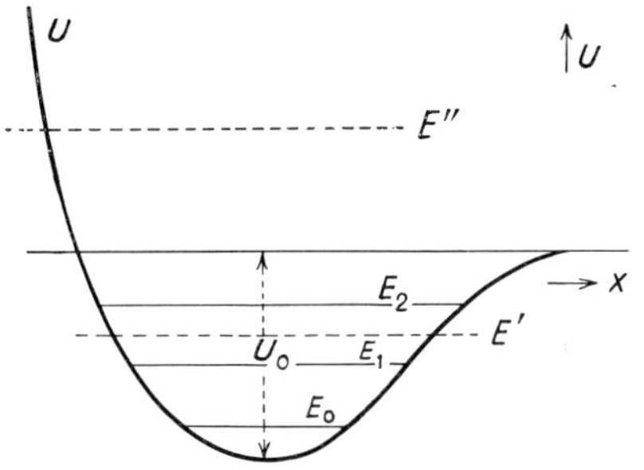

Fig. 9.

In the interval $(-U_0, 0)$ (E' in fig. 9) the situation is completely similar to the one discussed in § 17. There are two classical turning points, x_1 and x_2, where $E - U = 0$ and under certain circumstances one will find discrete eigenvalues, $E_0, E_1, E_2, \ldots$, which correspond to solutions of the Schrödinger equation which go to zero for $x \to \pm\infty$ and which have 0, 1, 2, . . . zeros. It depends on the exact form of U whether the number of these proper eigenvalues is zero, finite or infinite. For $E > 0$ (E'' in fig. 9) the situation is similar to the one discussed at the beginning of this section. There is one turning point and for each value of E there exists exactly one solution $\varphi(x, E)$ of equation (30) which is everywhere finite and which corresponds to an improper stationary state. We may assume that any normalisable function ψ_0 can be expanded in terms of both the proper and the improper eigenfunctions,

$$\psi_0(x) = \Sigma_k c_k \varphi_k(x) + \int c(E)\, \varphi(x, E)\, dE. \tag{84}$$

The series can be finite or infinite according to the number of discrete eigenvalues. Equation (84) can be considered to be the limiting case of a series expansion occurring for a potential which is infinite for $x \geqq G$ in the limit where $G \to \infty$. If the φ_k are normalised and the $\varphi(x, E)$ also chosen suitably, the following equations will hold

$$c_k = \int \varphi_k^* \psi_0 dx,\ c(E) = \int \varphi^*(x, E) \psi_0 dx,\ \int |\psi_0|^2 dx = \Sigma_k |c_k|^2 + |c(E)|^2 dE. \tag{85}$$

A physical situation characterised by ψ_0 at $t = 0$ will correspond to a probability $|c_k|^2$ for an energy value E_k (< 0) and a probability $|c(E)|^2 dE$ of finding the energy between E and $E+dE$ ($E > 0$).

A similar case where both proper and improper eigenvalues occur, but where degeneracy occurs as an additional complication occurs for U of the form given by figure 10. Here, $U \rightarrow U_1 (>0)$ for $x \rightarrow -\infty$, $U \rightarrow 0$ for $x \rightarrow +\infty$, and U possesses a minimum $-U_0$.

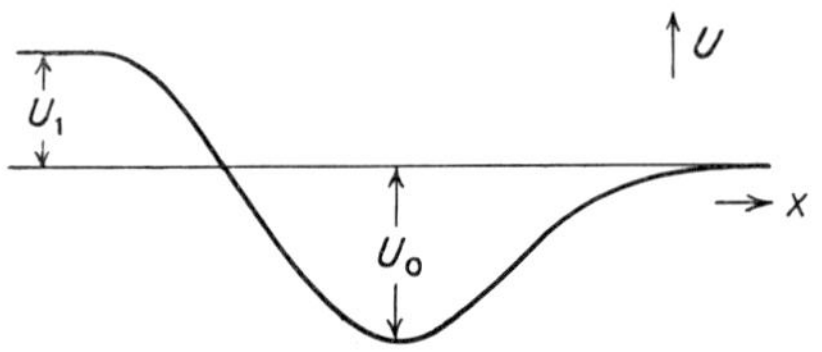

Fig. 10.

For $-U_0 < E < 0$ we may have discrete, non-degenerate eigenvalues, for $0 < E < U_1$ improper, non-degenerate eigenvalues and for $U_1 < E$ improper, two-fold degenerate eigenvalues.

§ 24. General discussion of eigenvalues and eigenfunctions

The totality of eigenvalues, proper and improper, which correspond to a given Schrödinger equation are called the *eigenvalue spectrum* and we have just seen that this spectrum can contain discrete "lines" as well as a "continuum." From our discussion it follows easily how one can construct three-dimensional one-particle problems with all kinds of eigenvalue spectra. The simplest method is to consider — as in § 17 — a potential function $U(x, y, z)$ which is a sum $U_1(x)+U_2(y)+U_3(z)$. The eigenvalues are thus the sum (37) of eigenvalues of three one-dimensional problems and the resulting spectrum may be a complicated combination of continua and discrete lines. The eigenfunctions are in each case the product of three one-dimensional eigenfunctions. An arbitrary normalisble function $\psi_0(x, y, z)$ can uniquely be expanded in terms of these eigenfunctions. * This means that an arbitrary physical situation of the system can be considered as a superposition of (proper and improper) stationary states.

* One first of all expands $\psi_0(x, y, z)$ in terms of the eigenfunctions of the x-coordinate. The coefficients will be functions of y and z and can be expanded in terms of the eigenfunctions in y with coefficients which are functions of z and can thus be expanded in terms of the eigenfunctions in z.

In actual applications it often occurs that although $U(\boldsymbol{x})$ can not be written in the form $U_1(x)+U_2(y)+U_3(z)$ the Schrödinger equation (28) can still be solved by a product of the form

$$\varphi = \varphi^{(1)}(q_1)\,\varphi^{(2)}(q_2)\,\varphi^{(3)}(q_3). \tag{86}$$

The three q_i are new coordinates which are functions of the Cartesian coordinates x, y, and z. Each $\varphi^{(k)}$ satisfies a linear homogeneous second order differential equation. In such a case one says that the Schrödinger equation can be solved by *separation of variables.*

Especially important is the case when a separation in polar coordinates is possible. This happens when U depends only on the distance r from the origin. We shall discuss this case which includes the theory of the hydrogen spectrum in detail in § 46. Separation in polar coordinates or in Cartesian coordinates are special cases of the separation in so-called elliptic coordinates which is the most general case of separation for a one-particle problem.

The one-dimensional eigenvalue problems to which the separation leads often have characteristics which are different from the ones we have just discussed and which are peculiar to the choice of variables. We shall encounter examples later on. We shall now consider the important general case where the potential U possesses cylindrical symmetry around the z-axis, that is, when U depends only on z and $\varrho = (x^2+y^2)^{1/2}$. In this case one introduces the azimuthal angle ζ by the equation $y = x \tan \zeta$ and this angle can then be separated, that is, there are solutions of the form

$$\varphi = f(\varrho, z)g(\zeta), \tag{87}$$

and one can easily check that $g(\zeta)$ and $f(\varrho, z)$ satisfy the differential equations

$$d^2g/d\zeta^2 - Kg = 0, \tag{88}$$

$$\left[\frac{\partial^2}{\partial z^2}+\frac{\partial^2}{\partial \varrho^2}+\frac{1}{\varrho}\frac{\partial}{\partial \varrho}\right]f+\frac{2m}{\hbar^2}\left[E-U+\frac{\hbar^2 K}{2m\varrho^2}\right]f = 0. \tag{89}$$

Using the relation $\nabla^2 \equiv \partial^2/\partial z^2+\partial^2/\partial\varrho^2+\varrho^{-1}\partial/\partial\varrho+\varrho^{-2}\partial^2/\partial\zeta^2$, one gets by substituting expression (87) for φ into the Schrödinger equation $\nabla^2\varphi+(2m/\hbar^2)(E-U)\varphi=0$ and dividing by φ

$$\frac{1}{f}\left[\frac{\partial^2 f}{\partial z^2}+\frac{\partial^2 f}{\partial\varrho^2}+\frac{1}{\varrho}\frac{\partial f}{\partial\varrho}+\frac{2m}{\hbar^2}(E-U)f\right]+\frac{1}{g\varrho^2}\frac{d^2g}{d\zeta^2} = 0.$$

It follows that $g^{-1}d^2g/d\zeta^2 = \text{constant} = K$ and that f must satisfy equation (89).

The most general solution of equation (88) is

$$g = A \exp\,[i\zeta(-K)^{1/2}]. \tag{90}$$

If φ is to be an eigenfunction, g should return to its original value, if ζ is increased by 2π, since otherwise φ would not be a one-valued function of $\boldsymbol{x}$. This means that $-K$ must be the square of an integer k,

$$K = -k^2, \; g = A \exp (ik\zeta), \; k = 0, +1, +2, \ldots \qquad (91)$$

The eigenvalues $K = -k^2$ of the differential equation (88) form an infinite discrete series and all of them, except $K = 0$, are two-fold degenerate. The problem of the eigenvalues and eigenfunctions of the system is reduced to the two-dimensional problem (89). We shall see in § 45 that the introduction of k corresponds to the "quantisation" of the z-component of the angular momentum of the particle.

We may make at this point a few general remarks on the existence of eigenvalues and eigenfunctions. In all practical applications we shall always assume that they exist and that any normalisable function can be expanded uniquely in terms of those eigenfunctions. This corresponds to the assumption that an arbitrary physical situation can always be considered to be a superpostion of stationary states. In using the terms "eigenvalues," "eigenfunctions" and "stationary states" we leave undecided the question whether we are dealing with proper or improper eigenvalues, . . . The question of what conditions must be satisfied by the potential U in order that our assumptions are justified is an extremely difficult mathematical problem. The fact that there is no simple general criterion for deciding whether a given non-normalisable solution of the Schrödinger equation can be considered to be an improper eigenfunction is connected with these difficulties. In the present section the criterion of being everywhere finite was sufficient. In applications one always assumes that the proper eigenvalues form a point set, and the improper ones a continuum. The justification of this assumption depends also on the form of the potential. Von Neumann and Wigner * have given a simple example of a potential for which all real values of E are eigenvalues which correspond to normalisable solutions. We must add, however, that up to now, as far as we can see, the theory has not encountered any logical difficulties which could be traced to purely mathematical complications.

The analytical problem of eigenvalues and expansion in eigenfunctions can always, and in manifold ways, be considered to be

* J. v. Neumann and E. Wigner, *Phys. Z.*, **30**, 465, 1929.

the limiting case of a purely algebraic problem. A simple procedure is to consider the lattice points on the faces of and inside a cube of edgelength $2\delta N$,

$$x = k\delta,\ y = l\delta,\ z = m\delta,\ k, l, m = 0,\ \pm 1,\ \pm 2, \ldots, \pm N. \quad (92)$$

The Schrödinger equation (28) is now the limiting case of a system of linear, homogeneous equations in the variables $\varphi_{k,l,m}$,

$$\nabla^2 \varphi_{k,l,m} + (2m/\hbar^2)(E - U_{klm})\varphi_{k,l,m} = 0, \quad (93)$$

where $U_{klm} = U(k\delta,\ l\delta,\ m\delta)$ and where $\nabla^2 \varphi_{k,l,m}$ is a short hand notation for the expression

$$\nabla^2 \varphi_{k,l,m} = \delta^{-2}[\varphi_{k+1,l,m} + \varphi_{k,l+1,m} + \varphi_{k,l,m+1} - 6\varphi_{k,l,m} + \varphi_{k-1,l,m} + \varphi_{k,l-1,m} + \varphi_{k,l,m-1}]. \quad (94)$$

The boundary conditions are that $\varphi_{k,l,m} = 0$ as soon as either k, or l, or m is equal to $+N$ or $-N$. We know from algebra (compare also the discussion in § 37) that the equation of n-th degree which according to the set of equations (93) must be satisfied by E possesses n real roots which are not necessarily all different (degeneracy). To each root $E^{(\nu)}$ there corresponds a set of numbers $\varphi_{klm}^{(\nu)}$ $(\nu = 1, 2, \ldots, n)$ which satisfy equations (93) with $E = E^{(\nu)}$ and which, moreover, satisfy the equations

$$\Sigma_{klm} \varphi_{klm}^{(\nu)*} \varphi_{klm}^{(\nu')} = \delta_{\nu\nu'} \quad (\delta_{\nu\nu'} = 0,\ \nu \neq \nu';\ \delta_{\nu\nu'} = 1,\ \nu = \nu'). \quad (95)$$

An arbitrary set of numbers ψ_{klm}^0 which satisfy the same boundary conditions as the $\varphi_{k,l,m}$ can always be written in the form

$$\psi_{klm}^0 = \Sigma_\nu c^{(\nu)} \varphi_{klm}^{(\nu)}, \quad (96)$$

and the $c^{(\nu)}$ are uniquely determined by the equations

$$c^{(\nu)} = \Sigma_{klm} \varphi_{klm}^{(\nu)*} \psi_{klm}^0. \quad (97)$$

If one now lets δ go to zero while keeping $N\delta = G$ constant, equations (93) will go over into the time-independent Schrödinger equation (28) for a system which is inside a cube of edgelength $2G$ with infinitely high potential walls and with a potential $U(\boldsymbol{x})$ existing inside. In this limit the $\varphi_{klm}^{(\nu)}$ form a discrete set of twice differentiable eigenfunctions with corresponding eigenvalues $E^{(\nu)}$. Equation (95) gives the orthonormality of the eigenfunctions and equation (96) shows how an arbitrary function can be expanded in terms of the eigenfunctions. If one now lets G to go ∞ one should obtain the proper and improper eigenvalues and eigenfunctions of the original problem. It is clear that one must investigate in detail in each case whether all limiting processes are allowed, whether the various limiting processes can be exchanged, and what will be the influence of possible singularities of U.

The algebraic problem (93) is often formulated in a geometric way as follows. One considers the $\varphi_{k,l,m}$ as the components of an n-dimensional vector. Writing the equations (93) in the form

$$\Sigma_{k'l'm'}\, a_{klm,\,k'l'm'}\, \varphi_{k',\,l',\,m'} = E\varphi_{k,\,l,\,m}\,, \tag{98}$$

with $a_{klm,\,klm} = (3\hbar^2/m\delta^2) + U_{klm}$, $a_{klm,\,k\pm 1lm} = a_{klm,\,kl\pm 1m} = a_{klm,\,klm\pm 1} = -\hbar^2/2m\delta^2$ and all other $a_{klm,\,k'l'm'} = 0$, we see that the coefficients $a_{klm,\,k'l'm'}$ define a hypersurface of the second degree in the n-dimensional space, the equation of which is $\Sigma_{klm\,k'l\,m'}\, a_{klm,\,k'l'm'}\, x_{klm}\, x_{k'l'm'} = 1$. The principal axes of this hypersurface are along the n unit vectors $\varphi^{(\nu)}_{klm}$ $(\nu = 1, 2, \ldots, n)$ and their lengths are equal to $[E^{(\nu)}]^{-\frac{1}{2}}$. The coefficients $c^{(\nu)}$ are now the components of an arbitrary vector when the axes of the system of coordinates are taken along the principal axes.

§ 25. Charged particles in an electromagnetic field

If an electrically charged particle is moving not only in an electric but also in a magnetic field, one is dealing with a problem which is in principle closely connected with the concepts of the theory of relativity. All the same it is possible to treat this problem in the framework of non-relativistic physics in a simple manner as long as one renounces some of the finer relativistic details. From classical relativistic electrodynamics we know that the electric vector $\mathfrak{E}$ and the magnetic vector $\mathfrak{H}$ can always be described through a scalar potential Φ and a vector potential $\mathfrak{A}$,

$$\mathfrak{E} = -\boldsymbol{\nabla}\Phi - c^{-1}\,\partial\mathfrak{A}/\partial t,\ \mathfrak{H} = \operatorname{curl}\mathfrak{A}, \tag{99}$$

where Φ and $\mathfrak{A}$ satisfy the equations

$$\begin{gathered}\boldsymbol{\nabla}^2\Phi - c^{-2}\,\partial^2\Phi/\partial t^2 = \Box\,\Phi = -4\pi\varrho,\quad (\boldsymbol{\nabla}\cdot\mathfrak{A}) + c^{-1}\,\partial\Phi/\partial t = 0,\\ \Box\,\mathfrak{A} = -4\pi\varrho\boldsymbol{v}/c,\end{gathered} \tag{100}$$

where ϱ is the charge density, $\varrho\boldsymbol{v}$ the current density, and c the velocity of light. The equations of motion of a charged particle of mass m and charge e follow from a Hamiltonian equation (compare §§ 11 and 1),

$$c^{-2}(E_{\mathrm{r}} - e\Phi)^2 - (\boldsymbol{P} - [e\mathfrak{A}/c]\cdot\boldsymbol{P} - [e\mathfrak{A}/c]) - m^2c^2 = 0, \tag{101}$$

where E_{r} and $\boldsymbol{P}$ are again the (relativistic) energy and the momentum of the particle. This equation follows from equation (1.13) by substituting $E - e\Phi$ for E and $\boldsymbol{P} - (e\mathfrak{A}/c)$ for $\boldsymbol{P}$. Just as one can call $e\Phi$ the potential energy and $E - e\Phi$ the kinetic energy, so one can call $e\mathfrak{A}/c$ the potential momentum and $\boldsymbol{P} - (e\mathfrak{A}/c)$ the kinetic momentum.

Writing as in equation (1.14) $E_{\mathrm{r}} = E_{\mathrm{n.r.}} + mc^2$ and neglecting a term $c^{-2}(E_{\mathrm{n.r.}} - e\Phi)^2$ one gets from equation (101)

$$(P^2/2m) - (e/mc)(\boldsymbol{P} \cdot \mathfrak{A}) + (e^2\mathfrak{A}^2/2mc^2) + e\Phi = E_{\mathrm{n.r.}}. \tag{102}$$

One checks easily that the Hamiltonian equations (3) which correspond to this energy function will lead to the sum of the electrical force $e\mathfrak{E}$ and the Lorentz-force $e[\boldsymbol{v} \wedge \mathfrak{H}]/c$ as the product of m and the acceleration of the particle.

The same prescription (5) which led earlier to the wave equation (8) can now be used to interpret the left hand side of equation (102) as an operator * and to obtain a wave equation. The result is

$$-(\hbar^2/2m)\nabla^2\psi + (e\hbar/2imc)[(\mathfrak{A} \cdot \nabla) + (\nabla \cdot \mathfrak{A})]\psi + [(e^2/2mc^2)\mathfrak{A}^2 + e\Phi]\psi = i\hbar\partial\psi/\partial t. \tag{103}$$

We may assume that equation (103) will describe in the non-relativistic approximation the change of the physical situation in time. From the considerations of § 27 it follows easily that the motion of a wave packet which satisfies equation (103) is to a first approximation the motion of a mass point in classical electrodynamics.

One can again interpret $|\psi|^2$ as the probability density J, but the probability current density is now given by the equation

$$\boldsymbol{S} = -i(\hbar/2m)[\psi^*\nabla\psi - \psi\nabla\psi^*] - (e/mc)\mathfrak{A}\psi^*\psi, \tag{104}$$

instead of by equation (21). One can easily check in the same way as before that the equation of continuity (20) is satisfied when $\boldsymbol{S}$ is given by equation (104) **.

If $\mathfrak{A}$ and Φ in equation (103) do not depend on the time, one can again investigate solutions of the form (27) to find stationary states in a stationary electromagnetic field. The situation is qualitatively the same as the one of a particle in a potential field and the

* The difficulty that the term $(e/mc)(\boldsymbol{P} \cdot \mathfrak{A})\psi = (e/mc)(\mathfrak{A} \cdot \boldsymbol{P})\psi$ can be interpreted both as $(e\hbar/imc)(\nabla \cdot \mathfrak{A})\psi$ and as $(e\hbar/imc)(\mathfrak{A} \cdot \nabla)\psi$ must be circumvented (see § 29) by taking half the sum of those two expressions. Note, though, that the difference of the two terms is equal to $(e\hbar/imc)\psi(\nabla \cdot \mathfrak{A}) = -(e\hbar/imc)\psi\partial\Phi/\partial t$ which is zero in the case of a stationary field.

** The choice of $\boldsymbol{S}$ follows immediately from equation (4.129).

existence of eigenvalues and eigenfunctions as well as the possibility of expanding an arbitrary function in terms of eigenfunctions can be made plausible by methods analogous to those of the previous section.

CHAPTER III

THE NON-RELATIVISTIC TREATMENT OF THE MANY-BODY PROBLEM

§ 26. The two-body problem

Before we consider the general quantum theory of interacting particles we shall illustrate in this section a few characteristic points of view by considering the case of two weakly interacting particles.

If during an experiment in which two particles are taking part one could assume that these particles did not interact at all, one could apply to both of them the considerations of the previous chapter. Indeed, those considerations could come to full fruition only when one considers experiments involving a large number of independent particles. At the moment, however, we want to consider the case where two particles can be treated as being free, although they may strongly interact during a possible collision. This situation is similar to the situation in kinetic theory when one considers the molecules in a perfect gas. If we want to investigate such a situation in more detail from the point of view of quantum theoretical concepts, we must find out in the first place how and in how far the conservation laws of energy and momentum can be taken into account. This will justify, by an appeal to the correspondence principle, the use of the concept "energy" and "momentum" even in the case of a one-particle problem (compare § 9).

Consider the case where the two particles approach each other in such a way that they may in the course of time collide. Classically both position and momentum of each particle could be well defined. The nearest approach to this classical situation is obtained in quantum theory by assigning to each of them a special wave packet (wave group) in such a way that the two packets are widely separated at $t = 0$ but that they are moving with such velocities and in such directions that they may partly or wholly overlap at a later time. The momenta $\boldsymbol{p}_1$ and $\boldsymbol{p}_2$ of the two particles are only approximately defined. If a collision occurs the momenta $\boldsymbol{p}_1'$ and $\boldsymbol{p}_2'$ after the collision may be very different from $\boldsymbol{p}_1$ and $\boldsymbol{p}_2$. However, according to the law of conservation of momentum the sum $\boldsymbol{p}_1'+\boldsymbol{p}_2'$ must be equal to $\boldsymbol{p}_1+\boldsymbol{p}_2$ within the same margin within which the latter sum was defined. The position at which the collision took place is indeterminate to the extent to which the two wave packets overlapped. To use the full accuracy with which the original momenta were defined one will assume that both packets have the same size — which corresponds to the same uncertainty for $\boldsymbol{p}_1$ and $\boldsymbol{p}_2$ — and that they occupy the same region of space during the "collision". In that way one obtains the closest analogy to the classical picture of a collision.

We now ask how we can describe mathematically the physical situation which was created by the "collision" of the two packets when we wish to preserve the analogy with the one-particle case. We may imagine that it is possible through suitable apparatus (gratings) to measure the momenta of the two particles after the collision. We may also assume that the experiment can be repeated as often as we want under identical conditions (by opening and shutting slits one can reproduce the original packets). If we find a wave vector $\boldsymbol{\sigma}_1$ for the first particle, we should find a wave vector $\boldsymbol{\sigma}_2$ for the second one such that $h(\boldsymbol{\sigma}_1+\boldsymbol{\sigma}_2)$ is approximately equal to the total momentum $h\boldsymbol{\sigma}$ *before* the collision, while energy conservation demands that $h^2[(\sigma_1^2/2m_1)+(\sigma_2^2/2m_2)]$ is equal to the total energy before the collision (we assume that we may use non-relativistic equations). We may, moreover, assume that it is possible to determine through suitable experiments a probability density and a probability current density for the second particle. This is done by considering only those results for which the momentum $h\boldsymbol{\sigma}_1$ of the first particle has a fixed value. We can thus describe the physical situation of the second particle by a wave function χ_2 which apart from depending on $\boldsymbol{x}_2$ and t_2 contains the wave number $\boldsymbol{\sigma}_1$ of the first particle as parameter,

$$\chi_2(\boldsymbol{\sigma}_1; \boldsymbol{x}_2, t_2) = \int A_2(\boldsymbol{\sigma}_1, \boldsymbol{\sigma}_2) \exp\{2\pi i[(\boldsymbol{\sigma}_2 \cdot \boldsymbol{x}_2) - \nu_2 t_2]\} d\boldsymbol{\sigma}_2. \tag{1}$$

We may assume that A_2 can be chosen, by suitably "normalising" χ_2, so that $|A_2|^2 d\boldsymbol{\sigma}_1 d\boldsymbol{\sigma}_2$ is the probability that the wave numbers of the two particles lie within the volume elements of $\boldsymbol{\sigma}$-space $d\boldsymbol{\sigma}_1$ and $d\boldsymbol{\sigma}_2$.

Similarly one reaches the conclusion that the physical situation of the first particle for a given momentum $h\boldsymbol{\sigma}_2$ of the second particle can be described by a wave function

$$\chi_1(\boldsymbol{x}_1, t_1; \boldsymbol{\sigma}_2) = \int A_1(\boldsymbol{\sigma}_1, \boldsymbol{\sigma}_2) \exp \{2\pi i[(\boldsymbol{\sigma}_1 \cdot \boldsymbol{x}_1) - \nu_1 t_1]\} \, d\boldsymbol{\sigma}_1, \qquad (2)$$

where clearly $|A_1|^2$ is the same function as $|A_2|^2$. These two functions will only differ appreciably from zero, if $h(\boldsymbol{\sigma}_1 + \boldsymbol{\sigma}_2)$ and $h^2[(\sigma_1^2/2m_1) + (\sigma_2^2/2m_2)]$ are practically equal to the total momentum and the total energy before the collision.

It is an obvious choice to take A_1 and A_2 identical and to put

$$A_1 \equiv A_2 \equiv A(\boldsymbol{\sigma}_1, \boldsymbol{\sigma}_2). \qquad (3)$$

We can now introduce a function χ by the equation

$$\chi(\boldsymbol{x}_1, t_1; \boldsymbol{x}_2, t_2) = \int A \exp\{2\pi i[(\boldsymbol{\sigma}_1 \cdot \boldsymbol{x}_1) + (\boldsymbol{\sigma}_2 \cdot \boldsymbol{x}_2) - \nu_1 t_1 - \nu_2 t_2]\} d\boldsymbol{\sigma}_1 d\boldsymbol{\sigma}_2. \qquad (4)$$

This function should enable us to characterise the physical situation and we see the analogy with the one-particle case. This function also enables us to state what the probability will be that the first particle at t_1 is within a volume element $d\boldsymbol{x}_1$ while the second particle at t_2 is within $d\boldsymbol{x}_2$. Such a probability gives us a possible experimental check. Denoting it by $J d\boldsymbol{x}_1 d\boldsymbol{x}_2$ we shall put

$$J = \chi^* \chi = |\chi|^2, \qquad (5)$$

since according to the theory of Fourier integrals it follows from equation (4) and the fact that $|A|^2$ is a probability density that

$$\int |\chi|^2 d\boldsymbol{x}_1 d\boldsymbol{x}_2 = \int |A|^2 d\boldsymbol{\sigma}_1 d\boldsymbol{\sigma}_2 = 1. \qquad (6)$$

If we now generalise this case, we shall assume that the general case involving two particles will be characterised by introducing an arbitrary (normalisable and according to equation (6) normalised) function A. This assumption is in accordance with the general quantum mechanical principle of the superposition of physical situations, mentioned in § 19. If we have two situations characterised by A' (or χ') and A'' (or χ''), there will also be a situation corresponding to a superposition of these two situations and characterised by $A''' = \alpha A' + \beta A''$ (or $\chi''' = \alpha\chi' + \beta\chi''$).

A special case occurs when A can be written as a product $A^{(1)}(\boldsymbol{\sigma}_1) \cdot A^{(2)}(\boldsymbol{\sigma}_2)$. The wave function χ will also be a product $\chi^{(1)}(\boldsymbol{x}_1, t_1)\chi^{(2)}(\boldsymbol{x}_2, t_2)$. The two particles can be considered to be independent as the result of an experiment on one particle will not depend on what one finds for the other one. In the case of a collision such as we just discussed, this case is realised before the collision. The most general situation can always be described by a superposition of such special situations. Equation (4) expresses especially that the general situation can be considered to be a superposition of situations where the momentum and energy of both particles is well defined.

The case where the two particles have not only the same mass but are completely indistinguishable is of the greatest physical importance. Examples are two electrons, two protons, or two α-particles. The physical equivalence of two such particles has as a consequence that two situations, characterised by A' and A'' can not be distinguished physically if A' can be obtained from A'' by exchanging the arguments belonging to the two particles,

$$A''(\boldsymbol{\sigma}_1, \boldsymbol{\sigma}_2) = A'(\boldsymbol{\sigma}_2, \boldsymbol{\sigma}_1), \ \chi''(\boldsymbol{x}_1, t_1;\ \boldsymbol{x}_2, t_2) = \chi'(\boldsymbol{x}_2, t_2;\ \boldsymbol{x}_1, t_1). \quad (7)$$

We shall return to the problems connected with this indistinguishability (Pauli exclusion principle) in Chapter VII.

It is well known that the fact that simultaneity has an absolute meaning in non-relativistic mechanics has important consequences. The interactions between particles separated in space may be considered to be due to instantaneously operating actions at distance which lead to a change in time of the instantaneous state of the system according to given laws. This instantaneous state is determined by giving the positions and momenta of all particles. In the example of two free particles one will characterise analogously the instantaneous physical situation by choosing in equation (4) $t_1 = t_2$ so that one obtains a wave function ψ which depends on the position coordinates of the two particles and the (absolute) time,

$$\psi(\boldsymbol{x}_1, \boldsymbol{x}_2, t) = \int A \exp\{2\pi i[(\boldsymbol{\sigma}_1 \cdot \boldsymbol{x}_1) + (\boldsymbol{\sigma}_2 \cdot \boldsymbol{x}_2) - (\nu_1 + \nu_2)t]\} d\boldsymbol{\sigma}_1 d\boldsymbol{\sigma}_2. \quad (8)$$

Since

$$\nu_1 = h\sigma_1^2/2m_1,\ \nu_2 = h\sigma_2^2/2m_2,\ \nu_1 + \nu_2 = h[(\sigma_1^2/2m_1) + (\sigma_2^2/2m_2)], \quad (9)$$

the integrand of equation (8) and thus also ψ will satisfy the linear homogeneous differential equation

$$\frac{1}{2m_1}\nabla_1^2\psi + \frac{1}{2m_2}\nabla_2^2\psi = -\frac{i}{\hbar}\frac{\partial\psi}{\partial t}, \quad \nabla_i^2 = \frac{\partial^2}{\partial x_i^2} + \frac{\partial^2}{\partial y_i^2} + \frac{\partial^2}{\partial z_i^2}. \quad (10)$$

This is the *Schrödinger equation for two free particles.* It can be obtained from the well-known relation between the momenta of the two particles and their total energy,

$$(p_1^2/2m_1)+(p_2^2/2m_2) = E, \tag{11}$$

by interpreting these equations as differential operators according to the rule

$$\boldsymbol{p}_1 = -i\hbar\nabla_1, \quad \boldsymbol{p}_2 = -i\hbar\nabla_2, \quad E = i\hbar\,\partial/\partial t, \tag{12}$$

and by letting equation (11), after substituting into it equations (12), operate on ψ.

If the instantaneous physical situation at $t=0$ is characterised by $\psi_0(\boldsymbol{x}_1, \boldsymbol{x}_2)$, it will be determined at all times according to equation (10). From equations (5) and (6) it follows that, if $J(\boldsymbol{x}_1, \boldsymbol{x}_2, t)$ is defined by

$$J = \psi^*\psi, \tag{13}$$

$Jd\boldsymbol{x}_1 d\boldsymbol{x}_2$ can be interpreted as being the probability that at t the first particle is within the volume element $d\boldsymbol{x}_1$ and the second one within $d\boldsymbol{x}_2$. It is necessary for this interpretation that $\int|\psi|^2 d\boldsymbol{x}_1 d\boldsymbol{x}_2$ does not vary in time and is equal to unity. That this integral is constant follows, provided it converges, either from equation (6) or from the differential equation (10) in the same way as equation (2.19) followed from equation (2.8). If it converges, one can always normalise ψ so that $\int|\psi|^2 d\boldsymbol{x}_1 d\boldsymbol{x}_2 = 1$.

By restricting ourselves to describing only the instantaneous situation of the two particles we have made it possible to describe in a simple manner the system even when there are all the time instantaneous interactions. Let the system be characterised by a wave function $\psi(\boldsymbol{x}_1, \boldsymbol{x}_2, t)$ such that if at $t=t_0$ all forces suddenly vanished the situation would be described by equation (10), that is, be characterised by a function of the form (8) which at $t = t_0$ was equal to $\psi(\boldsymbol{x}_1, \boldsymbol{x}_2, t_0)$.

Let $U(\boldsymbol{x}_1, \boldsymbol{x}_2)$ be the interaction potential in the classical description of the two-body system, so that instead of equation (11) we have classically

$$(p_1^2/2m_1)+(p_2^2/2m_2)+U = E. \tag{14}$$

If U were simply independent of $\boldsymbol{x}_1$ and $\boldsymbol{x}_2$, that is, if the two particles behaved as free particles, one could still use a superposition of free de Broglie waves as in equation (8), but equations (9) would no longer hold. The question of what one should write instead of the first two of these equations is obviously meaningless.

Instead of the third equation (9) one writes immediately

$$\nu_1+\nu_2 = h[\sigma_1^2/2m_1)+(\sigma_2^2/2m_2)]+U/h, \tag{15}$$

and instead of equation (10) we get a slightly more complicated equation,

$$(2m_1)^{-1}\nabla_1^2\psi+(2m_2)^{-1}\nabla_2^2\psi-\hbar^{-2}U\psi = -(i/\hbar)\partial\psi/\partial t. \tag{16}$$

The simplest way of extending this equation to the case where U depends on $\boldsymbol{x}_1$ and $\boldsymbol{x}_2$ is to declare equation (16) still to be valid. Indeed, one assumes in wave mechanics that equation (16), *the Schrödinger equation of the two-body problem*, will give us the correct change in time of the instantaneous physical situation. This way of describing the system is closely connected with the classical description of the interaction as we shall see in the next section.

§ 27. The Schrödinger equation of many interacting particles

Consider a system consisting of mass points which interact and are in external fields of force. Let $\boldsymbol{x}_i$ $(i = 1, 2, \ldots, n)$ be the position in Cartesian coordinates of the i-th particle and $\boldsymbol{F}_i$ the force acting on it so that the non-relativistic equations of motion are of the form

$$m_i\ddot{\boldsymbol{x}}_i = \dot{\boldsymbol{p}}_i = \boldsymbol{F}_i, \qquad i = 1, 2, \ldots, n. \tag{17}$$

In equations (17) m_i and $\boldsymbol{p}_i$ are the mass and momentum of the i-th particle.

If the forces can be derived from a potential $U(\boldsymbol{x}_1, \ldots, \boldsymbol{x}_n, t)$ according to

$$\boldsymbol{F}_i = -\nabla_i U, \tag{18}$$

the equations (17) can be written in the canonical form

$$\dot{p}_k = -\partial H/\partial q_k, \; \dot{q}_k = \partial H/\partial p_k, \qquad k = 1, 2, \ldots, 3n, \tag{19}$$

where the position coordinates and corresponding momenta are denoted by q_k and p_k while H is the energy E of the system as a function of the p_k and q_k,

$$E = H = \Sigma_i(\boldsymbol{p}_i^2/2m_i)+U. \tag{20}$$

If we denote as in § 11 the time with q_0 and introduce the conjugate momentum $p_0(= -E)$ we can derive the equations of motion from a Hamiltonian equation,

$$\mathscr{H} \equiv H(p_1, \ldots, p_{3n}, q_1, \ldots, q_{3n}, q_0) + p_0 = 0 \tag{21}$$

in the form

$$\dot{p}_k = -\partial\mathscr{H}/\partial q_k, \quad \dot{q}_k = \partial\mathscr{H}/\partial p_k, \quad k = 0, 1, 2, \ldots, 3n. \tag{22}$$

If we take into account the considerations of the previous section and the general treatment of the one-particle problem in chapter 2, it seems plausible to characterise the physical situation of our n-particle system by a wave function $\psi(\boldsymbol{x}_1, \ldots, \boldsymbol{x}_n, t) \equiv \psi(q_1, \ldots, q_{3n}, t)$ such that ψ at any time determines the physical situation of the system and that the change of ψ with time is governed by the following linear homogeneous differential equation,

$$\mathscr{H}_{\mathrm{op}}\,\psi = 0. \tag{23}$$

In equation (23) $\mathscr{H}_{\mathrm{op}}$ is a differential operator of the second order which is obtained from equation (21) by putting

$$p_k = -i\hbar\partial/\partial q_k, \qquad k = 0, 1, 2, \ldots, 3n, \tag{24}$$

or, in particular,

$$\boldsymbol{p}_i = -i\hbar\boldsymbol{\nabla}_i, \qquad -p_0 = E = i\hbar\partial/\partial q_0 = i\hbar\partial/\partial t. \tag{25}$$

From equations (20), (21), (23), and (25) we get the explicit form of the *Schrödinger equation of an n-particle system,*

$$-\sum_i \frac{\hbar^2}{2m_i}\boldsymbol{\nabla}_i^2\psi + U\psi = i\hbar\frac{\partial\psi}{\partial t}, \quad \boldsymbol{\nabla}_i^2 \equiv \frac{\partial^2}{\partial x_i^2} + \frac{\partial^2}{\partial y_i^2} + \frac{\partial^2}{\partial z_i^2}. \tag{26}$$

To discuss the wave equation (23) we consider the general case where $\mathscr{H}$ is an arbitrary finite polynomial in the p_k *

$$\mathscr{H}(p_0, p_1, \ldots) = \Sigma\, b\, p_0^{a_0} p_1^{a_1} \ldots, \tag{27}$$

where the coefficients b are functions of the q_k. If the b were constants, equation (23) would have solutions in the form

$$\psi = C \exp\,[2\pi i(\sigma_0 q_0 + \sigma_1 q_1 + \ldots)], \tag{28}$$

where the σ_k satisfy the equation $\mathscr{H}(h\sigma_0, h\sigma_1, \ldots) = 0$. Let us assume that we can find a solution of this equation where all σ_k are real so that ψ is periodic in all q_k. When the b are not constant, we assume ψ to be of the form

$$\psi = C(q_0, q_1, \ldots) \exp\,[2\pi i\chi(q_0, q_1, \ldots)], \tag{29}$$

* The discussion follows closely the treatment used by H. A. Lorentz in his lectures in 1927. I am grateful to Professor A. D. Fokker for communicating this discussion to me. Similar considerations can be found in § 40 of Dirac's *The principles of quantum mechanics* (Oxford University Press 1930).

where both C and χ are real functions. Near a point $\bar{q}_0, \bar{q}_1, \ldots$ in q-space we have

$$\chi = \chi(\bar{q}_0, \bar{q}_1, \ldots) + \Sigma_k [\partial\chi/\partial q_k]_{\bar{q}_0, \bar{q}_1, \ldots} (q_k - \bar{q}_k) + \ldots$$

Using the abbreviations

$$\partial\chi/\partial q_k = \sigma_k, \quad \partial^2\chi/\partial q_k \partial q_{k'} = \sigma_{kk'}, \tag{30}$$

we can write this last equation in the form

$$\chi = \bar{\chi} + \Sigma_k \bar{\sigma}_k (q_k - \bar{q}_k) + \ldots \tag{31}$$

The function ψ of equation (29) will still be approximately periodic in the q_k as long as both the wave numbers σ_k and the amplitude function C change very little inside a region of q-space of the order of magnitude of the wave lengths $(\bar{\sigma}_k)^{-1}$,

$$(\partial\sigma_k/\partial q_{k'})(\sigma_{k'})^{-1} \ll \sigma_k, \text{ or, } \sigma_{kk'} \ll \sigma_k \sigma_{k'}; \tag{32}$$

$$(\partial C/\partial q_k)(\sigma_k)^{-1} \ll C, \text{ or, } \partial C/\partial q_k \ll C\sigma_k. \tag{33}$$

We assume that C and χ are "smoothed" functions so that if equations (32) and (33) are differentiated a few times with respect to the q_k, the $\ll$-sign will still hold.

We now assume that there exist solutions (29) which satisfy all these conditions. If we consider p_k to be an operator according to equation (24) we get

$$p_{k\,\mathrm{op}}\psi = [Ch\sigma_k - i\hbar(\partial C/\partial q_k)] \exp(2\pi i\chi),$$

where the second term is small compared to the first one according to equation (33). By now applying $p_{k'\,\mathrm{op}}$ we get

$$p_{k'\,\mathrm{op}} p_{k\,\mathrm{op}}\psi = \{Ch\sigma_k h\sigma_{k'} - i\hbar[(\partial C/dq_{k'})h\sigma_k + (\partial C/\partial q_k)h\sigma_{k'} + Ch\sigma_{k'k}] \\ - \hbar^2(\partial^2 C/\partial q_k \partial q_{k'})\} \exp(2\pi i\chi). \tag{34}$$

The second term on the right hand side is small compared to the first one according to equations (32) and (33), and the third term is again small compared to the second one. This third term we shall neglect. Continuing we get

$$p_{k''\,\mathrm{op}} p_{k'\,\mathrm{op}} p_{k\,\mathrm{op}}\psi = \{Ch\sigma_k h\sigma_{k'} h\sigma_{k''} - i\hbar[(\partial C/\partial q_k)h\sigma_{k'} h\sigma_{k''} \\ + (\partial C/\partial q_{k'})h\sigma_k h\sigma_{k''} + (\partial C/\partial q_{k''})h\sigma_k h\sigma_{k'} + C(h\sigma_{kk'} h\sigma_{k''} + h\sigma_{kk''} h\sigma_{k'} \\ + h\sigma_{k'k''} h\sigma_k)]\} \exp(2\pi i\chi) + \ldots$$

The first term originated by operating with $p_{k''\,\mathrm{op}}$ on $\exp(2\pi i\chi)$ in the first term on the right hand side of equation (34), the second term by operating on exp $(2\pi i\chi)$ in the second term and on the factor of exp $(2\pi i\chi)$ in the first term. All other terms are small compared to the ones retained according to equations (32) and (33). Similarly we can operate on ψ with any number of factors p_k and we can retain only the dominant and subdominant terms. One can easily check that the final result of $\mathscr{H}_{\text{op}}$ acting on ψ will be of the form

$$\mathscr{H}_{\text{op}}\psi=\left\{C\mathscr{H}-i\hbar\left[\Sigma_k\frac{\partial C}{\partial q_k}\frac{\partial\mathscr{H}}{\partial p_k}+\tfrac{1}{2}\Sigma_{k,k'}h\sigma_{kk'}\frac{\partial^2\mathscr{H}}{\partial p_k\partial p_{k'}}\right]\right\}\exp(2\pi i\chi)+\cdots, \tag{35}$$

where everywhere the argument p_k in $\mathscr{H}$, $\partial\mathscr{H}/\partial p_k$, $\partial^2\mathscr{H}/\partial p_k\partial p_{k'}$ is replaced by $h\sigma_k$ and where the second summation is over both k and k'. As the ratio of the first to the second term on the right hand side of equation (35) is purely imaginary, we get from equation (23) two equations,

$$\mathscr{H}(h\sigma_0, h\sigma_1, \ldots)\equiv\mathscr{H}(\partial(h\chi)/\partial q_0, \partial(h\chi)/\partial q_1, \ldots)=0, \tag{36}$$

$$\sum_k\frac{1}{c}\frac{\partial C}{\partial q_k}\frac{\partial\mathscr{H}}{\partial p_k}=-\tfrac{1}{2}\sum_{k,k'}\frac{\partial^2\mathscr{H}}{\partial p_k\partial p_{k'}}h\sigma_{kk'}. \tag{37}$$

Equation (36) is a differential equation which $\chi(q_0, q_1, \ldots)$ must satisfy approximately. If one puts $h\chi = S$, it is identical with the Hamilton-Jacobi equation for S. The solution of the canonical Hamiltonian equation can in classical mechanics be reduced to a solution of this Hamilton-Jacobi equation.

Equation (37) is a condition which the amplitude function C must satisfy to a first approximation. If χ is a solution of equation (36) satisfying conditions (32), $\partial\mathscr{H}/\partial p_k$ and $\partial^2\mathscr{H}/\partial p_k\partial p_{k'}$ are known functions of the q_k and the solution of equation (37) can be interpreted as follows. We first of all define in q-space for each solution χ a set of curves satisfying the equations

$$dq_k/d\tau = \partial\mathscr{H}/\partial p_k\,, \tag{38}$$

where τ is a parameter and where the right hand side is a known function of the q_k for a given χ as we saw a moment ago. Consider now the change of C along such a curve. It is given by the equation

$$\frac{1}{C}\frac{dC}{d\tau}=\sum_k\frac{1}{C}\frac{\partial C}{\partial q_k}\frac{dq_k}{d\tau}=-\tfrac{1}{2}\sum_{k,k'}\frac{\partial^2\mathscr{H}}{\partial p_k\partial p_{k'}}h\sigma_{kk'}. \tag{39}$$

where we have used equations (37) and (38). Assume now that we may choose $C(q_0, q_1, \ldots)$ for a fixed value $\bar{q}_0$ of q_0 (which is again identical with the time t) as an arbitrary function of the other q_k, as long as it satisfies equation (33) *. The values of C for other values of $q_0(=t)$ follow from equation (39) by constructing in each point of the hyperspace $q_0 = \bar{q}_0$ the curves (38) and integrating along them according to equation (39). If we choose especially C such that C for $q_0 = \bar{q}_0$ is appreciably different from zero only in the neighbourhood of the point $q_1 = \bar{q}_1, q_2 = \bar{q}_2, \ldots$, we are dealing with a *wave group* which is moving in $q_1, q_2, \ldots$-space with "time" q_0 according to equation (38). The wave numbers $\bar{\sigma}_1, \bar{\sigma}_2, \ldots$ of this group [see equation (31)] are approximately defined and approximately equal to $[\partial\chi/\partial q_k]_{\bar{q}_0, \bar{q}_1, \ldots}$. The change of the wave numbers along the curves (38) is obtained from the equation

$$\frac{d\sigma_k}{d\tau} = \sum_{k'} \frac{\partial^2\chi}{\partial q_k \partial q_{k'}} \frac{dq_{k'}}{d\tau} = \sum_{k'} \frac{\partial\mathscr{H}}{\partial p_k} \frac{\partial^2\chi}{\partial q_k \partial q_{k'}} = -\frac{1}{h}\frac{\partial\mathscr{H}}{\partial q_k}, \tag{40}$$

where we have used equation (38) and the fact that χ satisfies equation (36). Indeed, by differentiating equation (36) with respect to q_k, taking into account that $\mathscr{H}$ depends explicitly on the q_k through the b, we find

$$0 = \frac{\partial\mathscr{H}}{\partial q_k} + \sum_{k'} \frac{\partial\mathscr{H}}{\partial p_{k'}} \frac{\partial p_{k'}}{\partial q_k} = \frac{\partial\mathscr{H}}{\partial q_k} + \sum_{k'} \frac{\partial\mathscr{H}}{\partial p_{k'}} \frac{\partial^2 h\chi}{\partial q_{k'} \partial q_k}.$$

Denoting the (approximately defined) values of $h\sigma_k$ along the curves (38) by p_k we find that these curves are characterised by the equations (38) and (40), or,

$$dp_k/d\tau = -\partial\mathscr{H}/\partial q_k, \quad dq_k/d\tau = \partial\mathscr{H}/\partial p_k, \quad k = 0, 1, 2, \ldots \tag{41}$$

The (approximately defined) motion of the wave group occurs thus just according to the same laws which govern in classical mechanics the change in time of a system the instantaneous state of which is determined by giving the values of the generalised coordinates $q_1, q_2, \ldots$ and the canonically conjugate momenta $p_1, p_2, \ldots$ while its energy $-p_0$ is determined as a function of $q_0, q_1, q_2, \ldots, p_1, p_2, \ldots$ by putting $\mathscr{H}$ of equation (27) equal to zero [compare equation (2.11) and (2.11′)]. A special case of this result is that of the motion of a wave group corresponding to a mass point in three-

* This is allowed if $dq_0/d\tau(=\partial\mathscr{H}/\partial p_0)$ is nowhere zero which is, for instance, the case for the $\mathscr{H}$ given by equation (21).

dimensional space. In the case of a free particle we get the uniform straight-line motion of Galilei's law of inertia which we discussed in detail in chapter 1. In the case of a particle in an external field of force we proved the correspondence between the motion of wave groups and Newton's laws of motion in § 12. To complete the connexion with those earlier cases we need only introduce the (approximately defined) frequency in time ν of the wave phenomenon described by equation (29) the absolute values of which is equal to $|\sigma_0|$ according to equation (31). We put $\nu = -\sigma_0$ and the (approximately defined) energy which corresponds to the motion of the wave group (29) is now given by the equation

$$E = -p_0 = -h\sigma_0 = h\nu. \tag{42}$$

§ 28. The interpretation of the wave function

If in accordance with the considerations of the previous section the physical situation of a system of several particles is characterised by a wave function $\psi(\boldsymbol{x}_1, \boldsymbol{x}_2, \ldots, \boldsymbol{x}_n, t)$ we can start the interpretation of this function by stating that at time t the value of a coordinate of one of the particles, say x_k, is approximately determined, if ψ at t is appreciably different from zero only in the neighbourhood of one particular value of x_k, independent of the values of all other coordinates. In order that one of the momentum coordinates, for instance p_{xk}, is approximately defined ψ should behave approximately periodically with regard to the corresponding x_k ($\psi \sim \exp[2\pi i\sigma_{xk}x_k]$) again independent of the values of all other coordinates; in that case $h\sigma_{xk}$ is the (approximate) value of p_{xk}. To get an approximate definition of the energy E of the system, ψ must be approximately periodic in the time ($\psi \sim \exp[-2\pi i\nu t]$), independent of the values of all coordinates, and we have $E \sim h\nu$. From these considerations it follows at once that there exists a complementarity of the definability of a coordinate and of its canonically conjugate momentum, that is, the validity of the Heisenberg relations (1.41) for each particle. Also we have the relation $\Delta E \,.\, \Delta t \sim h$ as in § 16.

We have seen in the last section that if we are dealing with a situation where at the same time the coordinates and the momenta of the particles are as well defined as possible, the change in time of these variables will take place according to the classical equations of motion. These equations can be obtained from a Hamiltonian equation $\mathscr{H} = 0$ by substituting in the wave equation (23) $ip_k/\hbar$

for $\partial/\partial q_k$. Therefore, if the Hamiltonian equation for a many particle system has been written down on the basis of purely classical considerations and if the prescription (25) leads uniquely to the construction of a quantum mechanical wave equation, then one would expect that such a description of the physical phenomena has been obtained which satisfies the requirements of both Bohr's correspondence principle (§ 4) and the basic de Broglie postulates of quantum theory.

Present day non-relativistic quantum mechanics, indeed, fulfills this expectation. However, it has become clear that the uniqueness of the construction of the wave equation and thus the possibility of a *quantitative formulation of probability statements* is clearly connected with the particular form of the Hamiltonian equations which occur in practice. We shall postpone until Chapter VI the complicating accompanying circumstances due to the fact that only a description which takes the *spin of the elementary particles* into account can be satisfactory. We now ask whether it is possible to introduce a real, positive function $J(\boldsymbol{x}_1, \ldots, \boldsymbol{x}_n, t)$ as probability density, that it, a function such that $J d\boldsymbol{x}_1 \ldots d\boldsymbol{x}_n$ is the probability that at t the first particle is within the volume element $d\boldsymbol{x}_1$, the second particle within $d\boldsymbol{x}_2$, and so on. The considerations of chapter 2 and § 26 lead us to the following general equation for J,

$$J = \psi^* \psi. \tag{43}$$

We require that J can be normalised,

$$\int \psi^* \psi \, d\tau = 1, \tag{44}$$

where the integral extends over all values of all the coordinates and where $d\tau \equiv d\boldsymbol{x}_1 \ldots d\boldsymbol{x}_n$.

The integral in equation (44) will not vary in time, if *two conditions* are satisfied. *First of all* in the wave equation (23) $\mathscr{H}_{\text{op}}$ the general form of which is given by equation (27), must be of the form (21), that is, *it must be possible to write the energy $E = -p_0$ of the system as a finite polynomial H in $p_1, p_2, \ldots$.* The coefficients in the polynomial may depend on the q_k and the time $t = q_0$. This condition is always fulfilled in the case of classical, non-relativistic many-particle systems [compare equation (20)]. The wave equation will be a differential equation of the first order in t of the form

$$H_{\text{op}} \psi = i\hbar \, \partial\psi/\partial t, \tag{45}$$

where H_{op} is obtained from the energy function H through the prescription (24).

Multiplying equation (45) by ψ^* and taking the imaginary part of the ensuing equation we have

$$\psi^* H_{\text{op}} \psi - \psi H^*_{\text{op}} \psi^* = i\hbar \partial(\psi^* \psi)/\partial t,$$

where H^*_{op} is obtained from H_{op} by everywhere substituting $i\hbar\partial/\partial q_k$ for $-i\hbar\partial/\partial q_k$. Integrating over all coordinates and assuming the necessary convergence we have

$$(d/dt) \int \psi^* \psi \, d\tau = -(i/\hbar) \int [\psi^* H_{\text{op}} \psi - \psi H^*_{\text{op}} \psi^*] \, d\tau. \tag{46}$$

The right hand side of equation (46) vanishes only if a *second condition* is fulfilled. This condition is that *the energy operator* H_{op} *must be a Hermitean operator.*

A Hermitean operator Ω_{Herm} is defined as an operator for which for any pair of functions f and g the following equation holds (compare also § 29)

$$\int f \Omega_{\text{Herm}} g \, d\tau = \int g \Omega^*_{\text{Herm}} f \, d\tau, \tag{47}$$

provided that the integrals converge. The integration is over all values of the coordinates $q_1, q_2, \ldots,$ and

$$d\tau = \varrho dq_1 dq_2 \ldots, \tag{48}$$

where $\varrho(q_1, q_2, \ldots)$ is a density function the form of which depends on the choice of the coordinates. In our case where the q_k are Cartesian coordinates, $\varrho = 1$ [compare equation (44)]. Introduction of new coordinates (for instance, polar coordinates) would at once lead to a ϱ which is different from one (compare § 34).

If we put in equation (47) $f = \psi^*$, $g = \psi$ and if H_{op} is Hermitean so that we can put $\Omega_{\text{Herm}} = H_{\text{op}}$, we see from equation (47) that the right hand side of equation (46) is equal to zero. The condition that H_{op} be Hermitean is actually satisfied for the Schrödinger equation (26) as can be shown by twice integrating by parts (compare also § 29). The expression (43) for the probability density is thus also justified for many-particle systems. If ψ is a solution of equation (45) for which $\int |\psi|^2 d\tau$ is finite, this integral can always be made equal to unity by suitable normalisation.

It is also simple to give a quantitative probability distribution law for the momenta of the particles in analogy with the treatment of one and two-particle systems. We write ψ in the form of a

Fourier integral,

$$\psi=\int A(\boldsymbol{\sigma}_1, \ldots, \boldsymbol{\sigma}_n, t) \exp\{2\pi i[(\boldsymbol{\sigma}_1 \cdot \boldsymbol{x}_1)+\cdots+(\boldsymbol{\sigma}_n \cdot \boldsymbol{x}_n)]\} d\boldsymbol{\sigma}_1 \cdots d\boldsymbol{\sigma}_n, \tag{49}$$

with

$$A=\int \psi \exp\{-2\pi i[(\boldsymbol{\sigma}_1 \cdot \boldsymbol{x}_1)+\cdots+(\boldsymbol{\sigma}_n \cdot \boldsymbol{x}_n)]\} d\boldsymbol{x}_1 \cdots d\boldsymbol{x}_n,$$
$$\int |A|^2 d\boldsymbol{\sigma}_1 \cdots d\boldsymbol{\sigma}_n = \int |\psi|^2 d\boldsymbol{x}_1 \cdots d\boldsymbol{x}_n. \tag{50}$$

We can clearly interpret $|A|^2 d\boldsymbol{\sigma}_1 \cdots d\boldsymbol{\sigma}_n = (|A|^2/h^{3n}) d\boldsymbol{p}_1 \cdots d\boldsymbol{p}_n$ as the probability that at t the momenta $\boldsymbol{p}_1, \ldots$ lie within the volume elements $d\boldsymbol{p}_1, \ldots$ of momentum space.

We shall discuss in Chapter IV how one can use the wave function to characterise the physical situation of a system by defining the instantaneous value of any mechanical quantity.

§ 29. Operators

Any rule which enables us to construct from a function f of some arguments a new function g of the same arguments defines an operator * and we write symbolically

$$g = \Omega f. \tag{51}$$

If for any two functions f_1 and f_2 and any two arbitrary constants a_1 and a_2 we have

$$\Omega(a_1 f_1 + a_2 f_2) = a_1 \Omega f_1 + a_2 \Omega f_2, \tag{52}$$

we call Ω a *linear operator* **. In the following we shall consider only linear operators. The sum Ω_3 of two operators Ω_1 and Ω_2 is defined by the equation

$$\Omega_3 f = (\Omega_1 + \Omega_2) f = \Omega_1 f + \Omega_2 f. \tag{53}$$

In expressions which contain linear operators as symbolical factors one can in general apply the distributive law of multiplication.

The product Ω_3 of two operators Ω_1 and Ω_2 is defined by the equation

$$\Omega_3 f = (\Omega_1 \Omega_2) f = \Omega_1(\Omega_2 f),$$

* In many cases it is necessary to impose on f some restrictions, for instance, differentiability, in order that equation (51) makes sense.

** Examples of such operators for the case of functions of one variable are

$$\Omega = x^2 \rightarrow [\Omega f(x) = x^2 f(x)], \quad \Omega = d/dx \rightarrow [\Omega f(x) = df/dx],$$
$$\Omega = \int_a^b dx' g(x', x) \ldots \rightarrow [\Omega f(x) = \int_a^b dx' g(x', x) f(x')].$$

where the last member of the equation indicates the function which is obtained by letting Ω_1 operate on $\Omega_2 f$. In general the commutative law of multiplication does not hold for the product of two operators, that is in general $\Omega^2\Omega_2 f \neq \Omega_2\Omega^2 f$. The associative law applies clearly to the product of three operators,

$$(\Omega_1\Omega_2)\Omega_3 f = \Omega_1(\Omega_2\Omega_3)f = \Omega_1\Omega_2\Omega_3 f.$$

If we denote the operator $\Omega_1\Omega_2-\Omega_2\Omega_1$ by $[\Omega_1, \Omega_2]_-$ so that we have

$$\Omega_1\Omega_2-\Omega_2\Omega_1 = [\Omega_1, \Omega_2]_- = -[\Omega_2, \Omega_1]_-, \tag{54}$$

then the following relations hold

$$\left.\begin{aligned} [\Omega_1, \Omega_3+\Omega_4+\cdots]_- &= [\Omega_1, \Omega_3]_-+[\Omega_1, \Omega_4]_-+\cdots; \\ [\Omega_1, \Omega_3\Omega_4\Omega_5\cdots]_- &= [\Omega_1, \Omega_3]_-\Omega_4\Omega_5\cdots \\ &\quad +\Omega_3[\Omega_1, \Omega_4]_-\Omega_5\cdots+\Omega_3\Omega_4[\Omega_1, \Omega_5]_-\cdots+\cdots \end{aligned}\right\} \tag{55}$$

The associative law is often not valid for expressions containing several operators and functions. For instance, in general

$$(\Omega_1 f_1)f_2 \neq \Omega_1(f_1 f_2), \quad (\Omega_1 f_1)(\Omega_2 f_2) \neq \Omega_1\{f_1(\Omega_2 f_2)\}.$$

In general one is also not allowed to exchange two factors of which one is an operator and the other a function. If all operators which occur in an expression are simply prescriptions to multiply by a given function, they can be treated in this expression in the same way as these functions and the associative and commutative laws hold again. An example of such an operator is the *unit operator* Ω_E which prescribes multiplication by 1,

$$\Omega_E f = f. \tag{56}$$

A simple and for quantum theory very important branch of operator theory is the one dealing with functions of a discrete variable. A function of such a variable is simply a (finite or infinite) series of numbers $f_{klm\ldots}$, which may be complex. The indices will in general be taken to have integer values. If one wants to, one can consider f to be a function of the variables $k, l, m, \ldots$ which then are subsequently restricted to a range of integer values. The number of values which the function takes on is in this case either finite or enumerably infinite. In principle, it is always possible to denote the values of the function by f_k where there is now only one index, k, which runs from 1 either to a finite number n or to infinity. In geometric language one can imagine the function represented by a point (or vector) with (complex) coordinates $f_1, f_2, \ldots$ in an n-dimensional or ∞-dimensional space.

A linear operator which acts on such a function of a discrete argument is determined by giving a set of numbers Ω_{km}, where the indices k and m independently take on the values 1, 2, ... By the equation $g = \Omega f$ is in this case understood the equation

$$g_k = \Sigma_m \Omega_{km} f_m . \tag{57}$$

Indeed, if $\delta^{(m)}$ is a function which is 1 when the index is equal to m, and zero otherwise and if we denote by Ω_{km} the values for index k of the function $\varepsilon^{(m)} = \Omega\delta^{(m)}$ we have, using equation (52),

$$(\Omega f)_k = (\Omega \Sigma_m f_m \delta^{(m)})_k = \Sigma_m \Omega_{km} f_m .$$

It must be noted that the f_m are numbers like the a_i in equation (52). In the last equation we denoted by $(\Omega f)_k$ the value of the function Ωf for index k. We shall always assume that all sums which we encounter will possess the necessary convergence properties.

We express the situation as follows. The operator Ω is characterised by the matrix elements Ω_{km}, or, each operator corresponds to a matrix *. If $\Omega^{(3)}$ is the product of $\Omega^{(1)}$ and $\Omega^{(2)}$ we find the matrix elements of $\Omega^{(3)}$ from the equation

$$(\Omega^{(3)} f)_k = \Sigma_m \Omega^{(3)}_{km} f_m = \Sigma_l \Omega^{(1)}_{kl} (\Omega^{(2)} f)_l = \Sigma_l \Omega^{(1)}_{kl} \Sigma_m \Omega^{(2)}_{lm} f_m = \Sigma_{l\,m} \Omega^{(1)}_{kl} \Omega^{(2)}_{lm} f_m .$$

Since the above equation must be valid for any f, and thus also for the functions $\delta^{(m)}$, it follows that

$$\Omega^{(3)}_{km} = \Sigma_l \Omega^{(1)}_{kl} \Omega^{(2)}_{lm} . \tag{58}$$

This equation expresses the fact that the matrix corresponding to $\Omega^{(3)}$ is obtained from those corresponding to $\Omega^{(1)}$ and $\Omega^{(2)}$ through the rules of *matrix multiplication.* This multiplication is in general non -commutative, that is, in general $(\Omega^{(1)}\Omega^{(2)})_{km} \neq (\Omega^{(2)}\Omega^{(1)})_{km}$.

The elements for which $k = l$ are called *diagonal elements.* A matrix is called a *diagonal matrix,* if all off-diagonal elements $(k \neq l)$ are equal to zero. In a product of two diagonal matrices one can change the factors.

The diagonal matrix of which all diagonal elements are equal to

* A matrix is the (finite or infinite) quadratic scheme of its elements:

$$= \begin{matrix} \Omega_{11} & \Omega_{12} & \Omega_{13} \ldots \\ \Omega_{21} & \Omega_{22} & \Omega_{23} \ldots \\ \ldots & \ldots & \ldots \end{matrix} =$$

1 is called the unit matrix and is denoted by the Weierstrass symbol δ_{kl},

$$\delta_{kl} = 1,\ k = l;\ \delta_{kl} = 0,\ k \neq l. \tag{59}$$

An operator Ω_E corresponding to the unit matrix will leave a function f on which it acts unchanged, $(\Omega_E f)_k = \Sigma_l \delta_{kl} f_l = f_k$. This operator corresponds thus to multiplication by 1 and is thus identical with the unit operator.

An arbitrary diagonal matrix D_{kl} can be represented by $D_{kl} = a_k \delta_{kl}$, where a_k is the value of a function a for index k. If an operator Ω_D corresponding to a diagonal matrix D acts on a function f, we have $(\Omega_D f)_k = \Sigma_l a_k \delta_{kl} f_l = a_k f_k$, and we see that Ω_D corresponds to "multiplying" f by the function a.

An operator is called *Hermitean,* if the corresponding matrix is Hermitean, that is, if the matrix elements Ω_{kl} satisfy the relation

$$\Omega_{kl} = \Omega^*_{lk}. \tag{60}$$

From this equation we see that the diagonal elements $(\Omega_{kk} = \Omega^*_{kk})$ must be real. If the off-diagonal elements also happen to be real so that we have $\Omega_{kl} = \Omega_{lk}$, we call the matrix *symmetrical.*

We shall denote by $\int f$ the sum of all values f_k of the function f, $\int f = \Sigma_k f_k$. For the sum of the product of two functions f and Ωg we have

$$\int f\Omega g = \Sigma_k f_k (\Omega g)_k = \Sigma_{kl} f_k \Omega_{kl} g_l = \Sigma_{kl} g_k \Omega_{lk} f_l = \int g\tilde{\Omega} f, \tag{61}$$

where $\tilde{\Omega}$ is the *adjoint operator* of Ω. Its matrix elements $\tilde{\Omega}_{kl}$ are obtained from those of Ω by transposition, that is, by exchanging the indices,

$$\tilde{\Omega}_{kl} = \Omega_{lk}. \tag{62}$$

The complex conjugated operator Ω^* of Ω is defined generally by the equation

$$\Omega^* f = (\Omega f^*)^*. \tag{63}$$

In the case of a discrete argument this means

$$\Sigma_l (\Omega^*)_{kl} f_l = (\Sigma_l \Omega_{kl} f_l^*)^* = \Sigma_l \Omega^*_{kl} f_l,$$

or, the matrix elements of the complex conjugated operator Ω^* are the complex conjugates of the corresponding matrix elements of Ω.

If an operator Ω_{Herm} is Hermitean this means according to equation (60) that the operator adjoint to Ω is identical with the complex conjugated operator, and for arbitrary functions f and g we have

$$\int f\Omega_{\text{Herm}}\, g = \int g\Omega^*_{\text{Herm}}\, f. \tag{64}$$

We can consider the case of continuous functions of one or more continuous variables $x, y, z, \ldots$ and (linear) operators which can act on these functions, to be the limit of the case of functions of a discrete argument. To do this we consider first the function only at selected points $x_k, y_l, z_m, \ldots$ $(k, l, m, \ldots = 1, 2. \ldots)$ in x-, y-, z-, . . . -space and then we let the density of these points increase, similarly to the procedure in § 24. Analogously with the representation of an operator in the discrete case by a matrix we can now immediately write down an explicit formula for $g = \Omega f$,

$$g(x, y, z, \ldots) = \int \Omega(x, y, z, \ldots; x', y', z', \ldots) f(x', y', z', \ldots) d\tau', \tag{65}$$

which equation takes the place of equation (57). The integration is over all possible values of $x', y', z', \ldots$, and $d\tau' = \varrho(x', y', z', \ldots) dx'dy'dz' \ldots$, where ϱ is a density function such as was mentioned in the discussion of equation (47). The density of points in the x-, y-, z-, . . . -space of the kind mentioned a moment ago will be proportional to ϱ in the limit of the continuous case. (In quantum mechanics one may put $\varrho = 1$, if the $x, y, z, \ldots$ are Cartesian coordinates).

In the discrete case there corresponded a matrix element Ω_{kl} to each pair of indices k, l. Now we have an *operator function* corresponding to each pair of points $x, y, z, \ldots$ and $x', y', z', \ldots$ in x-, y-, z-, . . . -space. One often speaks of a *continuous matrix*. The operator function corresponding to the sum Ω_3 of two operators Ω_1 and Ω_2 is according to equation (53) the sum of the operator functions of Ω_1 and Ω_2. To the product $\Omega_1\Omega_2$ there corresponds an operator function which in analogy with rule (58) for matrix multiplication is given by the equation

$$\Omega_3(x, y, \ldots; x', y', \ldots) = \int \Omega_1(x, y, \ldots; x'', y'', \ldots)\Omega_2(x'', y'', \ldots; x', y', \ldots)\varrho(x'', y'', \ldots)dx''dy'' \ldots \tag{66}$$

One sees easily by considering examples that the general representation (65) is mainly formal. If, for instance the operator Ω in the case of just one variable means multiplying by $F(x)$, $g(x) = \Omega f(x) = F(x)f(x)$, there will not be a function $\Omega(x, x')$ such that for all f, $F(x)f(x) = \int \Omega(x, x')f(x')dx'$, where for the sake of simplicity we have put $\varrho = 1$.

One can, however, define a series of functions $\Omega_n(x, x')$ such that for increasing n the expression $\int \Omega_n(x, x')f(x')\,dx'$ approaches the expression $F(x)f(x)$. One chooses, for instance, for $\Omega_n(x, x')$ a product of $F(x)$ and a function $\delta_n(\xi)$ of the argument $\xi = x-x'$,

$$\Omega_n(x, x') = F(x)\delta_n(\xi). \tag{67}$$

The function $\delta_n(\xi)$ must be such (compare a similar discussion in § 23) that it is appreciably different from zero only in a region of the order n^{-1} and that its absolute value is less than Cn, with C a positive constant, while finally the integral of $\delta_n(\xi)$ over ξ is always equal to 1 *,

$$\int_{-\infty}^{+\infty} \delta_n(\xi)\,d\xi = 1. \tag{68}$$

If all these conditions are satisfied, it follows that for increasing n the integral $\int_{-\infty}^{+\infty} \delta_n(x-x')f(x')dx'$ will converge to $f(x)$ and that the integral (65) with Ω given by equation (67) converges to $F(x)f(x)$.

We can express this as follows: the operator function δ_n corresponds in the limit $n \to \infty$ to the unit operator. If there are more than one variable we must take for the unit operator the limit for $n \to \infty$ of the operator function $\delta_n(x-x')\,\delta_n(y-y')\,\delta_n(z-z')\ldots$ — a product of δ-functions. It is often convenient and allowed to assign formally to the unit operator a real function,

$$\Omega_E(x, y, \ldots; x', y', \ldots) = \delta(x-x')\,\delta(y-y')\ldots \tag{69}$$

This symbolic function is usually called the Dirac δ-function, since Dirac was the first to introduce it in quantum mechanics **.

The formal treatment of differential operators is closely connected with this representation of the unit operator. There is clearly no operator $\Omega(x, x')$ such that $df(x)/dx = \int \Omega(x, x')\,f(x')\,dx'$ (ϱ is once again taken to be 1).

If we introduce the functions δ_n again, and now restrict ourselves to those which can be differentiated at least once (the third example in the footnote would therefore not qualify) we find if we in-

* Examples: $\delta_n = n \exp(-\pi n^2\xi^2)$; $\delta_n = (\sin n\xi)/\pi\xi$; $\delta_n = 0$ for $|\xi| \geqq (2n)^{-1}$ and $= n$ for $|\xi| < (2n)^{-1}$.

** P. A. M. Dirac, *Proc. Roy. Soc.*, **A113**, 625, 1927. We may draw attention to the important fact that $\delta(x)$ is an even function, that is, that $\delta(-x) = \delta(x)$; compare eq. (73). For the mathematical aspects of the δ-function we may refer to L. Schwartz, *Théorie des distributions*, Hermann, Paris, 1950.

tegrate by parts

$$\int_{-\infty}^{+\infty} \delta_n(x-x') \frac{df(x')}{dx'} dx' = -\int_{+\infty}^{+\infty} \frac{\partial \delta_n(x-x')}{\partial x'} f(x')\, dx'$$
$$= \int_{-\infty}^{+\infty} \delta'_n(x-x') f(x')\, dx', \tag{70}$$

$$\delta'_n(\xi) = d\delta_n/d\xi. \tag{71}$$

The first member of equation (70) converges as $n \to \infty$ to df/dx. The operator function $\delta'_n(\xi)$ represents thus in the limit $n \to \infty$ the differential operator d/dx. Following Dirac we assign now to this differential operator formally the operator function $\delta'(x-x')$ which in calculations may formally be treated as the derivative of the δ-function.

This representation can obviously be extended to higher derivatives and more variables. For instance, in the case of three variables x, y, z to the operator $\partial^3/\partial x^2 \partial y$ will formally be assigned the operator function $\delta''(x-x')\delta'(y-y')\delta(z-z')$.

Using equation (65) we can also easily give the adjoint operator $\tilde{\Omega}$ and the complex conjugated operator. The first one is, in analogy with equation (61), defined by $\int f\Omega g\, d\tau = \int g\tilde{\Omega} f\, d\tau$ and in analogy with equation (62) we find now $\tilde{\Omega}(x, y, \ldots; x', y', \ldots) \equiv \Omega(x', y', \ldots; x, y, \ldots)$.

The complex conjugated operator is defined by equation (63) and the corresponding operator function is obviously the function $\Omega^*(x, y, \ldots,; x', y', \ldots)$ which is the complex conjugate of the function Ω.

A *Hermitean* operator must again satisfy equation (47) which means that $\tilde{\Omega}$ must be identical with Ω^* so that the following condition must be fulfilled,

$$\Omega(x, y, \ldots; x', y', \ldots) = \Omega^*(x', y', \ldots; x, y, \ldots). \tag{72}$$

A non-Hermitean operator can always be written in the form $\Omega_1 + i\Omega_2$ where both Ω_1 and Ω_2 are Hermitean. This is done by putting $\Omega_1 = \frac{1}{2}(\Omega + \tilde{\Omega}^*)$, $\Omega_2 = -\frac{1}{2}i(\Omega - \tilde{\Omega}^*)$.

The unit operator is clearly Hermitean so that the δ-function which in the case of $\varrho = 1$ corresponds to the unit operator satisfies formally the equation

$$\delta(\xi) = \delta^*(-\xi), \tag{73}$$

although it is not necessarily true that $\delta_n(\xi) = \delta_n^*(-\xi)$. Multiplication by a coordinate or, more generally, by a real function F of the coordinates also clearly defines a Hermitean operator. In order to satisfy formally equation (72) one assigns to this operator instead of an operator function of the form (67) an operator function $F^{1/2}(x, y, \ldots)\delta(x-x')\delta(y-y')\ldots F^{1/2}(x', y', \ldots)$.

The operator $\partial/\partial x$ is non-Hermitean, but $i\partial/\partial x$ is Hermitean in the case where $\varrho = 1$, as one can see if one integrates by parts, $\int f(i\partial/\partial x)g\,dx\,dy \ldots = \int g(-i\partial/\partial x)f\,dx\,dy \ldots$.

Indeed, $-i\partial/\partial x$ is the complex conjugated operator of $i\partial/\partial x$. From equation (73) it follows formally that $\delta'(\xi) = -\delta'^*(-\xi)$ so that the formal operator function of $i\partial/\partial x$, which is $i\delta'(x-x')\delta(y-y')\cdots$, will satisfy equation (72).

If $\varrho \neq 1$ we have instead of $i\partial/\partial x$ the Hermitean operator $i\varrho^{-1/2}(\partial/\partial x)\varrho^{-1/2}$ with the symbolic operator function $i\varrho^{-1/2}(x, y, \ldots)$ $\delta'(x-x')\delta(y-y')\ldots\varrho^{-1/2}(x', y', \ldots)$ which satisfies equation (72). Similarly we have for the unit operator, $\varrho^{-1/2}(x, y, \ldots)\delta(x-x')$ $\delta(y-y')\ldots\varrho^{-1/2}(x', y', \ldots)$.

The sum of Hermitean operators is again Hermitean, but not in general the product. Indeed, when Ω_1 and Ω_2 are Hermitean, we have

$$\int f\Omega_1\Omega_2 g\,d\tau = \int f\Omega_1(\Omega_2 g)\,d\tau = \int (\Omega_2 g)(\Omega_1^* f)\,d\tau$$
$$= \int (\Omega_1^* f)\Omega_2 g\,d\tau = \int g\Omega_2^*\Omega_1^* f\,d\tau = \int g(\Omega_2\Omega_1^*)f\,d\tau.$$

Unless Ω_1 and Ω_2 commute, $\Omega_1\Omega_2$ is non-Hermitean [compare eqq. (58) and (66)]. The sum $\Omega_1\Omega_2 + \Omega_2\Omega_1$ is, however, again Hermitean.

Consider now an n-particle problem with Cartesian coordinates $q_1, \ldots, q_{3n}$ and conjugate momenta $p_1, \ldots, p_{3n}$, and consider a function F of the p_k and q_k. Let F be a finite polynomial in the p_k with real coefficients b which may depend on the q_k and possibly the time, $F \equiv \Sigma b p_1^{a_1} p_2^{a_2} \cdots$ We now construct the differential operator F_{op} which corresponds to F by using the prescription (24) ($p_k = -i\hbar\partial/\partial q_k$),

$$F_{\text{op}} = \Sigma b(-i\hbar)^{a_1+a_2+\cdots}\frac{\partial^{a_1+a_2+\cdots}}{\partial q_1^{a_1}\,\partial q_2^{a_2}\cdots}. \tag{74}$$

Each term consists of a product of two Hermitean operators b and $p_1^{a_1}p_2^{a_2}\cdots$, and in general this product is non-Hermitean. Simple examples where the terms are Hermitean are when b is constant, or when all the a_k are zero. We have an example of such an operator in the Hamiltonian operator given by equation (20). If we now write instead of equation (74)

$$F = \tfrac{1}{2}\Sigma[bp_1^{a_1}p_2^{a_2}\cdots + p_1^{a_1}p_2^{a_2}\cdots b], \tag{75}$$

the new F will be identical with the old one when the p_k are normal variables, but now F_{op} obtained through the prescription (24) from equation (75) will be Hermitean. By using the above procedure we can assign to each function F of this kind a Hermitean operator. The procedure is not unique. Classically one can write, for instance, $p_k b p_l$ instead of $b p_k p_l$. This means that we can write either $\frac{1}{2}(p_k b p_l + p_l b p_k)$ or $\frac{1}{2}(b p_k p_l + p_k p_l b)$ for the corresponding Hermitean operator, and this will in general, lead to two different operators. Our procedure is unique only when b is multiplied by only one factor p *.

Using such a procedure we are able to write the Schrödinger equation (45) of an n-particle system always in a form which satisfies the second condition of § 28, that is, to work always with a Hermitean energy operator, even if the classical energy function H is not of the simple type (20). Indeed, in classical physics one meets with a complicated H in the case of electrically charged particles, if one takes into account both the influence of an external magnetic field and of the magnetic field produced by the motion of the other particles. We have discussed in § 25 the influence of an external magnetic field. To a first approximation it is described by an additional term $-(e/mc)(\boldsymbol{p}\cdot\mathfrak{A})$ in the energy ($\boldsymbol{p}$: momentum of the particle; $\mathfrak{A}$: magnetic vector potential). It is impossible to describe the mutual magnetic interaction of the particles rigorously in a non-relativistic theory. However Darwin ** has shown that one can still describe the motion of the particles using a Hamiltonian energy function H, as long as one neglects terms of the order c^{-3} (c: velocity of light) and higher. The Hamiltonian is of the form [compare eq. (2.102)]

$$H = \Sigma_i p_i^2\, 2m_i - \Sigma_i p_i^4/8m_i^3c^2 + \Sigma_{ij} e_i e_j/r_{ij} - \Sigma_{ij}(e_i e_j/2m_i m_j c^2)\,. \tag{76}$$
$$\{[(\boldsymbol{p}_i\cdot\boldsymbol{p}_j)/r_{ij}] + [(\boldsymbol{p}_i\cdot\boldsymbol{x}_{ij})(\boldsymbol{p}_j\cdot\boldsymbol{x}_{ij})/r_{ij}^3]\} + \Sigma_i\{e_i\Phi_i - (e_i/m_i c)(\boldsymbol{p}_i\cdot\mathfrak{A}_i) + (e_i^2/2m_i c^2)\mathfrak{A}_i^2\}.$$

The first term is the non-relativistic kinetic energy, the second term is the first relativistic correction term [compare eqq. (1.13) and (1.14)], the third term is the electrostatic and the fourth term the magnetic interaction, while the last sum arises from external electromagnetic fields. In equation (76) $\boldsymbol{x}_{ij}$ is the radius vector from particle i to particle j with absolute magnitude r_{ij}. From our previous considerations it follows easily how one can obtain a Hermitean operator from H, for instance, by using equation (75). Apart from the difficulty of assigning uniquely a Hermitean operator to H (compare, however, the discussion at the end of § 34), there are many circumstances limiting the applicability of equation (76). We shall

* H. Weyl (*Z. Phys.*, **46**, 1, 1927; compare also M. H. McCoy, *Proc. Nat. Acad. Sc.*, **18**, 674, 1932) has developed a procedure to assign uniquely to each classical function $F(p, q)$ a Hermitean operator. His considerations have, however, up to now not been applied in any practical case.

** C. G. Darwin, *Phil. Mag.*, **39**, 537, 1920; see also G. Breit, *Phys. Rev.*, **34**, 553, 1929.

mention here only the fact that we have neglected the spin of the particles, which will give rise to other terms of the order c^{-2} (compare § 57).

The fact that quantum mechanical operators often do not commute is a consequence of the *non-commutability of the operators* p_k *and* q_k. From the equations

$$(q_k p_l - p_l q_k)_{\mathrm{op}} f = -i\hbar\{q_k(\partial f/\partial q_l) - \partial(q_k f)/\partial q_l\} = i\hbar\delta_{kl} f,$$
$$(q_k q_l - q_l q_k)_{\mathrm{op}} f = q_k q_l f - q_l q_k f = 0,$$
$$(p_k p_l - p_l p_k)_{\mathrm{op}} f = (-i\hbar)^2\{\partial^2 f/\partial q_k \partial q_l - \partial^2 f/\partial q_l \partial q_k\} = 0$$

we derive the fundamental *commutation relations* of the p and q operators,

$$[q_k, p_l]_- = q_k p_l - p_l q_k = i\hbar\delta_{kl}, \quad [q_k, q_l]_- = 0, \ [p_k, p_l]_- = 0. \qquad (77)$$

These equations which show a close formal connexion with the Heisenberg relations express the fact that all q_k (the Cartesian coordinates) commute with one another and also all p_k (the ordinary momentum components). Also a given q commutes with all p, except its own canonical conjugate. The operator $q_k p_k - p_k q_k$ is the only non-vanishing one and it corresponds to multiplication by $i\hbar$.

Not only two of the q_k, but any two functions of the q_k are commuting operators. In the same way two operators containing only the p_k commute.

The commutator of p_k with a function Q of the q_k satisfies the equation

$$\begin{aligned}[p_k, Q]_- f &= -i\hbar\{\partial(Qf)/\partial q_k - Q\partial f/\partial q_k\} \\ &= -i\hbar(\partial Q/\partial q_k) f, \text{ or, } [p_k, Q]_- = -i\hbar\partial Q/\partial q_k.\end{aligned} \qquad (78)$$

If we commute p_k with an operator F which is a sum of terms which are each products of factors containing only p_k or only q_k,

$$F = \Sigma Q_1 P_1 Q_2 P_2 \cdots, \qquad (79)$$

we get by applying equation (55)

$$\begin{aligned}[p_k, F]_- &= \Sigma\{[p_k, Q_1]_- P_1 Q_2 P_2 \cdots + Q_1 P_1 [p_k, Q_2]_- P_2 \cdots + \cdots\} \\ &= -i\hbar\Sigma\{(\partial Q_1/\partial q_k) P_1 Q_2 P_2 \cdots + Q_1 P_1 (\partial Q_2/\partial q_k) P_2 \cdots + \cdots\} \\ &= -i\hbar\partial F/\partial q_k,\end{aligned} \qquad (80)$$

where $\partial F/\partial q_k$ is defined by the preceding expression.

Using equation (55) twice one finds similarly for the commutator of q_k with a product P of p's,

$$[q_k, P]_- = [q_k, p_1^{a_1} p_2^{a_2} \cdots p_k^{a_k} \cdots]_- = p_1^{a_1} \cdots [q_k, p_k^{a_k}]_- \cdots$$
$$= i\hbar a_k p_1^{a_1} \cdots p_k^{a_k-1} \cdots = i\hbar \partial P/\partial p_k, \tag{81}$$

since from equations (55) and (77) it follows that

$$[q_k, p_k^{a_k}]_- = \{[q_k, p_k]_- p_k^{a_k-1} + p_k[q_k, p_k]_- p_k^{a_k-2} + \cdots\} = i\hbar a_k p_k^{a_k-1}.$$

Commutation of q_k with the operator F given by equation (79) gives the result

$$[q_k, F]_- = \Sigma\{Q_1[q_k, P_1]_- Q_2 P_2 \cdots + \cdots\}$$
$$= i\hbar\Sigma\{Q_1(\partial P_1/\partial p_k) Q_2 P_2 \cdots + \cdots\} = i\hbar\partial F/\partial p_k, \tag{82}$$

where $\partial F/\partial p_k$ is defined by the preceding expression. The only difference between differentiating an operator with respect to another operator and ordinary differentiation is that now one may never change the order of non-commuting factors.

Using equations (80) and (82) one can also prove that the commutator $[F, G]_-$ of two arbitrary operators F and G is given by the expression $i\hbar\Sigma_k \left[\frac{\partial F}{\partial q_k}\frac{\partial G}{\partial p_k} - \frac{\partial F}{\partial p_k}\frac{\partial G}{\partial q_k}\right]$ * which is just $i\hbar$ times the *Poisson bracket* ** of F and G.

§ 30. The generalised Ehrenfest theorem

In § 27 we discussed the close connexion between the Schrödinger equation (23) and the classical equations of motion (41) which correspond to the Hamiltonian equation. In quantum mechanics the Schrödinger equation has always the form (45), $H_{\rm op}\psi = i\hbar\partial\psi/\partial t$, with a Hermitean operator $H_{\rm op}$. This restriction enables us not only to introduce a simple definition for the probability density (43) but also to show in detail the close connexion with the classical equations of motion.

Let $F_{\rm op}$ be an operator obtained from a function of the p_k, q_k, and possibly t which is a polynomial in the p_k, and let ψ_1 and ψ_2 be two arbitrary solutions of equation (45). We now define a quantity F_{12} by the equation

$$F_{12} = \int \psi_1^* F_{\rm op} \psi_2 d\tau, \tag{83}$$

* In writing down this expression one treats the p_k and q_k as normal variables.

** See, for instance, E. T. Whittaker, *Analytical Dynamics*, Cambridge University Press, 1937, Ch. 11.

where $d\tau$ indicates integration over all spatial coordinates and where we assume the integral to converge. In general, F_{12} will depend on t, and using equations (47), (45) and its conjugate complex we find for its derivative with respect to t

$$\begin{aligned}\dot{F}_{12} &= \int \left[\psi_1^* F \frac{\partial \psi_2}{\partial t} + \frac{\partial \psi_1^*}{\partial t} F\psi_2 + \psi_1^* \frac{\partial F}{\partial t}\psi_2\right] d\tau \\ &= \frac{i}{\hbar}\int \{-\psi_1^*(FH\psi_2) + (F\psi_2)(H^*\psi_1^*)\} d\tau + \left(\frac{\partial F}{\partial t}\right)_{12} \\ &= \frac{i}{\hbar}\int \{-\psi_1^* FH\psi_2 + \psi_1^* H(F\psi_2)\} d\tau + \left(\frac{\partial F}{\partial t}\right)_{12} \\ &= \frac{i}{\hbar}[HF - FH]_{12} + \left(\frac{\partial F}{\partial t}\right)_{12} \end{aligned} \tag{84}$$

In special cases where F is either one of the p_k or one of the q_k we have, using equations (80) and (82),

$$(\dot{q}_k)_{12} = \frac{i}{\hbar}[H, q_k]_{-12} = \left(\frac{\partial H}{\partial p_k}\right)_{12}, \quad (\dot{p}_k)_{12} = \frac{i}{\hbar}[H, p_k]_{-12} = -\left(\frac{\partial H}{\partial q_k}\right)_{12}. \tag{85}$$

In contradistinction to the case of the approximately defined movement of wave groups which we discussed in § 27 equations (85) express an exact, quantitative relationship which has a complete formal analogy to the classical equations of motion (19). We refer to equations (85) as the *generalised Ehrenfest theorem.*

In the special case where ψ_1 and ψ_2 are the same normalised solution ψ of equation (45) we shall call the integral

$$\bar{F} = \int \psi^* F_{\text{op}}\, \psi\, d\tau \tag{86}$$

the expectation value $\bar{F}$ *of the operator* F_{op} *in the physical situation described by* ψ. The quantities $\overline{q_k}$ and $\overline{p_k}$ are clearly a measure for the approximately defined coordinate q_k or momentum p_k in the case of the wave packets discussed in § 27. Equations (85) now have the form

$$\dot{\overline{q_k}} = \overline{\partial H/\partial p_k}, \quad \dot{\overline{p_k}} = -\overline{\partial H/\partial q_k}, \tag{87}$$

and they give a quantitative relation for the (approximately defined) motion of the wave group as described by equations (41). For a particle in a three-dimensional field of force, where H is given by equation (2.4) we have

$$\dot{x} = \frac{\overline{p_x}}{m}, \quad \dot{y} = \frac{\overline{p_y}}{m}, \quad \dot{z} = \frac{\overline{p_z}}{m};$$

$$m\ddot{\bar{x}} = \dot{\bar{p}}_x = -\overline{\frac{\partial U}{\partial x}}, \quad m\ddot{\bar{y}} = \dot{\bar{p}}_y = -\overline{\frac{\partial U}{\partial y}}, \quad m\ddot{\bar{z}} = \dot{\bar{p}}_z = -\overline{\frac{\partial U}{\partial z}}.$$

These equations which show immediately the connexion with the Newtonian equations of motion were derived by Ehrenfest *.

§ 31. THE CONSERVATION OF MOMENTUM

The change in time of a function $F(p_k, q_k, t)$ is classically given by the equation

$$\dot{F} = \sum_k \left[\frac{\partial F}{\partial q_k}\dot{q}_k + \frac{\partial F}{\partial p_k}\dot{p}_k\right] + \frac{\partial F}{\partial t}$$

$$= \sum_k \left[\frac{\partial F}{\partial q_k}\frac{\partial H}{\partial p_k} - \frac{\partial F}{\partial p_k}\frac{\partial H}{\partial q_k}\right] + \frac{\partial F}{\partial t} = \{F, H\} + \frac{\partial F}{\partial t}. \quad (88)$$

The Poisson bracket of F and H which occurs in this equation is, as we noted in § 29, the classical counterpart of the quantum mechanical expression $(i/\hbar)[H, F]_-$ and we see that equations (84) and (88) are each others counterparts.

In particular, if an operator F_{op} which does not contain the time implicitly commutes with H_{op}, then the integrals of the type F_{12} in equation (83) will be time independent according to equation (84). One expresses this by saying that F_{op} is an *integral* of the quantum mechanical problem. This corresponds exactly to the classical case of a function of the p_k and q_k which does not change its value with time.

If the forces acting on the particles of the system depend only on the relative positions in space of the particles, we call the system a *free* system. In classical mechanics the total (linear) momentum of a free system is time independent. This law can be derived as follows. The Hamiltonian is invariant under an arbitrary translation, that is, under a transformation of the type

$$\boldsymbol{x}_i' = \boldsymbol{x}_i + \boldsymbol{\varepsilon}, \quad \boldsymbol{p}_i' = \boldsymbol{p}_i, \quad (89)$$

where $\boldsymbol{\varepsilon}$ is an arbitrary vector. We see then from the equation

$$H(\boldsymbol{x}_k', \boldsymbol{p}_k', t) = H(\boldsymbol{x}_k, \boldsymbol{p}_k, t) + (\boldsymbol{\varepsilon} \cdot \Sigma_i \boldsymbol{\nabla}_i H), \quad (90)$$

* P. Ehrenfest, *Z. Phys.*, **45**, 455, 1927.

which is true for any $\boldsymbol{\varepsilon}$, that the sum on the right hand side of this equation must be equal to zero,

$$\Sigma_i \boldsymbol{\nabla}_i H = 0. \tag{91}$$

From the equations of motion it follows that $\Sigma_i \boldsymbol{\nabla}_i H = -(d/dt)[\Sigma_i \boldsymbol{p}_i]$ and we see thus that the total momentum,

$$\boldsymbol{P} = \Sigma_i \boldsymbol{p}_i, \tag{92}$$

is a constant vector.

If H is of the form (20) we can immediately take over equation (91) in quantum mechanics. The left hand side is now an operator-vector with components $(i/\hbar)[P_x, H]_-$, $(i/\hbar)[P_y, H]_-$, $(i/\hbar)[P_z, H]_-$ when we use equation (80) and where P_x, P_y, P_z are the operators defined by equation (92), corresponding to the total momentum of the system. Equation (91) is thus equivalent to the equations

$$[H, P_x]_- = [H, P_y]_- = [H, P_z]_- = 0, \tag{93}$$

that is, the components of the total momentum commute with the energy operator and are thus integrals of motion in the quantum mechanical sense.

Equations (93) do not show immediately that one shall find a constant value for the momentum when one is interpreting physical experiments. That this is so, however, we can show as follows. First of all we remark that if F_{op} commutes with H_{op}, the same will hold for any power of F_{op}. According to § 30 this means that the expectation value $\overline{F_{\text{op}}^n}$ will be time independent, provided F_{op} does not contain the time implicitly. Using equations (49) and (50) we find for the expectation value of P_x^n.

$$\begin{aligned}\overline{P_x^n} &= \int \psi^* P_{x\,\text{op}}^n \psi \, d\tau = \iint \psi^* P_{x\,\text{op}}^n \exp\{2\pi i[(\boldsymbol{\sigma}_1 \,.\, \boldsymbol{x}_1) + \cdots]\}\, d\boldsymbol{\sigma}_1 \cdots d\tau \\ &= h^n \iint \psi^* (\Sigma_i \sigma_{xi})^n A \exp\{2\pi i\, [(\boldsymbol{\sigma}_1 \cdot \boldsymbol{x}_1) + \cdots]\}\, d\boldsymbol{\sigma}_1 \cdots d\tau \\ &= h^n \int (\Sigma_i \sigma_{xi})^n A^* A \, d\boldsymbol{\sigma}_1 \cdots\end{aligned}$$

The last integral is now according to our considerations time independent. Similarly we find that in general expressions of the kind

$$\overline{P_x^n P_y^m P_z^l} = h^{n+m+l} \int \sigma_x^n \sigma_y^m \sigma_z^l A^* A \, d\boldsymbol{\sigma}_1 \cdots \quad (\boldsymbol{\sigma} = \Sigma_i \boldsymbol{\sigma}_i)$$

will be time independent. It follows that for any arbitrary function

f of $\boldsymbol{\sigma}$ the integral

$$\int f(\boldsymbol{\sigma}) \, |A^2| \, d\boldsymbol{\sigma}_1 \cdots d\boldsymbol{\sigma}_n \tag{94}$$

will be time independent.

We saw in § 28 that $|A|^2 d\boldsymbol{\sigma}_1 \cdots d\boldsymbol{\sigma}_n$ was the probability that the momenta of the particles were lying within volume elements $h^3 d\boldsymbol{\sigma}_1, h^3 d\boldsymbol{\sigma}_2, \ldots, h^3 d\boldsymbol{\sigma}_n$. If we now in expression (94) for f choose a function which is 1 within a chosen volume element $d\boldsymbol{\sigma}$ of σ-space and 0 outside, we find that the fact that expression (94) does not change with time corresponds to the fact that *the probability that $h\boldsymbol{\sigma}$, that is, the total momentum, lies within a given interval is independent of time.* This is just the quantum mechanical counterpart of the law of the conservation of momentum. As we emphasised in §26 quantum mechanics is constructed in such a way that the fundamental laws such as those of the conservation of energy and momentum remain valid. The above considerations were concerned with a quantitative formulation of the law of the conservation of momentum.

To illustrate the general role of the conservation law we consider the problem posed by the straight cloud chamber tracks produced by α-particles (and fast β-particles). As we mentioned in § 2 these experiments supported very convincingly Rutherford's ideas about the corpuscular nature of the α-particles and the applicability of classical mechanics on charged elementary particles. After it was shown by experiments how fruitful the de Broglie hypothesis about the wave nature of those particles was, many people felt that the results of experiments with Wilson chambers were rather paradoxal and that the simple classical interpretation could be justified only by detailed quantum mechanical considerations. One would expect that according to quantum mechanics the emission of an α-particle by a nucleus should be described by an outgoing spherical wave field. It would thus not be expected that the air molecules which are ionised by the α-particle would lie on a straight line. However, this paradox does not really exist. This is seen most easily when one takes into account that the construction of quantum mechanics is always based on the possibility of defining phenomena in classical terms. This means that one can only describe the results of a physical experiment by using the concepts of classical physics. In the case of the α-particle one interprets the occurrence of a droplet as the consequence of the interaction of a charged particle with the vapour molecules at a more or less definite point in space-time. One can thus understand the straight orbit, if one assumes that the momentum and energy of the particle are very large compared to the exchange of momentum and energy at a single ionisation process. Quantum mechanics works also with the concepts both of localisation in space-time and of momentum and energy of particles, but under one restriction, namely, that a strict definition of one of them excludes that of the other. The droplets will mark those parts of the gas which are passed by the α-particle, but their marking is only a rough one. A rigorous definition of the position of the α-particle would be connected with a very large unpredictable exchange of momentum with the marking system and we would expect on the contrary a very small exchange of momentum from our knowledge of molecular structure. The explicit introduction of the quantum of action thus never enters and if there are droplets to mark the position of the α-particle, they should lie on a straight line in accordance with the law of conservation of momentum.

A consideration of the wave function which described the system consisting of the α-particle plus the vapour molecules must necessarily lead to the same result. If one knows that an (approximately) localised radioactive nucleus emits an α-particle, one will construct ψ such that initially the probability of finding a value of the momentum corresponding to a given direction is roughly independent of the direction. Any molecule in the surrounding space may be ionised at a later time. However, as soon as an experiment shows that such a process has taken place for one particular molecule it means that according to the quantum mechanical formulation of the law of conservation of momentum one particular direction is realised with more or less precision for the momentum of the α-particle. This means in turn that any other evidence of this kind which pertains to the behaviour of the system must be in agreement with this direction of the momentum.

The fact that the law of conservation of momentum is to a large extent valid in the interpretation of cloud chamber experiments because of the small amount of exchange of momentum (and energy) with the vapour molecules enables us to go a bit further in the wave mechanical description. One can describe the α-particle by means of a special wave packet which moves approximately according to the Galilean law of inertia which we discussed in § 6.

The approximate position of an ionised condensation nucleus which (at some distance from the nucleus) appeared at an approximately given time will make it possible to determine the wave function in sufficient detail so that we can describe a considerable part of the orbit of the α-particle. If the α-particle collides with a nucleus it is often impossible to predict quantum mechanically over what angle the α-particle will be deflected. The orbit after the collision can, however, again be described by a special wave packet, the actual form of which may well have to be settled by experimental evidence *.

§ 32. Stationary states

If it be possible to assign to a many particle system a definite value of its energy, the physical situation of the system must be characterised according to the basic ideas of quantum theory (compare § 16) by a wave function which is periodic in time. The frequency ν of the wave phenomenon is connected with the energy E through Planck's formula $E = h\nu$. This means that the wave function ψ must be of the form [compare §§ 16, 17 and § 27 eqq. (29), (31) and (42)]

$$\psi = \varphi \exp\,(-iEt/\hbar), \tag{95}$$

where φ depends on the spatial coordinates of the particles only. Substitution of equation (95) into the wave equation (45) gives us for φ the time independent Schrödinger equation.

$$H_{\text{op}}\,\varphi = E\varphi. \tag{96}$$

This equation has clearly a meaning only, of H_{op} does not contain the time explicitly. Similarly one can not talk about the energy of the system in classical physics, if the Hamiltonian contains the time explicitly.

The general discussion of equation (96) is completely analogous to the discussion in the case of one particle. We can therefore be short. In general not all solutions of equation (96) will have a physical meaning. If there is a value of E for which equation (96) possesses one or more normalisable solutions φ_k, that is, solutions for which $\int |\varphi_k|^2\,d\tau$ is finite with the integration extending over all possible values of all coordinates, we call this value of E an energy

* Compare, W. Heisenberg, *The physical principles of quantum theory*, University of Chicago Press, 1930.

eigenvalue and φ_k the corresponding eigenfunction of the differential equation (96). The corresponding wave function (95) characterises in this case a (proper) *stationary state* of the system.

In all applications in atomic physics the eigenvalues of this kind form a discrete set, that is, they lead to a line spectrum (compare § 24). The number of linearly independent eigenfunctions which correspond to the same eigenvalue is called the degree of degeneracy of the eigenvalue considered. Two eigenfunctions φ_1 and φ_2 belonging to different eigenvalues E_1 and E_2 are always orthogonal to each other, that is,

$$\int \varphi_2^* \varphi_1 d\tau = 0. \tag{97}$$

To prove this, we use equation (96) to write $\varphi_2^* H_{\text{op}} \varphi_1 = E_1 \varphi_2^* \varphi_1$, $\varphi_1 H_{\text{op}}^* \varphi_2^* = E_2^* \varphi_1 \varphi_2^*$, where we do not want to exclude at this moment the possibility that there exist complex eigenvalues. Integrating both equations over all possible values of all variables and subtracting the second equation from the first one, we get

$$\int (\varphi_2^* H_{\text{op}} \varphi_1 - \varphi_1 H_{\text{op}}^* \varphi_2^*) \, d\tau = (E_1 - E_2^*) \int \varphi_2^* \varphi_1 d\tau. \tag{98}$$

Since H_{op} is Hermitean the left hand side is equal to zero [compare eq. (47)]. Consider now first of all the case where φ_1 and φ_2 are the same, so that $E_1 = E_2$. It follows then from equation (98) that $E_1 = E_1^*$, that is, that the eigenvalues are real. This had been tacitly assumed when we wrote down equation (95). Consider now the case where $E_1 \neq E_2$, and thus also $\varphi_1 \neq \varphi_2$. We see that then equation (97) follows from equation (98).

By multiplying the φ by a suitable constant we can always make them normalised, $\int |\varphi|^2 d\tau = 1$. If the g eigenfunctions which correspond to the same eigenvalue are not orthogonal to each other, we can always (in an infinite variety of ways) replace them by g linear combinations in such a way that the new eigenfunctions are orthonormal (compare § 21).

As in the case of the one-particle problem there may also exist improper stationary states. These occur when there are solutions of the form (95) which are non-normalisable, but all the same correspond to the limiting case of a physical situation. The simplest case is given by a system consisting of free particles only. We can find for every positive value of E solutions of the kind

$$\psi = \exp\{2\pi i \, [\Sigma_i(\boldsymbol{\sigma}_i \cdot \boldsymbol{x}_i) - Et/h]\}, \; E = h^2 \Sigma_i(\boldsymbol{\sigma}_i^2/2m_i).$$

These solutions are the limiting case of the wave functions corresponding to the situation where the momenta of all the particles are defined with great precision. In this example the stationary states are infinitely degenerate, since for any E there are still infinitely many ways of choosing the $\boldsymbol{\sigma}_i$.

The improper eigenvalues E form a continuous set in the space of all real numbers. Together with possible discrete eigenvalues they form the total eigenvalue spectrum of the system.

An arbitrary physical situation of the system can always in one, and only one, way be considered to be a superposition of (proper and improper) stationary states,

$$\psi = \Sigma_k c_k \varphi_k \exp(-iE_k t/\hbar) + \int c(E)\varphi(E) \exp(-iEt/\hbar) dE, \quad (99)$$

where the summation is over all proper and the integration over all improper eigenvalues. Equation (99) with $t = 0$ expresses the fact that an arbitrary normalisable function of the coordinates ψ_0 can always be expanded in terms of eigenfunctions,

$$\psi_0 = \Sigma_k c_k \varphi_k + \int c(E)\varphi(E) dE. \quad (100)$$

It is possible to "normalise" the $\varphi(E)$ in such a way that always

$$1 = \int |\psi|^2 d\tau = \int |\psi_0|^2 d\tau = \Sigma_k |c_k|^2 + \int |c(E)|^2 dE. \quad (101)$$

This equation can be interpreted by stating that $|c_k|^2$ is the probability of finding E_k realised, and $|c(E)|^2 dE$ the probability of finding an energy between E and $E+dE$ realised in an experiment which measures the energy of the system. When the complete system of eigenfunctions is known we can express the coefficients c by using the (proper and improper) orthonormality conditions. If there is no degeneracy we have

$$c_k = \int \varphi_k^* \psi_0 d\tau, \; c(E) = \int \varphi^*(E)\psi_0 d\tau. \quad (102)$$

If the eigenvalues are degenerate the situation is much more complicated. The following equations illustrate an arbitrarily chosen, rather general case,

$$\begin{aligned} \varphi_k &= \Sigma_j c_{kj} \varphi_{kj}, \\ \varphi(E) &= \Sigma_j c_j(E) \varphi_j(E) \qquad\qquad (103) \\ &+ \Sigma_j \int c_j(E, r_1, r_2, \ldots)\varphi_j(E, r_1, r_2, \ldots)\varrho(r_1, r_2, \ldots) dr_1, dr_2, \ldots, \end{aligned}$$

$$\int |\varphi_k|^2 \, d\tau = \Sigma_j \, |c_{kj}|^2 = 1,$$
$$\Sigma_j \, |c_j(E)|^2 + \Sigma_j \int |c_j(E, r_1, r_2, \ldots)|^2 \, \varrho \, dr_1 \, dr_2 \ldots = 1. \tag{104}$$

Equation (103) expresses the fact that the most general eigenfunction corresponding to an energy value E_k can always be constructed as a linear combination of linearly independent eigenfunctions. The latter may be functions of one or more continuous parameters $r_1, r_2, \ldots$ and if this is the case one must introduce also a density function $\varrho(r_1, r_2, \ldots)$. The infinite number of possibilities of choosing the basic set of eigenfunctions may be restricted by requiring orthonormality. This restriction entails that equations (104) hold. We have now instead of equations (102),

$$c_k c_{kj} = \int \varphi_{kj}^* \, \psi_0 \, d\tau, \quad c(E) \, c_j(E) = \int \varphi_j^* (E) \, \psi_0 \, d\tau,$$
$$c(E) \, c_j(E, r_1, r_2, \ldots) = \int \varphi_j^* (E, r_1, r_2, \ldots) \, \psi_0 \, d\tau. \tag{105}$$

Using equation (104) and (105) we can calculate $|c_k|^2$ and $|c(E)|^2$. Our formulae are rather more general than those of § 23 in that we have considered the possibility that the improper eigenfunctions may be functions of continuous parameters *. Such a case can, for instance, occur in the case of a system of independent free particles.

In the limit where one can consider the different parts of the system to be independent or where one can neglect the interaction, at any rate to a first approximation (compare § 26), the discussion of the stationary states will be very simple. This independence means classically that the change in time of the coordinates and momenta of a particle depends only on the instantaneous values of these six quantities themselves, and that thus the Hamiltonian can be split up into parts which depend on the coordinates and momenta of only one particle,

$$H = H^{(1)} + H^{(2)} + \cdots + H^{(\varkappa)} + \cdots \tag{106}$$

We can now interpret the $H^{(\varkappa)}$ as operators in the usual quantum mechanical way and we shall denote by $\varphi_k^{(\varkappa)}$ the (proper and improper) eigenfunctions of the equation

$$H_{\text{op}}^{(\varkappa)} \, \varphi^{(\varkappa)} = E^{(\varkappa)} \, \varphi^{(\varkappa)}, \tag{107}$$

* We could have considered the even more general case where we have several sets of eigenfunctions which depend on one, on two, . . . continuous parameters respectively.

which involves the coordinates of the $\varkappa$-th particle only. It is easily seen that the product

$$\varphi_{k_1 k_2 \ldots} = \varphi_{k_1}^{(1)} \varphi_{k_2}^{(2)} \cdots \tag{108}$$

will be an eigenfunction of the Schrödinger equation (96) corresponding to an eigenvalue

$$E_{k_1 k_2 \ldots} = E_{k_1}^{(1)} + E_{k_2}^{(2)} + \cdots, \tag{109}$$

which is the sum of eigenvalues of equation (107).

The situation is clearly mathematically analogous to the case of a potential of the form $U_1(x)+U_2(y)+U_3(z)$ which we discussed in §§ 17 and 24. It is also easily checked that an arbitrary function ψ_0 can be expanded in terms of eigenfunctions of the form (108) (compare the discussion in § 24).

Even if equation (107) does not lead to degenerate eigenvalues, the eigenvalues (109) can easily be degenerate, since different combinations may lead to the same value of the sum. Linear combinations of the corresponding solutions (108) will also in this case again be eigenfunctions.

If we are dealing with systems where two or more particles are physically indistinguishable (for instance, a system, of a number of electrons whose interaction can either be neglected or be replaced by a universal potential, that is, a field which is the same for all electrons) such a degeneracy will always occur since the sum (109) will not change when we exchange the indices belonging to two equivalent particles. Such systems of equivalent particles show particular properties which are of the utmost importance for atomic physics (Pauli principle). We shall return to them later on.

§ 33. The law of conservation of energy; causality in quantum mechanics

From the considerations at the beginning of § 31 it follows easily that the energy function of a system, considered as a Hermitean operator, is an integral of the equations of motion, provided it does not contain the time explicitly. This follows since the energy operator commutes with itself. As in the case of the law of conservation of momentum this commutability does not mean immediately that the energy will come out to be constant in time when we interpret an experiment. The considerations of the previous section have, however, given the solution to this problem. The solution is completely analogous to the treatment of the case of the momentum conservation in § 31. We proved then that the probability distribution of the momenta is such that the probability of finding a given total momentum does not change in time. In § 32 we connected the probability distribution of the energy with $|c_k|^2$ and $|c(E)|^2 dE$ and these quantities do not depend on time.

In classical non-relativistic mechanics the law of conservation of energy was seen as a consequence of the special form of the interactions between the various particles (compare § 1). The instantaneous values of these interactions which determine the causal evolution of mechanical phenomena are derived by differentiation from a time independent potential. The far-going connexion which according to the theory of relativity exists between energy and momentum can not yet be fully unfolded in non-relativistic mechanics. This is due to the fact that this connexion is properly treated only in the detailed discussion of a system of charged particles, but this involves the explicit introduction of the electromagnetic field.

One can see this difference in the formal treatment of momentum and energy (in the non-relativistic theory) from the fact that while each particle has its own individual momentum, the energy is defined only for the total system and even then only apart from a possible additive constant. (This is also still true in the case where the electromagnetic interactions are treated approximately in the manner discussed in § 29). The same difference occurs also in the non-relativistic quantum mechanics of many particles since it follows as closely as possible classical mechanics — as can be seen most clearly from the considerations of § 26. The law of conservation of momentum which we discussed in § 31 refers to a statement about the sum of the individual momenta of the particles which might be assigned to the system when an experiment is interpreted while only the total energy of the system enters into the discussion of the quantum mechanical interpretation of the law of conservation of energy.

The laws of conservation of momentum and energy are the simplest examples of causal laws in physics. In the Introduction we mentioned that quantum theory leads to a peculiar restriction of the applicability of the causal laws to the interpretation of physical phenomena *. We start by noting that both the close connexion

* Compare N. Bohr, *Naturwiss.*, **16**, 245, 1928; **17**, 483, 1929; **18**, 73, 1930.

between the classical and the quantum mechanical description of the behaviour of a system and the possibility of defining the energy are guaranteed by the fact that we consider the quantum mechanical situation always during finite time intervals. We refer here not to actual observations on the systems but to the description, by means of a wave function, of a system which is not acted upon by an external agent. In this description the classical space-time concept plays an essential role — the wave function depends on the spatial coordinates and the time. We saw in chapter 2 that in the case of a free particle the casual law of inertia becomes an asymptotic law in quantum mechanics, and that this followed because the approximately defined energy of the particle could be connected through $E = h\nu$ with a wave phenomenon which was (approximately) periodic in time. Also in the case when several particles are interacting, it is possible to get a close connexion with the classical causal evolution of events (for instance, the possibility of defining continuously changing velocities) only because the situation is characterised by an (approximately defined) frequency (compare § 27). From such considerations one can draw the following general conclusion. *The interpretation of some event, that is, some interaction of a system with an external piece of apparatus, can be used to make predictions concerning the future behaviour of the system (in the sense of a classical causal description) only if the moment at which this event took place was inaccurately defined.* This is so because we can use only the usual classical concepts to interpret the event, and those classical quantities, which are needed to characterise (approximately) the situation of the system so that we can make some predictions about the future behaviour of the system in the sense of a causal law, are just those quantities which according to our considerations will not be defined if the moment at which the event took place was completely and rigorously defined. There exists, as Bohr expresses it, a complementarity between the possibility of defining events in a time sequence and the possibility of describing them causally (predictions). The interaction with a piece of apparatus which can teach us something about the properties of the system at a sharply defined moment will lead to an abrupt break in the causal chain of events. From the point of view of constructing physical concepts this interaction can be said to be in principle indescribable.

In the special case where we consider the conservation of energy, the Heisenberg relation $\Delta E \,.\, \Delta t \gtrapprox \hbar$ gives us a quantitative formulation of the complementary relation between the margin Δt within which the moment of interaction can be defined and the margin ΔE within which the energy after the interaction can be defined (compare § 16). A simple example is again the case of a free particle bombarding an opening in a rigid wall which is opened during a very short period (see § 5). The definition of its energy is the less, the shorter the hole is opened.

From the foregoing it follows that similar situations will occur not only in the conservation of energy, but also in the case of other causal laws, such as, for instance, the law of conservation of momentum. If one considers only the symbolical description of the situation by means of a wave function, one might argue that the probability distributions of momentum and energy are at any time well defined functions and that these functions do not change with time. One could even consider idealised experiments which repeat the same situation as often as required and which could be used to measure the momenta of all the particles at an arbitrarily chosen definite moment (for instance, by repeatedly suddenly letting the interactions vanish and investigating the interference patterns). Such experiments would show the fact that the momentum distribution is constant in time *. One sees, however, immediately that this does not give us a verification of the causal law of conservation of momentum, such as we were concerned with in the interpretation of the cloud chamber experiments with α-particles. In the latter case we were really dealing with an event which happened *only once* (the appearance of droplets) and which could be interpreted as a causal chain. We must require as a matter of principle that the observation of the droplets should never enable us to fix with complete accuracy the moment when the α-particle crossed a given geometrical plane (or the exact position where it crossed).

* Analogous idealised experiments to find the energy distribution function would require the explicit showing of a phenomenon which is periodic in time. The nature of the time coordinate seems to contradict the possibility of such experiments except when one bases the discussion on the kinematics of the theory of relativity. This is clearly closely connected with the occurrence of an undetermined constant in the expression of the non-relativistic energy.

CHAPTER IV

TRANSFORMATION THEORY

A. GENERAL THEORY

§ 34. Coordinate transformations

In classical mechanics it is often advisable to introduce for discussing the behaviour of the system new variables instead of the original coordinates and momenta. It has been found possible to develop corresponding methods in quantum mechanics. They form together what is usually called *quantum mechanical transformation theory.*

The simplest case is the pure coordinate transformation. Originally the positions of the n particles in the system were determined by the values of the Cartesian coordinates $q_1, q_2, \ldots, q_{3n}$. We now introduce instead generalised coordinates $q'_1, q'_2, \ldots, q'_{3n}$. We can treat the q'_k as functions of the q_k and vice versa. We assume that the Jacobians

$$\begin{aligned} \varrho &= \partial(q'_1, \ldots, q'_{3n})/\partial(q_1, \ldots, q_{3n}), \\ \varrho' &= \partial(q_1, \ldots, q_{3n})/\partial(q'_1, \ldots, q'_{3n}), \quad \varrho\varrho' = 1, \end{aligned} \tag{1}$$

are different from zero. The Schrödinger wave function $\psi(q_1, q_2, \ldots, t)$ goes over into a function of the q'_k and t through the transformation $q_k \to q'_k$ and we obtain the transformed wave function $\psi'(q'_1, q'_2, \ldots, t)$. Equation (3.44) goes over into the equation

$$1 = \int \psi^* \psi \, d\tau = \int \psi'^* \psi' \, d\tau', \tag{2}$$

$$d\tau = dq_1 \ldots dq_{3n} = \varrho' dq'_1 \ldots dq'_{3n} = d\tau'. \tag{3}$$

Clearly $\psi'^*\psi'\varrho'$ is the probability density in q'-space, that is, $\psi'^*\psi'\varrho' dq'_1 \ldots dq'_{3n}$ is the probability of finding a spatial configuration of the system corresponding to a point in the volume element $dq'_1 \ldots dq'_{3n}$.

The wave equation (3.45) can now be written in the form

$$H_{\text{op}}\psi = H'_{\text{op}}\psi' = i\hbar \, \partial\psi'/\partial t, \tag{4}$$

where H'_{op} is the *transformed energy operator*. In general the transformation of an operator $\Omega \to \Omega'$ corresponding to a coordinate transformation $q \to q'$ is defined by the condition that if

$$f_2 = \Omega f_1, \tag{5a}$$

where f_1 is an arbitrary function of the q_k and if f'_2 and f'_1 are obtained from f_2 and f_1 by the transformation $q \to q'$, the linear operator which transforms f'_1 into f'_2 is given by the equation

$$f'_2 = \Omega' f'_1. \tag{5b}$$

If H_{op} is Hermitean, H'_{op} will also be Hermitean,

$$\int f' H'_{\text{op}} g' \, d\tau' = \int f H_{\text{op}} g \, d\tau = \int g H^*_{\text{op}} f \, d\tau = \int g' H'^*_{\text{op}} f' \, d\tau'. \tag{6}$$

If H_{op} is of the simple type (3.20), which we can write in the form $H_{\text{op}} = \Sigma_k (2m_k)^{-1} (i\hbar)^2 \partial^2/\partial q_k^2$, where we leave out the term U which simply transforms into U', we find H'_{op} simply from the first equality of equation (6). We get

$$\int f H_{\text{op}} g \, d\tau = -\int \Sigma_k \frac{1}{2m_k} (i\hbar)^2 \frac{\partial f}{\partial q_k} \frac{\partial g}{\partial q_k} d\tau = -\tfrac{1}{2}(i\hbar)^2 \int \Sigma_{l,m} G^{lm} \frac{\partial f'}{\partial q'_l} \frac{\partial g'}{\partial q'_m} \varrho' dq \ldots$$

$$= \tfrac{1}{2}(i\hbar)^2 \int f' \Sigma_{l,m} \frac{1}{\varrho'} \frac{\partial}{\partial q'_l} \left[\varrho' G^{lm} \frac{\partial g'}{\partial q'_m} \right] d\tau',$$

where the first and the last equality follow by integrating by parts, where we have used the equation $\partial/\partial q_k = \Sigma_l (\partial q'_l/\partial q_k)(\partial/\partial q'_l)$ and where

$$G^{lm} = \sum_k \frac{1}{m_k} \frac{\partial q'_l}{\partial q_k} \frac{\partial q'_m}{\partial q_k} = G^{ml}. \tag{7}$$

We have thus

$$H'_{\text{op}} = \tfrac{1}{2}(i\hbar)^2 \frac{1}{\varrho'} \sum_{lm} \frac{\partial}{\partial q'_l} \left[\varrho' G^{lm} \frac{\partial}{\partial q'_m} \right] = \frac{1}{2\varrho'} \Sigma\, p'_l \varrho' G^{lm} p'_m, \tag{8}$$

if we use p'_l as an abbreviation for

$$p'_l = -i\hbar \partial/\partial q'_l. \tag{9}$$

From equation (7) it follows that the determinant $1/G$ of the G^{lm} is equal to ϱ^2 divided by all the m_k,

$$G^{-1} = |G^{lm}| = \varrho^2/m_1 m_2 \ldots = (\varrho'^2 m_1 m_2 \ldots)^{-1}. \tag{10}$$

The momenta p'_k which are canonically conjugate to the q'_k are found in classical mechanics by differentiating the kinetic energy T with respect to the generalised velocities $\dot{q}'_k$,

$$\begin{aligned} T &= \sum_k \tfrac{1}{2} m_k \dot{q}_k^2 = \sum_{lm} \tfrac{1}{2} G_{ml} \dot{q}'_l \dot{q}'_m, \\ G_{lm} &= \sum_k m_k \frac{\partial q_k}{\partial q'_l} \frac{\partial q_k}{\partial q'_m} = G_{ml}, \quad p'_l = \frac{\partial T}{\partial \dot{q}'_l} = \sum_m G_{lm} \dot{q}'_m . \end{aligned} \tag{11}$$

From equations (7) and (11) it follows that

$$\sum_m G^{lm} G_{mn} = \delta_{ln}, \quad \dot{q}'_l = \sum_m G^{lm} p'_m, \quad T = \tfrac{1}{2} \sum_l p'_l \dot{q}'_l = \tfrac{1}{2} \sum_{lm} G^{lm} p'_l p'_m . \tag{12}$$

The determinant of the G_{lm} is the reciprocal of the determinant of the G^{lm}, that is, it is equal to G,

$$|G_{lm}| = |G^{lm}|^{-1} = G = \varrho'^2 m_1 m_2 \ldots \tag{13}$$

From these formulae one sees that the operators $p'_{l\,\mathrm{op}}$ defined by equation (9) play formally the role of the classical generalised momenta. If we write equation (8) in the form

$$H'_{\mathrm{op}} = \tfrac{1}{2} \sum_{lm} G^{-\frac{1}{2}} p'_{l\,\mathrm{op}} G^m G^{\frac{1}{2}} p'_{m\,\mathrm{op}} , \tag{14}$$

we see that this expression would be identical with the classical energy expression (12) if we could neglect the non-commutability of the p'_l and q'_l. Conversely, equation (14) gives us the general prescription to find from the classical formula (12) the correct Hermitean operator in term of generalised coordinates for the case where the energy is quadratic in the p'_l *. We must draw attention to the fact that the $p'_{l\,\mathrm{op}}$ which we have introduced are non-Hermitean. We shall see in the following that the quantum mechanical operator corresponding to a classical quantity such as p'_l should always be Hermitean. The correct operator would here be given by the equation (compare the discussion in § 29)

$$(p_l)_{\mathrm{classical}} \rightarrow G^{-\frac{1}{4}} p'_{l\,\mathrm{op}} G^{\frac{1}{4}} = \varrho'^{-\frac{1}{2}} (i\hbar) (\partial/\partial q'_l) \varrho'^{\frac{1}{2}}. \tag{15}$$

§ 35. The definability of mechanical quantities

We can immediately extend to many particle systems the considerations of § 13 about the physical meaning of the wave function. If we define a mechanical quantity or an *observable* as any single-

* E. Schrödinger, *Ann. Phys.*, **79**, 748, 1926. It is possible through this prescription to solve the difficulty of the arbitrariness of the order of factors which occurred in equation (3.76).

valued real function of coordinates and momenta possibly containing the time explicitly such a quantity would in classical physics possess a well defined value at any moment and this value could in principle be determined by an experiment at one definite time. This will in general be impossible in quantum mechanics. We shall, however, assume that for any given physical situation we can always determine a quantitative probability law which expresses the instantaneous distribution of the quantity in question over all its possible values. In accordance with the usual practice of invoking idealised circumstances we may imagine that this distribution can be checked by instantaneous measurements — which may have to be repeated several times under the same circumstances at the same moment in the evolution of the physical process. We have earlier (see especially §§ 14 and 15) discussed examples of such measurements when the physical quantities involved were the spatial coordinates and the momenta. We must remember, however, that it is in general hardly possible to invent such an idealised experiment that the interaction between apparatus and system can give information about the instantaneous value of a physical quantity and this is true both in classical and in quantum physics. It is thus often preferable to consider the situation more formally and to make the following statement. In interpreting the physical situation of a system it must be possible to describe quantitatively by means of a distribution function the indefiniteness of the instantaneous value of a physical quantity.

If we use a complex wave function to describe the physical situation, we can find a unique way to construct such distribution functions. In principle we have solved this problem as it is not really different from the way by which in chapters 2 and 3 we found the distribution over possible energy values. That this should be so follows immediately from a consideration of the (completely unphysical) idealised experiment where suddenly at a given time the Hamiltonian is replaced by the mechanical quantity considered. The energy distribution function in the new system can be nothing but the distribution function we were looking for. Our original problem had, however, nothing to do with questions of energy and we shall discuss it independently in § 39. Before doing that we shall investigate in some detail the general problem of eigenvalues and eigenfunctions in quantum mechanics.

§ 36. Eigenvalues and eigenfunctions corresponding to an observable

Let $F(p_k, q_k, t)$ be a real mechanical quantity, where the q_k are Cartesian coordinates and p_k the corresponding momenta. We shall restrict our discussion to quantities which are finite polynomials in the p_k. Using the rule $p_k \to -i\hbar\partial/\partial q_k$ we can assign to F a Hermitean operator F_{op} (compare § 29). If the prescription for this procedure is not unique, the problem occurs whether a particular choice for F_{op} must be preferred on mathematical or physical grounds. Luckily, this point has up to now not led to any difficulties in the theory.

Consider now the differential equation

$$F_{\text{op}} S = F_{\text{e}} S, \tag{16}$$

where S is in general a complex function of the Cartesian coordinates of the particles and where F_{e} is a numerical constant. This equation if formally not different from equation (3.96) from which we found the eigenvalues and eigenfunctions of a Hermitean energy operator. The only difference is that F_{op} and thus also S may contain the time t as parameter. However, in most physical applications F_{op} does not depend explicitly on the time.

Exactly in the same way as in the case of the energy operator we now introduce *the eigenvalues F_e and the eigenfunctions of the operator* F_{op}. Equation (16) has as its solution one eigenfunction (non-degenerate case) or several linearly independent eigenfunctions (degenerate case), if F_{e} is an eigenvalue. An eigenfunction is defined as a one-valued function of the q_k which is either normalisable ($\int SS^* d\tau$ converges: proper eigenfunction and proper eigenvalue) or which can be considered to be the limiting case of such a normalisable function (improper eigenfunction and improper eigenvalue). The eigenvalues F_{e} are real. The set of all F_{e} are called the *eigenvalue spectrum of F_{op}. An arbitrary normalisable function of the coordinates can always be expanded in terms of the eigenfunctions of the operator* [compare eq. (3.100)].

The mathematical equations are simple in the case when the eigenvalue spectrum is a line spectrum and when each eigenfunction S corresponding to a degenerate eigenvalue F_k can be written as the linear combination of a finite number (g_k) of linearly independent eigenfunctions S_{kj} ($j = 1, 2, \ldots, g_k$). In this case we have

$$F_{\text{op}} S_{kj} = F_k S_{kj}. \tag{17}$$

From the considerations in § 32 it follows that the F_e must be real [replace in equation (3.98) φ by S, H_{op} by F_{op}, and E by F_e] and that two eigenfunctions corresponding to different eigenvalues are orthogonal to one another, that is

$$\int S^*_{k'j'} S_{kj}\, d\tau = 0, \quad k' \neq k, \quad d\tau = dq_1 dq_2 \ldots \tag{18}$$

We can, moreover, always choose the S_{kj} which belong to the same degenerate F_e in such a way that they are orthogonal to one another (compare § 21) and we can finally multiply each eigenfunction by a factor such that they are all normalised. If this is all done we have the general result

$$\int S^*_{k'j'} S_{kj}\, d\tau = \delta_{k'k}\,\delta_{j'j}. \tag{19}$$

If we write the expansion of an arbitrary normalisable function f in the form

$$f = \Sigma_k \left[\Sigma_{j=1}^{g_k} b_{kj} S_{kj}\right] = \Sigma_{kj} b_{kj} S_{kj}\,, \tag{20}$$

we can find the coefficients b_{kj} from the equations

$$b_{kj} = \int S^*_{kj} f\, d\tau, \tag{21}$$

which follows if we multiply equation (20) by S^*_{kj}, integrate over the whole of q-space and use equation (19). From equations (20) and (21) it also follows that

$$\int |f|^2\, d\tau = \Sigma_{kj} b_{kj} \int S_{kj} f^*\, d\tau = \Sigma_{kj} |b_{kj}|^2. \tag{22}$$

Substituting equation (21) into equation (20) and interchanging the order of summation and integration we get

$$f(q_1, q_2, \ldots) = \int \{\Sigma_{kj} S^*_{kj}(q_1', q_2', \ldots) S_{kj}(q_1, q_2, \ldots)\} f(q_1', q_2', \ldots)\, dq_1' dq_2' \ldots,$$

which shall be valid for any arbitrary function f. The expression within braces has formally the same properties as the Dirac δ-function which in § 29 was formally assigned to the unit operator,

$$\Sigma_{kj} S^*_{kj}(q_1', q_2', \ldots) S_{kj}(q_1, q_2, \ldots) = \delta(q_1 - q_1')\,\delta(q_2 - q_2') \ldots \tag{23}$$

As in § 32 we can easily generalise these equations to cover the case where the energy value spectrum is partly or wholly continuous and where the eigenfunctions belonging to a degenerate eigenvalue may depend on one or more parameters. In the case where there are only improper eigenvalues and where the degeneracy is fully covered by a number of continuous parameters $r_1, r_2, \ldots$ we get the following generalisations of equations (20) to (22),

$$f(q_1, q_2, \ldots) = \int b(F, r_1, r_2, \ldots) S(F, r_1, r_2, \ldots; q_1, q_2, \ldots)\, \varrho\, dF\, dr_1 dr_2 \ldots, \tag{24}$$

$$b(F, r_1, r_2, \ldots) = \int S^*(F, r_1, r_2, \ldots; q_1, q_2, \ldots)\, f(q_1, q_2, \ldots)\, d\tau, \tag{25}$$

$$\int |f|^2 d\tau = \int b(F, r_1, r_2, \ldots) \Big\{ \int S(F, r_1, \ldots; q_1, \ldots) f^*(q_1, \ldots) d\tau \Big\} \varrho\, dF\, dr_1 \ldots$$
$$= \int |b|^2 \varrho\, dF\, dr_1 dr_2 \ldots \tag{26}$$

We have written F to indicate the eigenvalues of F_{op}. In equations (24) to (26) the integrations extend over all possible values of $F, r_1, r_2, \ldots$ and the density function ϱ may depend on F as well as on $r_1, r_2, \ldots$ — which means a slight generalisation from the analogous case in § 32. Equation (24) expresses the general expansion of a function in terms of the eigenfunctions of F_{op}. If the eigenfunctions S are chosen in such a way that the coefficients b in equation (24) are determined by equation (25), we say that the eigenfunctions S are *orthonormal with respect to the density function* $\varrho(F, r_1, r_2, \ldots)$. Substitution of equation (24) into equation (25) gives

$$b(F, r_1, r_2, \ldots) = \iint S^*(F, r_1, \ldots; q_1, \ldots)\, S(F', r_1', \ldots; q_1, \ldots)$$
$$\varrho(F', r_1', \ldots)\, b(F', r_1', \ldots)\, d\tau\, dF'\, dr_1' \ldots,$$

which means that the integral $\int S^*(F, r_1, \ldots; q_1, \ldots) S(F', r_1', \ldots; q_1, \ldots) d\tau$ formally plays the role of the unit operator in the space of the variables $F, r_1, \ldots$ [compare eq. (3.65)]. Using the δ-function we can write

$$\int S^*(F, r_1, \ldots; q_1, \ldots) S(F', r_1', \ldots; q_1, \ldots)\, dq_1 \ldots$$
$$= \varrho^{-1/2} \delta(F - F')\, \delta(r_1 - r_1') \ldots \varrho'^{-1/2}. \tag{27}$$

This equation is the counterpart of equation (19) and expresses the (improper) orthonormality of the eigenfunctions.

If b in equation (24) is a function which is zero except within a volume element $F_0 < F < F_0 + \Delta F_0$, $r_{i0} < r_i < r_{i0} + \Delta r_{i0}$, where $b = 1$, then f is a so-called eigendifferential * and equations (25)

* H. Weyl, *Math. Ann.*, **68**, 220, 1910; Compare also J. R. Oppenheimer, *Z. Phys.*, **41**, 268, 1927.

and (27) state that $\int S^*(F, r_1, \ldots) f(F_0, r_{10}, \ldots)\, d\tau$ is equal to 1 or 0 according to whether the values of $F, r_1, r_2, \ldots$ lie within or outwith the chosen volume element.

The counterpart of equation (23) is obtained by substitution of equation (25) into equation (24),

$$\int S^*(F, r_1, \ldots;\, q_1', \ldots)\, S(F, r_1, \ldots;\, q_1, \ldots)\, \varrho\, dF\, dr_1\, dr_2 \ldots = \delta(q_1 - q_1')\,\delta(q_2 - q_2') \ldots \quad (28)$$

It is not difficult to write down the equations for a mixed case where the spectrum contains lines as well as a continuum, and where the degeneracy can be described partly by a discrete index and partly by continuous parameters. The equations are hybrids of equations (20) to (22) and (24) to (26). By using Stieltjes integrals one can always use equations of the type (24) to (26).

§ 37. Eigenvalues and eigenfunctions of finite Hermitean matrices

We have not tried to develop the theory of eigenvalues and eigenfunctions of Hermitean operators acting on functions of continuous variables (in our case the Cartesian coordinates of the particles) or the theorem about the expansion of functions in terms of eigenfunctions with full mathematical rigour *. The various theorems and equations become much more translucent if we consider them to be the limiting case of the theory of Hermitean operators acting on functions of a discrete argument (compare the discussion in § 24). We shall review this theory in this section. We may remark here that this theory is also important for its own sake.

In order to avoid possible convergence difficulties we consider the case where the argument of the functions can only take on N different values. This case can easily be treated with mathematical rigour. As before we indicate the N values of the argument by an index m which can take on the integral values $1, 2, \ldots, N$. An arbitrary function f is thus determined by giving the N (in general complex) numbers f_m. Geometrically speaking f is a vector with complex components f_m in an N-dimensional space.

A function f is called "normalised" if the "sum" of the product

* We refer the reader to J. v. Neumann, *Mathematical foundations of quantum mechanics*, Princeton University Press, 1955 where the mathematical aspects are given their full due.

f^*f is equal to 1, that is (compare § 29) if

$$\int f^* f = \Sigma_m f_m^* f_m = 1. \tag{29}$$

Two functions f and g are called orthogonal to one another if the sum of the product of f^* and g is zero, that is, if

$$\int f^* g = \Sigma_m f_m^* g_m = 0 = \Sigma_m g_m^* f_m = \int g^* f. \tag{30}$$

Geometrically speaking $\int f^*f$ is the square of the absolute magnitude of the vector f and $\int f^*g$ is the scalar product of the vectors f and g. The scalar product of f and g is thus complex conjugate to the scalar product of g and f. Orthogonality corresponds to a zero scalar product. In the case of real vectors these definitions are the same as those of Euclidean geometry, if the f_m refer to an orthogonal coordinate system.

If a set of N functions $T^{(\mu)}$ $(\mu = 1, 2, \ldots, N)$ is such that any function can be expanded in terms of the $T^{(\mu)}$, that is, that we can write for f

$$f = \Sigma_\mu \varphi_\mu T^{(\mu)}, \tag{31}$$

this set is called a complete set. Equation (31) is equivalent to the following N equations,

$$f_m = \Sigma_\mu \varphi_\mu T_m^{(\mu)}, \qquad m = 1, 2, \ldots, N. \tag{32}$$

In order that the φ_μ can be determined, it is necessary that the determinant of the $T_m^{(\mu)}$ is different from zero, that means that apart from the trivial solution $a_\mu = 0$ there is no solution for the a_μ of the N equations

$$\Sigma_\mu a_\mu T_m^{(\mu)} = 0. \tag{33}$$

This can also be expressed by requiring that the functions $T^{(\mu)}$ should be linearly independent, that is, that there should not exist relations of the following kind,

$$\Sigma_\mu a_\mu T^{(\mu)} = 0, \tag{34}$$

except when all a_μ vanish. Geometrically speaking, the vectors $T^{(\mu)}$ should span the whole of N-dimensional space, so that they can be used as new coordinate axes (which in general will not form an orthogonal system). The coefficients φ_μ are the components of f with respect to these axes expressed in suitable units.

In the language of the theory of functions one considers the φ_μ

in equation (32) to be the values of a new function φ of argument $\mu (\mu = 1, 2, \ldots, N)$ which is obtained from the original function f by means of the transformation matrix $T_m^{(\mu)}$. If we now consider a second transformation which changes φ into a function F of argument $M (M = 1, 2, \ldots, N)$,

$$\varphi_\mu = \Sigma_M F_M V_\mu^{(M)}, \tag{35}$$

where again the determinant of the $V_\mu^{(M)}$ must be different from zero, we see that there exists the following relation between f and F, $f_m = \Sigma_M F_M [\Sigma_\mu T_m^{(\mu)} V_\mu^{(M)}] = \Sigma_M F_M W_m^{(M)}$, so that we have for the transformation matrix $W_m^{(M)}$ the equation

$$W_m^{(M)} = \Sigma_\mu T_m^{(\mu)} V_\mu^{(M)}. \tag{36}$$

Formally this relation is the same as the multiplication law for matrices discussed, for instance, in § 29. There is, however, a difference. The matrices Ω_{km} which we considered earlier corresponded to operators which produced from one function f a new function g of the same argument [compare eq. (3.57); geometrically speaking the operator assigned to a given vector f another vector g]. Now, however, we use the transformation matrix $T_m^{(\mu)}$ to get a new *representation* of the original function f (geometrically speaking the vector f with components f_m is described in a new coordinate system by components φ_μ).

The transformation matrix which occurs when the φ_μ are expressed in the f_m and which can be obtained by solving equations (32) is called the *inverse matrix* T^{-1}. We have

$$\varphi_\mu = \Sigma_m f_m (T^{-1})_\mu^{(m)}. \tag{37}$$

Here, $(T^{-1})_\mu^{(m)}$ is the subdeterminant of $T_m^{(\mu)}$ in the transformation matrix, divided by the determinant of this matrix. We have $\Sigma_m (T^{-1})_\mu^{(m)} T_m^{(\mu')} = \delta_{\mu\mu'}$, and $\Sigma_\mu T_m^{(\mu)} (T^{-1})_\mu^{(m')} = \delta_{mm'}$ which can be written symbolically as $T^{-1} T = 1$ and $TT^{-1} = 1$, where 1 indicates the unit matrix in the μ- and in the m-representation respectively.

Consider N functions $U^{(\mu)}$ which are orthonormal, that is, for which the following equations hold,

$$\int U^{*(\mu')} U^{(\mu)} \equiv \Sigma_m U_m^{*(\mu')} U_m^{(\mu)} = \delta_{\mu'\mu}. \tag{38}$$

They will form a complete set since the square of the absolute value of the determinant of the $U_m^{(\mu)}$ will be equal to 1 according to equation (38). The $U^{(\mu)}$ are called an orthonormal set. One can immediately find the coefficients of the expansion

$$f = \Sigma \varphi_\mu U^{(\mu}, \text{ or, } f_m = \Sigma_\mu \varphi_\mu U_m^{(\mu)}, \tag{39}$$

by multiplying by $U_m^{*(\mu')}$ and summing over m,

$$\varphi_\mu = \Sigma_m f_m U_m^{*(\mu)}. \tag{40}$$

For the sum of the square of the absolute value of f we have

$$\int f^* f \equiv \Sigma_m f_m^* f_m = \Sigma_{m\mu\mu'} \varphi_\mu^* \varphi_{\mu'} U_m^{*(\mu)} U_m^{(\mu')} = \Sigma_\mu \varphi_\mu^* \varphi_\mu \equiv \int \varphi^* \varphi. \tag{41}$$

If we choose f to be the unit vector along the m'-axis so that $f_m = \delta_{mm'}$, we get $\varphi_\mu = U_{m'}^{*(\mu)}$ and from equation (39) we have

$$\Sigma_\mu U_m^{(\mu)} U_{m'}^{*(\mu)} = \delta_{mm'}. \tag{42}$$

We see from these equations that the representations of the given function either by the f_m or by the φ_μ are completely equivalent. Geometrically speaking we are describing one vector in two different, but equivalent coordinate systems. Expression (41) is the absolute square of the length of this vector for which we have analogous expressions in the original and in the "rotated" coordinate system. These expressions correspond to the Pythagorean theorem for real vectors in N dimensions.

The linear relations (39) and (40) between the φ_μ and the f_m are called a *unitary transformation* which changes the f_m into the φ_μ. According to equation (41) $\Sigma_m f_m^* f_m$ goes over into $\Sigma_\mu \varphi_\mu^* \varphi_\mu$. One could actually define a unitary transformation by this property. The N^2 numbers $U_m^{(\mu)}$ form a *unitary matrix*. Comparing equations (32) and (37) with equations (39) and (40) we see that $(U^{-1})_\mu^{(m)} = U_m^{*(\mu)}$, or,

$$U^{-1} = \tilde{U}^*, \ (U^{-1})^* = \tilde{U}, \tag{43}$$

that is, the inverse matrix of a unitary matrix is obtained by taking the complex conjugate of the transposed matrix (transposition, that is, the exchange of rows and columns, is indicated by a tilde ˜.) Equation (43) gives a necessary and sufficient condition for unitarity.

The determinant of a unitary matrix is always a complex number of modulus 1. The inverse of a unitary matrix is again unitary and so is the product of two unitary matrices, since the result of two successive unitary transformations is itself a unitary transformation.

An orthonormal system can be chosen in infinitely many ways.

Especially one can always choose $U^{(1)}$ to be the same as a given normalised function.

To see this we first of all note that $U^{(1)}$ can never be expressed as a linear combination of functions G which are orthogonal to $U^{(1)}$ so that among all possible G there can be at most $N-1$ linearly independent functions (geometrically speaking the G all lie in an $N-1$-dimensional subspace which is orthogonal to $U^{(1)}$). One can choose these $N-1$ functions as follows: $G_m^{(\mu)} = \delta_{m\mu} - U_m^{(1)} U_\mu^{*(1)}$, $\mu = 2, 3, \ldots, N$. Without restricting the general validity of our argument we can choose $U_1^{(1)}$ to be non-vanishing. In that case one can show that the $G^{(\mu)}$ are, indeed, linearly independent and orthogonal to $U^{(1)}$. We normalise $G^{(2)}$ and call it $U^{(2)}$. There can be at most $N-2$ linearly independent functions H among the functions which can be built up out of the $G^{(\mu)}$ and which are orthogonal to $U^{(2)}$ (geometrically speaking, the vectors H span an $N-2$-dimensional subspace which is orthogonal to both $U^{(1)}$ and $U^{(2)}$). We choose now the following $N-2$ functions, $H^{(\mu)} = G^{(\mu)} - U^{(2)} \int U^{*(2)} G^{(\mu)}$, $\mu = 3, 4, \ldots, N$. These are linearly independent, as any linear relation between them would involve such a relation between the $G^{(\mu)}$. Normalising $H^{(3)}$ we obtain $U^{(3)}$. Proceeding in this way we obtain an orthonormal system of the kind mentioned.

Using the concepts of orthonormal sets and unitary transformations we can easily discuss the eigenvalue and eigenfunction problem of a Hermitean operator. In our case such an operator F_{op} is defined by the elements of a finite Hermitean matrix ($F_{mn} = F_{nm}^*$; compare § 29). The counterpart of problem (16) is now to find a function S, that is, a set of numbers $S_1, S_2, \ldots, S_N$ such that for all values of m the following equation holds [compare eq. (3.57)],

$$\Sigma_n F_{mn} S_n = F S_m . \tag{44}$$

The number F is an eigenvalue of the Hermitean operator F_{op}, or, as one usually says in this case, an eigenvalue of the Hermitean matrix F_{mn}. The condition for the existence of non-trivial solutions of equation (44) is that the determinant of the system vanishes

$$\begin{vmatrix} F_{11}-F & F_{12} & F_{13} & \cdots \\ F_{21} & F_{22}-F & F_{23} & \cdots \\ \cdots & \cdots & \cdots & \cdots \end{vmatrix} = |F_{mn} - F\delta_{mn}| = 0. \tag{45}$$

We have thus an equation of the N-th degree, the so-called *secular equation* of the eigenvalue problem. The roots, the *eigenvalues*, are as we shall see in a moment all real. We shall denote them by F_μ ($\mu = 1, 2, \ldots, N$).

Let $T_m^{(\mu)}$ be a transformation matrix with non vanishing determinant which transforms the function S into a function σ,

$$S_m = \Sigma_\mu \sigma_\mu T_m^{(\mu)}, \quad \sigma_\mu = \Sigma_m S_m (T^{-1})_\mu^{(m)}. \tag{46}$$

We can transform equations (44) into a set of equations for the σ_μ,

$$F\sigma_\mu = \Sigma_m F S_m (T^{-1})_\mu^{(m)} = \Sigma_{mn} (T^{-1})_\mu^{(m)} F_{mn} S_n = \Sigma_{mn\nu} (T^{-1})_\mu^{(m)} F_{mn} T_n^{(\nu)} \sigma_\nu ,$$

or,

$$\Sigma_\nu \Phi_{\mu\nu} \sigma_\nu = F\sigma_\mu , \tag{47}$$

where the new matrix Φ is obtained from F by transformation,

$$\Phi_{\mu\nu} = \Sigma_{mn} (T^{-1})_\mu^{(m)} F_{mn} T_n^{(\nu)}, \quad F_{mn} = \Sigma_{\mu\nu} T_m^{(\mu)} \Phi_{\mu\nu} (T^{-1})_\nu^{(n)}. \tag{48}$$

Equation (47) is equivalent to equation (44) and the resulting secular equation must thus be identical to equation (45). If we choose for T a unitary matrix U equation (48) can be written in the form

$$\Phi_{\mu\nu} = \Sigma_{mn} U_m^{*(\mu)} F_{mn} U_n^{(\nu)}, \quad F_{mn} = \Sigma_{\mu\nu} U_m^{(\mu)} \Phi_{\mu\nu} U_n^{*(\nu)}. \tag{49}$$

The matrix Φ is Hermitean, since F is Hermitean. *It is always possible to find a unitary matrix U such that $\Phi_{\mu\nu}$ is a diagonal matrix* $F_\mu \delta_{\mu\nu}$. The F_μ are the eigenvalues of the original matrix and we have thus

$$\Phi_{\mu\nu} = F_\mu \delta_{\mu\nu} , \quad F_{mn} = \Sigma_\mu U_m^{(\mu)} F_\mu U_m^{*(\mu)}. \tag{50}$$

We can prove this theorem as follows *. Consider the expression $A = \Sigma_{mn} S_m^* F_{mn} S_n$, for the moment with arbitrary S. Since F is Hermitean, A will be real. There will be a set of S_m values, $S_m = S_m^{(1)}$, such that A is minimum under the condition that $\Sigma S_m^* S_m = 1$. The analytical condition for this minimum is: $\Sigma_{mn} S_m^{(1)*} (F_{mn} - \lambda\delta_{mn}) \delta S_n + \Sigma_{mn} (F_{mn} - \lambda\delta_{mn}) S_n^{(1)} \delta S_m^* = 0$, where λ is a constant and where the δS_m indicate infinitesimal changes in the S_m. As S_m and S_m^* are analytically independent, it follows that $\Sigma_n F_{mn} S_n^{(1)} = \lambda S_m^{(1)}$, $\Sigma_m F_{mn} S_m^{(1)*} = \lambda S_n^{(1)*}$, $A = \Sigma_m S_m^{(1)*} \lambda S_m^{(1)} = \lambda$. The $S_m^{(1)}$ satisfy thus equations (44) and λ is the smallest eigenvalue of F_{mn}, which we shall call F_1. We now introduce an orthonormal set by means of a unitary matrix $U_n^{(n')}$ which is chosen in such a way that $U^{(1')}$ is equal to $S^{(1)}$, $U_m^{(1')} = S_m^{(1)}$. The transformed matrix $F'_{m'n'}$ satisfies the equations [compare eq. (49)], $F'_{m'1'} = \Sigma_m U_m^{(m')*} F_1 U_m^{(1')} = \delta_{m'1'} F_1$, $F'_{1'n'} = \delta_{1'n'} F_1$. The N equations (47) are thus reduced to one equation, $F_1' S_{1'} = F S_{1'}$ which contains only $S_{1'}$, and a set of $N-1$ equations $\Sigma_{n'=2}^{N} F'_{m'n'} S_{n'} = F S_{m'}$, $m' = 2, \ldots, N$, which contain only $S_{2'}, \ldots, S_{N'}$.

We can reduce the problem one step further by another unitary transformation. We consider a set of values $S_{n'} = S_{n'}^{(2)}$ $(n' = 2, \ldots, N)$ such that the real expression $B = \Sigma_{m'n'} S^*_{m'} F'_{m'n'} S_{n'}$ has a minimum value F_2 under the condition $\Sigma_{n'} S^*_{n'} S_{n'} = 1$. Again, F_2 is an eigenvalue of our secular problem, and certainly $F_2 \geqq F_1$. We then introduce an orthonormal set $U^{(n'')}$ $(n'' = 2, \ldots, N)$ corresponding to a unitary matrix $U_{m'}^{(n'')}$ $(m' = 2, \ldots, N)$ such that $U^{(2'')}$ is the same as $S^{(2)}$, $U_{m'}^{(2'')} = S_{m'}^{(2)}$. This matrix is completed into a unitary matrix with N rows and columns by putting $U_{m'}^{(1'')} = \delta_{1''m'}$, $U_{1'}^{(n')} = \delta_{n''1'}$. Using this matrix one transforms $F'_{m'n'}$ into a matrix $F''_{m''n''}$, $F''_{m''n''} = \Sigma_{m'n'} U_{m'}^{(m'')*} F'_{m'n'} U_{n'}^{(n'')}$. One sees that all off-diagonal elements of the first two rows and columns of F'' are zero, and that

* Compare R. Courant and D. Hilbert, *Methods of Mathematical Physics*, Interscience Publishers, New York 1953, Vol. I.

$F''_{1''1''} = F_1$, $F''_{2''2''} = F_2$. This transformed matrix F'' is connected with the original matrix F by the equations $F''_{m''n''} = \Sigma_{mn} U_m^{(m'')*} F_{mn} U_n^{(n'')}$, $U_m^{(m'')} = \Sigma_{m'} U_m^{(m')} U_{m'}^{(m'')}$. The unitary matrix $U_m^{(m'')}$ is thus obtained by multiplying $U_m^{(m')}$ and $U_{m'}^{(m'')}$.

Proceeding in the same fashion we finally construct a unitary matrix $U_m^{(\mu)}$ which transforms F_{mn} into a diagonal matrix. The diagonal elements F_n of this matrix are the eigenvalues of our secular problem arranged in order of magnitude, $F_1 \leqq F_2 \leqq F_3 \leqq \ldots \leqq F_N$.

Applying now equations (46) to (49) to the case where we have a unitary matrix possessing this property, so that equations (50) hold, we get the secular equations (47) in the simple form

$$F_\mu \sigma_\mu = F \sigma_\mu , \quad \mu = 1, 2, \ldots, N. \tag{51}$$

If all eigenvalues F are different there are exactly N solutions

$$F = F_\mu , \quad \sigma_\mu \neq 0 \text{ and } \sigma_\nu = 0, \ \nu \neq \mu, \quad (\mu = 1, 2, \ldots, N)$$

and using equations (46) we find for the eigenfunctions S,

$$S_m = \sigma_\mu U_m^{(\mu)}, \quad \mu = 1, 2, \ldots, N. \tag{52}$$

The σ_μ are arbitrary numerical factors. Choosing for the sake of simplicity $\sigma_\mu = 1$ we see that *the orthonormal set $U^{(\mu)}$ forms a complete set of eigenfunctions of our Hermitean matrix.*

The most general set of normalised eigenfunctions is obtained by assigning arbitrarily to the σ_μ in equation (52) complex values with modulus 1. However, these factors can be included in the $U_m^{(\mu)}$ since the quantities $U'^{(\mu)}_m = \sigma_\mu U_m^{(\mu)}$ form also a unitary matrix which brings F_{mn} on diagonal form. In our case of eigenvalues which are all different one can determine the eigenfunctions immediately from the secular determinant (45), since the S_m have the same ratios as the subdeterminants of a column.

Consider now the case where some of the eigenvalues are the same, that is, the *degenerate* case. The different eigenvalues are now denoted by $F_k (F_1 < F_2 < F_3 \ldots)$ and the root F_k of the secular equation may be g_k-fold, that is, the eigenvalue F_k may be g_k-fold degenerate. The general solution of the transformed secular equation (51) is now that for $F = F_k$ all σ_i are equal to zero except those for which $i = g_1 + g_2 + \cdots + g_{k-1} + j_k$ with $j_k = 1, 2, \ldots, g_k$, while k takes on the values $1, 2, \ldots$ The eigenfunctions are now

$$S_m = \Sigma_\mu \sigma_\mu U_m^{(\mu)}, \tag{53}$$

where μ runs from $g_1 + g_2 + \cdots + g_{k-1} + 1$ to $g_1 + g_2 + \cdots + g_{k-1} + g_k$ $(k = 1, 2, \ldots)$. The most general eigenfunction corresponding to F_k contains g_k arbitrary parameters σ. Two eigenfunctions corresponding to different F_k are orthogonal to one another. One can always choose from the eigenfunctions belonging to the same k g_k linearly independent ones. A special choice is to choose these g_k functions by putting in turn all σ_μ in equation (53) equal to zero bar one which is put equal to 1. We thus have again the solution (52) and all eigenfunctions, even those belonging to the same eigenvalue, are orthonormal. One can obtain this last result in a more general way by choosing for the σ_μ in equation (53) the coefficients σ_μ^j $(j = 1, \ldots, g_k;$ μ from $g_1 + \cdots + g_{k-1} + 1$ to $g_1 + \cdots + g_{k-1} + g_k)$ of an arbitrary unitary matrix of rank g_k. This means that one writes

$$S_m^{kj} = \Sigma_\mu \sigma_\mu^j U_m^{(\mu)} = U'^{kj}_m. \tag{54}$$

The N^2 coefficients U'^{kj}_m $(m = 1, \ldots, N; j = 1, \ldots, g_k; k = 1, 2, \ldots)$ clearly form another unitary matrix. This matrix is obtained by multiplication of $U_m^{(\mu)}$ and a unitary matrix σ_μ^{kj} of rank N which has non-vanishing elements only within squares of rows and columns which are arranged along the diagonal *. One sees easily that the matrix U' transforms the original matrix F_{mn} into a diagonal matrix, just as U did. This means that the prescription which we gave earlier in this section to construct U by successive steps is not unique in the case of a degenerate eigenvalue **. A discussion of the solutions (53) based on the secular determinant is much more complicated in the degenerate case than in the non-degenerate case, as all subdeterminants of rank larger than $N - g$ will be equal to zero in the case of a g-fold root.

One can illustrate the construction of the eigenfunctions geometrically. The equation $A = \Sigma S_m{}^* F_{mn} S_n$ is in the case of real S and F a quadratic hypersurface. This equation is changed into an equation in a new system of coordinates, σ_μ, through the real unitary transformation U in such a way that all mixed terms disappear, $A = \Sigma \sigma_\mu{}^* F_\mu \sigma_\mu$. The unit vectors $U^{(\mu)}$ are along the principal axes of the hypersurface and the lengths of these principal axes are proportional to $F_\mu^{-\frac{1}{2}}$. One therefore often calls the problem (44) *the problem of reducing the Hermitean matrix F_{mn} to its principal axes*. In the case of a g-fold degenerate eigenvalue the hypersurface possesses g principal axes of equal length. They span a g-dimensional subspace and any set of g mutually orthogonal vectors in this subspace can be

* This matrix looks as follows: $\begin{array}{c:c:c} \sigma_1 & & 0 \\ \hdashline & \sigma_2 & \\ \hdashline 0 & & \end{array}$, where σ_k is the unitary matrix $\sigma_\mu{}^j (j = 1, \ldots, g_k;\ \mu = 1, 2, \ldots)$.

** One can, for instance, choose for $S^{(1)}$ any of the normalised functions S' corresponding to $k = 1$ which are give by equation (53).

considered to be the principal axes. Our construction of $U_m^{(\mu)}$ corresponds to the process where one first of all locates the longest principal axis (along the vector $S^{(1)}$ with a length proportional to $F_1^{-\frac{1}{2}}$), then one locates in the subspace orthogonal to this axis the second longest principal axis, and so on. If F_1 is g_1-fold degenerate we can choose $S^{(1)}$ along any direction in the subspace spanned by the g_1 principal axes of equal length.

One notices easily the close formal analogy between the equation discussed in the present section and those occuring in § 36 where Hermitean operators corresponding to observables were discussed. First of all we again draw attention to the fact that the differential equation (16) for the eigenfunctions of those operators are the counterpart of the linear homogeneous equations (44). The S_{kj} which according to equation (17) were the solutions of equation (16) correspond now for any k to some set of g_k linearly independent functions S of the form (53). Similarly the choice of the S_{kj} by the procedure outlined in § 21 such that equation (19) is satisfied corresponds to the choice of the S in equation (54) by using a unitary matrix σ_μ^j. A given set of S_{kj} can always be replaced by the linear combinations

$$S'_{kj'} = \Sigma_j U_j^{j'} S_{kj}, \tag{55}$$

where $U_j^{j'}$ is an arbitrary unitary matrix of g_k rows and columns.

The expansion of an arbitrary function f in equation (20) is the counterpart of equation (39) and one can consider the expansion coefficients b_{kj} to be a function of a discrete argument ($j=1, 2, \ldots, g_k$; $k = 1, 2, \ldots$) which is obtained from the original function f by means of a *generalised unitary transformation.* Expression (21) for the b_{kj} corresponds to the solution (40) of equation (39). Finally equation (22) corresponds to equation (41) while the formal orthogonality relation (23) corresponds to equation (42).

If we consider on the other hand the case where there are only improper eigenfunctions corresponding to a continuous spectrum, we can interpret equation (24) as a unitary transformation which transforms f into a function of the continuous variables $F, r_1, r_2, \ldots$ and equations (24) to (28) correspond to equations (39) to (41), (38), and (42). If there is degeneracy there are also here infinitely many ways of choosing the eigenvalues corresponding to one eigenvalue.

§ 38. The eigenfunctions of commuting Hermitean operators

Is it possible to assign the same set of eigenfunctions to two or more operators? Let F_k be the eigenvalues of F_{op} with the corresponding orthonormal set of eigenfunctions S_{kj}. According to equation (55) there are still infinitely many ways of choosing the S_{kj}. Would it be possible to choose the S_{kj} such that they are at the same time eigenfunctions of a second Hermitean operator G_{op}. The necessary and sufficient condition for this to be possible is that F_{op} *and* G_{op} *commute,* that is, that the following identity holds,

$$F_{op} G_{op} - G_{op} F_{op} \equiv 0. \tag{56}$$

First of all we shall prove that this condition is necessary. Let S_m be a complete set of eigenfunctions of both F_{op} and G_{op}, so that we have

$$F_{op} S_m = F_m S_m, \quad G_{op} S_m = G_m S_m . \tag{57}$$

The F_m and G_m are the eigenvalues*. From equations (57) it follows that $G_{op} F_{op} S_m = F_m G_{op} S_m = F_m G_m S_m = G_m F_{op} S_m = F_{op} G_{op} S_m$. As this last equation holds for every eigenfunction and as any arbitrary normalisable function f can be expanded in terms of the S_m we have for such an arbitrary function $G_{op} F_{op} f = F_{op} G_{op} f$ which means that $F_{op} G_{op}$ and $G_{op} F_{op}$ are identical operators. Equation (56) is thus a necessary condition.

To prove that it is sufficient we start from some complete set of eigenfunctions S_{kj} of F, so that equation (17) holds, that is, $F_{op} S_{kj} = F_k S_{kj}$. Assuming equation (56) to hold we have $F_{op} G_{op} S_{kj} = G_{op} F_{op} S_{kj} = F_k G_{op} S_{kj}$. From the equality of the first and last member of this equation it follows that all g_k functions $G_{op} S_{kj}$ $(j = 1, \ldots, g_k)$ are eigenfunctions of F_{op} corresponding to the eigenvalue F_k. They must thus all be linear combinations of the S_{kj} $(j = 1, \ldots, g_k)$ and we have, dropping the index k to simplify the equations,

$$G_{op} S_j = \Sigma_{j'} G_{j'j} S_{j'}, \quad j, j' = 1, \ldots, g, \tag{58}$$

$$\int S_{j'}^* G_{op} S_j \, d\tau = \Sigma_{j''} G_{j''j} \int S_{j'}^* S_{j''} d\tau = G_{j'j},$$

$$G_{jj'} = \int S_j^* G_{op} S_{j'} \, d\tau = \int S_{j'} G_{op}^* S_j^* \, d\tau = G_{j'j}^* . \tag{59}$$

We see that the $G_{jj'}$ form a Hermitean matrix. Let G_ι be their eigenvalues and let u_j^ι be a unitary matrix which reduces $G_{jj'}$ to principal axes [compare eqq. (49) and (50)]

$$G_{j'j} = \Sigma_\iota u_{j'}^\iota G_\iota u_j^{\iota *} . \tag{60}$$

* For this part of the proof we do not need to take degeneracy explicitly into account.

Substitution of equation (60) into equation (58), multiplication of the resulting equation by $u_j^{\iota'}$ and summation over j leads to the result

$$G_{\text{op}}\,[\Sigma_j u_j^{\iota'} S_j] = \Sigma_{j'j\iota} u_{j'}^{\iota} G_\iota u_j^{\iota *} S_{j'} u_j^{\iota'} = G_{\iota'}\,[\Sigma_{j'} u_{j'}^{\iota'} S_{j'}].$$

We see thus that the g orthonormal functions

$$\sigma_\iota = \Sigma_j u_j^\iota S_j\,,\ \iota = 1, 2, \ldots, g, \tag{61}$$

are eigenfunctions of the operator G_{op} corresponding to the eigenvalue G_ι. As they are linear combinations of the S_j they are also eigenfunctions of F_{op}. This concludes the proof.

The analogous case of Hermitean operators acting on functions of a discrete argument, that is of Hermitean matrices, is very easily treated. Its consideration illiminates the previous case. The fact that the eigenfunctions of F and G are the same means that their matrix elements can be written in the form $F_{mn} = \Sigma_\mu U_m^\mu F_\mu U_n^{\mu *}$, $G_{mn} = \Sigma_\mu U_m^\mu G_\mu U_n^{\mu *}$, where U_m^μ is a unitary matrix and where the F_μ and G_μ are the eigenvalues. One now sees easily that the matrices FG and GF are identical, $(FG)_{mn} = (GF)_{mn} = \Sigma_\mu U_m^\mu F_\mu G_\mu U_n^{\mu *}$. If the two matrices commute, they will still commute if they are transformed by the same matrix U_m^μ. Choosing for this matrix one which reduces F to principal axes, it follows from their commutability that $F_\mu G_{\mu\nu} - G_{\mu\nu} F_\nu = (F_\mu - F_\nu)\, G_{\mu\nu} = 0$. If the F_k are g_k-fold degenerate eigenvalues, $G_{\mu\nu}$ will reduce to smaller Hermitean matrices of rank g_k which are grouped along the diagonal, with all the other $G_{\mu\nu}$ equal to zero. By another unitary transformation by a unitary matrix σ_μ^j of the kind discussed earlier we can reduce these smaller matrices also to principal axes without disturbing the diagonal form of the F matrix.

§ 39. The distribution function of an observable; probability amplitudes

Let the physical situation of a given system be described by the normalised Schrödinger function $\psi(q_1, q_2, \ldots, t)$. We can expand the function at any time t in terms of the eigenfunctions of a Hermitean operator F_{op}. Restricting our discussion for the time being to the case of a discrete spectrum we have [see eq. (20)]

$$\psi(q_1, \ldots, t) = \Sigma_{kj}\, a_{kj}(t)\, S_{kj}(q_1, \ldots). \tag{62}$$

The expansion coefficients a_{kj} depend on t, since ψ depends on t. They are given according to equation (21) by the relations

$$a_{kj} = \int S_{kj}^* \psi\, d\tau. \tag{63}$$

It follows from equation (22) that

$$1 = \int |\psi|^2 \, d\tau = \Sigma_k \, [\Sigma_j |a_{kj}|^2] \, . \tag{64}$$

We now introduce the following basic postulate. *The expression* $\Sigma_j |a_{kj}|^2$ *may be considered to be the probability that the mechanical quantity* F *is equal to its eigenvalue* F_k; *other values of the quantity* F *do not have to be considered* *. If we are dealing with a continuous spectrum and degeneracy described by continuous parameters we get from equations (24) to (26) instead of equations (62) to (64) the equations

$$\psi(q_1, \ldots, t) = \int a(F, r_1, \ldots, t) \, S(F, r_1, \ldots; q_1, \ldots) \, \varrho \, dF \, dr_1 \ldots, \tag{65}$$

$$a = \int S^* \, \psi \, d\tau, \tag{66}$$

$$1 = \int |\psi|^2 \, d\tau = \int |a|^2 \, \varrho \, dF \, dr_1 \ldots, \tag{67}$$

and $dF \int |a|^2 \varrho \, dr_1 \ldots$ *is the probability that the mechanical quantity* F *has a value between* F *and* $F + dF$.

If we consider ψ as the limiting case of a function of a discrete argument with values $\psi_1, \psi_2, \ldots, \psi_m, \ldots$, the operator F_{op} will correspond to a Hermitean matrix F_{mn}. A unitary transformation which reduces F to its principal axes will transform ψ into a function a_{kj} ($\psi_m = \Sigma_{kj} \, a_{kj} \, S_m{}^{kj}$). Equations (64) and (67) can thus be considered to be the limiting cases of the equation $1 = \int \psi^* \psi = \Sigma_m \psi_m{}^* \, \psi_m = \Sigma_{kj} \, a_{kj}{}^* \, a_{kj} = \Sigma_k \{\Sigma_j \, |a_{kj}|^2\}$ [compare eq. (41)].

Several authors (Weyl, Dirac) have used the above probability postulate—or a very similar one—as the basic axiom of quantum theory together with the general principle of superposition (compare § 19). This has turned out to be fully justified in the non-relativistic description of many particle systems. We shall show by using the correspondence principle how it is closely connected with the classical description of such systems and also that it contains the previously discussed distribution laws for coordinates and momenta as special cases.

This probability law enables us to calculate the *average value of the n-th power* (n: integer) of our mechanical quantity F in a given physical situation. First of all we note that it follows from equation

* If F is the energy operator of the system, this postulate corresponds to the energy distribution function discussed in § 32 (compare also § 35). Instead of writing [compare eqq. (3.99), (3.103), (3.105)] $\psi = \Sigma_{kj} \, c_k \, c_{kj} \exp{[-i \, E_k t/\hbar]} \, \varphi_{kj}$ we now write $\psi = \Sigma_{kj} \, a_{kj}(t) \, \varphi_{kj}$, so that $a_{kj} = c_k c_{kj} \exp{[-i \, E_k t/\hbar]}$ and $\Sigma_j |a_{kj}|^2 = |c_k|^2 \Sigma_j |c_{kj}|^2 = |c_k|^2$.

(17) that $F_k^n S_{kj} = F_{\text{op}}^n S_{kj}$. Using the complete conjugate of equation (63) we find

$$\overline{F^n} = \Sigma_k [\Sigma_j |a_{kj}|^2] F_k^n = \Sigma_{kj} a_{kj} \int \psi^* F_k^n S_{kj} d\tau = \int \psi^* F_{\text{op}}^n [\Sigma_{kj} a_{kj} S_{kj}] d\tau,$$

$$\text{or,} \quad \overline{F^n} = \int \psi^* F_{\text{op}}^n \psi \, d\tau \tag{68}$$

It is easily checked that this last equation holds also in the case of a continuous spectrum, or in general in the case of any complicated eigenvalue spectrum.

As in § 30 expression (68) is called the expectation value of the n-th power of the operator F_{op}. At that time we were not concerned with a distribution function for F, but we see now that the definition (3.86) can be connected without inner inconsistencies with the existence of a distribution function. One could actually have defined this function using equation (68), since according to a well-known result of probability theory the distribution function of a quantity is determined by the average values of all its integer powers *. The average value $\overline{F}$, that is, the integral $\int \varphi^* F_{\text{op}} \varphi d\tau$, is usually called the *expectation value* of the mechanical quantity considered.

Let now the situation at a given time be represented approximately by the normalised wave function $\psi_0 \doteqdot C(q_1, q_2, \ldots) \times \exp[2\pi i \, \Sigma \sigma_k q_k]$, where C is a "smooth" function of the q_k which varies but slowly compared with the exponential and which is different from zero only in the neighbourhood of $q_k = q_k^0$. The situation approximates to the classical case where the q_k are equal to q_k^0 and simultaneously the p_k are equal to $h\sigma_k = p_k^0$ (compare § 27). One sees easily that approximately $F_{\text{op}}(p_{k\,\text{op}}, q_k) \psi_0 \doteqdot F(p_k^0, q_k) \psi_0$, since to a first approximation the prescription $p_{k\,\text{op}} = -i\hbar \partial/\partial q_k$ means multiplication by $h\sigma_k$. We have thus from equation (68)

$$\overline{F^n} \doteqdot \int \psi_0^* F^n (p_k^0, q_k) \psi_0 d\tau \doteqdot F^n(p_k^0, q_k^0) \int \psi_0^* \psi_0 d\tau = F^n(p_k^0, q_k^0).$$

The mechanical quantity F has thus in this approximation the well defined value $F(p_k^0, q_k^0)$ in agreement with the classical definition of this quantity.

Let now F be a function $F(q_1, q_2, \ldots)$ of the coordinates only. According to equation (68) its distribution function with respect to an arbitrary solution ψ of the Schrödinger equation will satisfy the relations

* Since F_{op} is a Hermitean operator, the integral (68) is real. We have assumed it to converge, although that will often not be the case in actual cases.

$$\overline{F^n} = \int F^n(q_k)\,|\psi|^2\,d\tau. \tag{69}$$

This is in complete agreement with the general interpretation of the wave function (§ 28) according to which $|\psi|^2 d\tau$ is the probability that the q_k have values lying within the volume element $d\tau$ of $q_1, q_2, \ldots$-space.

Let now F be a polynomial $F(p_1, p_2, \ldots)$ which is a function of the p_k only. Using the Fourier integral of ψ,

$$\begin{aligned} \psi &= \int A(\sigma_1, \sigma_2, \ldots, t) \exp\,[2\pi i\,\Sigma\sigma_k q_k]\,d\sigma_1 d\sigma_2 \ldots, \\ A &= \int \psi \exp\,[-2\pi i\,\Sigma\sigma_k q_k]\,d\tau, \end{aligned} \tag{70}$$

we get from equation (68)

$$\begin{aligned} \overline{F^n} &= \int d\tau \int d\sigma_1 \ldots \psi^* F^n_{\rm op} A \exp[2\pi i\,\Sigma\sigma_k q_k] = \\ &\int d\sigma_1 \ldots F^n(h\sigma_1, \ldots)\,A \int \psi^* \exp[2\pi i\,\Sigma\sigma_k q_k]\,d\tau = \int F^n(h\sigma_1, \ldots)\,|A|^2 d\sigma_1, \ldots \end{aligned} \tag{71}$$

The distribution function of F corresponds thus to a situation where the probability that the momenta $p_k = h\sigma_k$ lie in the intervals $[h\sigma_k, h(\sigma_k + d\sigma_k)]$ $(k = 1, 2, \ldots)$ is given by the expression $|A|^2 d\sigma_1 d\sigma_2 \ldots$. This is, however, exactly the distribution law which we met with in the treatment of one-particle and many-particle systems (§§ 15, 28).

Equation (68) is thus a natural generalisation for the conditions which must be satisfied by the distribution function of an observable. We discussed earlier the general prescription for constructing this function by means of the eigenvalues and eigenfunctions of the corresponding operator. Characteristic of this prescription was that the distribution function could always be represented by the absolute square of a complex function of a discrete continuous (or mixed) argument. This function is called after Pauli * the *probability amplitude of the given physical situation with respect to the mechanical quantity under consideration.* The determination of the probability function is completely analogous to the calculation of the energy of a harmonic vibration in terms of its amplitude.

A special case occurs when at a time t_0 the physical situation is such that the wave function is identical with an eigenfunction of $F_{\rm op}$, so that $F_{\rm op}\,\psi(t_0) = F_{\rm e}\,\psi(t_0)$. In this case the value of the observable in the given situation is clearly well defined and equal to $F_{\rm e}$.

* Compare P. Jordan, *Z. Phys.*, **40**, 811, 1927.

This case can only occur exactly, if F_e is a proper eigenvalue corresponding to a normalisable eigenfunction. If F_e belongs to the continuous part of the eigenvalue spectrum there are physical situations such that the probability amplitudes are appreciably different from zero only in an (arbitrary small) interval [$\psi(t_0)$ may, for instance, be an eigendifferential; see § 36]. The situation is clearly completely analogous to that of the improper stationary states.

So far we have considered only *one* mechanical quantity. We saw, however, in § 38 that one can assign the same set of eigenfunctions to a number of quantities, provided the corresponding operators commute. This fact can often be used to avoid partly or completely the complications occurring when there are degenerate eigenvalues (arbitrariness in the choice of eigenfunctions). Indeed, one can often choose a number of commuting operators [$F^{(1)}$, $F^{(2)}$, . . .] in such a way that the set of corresponding eigenfunctions is uniquely determined (apart from possible phase factors) by the eigenvalues. For instance, the eigenvalue $F^{(1)}_{k_1}$ may be degenerate, and also $F^{(2)}_{k_2}$, $F^{(3)}_{k_3}$ and so on. However, it may happen that there is only one eigenfunction $S_{k_1, k_2, k_3, \ldots}$ which is simultaneously an eigenfunction of $F^{(1)}_{\text{op}}$, $F^{(2)}_{\text{op}}$, $F^{(3)}_{\text{op}}$, . . . ,

$$F^{(r)}_{\text{op}} S_{k_1, k_2, \ldots} = F^{(r)}_{k_r} S_{k_1, k_2, \ldots}, \quad r = 1, 2, \ldots \tag{72}$$

The probability amplitudes will now form a set of (time dependent) coefficients $a_{k_1, k_2, \ldots}(t)$, the indices of which refer to the eigenvalues of $F^{(1)}_{\text{op}}$, $F^{(2)}_{\text{op}}$, . . . and instead of equations (62) to (64) we have

$$\begin{aligned} &\psi(q_k, t) = \Sigma_{k_1, k_2, \ldots} a_{k_1, k_2, \ldots}(t) S_{k_1, k_2, \ldots}(q_k), \\ &a_{k_1, k_2, \ldots}(t) = \int S^*_{k_1, k_2, \ldots} \psi \, d\tau, \; 1 = \int |\psi|^2 d\tau = \Sigma_{k_1, k_2, \ldots} |a_{k_1, k_2, \ldots}|^2. \end{aligned} \tag{73}$$

One should be careful not to introduce more operators than are necessary. If, for instance, the eigenfunctions are completely fixed by giving k_1, k_2, and k_3 it is unnecessary to introduce a fourth operator $F^{(4)}_{\text{op}}$ corresponding to an index k_4.

Let us finally consider the case where each of the operators has only a continuous spectrum and where the eigenfunctions are determined by giving the set of the corresponding eigenvalues of all the operators. Instead of equations (65) to (67) we get now the following equations from which the parameters $r_1, r_2, \ldots$ which described the degeneracy have disappeared, and where $F^{(1)}$, $F^{(2)}$, . . . indicate the eigenvalues.

$$\left.\begin{aligned}&\psi(q_k, t)\\&=\int a(F^{(1)}, F^{(2)}, \ldots, t) S(F^{(1)}, F^{(2)}, \ldots; q_k)\, \varrho(F^{(1)}, F^{(2)}, \ldots)\, dF^{(1)} dF^{(2)} \ldots\\&\qquad a = \int S^* \psi\, d\tau,\ 1 = \int |\psi|^2 d\tau = \int |a|^2 \varrho\, dF^{(1)} dF^{(2)} \ldots\end{aligned}\right\} \quad (74)$$

One can also use finally equation (74) in the case where the spectrum is partly discrete and partly continuous as long as one takes into account that the $F^{(1)}, F^{(2)}, \ldots$ are partly discrete and partly continuous variables and that the integration sign indicates partly integration and partly summation over discrete eigenvalues.

The absolute square of a probability amplitude, which may be a function of eigenvalues of several commuting operators, gives us information about the probability distribution of these quantities. In particular, if the wave function at t_0 is an eigenfunction of both $F^{(1)}$ and $F^{(2)}$, $F^{(1)}_{\text{op}} \psi(t_0) = F^{(1)}_{\text{e}} \psi(t_0)$, $F^{(2)}_{\text{op}} \psi(t_0) = F^{(2)}_{\text{e}} \psi(t_0)$, then both the values of $F^{(1)}$ and that of $F^{(2)}$ are well defined at that time. If $F^{(1)}$ and $F^{(2)}$ are two non-commuting quantities it is in general impossible to find an eigenfunction corresponding to the eigenvalue $F^{(1)}_{\text{e}}$ which satisfies the above equations and which at the same time corresponds to an eigenvalue $F^{(2)}_{\text{e}}$ of $F^{(2)}_{\text{op}}$. This means that *in general only commuting operators can have simultaneously well defined values.* It is only by accident that non-commuting quantites can be simultaneously well defined *.

§ 40. Transformation of functions

In describing the physical situation of a system we can replace the Schrödinger function ψ by the probability amplitudes a with respect to a mechanical quantity or to a set of quantities corresponding to commuting operators since the a follow from ψ just as ψ follows from the a. This means that the equations (73) and (74) produce a *transformation of the wave function* into an equivalent *representation of the physical situation.* The causal evolution of events in classical physics corresponds in quantum mechanics to the Schrödinger equation (3.45) which describes the change in time of the wave function. We can find an equivalent law which describes the change in time of the a.

* Since it follows from our equations that $F_{\text{op}}^{(1)} F_{\text{op}}^{(2)} \psi = F_{\text{op}}^{(2)} F_{\text{op}}^{(1)} \psi = F_{\text{e}}^{(1)} F_{\text{e}}^{(2)} \psi$, it is necessary that the Hermitean operator $iF_{\text{op}}^{(1)} F_{\text{op}}^{(2)} - iF_{\text{op}}^{(2)} F_{\text{op}}^{(1)}$ possesses the eigenvalue zero. An example of this is the case of an atom in a stationary state in which the angular momentum is zero in all directions (compare § 45).

The construction of this equation is called the *transformation of the Schrödinger equation* and it follows from equation (74). We shall denote the coordinates $q_1, q_2, \ldots$ and the eigenvalues $F^{(1)}, F^{(2)}, \ldots$ by q and F in order to simplify the formulae. We have now

$$i\hbar\,\partial a(F,t)/\partial t = i\hbar\int S^*(F,q)(\partial\psi/\partial t)\partial q = \int S^*(F,q)H_{\mathrm{op}}\,\psi\,dq$$
$$= \int S^*(F,q)H_{\mathrm{op}}\left\{\int a(F',t)\,S(F',q)\,\varrho(F')\,dF'\right\}dq.$$

The last member of this equation is a function of F and t which is obtained from $a(F, t)$ by a prescription which obviously corresponds to a linear operator. We write this in the form

$$i\hbar\partial a(F,t)/\partial t = H'_{\mathrm{op}}\,a(F,t), \tag{75}$$

and we call H'_{op} the *transformed energy operator* of the system with respect to the observables F—or with respect to the eigenvalues of these quantities.

We want to discuss the general problem of the transformation of functions and operators without referring specifically to the wave function and the energy. An arbitrary normalisable function of the coordinates f can be expanded in terms of the eigenfunctions [compare eqq. (73) and (74)]. The set of coefficients or the coefficient-function b which occurs in this expansion is the function produced by the transformation of f. In analogy with equation (74) [or eqq. (24) to (26)] we write

$$f(q) = \int b(F)S(F,q)\varrho(F)dF, \quad (76\text{a}); \qquad b = \int S^* f\,dq, \quad (76\text{b});$$
$$\int f^* f\,dq = \int b^*\,b\,\varrho d\,F. \qquad (76\text{c})$$

We call $S(F, q)$ the transformation function; it can be considered to be the limiting case of a unitary matrix (compare § 37). The equations are general, if one allows the integration sign to include possible summation over discrete F-values. A linear combination $\Sigma_\lambda c_\lambda f_\lambda$ of functions $f_1, f_2, \ldots$ with constant coefficients $c_1, c_2, \ldots$ is transformed into the corresponding linear combination of the b's,

$$\Sigma_\lambda c_\lambda f_\lambda \rightarrow \Sigma_\lambda c_\lambda b_\lambda. \tag{77}$$

We can generalise equation (76c) by the introduction of two functions f_1 and f_2 as follows,

$$\int f_2^* f_1\,dq = \int b_2^*\,b_1\,\varrho\,dF \tag{78}$$

The equation expresses the invariance of the so-called "scalar product" of two functions. It is characteristic for the unitary character (in the generalised sense) of the transformation (compare § 37).

It is typical of the nature of these transformations that one can immediately go from one transformed function of f to another one. Let a second transformation of f with respect to the quantities G be given by the equations

$$f(q) = \int c(G)\, T(G, q)\, \sigma(G)\, dG, \quad (79a); \quad c = \int T^* f\, dq, \quad (79b);$$

$$\int f_1^* f_2\, dq = \int c_1^* c_2\, \sigma\, dG. \quad (79c)$$

One finds easily that the following equations hold

$$c(G) = \int b(F)\, U(F, G)\, \varrho(F)\, dF, \quad (80a);$$

$$b(F) = \int U^*(F, G) c(G) \sigma(G) dG, \quad (80b); \quad \int c_1^* c_2 \sigma\, dG = \int b_1^* b_2 \varrho\, dF \quad (80c).$$

Equation (80a) is obtained by substituting equation (76a) into equation (76b) and the transformation function $U(F, G)$ is given by the equation

$$U(F, G) = \int T^*(G, q)\, S(F, q)\, dq. \tag{81}$$

The complete reciprocity of $b(F)$ and $c(G)$ is seen even better, if we write equation (80b) in the form

$$b\,(F) = \int c(G)\, V(G, F)\, \sigma(G)\, dG, \tag{82}$$

where U is obtained from V, and also V from U, by interchanging the two arguments and taking the complex conjugate. Also $f(q)$ is completely equivalent to b and c. This is not immediately clear since we had arranged things in such a way that there was no density function involved in the integration over the (Cartesian) coordinates q. We can, however, always arrange everything in such a way that the density function is equal to 1, at any rate, as long as we are dealing with real integrations over continuous values of F or G. This procedure is not always advantageous, but consists in replacing b, c, S, T, and U by b', c', S', T', and U' which are given by the equations $b' = b\varrho^{1/2}$, $c' = c\sigma^{1/2}$, $S' = S\varrho^{1/2}$, $T' = T\sigma^{1/2}$, $U' = U\varrho^{1/2}\sigma^{1/2}$.

If both F and G are discrete variables, the values of which are distinguished by indices k and l, the function $U(F, G)$ becomes a matrix (which may be infinite) with elements U_l^k and equation (80) is now as follows

$$\left.\begin{aligned} c_l = \Sigma_k U_l^k b_k,\ c_l^* = \Sigma_k b_k^* U_l^{k*},\ b_k = \Sigma_l U_l^{k*} c_l,\ b_k^* = \Sigma_l c_l^* U_l^{k*}, \\ \Sigma_l U_l^{k*} U_l^{k'} = \delta_{kk'},\ \Sigma_k U_l^k U_{l'}^{k*} = \delta_{ll'},\ \Sigma_l c_l^* c_l = \Sigma_k b_k^* b_k. \end{aligned}\right\} \qquad (83)$$

These equations corrrespond to equations (38) to (42). The order in which the factors occur is such that the indices over which the summation occurs are next to each other, if we read U_l^k as U_{lk} and U_l^{k*} as U^*_{kl}. Using matrix notation and the symbol $\int$ for the scalar product we can write equations (83) in the form

$$c = Ub,\ c^* = b^* U^*,\ b = U^*c,\ b^* = c^*U,\ U^* U = 1,\ UU^* = 1,\ \int c^*c = \int b^*b. \qquad (84)$$

This symbolism can also be used to write equations (80) in a simpler form. We consider U then to be a unitary matrix in the generalised sense. The combination of two transformations leads to the equations [compare eq. (81)],

$$U = T^* S,\ U^* = S^* T,\ S = TU,\ S^* = U^* T^*. \qquad (85)$$

Here T transforms from the q- into the G-representation, U from the G-into the F-representation, and their product S from the q- into the F-representation.

Dirac's notation * Dirac has introduced a notation which expresses fully the equivalence of the various ways of describing a function. It may be remarked also that in the case of functions of more than one variable the Dirac notation has great advantages, and it is used extensively in the modern literature. Dirac uses the "bra" notation $\langle q'|$, $\langle F'|$, $\langle G'|$ for what we called $f(q)$, $b(F)$, and $c(G)$, and the "ket" notation $|q'\rangle$, $|F'\rangle$, $|G'\rangle$ for their complex conjugate. This indicates that one and the same entity is described in the q-"language", the F-"language", or the G-"language". The primes indicate that we are dealing with eigenvalues while Dirac uses unprimed symbols for the general description of mechanical quantities. For the transformation functions S and S^* he writes $\langle q'|F'\rangle$ and $\langle F'|q'\rangle$, for T and T^* $\langle q'|G'\rangle$ and $\langle G'|q'\rangle$, and for U and U^* $\langle G'|F'\rangle$ and $\langle F'|G'\rangle$. Equations (80) are now in the form

$$\left.\begin{aligned} \langle G'| = \int \langle G'|F'\rangle\, dF' \langle F'|,\ |G'\rangle = \int |F' \langle dF'\rangle\, F'|G'\rangle, \\ \langle F'| = \int \langle F'|G'\rangle\, dG' \langle G'|,\ F|'\rangle = \int |G'\rangle\, dG' \langle G'|F'\rangle, \\ \int |G'\rangle\, dG' \langle G'| = \int |F'\rangle\, dF' \langle F'|. \end{aligned}\right\} \qquad (86)$$

The unitary character of the transformation $\langle F'|G'\rangle$ is expressed by the equations

$$\int \langle G'|F'\rangle\, dF' \langle F'|G''\rangle = \delta_{G'G''},\qquad \int \langle F'|G'\rangle\, dG' \langle G'|F''\rangle = \delta_{F'F''}, \qquad (87)$$

where $\delta_{G'G''}$ and $\delta_{F'F''}$ are the operator functions of the unit operator in the G- and F-representation, which correspond to the unit matrix in the case of discrete arguments. The combination of two transformations leads to [compare eq. (81)]

$$\langle G'|F'\rangle = \int \langle G'|q'\rangle\, dq' \langle q'|F'\rangle. \qquad (88)$$

§ 41. Transformation of operators; matrix representation of an observable

Let the linear operator Ω change an arbitrary function f_1 into another function f_2,

$$f_2 = \Omega f_1. \qquad (89)$$

* P. A. M. Dirac, *The Principles of quantum mechanics*, Oxford University Press, especially 3rd edition 1947.

Using a complete orthonormal system $S(F, q)$ we can transform f_1 and f_2 into b_1 and b_2 [compare eq. (76)]. From equations (76) and (77) it follows easily that b_1 and b_2 are connected by means of a linear operator,

$$b_2 = \Omega' b_1. \tag{90}$$

The new operator Ω' is obtained from the operator Ω by a *transformation*. This operator Ω' acts on the (continuous, discrete, or mixed) arguments $F^{(1)}, F^{(2)}, \ldots$ This is called the *representation of the operator with respect to the* F, or in the "F-language".

Without writing down Ω' explicitly we can see immediately the following properties. The unit operator transforms into the unit operator. The sum or product of two operators transforms into the sum or product of the transformed operators. The order of the factors in a product remains unchanged. In general any operator which is in the form of a polynomial of operators transforms in the same polynomial of transformed operators. Especially commutability and the commutation rules (3.77) are invariant under a transformation. The transformed operators of two operators which were adjoint or complex conjugate are not necessarily each others adjoint or complete conjugate. However, the operator which is both the adjoint and the complex conjugate of a given operator will transform into the adjoint and complex conjugate of the transformed operator. This means that a transformed Hermitean operator ($\overset{\frown}{\Omega}{}^* = \Omega$) is again Hermitean.

Proof: If f_1 and f_2 are two arbitrary functions, we have

$$\left.\begin{aligned} \int f_3{}^* \Omega f_1\, d\tau &= \int b_3{}^* \Omega' b_1\, \varrho\, dF = \int b_1 \tilde{\Omega}' b_3{}^*\, \varrho\, dF \\ &= \left\{ \int b_1{}^* \tilde{\Omega}'{}^* b_3\, \varrho\, dF \right\}^* \\ \int f_3{}^* \Omega f_1\, d\tau &= \left\{ \int f_1{}^* \tilde{\Omega}^* f_3\, d\tau \right\}^* = \left\{ \int b_1{}^* (\tilde{\Omega}^*)' b_3\, \varrho\, dF \right\}^* \end{aligned}\right\} \to (\overset{\frown}{\Omega}{}^*)' = (\overset{\frown}{\Omega}{}')^* \tag{91}$$

An explicit formal expression for the transformed operator follows easily from equations (76) and (89)

$$\begin{aligned} b_2(F) &= \int S^*(F, q) f_2(q)\, dq = \int S^*(F, q)\, \Omega f_1(q)\, dq \\ &= \int S^*(F, q)\, \Omega \left\{ \int b_1(F') S(F', q)\, \varrho(F')\, dF' \right\} dq \\ &= \int \left\{ \int S^*(F, q)\, \Omega S(F', q)\, dq \right\} b_1(F')\, \varrho(F')\, dF', \end{aligned}$$

or,

$$b_2(F) = \int \Omega(F, F')\, b_1(F')\, \varrho(F')\, dF', \tag{92}$$

$$\Omega(F, F') = S^*(F, q)\, \Omega S(F', q)\, dq. \tag{93}$$

The function $\Omega(F, F')$ which occurs in equation (92) is the operator function, or continuous matrix, which corresponds to the transformed operator of equation (90); it is given explicitly by equation (93). This last equation shows that one can consider $\Omega(F, F')$ to be the coefficient in the expansion of $\Omega S(F, q)$ in terms of eigenfunctions [compare eq. (76b)],

$$\Omega S(F', q) = \int \Omega(F', F) S(F, q) \varrho(F) dF. \tag{94}$$

If the eigenvalue spectrum of F is discrete, so that we can write the $S(F, q)$ in the form $S_k(q)$ the integral over F is replaced by a summation over the k. Instead of $\Omega(F, F')$ we get an ordinary matrix and we have

$$\left.\begin{aligned} &f(q) = \Sigma_k b_k S_k(q),\ f^{(2)} = \Omega f^{(1)},\ \Omega_{kk'} = \int S_k^*(q)\, \Omega\, S_{k'}(q)\, dq, \\ &b_k = \int S_k^* f\, dq, b_k^{(2)} = \Sigma_{k'} \Omega_{kk'} b_{k'}^{(1)},\ \Omega S_{k'}(q) = \Sigma_k \Omega_{kk'} S_k(q), \\ &\int f^* f\, dq = \Sigma_k b_k^* b_k. \end{aligned}\right\} \tag{95}$$

The $\Omega_{kk'}$ is called the matrix (or matrix representation) of the operator considered with respect to the given set of eigenfunctions (or with respect to the mechanical quantity (quantities) corresponding to these eigenfunctions).

One can easily generalise equations (92) and (93) to include the case where the F are partly discrete and partly continuous. If F is a continuous variable, it may happen that $\Omega(F, F')$ will be an improper mathematical function (compare § 29).

As in the case of functions we can also now immediately go from one transformed form Ω' to another one Ω''. Consider, for instance, equations (80) which give the transformation of a function from the G-representation to the F-representation. Let the corresponding representations of an operator Ω be denoted by $\Omega(G, G')$ and $\Omega(F, F')$. By following the same line of reasoning as the one which led to equations (92) and (93) we find

$$\begin{aligned} \Omega(F, F') &= \iint U^*(F, G)\, \Omega(G, G')\, U(F', G')\, \sigma(G)\, \sigma(G')\, dG\, dG', \\ \Omega(G, G') &= \iint U(F, G)\, \Omega(F, F')\, U^*(F', G')\, \varrho(F)\, \varrho(F')\, dF\, dF'. \end{aligned} \tag{96}$$

Here $U(F, G)$ is again the transformation function (81) which transforms from the G- to the F-representation. If both F and G are discrete variables equations (96) give us the transformation of one matrix into another,

$$\Omega_{kk'} = \Sigma_{ll'} U_l^{k*} \Omega_{ll'} U_{l'}^{k'}, \Omega_{ll'} = \Sigma_{kk'} U_l^k \Omega_{kk'} U_{l'}^{k'*}. \tag{97}$$

We have met with the same equations in the problems of transformation of the eigenvalue problems of a finite matrix [compare eq. (49)]. It is immaterial for the present argument that we were then dealing with a Hermitean matrix.

In equation (97) U_l^k is a unitary matrix with infinitely many rows and columns and using the matrix notation we can write

$$\Omega' = U^* \Omega'' U, \quad \Omega'' = U \Omega' U^*. \tag{98}$$

If U is considered to be a unitary matrix in the general sense, equation (98) can also be seen as the short hand notation of equation (96).

Dirac denotes the operator functions by $\langle q'|\Omega|q''\rangle$, $\langle F'|\Omega|F''\rangle$. The primes again indicate the eigenvalues of the corresponding mechanical quantities. Using the notation introduced at the end of § 40 we can write equations (89) and (90) in the form

$$\langle q'|_2 = \int \langle q'|\Omega|q''\rangle \, dq'' \langle q''|_1, \langle F'|_2 = \int \langle F'|\Omega|F''\rangle \, dF'' \langle F''|_1. \tag{99}$$

The transformation of an operator can be written in the form

$$\langle F'|\Omega|F''\rangle = \int \langle F'|q'\rangle \, dq' \langle q'|\Omega|q''\rangle \, dq'' \langle q''|F''\rangle. \tag{100}$$

There exists clearly a close connexion between the Dirac notation and the matrix notation (84), (85), and (98). A further advantage of his notation consists in that it can be applied immediately to the case of the most general linear transformations and not only to the class of (generalised) unitary transformations. These linear transformations correspond to the transformations discussed in § 37 by means of an arbitrary matrix T (with non-vanishing determinant) and they occur when a function is expanded in terms of an arbitrary (that is, not orthonormal) complete set of functions. One can still retain equations (84), (85) and (98), but the asterisk does no longer indicate complex conjugation. Instead U^* corresponds to the inverse matrix U^{-1}, and the $b^*(c^*)$ are no longer the complex conjugate of the $b(c)$ but the so-called complementary functions which are defined by their transformation from the original function f^*. For functions of the coordinates q, the complementary function f^* is identical with the complex conjugate function. These general transformations are of advantage for certain quantum mechanical applications.

An eigenfunction $S(q)$ of a Hermitean operator F_{op} satisfies equation (16). Let us transform that equation by means of an orthonormal set of functions $T(G, q)$ which refers to a mechanical quantity G or to a set of such quantities. Let $S(q)$ be transformed into $U(G)$ and F_{op} into F'_{op}. From equation (16) it then follows that

$$F'_{\text{op}} U(G) = F_e U(G), \; U(G) = \int T^*(G,q) \, S(q) \, dq. \tag{101}$$

or, if we introduce explicitly the operator functions, we have

$$\begin{aligned} &\int F(G, G') \, U(G') \, \sigma(G') \, dG' = F_e U(G), \\ &F(G, G') = \int T^*(G,q) \, F_{\text{op}} T(G', q) \, dq. \end{aligned} \tag{102}$$

The eigenvalue and eigenfunction problem (101) is equivalent to the original problem (16). It is formally a Fredholm integral equation * with Hermitean kernel $F(G, G') = F^*(G', G)$. One can discuss the solutions of equation (101) formally in the same way as those of equation (16). If $F(G, G')$ is a proper mathematical function of its argument one can also appeal to the theory of integral equations. If the eigenvalues of G_{op} are all discrete corresponding to eigenfunctions $T_l(q)$ we have instead of equation (102)

$$\Sigma_{l'} F_{ll'} U_{l'} = F_{\text{e}} U_l, \; F_{ll'} = \int T_l^*(q)\, F_{\text{op}}\, T_{l'}(q)\, dq, \qquad (103)$$

and the problem is exactly analogous to the algebraic problem (44). The only difference is that we are now dealing with infinitely many equations with infinitely many unknowns and a secular equation of infinitely high degree.

The solutions of equation (101) we denote by $U(F, G)$ indicating that they may depend on the eigenvalues F. It is possible to choose the U in such a way that they form a complete orthonormal set, even if the eigenvalues F are degenerate. In this case F stands both for the eigenvalues of F_{op} and for the parameters $r_1, r_2, \ldots$. One can also consider several commuting operators $F^{(1)}_{\text{op}}, F^{(2)}_{\text{op}}, \ldots$ instead of *one* operator F_{op} and see to it that by giving their eigenvalues the eigenfunctions $U(F, G)$ are completely determined apart from phase factors. If we now identify F_{op} and G_{op} with the quantities of the same name which occurred in equations (76) and (79), $U(F,G)$ is clearly identical with the transformation function in equation (80) which transformed from the G- into the F-representation. The second equation (101) goes, indeed, over into equation (81).

If the observable F_{op} commutes with the quantities $G^{(1)}, G^{(2)}, \ldots$ the transformed operator will be very simple. Its meaning is multiplication by a function of the G-eigenvalues. Indeed, the eigenfunctions of the $G^{(r)}$ are in this case also eigenfunctions of F_{op} (compare § 38) and the eigenvalues of F_{op} can be denoted by $F(G^{(1)}, G^{(2)}, \ldots)$ since there corresponds an eigenvalue of F to each common eigenfunction $T(G^{(r)}, q)$ of the operators $G^{(1)}_{\text{op}}, G^{(2)}_{\text{op}}, \ldots$,

$$F_{\text{op}}\, T(G^r, q) = F(G^{(r)})\, T(G^{(r)}, q). \qquad (104)$$

From equation (79a) it now follows that

$$f^{(2)}(q) = F_{\text{op}} f^{(1)}(q) = F_{\text{op}} \int c^{(1)}(G^{(r)})\, T(G^{(r)}, q)\, \sigma\, dG$$
$$= \int F(G^{(r)})\, c^{(1)}(G^{(r)})\, T(G^{(r)}, q)\, \sigma\, dG.$$

* Compare, for instance, R. Courant and D. Hilbert, *Methods of mathematical physics*, Interscience Publishers, New York, 1953, Ch. 3.

It follows thus that the function $c^{(2)}(G^{(r)})$ obtained by transforming $f^{(2)}$ satisfies the equation

$$c^{(2)}(G^{(r)}) = F(G^{(r)})\, c^{(1)}(G^{(r)}), \tag{105}$$

which completes the proof. In the case of discrete $G^{(r)}$ eigenvalues equation (105) expresses the fact that a matrix corresponding to a quantity which commutes with all the $G^{(r)}_{\text{op}}$ will be a diagonal matrix in the G-representation. The diagonal elements are the corresponding eigenvalues of F_{op}.

This result could have been obtained directly from the expression (93) for the operator function which corresponds to the transformed operator. We have [compare eq. (27)], $F(G^{(r)}, G^{(r)\prime}) = F(G^{(r)\prime}) \int T^*(G^{(r)}, q) T(G^{(r)\prime}, q)\, dq = F(G^{(r)\prime})\, \delta_{GG'}$, where $\delta_{GG'}$ is here as in equation (87) the operator function corresponding to the unit operator. If the $G^{(r)}$ are discrete, $\delta_{GG'}$ is the unit matrix and $F(G^{(r)}, G^{(r)\prime})$ is a diagonal matrix. If they are continuous, $\delta_{GG'}$ and $F(G^{(r)}, G^{(r)\prime})$ are improper mathematical functions. All the same one says that the operator is brought on diagonal form by this transformation. This reminds us of the fact that we must expect that we may encounter such improper functions, if we apply the formalism of the present chapter *. This formalism must be considered, as we have stressed before, to be a generalisation of or analogy to the pure algebraic theory of unitary transformations in a "space" of a finite number of dimensions. The following example may show how careful one must be with such generalisations.

The quantities which occur in quantum mechanics are always constructed from the basic operators q and p—the coordinates and their canonically conjugate momenta. These operators satisfy the commutation rules (3.77). If one transforms the p and q these relations remain valid. If the eigenfunctions which lead to this unitary transformation form a discrete set so that they can be characterised by indices $k_1, k_2, \ldots$ (k for short), we obtain for the q and p infinite matrices $q_{kk'}, p_{kk'}$. Let in particular p and q be each other's canonical conjugate, so that we have

$$qp - pq = i\hbar 1. \qquad \text{(1: unit matrix)} \tag{106}$$

This relation is in the k-representation $\Sigma_{k'} q_{kk'} p_{k'k''} - \Sigma_{k'} p_{kk'} q_{k'k''} = i\hbar \delta_{kk''}$. If we were dealing with finite matrices of N rows and

* Von Neumann (*Mathematical foundations of quantum mechanics,* Princeton University Press, 1955) has shown how one can discuss the transformation of normalised functions without introducing improper functions.

columns this equation would be inconsistent. Putting $k = k''$ and summing over k leads to the equation

$$\Sigma_{kk'} q_{kk'} p_{k'k} - \Sigma_{kk'} p_{kk'} q_{k'k} = i\hbar N. \tag{107}$$

The left hand side is equal to zero, but not the right hand side. That equation (106) is all the same valid in quantum mechanics, is due to the fact that we are always dealing with infinite matrices. The double sums in equation (107) do not converge and we can thus not interchange the order of summation. In a space of finite dimensions there are no operators which satisfy equation 106 *.

§ 42. The transformed Schrödinger equation

We return now to the transformation of the Schrödinger equation (§ 40). Let $S(F, q)$ be an orthonormal set of eigenfunctions (F indicates the eigenvalues of a Hermitean operator F_{op} and possible degeneracy parameters $r_1, r_2, \ldots$ or it indicates the eigenvalues of a number of commuting Hermitean operators $F^{(1)}_{(\mathrm{op})}, F^{(2)}_{(\mathrm{op})}, \ldots$). We use the $S(F, q)$ to transform the Schrödinger equation (3.45) and we obtain the transformed Schrödinger equation (75), $H'_{\mathrm{op}} a(F,t) = i\hbar\, \partial a(F, t)/\partial t$, where a is the probability amplitude of the given situation with respect to the $S(F, q)$ while H'_{op} is the energy operator in the corresponding representation.

If the F-eigenvalue spectrum is continuous we have

$$\begin{gathered} a(F,t) = \int S^*(F,q)\,\psi(q,t)\,d\tau, H'_{\mathrm{op}} a = \int H(F,F')\,a(F',t)\,\varrho(F')\,dF', \\ H(F, F') = \int S^*(F, q)\, H_{\mathrm{op}}\, S(F', q)\, d\tau \end{gathered} \tag{108}$$

If the spectrum is discrete, however, we have

$$\begin{gathered} a_k(t) = \int S^*_k(q)\,\psi(q,t)\,d\tau,\ H'_{\mathrm{op}} a_k = \Sigma_{k'} H_{kk'} a_{k'}, \\ H_{kk'} = \int S^*_k(q)\, H_{\mathrm{op}}\, S_{k'}(q)\, d\tau, \end{gathered} \tag{109}$$

where $H_{kk'}$ is the *energy matrix* in the k-representation.

If the transformation brings the energy on diagonal form [eigenvalues $E(F)$ or E_k], we can immediately find a solution of the transformed equation, which now has the form

* H. Weyl (*Z. Phys.*, **46**, 1, 1927) has shown how one can define operators q_i and p_i in a space of finite dimensions in such a way that they satisfy the commutation relations (3.77) in the limit where the number of dimensions goes to infinity.

$$E(F)\,a(F,t) = i\hbar\,\partial a(F,t)/\partial t, \text{ or, } E_k a_k(t) = i\hbar\,\dot{a}_k(t). \quad (110)$$

The solution is

$$\left.\begin{aligned} a(F,t) &= c(F)\exp[-iE(F)t/\hbar], \quad \int |c(F)|^2 \varrho(F)\,dF = 1, \\ \text{or,} \quad a_k(t) &= c_k \exp(-iE_k t/\hbar), \quad \Sigma_k |c_k|^2 = 1, \end{aligned}\right\} \quad (111)$$

where the c's are time independent integration constants. In this way we have come back to the solution of the Schrödinger equation which corresponds to a superposition of stationary states (compare § 32). It is immaterial whether one describes the distribution function of the energy by means of the $|a|^2$ or by means of the $|c|^2$ since the $|a|^2$ do not depend on the time and are equal to the $|c|^2$.

One can arrange it in such a way that the a in equation (111) are not only probability amplitudes for the energy itself, but also for any observable which commutes with the energy that is, for any integral of the given quantum mechanical system (compare § 31). The probability distribution of all these integrals is thus constant. This corresponds to the fact in classical physics that an integral of the equations of motion is constant in time [compare the law of conservation of momentum (§ 31) and the law of conservation of energy (§ 33)].

If an observable does not commute with the energy, its probability distribution will in general change with time in accordance with equation (75). Only if the system is in a stationary state, will the distribution function of any mechanical quantity F be constant in time, provided F does not contain the time explicitly. To see this one considers expression (68) for the average value $\overline{F^n}$. This expression is constant in time, as ψ contains the time only in the form of a phase factor.

The transformation of the Schrödinger equation is the quantum mechanical counterpart of the canonical transformation of the classical equations of motion. For this reason one sometimes calls the generalised unitary transformations (76) and (84) canonical transformations. In both cases one is concerned with the introduction of new mechanical quantities, functions of the original coordinates and momenta which form the basis of the new description of the behaviour of the system in time. Classically the transformation is described as follows *. Let $F^{(1)}, F^{(2)}, \ldots, F^{(s)}$ (s: number of degrees

* See, for instance, E. T. Whittaker, *Analytical Dynamics*, Cambridge University Press, 1937, Ch. 11.

of freedom) be s independent functions of the q and the p. If, and only if, the $F^{(r)}$ are in involution, that is, if all Poisson brackets $\{F^{(r')}, F^{(r)}\}$ are identically zero,

$$\{F^{(r')}, F^{(r)}\} = \sum_i \left[\frac{\partial F^{(r)}}{\partial p_i}\frac{\partial F^{(r')}}{\partial q_i} - \frac{\partial F^{(r)}}{\partial q_i}\frac{\partial F^{(r')}}{\partial p_i}\right] = 0, \quad (112)$$

is it possible to introduce another s functions $J^{(r)}$ $(r = 1, 2, \ldots, s)$ of the p, q which are canonically conjugate to the $F^{(r)}$. The equations of motions are now $\dot{F}^{(r)} = \partial H/\partial J^{(r)}$, $\dot{J}^{(r)} = -\partial H/\partial F^{(r)}$, where H is the energy of the system expressed in the F and the J. The following relations hold,

$$\{J^{(r')}, J^{(r)}\} = 0, \quad \{F^{(r')}, J^{(r)}\} = \delta_{rr'} \quad (113)$$

In quantum mechanics one also introduces a number of mechanical quantities $F^{(r)}_{\text{op}}$. They must commute,

$$[F^{(r)}_{\text{op}}, F^{(r')}_{\text{op}}]_- \equiv F^{(r)}_{\text{op}} F^{(r')}_{\text{op}} - F^{(r')}_{\text{op}} F^{(r)}_{\text{op}} = 0, \quad (114)$$

and the physical situation is described by a function a of the eigenvalues of the F, the probability amplitude. This function satisfies the transformed Schrödinger equation. Equations (114) and (112) are each others counterpart (compare § 29).

One might expect that it would be possible to introduce also in quantum mechanics a set of quantities $J^{(r)}$ which would satisfy the relations

$$[J^{(r)}_{\text{op}}, J^{(r')}_{\text{op}}]_- = 0, \quad [F^{(r)}_{\text{op}}, J^{(r')}_{\text{op}}]_- = i\hbar\,\delta_{rr'}, \quad (115)$$

so that we can consider for each $F^{(r)}$ its canonical conjugate $J^{(r)}$. Putting in equations (114) and (115) $J^{(r)} = p_r$ and $F^{(r)} = q_r$ one obtains the commutation relations (3.77). Also in the case where the F are simply functions, q', of the coordinates only (compare § 34) one can introduce differential operators [compare eq. (15)] which correspond to the classical generalised momenta and which satisfy the commutation relations. In general, however, the analogy with classical mechanics is not present. It usually impossible to define in a natural manner quantities $J^{(r)}$ which are canonically conjugate to the $F^{(r)}$ that is, which satisfy equations (115)*. The precise mathem-

* In Dirac's and Jordan's original papers (*Proc. Roy. Soc.*, **A 113**, 621, 1927; *Z. Phys.*, **40**, 809, 1927) in which the transformation theory was first developed, the formal introduction of canonically conjugate variables played a large part. They never, however, played a role in any applications. Jordan's attempt (*Z. Phys.*, **44**, 6, 1927) to define canonical conjugation of two variables by means of the function which transforms from F to J has proved to be unusable.

atical treatment of these problems is not easy *. We shall draw attention only to the fact that in general the $J^{(r)}$ which are canonically conjugate to the $F^{(r)}$ depend in such a way on the p and q that it seems impossible to find a quantum mechanical analogy. If the $F^{(r)}$ are polynomials in the p so that one can easily define $F^{(r)}_{\mathrm{op}}$, the $J^{(r)}$ depend on the p in a complicated manner **. Even without maintaining that one might derive immediately from the classical $J^{(r)}$ operators $J^{(r)}_{\mathrm{op}}$ which are canonically conjugate to $F^{(r)}_{\mathrm{op}}$, this circumstance shows all the same that one can not define $J^{(r)}_{\mathrm{op}}$ in a sensible way.

This peculiar difference between classical mechanics and quantum mechanics is closely connected to the fact that differential calculus is an essential tool in classical mechanics, but that it is a strange element in operator calculus and that it can only be applied in some simple cases and even then in a highly artificial manner (compare § 29). The idea of infinitesimal changes is foreign to the character of an operator. The state of affairs is rather such that expressions containing derivatives in the classical theory have a quantum mechanical counterpart which is constructed purely algebraically and in which non-commutability plays an important part. We have especially in mind the analogy between the Poisson bracket $\{G, F\}$ and the operator $[G_{\mathrm{op}}, F_{\mathrm{op}}]_-$ (see § 29). This restriction in the choice of mechanical quantities which can be considered in quantum mechanics as compared to the infinite choice of canonical transformations in classical mechanics simplifies to some extent the quantum mechanical discussion.

§ 43. The time dependence of observables

Up to now we have only discussed the instantaneous values of a mechanical quantity F. We showed in § 39 how one could talk at any time t of a probability distribution of F over its eigenvalues. In this section we shall show how one can extend these consideration in such a way that we obtain the *quantum mechanical counterpart of those classical quantities whose definition depends on the variation of the p and q in time.*

* A large part is played here by the questions which operators can be allowed to occur and in how far one can retain the general framework of the transformation theory.

** For instance, in the case of one degree of freedom the canonical conjugate of $F = \frac{1}{2}(p^2+q^2)$ is $J = \tan^{-1}(p/q)$.

In the preceding sections F_{op} meant approximately "multiplication of the wave function $\psi(t)$ by the value $f(t)$ which the quantity F possesses at the instant t". This would be rigorously correct, if $\psi(t)$ were an eigenfunction of F_{op}. We now define preliminarily an operator $F(t')'_{\text{op}}$ which to the same approximation will mean multiplication of the wave function $\psi(t)$ by the value which the quantity F is going to possess at an instant t' later, that is, at $t_1 = t+t'$. The value $f(t+t')$ which F will possess at t_1 could also have been obtained by operating with F_{op} on $\psi(t+t')$. We now introduce a wave function χ by the equation

$$\chi(t_1) = F_{\text{op}}\,\psi(t_1). \tag{116}$$

We consider χ to satisfy the Schrödinger equation (3.45), that is, χ is that solution of the Schrödinger equation whose value at $t = t_1$ is given by equation (116). If ψ is also a solution of the Schrödinger equation, we see that χ and ψ are approximately the same solution, namely, χ is the solution obtained by multiplying the solution ψ by a constant, the constant being the value of F at $t = t_1$, that is, $f(t_1)$. This means that approximately $\chi(t) \doteqdot f(t_1)\psi(t)$ for all values of t. From our preliminary definition of $F(t')'_{\text{op}}$ it follows that we can write this equation in the form $\chi(t) \doteqdot F(t_1-t)'_{\text{op}}\,\psi(t)$.

We can now introduce an operator $F(t)_{\text{op}}$ which for any wave function, that is, solution of the Schrödinger equation, ψ will satisfy rigorously the equation

$$\chi(t_2) = F(t_1-t_2)_{\text{op}}\,\psi(t_2), \tag{117}$$

for any values of t_1 and t_2, where $\chi(t)$ is that solution of the Schrödinger equation which satisfies equation (116). We see that $F(0)_{\text{op}}$ is the same as F_{op}.

If we now denote the derivative of the operator $F(t)_{\text{op}}$ with respect to its argument by $\dot{F}(t)_{\text{op}}$, we find by taking the derivative of equation (117) with respect to t_2,

$$\left(\frac{\partial\chi}{\partial t}\right)_{t=t_2} = -\dot{F}(t_1-t_2)_{\text{op}}\,\psi(t_2) + F(t_1-t_2)_{\text{op}}\left(\frac{\partial\psi}{\partial t}\right)_{t=t_2},$$

which can be written in the form

$$-(i/\hbar)H_{\text{op}}\,\chi(t_2) = -\dot{F}(t_1-t_2)_{\text{op}}\,\psi(t_2) - F(t_1-t_2)_{\text{op}}\,(i/\hbar)H_{\text{op}}\,\psi(t_2),$$

since ψ and χ both satisfy the Schrödinger equation. Using equation (117) to replace $\chi(t_2)$ by $F(t_1-t_2)_{\text{op}}\psi(t_2)$ and remembering that $\psi(t_2)$ is an arbitrary space function, we see that

$$i\hbar\, \dot{F}(t)_{\rm op} = F(t)_{\rm op} H_{\rm op} - H_{\rm op} F(t)_{\rm op}, \tag{118}$$

where we have put $t_1 - t_2 = t$.

The operator $F(t)_{\rm op}$ *is the quantum mechanical counterpart of a classical observable whose value is studied as a function of time.* This follows from our considerations regarding $F(t')'_{\rm op}$, but we can improve on this by using the generalised Ehrenfest theorem. We saw in § 30 that the expectation value of the right hand side of equation (118) is at $t = 0$ equal to the time derivative of the expectation value of $F_{\rm op}$.

We can express $F(t)_{\rm op}$ as follows explicitly in $F_{\rm op}$ [$= F(0)_{\rm op}$] and $H_{\rm op}$,

$$F(t)_{\rm op} = \exp[it\, H_{\rm op}/\hbar]\, F_{\rm op} \exp[-it\, H_{\rm op}/\hbar], \tag{119}$$

where $\exp[\pm it\, H_{\rm op}/\hbar]$ is defined by the exponential series $\Sigma(\pm it\, H_{\rm op}/\hbar)^n/n!$. We have assumed that the double sums occuring in equation (119) all converge. It is easily checked that expression (119) satisfies equation (118). We could have derived equation (119) from equations (116) and (117) by noting that the Taylor series $\psi(t_2) = \Sigma\,[(t_2 - t_1)^n/n!](\partial/\partial t)^n\, \psi(t_1)$ could be written in the form $\psi(t_2) = \Sigma\,[(t_1 - t_2)^n/n!](i/\hbar)^n H^n_{\rm op}\, \psi(t_1) = \exp[it\, H_{\rm op}/\hbar]\, \psi(t_1)$ and that a similar expression holds for χ so that equation (117) goes over into $\exp[i(t_1 - t_2)\, H_{\rm op}/\hbar] F_{\rm op}\, \psi(t_1) = F(t_1 - t_2)_{\rm op} \exp[i(t_1 - t_2)\, H_{\rm op}/\hbar]\, \psi(t_1)$ from which equation (119) follows since ψ and t_2 are arbitrary. It is easily checked that $F(t)_{\rm op}$ is Hermitean, since

$$\int \psi_1^* \exp[it\, H_{\rm op}/\hbar]\, \psi_2\, d\tau = \int \psi_2 \exp[it\, H^*_{\rm op}/\hbar]\, \psi_1^*\, d\tau.$$

Note that the commutation relations (3.77) of the p and q remain the same when we consider the p and q to be time dependent operators,

$$\begin{gathered} q_i(t)\, q_j(t) - q_j(t)\, q_i(t) = 0, \quad p_i(t)\, p_j(t) - p_j(t)\, p_i(t) = 0, \\ q_i(t)\, p_j(t) - p_j(t)\, q_i(t) = i\hbar\, \delta_{ij}. \end{gathered} \tag{120}$$

The eigenvalues of $F(t)_{\rm op}$ are the same as those of $F_{\rm op}$, while its eigenfunctions $S(t)$ derive from those of $F_{\rm op}$ — which we shall denote by $S(0)$ — through the equation $S(t) = \exp(it\, H_{\rm op}/\hbar)\, S(0)$.

These considerations enable us in principle to extend the quantum mechanical concepts of distribution functions and probability amplitudes to those classical quantities which refer to the change in time of a system rather than its instantaneous situation.

The matrix elements of $F(t)$ with respect to some orthonormal set

$T_k(q)$ are time dependent and given by the equations [compare eq. (95)]

$$\{F(t)\}_{ll'} = \int T_l^*(q)\, F(t)_{\rm op}\, T_{l'}(q)\, d\tau = \int T_l^*(q, t)\, F_{\rm op}\, T_{l'}(q, t)\, d\tau. \quad (121)$$

In this equation the $T_l(q, t)$ satisfy the Schrödinger equation $H_{\rm op} T_l(q, t) = i\hbar \partial T_l(q, t)/\partial t$ with the boundary condition that at $t = 0$ the $T_l(q, t)$ are identical with the given orthonormal set, $T_l(q, 0) \equiv T_l(q)$. As $(d/dt) \int T_l^*(q, t)\, T_{l'}(q, t)\, d\tau = 0$ [a consequence of eq. (3.84)], the $T_l(q, t)$ themselves will form an orthonormal set at all times t.

If especially the $T_l(q)$ are a set of eigenfunctions, $\varphi_l(q)$, of the energy operator we have [compare eq. (111)] $T_l(q, t) = \varphi_l(q) \exp[-iE_l t/\hbar]$ and the matrix elements of $F(t)_{\rm op}$ become simply

$$\begin{aligned}\{F(t)\}_{ll'} &= \int \varphi_l^*\, F_{\rm op}\, \varphi_{l'}\, d\tau \cdot \exp[i(E_l - E_{l'})\, t/\hbar] \\ &= F_{ll'} \exp(2\pi i \nu_{ll'} t).\end{aligned} \quad (122)$$

Here $\nu_{ll'}$ is the Bohr frequency of the $l \to l'$ transition (compare §18).

The observables $F(t)$ correspond always to Hermitean operators and one can obtain new observables by adding them one to another or by multiplying them by each other and one does not have to worry on which argument they act. This means that the possibility of forming an analytic expression from them is independent of the representation. The operator equations (118) are also independent of the representation. These equations describe the variation in time of the $F(t)$ and might be called the general "quantum mechanical equations of motion". A similar situation occurs in classical mechanics where it is immaterial in which variables the observables are expressed and where one can write the equations of motion (3.19) in the form $\dot{F} = \{F, H\}$ which is analogous to equation (118). A big difference arises, however, in that in quantum mechanics one can in general not assign a definite numerical value to the observables. This is connected to the fact that these quantities often do not commute. Dirac * has introduced the nomenclature "q-numbers" (q: quantum) for such quantities and has called quantities which possess a definite numerical value and which can occur at any place in a product without raising questions of commutability "c-numbers" (c: commuting). Apart from the electronic charge and mass, the quantum of action, and so on, the eigenvalues of the observables and (at any rate in non-relativistic quantum mechanics) the time are c-numbers.

The quantities (122) played a leading role in Heisenberg's first paper on matrix mechanics ** and we shall call them therefore *Heisenberg matrices*. In many respects they are the counterparts of the Fourier components in which one can expand a quantity in classical mechanics (compare Ch. VIII). In the original matrix

* P. A. M. Dirac, *Proc. Roy. Soc.* A., **110**, 562, 1926.

** W. Heisenberg, *Z. Phys.*, **33**, 879, 1925.

mechanics of Born, Jordan and Heisenberg * (see § 18) the main problem of quantum mechanics was formulated as follows. Find solutions of the quantum mechanical equations of motion (118) such that all observable are represented by Heisenberg matrices and that the matrices of the p and q satisfy the commutation relations (120). The energy becomes a diagonal matrix $E_l \delta_{ll'}$.

Although in principle many problems of quantum theory can be solved on this basis, it is hardly possible to derive in this way a theory of atomic phenomena which obeys clearly the correspondence principle. For this purpose one uses the Schrödinger equation. In the Schrödinger theory the concept of a physical situation which is described by the wave function, and the quantum mechanical superposition principle are of the greatest importance.

We may add a few words here about the difference between the Heisenberg picture and the Schrödinger picture. In the latter the physical situation is described by a wave function which changes in time while to each observable there is assigned an operator which is constant. The expectation value, $\overline{F}$, of an observable F is obtained in this picture from the equation $\overline{F} = \int \psi^*(t) F_{op} \psi(t)\, d\tau$, and the change of ψ is governed by the Schrödinger equation $i\hbar\dot{\psi} = H_{op}\psi$. In the Heisenberg picture, one describes the situation by a constant wave function, say, $\psi(0)$, and there are time dependent operators $F(t)_{op}$ which correspond to the observables. These operators satisfy equation (118). The expectation value $\overline{F}$ is now given by the equation $\overline{F} = \int \psi^*(0)\, F(t)_{op}\psi(0)\, d\tau$. It can be checked that the expectation value is the same in the two pictures. This follows from equation (119) and from the fact that (i) $\psi(t) = \exp\,[-itH_{op}/\hbar]\,\psi(0)$ is the solution of the Schrodinger equation with boundary condition $\psi = \psi(0)$ at $t = 0$ and (ii) $\int \psi^*(0) \exp\,[itH_{op}/\hbar]\, F_{op} \exp\,[-itH_{op}/\hbar]\,\psi(0)\, d\tau = \int \psi^*(0) \exp\,(itH_{op}/\hbar)\, F_{op}\, \psi(t)\, d\tau = \int [F_{op}\,\psi(t)] \exp\,[itH_{op}/\hbar]\,\psi^*(0)\, d\tau = \int [F_{op}\,\psi(t)]\,\psi^*(t)\, d\tau$, because $\exp\,[itH_{op}/\hbar]$ is Hermitean.

In recent quantum mechanical literature an intermediate picture is often used. One splits the Hamiltonian into two parts, $H_{op} = H_{op}^{(1)} + H_{op}^{(2)}$ and works with a wave function satisfying the equation $i\hbar\dot{\psi}'(t) = H_{op}^{(2)}\,\psi'(t)$ and operators satisfying the equation $i\hbar\,\dot{F}'(t)_{op} = [F'(t)_{op}, H_{op}']_-$. The expectation value $\overline{F}$ now satisfies the equation $\overline{F} = \int \psi'^*(t)\; F'(t)_{op}\,\psi'(t)\, d\tau$ with the same result as before. In many cases one chooses for $H_{op}^{(2)}$ the interaction energy and this picture is then called the *interaction picture*. The wave functions $\psi(t)$, $\psi(0)$, and $\psi'(t)$ are called respectively *the Schrödinger, the Heisenberg and the interaction representation* of the physical situation.

B. EXAMPLES

§ 44. The probability distribution of coordinates and momenta; the probability current density

Consider a system with one degree of freedom. Let q be the coord-

* M. Born, P. Jordan, and W. Heisenberg, *Z. Phys*l, **34**, 858, 1925; **35**, 557, 1925.

inate, that is, q defines the position of a particle on a straight line $(-\infty < q < \infty)$, let p be the momentum of the particle, and let the normalised wave function $\psi(q, t)$ describe the physical situation of the system. What will be the eigenvalues, eigenfunctions and the probability distribution of the following observable F. This quantity is a function of q only and satisfies the equations

$$\begin{aligned} F(q) &= 1, \qquad q' \leqq q \leqq q''; \\ F(q) &= 0, \qquad -\infty < q < q' \text{ and } q'' < q < \infty. \end{aligned} \tag{123}$$

If F_e is an eigenvalue of F corresponding to the eigenfunction $S(q)$, we have according to equation (16) $F_{op}(q)\,S(q) = F_e S\,(q)$. There are obviously two kinds of solution, namely, those for which $S(q)$ is different from zero only for $q' \leqq q \leqq q''$ (eigenvalue 1) and those for which $S(q)$ is different from zero only outwith the interval (q', q'') (eigenvalue 0),

$$\begin{aligned} F_e = 1,\ S_1(q) &= 0, \qquad q < q' \text{ or } q'' < q; \\ F_e = 0, \quad S_0(q) &= 0, \qquad q' \leqq q \leqq q''. \end{aligned} \tag{124}$$

We see that there is still an infinitely large amount of freedom in the choice of the S_1 and the S_0. Both eigenvalues are infinitely degenerate and one can easily construct sets of orthonormal functions $S_{1k}(q)$ and $S_{0k}(q)$ such that any function can be expanded in terms of them. This is, however, not necessary in order to find the probability distribution of F. We can introduce two functions ψ_0 and ψ_1 such that $\psi_0 = 0$ for $q' \leqq q \leqq q''$ and $\psi_0 = \psi(q, t)$ for $q < q'$ or $q > q''$ and $\psi_1 = \psi - \psi_0$. The ψ_1 and ψ_0 are both eigenfunctions of F_{op} and the equation $\psi = \psi_1 + \psi_0$ is already an expansion in terms of eigenfunctions. We can now get the following expansions in terms of the normalised eigenfunctions ψ_1' and ψ_0',

$$\begin{aligned} \psi = b_1\psi_1' + b_0\psi_0', \qquad \psi_1' &= \psi_1/b_1, \qquad \psi_0' = \psi_0/b_0, \\ b_1(t)^2 = \int |\psi_1|^2\,dq, \qquad b_0(t)^2 &= \int |\psi_0|^2\,dq, \qquad b_1^2 + b_0^2 = 1. \end{aligned} \tag{125}$$

In accordance with the considerations of § 39 we can consider b_1^2 and b_0^2 to be the probabilities that $F = 1$ or $F = 0$. The equation $F = 1$ expresses the fact that the particle is within the interval (q', q'') and the probability for this to happen is thus equal to $b_1^2 = \int_{q'}^{q''} |\psi|^2\,dq$, in complete agreement with the interpretation of ψ as the probability amplitude of the coordinates; it is an illustration of the considerations of § 39.

We have chosen here the simplest case of a function of the coordinates whose eigenfunctions are proper mathematical functions. As soon as $F(q)$ varies continuously within a finite interval, we must introduce improper functions to describe the eigenfunctions. One can always write formally $F_{op}(q)\,\delta(q-q') = F(q')\,\delta(q-q')$, where $\delta(q-q')$ is the Dirac delta-function. In this equation q' is arbitrary and the value $F(q')$ occurs as an eigenvalue of $F_{op}(q)$. One can thus treat the functions $\delta(q-q')$ $(-\infty < q < \infty)$ as an orthonormal set in terms of which any function of q can be expanded: $f(q) = \int_{-\infty}^{\infty} f(q')\,\delta(q-q')\,dq'$, $f(q') = \int_{-\infty}^{\infty} f(q)\,\delta^*(q-q')dq$. (As δ is a real function, $\delta = \delta^*$). We have formally assigned the operator function $F(q')\,\delta(q-q')$ to the operator $F_{op}(q)$. The eigenvalues of the coordinate itself, considered as an operator q_{op}, is simply the set of all real values which this coordinate can take on according to its classical definition. Its operator function is $q'\delta(q-q')$.

We see that the use of the general transformation theory requires the introduction of improper functions, especially for the simplest possible operators. The coordinate transformation $q \to Q(q)$ (compare § 34) requires in this general framework the existence of such a transformation function $S(Q, q)$ that $f(q) = \int b(Q)\,S(Q, q)\varrho(Q)\,dQ$, $b(Q) = \int S^*(Q, q)\,f(q)\,dq$. These equations are satisfied by the relations $S(Q, q) = S^*(Q, q) = \delta(q'(Q)-q) = \varrho^{-1/2}(Q)\,\delta(Q-Q'(q))\,\varrho^{1/2}(Q')$, $\varrho(Q) = dq/dQ$. This is clearly a very complicated way to express the fact that $b(Q)$ is derived from $f(q)$ by expressing the q in terms of Q. It implies the formal transformation of the delta-function from q to Q [$q'(Q)$ and $Q'(q)$ denote q expressed in terms of Q and Q expressed in terms of q].

One can easily extend these formulae to the case of several degrees of freedom. It is also not difficult to perform the transformation of the Schrödinger equation, discussed in § 34, according to this general theory. A simple coordinate transformation corresponds clearly in the case of discrete arguments to the trivial process of a renumbering of the arguments.

The *momentum eigenvalue problem* requires for the case of our one-dimensional system the solution of the differential equation $p_{op}S(q) = -i\hbar\partial S/\partial q = p_e S$. The solution is $S(q) = c\exp(ip_e q/\hbar)$, where c is a constant. One obtains functions in terms of which any normalisable function can be expanded only, if $p_e(-\infty < p_e < \infty)$ is real. This means that, as expected, the eigenvalues of p_{op} are real and we are back to expressing the wave function as a Fourier integral as in Chapter 1 and to our original interpretation of this Fourier integral (compare also § 15). Without introducing the wave numbers and writing simply p for the eigenvalues of p_{op} we have $\psi(q, t) = \int a(p, t)\exp(ipq/\hbar)[dp/h]$, $a(p, t) = \int \psi(q, t)\exp(-ipq/\hbar)dq$. We shall choose the eigenfunctions as follows,

$$S(p, q) = \exp(ipq/\hbar), \tag{126}$$

so that the $S(p, q)$ are normalised with respect to a density function h^{-1} and $|a|^2 dp/h$ is the probability that the value of the momentum lies in the interval $(p, p+dp)$ (compare § 15). One can extend the discussion easily to the case of several degrees of freedom (many particles in three-dimensional space).

One often introduces with advantage the probability amplitude $a(p_x, p_y, \ldots, t)$ with respect to the momenta of the particles, when applying the theory. Many experiments are concerned with collisions of free particles with each other or with other systems. For instance, one measures the energies and angular distribution of elastically and inelastically scattered electrons which initially were moving in a known direction with a known energy. A description of such a situation by means of the probability amplitudes with respect to the momenta is clearly more suitable to the experimental set-up. On the other hand the transformed Schrödinger equation is not as tractable as the original one. The relation $f_2(q) = \Omega f_1(q)$, for instance, is in the one-dimensional case transformed into [compare eq. (93)]

$$b_2(p) = \iint \exp(-ipq/\hbar)\, \Omega \exp(ip'q/\hbar)\, b_1(p')\, (dp'/h)\, dq, \qquad (127)$$

corresponding to an operator function $\Omega(p, p') = \int \exp(-ipq/\hbar)$ $\Omega \exp(ip'q/\hbar) dq$. If Ω is equal to $p_{\text{op}} = -i\hbar\partial/\partial q$, we get from equation (127) simply

$$b_2(p) = \iint \exp[iq(p'-p)/\hbar]\, p' b_1(p')\, (dp'/h)\, dq = p b_1(p)$$

where we have used the fact that exp (ikx) is the Fourier transform of the delta function, which means that $\int_{-\infty}^{+\infty} \exp(ikx)\, dx = \delta(k)$. If, however, $\Omega = q_{\text{op}}$ we find

$$\begin{aligned} b_2(p) &= \iint q \exp[iq(p'-p)/\hbar]\, b_1(p')\, (dp'/h)\, dq \\ &= -i\hbar \iint [(\partial/\partial p') \exp\{iq(p'-p)/\hbar\}]\, b_1(p')\, (dp'/h)\, dq \\ &= (i/2\pi) \iint \exp[iq(p'-p)/\hbar][\partial b_1(p')/\partial p']\, dp' dq = i\hbar \partial b_1(p)/\partial p. \end{aligned}$$

We see now that p_{op} means multiplying by p and q_{op} differentiation with respect to p and multiplying by $i\hbar$. The commutation relations (3.77) remain valid. We see also that the connexion between b_1 and b_2 through a linear operator is mathematically complicated as soon as Ω can not be written as a polynomial in the q. In general the energy operator, which is always a polynomial in the momenta, will not be a polynomial in the q.

Let R be the distance of a particle at $\boldsymbol{x}$ to a fixed point $\boldsymbol{x}_0[R^2 = (\boldsymbol{x}-\boldsymbol{x}_0 \cdot \boldsymbol{x}-\boldsymbol{x}_0)]$. The operator function $\Omega(\boldsymbol{p}, \boldsymbol{p}')$ in the momentum representation corresponding to the operator $e^{-\lambda R}/R$ is given by the equation

$$\Omega(\boldsymbol{p},\boldsymbol{p}')=\int[e^{-\lambda R}/R]\exp[i(\boldsymbol{p}'-\boldsymbol{p}\,.\boldsymbol{x})/\hbar]d\boldsymbol{x}=4\pi\exp[i(\boldsymbol{p}-\boldsymbol{p}\cdot\boldsymbol{x}_0)/\hbar]/[\lambda^2+\hbar^{-2}|\boldsymbol{p}'-\boldsymbol{p}|^2].$$

This equation goes over into the following one for $\lambda=0$:

$$R^{-1}(\boldsymbol{p},\,\boldsymbol{p}')=[h^2/\pi\,|\boldsymbol{p}'-\boldsymbol{p}|^2]\exp[i(\boldsymbol{p}'-\boldsymbol{p}\cdot\boldsymbol{x}_0)/\hbar].$$

The density function in this case is h^{-3}.

Let now $\boldsymbol{x}_0$ be the position of a second particle with momentum $\boldsymbol{p}_0$ and consider the operator function $R^{-1}(\boldsymbol{p},\,\boldsymbol{p}_0;\,\boldsymbol{p}'\,\boldsymbol{p}_0')$ of $1/R$ with respect to the momenta of the two particles (density function h^{-6}).

$$\begin{aligned}R^{-1}(\boldsymbol{p},\boldsymbol{p}_0;\boldsymbol{p}',\boldsymbol{p}_0')&=[h^2/\pi|\boldsymbol{p}'-\boldsymbol{p}|^2]\int\exp[i(\boldsymbol{p}'-\boldsymbol{p}\cdot\boldsymbol{x}_0)/\hbar]\,.\,\exp.\,[i(\boldsymbol{p}_0'-\boldsymbol{p}_0\cdot\boldsymbol{x}_0)/\hbar]\,d\boldsymbol{x}\\&=[h^5/\pi|\boldsymbol{p}'-\boldsymbol{p}|^2]\delta(\boldsymbol{p}'+\boldsymbol{p}_0'-\boldsymbol{p}-\boldsymbol{p}_0),\end{aligned}\tag{128}$$

where the three dimensional delta function $\delta(\boldsymbol{x})$ is defined as the product of three delta functions, $\delta(x)\delta(y)\delta(z)$. The expression for R^{-1} is symmetrical in the two particles, as it is different from zero only if $\boldsymbol{p}'+\boldsymbol{p}_0'-\boldsymbol{p}-\boldsymbol{p}_0=0$, that is, when $|\boldsymbol{p}'-\boldsymbol{p}|=|\boldsymbol{p}_0'-\boldsymbol{p}_0|$.

These equations are useful in the computation of matrix elements which occur in the perturbation treatment of scattering problems *.

We can connect the expression derived in chapters 1 and 2 for the *probability current density* as follows with transformation theory. Let $F(q)$ again be the function defined by equation (123). Consider now the observable pF/m. This function is in the classical theory equal to the velocity v of the particle if it is inside the interval $(q',\,q'')$, and otherwise equal to zero. In the first case the particle will certainly leave the interval $(q',\,q'')$ within a period of time $(q''-q')/|v|$ either at q'' (v positive) or at q' (v negative).

Quantum mechanically p and q can not be defined simultaneously, as they do not commute. We first of all make pF/m Hermitean by writing $(pF+Fp)/2m$ which is thus the analogy to the real classical quantity pF/m (compare § 29). It is clearly difficult to imagine a sensible experiment which enables us to measure the value of this expression. Its expectation value will, however, correspond to the classical value pF/m, averaged over a large number of cases, provided $q'-q''$ is sufficiently small. A moment's reflexion shows that this expectation value can thus be considered to be the probability that during a time interval $q''-q'$ the particle passes through the interval in the positive direction minus the probability that it passes through the interval in the opposite direction. In the limit where $q''-q'\to 0$ the average $\overline{(pF+Fp)/2m(q''-q')}$ must be equal to the previously defined probability current density S. We have

* Compare H. Bethe, *Ann. Phys.*, **5**, 325, 1930.

$$\overline{(pF+Fp)/2m} = (2m)^{-1}\int \psi^*(pF+Fp)\psi\,dq$$
$$= (2m)^{-1}\int (\psi^*Fp\psi+\psi F^*p^*\psi^*)\,dq$$
$$= (\hbar/2mi)\int [\psi^*F(\partial\psi/\partial q)-\psi F(\partial\psi^*/\partial q)]\,dq$$
$$= (\hbar/2mi)\int_{q'}^{q''} [\psi^*(\partial\psi/\partial q)-\psi(\partial\psi^*/\partial q)]\,dq,$$
$$S = \mathrm{Lim}_{q''-q'\to 0}\,\overline{(pF+Fp)/2m(q''-q')}$$
$$= (\hbar/2mi)[\psi^*(\partial\psi/\partial q)-\psi(\partial\psi^*/\partial q)].$$

This expression agrees indeed with our previous formulae (compare § 14), with a slight difference in that earlier we were dealing with the three-dimensional case. It is, however, easy to extend the above analysis to the case of three-dimensions.

Without having recourse to the special form of the energy operator we could try to write the probability current density $\boldsymbol{S}$ of a particle in three dimensions (position $\boldsymbol{x}$) in the form

$$\boldsymbol{S} = \tfrac{1}{2}[\psi^*\dot{\boldsymbol{x}}\psi+\psi\dot{\boldsymbol{x}}^*\psi^*], \tag{129}$$

where the vector operator $\dot{\boldsymbol{x}}$ is defined by the three equations (see § 43) $i\hbar\dot{x} = xH-Hx$, $i\hbar\dot{y} = yH-Hy$, $i\hbar\dot{z} = zH-Hz$. If $F(\boldsymbol{x})$ be an arbitrary function one finds by a straightforward calculation $i\hbar\int F(\nabla\cdot\boldsymbol{S})\,d\tau = \frac{1}{2}\int\psi^*\{([\boldsymbol{p}F-F\boldsymbol{p}]\cdot\dot{\boldsymbol{x}})+(\dot{\boldsymbol{x}}\cdot[\boldsymbol{p}F-F\boldsymbol{p}])\}\psi\,d\tau$. If $J = \psi^*\psi$ is the probability density we have $i\hbar\int F(\partial J/\partial t)\,d\tau = \int\psi^*(FH-HF)\psi\,d\tau$. The validity of the equation of continuity (see § 7) $\partial J/\partial t+(\nabla\cdot\boldsymbol{S}) = 0$ requires thus that the following identity holds for arbitrary F,

$$FH-HF = \tfrac{1}{2}\{([F\boldsymbol{p}-\boldsymbol{p}F]\cdot\dot{\boldsymbol{x}})+(\dot{\boldsymbol{x}}\cdot[F\boldsymbol{p}-\boldsymbol{p}F])\}. \tag{130}$$

This equation is the quantum mechanical analogy of the classical equation, $dF/dt = (\nabla F\cdot\dot{\boldsymbol{x}})$. One sees easily that equation (130) in general holds only, if H considered as a polynomial in the $\boldsymbol{p}$ does not contain terms of a higher degree than the second. This is connected with the fact that the prescription $\dot{\boldsymbol{x}}F\to\frac{1}{2}(\dot{\boldsymbol{x}}F+F\dot{\boldsymbol{x}})$ for constructing a Hermitean operator, which is the basis of expression (129) is only one out of several possibilities as soon as $\dot{\boldsymbol{x}}$ contains the $\boldsymbol{p}$ to a higher degree than the second. It is not difficult to settle the problem of how to construct a probability current density which satisfies the equation of continuity in the case of an arbitrary energy function H; this problem is, however, not relevant to any applications of quantum theory.

§ 45. The eigenvalues and eigenfunctions of the angular momentum

Of the utmost importance for quantum mechanics is the eigenvalue problem of the angular momentum of a particle in three-dimensional space and the corresponding transformations. If $\boldsymbol{M}$ be the angular momentum vector taken from the origin we have

$$M_x = yp_z - zp_y,\ M_y = zp_x - xp_z,\ M_z = xp_y - yp_z\,;\qquad \boldsymbol{M} = [\boldsymbol{x}\times\boldsymbol{p}]. \tag{131}$$

If we consider M_x, M_y, and M_z to be operators, we see from equations (3.77) that the order of the factors in each of the terms on the right hand side of equations (131) is immaterial (for instance, $[yp_z - zp_y]_{\text{op}} = [p_z y - p_y z]_{\text{op}}$). It follows then that these operators are Hermitean. The operators M_x, M_y, and M_z do not commute and they can thus in general not possess simultaneously well defined values. Using the commutation relations (3.77) we find

$$M_xM_y - M_yM_x = (yp_z - zp_y)(zp_x - xp_z) - (zp_x - xp_z)(yp_z - zp_y)$$
$$= yp_x(p_z z - zp_z) + p_y x(zp_z - p_z z),\ \text{or,}$$

$$[M_x, M_y]_- = i\hbar M_z,\ \text{and similarly,}$$
$$[M_y, M_z]_- = i\hbar M_x,\ [M_z, M_x]_- = i\hbar M_y. \tag{132}$$

The left hand sides of equations (132) have the same form as the cross product of $\boldsymbol{M}$ with itself, and we can write equations (132) symbolically in the form $[\boldsymbol{M}\times\boldsymbol{M}] = i\hbar\boldsymbol{M}$. The components of the angular momentum are also in general not commuting with the coordinates and momenta. It is easily checked that the following equations hold

$$\left.\begin{aligned} &[M_x, x]_- = 0, && [M_x, y]_- = i\hbar z, && [M_x, z]_- = -i\hbar y,\\ &[M_x p_x]_- = 0, && [M_x, p_y]_- = i\hbar p_z, && [M_x, p_z]_- = -i\hbar p_y, \end{aligned}\right\} \tag{133}$$

and similar equations for the M_y and M_z. The square of the angular momentum M^2 $(= M_x^2 + M_y^2 + M_z^2)$ considered as an operator commutes with M_x, M_y, and M_z. Using equations (132) we find

$$[M_x^2, M_x]_- = 0,$$

$$[M_y^2, M_x]_- = [M_y, M_x]_- M_y + M_y[M_y, M_x]_- = -i\hbar[M_zM_y + M_yM_z],$$

$$[M_z^2, M_x]_- = [M_z, M_x]_- M_z + M_z[M_z, M_x]_- = i\hbar[M_yM_z + M_zM_y],$$

$$[M^2, M_x]_- = [M^2, M_y]_- = [M^2, M_z]_- = 0. \tag{134}$$

Using equations (133) one sees that M_x, M_y, and M_z commute both with r^2 $(= x^2 + y^2 + z^2)$ and with p^2 $(= p_x^2 + p_y^2 + p_z^2)$,

$$\left.\begin{aligned} &[M_x, r^2]_- = [M_y, r^2]_- = [M_z, r^2]_- = 0, \\ &[M_x, p^2]_- = [M_y, p^2]_- = [M_z, p^2]_- = 0, \\ &[M^2, r^2]_- = 0, \ [M^2, p^2]_- = 0. \end{aligned}\right\} \tag{135}$$

The eigenfunctions of the z-component of the angular momentum, M_z, follow from the differential equation $M_{z\text{op}} S = -i\hbar[x(\partial/\partial y)-y(\partial/\partial x)] S = (M_z)_e S$. Introducing the azimuth φ and the distance ϱ from the z-axis, $x = \varrho \cos \varphi$, $y = \varrho \sin \varphi$, we can consider S to be a function of z, ϱ, and φ. The operator $x(\partial/\partial y)-y(\partial/\partial x)$ becomes $\partial/\partial\varphi$ and the differential equation is of the form $-i\hbar\partial S(z, \varrho, \varphi)/\partial\varphi = (M_z)_e S$ with the solution $S = \sigma(z, \varrho) \exp [i(M_z)_e\varphi/\hbar]$. We require the eigenfunctions to be single-valued functions of the space coordinates, since the wave functions which can be expanded in terms of them are single-valued because of their physical meaning. This means that the exponent must be an integral multiple of $i\varphi$. We have thus for the eigenvalues the equations

$$(M_z)_e = m\hbar, \qquad m = 0, \pm 1, \pm 2, \ldots, \tag{136}$$

while the eigenfunctions satisfy the equations

$$S = \sigma(z, \varrho)e^{im\varphi}, \qquad m = 0, \pm 1, \pm 2, \ldots \tag{137}$$

One calls m the *quantum number of the z-component of the angular momentum or the magnetic quantum number* (see § 76). In order that S be normalised σ must satisfy the equation $\int S^*S\, d\mathbf{x} = 1 = 2\pi \int \sigma^*\sigma\varrho\, dz d\varrho$. Otherwise σ is arbitrary and the eigenvalues are infinitely degenerate.

The expression of a function in terms of these eigenfunctions means nothing more than that they as functions of z, ϱ, and φ can be expanded in a Fourier series of the angle φ. The wave function of a particle in three-dimensional space can be written as follows

$$\psi(z, \varrho, \varphi, t) = \Sigma_{m=-\infty}^{+\infty} a_m(z, \varrho, t)e^{im\varphi}, \ a_m = (2\pi)^{-1} \int_0^{2\pi} e^{-im\varphi}\, \psi\, d\varphi. \tag{138}$$

We can interpret this expansion differently from the point of view of transformation theory. We write thereto $\psi = \Sigma_m N_m^{1/2}[a_m N_m^{-1/2} e^{im\varphi}]$, $N_m = 2\pi \int|a_m|^2\varrho\, dz d\varrho$. The function $a_m N_m^{-1/2} e^{im\varphi}$ is a normalised eigenfunction, and $N_m^{1/2}$ plays the part of a probability amplitude while N_m is the probability that M_z has the value $m\hbar$. One can also consider the $a(z, \varrho, t)$ to be probability amplitudes; in that case z and ϱ are considered to be parameters like the r in equation (24) and the eigenfunctions are the $e^{im\varphi}$ which are normalised with respect to a density function $2\pi\varrho^*$. We are dealing with a quantum mechanical transformation which leaves z and ϱ unchanged and only exchanges the angle φ for the index m. In this interpretation N_m is again the probability that M_z has the value $m\hbar$ [see eq. (67)].

If H and M_z commute, that is, if M_z is an integral of the problem (classically M_z is constant), N_m will be time independent. For instance, N_m may be zero for all values of m except $m = m_0$ ($N_{m_0} = 1$); in that case M_z has the well defined value $m_0\hbar$. This corresponds to the quantisation of the angular momentum with respect to a chosen axis in the old quantum theory.

The transformed Schrödinger equation is of the form (compare § 42)

$$\Sigma_{m'} H_{mm'} a_{m'}(z, \varrho, t) = i\hbar\, \partial a_m(z, \varrho, t)/\partial t, \tag{139}$$

where the $H_{mm'}$ form an infinite set of operators which act on functions of z and ϱ. If H is the original energy function acting on x, y, z, or on z, ϱ, φ we have using equations (93) and (138)

$$\left.\begin{aligned} i\hbar\partial a_m/\partial t &= (2\pi)^{-1}\int e^{-im\varphi}H\psi d\varphi = (2\pi)^{-1}\Sigma_{m'}\int e^{-im\varphi}H a_{m'} e^{im'\varphi}d\varphi, \\ H_{mm'} &= (2\pi)^{-1}\int e^{-im\varphi}\, He^{im\varphi}\, d\varphi. \end{aligned}\right\} \tag{140}$$

Since M_z does not commute with M_x and M_y [see eq. (132)], there does not exist a set of eigenfunctions common to M_z and M_x or to M_z and M_y. However, it should be possible to find a set of eigenfunctions common to M_z and M^2 as they do commute.

We shall now consider the *eigenfunctions of the square of the angular momentum. These are the Laplacian spherical harmonics multiplied by an arbitrary function of r*. We shall briefly discuss the definition and properties of spherical harmonics in view of applications later in the book **. A point in space is determined by the absolute value r and the direction of its radius vector. The direction depends on two angles, say the polar angle ϑ and azimuth φ. We denote the direction by the symbol $\boldsymbol{\omega}$, which can be considered to determine a point on the unit sphere. An element of solid angle is denoted by $d\boldsymbol{\omega}$ and integration over all possible directions by

* To retain equations (24) to (26) one should take $S_m(z', \varrho'; z, \varrho; \varphi) = (2\pi\varrho)^{-1}\delta(z-z')\delta(\varrho-\varrho')\exp(im\varphi)$ for the eigenfunctions and take instead of the expansion (138) the equivalent expansion

$$\psi = \Sigma_m \iint a_m(z', \varrho', t) S_m(z', \varrho'; z, \varrho; \varphi)\, 2\pi\varrho' d\varrho' d\varrho' dz', \quad a_m = \int S_m{}^* \psi d\tau.$$

** See for instance, R. Courant and D. Hilbert, *Methods of Mathematical Physics*, Interscience Publ., New York, 1953, Ch. 7; E. T. Whittaker and G. N. Watson, *Modern Analysis*, Cambridge University Press, 1927, Ch. 18.

$\int \ldots d\boldsymbol{\omega}$. A (Laplacian) spherical harmonic, $Y_l(\boldsymbol{\omega})$, is defined to be r^{-l} times a homogeneous polynomial of the l-th degree in x, y, z which satisfies the Laplacian differential equation ("harmonic polynomial" of the l-th degree),

$$Y_l(\boldsymbol{\omega}) = r^{-l} H_l(x, y, z), \qquad \nabla^2 H_l = 0. \tag{141}$$

An arbitrary homogeneous polynomial H_l of the l-th degree contains $\frac{1}{2}(l+1)(l+2)$ arbitrary coefficient, and $\nabla^2 H_l$ contains $\frac{1}{2}(l-1)l$ coefficients which are linear combinations of the coefficients of H_l. Since $\nabla^2 H_l = 0$, we can choose $\frac{1}{2}(l+1)(l+2) - \frac{1}{2}l(l-1) = 2l+1$ coefficients arbitrarily. This means that there are $2l+1$ linearly independent harmonic polynomials (and the same number of spherical harmonics) of the l-th degree.

Clearly we can obtain a solution of $\nabla^2 H_l = 0$ as follows

$$H_l = T^l = (ax+by+cz)^l, \tag{142}$$

$$a^2+b^2+c^2 = 0. \tag{143}$$

The quantities a, b, and c cannot be all three real because of equation (143). We can express them as follows in terms of two arbitrary complex numbers ξ and η,

$$a = \tfrac{1}{2}(\xi^2-\eta^2), \qquad b = -\tfrac{1}{2}i(\xi^2+\eta^2), \qquad c = -\xi\eta. \tag{144}$$

Substituting equations (144) into equations (142) gives us

$$\begin{aligned} H_l &= 2^{-l}\{\xi^2(x-iy) - 2\xi\eta z - \eta^2(x+iy)\}^l \\ &= \Sigma_{m=-l}^{+l}\, \xi^{l-m}\, \eta^{l+m}\, Q_l^m(x, y, z). \end{aligned} \tag{145}$$

Substituting expression (145) into equation (141) we get $\Sigma_m \xi^{l-m} \eta^{l+m} \nabla^2 Q_l^m = 0$, and we see that each of the Q_l^m is harmonic, as the ξ and η are arbitrary. We have in this way obtained a particular representation of the $2l+1$ linearly independent harmonic polynomials. Introducing polar coordinates,

$$x \pm iy = r \sin \vartheta e^{i\varphi}, \qquad z = r \cos \vartheta, \tag{146}$$

and using the substitution $\xi = Xe^{\frac{1}{2}i\varphi}$, $\eta = Ye^{-\frac{1}{2}i\varphi}$, $\sigma = \cos\vartheta + Y \sin\vartheta/X$, $\tau = \cos\vartheta - X \sin\vartheta/Y$, we get from equation (145) the following expression for T^l,

$$\begin{aligned} T^l &= (\tfrac{1}{2}r)^l[X^2 \sin\vartheta - 2XY \cos\vartheta - Y^2 \sin\vartheta]^l \\ &= (\tfrac{1}{2}r)^l X^{2l}[(1-\sigma^2)/\sin\vartheta]^l = (\tfrac{1}{2}r)^l Y^{2l}[(\tau^2-1)/\sin\vartheta]^l. \end{aligned}$$

Expanding T^l in a Taylor series, once in terms of Y/X and once in terms of X/Y we get

$$T^l = \Sigma_m (\tfrac{1}{2}r)^l X^{l-m} Y^{l+m} [(l+m)!]^{-1} [\sin\vartheta\, d/d\sigma]^{l+m} [(1-\sigma^2)/\sin\vartheta]^l_{\sigma=\cos\vartheta}$$
$$= (\tfrac{1}{2}r)^l \Sigma_m \xi^{l-m} \eta^{l+m} [(l+m)!]^{-1} \sin^m\vartheta\, e^{im\varphi} [d/d(\cos\vartheta)]^{l+m} (1-\cos^2\vartheta)^l$$
$$= \Sigma_m (\tfrac{1}{2}r)^l X^{l-m} Y^{l+m} (-1)^l [(l-m)!]^{-1} [-\sin\vartheta\, d/d\tau]^{l-m} [(1-\tau^2)/\sin\vartheta]^l_{\tau=\cos\vartheta}$$
$$= (\tfrac{1}{2}r)^l \Sigma_m \xi^{l-m} \eta^{l+m} [(l-m)!]^{-1} (-\sin\vartheta)^{-m} e^{im\varphi} [d/d(\cos\vartheta)]^{l-m} (1-\cos^2\vartheta)^l.$$

Comparing these equations with equation (145) we find two equivalent representations of the $Q_l^m(x, y, z)$. We shall denote by $P_l^m(\vartheta, \varphi)$ [in the literature our P_l^m are often denoted by $P_l^m(\cos\vartheta)\, e^{im\varphi}$] the corresponding spherical harmonics obtained by dividing the Q_l by $(-r)^l$,

$$T^l = (-r)^l \Sigma_m \xi^{l-m} \eta^{l+m} P_l^m. \tag{147}$$

We find then for the P_l^m the equations

$$\left.\begin{aligned} P_l^m(\vartheta, \varphi) &= \sin^m\vartheta\, e^{im\varphi} [2^l (l+m)!]^{-1} [d/d(\cos\vartheta)]^{l+m} (\cos^2\vartheta - 1)^l \\ &= (-\sin\vartheta)^m e^{im\varphi} [2^l (l-m)!]^{-1} [d/d(\cos\vartheta)]^{l-m} (\cos^2\vartheta - 1)^l, \\ m &= 0, \pm 1, \pm 2, \ldots, +l; \quad P_l^{m*}(\vartheta, \varphi) = (-1)^m P_l^{-m}(\vartheta, \varphi). \end{aligned}\right\} \tag{148}$$

If $m = 0$, $P_l^0(\vartheta, \varphi)$ depends on ϑ only and it is a polynomial in $\cos\vartheta$. The functions $P_l^0(\vartheta, \varphi)$ or $P_l(\cos\vartheta)$ are the *Legendre functions* or *zonal spherical harmonics*; the P_l^m $(m \neq 0)$ are the *tesseral spherical harmonics* and all of the P_l^m are called the *polar spherical harmonics.* An arbitrary spherical harmonic of the l-th degree can always be expressed as a linear combination of the P_l^m, $Y_l(\boldsymbol{\omega}) = \Sigma_m b_m P_l^m(\vartheta, \varphi)$.

Spherical harmonics satisfy the following orthogonality relation

$$Y_l^*(\boldsymbol{\omega})\, Y_{l'}(\boldsymbol{\omega})\, d\boldsymbol{\omega} = 0, \qquad \text{if } l \neq l'. \tag{149}$$

The * in this equation is superfluous as the Y^* are themselves spherical harmonics. Equation (149) is proved by using Green's theorem *. We take over from mathematics the result that any single-valued function $f(\boldsymbol{\omega})$ of $\boldsymbol{\omega}$ can always be expanded in terms of spherical harmonics,

$$f(\boldsymbol{\omega}) = \Sigma_l b_l Y_l(\boldsymbol{\omega}) = \Sigma_{lm} b_{lm} P_l^m. \tag{150}$$

The Y_l do thus for the sphere what the functions $\cos m\varphi$, $\sin m\varphi$,

* This theorem gives us

$$\int [H_l^* \nabla^2 H_{l'} - H_{l'} \nabla^2 H_l^*] d\boldsymbol{x} = \oint [H_l^* (\partial H_{l'}/\partial r) - H_{l'} (\partial H_l^*/\partial r)] dS.$$

The integrals are over the volume and the surface of a sphere respectively. The left hand side is zero because of equation (141) and the right hand side is equal to $(l'-l) r^{l+l'+1} \int Y_l^* Y_{l'}\, d\boldsymbol{\omega}$.

or $e^{im\varphi}$ (m: integral) do for the circle (any single-valued function on the circle, that is, any function of φ with period 2π, can be expressed as a Fourier series).

We now return to the operator M^2. The harmonic polynomials (142) satisfy the differential equation*

$$M^2_{\text{op}} H_l = \hbar^2 l(l+1) H_l. \tag{151}$$

From the physical meaning of M_{op} it follows that $M^2_{\text{op}} f(x, y, z)$ can depend only on the change of f on the sphere which passes through the point x, y, z **. We can thus multiply H_l on both sides of equation (151) by an arbitrary function $f(r)$. That this is possible also follows from equation (135). We have thus

$$f(r)\, M^2_{\text{op}} H_l = M^2_{\text{op}} f(r) H_l = \hbar^2 l(l+1) f(r)\, H_l,$$

or,

$$M^2_{\text{op}}\, g(r) Y_l(\boldsymbol{\omega}) = \hbar^2 l(l+1)\, g(r)\, Y_l(\boldsymbol{\omega}). \tag{152}$$

Since according to equation (150) the $Y_l(\boldsymbol{\omega})$ form a complete set, in terms of which any function of $\boldsymbol{\omega}$ can be expanded, we have found that the spherical harmonics multiplied by arbitrary functions of r form the complete set of eigenfunctions of M^2 with eigenvalues

$$[M^2_{\text{op}}]_e = l(l+1)\hbar^2, \qquad l = 0, 1, 2, \ldots \tag{153}$$

One calls l the *total angular momentum quantum number*. Equation (153) corresponds to the *quantisation of the total angular momentnm* in the old quantum theory. In that case one had the rule that if the absolute value of the total angular momentum of a particle were constant in time, it would be equal to an integral multiple of $\hbar$, corresponding to a value $l^2\hbar^2$ for its square. Equation (153) gives instead the correct quantum mechanical expression $l(l+1)\hbar^2$.

Equation (149) expresses simply the general orthogonality of the eigenfunctions of M^2. These are $2l+1$-fold degenerate with respect to their $\boldsymbol{\omega}$-dependence and infinitely degenerate with respect to their r-dependence. One can in infinitely many ways construct $2l+1$ mutually orthogonal spherical harmonics $Y_l(\boldsymbol{\omega})$. The simplest example of such a set of orthogonal functions are the polar spherical harmonics P_l^m given by equation (148), since it follows easily that $[d\boldsymbol{\omega} = d\varphi d(\cos\vartheta)]$

* $M_x T^l = -i\hbar l(yc-zb)\, T^{l-1}$, $M_x{}^2 T^l = \hbar^2\{l(yb+zc)\, T^{l-1} - l(l-1)\, (yc-zb)^2 T^{l-2}\}$, $M^2 T^l = \hbar^2\{2lT^l + l(l-1)\, [T^2 - r^2(a^2+b^2+c^2)\,] T^{l-2}\} = \hbar^2 l(l+1) T^l$.

** One could write $M_{\text{op}}{}^2 = (-i\hbar)^2[(\partial^2/\partial\varphi_x{}^2) + (\partial^2/\partial\varphi_x{}^2) + (\partial^2/\partial\varphi_z{}^2)]$ where $\varphi_x, \varphi_y, \varphi_z$ are the azimuths with respect to the x-, y-, and z-axes. These φ's are *not* independent coordinates.

$$\int P_l^{m*} P_l^{m'} d\boldsymbol{\omega}$$
$$= \left[\int_{-1}^{+1} P_l^{m*}(\cos\vartheta)\, P_l^{m'}\, (\cos\vartheta)\, d(\cos\vartheta)\right]\left[\int_0^{2\pi} e^{i(m-m')\varphi} d\varphi\right] = 0,$$
$$\text{if } m \neq m'.$$

Comparing equations (137) and (148) we see that *the P_l^m are also eigenfunctions of the z-component of the angular momentum*. For given (non-negative integral) l and integral m we have a P_l^m only provided $|m| \leqq l$. This corresponds to the classical fact that the angular momentum in one direction can never be larger than the total angular momentum.

If we now expand an arbitrary function $f(x, y, z)$ or $f(r, \vartheta, \varphi)$ in terms of the P_l^m we have obtained a *transformation to the r, l, m-representation, where l and m classify the eigenvalues of two commuting observables M^2 and M_z,*

$$f(r, \vartheta, \varphi) = \Sigma_l \Sigma_{|m| \leqq l}\, b_{lm}(r) P_l^m(\vartheta, \varphi). \tag{154}$$

The coefficients can be obtained from the equation

$$\int P_l^{m*}(\vartheta, \varphi) f(r, \vartheta, \varphi)\, d\boldsymbol{\omega} = b_{lm}(r) \int P_l^{m*}(\vartheta, \varphi)\, P_l^m\, (\vartheta, \varphi)\, d\boldsymbol{\omega}.$$

The last integral can be evaluated by using equation (148) (we write $\cos\vartheta = \tau$, $d\boldsymbol{\omega} = d\tau\, d\varphi$)

$$\int |P_l^m|^2\, d\boldsymbol{\omega}$$
$$= 2\pi(-1)^m[2^{2l}(l+m)!(l-m)!]^{-1} \int_{-1}^{+1} (d/d\tau)^{l+m}(\tau^2-1)^l (d/d\tau)^{l-m}(\tau^2-1)^l\, d\tau$$
$$= 2\pi(2l)![2^{2l}(l+m)!(l-m)!]^{-1} \int_{-1}^{+1} (1-\tau^2)^l\, d\tau$$
$$= [4\pi/(2l+1)][(l!)^2/(l+m)!(l-m)!] = C_{lm}. \tag{155}$$

We have thus

$$b_{lm} = C_{lm}^{-1} \int P_l^{m*}\, f\, d\boldsymbol{\omega}. \tag{156}$$

From equations (154) and (156) it follows that

$$\int |f|^2 d\mathbf{x} = \int |f|^2 r^2\, dr\, d\boldsymbol{\omega} = \Sigma_{lm} \int_0^\infty |b_{lm}(r)|^2\, C_{lm} r^2\, dr.$$

We can see thus that the P_l^m form an orthogonal set of eigenfunctions of M^2 and M_z which transform r, ϑ, φ into r, l, m and which are normalised with respect to a density function $C_{lm} r^2$ *.

* If we want to retain equation (76) we should write for the eigenfunctions $S_{lm}(r'; r, \vartheta, \varphi) = (r^2 C_{lm})^{-1} \delta(r'-r)\, P_l{}^m(\vartheta, \varphi)$, and instead of equations (154) and (156) we would have $f(r, \vartheta, \varphi) = \Sigma_{lm} \int b_{lm}(r')\, S_{lm}(r'; r, \vartheta, \varphi) r'^2 C_{lm}\, dr'$, $b_{lm} = \int S_{lm}{}^* f d\tau$.

If the operator Ω acting on $f^{(1)}$ produces $f^{(2)}$, $f^{(2)} = \Omega f^{(1)}$ we get after the transformation to r, l, m,

$$\begin{aligned} b^{(2)}_{lm} &= C^{-1}_{lm} \int P^{m*}_l \, \Omega f^{(1)} \, d\boldsymbol{\omega} = C^{-1}_{lm} \Sigma_{l'm'} \int P^{m*}_l \, \Omega \, b^{(1)}_{l'm'} P^{m'}_{l'} \, d\boldsymbol{\omega} \\ &= \Sigma_{l'm'} \Omega_{lm,\, l'm'} b^{(1)}_{l'm'}, \; \Omega_{lm,\, l'm'} = C^{-1}_{lm} \int P^{m*}_l \, \Omega P^{m'}_{l'} \, d\boldsymbol{\omega}. \end{aligned} \tag{157}$$

The $\Omega_{lm,\, l'm'}$ form an infinite matrix with elements which are operators acting on r. In some simple cases they can be reduced to functions of r or even to constants.

We want to discuss in some detail a few operators in the r, l, m-representation. First of all the operators $M_{x\text{op}}$, $M_{y\text{op}}$, $M_{z\text{op}}$ and $M^2{}_{\text{op}}$. According to the theory developed in § 41 M_z and M^2 become diagonal matrices. From equations (153) and (136) we have

$$(M^2)_{lm,\, l'm'} = \hbar^2 l(l+1)\, \delta_{ll'} \delta_{mm'}, \qquad (M_z)_{lm,\, l'm'} = \hbar m \delta_{ll'} \delta_{mm'}. \tag{150}$$

Since M_x and M_y commute with M^2 they will be diagonal with respect to l, but as they do not commute with M_z they cannot be diagonal with respect to m. Since they commute with M^2 it follows that $M_x Y_l(\boldsymbol{\omega})$ and $M_y Y_l(\boldsymbol{\omega})$ are themselves spherical harmonics of the l-th degree [compare the derivation of equations (152) from equation (135)].

We have $M_x T^l = l T^{l-1} (y p_z - z p_y)_{\text{op}} \, (ax+by+cz) = -i\hbar l (yc - zb) T^{l-1}$, $M_y T^l = -i\hbar l (za - xc) T^{l-1}$, $(M_x \pm i M_y) \, T^l = -i\hbar l [c(y \mp ix) + z(-b \pm ia)] \, T^{l-1}$. Using equation (144) we get $(M_x + iM_y) T^l = -i\hbar l [-\xi\eta(-ix+y) + i\xi^2 z] T^{l-1} = -\hbar l \xi (\partial T/\partial\eta) T^{l-1} = -\hbar\xi [\partial(T^l)/\partial\eta]$, $(M_x - iM_y) T^l = -\hbar y [\partial(T^l)/\partial\xi]$. Using equation (147) we get $\Sigma_m (M_x \pm iM_y) \, \xi^{l-m} \eta^{l+m} P_l{}^m = -\Sigma_m \hbar \xi^{l-m\pm 1} \eta^{l+m\mp 1} P_l{}^m (l \pm m)$.

By comparing the coefficients of $\xi^{l-m} \eta^{l+m}$ on both sides of this equation we get the following relations,

$$\begin{aligned} (M_x + iM_y)_{\text{op}} P^{m-1}_l &= -\hbar(l+m) \, P^m_l, \\ (M_x - iM_y)_{\text{op}} P^{m+1}_l &= -\hbar(l-m) \, P^m_l. \quad (P^m_l = 0, \text{ if } |m| > l) \end{aligned} \tag{159}$$

From equations (157) and (159) it follows that all matrix elements are zero except the following ones

$$(M_x \pm iM_y)_{l,m;\, l,m\mp 1} = -\hbar(l \pm m). \tag{160}$$

From equation (160) we can obtain the matrices for M_x and M_y. These are, however, not Hermitean. This is due to the fact that we have normalised the functions and operators in the r, l, m-representation with respect to a density function $C_{lm} r^2$ which depends on m. It is often convenient to take for the density function simply r^2.

This means a slightly different normalisation of the polar spherical harmonics. We do this by introducing $\mathscr{P}_l^m$ by the equations [compare eq. (155)]

$$\mathscr{P}_l^m = (-1)^m C_{lm}^{-1/2} P_l^m, \qquad \int |\mathscr{P}_l^m|^2 d\boldsymbol{\omega} = 1. \tag{161}$$

The factor $(-1)^m$ is introduced to get the equations in their usual form. Using the $\mathscr{P}_l^m$ we find that the factor C_{lm} in equations (156) and (157) disappears; equations (158) remain unchanged and we get instead of equations (159) and (160),

$$\left.\begin{aligned} &(M_x \pm iM_y)_{\text{op}} C_{l,m\mp1}^{1/2} \mathscr{P}_l^{m\mp1} = \hbar(l\pm m) C_{lm}^{1/2} \mathscr{P}_l^m, \\ &(M_x \pm iM_y)_{l,m;\,l,m\mp1} = \hbar C_{l,m}^{1/2} C_{l,m\mp1}^{-1/2} (l\pm m) \\ &= \hbar[(l\pm m)(l\mp m+1)]^{1/2} = \hbar[l(l+1)-m(m\mp1)]^{1/2}. \end{aligned}\right\} \tag{162}$$

$m \backslash m'$	l	$l-1$	$l-2$	...	$-l+1$	$-l$
l	0	$\frac{1}{2}\sqrt{2l\cdot 1}$	0	...	0	0
$l-1$	$\frac{1}{2}\sqrt{2l\cdot 1}$	0	$\frac{1}{2}\sqrt{(2l-1)2}$	...	0	0
$l-2$	0	$\frac{1}{2}\sqrt{(2l-1)2}$	0	...	0	0
.	.	.	.	.	.	.
.	.	.	.	.	.	.
.	.	.	.	.	.	.
$-l+1$	0	0	0	...	0	$\frac{1}{2}\sqrt{1\cdot 2l}$
$-l$	0	0	0	...	$\frac{1}{2}\sqrt{1\cdot 2l}$	0

Fig. 11a.
The matrix representation of $M_x/\hbar$.

From equation (162) we get Hermitean matrices in m, m' for M_x and M_y. In figure 11 we have illustrated the m, m'-matrices of M_x, M_y, and M_z. Using these matrices one sees that the commutation relations (132) hold.

The transformation of the operators $x_{\text{op}}, y_{\text{op}}, z_{\text{op}}, p_{x\text{op}}, p_{y\text{op}}, p_{z\text{op}}$ is obtained from the equations [compare eq. (157)]

$m \backslash m'$	l	$l-1$	$l-2$	$\dots$	$-l+1$	$-l$
l	0	$-\frac{i}{2}\sqrt{2l \cdot 1}$	0	$\dots$	0	0
$l-1$	$\frac{i}{2}\sqrt{2l \cdot 1}$	0	$-\frac{i}{2}\sqrt{(2l-1)2}$	$\dots$	0	0
$l-2$	0	$\frac{i}{2}\sqrt{(2l-1)2}$	0	$\dots$	0	0
.	.	.	.	.	.	.
.	.	.	.	.	.	.
.	.	.	.	.	.	.
$-l+1$	0	0	0	$\dots$	0	$-\frac{i}{2}\sqrt{1 \cdot 2l}$
$-l$	0	0	0	$\dots$	$\frac{i}{2}\sqrt{1 \cdot 2l}$	0

Fig. 11b.
The matrix representation of $M_y/\hbar$.

$m \backslash m'$	l	$l-1$	$l-2$	$\dots$	$-l+1$	$-l$
l	l	0	0	$\dots$	0	0
$l-1$	0	$l-1$	0	$\dots$	0	0
$l-2$	0	0	$l-2$	$\dots$	0	0
.	.	.	.	.	.	.
.	.	.	.	.	.	.
.	.	.	.	.	.	.
$-l+1$	0	0	0	$\dots$	$-l+1$	0
$-l$	0	0	0	$\dots$	0	$-l$

Fig. 11c.
The matrix representation of $M_z/\hbar$.

$$\left.\begin{aligned} zb(r)P_{l-1}^m &= \{[(l-1)/(2l-1)]P_{l-2}^m \\ &\quad +[(l+m)(l-m)/l(2l-1)]P_l^m\}b(r)r, \\ (x\pm iy)b(r)P_{l-1}^m &= \mp\{[(l-1)/(2l-1)]P_{l-2}^{m\pm1} \\ &\quad -[(l\pm m+1)(l\pm m)/l(2l-1)]P_l^{m\pm1}\}b(r)r; \end{aligned}\right\} \quad (163)$$

$$\left.\begin{aligned} (i/\hbar)p_z bP_{l-1}^m &= [(l-1)/(2l-1)]P_{l-2}^m r^{-l}[d(br^l)/dr] \\ &\quad +[(l+m)(l-m)/l(2l-1)]P_l^m r^{l-1}[d(br^{-l+1})/dr], \\ (i/\hbar)(p_x\pm ip_y)b\,P_{l-1}^m &= \mp\{[(l-1)/(2l-1)]P_{l-2}^{m\pm1}r^{-l}[d(br^l)/dr] \\ &\quad -[(l\pm m+1)(l\pm m)/l(2l-1)]P_l^{m\pm1}r^{l-1}[d(br^{-l+1})/dr]\}. \end{aligned}\right\} \quad (164)$$

We derive these equations as follows. Let A, B, and C be three arbitrary numbers and let $V = (Ax+By+Cz)\,T^l$, $W = (Aa+Bb+Cc)r^2\,T^{l-1}$. We then have $\nabla^2 V = 2l(Aa+Bb+Cc)\,T^{l-1}$, and $\nabla^2 W = 2(2l+1)(Aa+Bb+Cc)T^{l-1}$. We see thus that $V-[l/(2l+1)]W$ is a spherical harmonic of the $l+1$-st degree. Introducing ξ and η through equations (144) we have

$$(Ax+By+Cz)T^l = [l/(2l+1)](Aa+Bb+Cc)r^2\,T^{l-1}+[2(l+1)(2l+1)]^{-1}\,DT^{l+1},$$

where D is the differential operator $D = (A+iB)(\partial^2/\partial\xi^2)-2C(\partial^2/\partial\xi\partial\eta) - (A-iB)(\partial^2/\partial\eta^2)$. We now use equation (147) and by comparing the coefficients of A, B, and C and of $\xi^{l-m}\eta^{l+m}$ we find equation (163). If one then uses the formulae $(i/\hbar)p_z f(r)T^l = (\partial/\partial z)f(r)T^l = r^{-1}(df/dr)zT^l+lf(r)cT^{l-1}$, and so on, one obtains equation (164).

From equations (163) and (164) we can immediately calculate the matrix elements of x_{op}, y_{op}, z_{op}, $p_{x\text{op}}$, $p_{y\text{op}}$, $p_{z\text{op}}$ with respect to the P_l^m or the $\mathscr{P}_l^m$. One sees that for given l and l' the dependence on m and m' is the same for z and p_z, for y and p_y, and for x and p_x. We do not want to write down these matrix elements explicitly.

Let us finally consider the transformation of the operator $p_{\text{op}}^2 = -\hbar^2\nabla^2$. From equation (157) it follows that we must evaluate $p_{\text{op}}^2\,b(r)\,P_l^m$. We get

$$\begin{aligned} p_{\text{op}}^2\,b(r)\,P_l^m &= \hbar^2\left\{-\frac{1}{r}\frac{d^2br}{dr^2}+\frac{l(l+1)b}{r^2}\right\}P_l^m \\ &= -\hbar^2\left\{\frac{d^2br^{-l}}{dr^2}+\frac{2(l+1)}{r}\frac{dbr^{-l}}{dr}\right\}r^l\,P_l^m. \end{aligned} \quad (165)$$

Equation (165) is derived as follows

$$\begin{aligned} \nabla^2(f(r)T^l) &= T^l\nabla^2 f(r)+2(\nabla f(r)\cdot\nabla T^l) = [(d^2f/dr^2)+2r^{-1}(df/dr)+2lr^{-1}(df/dr)]T^l \\ &= [(d^2/dr^2)+2(l+1)r^{-1}(d/dr)]fT^l. \end{aligned} \quad (165a)$$

Substituting for f, br^{-l} and T^l expression (147) we get equation (165).

Comparing equation (165) with equation (157) we see that p^2 is diagonal with respect to l and m in the r, l, m-representation, as was to be expected, since p^2 commutes with M^2 and M_z, and we have $(p^2)_{lm,l'm'} = \hbar^2[-r^{-1}(d^2/dr^2)r+l(l+1)r^{-2}]\,\delta_{ll'}\,\delta_{mm'}$. From this formula we get the following identity,

$$p^2_{\rm op} = (-i\hbar)^2 r^{-1}(\partial^2/\partial r^2)r + r^{-2}M^2_{\rm op}, \tag{166}$$

where $\partial/\partial r$ denotes differentiation in the direction of the radius vector. This identity could have been derived elementarily from equation (131) if we remember that $xp_x+yp_y+zp_z = -i\hbar r(\partial/\partial r)$.

If we now express the Laplacian in polar coordinates [for instance, by using eq. (14)],

$$\nabla^2 \equiv \frac{1}{r}\frac{\partial^2}{\partial r^2}r + \frac{1}{r^2}\left[\frac{1}{\sin\vartheta}\frac{\partial}{\partial\vartheta}\sin\vartheta\frac{\vartheta}{\partial\partial} + \frac{1}{\sin^2\vartheta}\frac{\partial^2}{\partial\varphi^2}\right], \tag{167}$$

we find by comparing equations (166) and (167)

$$M^2_{\rm op} = (-i\hbar)^2\left[\frac{1}{\sin\vartheta}\frac{\partial}{\partial\vartheta}\left(\sin\vartheta\frac{\partial}{\partial\vartheta}\right) + \frac{1}{\sin^2\vartheta}\frac{\partial^2}{\partial\varphi^2}\right]. \tag{168}$$

We could have obtained this equation by expressing M_x, M_y, and M_z in polar coordinates,

$$M_x = -i\hbar\left[-\sin\varphi\frac{\partial}{\partial\vartheta} - \cot\vartheta\cos\varphi\frac{\partial}{\partial\varphi}\right],$$

$$M_y = -i\hbar\left[\cos\varphi\frac{\partial}{\partial\vartheta} - \cot\vartheta\sin\varphi\frac{\partial}{\partial\varphi}\right],\quad M_z = -i\hbar\frac{\partial}{\partial\varphi}.$$

The sum of the squares of these operators gives us expression (168). The eigenfunctions of M^2 could thus have been obtained from the differential equation

$$(-i\hbar)^2\left[\frac{1}{\sin\vartheta}\frac{\partial}{\partial\vartheta}\left(\sin\vartheta\frac{\partial S}{\partial\vartheta}\right) + \frac{1}{\sin^2\vartheta}\frac{\partial^2 S}{\partial\varphi^2}\right] = (M^2)_{\rm e}\,S,$$

by using the method of the separation of coordinates,

$$S = f(\vartheta)e^{im\varphi},\quad \frac{1}{\sin\vartheta}\frac{d}{d\vartheta}\left(\sin\vartheta\frac{df}{d\vartheta}\right) - \frac{m^2}{\sin^2\vartheta}f = -\frac{1}{\hbar^2}(M^2)_{\rm e}f.$$

A discussion of this equation shows that it has solutions which are everywhere finite, only if M^2 is equal to one of its eigenvalues $\hbar^2 l(l+1)$ $(l = |m|, |m|+1, \ldots, \infty)$. The solutions S which we obtain in this way are just the functions $P_l^m(\vartheta, \varphi)$ defined by equation (148).

§ 46. A PARTICLE IN A CENTRAL FIELD OF FORCE; THE HYDROGEN ATOM

Consider a particle in a three-dimensional field of force, derivable from a potential $U(\boldsymbol{x})$. We can introduce the probability amplitude $a_{lm}(r, t)$ with respect to the polar spherical harmonics [see eq. (154)],

$$\begin{aligned} \psi(r, \vartheta, \varphi, t) &= \Sigma_{lm} a_{lm}(r, t)\, P_l^m(\vartheta, \varphi), \\ 1 = \int |\psi|^2 d\tau &= \Sigma_{lm} \int_0^\infty |a_{lm}|^2\, C_{lm} r^2\, dr. \end{aligned} \tag{169}$$

We find that $|a_{lm}|^2 C_{lm} r^2 dr$ is the probability that the particle is at a distance between r and $r+dr$ from the origin, while M^2 has the value $\hbar^2 l(l+1)$ and M_z the value $m\hbar$. The energy operator is $H_{\mathrm{op}} = (2m)^{-1} p_{\mathrm{op}}^2 + U$ and the transformed Schrödinger equation is in the form of an infinite set of simultaneous equations. From equations (166), (158), and (157) we find

$$\begin{aligned} (2m)^{-1}\{(-i\hbar)^2 r^{-1}[\partial^2(a_{lm} r)/\partial r^2] + \hbar^2 l(l+1) r^{-2} a_{lm}\} \\ + \Sigma_{l'm'} U_{lm,l'm'}\, a_{l'm'} = i\hbar \partial a_{lm}/\partial t. \end{aligned} \tag{170}$$

If we are dealing with central forces, that is, if U is a function of r only, $U_{lm,\, l'm'}$ is a diagonal matrix $U(r)\delta_{ll'}\, \delta_{mm'}$ and we have to solve only one equation for each l,

$$\begin{aligned} -(\hbar^2/2m)\, r^{-1}[\partial^2(a_{lm}\, r)/\partial r^2] \\ + (\hbar^2/2m) l(l+1)\, r^{-2}\, a_{lm} + U a_{lm} = i\hbar \partial a_{lm}/\partial t. \end{aligned} \tag{171}$$

This equation has the form of a one-dimensional Schrödinger equation which contains the total angular momentum quantum number l as a parameter.

This procedure is completely analogous to the classical reduction of the motion in a central field of force to the one-dimensional problem of radial motion. If we denote the probability amplitude multiplied by r by R_{lm}, $R_{lm}(r, t) = a_{lm}(r, t) r$, we get the following equation for $R_{lm}(r, t)$,

$$\begin{aligned} i\hbar \partial R/\partial t &= H_{\mathrm{op}}\, R \\ &= (2m)^{-1}\{(-i\hbar)(\partial^2/\partial r^2) + \hbar^2 l(l+1) r^{-2}\} R + U(r) R. \end{aligned} \tag{172}$$

In classical mechanics the energy has the following form in polar coordinates

$$H = (2m)^{-1}(p_r^2 + M^2 r^{-2}) + U, \tag{173}$$

where p_r is the momentum which is canonically conjugate to r

($p_r = m\dot{r}$) and M^2r^{-2} is the potential energy of the centripetal force. One can thus obtain the Schrödinger equation (172) from the classical energy (173) by using the same rule as in the one-dimensional case ($H \to H_{\text{op}}$; $p_r \to -i\hbar\partial/\partial r$) provided one substitutes for M^2 its correct quantum mechanical eigenvalue.

Equation (172) can be used to discuss the eigenvalues and eigenfunctions of the problem of a particle in a central field of force. We could have considered this problem also without introducing explicitly the transformed Schrödinger equation. The original Schrödinger equation has the form

$$\nabla^2\varphi + (2m/\hbar^2)[E - U(r)]\varphi = 0. \tag{174}$$

The energy operator commutes with M^2 (since U depends on r only) and one can look for solutions of the form

$$\varphi = b_l(r)Y_l(\boldsymbol{\omega}),$$

$$\text{or, } \varphi = f_l(r)(ax+by+cz)^l, \quad b_l(r) = r^l f_l(r), \quad a^2+b^2+c^2 = 0. \tag{175}$$

Substitution of expression (175) into equation (174) leads to a differential equation for b_l or f_l. If we use the expression involving f_l and use equation (165a) we find

$$(d^2f_l/dr^2) + 2(l+1)r^{-1}(df_l/dr) + (2m/\hbar^2)[E - U(r)]f_l = 0. \tag{176}$$

From equation (176) it follows that the „radial part" of the wave function b_l satisfies the equation

$$[d^2(b_l r)/dr^2] + \{(2m/\hbar^2)[E - U(r)] - l(l+1)r^{-2}\}b_l r = 0, \tag{177}$$

which is the eigenvalue problem (172).

We can discuss equation (177) qualitatively in the same way as we discussed equation (2.30) in § 17. We must bear in mind that r varies only between 0 and ∞. Each eigenvalue E of equation (177) corresponds to a $2l+1$-fold degenerate eigenvalue of the original problem, because of the properties of the spherical harmonics. It corresponds to a stationary state of the particle in a central field of force with the total angular momentum quantum number equal to l. In the context of the correspondence principle we can compare this state with a classical motion having an angular momentum $l\hbar$ perpendicular to the orbital plane. The $2l+1$-fold degeneracy corresponds to the different possible orientations of the orbital plane. If we choose one of the polar spherical harmonics $P_l^m(\vartheta, \varphi)$ ($m = l, l-1, \ldots, -l$) for the $Y_l(\boldsymbol{\omega})$ in equation (175) we have a situation where the angular momentum with respect to the z-axis is equal to $m\hbar$ which means classically a definite orientation of the orbital plane with respect to the z-axis. If we had chosen for $Y_l(\boldsymbol{\omega})$ an arbitrary spherical harmonic we could no longer have shown the analogy between the given situation and a classical, definitely orientated orbital plane.

The case of an attractive force inversely proportional to the square of the distance to the origin corresponds to the *problem of an electron in the field of a positively charged nucleus* (hydrogen atom, helium ion, and so on), where the nucleus is taken as the origin. The potential energy in this case the *Coulomb potential*,

$$U(r) = -Ze^2/r. \tag{178}$$

In equation (178) Ze is the nuclear charge and $-e$ the electronic charge, while m is now the electronic mass. If we consider the problem rigorously as a two-body problem we also obtain equation (174), but then m is the reduced mass,

$$m = m_{\text{nucl}}\, m_{\text{el}}/(m_{\text{nucl}} + m_{\text{el}}). \tag{179}$$

and x, y, z are the relative coordinates ($\boldsymbol{x} = \boldsymbol{x}_{\text{el}} - \boldsymbol{x}_{\text{nucl}}$).

From equation (3.26) it follows that the two-body Schrödinger equation has the form

$$(-i\hbar)^2[(2m_1)^{-1}\,\nabla_1{}^2\psi + (2m_2)^{-1}\,\nabla_2{}^2\psi] + U(r)\psi = i\hbar\partial\psi/\partial t;\ r^2 = (\boldsymbol{x}_1 - \boldsymbol{x}_2 \cdot \boldsymbol{x}_1 - \boldsymbol{x}_2), \tag{180}$$

where $\boldsymbol{x}_1$, m_1 and $\boldsymbol{x}_2$, m_2 are the positions and masses of the electron and the nucleus respectively. Introducing centre of gravity coordinates $(m_1+m_2)\boldsymbol{X} = m_1\boldsymbol{x}_1 + m_2\boldsymbol{x}_2$ and relative coordinates $\boldsymbol{x} = \boldsymbol{x}_1 - \boldsymbol{x}_2$ we get instead of equation (180)

$$\tfrac{1}{2}(-i\hbar)^2[(m_1+m_2)^{-1}\,\nabla_{\text{cog}}{}^2\,\psi + \{(m_1+m_2)/m_1m_2\}\,\nabla_{\text{r}}{}^2\,\psi] + U(r)\psi = i\hbar\partial\psi/\partial t,$$

$$\nabla_{\text{cog}}{}^2 = (\partial^2/\partial X^2) + (\partial^2/\partial Y^2) + (\partial^2/\partial Z^2), \quad \nabla_{\text{r}}{}^2 = (\partial^2/\partial x^2) + (\partial^2/\partial y^2) + (\partial^2/\partial z^2).$$

Using the method of separation of variables, $\psi = \psi_{\text{cog}}(\boldsymbol{X}, t)\psi_r(\boldsymbol{x}, t)$, we get

$$\frac{(-i\hbar)^2}{2(m_1+m_2)}\,\nabla_{\text{cog}}{}^2\,\psi_{\text{cog}} = i\hbar\,\frac{\partial\psi_{\text{cog}}}{\partial t},\quad \frac{2m_1m_2}{m_1+m_2}(-i\hbar)^2\nabla_{\text{r}}{}^2\psi_{\text{r}} + U(r)\,\psi_{\text{r}} = i\hbar\,\frac{\partial\psi_{\text{r}}}{\partial t.} \tag{181}$$

The first equation is the Schrödinger equation of a free particle of mass m_1+m_2. The second equation is the Schrödinger equation of a particle with the reduced mass $m = m_1m_2/(m_1+m_2)$ in the potential $U(r)$. We have here the analogy to the classical *elimination of the motion of the centre of gravity*.

We write as in § 17 equation (177) in the form

$$\frac{d^2R}{dr^2} + C(E-V)\,R = 0,\ R = rb(r),\ C = \frac{2m}{\hbar^2},\ V = -\frac{Ze^2}{r} + \frac{l(l+1)}{Cr^2.} \tag{182}$$

The qualitative behaviour of $C(E-V)$ is indicated in figure 12.

If l is positive the considerations of §§ 17 and 23 give the following results. *If E is negative there will be a set of discrete energy values,* which are all larger than the minimum value V_m of V. The corresponding eigenfunctions go to zero both as $r \to 0$ and as $r \to \infty$.

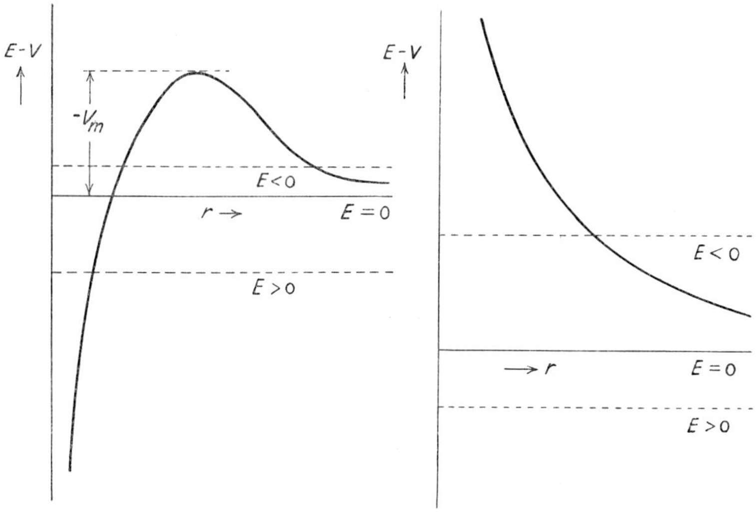

Fig. 12

These eigenfunctions can be distinguished by the number of zeros of R. *Any positive value of E, however, is an improper eigenvalue.* The corresponding eigenfunctions go to zero as $r \to 0$ and show oscillatory behaviour as $r \to \infty$.

The case $l = 0$ must be considered separately since $E - V \to \infty$ as $r \to 0$. Trying to solve equation (182) by a series expansion, $R = \lambda^{\lambda}(a_0 + a_1 r + \ldots)$ gives us for λ the equation $\lambda(\lambda - 1) = l(l+1)$, or, $\lambda = l+1$ or $-l$ *. If $\lambda = l+1$ all a_k can be expressed in terms of a_0 and one obtains for each value of E a solution which vanishes as r^{l+1} for $r \to 0$. There is thus also for $l = 0$ a solution which goes to zero as $r \to 0$. We must assume that the eigenfunctions correspond to this solution. We see that again the deliberations of §§ 17 and 23 lead to the result that for $E < 0$ there exists a discrete set of eigenvalues and that any positive E is an improper eigenvalue.

If $R_1 = r(a_0 + a_1 r + \ldots)$ is this particular solution of equation (182) for $l = 0$, the most general solution is $R = R_1(r) \int^r dr' [R_1(r')]^{-2} = R_1(r) \int^r r'^{-2} \times [a_0^{-2} - 2a_1 a_0^{-3} r' + \ldots] dr' = -2a_1 a_0^{-3} R_1 [\ln r + \text{constant}] + [-a_0^{-1} + b_1 r + b_2 r^2 + \ldots]$. This solution is finite for $r = 0$ and for any negative value of E one can construct a

* This is true as long as $rU(r)$ does not go to infinity as $r \to 0$.

solution which goes exponentially to zero as $r \to \infty$ and which remains finite as $r \to 0$. These solutions correspond to normalisable solutions of the three-dimensional problem, since we have from equation (175) $\int |\varphi|^2 d\tau = \int_0^\infty |b_l(r)|^2 r^2 dr . \int (Y_l(\boldsymbol{\omega})|^2 d\boldsymbol{\omega} = \int_0^\infty R^2(r)\, dr \cdot \int |Y_l(\boldsymbol{\omega})|^2 d\boldsymbol{\omega}$ and the integral $\int |Y_l(\boldsymbol{\omega})|^2 d\boldsymbol{\omega}$ is always finite. However, there are several reasons to discard these solutions in general. The circumstance that the wave function itself, $\varphi = bY = RY/r$, goes to infinity as r^{-1} is by itself not a sufficient physical reason. However, we can consider the potential (178) to be the limiting case $(r_0 \to 0)$ of the potential

$$U(r) = +\infty \qquad (r \leqq r_0), \qquad U(r) = -Ze^2/r \qquad (r > r_0) \tag{183}$$

corresponding to the nucleus being a small billiard ball of radius r_0. In the case of the potential (183) $R(r)$ must be equal to zero for $r = r_0$ and thus in the limit go to zero as $r \to 0$. This is still not a satisfactory argument and neither is the argument that the general solution $R(r)$ does not form part of an orthonormal set, as one can show (compare the proof of orthonormality in § 19). If we choose for $U(r)$ $(r < r_0)$ a different behaviour than in equation (183) [for instance, $U(r) = -C$, with C a large and positive constant and $C \to \infty$ as $r_0 \to 0$] we can arrange it so that the eigenfunctions for $r > r_0$, $l = 0$ do not coincide with R_1 in the limit $r_0 \to 0$. The assumption that they should behave like R_1 is mathematically the most natural one and gives the right result, namely the one agreeing with observations. We may add that this solution also goes over into the corresponding solution of the case of a particle in a central field of force in the Dirac theory.

The quantitative solution of equation (182) we obtain in the following way. We first of all define

$$r = \varrho a_1, \quad a_1 = \hbar^2/Ze^2 m, \quad E = \varepsilon W_1, \quad W_1 = Z^2 e^4 m/2\hbar^2. \tag{184}$$

We then get instead of equation (182)

$$(d^2R/d\varrho^2) + [\varepsilon + 2\varrho^{-1} - l(l+1)\,\varrho^{-2}]R = 0. \tag{185}$$

Substitution of

$$R = \varrho^{l+1} \exp\,[\pm\varrho\sqrt{(-\varepsilon)}]\; S(\varrho), \tag{186}$$

where we require $S(\varrho)$ to remain finite as $\varrho \to 0$, gives us the following equation for $S(\varrho)$,

$$S'' + [\pm 2(-\varepsilon)^{1/2} + 2(l+1)\varrho^{-1}]S' + [2 \pm 2l(l+1)(-\varepsilon)^{1/2}]\varrho^{-1}S = 0. \tag{187}$$

We now define a function ${}_1F_1(a, c; x)$ by the following series which is convergent for all values of x,

$${}_1F_1 = 1 + ax/1\cdot c + [a(a+1)/1\cdot 2\cdot c(c+1)]x^2 + \ldots \tag{188}$$

If the real part of x goes to $+\infty$, ${}_1F_1$ behaves asymptotically as $x^{a-c}e^x$, except when a is a non-positive integer. It satisfies the differential equation

$${}_1F_1'' + (cx^{-1} - 1)\,{}_1F_1' - ax^{-1}\,{}_1F_1 = 0. \tag{189}$$

Comparing equations (187) and (189) we see that the solution we were looking for is given by the equation

$$S = \text{constant}\ {}_1F_1(\pm(-\varepsilon)^{-1/2}+l+1,\ 2(l+1);\ \mp 2\varrho(-\varepsilon)^{1/2}). \tag{190}$$

If we now take ε to be negative and if we choose in equations (186), (187) and (190) the lower sign, we find that in general R will go to infinity as $\varrho^{-(-\varepsilon)^{-\frac{1}{2}}}$ exp $(\varrho(-\varepsilon)^{1/2})$ as $\varrho \to \infty$, except when $-(-\varepsilon)^{-1/2}+l+1$ is a non-positive integer. In that case, the series for S breaks off and becomes a finite polynomial. The function R goes exponentially to zero as $r \to \infty$ and is thus an eigenfunction.

If we put

$$(-\varepsilon)^{-1/2} = n, \quad n = l+1, l+2, \ldots, \tag{191}$$

n is the *principal or Bohr quantum number* of the corresponding stationary state. It is at least one larger than the angular momentum quantum number l, and $n-l-1$ is the number of zeros of the radial part of the wave function. We have for the energy

$$\varepsilon_n = -n^{-2},\ E_n = -W_1 n^{-2} = -Z^2e^4m/2n^2\hbar^2. \tag{192}$$

This formula was given by Bohr * in his theory of the hydrogen spectrum. The negative sign means that it takes a positive amount of work to remove the electron to infinity. The electron is bound in the field of the nucleus and the situation is analogous to the classical situation where the particle is moving in a Kepler ellipse. The values $n = 1$, $E_1 = -W_1$ correspond to the lowest state of the system. This state is called the *ground state* of the system.

An essential point is that the energy does not depend on l. For a given value of n, l can take on the values $0, 1, \ldots, n-1$, all corresponding to different wave functions but the same energy. The negative energy eigenvalues of a particle in a Coulomb field are thus even worse degenerate than we would expect in the general case of a central force problem (usual degeneracy $2l+1$). The total degree of degeneracy g_n of the state with principal quantum number n is given by the equation

$$g_n = \Sigma_{l=0}^{n-1}(2l+1) = n^2. \tag{193}$$

According to classical mechanics the orbit of a particle moving in the Coulomb field (178) will be a conic section with the centre of attraction in one of the foci. A situation with negative energy cor-

* N. Bohr, *Phil. Mag.*, **26**, 1, 1913; *Nature*, **92**, 231, 1913.

responds in particular to an elliptical *Kepler orbit.* The major axis of such an orbit and its period are fixed by the value of the energy, and do not depend on the value of the angular momentum. From classical mechanics we have

$$a = Ze^2/2|E|, \quad a(1-\eta^2) = M^2/Ze^2m, \tag{194}$$

where a is the semi-major axis and η the eccentricity ($\eta = 0$: circular orbit). In this case M^2 could take on all values from 0 to $Z^2e^4m/2|E|$. If we illustrate, as Bohr did in 1913, a stationary state with principal quantum number n by a classical orbit of the same energy (*Bohr orbit*) this maximum value of M^2 is equal to $(n\hbar)^2$ [see eq. (192)]. This is clearly the classical analogy of the restriction of l-values for given n.

The semi-major axis of a classical orbit with energy E_n is according to equations (194) and (154) equal to

$$a_n = n^2\hbar^2/Ze^2m = n^2a_1. \tag{195}$$

In the case of a circular orbit a_n is equal to the radius of the circle, and in particular a_1 *is equal to the radius of the first circular Bohr orbit.* From equation (196) below one can easily check that n^2a_1 gives us the order of magnitude of the region in space in which the wave function is appreciably different from zero.

Substituting expression (191) into equation (190) we find the following expression for the radial part R_{nl} of the wave function,

$$\begin{aligned} R_{nl} &= -(\tfrac{1}{2}n)^{2l+1}\,\varrho^{l+1}\,e^{\varrho/n}\left(\frac{d}{d\varrho}\right)^{n+l}\varrho^{n-l-1}\,e^{-2\varrho/n} \\ &= \varrho^{-l}\,e^{\varrho/n}\left(\frac{d}{d\varrho}\right)^{n-l-1}\varrho^{n+l}\,e^{-2\varrho/n}, \quad \varrho = \frac{r}{a_1}. \end{aligned} \tag{196}$$

From this equation we see immediately that R_{nl} has $n-l-1$ zeros. It is often necessary to know the normalisation factor. We have

$$\int_0^\infty |R_{nl}|^2\,d\varrho = 4(\tfrac{1}{2}n)^{2l+4}\{(n-l-1)![(2l+1)!]^2/(n+l)!\}. \tag{197}$$

If we now choose ε to be positive, we are dealing with the continuum part of the spectrum ($E=\varepsilon W_1$). The electron is "free" and the situation is analogous to the classical hyperbolic orbit, in which the particle approaches the nucleus with a finite velocity, coming from infinity, and disappearing with the same velocity into infinity. The de Broglie wave number of a free particle with energy εW_1 will be denoted by $\varkappa/2\pi$, and we have

$$\varkappa = (2m\varepsilon W_1)^{1/2}/\hbar = \varepsilon^{1/2}/a_1. \tag{198}$$

We see from this equation that $\varkappa/2\pi$ is the wave number which characterises the Schrödinger wave functions at large distances from the attracting centre. The argument $\varrho(-\varepsilon)^{1/2}$ in equation (190) becomes $i\varkappa r$ and using equation (190) we can write for the radial part of the wave function

$$R_l(\varkappa, r) = (\varkappa r)^{l+1} e^{\pm i\varkappa r} {}_1F_1(\mp i(\varkappa a_1)^{-1}+l+1,\ 2l+2;\ \mp 2i\varkappa r). \tag{199}$$

This expression is a function of r which, apart from a (possibly complex) constant factor, should be real for all values of r between 0 and ∞ and which should behave asymptotically for large values of r as $\cos(\varkappa r-\beta)$ (compare § 23). One finds, indeed, that the first term in an asymptotic series expansion in inverse powers of r is given by the equation (Γ^{-1} is the reciprocal gamma function),

$$R_l(\varkappa, r) \to 2^{-l} e^{-\frac{1}{2}\pi\sigma}(2l+1)!\,\mathrm{Re}\{\Gamma^{-1}(i\sigma+l+1)\exp i[\varkappa r+\sigma\ln(2\varkappa r)-\tfrac{1}{2}\pi(l+1)]\},$$
$$\sigma^{-1} = \varkappa a_1 = \varepsilon^{1/2}. \tag{200}$$

One obtains equation (200) by using the integral representation of ${}_1F_1$,*

$${}_1F_1(a,\ c;\ x) = \text{const.} \oint z^{a-1}(1-z)^{c-a-1}e^{xz}\,dz, \tag{201}$$

where the integration is for integral c over a closed contour which encircles 0 and 1. The constant is obtained from the requirement ${}_1F_1 \to 1$ as $x \to 0$. To obtain equation (200) (x large and positive imaginary) one changes the contour until it consists of two loops both going to infinity in the direction of the positive imaginary axis and encircling respectively the points 0 and 1. One then expands $(1-z)^{c-a-1}$ along the first loop in powers of z and one expands z^{a-1} along the second loop in powers of $1-z$.

The occurrence of the term $\sigma\ln(2\varkappa r)$ in the exponent means that even for large values of r the phase increases more strongly than $\varkappa r$.

The states of positive energy are clearly infinitely degenerate, as for a given value of $\varkappa$ l can still take on any value from 0 to ∞.

The special case where the energy of the particle is zero corresponds to the parabolic orbits of classical mechanics. In this case the radial part of the wave function contains a Bessel function. Equation (185) is now of the form $R''+[2\varrho^{-1}-l(l+1)\varrho^{-2}]R = 0$ with the solution **

$$R = \varrho^{1/2} J_{2l+1}[(8\varrho)^{1/2}], \tag{202}$$

which for large values of ϱ behaves as follows,

$$R \to (\varrho/2\pi^2)^{1/4} \cos[(8\varrho)^{1/2}-(l+\tfrac{3}{4})\pi]. \tag{203}$$

* See, for instance, H. and B. S. Jeffreys, *Methods of Mathematical Physics*, Cambridge University Press, 1946, p. 576.

** See, for instance, E. Jahnke and F. Emde, *Tables of functions*, Dover, New York, 1945.

CHAPTER V

PERTURBATION THEORY

§ 47. Introduction

Once we know an approximate solution of a quantum mechanical problem, there are several methods which enable us to obtain a more accurate solution. These methods which vary from problem to problem are collectively called *perturbation theory*. Sometimes we are dealing with an expansion of the eigenfunctions and eigenvalues of a system in powers of one or several small parameters which occur in the energy (for instance, in the case of atoms or molecules in homogeneous magnetic or electric fields, one expands in powers of the field strengths). At other times we want to describe the gradual change of a physical situation which occurs because of a small interaction ("coupling") with another system [for instance, collisions of atoms with fast particles (stopping power problems) or the interaction between atoms and radiation]. Again at other times we want to investigate the interaction within a system consisting of many particles (for instance, the theory of imperfect gases.) We could go on giving more examples. In all cases we are interested as far as the mathematics of the perturbation problems are concerned in finding an approximate solution of a Schrödinger equation, the exact solution of which is either unobtainable or too complicated to interpret. Both in the choice of the kind of method to be used as in the interpretation of the formulae obtained physical points of view are the deciding factor. This one can see, for instance, from the fact that one can often use with great advantage expansions in terms of a small parameter which are either not convergent, or the second term of which is already large. In most cases the perturbation methods of quantum theory have their immediate counterparts in the perturbation theory of classical mechanics, even if the individual steps in the calculations may be completely different.

In the present chapter we shall give a general discussion of the more important perturbation methods. At the same time we shall use the opportunity to discuss a few important quantum mechanical problems, such as the influence of „adiabatic" processes and the question of the time average of an observable.

§ 48. The perturbation of a non-degenerate discrete stationary state *

We want to find the stationary states and the energy eigenvalues of a system with an energy function

$$H = H^{(0)} + \lambda H^{(1)} + \lambda^2 H^{(2)} + \ldots \tag{1}$$

The right hand side of this equation is a finite or infinite series and λ is a "small" real parameter of such a nature that each term can be considered to be large compared to the next one. The system is called "unperturbed" when $\lambda = 0$, and the terms with λ, λ^2, ... are called the "perturbation energy". The $H^{(n)}$ are functions of the coordinates and momenta and they correspond to Hermitean operators $H^{(n)}_{\text{op}}$ which we shall simply denote by $H^{(n)}$. The coordinates q_k which are the argument of the wave functions will be generalised Lagrangian coordinates. If H is originally expressed in the Cartesian coordinates and momenta we can transform the $H^{(n)}$ by the methods described in § 34 in such a way that they operate on functions of the q_k.

We expand the eigenfunctions φ and eigenvalues E in terms of λ,

$$\varphi = \varphi^{(0)} + \lambda\varphi^{(1)} + \lambda^2 \varphi^{(2)} + \ldots, \qquad E = E^{(0)} + \lambda E^{(1)} + \lambda^2 E^{(2)} + \ldots \tag{2}$$

The time independent Schrödinger equation is of the form

$$(H-E)\varphi \equiv \{\Sigma_n \lambda^n (H^{(n)} - E^{(n)})\} \{\Sigma_m \lambda^m \varphi^{(m)}\} = 0.$$

Equating the coefficients of λ^n on both sides we get the following set of differential equations,

$$[H^{(0)} - E^{(0)}]\,\varphi^{(0)} = 0; \tag{3a}$$

$$[H^{(0)} - E^{(0)}]\,\varphi^{(1)} = -[H^{(1)} - E^{(1)}]\,\varphi^{(0)}; \tag{3b}$$

$$[H^{(0)} - E^{(0)}]\,\varphi^{(2)} = -[H^{(1)} - E^{(1)}]\,\varphi^{(1)} - [H^{(2)} - E^{(2)}]\,\varphi^{(0)}; \tag{3c}$$

. . . .

$$[H^{0)} - E^{(0)}]\,\varphi^{(n)} = -\,\Sigma_{m=1}^{n}\,[H^{(m)} - E^{(m)}]\,\varphi^{(n-m)}. \tag{3d}$$

Equation (3a) is a homogeneous differential equation which is the Schrödinger equation of the "unperturbed" system. We shall to begin with assume that this system possesses only discrete non-de-

* See, E. Schrödinger, *Ann. Phys.*, **80**, 437, 1926.

generate energy eigenvalues $E_k^{(0)}$ with corresponding normalised eigenfunctions $\varphi_k^{(0)}$. We shall further assume that the $E_k^{(0)}$ and the $\varphi_k^{(0)}$ are known.

As an example we can consider the case of a one-dimensional system of a particle of mass m in the potential field

$$U = ax^2 + \lambda bx^3 + \lambda^2 cx^4 + \ldots \qquad (a > 0) \tag{4}$$

The unperturbed system corresponds to an energy function $(p^2/2m) + ax^2$, that is to a linear harmonic oscillator with a discrete eigenvalue spectrum, the eigenvalues and eigenfunctions of which are known (§ 17).

Equations (3b), (3c), . . . are inhomogeneous differential equations which can be used to determine successively higher terms in equation (2). We first of all multiply equation (3b), where $E^{(0)}$ and $\varphi^{(0)}$ refer to a definitely given stationary state, by $\varphi^{(0)*}$ and integrate over all coordinates (if necessary we also multiply by a suitable density function),

$$\int \varphi^{(0)*}[H^{(0)} - E^{(0)}]\varphi^{(1)}\,d\tau = -\int \varphi^{(0)*}[H^{(1)} - E^{(1)}]\varphi^{(0)}d\tau, \tag{5}$$
$$d\tau = \varrho\, dq_1 dq_2 \ldots$$

Since $H^{(0)}$ is Hermitean and $\varphi^{(0)}$ normalised we can write this equation in the form

$$\int \varphi^{(1)}[\, H^{(0)*} - E^{(0)}]\, \varphi^{(0)*}\, d\tau = -\overline{H^{(1)}} + E^{(1)},$$

where $\overline{H^{(1)}}$ is the expectation value of $H^{(1)}$ in the unperturbed stationary state (§ 30). The left hand side is equal to zero [see eq. (3a)] and we get

$$E^{(1)} = \overline{H^{(1)}}, \tag{6}$$

or, *the change in the energy eigenvalue is to a first approximation equal to the expectation value of the first term of the perturbation energy in the unperturbed state.*

We know now the right hand side of the inhomogeneous equation (3b). Its solution can thus be obtained apart from an additive term which is a solution of the homogeneous equation, that is, apart from a term $a\varphi^{(0)}$. We shall fix $\varphi^{(1)}$ by requiring it to be orthogonal to $\varphi^{(0)}$,

$$\int \varphi^{(0)*}\, \varphi^{(1)}\, d\tau = 0. \tag{7}$$

This requirement has the advantage that $\varphi^{(0)} + \lambda\varphi^{(1)}$ is normalised not only in the zeroth, but also in the first approximation. This would also be the case, if we had taken $\varphi = \varphi^{(0)}+\lambda(a\varphi^{(0)}+\varphi^{(1)})$ with a being purely imaginary, but in that case $\varphi^{(0)}+a\lambda\varphi^{(0)}$ would be the same as $\varphi^{(0)}$, but for a phase factor, and we are essentially dealing with a wave function $\varphi^{(0)}+\lambda\varphi^{(1)}$.

We shall discuss later on how we can determine $\varphi^{(1)}$ from equation (3b). Let us for the moment assume that it has been determined and let us consider the solution of equation (3c). First of all we determine $E^{(2)}$ by the same method which was used to determine $E^{(1)}$,

$$\int \varphi^{(0)*}[H^{(0)}-E^{(0)}]\,\varphi^{(2)}\,d\tau$$
$$= -\int \varphi^{(0)*}[H^{(1)}-E^{(1)}]\,\varphi^{(1)}\,d\tau - \int \varphi^{(0)*}[H^{(2)}-E^{(2)}]\varphi^{(0)}\,d\tau.$$

The left hand side of this equation is zero since $H^{(0)}$ is Hermitian. As $\varphi^{(1)}$ is orthogonal to $\varphi^{(0)}$, the term with $E^{(1)}$ vanishes and we find

$$E^{(2)} = \int \varphi^{(0)*}\,H^{(1)}\,\varphi^{(1)}\,d\tau + \overline{H^{(2)}}. \tag{8}$$

The second term is the expectation value of the second term of the perturbation energy in the unperturbed state. The first term is equal to half of the factor of λ in the expectation value of $H^{(1)}$ in the perturbed state *.

We now know the right hand side of equation (3c) and we may assume that we have found a solution $\varphi^{(2)}$. This solution is determined, but for an additive term $a\varphi^{(0)}$. Once again we shall require $\varphi^{(2)}$ to be orthogonal to $\varphi^{(0)}$,

$$\int \varphi^{(0)*}\,\varphi^{(2)}\,d\tau = 0. \tag{9}$$

The normalising integral of the approximate expression $\varphi = \varphi^{(0)} + \lambda\varphi^{(1)}+\lambda^2\varphi^{(2)}$ will then up to terms in λ^2 be equal to $1+\lambda^2\int\varphi^{(1)*}\varphi^{(1)}\,d\tau$.

We can obviously continue this procedure as long as we want to. We first determine from equation (3d) $E^{(n)}$ by multiplying it by $\varphi^{(0)*}$ and integrating. We then solve for $\varphi^{(n)}$ and require $\varphi^{(n)}$ to be orthogonal to $\varphi^{(0)}$.

The differential equations (3b), (3c), . . . are of the form

$$\int [H^{(0)}-E^{(0)}]\chi = K, \tag{10}$$

where K is a known function of the coordinates which is orthogonal to the (quantum mechanical) solution $\varphi^{(0)}$ of the homogeneous solution.

* This expectation value is equal to $\int [\varphi^{(0)*}+\lambda\varphi^{(1)*}]\,H^{(1)}[\varphi^{(0)}+\lambda\varphi^{(1)}]d\tau$ and the factor of λ is $[\int\varphi^{(1)*}\,H^{(1)}\,\varphi^{(0)}\,d\tau + \int\varphi^{(0)*}\,H^{(1)}\,\varphi^{(1)}\,d\tau]$. Both integrals are real and they have the same value, since they are equal to $-\int\varphi^{(1)*}[H^{(0)}-E^{(0)}]\,\varphi^{(1)}\,d\tau$ [see equation (3b)].

Only in special cases is it possible to obtain a solution of equation (10) in closed form, say, by quadrature. An example where this is possible is when $H^{)0)} = (2m)^{-1}(-i\hbar)^2(d^2/dq^2)+U(q)$ (one-dimensional case), where we have the following solution,

$$\chi = -(2m/\hbar^2)\,\varphi\int^q \varphi^{-2}dq'\int_{-\infty}^{q'} K\varphi\, dq'', \tag{11}$$

where φ is the (quantum mechanical) solution of the homogeneous equation. One must be extremely careful in using equation (11) because of possible zeros of φ.

The most common method of finding a solution of equation (10) is the following one. Let $E^{(0)}$ be the eigenvalue $E_k^{(0)}$ corresponding to the k-th stationary state of the unperturbed system,

$$[H^{(0)}-E_k^{(0)}]\chi = K. \tag{12}$$

Expand now χ in terms of the eigenfunctions $\varphi_l^{(0)}$ of $H^{(0)}$,

$$\chi = \Sigma_l\, a_{lk}\, \varphi_l^{(0)}. \tag{13}$$

Multiplying equation (12) by $\varphi_l^{(0)*}$ and integrating we have

$$\begin{gathered}\Sigma_{l'}\, a_{l'k}\int \varphi_l^{(0)*}[H^{(0)}-E_k^{(0)}]\varphi_{l'}^{(0)}\, d\tau = \int \varphi_l^{(0)*}K\, d\tau,\\ a_{lk}[E_l^{(0)}-E_l^{(0)}] = \int\varphi_l^{(0)*}K\, d\tau = K_l.\end{gathered} \tag{14}$$

We see that $K_k = 0$.

From equation (14) we see that the a_{lk} $(k \neq l)$ are determined. By requiring the solution of equation (12) to be orthogonal to $\varphi_k^{(0)}$ we determine a_{kk} which is equal to zero.

We write now

$$E_l^{(0)}-E_k^{(0)} = h\nu_{lk}, \quad (\nu_{lk} = -\nu_{kl}) \tag{15}$$

where ν_{lk} is thus the frequency of the light emitted during a radiative transition from the l-th to the k-th state of the unperturbed system according to Bohr's frequency relation (see § 18; ν_{lk} is negative, if $E_l^{(0)} < E_k^{(0)}$). From equations (13), (14) and (15) we get

$$\chi = \Sigma_{l(\neq k)}\, K_l\varphi_l^{(0)}/h\nu_{lk}. \tag{16}$$

Equation (10) is a special case of the problem of finding from the equation

$$g = \Omega f \tag{17}$$

the inverse operator Ω^{-1}, that is, the operator which produces f out of g, $f = \Omega^{-1}g$. If Ω is a linear operator, Ω^{-1} will also be linear, provided it exists. In the case of functions of a discrete argument Ω^{-1} can be determined as a solution of a set of inhomogeneous linear equations. If Ω is Hermitian, the same will be true for Ω^{-1}.*

* The equation $\int f_1{}^*\,\Omega f_2\, d\tau = \int f_2\,\Omega^* f_1\, d\tau$ is equivalent to the equation $\int g_2\,\Omega^{-1*}\, g_1{}^*\, d\tau = \int g_1{}^*\,\Omega^{-1} g_2\, d\tau$.

If there are two different solutions f_1 and f_2 of equation (17), their difference $f_1 - f_2$ will satisfy the homogeneous equation. This case will thus only occur, when Ω possesses the eigenvalue zero. If in this case φ is an eigenfunction belonging to the eigenvalue 0, g must be orthogonal to φ, in order that equation (17) possesses a solution.* This corresponds to the solution of a set of linear equations with zero determinant. In equation (10) we are in fact dealing with this case.

If 0 is not an eigenvalue of Ω, and if we expand f in terms of the eigenfunctions φ_k of Ω (eigenvalues λ_k) we find

$$\left.\begin{aligned} f &= \Sigma_k a_k \varphi_k, \quad g = \Sigma_k a_k \lambda_k \varphi_k, \quad a_k = \lambda_k^{-1} \int \varphi_k^* g \, d\tau, \\ f(q) &= \int G(q, q') \, g(q') \, d\tau, \quad G(q, q') = \Sigma_l \lambda_l^{-1} \varphi_l(q) \varphi_l^*(q'). \end{aligned}\right\} \tag{18}$$

The so-called *Green function* $G(q, q')$ is simply the Hermitean operator function corresponding to the inverse operator. If g is orthogonal to $\varphi_k^{(0)}$, the term with $l = k$ in the series for $G(q, q')$ is omitted. If we consider the transition to the case where $\lambda_k = 0$, all terms in the series for $G(q, q')$ remain finite and equation (18) is identical with the solution (16) of equation (10). One can, indeed, write equation (16) in the form $\chi = \int \Sigma_{l(\neq k)} \varphi_l^{(0)}(q) \, \varphi_l^{(0)*}(q') K(q') \, d\tau / h\nu_{lk}$, where the denominators $h\nu_{lk}$ are just the eigenvalues of $H^{(0)} - E_k^{(0)}$.

We can use equations (14) and (16) to write down explicit expressions for the eigenfunctions and eigenvalues of the perturbed k-th state. Using equations (6) and (8) and denoting by Ω_{kl} the matrix elements of an operator Ω with respect to the eigenfunctions of $H^{(0)}$, $\Omega_{kl} = \int \varphi_k^{(0)*} \Omega \varphi_l^{(0)} \, d\tau$, we find

$$\varphi_k = \varphi_k^{(0)} + \lambda \varphi_k^{(1)} + \lambda^2 \varphi_k^{(2)} + \ldots, \quad E_k = E_k^{(0)} + \lambda E_k^{(1)} + \lambda^2 E_k^{(2)} + \ldots, \tag{19}$$

$$E_k^{(1)} = H_{kk}^{(1)}, \quad E_k^{(2)} = -\sum_{l(\neq k)} \frac{H_{kl}^{(1)} H_{lk}^{(1)}}{h\nu_{lk}} + H_{kk}^{(2)}, \ \ldots \tag{20a}$$

$$\left.\begin{aligned} \varphi_k^{(1)} &= -\sum_{l(\neq k)} \frac{H_{lk}^{(1)} \varphi_l^{(0)}}{h\nu_{lk}}, \\ \varphi_k^{(2)} &= \sum_{l(\neq k)} \frac{1}{h\nu_{lk}} \left\{ \sum_{m(\neq k)} \frac{H_{lm}^{(1)} H_{mk}^{(1)}}{h\nu_{mk}} - \frac{H_{lk}^{(1)} H_{kk}^{(1)}}{h\nu_{lk}} - H_{lk}^{(2)} \right\} \varphi_l^{(0)}, \ \ldots \end{aligned}\right\} \tag{20b}$$

These expansions are formal, and even when the series in equations (20) converge, it is questionable whether the series (19) will also converge. If we know, however, from considering the complete energy function H of equation (1) that the perturbed system also possesses a discrete energy spectrum, we may assume that *the perturbed energy levels will lie close to the unperturbed levels, provided λ is sufficiently small,* and that equation (19) enables us to calculate the deviations from the unperturbed state. Even so we cannot be sure that the series converge.

* $\int \varphi^* g \, d\tau = \int \varphi^* \Omega f \, d\tau = \int f \Omega^* \varphi^* \, d\tau = 0.$

One can easily find cases where the series (19) do not converge, and these cases often occur in applications. For instance, if the harmonic oscillator is perturbed by a term λx^3 [compare eq. (4)], the potential energy in the neighbourhood of $x = 0$ will still be approximately equal to ax^2, but for sufficiently large values of $|x|$, the term λx^3 will dominate in such a way that U will first show a maximum at a large negative x-value (absolute value of the order of λ^{-1}) and then go to $-\infty$ as $x \to -\infty$. From the considerations of § 17 it follows that there can no longer exist a discrete eigenvalue spectrum *. All the same equations (19) and (20) have a certain "asymptotic" meaning for those values of x where the perturbation term in U is small compared to the main term. They correspond to a physical situation where the particle stays a very long period in the „potential trough" near $x = 0$, and "oscillates up and down" if we use the classical correspondence picture.

A similar case occurs when in the case of a harmonic oscillator the perturbed term is λx^4. If $\lambda < 0$, U has two maxima which are far away from the origin, and $U \to -\infty$ as $|x| \to \infty$. There are no discrete levels and the series (19) will certainly not converge. The same will then be true for $\lambda > 0$ even though this leads to a discrete spectrum.

We want to make two remarks about equations (19) and (20). First of all we note that the occurrence of $h\nu_{lk} = E_l^{(0)} - E_k^{(0)}$ in the denominators entails that $\varphi_k^{(1)}$, $E_k^{(2)}$, . . . will be large if there exist unperturbed eigenvalues $E_l^{(0)}$ which lie close to $E_k^{(0)}$. If we assume for a moment that $H_{kl}^{(1)}$ is of the same order of magnitude as $H_{kk}^{(1)}$, we see that $\lambda^2 E_k^{(2)}$ in equation (19) will become of the same order of magnitude as $\lambda E_k^{(1)}$, if the energy difference $E_l^{(0)} - E_k^{(0)}$ between the k-th level and its nearest neighbour is of the same order as $\lambda E_k^{(1)}$, that is, as the perturbation of the k-th level. We would expect that we can no longer use equations (19) and (20). We shall see in § 50 how we must proceed in such a case.

Secondly, it may occur that the eigenvalue spectrum of $H^{(0)}$ is partly continuous. If this is the case, one can still use equations (19) and (20), provided $E_k^{(0)}$ and $\varphi_k^{(0)}$ refer to a discrete stationary state. The only difference is that each sum in equations (20) is replaced

* One finds that there exists a normalisable solution of the time-independent Schrödinger equation for any positive value of E (compare § 24). One can thus not discuss the question of eigenvalues and eigenfunctions in the framework of § 36. The time-dependent Schrödinger equation has, however, still a well-defined meaning.

by a sum and an integral. The integrals are over the continuous parameters which characterise the stationary states of the unperturbed system.

§ 49. THE PERTURBATION OF A DEGENERATE DISCRETE STATIONARY STATE *

Let equation (1) again give us the energy function of a perturbed system. We use equations (2) again to obtain the differential equations (3). Let now, however, the energy eigenvalue $E^{(0)}$ of the operator $H^{(0)}$ correspond to a degenerate stationary state so that the general solution of equation (3a) for this $E^{(0)}$-value is given by the equation (see § 21),

$$\varphi^{(0)} = \Sigma_{j=1}^{g} c_j \varphi_j^{(0)} . \tag{21}$$

We have assumed the degree of degeneracy g to be finite. Furthermore the $\varphi_j^{(0)}$ are chosen to be orthonormal, and $\varphi^{(0)}$ is normalised, provided

$$\Sigma_j |c_j|^2 = 1. \tag{22}$$

As in § 48 we multiply equation (3b) by $\varphi_j^{(0)*}$ and integrate over coordinate space. We thus obtain the following set of g equations,

$$\int \varphi_j^{(0)*} [H^{(0)} - E^{(0)}] \varphi^{(1)} \, d\tau = - \int \varphi_j^{(0)*} [H^{(1)} - E^{(1)}] [\Sigma_{j'} c_{j'} \varphi_{j'}^{(0)}] \, d\tau,$$

$$j = 1, 2, \ldots, g.$$

As $H^{(0)}$ is Hermitean the left hand side can be written in the form $\int \varphi^{(1)} [H^{(0)*} - E^{(0)}] \varphi_j^{(0)*} \, d\tau$ which is zero in view of equation (3a). Introducing the notation

$$\int \varphi_j^{(0)*} H^{(1)} \varphi_{j'}^{(0)} \, d\tau = H_{jj'}^{(1)} = H_{j'j}^{(1)*}, \tag{23}$$

we get the equations

$$\Sigma_{j'} H_{jj'}^{(1)} c_{j'} = E^{(1)} c_j , \quad j = 1, 2, \ldots, g,$$

which correspond to a finite secular problem of the kind discussed in § 37 [see eq. (4.44)].

There are g different solutions, which we shall distinguish by an index ι,

$$\Sigma_{j'} H_{jj'}^{(1)} c_{j'}^{\iota} = E_{\iota}^{(1)} c_{j'}^{\iota}, \quad \iota = 1, 2, \ldots, g. \tag{24}$$

* See, E. Schrödinger, *Ann. Phys.*, **80**, 437, 1926.

The eigenvalues $E_\iota^{(1)}$ are the roots of the secular equation

$$|H_{jj'}^{(1)} - E^{(1)}\delta_{jj'}| = 0. \tag{25}$$

If the roots are all different, the c_j^ι are uniquely determined but for a factor which depends on ι only. These factors can be chosen in such a way that the c_j^ι form a unitary matrix,

$$\Sigma_j c_j^{\iota *} c_j^{\iota'} = \delta_{\iota\iota'}, \qquad \Sigma_\iota c_{j'}^\iota c_j^{\iota *} = \delta_{j'j}. \tag{26}$$

We have now the choice between g orthonormal functions

$$\varphi_\iota^{(0)} = \Sigma_j c_j^\iota \varphi_j^{(0)}, \qquad \iota = 1, 2, \ldots, g, \tag{27}$$

for the eigenfunctions (21) which according to equation (2) describe the zeroth approximation of the perturbed state. These functions correspond to the following first approximation of the energy eigenvalue,

$$E_\iota = E^{(0)} + \lambda E_\iota^{(1)}. \tag{28}$$

After this we can solve successively equations (3b), (3c), . . . for each value of ι. We see that *under the influence of the perturbation the g-fold degenerate eigenvalue $E^{(0)}$ (stationary state) is split up into g non-degenerate eigenvalues (stationary states)*. The $\varphi^{(0)}$ are the eigenfunctions of the unperturbed energy operator which correspond to the perturbation $\lambda H^{(1)}$; they are called the zeroth approximation eigenfunctions.

If equation (25) possesses multiple roots, there is still a large amount of freedom in the choice of the c_j^ι. One can, however, always make the choice in such a way that the c_j^ι form a unitary matrix and the corresponding eigenfunctions an orthonormal set. According to equation (28) the eigenvalue $E^{(0)}$ is split up in less than g states. The original degeneracy is in the first approximation not completely lifted. One can decide whether higher approximations will remove the last degeneracies only by solving equation (3b) and discussing equation (3c) and so on.

We only considered the first step in the solution of our problem. Let us now discuss the next step for the case where in the first approximation the degeneracy is lifted. We first of all write instead of equations (2),

$$E_\iota = E^{(0)} + \lambda E_\iota^{(1)} + \lambda^2 E_\iota^{(2)} + \ldots, \quad \varphi_\iota = \varphi_\iota^{(0)} + \lambda\varphi_\iota^{(1)} + \lambda^2 \varphi_\iota^{(2)} + \ldots, \tag{29}$$

and we have now instead of equations (3) the equations

$$[H^{(0)}-E^{(0)}]\varphi_\iota^{(0)} = 0, \tag{30a}$$

$$[H^{(0)}-E^{(0)}]\varphi_\iota^{(1)} = -[H^{(1)}-E_\iota^{(1)}]\varphi_\iota^{(0)}, \tag{30b}$$

$$[H^{(0)}-E^{(0)}]\varphi_\iota^{(2)} = -[H^{(1)}-E_\iota^{(1)}]\varphi_\iota^{(1)}-[H^{(2)}-E_\iota^{(2)}]\varphi_\iota^{(0)}, \ldots \tag{30c}$$

Equations (30b), (30c), . . . are inhomogeneous equations of the kind $[H^{(0)}-E^{(0)}]\chi_\iota = K$, with K a function which is orthogonal to every $\varphi_{\iota'}^{(0)}$, $\int \varphi_{\iota'}^{(0)*} K\,d\tau = 0$, $(\iota' = 1, 2, \ldots, g)$. The solution of these inhomogeneous equations is determined but for the addition of a solution of the homogeneous equation. If χ_ι' is that solution which is orthogonal to all the $\varphi_\iota^{(0)}$ $(\iota = 1, 2, \ldots, g)$, we have in general $\chi_\iota = \chi_\iota' + \Sigma_{\iota'} f_{\iota'} \varphi_{\iota'}^{(0)}$. The functions χ_ι' may be thought to be expanded in terms of the eigenfunctions of $H^{(0)}$ corresponding to eigenvalues different from $E^{(0)}$. It turns out that in solving equations (30) by successive approximations one can not put all the $f_{\iota'}$ except f_ι equal to zero at each step. Let, for instance

$$\varphi_\iota^{(1)} = \varphi_\iota^{(1)'} + \Sigma_{\iota'} f_{\iota'\iota}\varphi_{\iota'}^{(0)} \tag{31}$$

be a solution of equation (30b). Substitute now this solution into equation (30c) and observe that since $\int\varphi_{\iota'}^{(0)*}[H^{(0)}-E^{(0)}]\varphi_\iota^{(2)}\,d\tau = \int \varphi_\iota^{(2)}[H^{(0)*}-E^{(0)}]\varphi_{\iota'}^{(0)*}\,d\tau = 0$, the right hand side of equation (30c) must be orthogonal to *each* of the $\varphi_\iota^{(0)}$. This last condition leads to the following set of equations $(\iota' = 1, 2, \ldots, g)$

$$\int\varphi_{\iota'}^{(0)*}[H^{(1)}-E_\iota^{(1)}]\varphi_\iota^{(1)}d\tau + f_{\iota'\iota}[E_{\iota'}^{(1)}-E_\iota^{(1)}] + \int\varphi_{\iota'}^{(0)*}[H^{(2)}-E_\iota^{(2)}]\varphi_\iota^{(0)}d\tau = 0.$$

For $\iota' = \iota$ we get an equation for $E_\iota^{(2)}$ which is analogous to equation (8),

$$E_\iota^{(2)} = \int\varphi_\iota^{(0)*}H^{(1)}\varphi_\iota^{(1)'}d\tau + \int\varphi_\iota^{(0)*}H^{(2)}\varphi_\iota^{(0)}\,d\tau, \tag{32}$$

and which can be interpreted in the same way as equation (8). For $\iota' \neq \iota$ we find the values for $f_{\iota'\iota}$,

$$f_{\iota'\iota} = -\left[\int\varphi_{\iota'}^{(0)*}H^{(1)}\varphi_\iota^{(1)'}\,d\tau + \int\varphi_{\iota'}^{(0)*}H^{(2)}\varphi_\iota^{(0)}\,d\tau\right]/(E_{\iota'}^{(1)}-E_\iota^{(1)}). \tag{33}$$

The coefficient $f_{\iota\iota}$ can be chosen arbitrarily and may, for instance, be put equal to zero.

The occurrence of the second term in the numerator shows that *it is necessary to know the perturbing energy in the second approximation in order to determine the eigenfunctions of the perturbed system in the first approximation* [using eq. (31)]. This is analogous to the fact that the eigenfunctions in the zeroth approximation depended on the first term of the energy perturbation.

If the degeneracy is not lifted in the first approximation, equations (32) and (33) can not be used. Equation (30c) must first of all be used to determine those $\varphi_\iota^{(0)}$ which correspond to the perturbation of the second order and this is done by solving another secular problem.

If the $\varphi_\iota^{(1)\prime}$ are expanded in terms of the eigenfunctions of $H^{(0)}$, equations (31) to (33) can be expressed in terms of the matrix elements of $H^{(1)}$ and $H^{(2)}$ corresponding to these functions. We shall, however, not write down the resulting equations explicitly.

§ 50. Perturbation theory and infinitesimal transformations *

In the previous two sections we discussed perturbations of the eigenvalues and eigenfunctions of the energy operator. The equations can, however, also be applied to the case when another observable instead of the energy is involved. From the point of view of quantum mechanical transformation theory we are dealing with the following problem (see § 37). Find a (generalised unitary) transformation S which brings the Hermitean operator F_{op} on diagonal form. Let F_{op} be the form

$$F_{\text{op}} = F_{\text{op}}^{(0)} + \lambda F_{\text{op}}^{(1)} + \lambda^2 F_{\text{op}}^{(2)} + \ldots, \tag{34}$$

where λ is a small real parameter, and let $S^{(0)}$ which brings $F_{\text{op}}^{(0)}$ on diagonal form be known. To obtain S from $S^{(0)}$ we need an additional transformation $U = S^{(0)*}S$ [see eq. (4.85)]. We expect that this transformation will not be very different from the identical transformation and that we can express U in a power series in λ.

The transformation $S^{(0)}$ transforms arbitrary functions in the q-representation into the $F^{(0)}$-representation. Let us assume that the eigenvalues of $F_{\text{op}}^{(0)}$ are discrete ($F_k^{(0)}$, $k = 1, 2, \ldots$) with corresponding normalised eigenfunctions $S_k^{(0)}(q)$. The transformed operator will be an infinite matrix [compare eq. (4.95)],

$$\begin{aligned} F_{kk'} &= \int S_k^{(0)*}(q)\, F_{\text{op}}\, S_{k'}^{(0)}(q)\, d\tau = F_{kk'}^{(0)} + \lambda F_{kk'}^{(1)} + \lambda^2 F_{kk'}^{(2)} + \ldots \\ &= F_k^{(0)}\, \delta_{kk'} + \lambda F_{kk'}^{(1)} + \lambda^2 F_{kk'}^{(2)} + \ldots \end{aligned} \tag{35}$$

The eigenvalue problem of F_{op} in the $F^{(0)}$-representation is of the form [see eq. (4.103)]

* See M. Born, W. Heisenberg, and P. Jordan, *Z. Phys.*, **35**, 557, 1926.

$$\Sigma_{k'} F_{kk'} U_{k'} = F U_k, \qquad k = 1, 2, \ldots \tag{36a}$$

$$\lambda \Sigma_{k'} F^{(1)}_{kk'} U_{k'} + \lambda^2 \Sigma_{k'} F^{(2)}_{kk'} U_{k'} + \ldots = [F - F^{(0)}_k] U_k, \tag{36b}$$

where F is an eigenvalue of $F_{kk'}$, that is, of F_{op}. The eigenfunctions of the perturbed operator are given by the equations,

$$S(q) = \Sigma_k U_k S^{(0)}_k(q). \tag{37}$$

By assuming that F_{op} possesses only discrete eigenvalues $F\mu$ ($\mu = 1, 2, \ldots$) we find that for any μ equations (36) have a solution U^μ_k which can be chosen in such a way that the U^μ_k form an (infinite) unitary matrix [see eq. (4.83)],

$$U^{-1} = U^*, \quad \Sigma_k U^{\mu *}_k U^{\mu'}_k = \delta_{\mu\mu'}, \quad \Sigma_\mu U^{\mu *}_k U^{\mu}_{k'} = \delta_{kk'}. \tag{38}$$

For $\lambda = 0$ we see that $U^\mu_k = \delta_{k\mu}$, $F_\mu = F^{(0)}_\mu$ is clearly a solution of equations (36). This solution corresponds to the identical transformation. If the eigenvalue $F^{(0)}_k$ is non-degenerate we try for $\lambda \neq 0$ the following series expansions

$$U^\mu_k = \delta_{k\mu} + \lambda U^{(1)}_{k\mu} + \lambda^2 U^{(2)}_{k\mu} + \ldots, \tag{39a}$$

$$F_\mu = F^{(0)}_\mu + \lambda F^{(1)}_\mu + \lambda^2 F^{(2)}_\mu + \ldots \tag{39b}$$

From equation (38) we get

$$\delta_{\mu\mu'} = \Sigma_k [\delta_{k\mu} + \lambda U^{(1)*}_{k\mu} + \lambda^2 U^{(2)*}_{k\mu} + \ldots][\delta_{k\mu'} + \lambda U^{(1)}_{k\mu'} + \lambda^2 U^{(2)}_{k\mu'} + \ldots]$$

which leads to the equations

$$U^{(1)*}_{\mu'\mu} + U^{(1)}_{\mu\mu'} = 0, \ (40\text{a}); \quad U^{(2)*}_{\mu'\mu} + U^{(2)}_{\mu'\mu} + \Sigma_k U^{(1)*}_{k\mu} U^{(1)}_{k\mu'} = 0, (40\text{b}); \ \ldots$$

From equation (40a) it follows that $u_{\mu\mu'} = -iU^{(1)}_{\mu\mu'}$ is Hermitean*. If we let λ go to zero so that we can neglect all but the first term in λ in equation (39a) we are dealing with a so-called *infinitesimal unitary transformation. Such a transformation is thus always characterised by a Hermitean matrix u* through the equation

$$U^\mu_k = \delta_{k\mu} + i\lambda u_{k\mu}. \tag{41}$$

The matrix $U^{(2)}$ is in general neither Hermitean nor anti-Hermitean.

If we substitute equations (39) into equation (36b), equations (40) will be satisfied automatically. This substitution leads, if we equate the coefficients of the various powers of λ on both sides of the equation, to the equations

* The matrix $U_{\mu\mu'}{}^{(1)}$ is called an anti-Hermitean matrix, as transposition produces the negative of its complex conjugate.

$0 = [F_\mu^{(0)} - F_k^{(0)}]\delta_{k\mu}$, (42a); $F_{k\mu}^{(1)} = [F_\mu^{(0)} - F_k^{(0)}]U_k^{(1)} + F_\mu^{(1)}\delta_{k\mu}$, (42b); ...

Equation (42a) is an identity. From equation (42b) we get

$$k = \mu \rightarrow F_k^{(1)} = F_{kk}^{(1)}, \qquad (43a);$$

$$k \neq \mu \rightarrow U_{k\mu}^{(1)} = -F_{k\mu}^{(1)}/[F_k^{(0)} - F_\mu^{(0)}]. \qquad (43b)$$

The diagonal element $U_{kk}^{(1)}$ must be purely imaginary, but is otherwise arbitrary. Equation (43b) shows clearly the anti-Hermitean character of $U^{(1)}$. Equation (43a) expresses the fact that in the first approximation the perturbation of an eigenvalue is equal to the expectation value of $F^{(1)}$ in the physical situation corresponding to the corresponding eigenfunction. This is analogous to the result of equation (6). Substituting from equations (43b) and (39a) into equation (37) leads to an expression for the first approximation of the eigenfunction which is equivalent to equation (20b).

The form (36) of the perturbation problem is particularly suited to deal with the case where the perturbation of an eigenvalue is in the first approximation already of the same order of magnitude as the difference between $F_k^{(0)}$ and one or more of the neighbouring eigenvalues of the unperturbed operator (compare the discussion in § 48).

In this case we write generally for the solution U_k^μ of equation (36) instead of equation (39a)

$$U_k^\mu = U_k^{(0)\mu} + \lambda U_k^{(1)\mu} + \ldots \qquad (44)$$

This is done, because substitution of equations (44) and (39b) into equation (36) gives in the zeroth approximation $0 = [F_\mu^{(0)} - F_k^{(0)}]U_k^{(0)\mu}$ and it follows that we may only put $U_k^{(0)\mu}$ equal to zero, if $F_\mu^{(0} - F_k^{(0)}$ is different from zero in the zeroth approximation, that is, provided the difference $F_\mu^{0)} - F_k^{(0)}$ is not itself of the order of magnitude of the higher order perturbation terms. Let us assume that the eigenvalues are numbered in such a way that the g eigenvalues $F_j^{(0)}$ ($j = 1, 2, \ldots, g$) from a narrow group which lies well away from the other eigenvalues of $F_{\text{op}}^{(0)}$. The elements of that part of the matrix $U_k^{(0)\mu}$ where both k and μ vary between 1 and g will form a unitary matrix with g rows and columns, while $U_k^{(0)\mu}$ will be equal to zero if only one of the pair k and μ possesses such a value. This unitary matrix can be obtained from a discussion of those terms in equation (36b) which are proportional to λ, and we find

$$\Sigma_{k'=1}^{g} \lambda F_{kk'}^{(1)} U_{k'}^{(0)\mu} = [F_{\mu}^{(0)} - F_{k}^{(0)}] U_{k}^{(0)\mu} + \lambda F_{\mu}^{(1)} U_{k}^{(0)\mu}, \quad \mu, k = 1, 2, \ldots, g. \tag{45}$$

These equations lead to a finite secular problem of the usual kind and equation (45) possesses non-trivial solutions, only if its determinant (of order g) vanishes,

$$\begin{vmatrix} \lambda F_{11}^{(1)} + F_{1}^{(0)} - F & \lambda F_{12}^{(1)} & \lambda F_{13}^{(1)} \ldots \\ \lambda F_{21}^{(1)} & \lambda F_{22}^{(1)} + F_{2}^{(0)} - F & \lambda F_{23}^{(1)} \ldots \\ \ldots & \ldots & \ldots \end{vmatrix} = 0, \quad F = F_{\mu}^{(0)} + \lambda F_{\mu}^{(1)} \tag{46}$$

The roots of this equation are the perturbed eigenvalues.

Apparently the diagonal elements contain terms of different orders of magnitude. However, if we replace the $F_j^{(0)}$ by $F_j^{(0)} - \Delta$ and F by $F - \Delta$, where Δ is, for instance, the arithmetic mean of the g $F_j^{(0)}$'s, we see that only terms of the order λ occur.

The case of the perturbation of a degenerate eigenvalue follows a special case of equation (45) which occurs if the $F_j^{(0)}$ are exactly equal. In that case equation (46) reduces to an equation which is equivalent to equation (25).

Up to now we have expanded the solution of the perturbation problem in terms of λ, and we could have considered the solution of equation (46) in order to obtain higher terms in the series (44) and (39b), after having written purely formally $F_j^{(0)} = \Delta + \lambda f_j$, say. For many applications it is sufficient to consider the solution of equations (45) and (46) and to discuss the perturbation problem in the following simple way. We know that in the infinite secular equation

$$|F_{kk'} - F\delta_{kk'}| = 0, \tag{47}$$

which governs the solution of equation (36a), all off-diagonal elements $F_{kk'}$ $(k \neq k')$ can be considered to be small compared to the diagonal terms F_{kk}, since the k-representation is obtained by using the eigenfunctions $S_k^{(0)}$ of an operator $F_{\text{op}}^{(0)}$ which approximates the given operator F_{op}. If the differences between the different F_{kk} are all large compared to the off-diagonal elements, the perturbed eigenvalues will in first approximation simply be equal to F_{kk} and the $S_k^{(0)}(q)$ are the corresponding eigenfunctions in zeroth approximation. If there is, however, a group of elements F_{kk} $(k = 1, 2, \ldots, g)$ whose differences are not large compared to the $F_{kk'}$, while their differences with the other F_{kk} are large compared to the $F_{kk'}$, one must solve the g-th order secular problem

$$|F_{jj'} - F\delta_{jj'}| = 0, \qquad j, j' = 1, 2, \ldots, g. \tag{48}$$

The roots of this equation F_ι $(\iota = 1, 2, \ldots, g)$ give us the perturbed eigenvalues in first approximation. The corresponding unitary matrix U_j^ι then leads to the following expressions for the corresponding eigenfunctions in the zeroth approximation,

$$S_\iota = \Sigma_{j=1}^{g} U_j^\iota S_j^{(0)}. \tag{49}$$

Geometrically speaking one is dealing with the problem of finding the directions and lengths of the principal axes of a hypersurface of the second degree which is not very different from a known hypersurface of this kind. If the principal axes of $H_{\text{op}}^{(0)}$ possess very different lengths, the principal axes of H_{op} will in the zeroth approximation fall along those of $H_{\text{op}}^{(0)}$; if this is not the case, the principal axes of $H_{\text{op}}^{(0)}$ may be rotated by an appreciable amount with respect to the principal axes of $H_{\text{op}}^{(0)}$.

§ 51. METHOD OF APPROXIMATE SOLUTIONS; THE VARIATIONAL PRINCIPLE

It often happens that it is possible to find an approximately correct expression for the Schrödinger eigenfunctions corresponding to one or more of the discrete stationary states of a system, even though these functions are not eigenfunctions of an operator which approximates the energy operator H_{op}.

We consider, for instance, the case of a differential equation of the type [see eq. (2.30)], $\varphi'' + C(E-U)\varphi = 0$. It may be impossible to find the exact solution of this equation. However, it may well be that a qualitative discussion (for instance, about number and position of zeros, behaviour at infinity, and so on) enables us to construct a simple mathematical expression — which may contain adjustable parameters — which we expect to give an approximate solution of the problem.

Another typical example is the case of an electron moving in the field of two or more nuclei. Let $\varphi_k^{(i)}$ be the wave-function of the k-th stationary state (energy $E_k^{(i)}$) in the field of the i-th nucleus alone. Any function $\varphi_k^{(i)}$ which is appreciably different from zero only in those regions where there is no other nucleus, is clearly an approximate solution of the Schrödinger equation corresponding to the approximate eigenvalue $E_k^{(i)}$. There is, however no "unperturbed energy operator" of which the $\varphi_k^{(i)}$ are the exact eigenfunctions (two $\varphi_k^{(i)}$ with different values of i are, for instance, not orthogonal to one another). Of particular importance is the case where the $E_k^{(i)}$-values may be the same for different values of i which occurs, for instance, when all nuclei have the same charge. The problem is then degenerate in zeroth approximation and the Schrödinger equation is also solved approximately by any linear combination of the $\varphi_k^{(i)}$. In general, however, the exact solution of the problem will lead to a splitting up of the degenerate eigenvalue $E_k^{(i)}$ into non-degenerate eigenvalues.

The following procedure can be used in such cases to obtain as exact an energy and a wave function as possible. We first of all normalise the approximate wave function φ_0,

$$\int \varphi_0^* \varphi_0 d\tau = 1. \tag{50}$$

We then try to determine both the energy E and the values of the adjustable parameters in φ_0 in such a way that the expression *

$$\int |(H_{\text{op}} - E)\varphi_0|^2 d\tau = \int \varphi_0^* (H_{\text{op}} - E)^2 \varphi_0 d\tau, \tag{51}$$

which would be equal to zero, if E and φ_0 corresponded to the exact solution, is a minimum,

$$\delta \int \varphi_0^* (H_{\text{op}} - E)^2 \varphi_0 d\tau = 0. \tag{52}$$

If there are no adjustable parameters we obtain the "best" value E_0 of E by differentiating expression (51) with respect to the energy and putting the resulting expression equal to zero, $\int \varphi_0^* (H_{\text{op}} - E)\varphi_0 d\tau = 0$, which leads to

$$E_0 = \int \varphi_0^* H_{\text{op}} \varphi_0 d\tau. \tag{53}$$

If the exact wave function is $\varphi_0 + \varphi_1$ and if we may assume φ_1 to be small of the *first* order compared to φ_0, it follows that E_0 is up to correction terms of the *second* order equal to the energy of the corresponding stationary state. Indeed, from the Schrödinger equation $(H_{\text{op}} - E)(\varphi_0 + \varphi_1) = 0$ it follows that

$$\begin{aligned} 0 = \int \varphi_0^* (H_{\text{op}} - E)(\varphi_0 + \varphi_1) d\tau &= E_0 - E + \int \varphi_1 (H_{\text{op}}^* - E)\varphi_0^* d\tau \\ &= E_0 - E - \int \varphi_1 (H_{\text{op}}^* - E)\varphi_1^* d\tau, \end{aligned}$$

and the last integral is clearly of the second order.

If the normalised φ_0 depends on adjustable parameters $a_1, a_2, \ldots$, E_0 will itself be a function of these parameters. If we know that we are looking for the value of the energy of just one discrete stationary state — as is the case, for instance, in the first example given above

* Equation (51) follows since H_{op} is Hermitean and E real.

— we can minimise the integral (51) with respect to the a_i and obtain in that way the "best" value of the energy.

The method which is usually applied in the literature is a different one. It consists of choosing the a_i in such a way that the expression (53) for E_0 itself is minimum. If we are looking for the *ground state*, that is, the lowest energy level — and this will be the case in many applications — one will certainly obtain a lower value of the energy than the one obtained by the first method. This method is best understood from the remark that *the correct wave functions satisfy the variational principle* *

$$\delta \int \varphi^* H_{\text{op}} \varphi \, d\tau = 0 \,, \qquad \int \varphi^* \varphi \, d\tau = 1 \,, \tag{54}$$

where the φ can be varied, provided they stay normalised. The value of the integral which is obtained by substituting for φ a solution of equations (54) *is the corresponding eigenvalue,*

$$\int \varphi^* H_{\text{op}} \varphi \, d\tau = E \,. \tag{55}$$

This follows by the usual method of variational calculus. Introducing a (real) Lagrangian multiplier λ we have from equations (54) the condition $\int \{\delta\varphi^* (H_{\text{op}} \varphi - \lambda\varphi) + \partial\varphi (H^*_{\text{op}} \varphi^* - \lambda\varphi^*)\} \, d\tau = 0$. Since $\delta\varphi^*$ and $\delta\varphi$ are analytically independent the Schrödinger equation, $(H_{\text{op}} - \lambda)\varphi = 0$, $(H^*_{\text{op}} - \lambda)\varphi^* = 0$, follows and λ is an eigenvalue E of H_{op} which — as can easily be checked — is equal to the extremum value of $\int \varphi^* H_{\text{op}} \varphi \, d\tau = \lambda \int \varphi^* \varphi \, d\tau$. We met a similar variational principle in § 37 when we were dealing with the eigenvalue problem of finite Hermitean matrices.

If we consider φ_0 in equation (52) to be an arbitrary normalised function, the variational principle will also here lead to the Schrödinger equation, provided E is also varied. The form of this variational problem is more complicated than equations (54), and calculations are correspondingly lengthier in the case of actual applications. There is, however, the advantage that we are always dealing with a minimum, and with a minimum which is equal to zero, if we obtain the correct solution.

If the Hamiltonian operator is quadratic in the momenta, which will be the case in most quantum mechanical problems (see § 27),

* One sometimes calls the use of this variational principle in quantum mechanics the Ritz-method. This nomenclature is historically hardly justified as Ritz used expansions of unknown functions in terms of known functions in his consideration of variational problems. (Compare W. Ritz, *Gesammelte Werke*, Gauthiers-Villars, Paris, 1911, pp. 192, 251.)

one can change the expression $\int \varphi^* H_{\text{op}} \varphi \, d\tau$ by integrating by parts in such a way that only first derivatives of φ and φ^* with respect to the coordinates occur. The calculations are accordingly simplified. If, for instance, H is of the form $\Sigma_k (p_k^2/2m_k) + U(q)$, one may assume φ to be real and the variational principle is of the form *

$$\delta \int \{\Sigma_k (\hbar^2/2m_k)(\partial\varphi/\partial q_k)^2 + U(q)\,\varphi^2\}\, d\tau = 0, \qquad \int \varphi^2 \, d\tau = 1. \tag{56}$$

The absolute minimum value of $\int \varphi^* H_{\text{op}} \varphi \, d\tau$ is clearly the energy of the ground state of the system. Thus, if $\varphi_0(a_1, a_2, \ldots)$ is an approximate expression for the wave function of this state, $E_0(a_1, a_2, \ldots)$ given by equation (53) will always be larger than the exact energy value. The most accurate energy value which we can obtain from $\varphi_0(a_1, a_2, \ldots)$ is thus, as we mentioned before, the minimum value of $E_0(a_1, a_2, \ldots)$. The corresponding values of the a_i are in general different from those obtained from the variational principle (52), but, up to a point, it is justified to claim that the variational principle (52) leads to a more accurate determination of the eigenfunction.

If the stationary state considered is not the ground state, one can not be sure a priori that the extremum of $E_0(a_1, a_2, \ldots)$ is closer to the correct energy value than the energy obtained from equation (52).

If the φ_0 are of the form (compare the case of an electron in the field of several nuclei discussed earlier)

$$\varphi_0 = \Sigma_{j=1}^{g} c_j \varphi_0^{(j)}, \tag{57}$$

where in the zeroth approximation all $\varphi_0^{(j)}$ correspond to the same or to practically the same energy of the system we are dealing with a case of g-fold degeneracy **. The $\varphi_0^{(j)}$ may depend on parameters a_i, but we shall not discuss this complication.

We now want to find the values of the c_j which lead to a φ_0 which approximates a stationary state most closely. Let $\varphi_0 + \varphi_1$ be a wave function which is an exact solution (φ_1 small of the first order compared to φ_0) and let E be the corresponding energy. We then have the following g equations,

$$\begin{aligned} 0 &= \Sigma_j \varphi_0^{(j)*} (H_{\text{op}} - E) [\Sigma_{j'} c_{j'} \varphi_0^{(j')} + \varphi_1] \, d\tau \\ &= \Sigma_{j'} (H_{jj'} - E I_{jj'}) + \int \varphi_1 (H_{\text{op}}^* - E) \varphi_0^{(j)*} \, d\tau, \end{aligned} \tag{58a}$$

* See, E. Schrödinger, *Ann. Phys.*, **79**, 361, 1926.

** The terminology "degeneracy" is slightly generalised for this case.

where

$$H_{jj'} = \int \varphi_0^{(j)*} H_{\rm op} \varphi_0^{(j')} d\tau = H_{j'j}^*, \qquad I_{jj'} = \int \varphi_0^{(j)*} \varphi_0^{(j')} d\tau = I_{j'j}^*. \quad (58b)$$

As $(H_{\rm op}^* - E)\varphi_0^{(j)*}$ is small of the first order, the last integral in equation (58a) is small of the second order and may be neglected. The c_j in equation (57) follow thus from the linear homogeneous equations

$$\Sigma_{j'} c_{j'}(H_{jj'} - E I_{jj'}) = 0, \qquad j = 1, 2, \ldots, g. \quad (59)$$

These equations are somewhat more complicated than the eigenvalue problems considered up to now, as $I_{jj'}$ is not zero for $j \neq j'$. All the same one can show that the determinant $|H_{jj'} - EI_{jj'}|$ possesses only real roots E and that one can find for every root a set of c_j which solve equation (59) *. In this way we find to a first approximation a splitting up into g energy values and we can determine the corresponding eigenfunctions.

Of course, we could have replaced at the beginning the $\varphi_0^{(j)}$ of equation (57) by an orthonormal set of linear combinations of these functions. In most actual applications this will involve an unnecessary complication of the calculations.

It can be checked easily that equations (59) could also have been obtained from expression (57) by applying the variational principle (52) or (54) and varying the c_j, keeping φ_0 normalised.

A drawback of the variational principle (54) is that there is no known practical method of estimating the error in the energy value obtained.

If we want to find a higher approximation of a problem of the kind discussed in this section, the general method for this is to represent again the wave function by a sum of the kind (57). In this case we take for the $\varphi_0^{(j)}$ also some of those functions which in the zeroth approximation correspond to states with energies at some distance from the energy of the state investigated. We shall find that in general the c_j in equation (57) will be the smaller, the larger the corresponding energy differences are. We can clearly consider the methods discussed in §§ 48 to 50 to be special cases of the application of this general method. In those §§ we were expanding the eigen-

* A unitary transformation which brings $I_{jj'}$ on principal axes (only positive eigenvalues I_j) will bring the matrix $H_{jj'} - E I_{jj'}$ on the form $H_{jj'}' - E I_j \delta_{jj'}$. If we then write $H_{jj'}' = H_{jj'}''(I_j I_{j'})^{1/2}$ our problem is reduced to the eigenvalue problem of the Hermitean matrix $H_{jj'}''$

functions we wanted to determine in terms of a complete orthonormal set $\varphi_0^{(j)}$ (the eigenfunctions of the unperturbed operator).

The theory of the London-van der Waals forces between neutral atoms is a simple case of the application of the general method outlined here.

§ 52. Expectation values and time averages

The formalism of quantum theory and the results of the calculations performed with this formalism derive their meaning in the last resort always from the interpretation of certain experimental data using classical concepts. In particular if we want to understand quantum mechanically the process of "observing" an atomic system, that is, the interaction of such a system with an "observational apparatus", those considerations have a meaning in the physical sense only when we can connect the characteristics with which we describe the apparatus with certain observational data. The examples of perturbation theory which we have considered enable us to discuss a certain type of observation in this way and to indicate how the observations can lead to quantitative statements about the properties of the system. Let an apparatus be situated in the neighbourhood of an atomic system. Let the interaction between the system and the apparatus be described classically by an expression of the form $\lambda F(p_k, q_k)$. The p and q are here the momenta and coordinates of the particles in the system, and λ is a parameter which is connected with the state of the apparatus and whose value can be "measured". To fix the ideas we can imagine an atom at a fixed position in space between the plates of an electrical condenser. In this case $F(p_k, q_k)$ is the electrical moment of the atom in the direction of the electrical field, and $-\lambda$ is the field strength the value of which can be obtained by measuring the position of a needle or some other classical quantity. Let the atom be in its ground state which we shall assume to be non-degenerate. If the interaction energy is small, its presence will mean, as we saw in § 48, that the energy of the total system is increased by $\lambda\overline{F}$, where $\overline{F}$ is the expectation value of F_{op} in the unperturbed state of the atom. This change of energy will lead to an observable "force" on the parameter λ, since an increase by $\delta\lambda$ of the parameter λ needs an energy $\overline{F}\delta\lambda$. It is only because we used a classical description of the apparatus that we were able to discuss at all the measurement of $\overline{F}$.

If we are dealing with an apparatus with an interaction of the form $\lambda F + \mu F^2 + \ldots$, we could measure not only $\overline{F}$, but also $\overline{F^2}, \ldots$, and we could obtain a picture of the distribution function of the observable F in the ground state of the atom.

In classical mechanics the p and q in the atomic system possess well defined values at all times. At any moment $F(p, q)$ is the force on the parameter λ and the measured force wuld be considered to be the time average $\breve{F}$ of the quantity F over a, relatively long, period of time (t_1, t_2), $\breve{F} = (t_2 - t_1)^{-1} \int_{t_1}^{t_2} F\, dt$. This integral corresponds to a Hermitean operator in quantum mechanics where we now take for F the operator $F(t)_{\text{op}}$ defined in § 43. Dropping the index "op" for the sake of simplicity and putting $t_1 = 0$, $t_2 = t$ we have

$$\breve{F} = t^{-1} \int_0^t F(t)\, dt. \tag{60}$$

The expectation value of the n-th power of $\breve{F}$ in a physical situation characterised by $\psi(t)$ is given by the equation

$$\overline{\breve{F}^n} = t^{-n} \int \psi^*(0) \left\{ \int_0^t F(t)\, dt \right\}^n \psi(0)\, d\tau. \tag{61}$$

We shall calculate this expectation value for the cases $n = 1$ and $n = 2$ in the case where $\psi(t)$ corresponds to a stationary state with eigenvalue E_k so that $\psi(0)$ is the normalised eigenfunction φ_k. From $H\varphi_k = E_k\varphi_k$ it follows that $\exp[\pm iHt/\hbar]\varphi_k = \exp[\pm iE_k t/\hbar]\varphi_k$. We find then by changing the order of integration and also by changing the order of the factors in the integrand [compare eq. (4.119)]

$$\overline{\breve{F}} = t^{-1}\int_0^t dt \int \varphi_k^* \exp(iHt/\hbar) F(0) \exp(-iHt/\hbar)\varphi_k d\tau = t^{-1}\int_0^t dt \int \varphi_k^* F \varphi_k d\tau = F_{kk}, \quad (6$$

$$\overline{\breve{F}^2} = t^{-2}\int_0^t dt_1 \int_0^t dt_2 \int \varphi_k^* \exp(iHt_1/\hbar) F(0) \exp[iH(t_2 - t_1)/\hbar] F(0) \exp(-iHt_2/\hbar)\varphi_k d\tau$$

$$= t^{-2}\int_0^s dt_1 \int_0^t dt_2 \exp[iE_k(t_1 - t_2)/\hbar] \int \varphi_k^* F \exp[iH(t_2 - t_1)/\hbar] F \varphi_k d\tau$$

$$= t^{-2}\int_0^t dt_1 \int_0^t dt_2 \exp[iE_k(t_1 - t_2)/\hbar] \Sigma_l F_{lk} \int \varphi_k^* F \exp[iE_l(t_2 - t_1)/\hbar] \varphi_l d\tau$$

$$= \Sigma_l F_{kl} F_{lk} \left| t^{-1} \int_0^t \exp[i(E_k - E_l)t_1/\hbar] dt_1 \right|^2, \quad \text{or,}$$

$$\overline{\breve{F}^2} = F_{kk}^2 + \Sigma_{l(\neq k)} F_{kl} F_{lk} 2(2\pi\nu_{kl}t)^{-2}(1 - \cos 2\pi\nu_{kl}t). \tag{63}$$

The F_{kl} are the matrix elements of F [$\equiv F(0)$] with respect to the energy eigenfunctions while the ν_{kl} are the Bohr frequencies as in equation (15), $\nu_{kl} = (E_k - E_l)/h$.

Equation (62) shows that the expectation value of the time average is simply equal to the expectation value of the operator itself. From equation (63) we see, however, that the expectation value of the square of the time average approaches the square of F_{kk} with increasing t, if E_k is a non-degenerate eigenvalue. As soon as t is only a few times larger than the period $|\nu_{kl}|^{-1}$ corresponding to the smallest of the ν_{kl}, $\overline{\tilde{F}^2}$ will practically be equal to F^2_{kk}. *If we consider the time average of an observable over a long time interval, it will possess a well defined value, just as the energy, and this value is the expectation value of the observable in the state considered, which is assumed to be non-degenerate* *. We can thus justifiably state that also from the quantum mechanical point of view measurements of the kind discussed at the beginning of this section are really measurements of time averages.

In the limit where t is very small, equation (63) leads to $\overline{\tilde{F}^2} = \Sigma_l F_{kl} F_{lk} = (F^2)_{kk}$. We see that we now are dealing with the instantaneous value of F, as we could have expected.

Equations (62) and (63) could also have been derived by integration over t of the Heisenberg matrix elements of $F(t)$ [compare eq. (4.122)].

§ 53. THE METHOD OF THE VARIATION OF CONSTANTS **

The application of perturbation theory which we discussed up to now were practically all concerned with a more accurate determination of discrete stationary states. In many cases, however, the physical problem is such that we are interested not so much in possible stationary states, but in the changes in time which are produced by the perturbation. If the energy operator of the system contains the time explicitly, there is no sense even in looking for stationary states.

In such cases we use the so-called method of the *variation of constants*, which is closely connected with the method in classical mechanics with the same name. It can in many respects be considered to be a particular application of the general transformation theory, as in the parallel case in classical mechanics.

Let us, as in § 48, consider a system with an energy operator H_{op} which can be expanded in terms of a small parameter λ, $H_{\text{op}} = H^{(0)}_{\text{op}} + \lambda H^{(1)}_{\text{op}} + \ldots$

* We do not wish to discuss the complications which enter in the case of degeneracy.

** See, P. A. M. Dirac, *Proc. Roy. Soc.*, **A 112**, 673, 1926.

We assume that the unperturbed operator $H_{\text{op}}^{(0)}$ does not contain the time explicitly and that we know its eigenvalues E_k and the orthonormal set of its eigenfunctions φ_k [we have dropped the superscript "(0)"]. We have used the notation which refers to a discrete spectrum, but the discussion can easily be generalised so as to include a continuous spectrum. We do not introduce degeneracy explicitly, so that several E_k may be the same. We now transform the Schrödinger equation $H_{\text{op}}\psi = i\hbar\partial\psi/\partial t$ in such a way that the physical situation is described in terms of the probability amplitudes $a_k(t)$ with respect to the eigenfunctions of the operator $H_{\text{op}}^{(0)}$ (compare § 42),

$$\psi = \Sigma_k a_k(t)\,\varphi_k, \qquad a_k = \int \varphi_k^*\,\psi\,d\tau, \qquad \Sigma_k|a_k(t)|^2 = 1, \tag{64}$$

$$H_{kl} = \int \varphi_k^*\,H_{\text{op}}\,\varphi_l\,d\tau = E_k\,\delta_{kl} + \lambda H_{kl}^{(1)} + \lambda^2 H_{kl}^{(2)} + \ldots,$$

$$H_{kl}^{(r)} = \int \varphi_k^*\,H_{\text{op}}^{(r)}\,\varphi_l\,d\tau,\; r = 1, 2, \ldots \tag{65}$$

The transformed Schrödinger equation

$$i\hbar\dot{a}_k = \Sigma_l H_{kl}\,a_l \tag{66}$$

can be written in the form $i\hbar\dot{a}_k - E_k a_k = \lambda\Sigma_l H_{kl}^{(1)} a_l + \lambda^2\Sigma_l H_{kl}^{(2)}\,a_l + \ldots$ where we used equation (65). Introducing the Bohr frequencies $\nu_{kl} = (E_k - E_l)/h$ and a set of quantities γ_k by the equations

$$a_k = \gamma_k \exp\left(-iE_k t/\hbar\right), \qquad \Sigma_k|\gamma_k|^2 = 1, \tag{67}$$

we find the following simultaneous differential equations for the γ_k *,

$$i\hbar\dot{\gamma}_k = \lambda\Sigma_l H_{kl}^{(1)} \exp\left(2\pi i\nu_{kt}\right)\gamma_l + \ldots \tag{68}$$

If there were no perturbation terms, the γ_k would be constant and the Schrödinger wave function

$$\psi = \Sigma_k \gamma_k \varphi_k \exp\left(-iE_k t/\hbar\right) \tag{69}$$

would correspond simply to a superposition of stationary states of the unperturbed problems. If there is a perturbation we get from equation (68) γ_k-values which vary in time; hence the name of this method.

* The quantites $H_{kl}{}^{(r)} \exp\left(2\pi i\nu_{kl}t\right)$ are the (Heisenberg) matrix elements of the operator $H^{(r)}(t)_{\text{op}}$ with respect to the φ_k, if $H^{(r)}(t)_{\text{op}}$ is the time dependent operator corresponding to $H^{(r)}$ (compare § 43).

Equations (68) are similar to the perturbation equations in celestial mechanics and their discussion runs along the same line as there, as long as E_k is a discrete spectrum. Apart from time-dependent terms there may also be time-independent terms on the right hand side of equation (68). If the $H_{\mathrm{op}}^{(r)}$ [and thus the $H_{kl}^{(r)}$] themselves do not depend on the time, these time-independent terms will be both the expectation values $\lambda^r H_{kk}^{(r)}$ and all terms for which $\nu_{kl} = 0$, that is, terms due to degeneracy of E_k. If we know the values of the γ_k at $t = 0$ ($\gamma_k = \gamma_k^0$ at $t = 0$) we can immediately calculate the change in the γ_k over a sufficiently small time interval $(0, t)$ by substituting on the right hand side of equations (68) for γ_k their initial values γ_k^0. The time-dependent terms give rise to the so-called *periodic perturbations* and the time-independent terms to the *secular perturbations*, which to a first approximation lead to a linear term in t. If all $H_{kl}^{(1)}$ are time-independent and if E_k is non-degenerate, by integrating equation (68) we find up to the first order in λ for γ_k the expression

$$\gamma_k - \gamma_k^0 = -(\lambda/h)\{2\pi i\, H_{kk}^{(1)} \gamma_k^0 t + \Sigma_{l(\neq k)}\, H_{kl}^{(1)} \gamma_l^0 \,[\exp(2\pi i \nu_{kl} t) - 1]/\nu_{kl}\},$$

where the first term on the right hand side corresponds to the secular perturbation and the sum to the periodic perturbations.

If the expressions $\lambda H_{kl}^{(1)}/h\nu_{kl}$ are small, the periodic perturbations have small amplitudes and only the secular perturbations change the results for large values of t. This difficulty is avoided easily — as it is in classical mechanics * — by solving first of all equations (68) neglecting the periodic terms (solution of the so-called secular perturbation equations). After that we can vary the constants which occur in this solution. It turns out that these new constants are subject only to periodic perturbations.

If the E_k are non-degenerate, the secular perturbation equations are simply $i\hbar\dot{\gamma}_k = \lambda H_{kk}^{(1)} \gamma_k$ with the solution $\gamma_k = \gamma_k' \exp[-i\lambda H_{kk}^{(1)} t/\hbar]$ ** Expressed in the new constants γ_k' we have the wave function

$$\psi = \Sigma_k \gamma_k' \varphi_k \exp\,[-i(E_k + \lambda H_{kk}^{(1)})t/\hbar], \tag{70}$$

and from equations (68) we get for the γ_k' the equations

$$\begin{aligned} i\hbar\dot{\gamma}_k' &= \lambda \Sigma_{l(\neq k)}\, H_{kl}^{(1)} \exp(2\pi i \nu_{kl}' t)\gamma_l' + \ldots, \\ h\nu_{kl}' &= (E_k + \lambda H_{kk}^{(1)}) - (E_l + \lambda H_{ll}^{(1)}). \end{aligned} \tag{71}$$

* Compare Pauli's article on "Störungstheorie" in *Physikalisches Handwörterbuch*, Springer, Berlin, 1931.

** We see that for small values of t we get the linear term in t as before.

Up to the first order in λ equations (71) are solved by the expressions

$$\gamma'_k - \gamma'^0_k = -\lambda \Sigma_{l(\neq k)} (H^{(1)}_{kl}/h\nu'_{kl})\, \gamma'^0_l \exp (2\pi i\, \nu'_{kl} t), \tag{72}$$

where now the γ'^0_k occur as the time averages of the γ'_k.

Substituting expressions (72) into equation (70) we see that the wave function (70) corresponds to a superposition of stationary states with energies $E_k + \lambda H^{(1)}_{kk}$ and eigenfunctions $\varphi_k - \lambda \Sigma_{l(\neq k)} (H^{(1)}_{lk}/h\nu'_{lk})\varphi_l$. This agrees exactly with equations (19) and (20) apart from the occurrence of ν'_{lk} instead of ν_{lk}. This last difference is, however, one of the second order in λ.

If we would proceed to the next approximation of the solution of equation (71) we would, of course, again find a result which is equivalent to equation (20). If we are interested in stationary states it would have been easier to look for solutions $a_k(t)$ of equations (66) of the form $a^{(0)}_k \exp (-iEt/\hbar)$ which would have led to the equations (36) which do no longer contain the time. The same holds for the case where the unperturbed system possesses degenerate eigenvalues or groups of relatively close lying energy values (compare § 50). All the same it is instructive to discuss also in that case the effect of the perturbation by studying equations (68).

Let the energy levels $E_1, E_2, \ldots, E_g$ form a close lying group so that the corresponding $h\nu_{kl}$ $(k, l = 1, 2, \ldots, g)$ cannot be considered to be large compared to the matrix elements of the perturbation energy. A special case would be exact degeneracy when some ν_{kl} would be exactly equal to zero. To simplify the equations we shall not use a series expansion in terms of λ, but put

$$H_{\text{op}} = H^{(0)}_{\text{op}} + G_{\text{op}}, \tag{73}$$

where G_{op} is small compared to $H^{(0)}_{\text{op}}$. For the time being we shall assume G_{op} not to contain the time explicitly. In the equations which correspond to equations (68),

$$i\hbar\dot{\gamma}_k = \Sigma_l\, G_{kl}\, \gamma_l \exp (2\pi i \nu_{kl} t), \tag{74}$$

we drop all terms for which $h\nu_{kl}$ is large compared to G_{kl}, in the cases where $k = 1, 2, \ldots, g$. These terms lead to fast oscillations of the γ with small amplitudes. We are left with a finite set of linear equations $(k, l = 1, 2, \ldots, g)$ from which we can compute the finite changes in the γ_k. As the G_{kl} are time-independent, this set of equations possesses solutions of the form $\gamma_k = \gamma'_k \exp [i(E_k - \Delta)t/\hbar]$ where the γ'_k are constants satisfying the set of homogeneous equations $\gamma'_k(\Delta - E_k) = \Sigma_l G_{kl}\gamma'_l$. This set of equations can be solved only provided the secular determinant is equal to zero, $|(E_k - \Delta)\delta_{kl} + G_{kl}| = 0$. We are back at the discussion of the perturbation of degenerate or nearly degenerate stationary states which was the subject of § 50 [see, for instance, eq. (46)].

The importance of the method of the variations of constants lies especially in its application to problems either where the perturbation energy operator depends explicitly on the time or where we are dealing with continuous spectra. In the first category falls the problem of the influence of slowly varying fields of force on atomic systems and in that connexion the *adiabatic theorem*. Also in this category are some collision- and radiation problems where the reaction of the bombarded (or irradiated) atomic system on the bombarding system (or radiation field) is neglected. The occurrence of continuum eigenvalues is typical of many quantum mechanical perturbation problems. It is basic for an explanation of the fact that many phenomena can be interpreted by introducing the concept of *transition probability per unit time* (probability for a "quantum jump" per unit time).

Before we discuss these problems we draw attention to the fact that one can also apply the method of the variation of constants to the case where we know approximate solutions, which are not necessarily solutions corresponding to an unperturbed energy operator (compare § 51) *. Let $\psi_1, \psi_2, \ldots, \psi_k, \ldots$ be approximate solutions of the Schrödinger equation $[H_{\text{op}} - i\hbar(\partial/\partial t)]\psi = 0$, where H_{op} may contain the time. The ψ_k are functions of the coordinates and the time which are linearly independent, but which do not necessarily form a closed set. Let

$$\psi = \Sigma_k \gamma_k(t)\,\psi_k + \xi \tag{75}$$

be an exact solution, where ξ is always orthogonal to all the ψ_k and may be considered to be small at least of the first order compared to the ψ_k (at any rate during the finite span of time). Let p_t denote $-i\hbar\partial/\partial t$ [compare eq. (3.25)]. Multiply the Schrödinger equation by ψ_l^* and integrate over all coordinates (compare § 51), $\int \psi_l^* (H_{\text{op}} + p_t)(\Sigma\gamma_k\psi_k + \xi)\,d\tau = 0$. Neglecting second order terms there are no terms involving ξ left ** and we have

* J. R. Oppenheimer, *Phys. Rev.*, **31**, 74, 1928.

** As $(H_{\text{op}} + p_t)\psi_l$ is a first order term, we find that $\int \psi_l^* H_{\text{op}}\,\xi\,d\tau = \int \xi H_{\text{op}}^* \psi_l^* d\tau$ is equal to $-\int \xi p_t^* \psi_l^*\,d\tau = i\hbar \int \xi \dot{\psi}_l^*\,d\tau$, up to terms of the second order. Up to this order, the terms involving ξ reduce to $-i\hbar\{\int \xi\dot{\psi}_l^* d\tau + \int \psi_l^* \dot{\xi} d\tau\} = -i\hbar(d/dt)\int \psi_l^* \xi d\tau = 0$

$$i\hbar\Sigma_k\dot{\gamma}_k\int\psi_l^*\psi_k\,d\tau=\Sigma_k\gamma_k\int\psi_l^*(H_{\rm op}+p_t)\,\psi_k\,d\tau. \tag{76}$$

These equations which show a great similarity to equations (59) but which are more general, enable us in principle to compute the time-dependence of the γ_k for given initial values of these quantities. If we solve these equations, we must first of all investigate whether it is possible to separate periodic and secular perturbations.

Equations (76) form the basis of many investigations of processes which occur when atoms or molecules collide, such as, electron capture by α-particles or chemical reactions.

§ 54. Variable fields of force; adiabatic theorem

Consider a system with a time-independent energy operator $H_{\rm op}^{(0)}$ all the eigenvalues of which we shall assume for the time being to be discrete. Let it be perturbed by a weak time-dependent field of force (corresponding to a term $G_{\rm op}$ in the energy operator). The perturbation will be described by equations (74) with time-dependent G_{kl}. In solving these equations we must investigate whether there are on the right hand side of these equations apart from fast oscillating terms slowly varying ones which would lead to secular perturbations of the γ_k. Such terms need not come only from small or vanishing ν_{kl}-values. If, for instance, G_{kl} had a component which behaved as $\exp(-2\pi i\nu t)$ (for instance, an atom in an oscillating radiation field), the term $G_{kl}\exp(2\pi i\nu_{kl}t)$ will give rise to secular perturbations if $\nu_{kl}-\nu$ is small or vanishing. We can try to solve equations (74) in the following way. First of all we retain only the constant terms and those which vary only slowly. If this leads in a natural way to a decomposition of equations (74) into finite sets of simultaneous equations, we can use these equations to obtain the secular perturbations and afterwards we can take care of the periodic perturbation by integrating the resultant equations. In general, however, the complete solution of the secular perturbations will encounter great mathematical difficulties. Sometimes the problem is such that we can use solutions which are easily obtained, but which are valid only during a relatively short time interval (power series in t).

If the perturbation changes slowly in time, that is, if the frequencies characterising the time dependence of the G_{kl} are small compared to the ν_{kl}, we can easily determine the secular perturbations as long as we are dealing with non-degenerate states. If the E_k

are non-degenerate eigenvalues of $H_{\mathrm{op}}^{(0)}$ the secular perturbation of the γ_k follows from the equations $i\hbar\dot{\gamma}_k = G_{kk}(t)\gamma_k$ with the solution $\gamma_k = \gamma_k' \exp[-2\pi i \Phi_k(t)]$, where the γ_k' are constant and where the $\Phi_k(t)$ are defined by the equations

$$\Phi_k(t) = h^{-1}\int_0^t G_{kk}(t)\,dt. \tag{77}$$

The periodic perturbations of the γ_k' are determined from the equations $i\hbar\dot{\gamma}_k' = \Sigma_{l(\neq k)} G_{kl}\gamma_l' \exp[2\pi i(\nu_{kl}t+\Phi_k-\Phi_l)]$. These equations are solved by putting in the right hand side the γ_l' equal to their values $\gamma_l'^0$ at a given time,

$$\gamma_k'(t) = \gamma_k'^0 - \Sigma_{l(\neq k)}(i/\hbar)\,\gamma_l'(0)\int_0^t G_{kl}\exp[2\pi i(\nu_{kl}t+\Phi_k-\Phi_l)]\,dt.$$

If we are interested in a physical situation which differs only very little from an unperturbed stationary state (index k_0), and if E_{k_0} is non-degenerate, only the secular perturbation of γ_{k_0} is of any importance, and we do not have to worry whether the other eigenvalues are degenerate, nearly degenerate (or even continuous) when we determine the variation in time of the situation. If, for instance, at $t = 0$ we had $\gamma_{k_0} = 1$, $\gamma_k = 0$ $(k \neq k_0)$, we substitute in the expression on the right hand side of equations (74)

$$\gamma_{k_0} = \exp[-2\pi i\,\Phi_{k_0}(t)],\quad \gamma_k = 0\ \ (k \neq k_0)$$

[compare eq. (77)]. As a first approximation we then get

$$\begin{aligned}\gamma_{k_0} &= \exp[-2\pi i\,\Phi_{k_0}(t)],\\ \gamma_k(t) &= -(i/\hbar)\int_0^t G_{kk_0}\exp\{2\pi i[\nu_{kk_0}t-\Phi_{k_0}(t)]\}dt \quad (k\neq k_0).\end{aligned} \tag{78}$$

The quantities $|\gamma_k(t)|^2$ are small of the second order so that these expressions satisfy, indeed, the relation $\Sigma|\gamma_k|^2 = 1$ to a first approximation.

We shall use equation (78) to discuss the *adiabatic theorem.*

Ehrenfest * introduced the adiabatic theorem in the old quantum theory where one illustrated the stationary states of an atom by mechanical orbits. The situation was the following one. A Hamiltonian energy function $H(p_k, q_k; a_1, a_2, \ldots)$ governed the mechanical properties of the system considered. This function depended not only on the coordinates, q_k, and momenta, p_k, but also on a number of parameters $a_1, a_2, \ldots$ which could vary continuously within certain intervals. To simplify the discussion we shall assume that there is only one parameter a. Let the system possess discrete stationary states for all values of a. The corresponding mechanical orbits are selected out of the continuum of all possible orbits by means of certain

* P. Ehrenfest, *Proc. Acad. Sc. Amsterdam,* **16**, 591, 1914.

„quantisation rules". A continuous set of a-values corresponds to a continuous set of "stationary orbits". If through some agent a changes with time, we are in general not allowed to describe the time-dependence of the system by the classical equations of motion. If, for instance for $t < 0$ the parameter a were constant and the system were in a stationary state, and if $a(t)$ were again constant for $t > T$, it would be fortuitous if the state of the system for $t \geqq T$ would again correspond to a stationary state for any choice of $a(t)$ between $t = 0$ and $t = T$. According to Ehrenfest's adiabatic theorem, in the limit where a changes infinitely slowly ($\dot{a} \to 0$; adiabatic process) one can, indeed, describe the change in the system by using the classical equations of motion. Bohr called this theorem therefore *the principle of the possibility of mechanical transformations of stationary states.* Using this principle it was possible to show* that the quantities $J_1, J_2, \ldots$ which are the periodicity moduli of the action integral and which are used to write down the quantisation rules ($J_1 = n_1 h$, $J_2 = n_2 h, \ldots; n_1, n_2, \ldots$ quantum numbers) do not change their value during an adiabatic process (adiabatic invariants).

The situation is, of course, completely different in quantum mechanics. There is no longer the question whether classical mechanics can be used and we may assume that in the case of time-dependent parameters the change of the physical situation in time is described by the Schrödinger equation with a time-dependent energy operator. All the same there exists still a theorem which is closely connected to the principle of the possibility of mechanical transformations and which is called the *quantum mechanical adiabatic theorem.***

Let us consider a system with an energy operator $H(p_k, q_k, a)$ which is such that for certain ranges of a-values the (discrete) energy eigenvalues and eigenfunctions can be considered to be continuous functions of a. If a changes in time, the situation corresponds at every moment to a certain probability distribution which refers to the stationary state corresponding to the value of a at that moment. In general this distribution will change in time. The adiabatic theorem states, however, that it will not change in the limit where a is changing infinitely slowly. We can lift the restriction to only discrete eigenvalues for all values of a by expressing the theorem in the following form. If there exist for all values of a a discrete energy eigenvalue $E_{k_0}(a)$ corresponding to an eigenfunction $\varphi_{k_0}(q; a)$, both of which are continuous functions of a, and if the physical situation at a given time, when $a = a_0$, is represented by $\varphi_{k_0}(q; a_0)$, then the situation will always be represented by $c_{k_0}(a)\varphi_{k_0}(q; a)$ $[|c_{k_0}(a)| = 1]$ if a is changed infinitely slowly (adiabatically). In short *a system which is in a stationary state, will stay in a stationary state during adiabatic processes.*

We can prove this theorem using equations (78), dividing the change of a up into small steps and proving the correctness of the

* J. M. Burgers, *Ann. Phys.*, **52**, 195, 1917; N. Bohr, *Z. Phys.*, **13**, 132, 1923.
** M. Born. *Z. Phys.*, **40**, 167, 1927.

theorem for one of these steps *. Let a_0 correspond to the energy operator $H^{(0)}_{\text{op}}$ and $a_0+\lambda$ (λ infinitesimal) to the energy operator $H^{(0)}_{\text{op}}+\lambda(\partial H_{\text{op}}/\partial a)_0 = H^{(0)}_{\text{op}}+\lambda H^{(1)}_{\text{op}}$, where $H^{(1)}_{\text{op}}$ depends on the time. Let the change in a be described by the equation

$$a = a_0+\lambda t/T, \qquad 0 \leqq t \leqq T. \tag{79}$$

The limit $T \to \infty$ corresponds to an adiabatic process. To apply equations (78) we write $G_{\text{op}} = \lambda t H^{(1)}_{\text{op}}/T$ and using equation (77) we find $\Phi_{k_0}(t) = \lambda H^{(1)}_{k_0k_0}t^2/2hT$, $G_{kk_0} = \lambda t H^{(1)}_{kk_0}/T$. Integrating by parts and neglecting terms of the order λ^2 we find from equations (78)

$$\gamma_{k_0}(T) = \exp[-i\lambda T H^{(1)}_{k_0k_0}/2\hbar),$$

$$\gamma_k(T) = -(\lambda H^{(1)}_{kk_0}/h\nu_{kk_0})[1-(2\pi i\nu_{kk_0}T)^{-1}]\cdot\exp\{2\pi i[\nu_{kk_0}T-(\lambda H^{(1)}_{k_0k_0}T/2h)]\},$$
$$k \neq k_0.$$

From these equations and equation (69) it follows that the wave function at $t = T$ is given by the equation

$$\begin{aligned}\psi(T) &= \Sigma_k \gamma_k(T)\varphi_k \exp(-iE_k t/\hbar)\\ &= \exp[-i(E_{k_0}+\tfrac{1}{2}\lambda H^{(1)}_{k_0k_0})T/\hbar]\\ &\qquad \{\varphi_{k_0}+\Sigma_{k(\neq k_0)}(\lambda H^{(1)}_{kk_0}/h\nu_{kk_0})[1-(2\pi i\nu_{kk_0}T)^{-1}]\varphi_k\}.\end{aligned} \tag{80}$$

As $T \to \infty$, the expression in braces goes over into that eigenfunction of the perturbed operator $H^{(0)}+\lambda H^{(1)}$ which correspond to the energy eigenvalue $E_{k_0}+\lambda H^{(1)}_{k_0k_0}$ [compare eq. (20)]. The system has stayed, so to say, the whole time during the adiabatic process in the same stationary state **. If T is finite, but $2\pi\nu_{kk_0}T \gg 1$ we see that the expression

$$\lambda^2|H^{(1)}_{kk_0}|^2/(2\pi h\nu_{kk_0}T)^2 \tag{81}$$

is the probability that during the process a "quantum jump" to the k-th state ($k \neq k_0$) has taken place. This probability is clearly proportional to the square of the speed with which the parameter a changes.

In general the adiabatic theorem loses its meaning when we are

* Compare N. Bohr, *Z. Phys.*, **13**, 132, 1923, and *Über die Quantentheorie der Linienspektren*, Vieweg, Brunswick, 1923, p. 14.

** If the system passes a degenerate state during the adiabatic process, which means that one or more of the ν_{kk_0} may vanish, the adiabatic theorem may still be valid, provided the derivative $\partial H_{\text{op}}(a)/\partial a$ is uniquely defined for the a-value for which the degeneracy occurred. This condition is necessary in order that H_{kk_0} be zero when ν_{kk_0} is zero. This condition is not necessary otherwise. We refer to a paper by Born and Fock (*Z. Phys.*, **51**, 165, 1928) where the adiabatic theorem is proved by more elegant methods.

dealing with a situation where improper stationary states, corresponding to a continuous spectrum, are excited. This is connected with the fact that one can interpret the eigenvalues corresponding to the continuum and their eigenfunctions in infinitely many ways as continuous functions of the parameter a.

§ 55. Time proportional transition probabilities

Let the eigenvalue spectrum of an unperturbed system be such that there exists a discrete energy eigenvalue E_0 in an interval which also contains a continuum of energy eigenvalues. The simplest example of such a situation is given by a particle in a two-dimensional field of force where the potential energy depends on the Cartesian coordinates as follows, $U = U^{(1)}(x)+U^{(2)}(y)$. The problem of finding the energy eigenvalues and eigenfunctions splits into two one-dimensional problems which refer only to the x- or to the y-coordinate (see § 17). Any eigenvalue E can be written in the form $E = E^{(1)}+E^{(2)}$ where $E^{(1)}$ corresponds to the x-problem and $E^{(2)}$ to the y-problem. The corresponding eigenfunction φ is the product of the x- and y-eigenfunctions, $\varphi = \varphi^{(1)}(x)\varphi^{(2)}(y)$. If now, for instance, the eigenvalue spectrum $E^{(1)}$ is discrete (for instance, $U^{(1)}(x) = ax^2$, the potential of a harmonic oscillator) and if the eigenvalue spectrum $E^{(2)}$ is partly discrete and partly continuous [$U^{(2)}(y)$ is of the form illustrated by fig. 9], a case such as we mentioned a moment ago may occur.

We assume that there is, apart from this complex unperturbed potential energy, a perturbation energy which may be due to a weak field of force with a potential which depends on both x and y.

Let us denote the unperturbed energy operator by $H_{\text{op}}^{(0)}$ and the operator of the perturbing energy by G_{op}, $H_{\text{op}} = H_{\text{op}}^{(0)}+G_{\text{op}}$. We shall assume that G_{op} does not contain the time explicitly. It is a very complicated problem to determine the exact eigenvalue spectrum of the perturbed system, since such a discrete eigenvalue which is embedded in a continuum will in general disappear due to the perburbation. We shall, however, describe the physical situation in terms of the unperturbed eigenfunctions and try to solve the problem of what will happen when at $t = 0$ the system were just in the stationary state corresponding to E_0.

Using the method of the variation of constants we write the wave function in the form [see eq. (69)]

$$\psi = \Sigma\gamma\varphi \exp(-iEt/\hbar), \qquad \Sigma|\gamma|^2 = 1, \tag{82}$$

where φ and E denote the eigenfunctions and eigenvalues of the unperturbed system, and where the summation sign indicates both summation over the discrete and integration (with a suitable density function) over the continuum stationary states. In the unperturbed system they will be constants while in the perturbed system equations similar to equations (74) hold.

Our system shows a peculiar kind of degeneracy. Because of the existence of a continuum the ν_{kl}-values which occur in equations (74) can take all values in the neighbourhood of 0, if we let k refer to E_0 and l to a level of the embedding continuum. We can therefore no longer effect a sharp separation of secular and periodic perturbations.

Let φ_0 be the eigenfunction of $H^{(0)}$ belonging to the discrete eigenvalue E_0, and let $\varphi(E)$ be the eigenfunctions belonging to the continuum in the neighbourhood of E_0. As a first approximation we have now instead of equations (82)

$$\psi = \gamma_0(t)\varphi_0 \exp(-iE_0t/\hbar) + \int \gamma(E,t)\varphi(E)\exp(-iEt/\hbar)\varrho(E)dE, \quad (83a)$$

$$|\gamma_0|^2 + \int |\gamma|^2 \varrho \, dE = 1, \qquad \gamma_0(0) = 1. \quad (83b)$$

These equations express the fact that the change in time of γ_0 — which at $t = 0$ was equal to 1, corresponding to the circumstance that at $t = 0$ only E_0 was excited — to a first approximation is governed only by the interaction with the neighbouring levels of the unperturbed system (secular perturbations) so that to that approximation only those $\gamma(E)$ will be appreciably different from zero. The integrals extend over intervals which contain E_0. The function $\varrho(E)$ is the density function corresponding to the choice of the (improper) eigenfunctions $\varphi(E)$. We could have arranged ϱ to be equal to unity by a suitable transformation (see § 40). The perturbation equations (74) are now of the form

$$i\hbar\dot{\gamma}_0 = G_{00}\gamma_0 + \int G_0^*(E)\gamma(E)\,\varrho(E)\exp[-i(E-E_0)t/\hbar]\,dE, \quad (84a)$$

$$\begin{aligned} i\hbar\dot{\gamma}(E) &= G_0(E)\,\gamma_0 \exp[i(E-E_0)t/\hbar] + \\ &\quad + \int G(E,E')\,\gamma(E')\,\varrho(E')\exp[i(E-E')t/\hbar]\,dE', \end{aligned} \quad (84b)$$

$$\left.\begin{aligned} G_{00} &= \int \varphi_0^* G_{\mathrm{op}}\varphi_0 \, d\tau, \qquad G(E,E') = \int \varphi^*(E)\, G_{\mathrm{op}}\varphi(E')\, d\tau, \\ G_0(E) &= \int \varphi^*(E)\, G_{\mathrm{op}}\varphi_0 \, d\tau, \qquad G_0^*(E) = \int \varphi_0^* G_{\mathrm{op}}\varphi(E)\, d\tau. \end{aligned}\right\} \quad (85)$$

Equations (84) could have been obtained by substituting expression (83) into the Schrödinger equation, multiplying the Schrödinger equation by φ_0^* or $\varphi^*(E)$ and integrating over all coordinates.

Without losing any generality we can put $G_{00} = 0$. The secular perturbation which γ_0 exerts on itself can always be eliminated by replacing γ_0 by $\gamma_0' = \gamma_0 \exp(-iG_{00}t/\hbar)$ and E_0 by $E_0' = E_0 + G_{00}$. Secondly we know that the following boundary condition must be satisfied,

$$t = 0 \rightarrow \gamma_0 = 1, \qquad \gamma(E) = 0, \tag{86}$$

so that $\gamma(E)$ will be small compared to γ_0 as long as t is not too large. This means that to a first approximation we may replace on the right hand side of equation (84) γ_0 and $\gamma(E)$ by their values at $t = 0$. This leads to the equations

$$i\hbar\dot{\gamma}(E) = G_0(E) \exp[i(E-E_0)t/\hbar], \tag{87}$$

with the solution

$$\gamma(E, t) = G_0(E)\{1 - \exp[i(E-E_0)t/\hbar]\}/(E-E_0). \tag{88}$$

The expression $|\gamma(E, t)|^2 \varrho \, dE$ is the probability that at t the states of the continuum for which the energy lies between E and $E+dE$ are excited. We can calculate this probability from equation (88) as long as the integral

$$J = \int |\gamma(E, t)|^2 \varrho \, dE \tag{89}$$

is small compared to unity. We introduce the notation

$$x = (E-E_0)/\hbar, \qquad E = E_0 + \hbar x, \tag{90}$$

and find for J,

$$J = \hbar^{-1} \int |G_0(E_0+\hbar x)|^2 \, 2(1-\cos xt) \, x^{-2} \, \varrho(E_0+\hbar x) \, dx.$$

The larger t is, the smaller is the range of x-values which contribute appreciably to the integral. Assuming $G_0(E)$ and $\varrho(E)$ to be smooth functions of E in the neighbourhood of $E = E_0$, we may for sufficiently large t-values replace the argument $E_0+\hbar x$ by E_0. Since $\int x^{-2}(1-\cos xt) \, dx \doteq t \int_{-\infty}^{+\infty} (1-\cos y) dy/y^2 = \pi t$, we find for J,

$$J = (2\pi/\hbar)|G_0(E_0)|^2 \varrho(E_0) t. \tag{91}$$

We have just seen that this expression is only valid if t is sufficiently large. If for such values of t expression (91) is still small compared to unity, due to the smallness of the perturbation, there exists, indeed, an interval of t-values within which J increases linearly with time.

From equation (83b) we see that $|\gamma_0|^2$ in the same time interval has decreased to $1-J$. This is not in disagreement with the fact that according to equations (84a) and (86) $\dot{\gamma}_0$ is zero to a first approximation, since J is small of the second order, if we assume G_{op} to be small of the first order. If we solve equation (84a) including second order terms, by substituting expression (88) for $\gamma(E)$ into the right hand side, we find that, indeed, $|\gamma_0|^2$ has decreased by an amount J.

Physically speaking we have obtained the following result. *There exists a certain probability per unit time, A,*

$$A = (2\pi/\hbar)|G_0(E_0)|^2\varrho(E_0), \tag{92}$$

that the system makes a transition from the discrete energy level E_0 to a continuum state with approximately the same energy.

Time proportional transition probabilities were introduced into quantum theory first of all by Einstein * in his considerations on the interaction between radiation and atoms. Dirac ** was the first to give the quantum mechanical interpretation of these probabilities, again for the case of radiative transitions. As well as in radiation theory, they play an important part in the discussion of collision problems, and also in the theory of spontaneous decay processes of atoms (Rosseland-Auger-effect), of molecules (predissociation) and of radioactive nuclei (α-decay).

In all these cases the situation is often more complicated than the one we discussed, although there is no basic difference. One of the complications which occurs sometimes is that the continuous spectrum itself is degenerate. In these cases one must replace $\varphi(E)$ and $\varrho(E)$ by $\varphi(E, r_1, r_2, \ldots)$ and $\varrho(E, r_1, r_2, \ldots)$ where the r_i are degeneracy parameters which we shall assume to be continuous [compare eq. (4.24)]. Expression (92) must now be replaced by the expression

$$A = (2\pi/\hbar) \int |G_0(E_0, r_1, r_2, \ldots)|^2 \varrho(E_0, r_1, r_2, \ldots)\, dr_1 dr_2 \ldots \tag{93}$$

In the theory of decay processes there is another complication. There we meet with the case (mentioned, for instance, at the begin-

* A. Einstein, *Phys. Z.*, **18**, 121, 1917.
** P. A. M. Dirac, *Proc. Roy. Soc.*, **A114**, 243, 1927.

ning of § 51) that we know certain approximate solutions of the Schrödinger equation which, however, do not correspond to an unperturbed energy operator.

Let us consider a system which possesses an approximately stationary state with wave function φ_0 and energy $E_0 = \int \varphi_0^* H_{\rm op} \varphi_0 d\tau$ (see § 51) and which also possesses a set of improper stationary states the wave functions of which are approximately equal to $\varphi(E)$. Let E_0 lie within the interval of the corresponding energy values E. Once again equation (83) will give us an approximate solution of the Schrödinger equation. Substituting this solution into the Schrödinger equation, multiplying this equation by $\varphi^*(E)$ and integrating over all coordinates we find apart from terms which certainly correspond only to fast, periodic perturbations [compare the analogous derivation of eq. (76)],

$$\int \varphi^*(E)[H_{\rm op} - i\hbar(\partial/\partial t)]\{\gamma_0(t)\,\varphi_0 \exp(-iE_0 t/\hbar) + \\ + \int \gamma(E', t)\varphi(E') \exp(-iE't/\hbar)\,\varrho(E')\,dE'\}\,d\tau = 0.$$

To simplify the considerations we shall assume that φ_0 is orthogonal to all $\varphi(E)$, at any rate to a good approximation, and that also the $\varphi(E)$ are approximately orthogonal to one another [compare eq. (4.27)], $\int \varphi^*(E)\varphi(E')\,d\tau = \varrho^{-1}\delta(E-E')$. Substituting for the γ their approximate values (86) we find

$$\exp(-iE_0 t/\hbar) \int \varphi^*(E)(H_{\rm op} - E_0)\,\varphi_0\,d\tau - i\hbar\dot{\gamma}(E) \exp(-iEt/\hbar) = 0.$$

This equation is the same as equation (87), if we put

$$G_0(E) = \int \varphi^*(E)(H_{\rm op} - E_0)\,\varphi_0\,d\tau, \tag{94}$$

and we can thus proceed as before.

We may draw attention to one particular point in connexion with equation (91). The probability distribution of the perturbed system over the continuous E-values is at t given by the expression $|\gamma(E, t)|^2 \varrho(E) dE$. If we introduce x by equation (90) this equation is, for sufficiently large values of t, equal to $W(x)dx = (A/\pi) \cdot (1 - \cos xt)dx/x^2$. The fact that such a distribution exists is not in contradiction to the principle of the conservation of energy, since the original situation referred to a stationary state of the unperturbed system and the same is true for the states characterised by E. The normalised distribution function $W'(x)$,

$$W'(x) = [(1 - \cos xt)/\pi t x^2]dx, \qquad \int W'(x)dx = 1, \tag{95}$$

depends on t. The larger t is, the larger is $W'(0) = t/2\pi$ and the more often will $W'(x)$ oscillate in a given interval. For very large values of t, equation (95) will no longer be valid, as J in expression (91) is no longer small compared to unity. Presumably $|\gamma_0|^2$ will decrease and approach zero for increasing values of t and we may ask whether there will exist a definite distribution of the system over E-values (or x-values) as $t \to \infty$. It is, indeed, possible to derive such an asymptotic distribution from equation (84), if one may assume that the matrix elements $G(E, E')$ all vanish, that is, that the continuous E-states do not lead to secular perturbations of each other. In many applications (for instance, radiation processes) the $G(E, E')$ are in fact equal to zero.

In this case we have instead of equations (84) the equations

$$i\dot{\gamma}_0(t) = G_0^*(E_0)\,\varrho(E_0)\int_{-\infty}^{+\infty}\gamma(x, t)e^{-ixt}\,dx, \qquad (84a');$$

$$i\hbar\dot{\gamma}(x, t) = G_0(E_0)\,\gamma_0(t)e^{ixt}. \qquad (84b')$$

In equations (84′) we have replaced the argument E of G_0 and ϱ by E_0 and we have written $\gamma(x)$ for $\gamma(E_0+\hbar x)$. If we take $\int_{-\infty}^{+\infty}$ to mean $\lim_{M\to\infty}\int_{-M}^{+M}$ we see that equations (84′) are solved by the expressions*

$$\gamma_0(t) = \exp\left(-\tfrac{1}{2}At\right), \qquad |\gamma_0|^2 = e^{-At}; \qquad (96a)$$

$$\left.\begin{aligned}\gamma(x, t) &= (G_0/\hbar)\{[1-\exp\left(-\tfrac{1}{2}At+ixt\right)]/(x+\tfrac{1}{2}iA)\},\\ |\gamma|^2 &= (|G_0|^2/\hbar^2)[(1-2\cos xt\,e^{-\frac{1}{2}At}+e^{At})/(x^2+\tfrac{1}{4}A^2)].\end{aligned}\right\} \qquad (96b)$$

We see that $|\gamma_0|^2$ approaches zero exponentially as we would expect from the definition of a transition probability, while in the limit $t\to\infty$ $|\gamma|^2$ corresponds to the probability distribution,

$$W'(x)dx = (A/2\pi)[dx/(x^2+\tfrac{1}{4}A^2)]. \qquad (97)$$

* Equation (84b′) is easily seen to be satisfied by equation (96), but to see that these equations also satisfy equations (84a′) we must prove that

$$\int_{-\infty}^{+\infty}[\exp(\tfrac{1}{2}At-ixt)/(x+\tfrac{1}{2}iA)]dx - \int_{-\infty}^{+\infty}[dx/(x+\tfrac{1}{2}iA)] = -\pi i.$$

The first integral is evaluated by integrating in the complex x-plane and completing the contour by a semi-circle of radius M in the lower half of the plane. The result is $-2\pi i$. The second integral is equal to $-\lim_{M\to\infty}\ln[(\tfrac{1}{2}iA+M)/(\tfrac{1}{2}iA-M)]$ $= \lim_{M\to\infty}\ln[-1+(iA/M)+\ldots] = +\pi i$, and the sum is, indeed, equal to $-\pi i$.

CHAPTER VI

THE SPINNING ELECTRON

A. Non-relativistic Spin Theory

§ 56. Uhlenbeck and Goudsmit's hypothesis of the rotating magnetic electron

In the first decade after its birth the Bohr theory of atomic structure turned out to be a very reliable guide for studying many atomic properties. Its strong point was especially its ability to explain spectral regularities and it thus led to new developments in spectroscopic investigations. In the early twenties it became gradually clear that the old quantum theory had its limitations. First of all it turned out that although many regularities could be interpreted by using models * and the correspondence principle they could not be treated quantitatively. Secondly there existed empirical laws — some of them known for a long time: for instance, the anomalous Zeeman effect — which could not be interpreted by the old quantum theory, however much one tried. In those cases one had to be satisfied with a suggestive theoretical terminology. These difficulties looked at least partially overcome when in 1925 Uhlenbeck and Goudsmit put forward the hypothesis of the rotating magnetic electron **. By extending the classical ideas about the nature of the electron they enriched the models of the atom in such a way that several spectroscopic regularities which were under discussion could be interpreted using the correspondence principle.

Uhlenbeck and Goudsmit were first of all concerned with an interpretation of the anomalous Zeeman effect and the "branching law" which expresses a fundamental property of the multiplet structure of atomic spectra. In their considerations the interpretation of the fourth quantum number, which Pauli had introduced for an electron

* The use of models consisted mainly in describing the physical situation of the atomic system, and especially the stationary states, by identifying it with classically possible orbits and by calculating the properties of the system in that way.

** G. E. Uhlenbeck and S. Goudsmit, *Naturwiss.*, **13**, 953, 1925; *Nature*, **117**, 264, 1926.

in a central field of force, played a prominent part. We can not enter into a discussion of historical details, but refer the reader to the literature *. At this point we are interested only in the formulation of the Uhlenbeck-Goudsmit hypothesis: *An electron behaves as a charged particle which is rotating around an axis in such a way, that it possesses an angular momentum of constant absolute magnitude* $\frac{1}{2}\hbar$ *around that axis and that it possesses at the same time a magnetic moment* $e\hbar/2mc$ *which has a direction opposite to the direction of the angular momentum.* In general the direction of the axis of the electron will change in time due to external forces. This change will be similar to the precession of the earth's axis. One nowadays calls these properties "electron spin", which immediately shows the analogy with rotations.

The occurrence of $\hbar$ in the hypothesis introduces from the beginning a quantum theoretical aspect and the hypothesis can be split up into two parts. The first part is purely classical: the electron possesses apart from a charge and a mass also an angular momentum of constant magnitude and a corresponding magnetic moment which is equal to $-e/mc$ times the angular momentum. The second part is quantum theoretical: the absolute magnitude of the angular momentum is always equal to $\frac{1}{2}\hbar$.

The meaning of these statements can be illustrated by their applications in the old quantum theory such as were made by Uhlenbeck and Goudsmit. There are two kinds of problem to be considered. Firstly, how can we complete the classical description of the motion of an electron or of a system of electrons so as to take the spin properties into account? Secondly, how can we "quantise" this classical theory so as to arrive at a more or less useful model of stationary states?

The first problem will be discussed in § 57. The exact treatment of the second problem is only of historical interest and we shall not give it, but we shall in § 58 immediately start the discussion of the modern quantum mechanical treatment of the spinning electron. It has turned out to be possible to include the spin in a sensible, logical way in quantum mechanics. This was first of all done in a paper by Heisenberg and Jordan** on the anomalous Zeeman-effect where the

* See, for instance, H. Kuhn, *Atomspektren, Hand- u. Jahrb. Chem. Phys.*, Vol 9/I, Akademische Verlag, Leipzig 1934 or L. Pauling and S. Goudsmit, *The structure of line spectra*, McGraw-Hill, New York, 1930.

** W. Heisenberg and P. Jordan, *Z. Phys.*, **37**, 263, 1926.

spin was included in the framework of matrix mechanics. Soon after Schrödinger's first papers had appeared Pauli * showed how one could treat the spin in the framework of wave mechanics, that is, how one had to modify the wave function description of atomic phenomena. It turned out that Heisenberg and Jordan's paper contained already all essential points. However, the wave mechanical approach illuminated certain new aspects of the quantum mechanical spin theory. We have especially in mind the impossibility of using continuous variables to describe the spin and the two-valued character of the wave functions. We shall return to these points.

An essential extension of the quantum mechanical treatment of the electron spin was obtained when Dirac's papers appeared in 1928 **. Pauli's treatment of the spin could satisfactorily explain a large number of phenomena, but it was unsatisfactory since it did not satisfy the requirement of the relativistic invariance of the laws of nature. This failing it had in common with the spinless non-relativistic quantum mechanical treatment of one particle or many particle systems which was the basis of Pauli's theory of the electron spin. Dirac investigated the case of one electron in an external electromagnetic field — without introducing the spin explicitly. He discovered that the requirement of relativistic invariance could be met by generalising the Schrödinger equation to a set of four simultaneous partial differential equations of the first order, *the so-called linear Dirac equations,* while still retaining the general physical aspects of the wave mechanical description. It was then found that the spin properties are automatically contained in these equations and that Pauli's treatment is equivalent to the limiting case of small velocities.

§ 57. The classical description of the motion of a spinning electron

According to the Dirac theory of the electron the spin properties are so closely connected with the (relativistic) quantum mechanical description of nature that it is rather artificial and hardly relevant to discuss first a "classical" spin theory. However, classical considerations played an important role in the historical development. Also, they are a powerful help if one wants to develop the theory gradually using heuristic points of view. On top of it there does not exist at the

* W. Pauli, *Z. Phys.*, **43** 601, 1927.
** P. A. M. Dirac, *Proc. Roy. Soc.*, **A117**, 610, 1928; **A118**, 351, 1928.

moment * a consistent theory of the electron and its spin which satisfies all physical requirements and at the discussion of which one could aim one's development of the theory.

The classical theory of the spinning electron consists of a set of equations of motion which closely resemble the equations of motion of a rigid body and in which the spin enters as a *new degree of freedom* in addition to the three translational degrees of freedom. It is true that according to classical mechanics a rigid body possesses three rotational degrees of freedom, that is, the state of motion with respect to the centre of gravity depends on six variables. The spinning electron must, however, be compared with a limiting or special case where only two variables (apart from the position and velocity of the centre of gravity) are needed to describe the state of motion. These two variables determine the spin direction, since according to Uhlenbeck and Goudsmit only the direction of the angular momentum of the electron with respect to the centre of gravity can change.

Such a limiting case occurs, for instance, when a rigid body is rotating so rapidly that the influence of external forces leads only to a small perturbation of the rotational motion and when, furthermore, the secular effect of this perturbation is such that only the direction of the total angular momentum changes, while the other coordinates describing the rotational motion remain unchanged. A well-known example is that of a rotationally symmetric body which is rotating rapidly round an axis close to its axis of symmetry while the total moment of the external forces with respect to the centre of gravity is always perpendicular to the axis of symmetry (top, precession of the earth).

The equations of motion of the spinning electron must express the fact that the rate of change of the momentum (or of the angular momentum) is equal to the resultant external force (or the resultant moment of the external forces). These external forces derive from the electromagnetic fields in which the electron is situated, and their action depends on the charge and current distribution inside the electron.

The consistent definition of field quantities and of charge and current densities is closely connected with the formalism of the theory of special relativity. If we go over from one inertial system to

* This was written in 1938, but it seems to the translator that this is still true, even though the recent renormalisation theories have proved to be very successful, at least as far as the theory of the electron is concerned.

another by means of a Lorentz transformation, these quantities transform according to well defined formulae. However, the idea of a rigid body whose state of rotation changes gradually is on purely kinematic grounds irreconcilable with the theory of special relativity*. We shall thus expect to run into trouble when we try to formulate equations of motion for the spinning electron which are Lorentz invariant. These difficulties are particularly serious if one tries — as Lorentz once did ** — to discuss the situation using a definite model of the electron which satisfies an ingenious definition of rigidity. Fortunately it turns out that one can derive a simple relativistic set of equations of motion which satisfies the requirements of atomic physics by formally extending the non-relativistic formulae without introducing any special model.

To begin with let us assume with Uhlenbeck and Goudsmit that the electron possesses apart from a charge $-e$ a magnetic dipole moment $\boldsymbol{\mu}$ in the direction of the spin. We shall denote the ratio of the magnetic moment and the angular momentum (or spin) $\boldsymbol{A}$ by α,

$$\boldsymbol{\mu} = \alpha \boldsymbol{A}. \tag{1}$$

According to Uhlenbeck and Goudsmit α is equal to $-e/mc$, but we shall leave its value open for the time being. We emphasise here that the electron does not possess an electrical dipole moment, quadrupole moment, . . ., or a magnetic quadrupole moment, octupole moment, ... We could express this as follows. The electromagnetic field due to the electron is the same as a field generated by a uniformly rotating sphere which is covered uniformly with charge.

If we may treat the electron as a point particle, the moment of the forces which will act on an electron in an external electromagnetic field will be equal to $[\boldsymbol{\mu} \wedge \mathfrak{H}]$ and the rate of change of its spin is given by the equation

$$\dot{\boldsymbol{A}} = \alpha[\boldsymbol{A} \wedge \mathfrak{H}], \tag{2}$$

where we have used equation (1). From equation (2) it follows that A^2 is constant $[(\boldsymbol{A} \cdot \dot{\boldsymbol{A}}) = 0]$.

If the electron is moving with a velocity $\boldsymbol{v}$, we should expect that also the electrical field will affect $\boldsymbol{A}$. The magnetic field $\mathfrak{H}'$ in the inertial system which moves with the electron is equal to $\mathfrak{H}' =$

* See, for instance, P. Ehrenfest, *Phys. Z.*, **10**, 918, 1909; **11**, 1127, 1910.

** H. A. Lorentz, Volta congress 1927; see: *Collected papers*, **7**, 178, Nijhoff, The Hague, 1934.

$\{\mathfrak{H}+c^{-1}[\mathfrak{E}\wedge \boldsymbol{v}]\}[(1-(v^2/c^2)]^{-1/2}$. If the assumption that the spin vector would be the same in both inertial systems were justified — in Newtonian mechanics the angular momentum of a rigid body with respect to the centre of gravity is the same in all inertial systems — we would get the following generalisation of equation (2) for a moving electron, neglecting terms of the order v^2/c^2,

$$\dot{\boldsymbol{A}} = \alpha[\boldsymbol{A} \wedge \{\mathfrak{H}+c^{-1}[\mathfrak{E} \wedge \boldsymbol{v}]\}]. \tag{3}$$

However, equation (3) is incorrect even if we may neglect terms of the order v^2/c^2. This is so, since the existence of a spatial vector $\boldsymbol{A}$ which would be the same in all inertial systems is in contradiction to the theory of special relativity. We must renounce the Newtonian concept of a rigid body and try to find a formal generalisation of equation (2) which is Lorentz invariant. A simple method and the most convenient one for the further development of the theory is the following one. Introduce apart from $\boldsymbol{A}$ a second spatial vector $\boldsymbol{B}$ to characterise the spin. This vector must satisfy the following requirements *.

1. The vector $\boldsymbol{B}$ is equal to zero when the electron is at rest.
2. The components of $\boldsymbol{A}$ and $\boldsymbol{B}$ form a six-vector, that is, they transform under a Lorentz transformation in the same way as the components of $\mathfrak{H}$ and $\mathfrak{E}$.

The simplest relativistically invariant generalisation of equation (2) is given by

$$d\boldsymbol{S}/d\tau = \alpha[\boldsymbol{S} \wedge \mathfrak{G}], \tag{4}$$

where the complex vectors $\boldsymbol{S}$ and $\mathfrak{G}$ are defined by the equations

$$\boldsymbol{S} = \boldsymbol{A}-i\boldsymbol{B}, \tag{5}$$

$$\mathfrak{G} = \mathfrak{H}-i\mathfrak{E}, \tag{6}$$

and where $d\tau$ is an element of eigentime. Equating both the real and the imaginary part of equation (4) leads to the equations

$$d\boldsymbol{A}/d\tau=\alpha\{[\boldsymbol{A}\wedge\mathfrak{H}]-[\boldsymbol{B}\wedge\mathfrak{E}]\}, \text{ (7a)}; \quad d\boldsymbol{B}/d\tau=\alpha\{[\boldsymbol{B}\wedge\mathfrak{H}]+[\boldsymbol{A}\wedge\mathfrak{E}]\} \text{ (7b)}$$

From equation (4) it follows that $S^2 = (\boldsymbol{A}-i\boldsymbol{B} \cdot \boldsymbol{A}-i\boldsymbol{B}) = A^2-B^2-2i(\boldsymbol{A} \cdot \boldsymbol{B})$ does not change along the world line. The real Lorentz invariant quantity A^2-B^2 can thus be considered to be the constant absolute square of the spin.

* J. Frenkel, *Z. Phys.*, **37**, 243, 1926.

We can see as follows that equation (4) is Lorentz invariant. The components of the vector $\boldsymbol{S}$ and $\mathfrak{G}$ are subject to an orthogonal transformation when we perform a general Lorentz transformation. This means that the components $\mathfrak{G}_x'$, $\mathfrak{G}_y'$, $\mathfrak{G}_z'$ in the transformed inertial system are linear homogeneous expressions in the components $\mathfrak{G}_x$, $\mathfrak{G}_y$, $\mathfrak{G}_z$ in the original inertial system and

$$\mathfrak{G}'^2 \equiv \mathfrak{G}_x'^2+\mathfrak{G}_y'^2+\mathfrak{G}_z'^2 = \mathfrak{G}_x^2+\mathfrak{G}_y^2+\mathfrak{G}_z^2 \equiv \mathfrak{G}^2. \tag{8}$$

If the two inertial systems move with respect to one another, the transformation coefficients will be complex. For instance, if we consider the well-known special case of the transformation

$$x' = \beta(x-vt), \quad y' = y, \quad z' = z, \quad t' = \beta[t-(xv/c^2)], \quad \beta = [1-(v^2/c^2)]^{-\frac{1}{2}},$$

we get for $\mathfrak{H}$ and $\mathfrak{E}$ the transformation

$$\mathfrak{H}_x' = \mathfrak{H}_x, \qquad \mathfrak{H}_y' = \beta[\mathfrak{H}_y+(v/c)\,\mathfrak{E}_z], \qquad \mathfrak{H}_z' = \beta[\mathfrak{H}_z-(v/c)\,\mathfrak{E}_y];$$
$$\mathfrak{E}_x' = \mathfrak{E}_x, \qquad \mathfrak{E}_y' = \beta[\mathfrak{E}_y-(v/c)\,\mathfrak{H}_z], \qquad \mathfrak{E}_z' = \beta[\mathfrak{E}_z+(v/c)\,\mathfrak{H}_y].$$

The components of $\mathfrak{G}$ are then transformed according to the equations

$$\mathfrak{G}_x' = \mathfrak{G}_x, \qquad \mathfrak{G}_x' = \beta[\mathfrak{G}_y+i(v/c)\,\mathfrak{G}_z], \qquad \mathfrak{G}_z' = \beta[\mathfrak{G}_z-i(v/c)\,\mathfrak{G}_y], \tag{9}$$

corresponding to a rotation round the x-axis over an angle $\arctan(iv/c)$. The invariance of $\mathfrak{G}^2 = (\mathfrak{H}-i\mathfrak{E}\cdot\mathfrak{H}-i\mathfrak{E})$ expresses the wellknown invariance of the expressions $\mathfrak{E}^2-\mathfrak{H}^2$ and $(\mathfrak{E}\cdot\mathfrak{H})$.

We see that the six-vectors of the theory of relativity correspond to complex vectors in a three-dimensional space and that any relation between a number of such complex vectors leads to two Lorentz invariant relations between six-vectors when we separate the real and imaginary parts. Equation (4) is an example of such a relation. The left hand side is a six-vector, since we differentiate with respect to the Lorentz invariant eigentime. The right hand side is also a six-vector as the components of a cross product transform as the components of the factors under a complex orthogonal transformation. We have restricted ourselves to those transformations which can be brought back to the identical transformation by continuous changes of the coefficients, that is, we have excluded inversions.

Equation (7) describes the rate of change of $\boldsymbol{A}$ and $\boldsymbol{B}$ along the world line. We must investigate whether, indeed, these equations satisfy the condition that $\boldsymbol{B} = 0$ in the inertial system moving with the electron, because only in that case can we consider equation (4) to be a justifiable generalisation of equation (2). Mathematically speaking this means that $\boldsymbol{B}$ must satisfy along the world line the condition

$$\boldsymbol{B} = c^{-1}[\boldsymbol{A} \wedge \boldsymbol{v}], \tag{10}$$

since this equation is Lorentz invariant and must be satisfied for $\boldsymbol{v} = 0$.

To prove equation (10) is Lorentz invariant we note that $\beta\{\mathfrak{E}+c^{-1}[\boldsymbol{v}\wedge\mathfrak{H}]\}$ and $\beta c^{-2}(\mathfrak{E}\cdot\boldsymbol{v})$ form a four-vector. Replacing $\mathfrak{E}$ by $\boldsymbol{B}$ and $\mathfrak{H}$ by $\boldsymbol{A}$ we still have a four-vector. This four-vector is zero for $\boldsymbol{v} = 0$ and must thus be equal to zero in every inertial system. This means that $\beta\{\boldsymbol{B}+c^{-1}[\boldsymbol{v}\wedge\boldsymbol{A}]\} = 0$, and equation (10) follows.

If equations (10) and (7) are both satisfied we have

$$d\boldsymbol{B}/d\tau = c^{-1}[(d\boldsymbol{A}/d\tau) \wedge \boldsymbol{v}] + c^{-1}[\boldsymbol{A} \wedge (d/\boldsymbol{v}d\tau)] = \alpha\{[\boldsymbol{B} \wedge \mathfrak{H}] + [\boldsymbol{A} \wedge \mathfrak{E}]\}.$$

In an inertial system in which $\boldsymbol{v} = 0$, $\boldsymbol{B}$ also vanishes and in this inertial system the eigentime is the same as the ordinary time so that we get

$$[\boldsymbol{A} \wedge (d\boldsymbol{v}/dt)] = \alpha c[\boldsymbol{A} \wedge \mathfrak{E}]. \tag{11}$$

We can compare this equation with the law which governs the rate of change of the position of the electron. According to the model we are using the force on the electron in the inertial system moving with the electron would be $-e\mathfrak{E} + (\boldsymbol{\mu} \cdot \nabla)\mathfrak{H}$ (the second term is due to a possible inhomogeneity of the magnetic field). If we use equation (1) we get for the equation of motion

$$m d\boldsymbol{v}/dt = -e\mathfrak{E} + \alpha(\boldsymbol{A} \cdot \nabla)\mathfrak{H}. \tag{12}$$

The last term on the right hand side expresses a spin-orbit interaction.

Substituting expression (12) into equation (11) and neglecting terms of the second degree in $\boldsymbol{A}$ we get for the constant α the following value:

$$\alpha = \text{ratio of spin magnetic moment to spin angular momentum} = -e/mc. \tag{13}$$

It is interesting to note that the simple relativistic equation (4) immediately leads to the value of α which was postulated by Uhlenbeck and Goudsmit. That it is necessary to neglect the spin-orbit interaction term means that a simple classical relativistic spin theory is possible only in the limit where the spin is so small that the reaction of the spin magnetic moment on the orbital motion of the electron can be neglected.

Equation (4) was a special choice. More generally, one could add to $\mathfrak{G}$ a vector $\mathfrak{G}'$ which transforms as $\mathfrak{G}$ and $\boldsymbol{S}$ and whose real part vanishes in the inertial system moving with the electron.* If we demand that $\mathfrak{G}'$ depends only on the Minkowski velocity and on $\mathfrak{E}$ and $\mathfrak{H}$ we have *

$$\mathfrak{G}' = k\left\{i\mathfrak{E} + \frac{i}{c}\frac{dt}{d\tau}\left[\frac{d\boldsymbol{x}}{d\tau} \wedge \mathfrak{G}^*\right] - \frac{1}{c^2}\left[\frac{d\boldsymbol{x}}{d\tau} \wedge \left[\frac{d\boldsymbol{x}}{d\tau} \wedge \mathfrak{G}^*\right]\right]\right\}, \qquad \mathfrak{G}^* = \mathfrak{H} + i\mathfrak{E}.$$

Condition (10) would lead to a value of α which depends on $k : \alpha = -e/mc(1-k)$. Only in the simple case where $k = 0$ do we get the value (13) which is considered in accordance with experimental data.

We might try to get rid of the restriction of negligible spin-orbit interaction by assuming that the electron at rest possesses apart from a charge and a magnetic

* See J. Frenkel, *Z. Phys.*, **37**, 243, 1926.

dipole moment also higher multipole moments which are all symmetric with respect to the spin axis. Equation (4) would then be the first term in a series expansion in powers of $\boldsymbol{S}$. We do not have the space here to enter into a discussion of this problem. Apart from the question of whether it would be at all possible to find a solution, we must draw attention to the fact that these classical multipoles will vanish as soon as the theory is quantised and the electron is given an angular momentum $\frac{1}{2}\hbar$. If the particle possessed an angular momentum $1 \cdot \hbar$ there would be a quadrupole moment; if its angular momentum were $\frac{3}{2}\hbar$ there would also be an octupole moment, and so on.

If a charged mass point or a system of identical charged mass points (mass m, charge $-e$) moves in an attractive central field according to the laws of Newtonian mechanics, the system under consideration possesses first of all an angular momentum which is constant in magnitude and direction. Secondly, it produces a magnetic field the time average of which to a first approximation is the field of a constant magnetic dipole. The direction of this orbital magnetic moment is the same as that of the orbital angular momentum and the ratio of the two is in the non-relativistic approximation equal to $-e/2mc$, or,

$$\text{ratio of orbital magnetic moment to orbital angular momentum} = -e/2mc. \tag{14}$$

Proof: In the non-relativistic approximation the magnetic vector potential $\mathfrak{A}$ produced by the movement of the charged mass points at the point $\boldsymbol{X}$ ($|\boldsymbol{X}| = R$) is given by the expression

$$\mathfrak{A} = -(e/c)\,\Sigma_i[\dot{\boldsymbol{x}}_i/|\boldsymbol{X}-\boldsymbol{x}_i|] = -(e/c)\Sigma_i\,\dot{\boldsymbol{x}}_i[R^2-2(\boldsymbol{X}\cdot\boldsymbol{x}_i)+r_i^2]^{-1/2},$$

where $\boldsymbol{x}_i$ is the position of the i-th particles ($|\boldsymbol{x}_i| = r_i$), and where the attractive centre is at the origin.

Expanding in powers of R^{-1} leads to $\mathfrak{A} = -(e/c)\{R^{-1}\Sigma_i\dot{\boldsymbol{x}}_i+R^{-3}\Sigma_i\dot{\boldsymbol{x}}_i(\boldsymbol{X}\cdot\boldsymbol{x}_i)+\ldots\}$ $= -(e/c)\{R^{-1}(d/dt)\Sigma_i\boldsymbol{x}_i+R^{-3}\Sigma_i\frac{1}{2}[[\boldsymbol{x}_i\wedge\dot{\boldsymbol{x}}_i]\wedge\boldsymbol{X}]+R^{-3}(d/dt)\,\Sigma_i\,\frac{1}{2}\boldsymbol{x}_i(\boldsymbol{X}\cdot\boldsymbol{x}_i)+\ldots\}$. If the particles form a system which is confined to a finite volume of space—and otherwise it has no sense to talk about a magnetic moment—the time average of the terms with d/dt is equal to zero and we have

$$\mathfrak{A} = R^{-3}[\boldsymbol{\mu}\wedge\boldsymbol{X}],\quad \boldsymbol{\mu} = -(e/2mc)\,\Sigma_i\,m[\boldsymbol{x}_i\wedge\dot{\boldsymbol{x}}_i].$$

This expression for $\mathfrak{A}$ is just the vector potential of a magnetic dipole which as we have just derived is equal to the total angular momentum multiplied by $-e/2mc$.

A different derivation and one which is more closely connected with the applications of the concept of magnetic moments and of their measurement consists in investigating what terms must be added to the Hamiltonian when an external magnetic field $\mathfrak{H}$ is applied. The extra term due to the i-th particle is $(e/mc)(\boldsymbol{p}_i\cdot\mathfrak{A}_i)$ [compare eqq. (2.101) and (3.76) with $-e$ replacing e], where $\boldsymbol{p}_i$ is the momentum of the i-th particle and $\mathfrak{A}_i$ the vector potential of $\mathfrak{H}$ at the position of the i-th particle. If we are dealing with a homogeneous magnetic field, we have $\mathfrak{A}_i = \frac{1}{2}[\mathfrak{H}\wedge\boldsymbol{x}_i]$, and we get for the total additional term in the Hamiltonian $(e/mc)\Sigma_i(\boldsymbol{p}_i\cdot[\mathfrak{H}\wedge\boldsymbol{x}_i]) = (\{e/2mc\}\Sigma_i[\boldsymbol{x}_i\wedge\boldsymbol{p}_i]\cdot\mathfrak{H})$. This corresponds to the energy of a magnetic moment $\boldsymbol{\mu} = -(e/2mc)\Sigma_i[\boldsymbol{x}_i\wedge\boldsymbol{p}_i]$ in a homogeneous magnetic field $\mathfrak{H}$.

If we are in particular dealing with electrons, the ratio (14) is half of the corresponding ratio (13). This difference is responsible for the anomalous Zeeman effect as Uhlenbeck and Goudsmit showed.

Uhlenbeck and Goudsmit discussed the possibility of interpreting classically the ratio (13). In particular they calculated the ratio of the magnetic moment of a uniformly rotating spherical shell which is uniformly covered with charge to that part of the angular momentum which according to electron theory is due to the electromagnetic field. They found for this ratio $-e/m_{em}c$, where m_{em} is the electromagnetic mass. In many ways this kind of consideration, based on a definite model, is unsatisfactory. To begin with one neglects the forces which are not electromagnetic of origin. Secondly, the result is no longer correct if the surface charge distribution is replaced by an arbitrary radially symmetric charge distribution. Finally one would expect that the magnetic moment of such a model would at most be of the order of magnitude er, where r is a measure for the linear dimensions of the electron. If one takes the classical electron radius e^2/mc^2 for r we get an expression for the magnetic moment which is smaller than the quantum mechanical value $e\hbar/2mc$ by a factor of the order of $2e^2/\hbar c \doteq 0.015$. In our treatment we have used the equations of motion of a rotating body [see, for instance, eq. (2)], but otherwise not introduced any model. We emphasised only the need for a consistent set of relativistically invariant equations of motion for a mass point.

It is perhaps not justified to call our theory "classical", but it is definitely not quantum mechanical.

We shall now investigate the problem of *energy* and *classical equations of motion*. We shall use, to begin with, the non-relativistic equation (2) and substitute for α its value (13),

$$\dot{\boldsymbol{A}} = -(e/mc)[\boldsymbol{A} \wedge \mathfrak{H}]. \tag{15}$$

In classical electromagnetic theory the energy of a magnetic dipole $-(e/mc)\boldsymbol{A}$ in a field $\mathfrak{H}$ would be equal to

$$H_s = (e/mc)(\boldsymbol{A} \cdot \mathfrak{H}). \tag{16}$$

If $\mathfrak{H}$ is constant, we see from equations (15) and (16) that H_s is also constant. We now introduce canonical spin coordinates p_s and q_s through the equations

$$p_s = A_z, \qquad q_s = \arctan(A_y/A_x), \tag{17}$$

that is, p_s is the spin z-component and q_s the azimuth of the direction of the spin. It is easily verified, that equation (15) is equivalent to

$$\dot{p}_s = -\partial H/\partial q_s, \qquad \dot{q}_s = \partial H/\partial p_s, \tag{18}$$

if we write expression (16) for H_s in the form

$$H_s = (e/mc)\,\{[A^2-p_s^2]^{1/2}[\mathfrak{H}_x \cos q_s + \mathfrak{H}_y \sin q_s] + p_s\mathfrak{H}_z\}.$$

We can now take into account the relativistic details which are embodied in equations (7) and (10) without changing equations (15) to (18) too much. Substituting equation (10) into equation (7a) we have

$$dA/d\tau = -(e/mc)\{[A \wedge \mathfrak{H}] - c^{-1}[[A \wedge v] \wedge \mathfrak{E}]\}. \tag{19}$$

It is impossible to find a pair of variables p_s and q_s such that equation (19) is equivalent to a canonical set of equations. However, we know that to a first, non-relativistic approximation $-e\mathfrak{E} = m\dot{v}$ so that $-[[A \wedge v] \wedge \mathfrak{E}] = \frac{1}{2}[A \wedge [\mathfrak{E} \wedge v]] + A(\mathfrak{E} \cdot v) - \frac{1}{2}v(A \cdot \mathfrak{E}) = \frac{1}{2}[A \wedge [\mathfrak{E} \wedge v]] - (m/2e)dV/dt$, where $V = Av^2 - v(A \cdot v) = [v \wedge [A \wedge v]]$. This means that we can write equation (19) in the form $\dot{A} = -(e/mc)[A \wedge \mathfrak{H}] - (e/2mc^2)[A \wedge [\mathfrak{E} \wedge v]] + \frac{1}{2}c^{-2}\dot{V}$. This equation is correct up to terms which are of the order v^2/c^2 compared to the first term (this is the order of the terms neglected in the equation $-e\mathfrak{E} = m\dot{v}$), that is, up to terms including c^{-3}. If we are dealing with a case where the direction of the spin changes only slowly as compared to the change in velocity — as will happen in most atomic problems — and where the velocity will periodically take on the same values, we can neglect the last term in the expression for $\dot{A}$ as it will not contribute to a secular change of A, since its time average, taken over period of the motion, is equal to zero. Hence, if we neglect the short periodic perturbations on the spin direction, we find for the rate of change of the spin the equation

$$\dot{A} = -(e/mc)[A \wedge (\mathfrak{H} + \tfrac{1}{2}c^{-1}[\mathfrak{E} \wedge v])]. \tag{20}$$

This equation of motion differs from equation (3) by the occurrence of a factor $\frac{1}{2}$, the so-called *Thomas factor*. This equation is formally of the same type as equation (15) and can thus immediately be written in the canonical form (15), provided we take for the energy instead of equation (16) the expression

$$H_s = (e/mc)(A \cdot \{\mathfrak{H} + \tfrac{1}{2}c^{-1}[\mathfrak{E} \wedge v]\}). \tag{21}$$

This expression is correct up to terms of order v^2/c^2. We have obtained the so-called *non-relativistic expression for the spin energy.*

In a hydrogen atom the term with $[\mathfrak{E} \wedge v]$ is the only one which can produce a change in the direction of the spin, if there is no external magnetic field. It played an important role in the discussion of the hydrogen spectrum. Originally one used equation (3) and the corresponding expression for the energy which differed from expression (21) by the absence of the factor $\frac{1}{2}$. One noticed then that one could obtain agreement with observational data only if this factor were introduced. Soon afterwards Thomas * showed that, indeed, relativistic kinematics required a change in equation (3). If a small solid ** non-rotating body, which is not subject to a torque, is moving in a periodic orbit, and if one assumes that the accelerations of all parts of the body in the systems in which the body is at rest are always parallel, one finds that any vector $\boldsymbol{A}$ which is fixed to the body, will be subject to a secular change which to a first approximation is governed by the equation

$$\boldsymbol{A} = -\tfrac{1}{2}c^{-2}[\boldsymbol{A}\wedge\dot{\boldsymbol{v}}[\wedge\boldsymbol{v}]]. \tag{22}$$

One can prove that a Lorentz transformation (without rotation) to an inertial system moving with a velocity $\boldsymbol{v}+\boldsymbol{dv}$ does not lead to the same results as the successive application of two Lorentz transformations (both without rotation), the first transforming to an inertial system moving with a velocity $\boldsymbol{v}$ and the second one transforming from this latter inertial system to an inertial system moving with a velocity $\boldsymbol{v}+\boldsymbol{dv}$. The two coordinate systems which one obtains through those two different operations are rotated over an angle $\frac{1}{2}c^{-2}[\boldsymbol{dv}\wedge\boldsymbol{v}]$ with respect to each other. The secular change of $\boldsymbol{A}$ must thus be described by an equation which is obtained from equation (3) by adding to its right hand side the right hand side of equation (22). Using the equations of motion (12) one is led to equation (20) and to expression (21) for the energy.

The Hamiltonian energy function which governs in the non-relativistic approximation the motion of a spinning electron in an electromagnetic field is obtained by simply adding to the old equation for the energy of an electron [see eq. (3.76)] expression (21) for H_s. We get thus

$$H = \frac{p^2}{2m} - \frac{p^4}{8m^3c^2} - e\Phi + \frac{e}{mc}(\boldsymbol{p}\cdot\mathfrak{A}) + \frac{e^2}{2mc^2}\mathfrak{A}^2 + \frac{e}{em}\left(\boldsymbol{S}\cdot\mathfrak{H} + \frac{[\mathfrak{E}\wedge\boldsymbol{p}]}{2mc}\right), \tag{23}$$

where we have denoted the spin vector by $\boldsymbol{S}$ for reasons which will become clear later on.

Equation (23) is correct if we may neglect terms of the order of c^{-3} ***. The energy function (23) cannot be used to discuss certain periodic fluctuations in the spin-direction, because of the terms neglected when we derived expression (21) for H_s. Furthermore $\boldsymbol{S}$ is considered to be so small that terms in S^2 may be neglected [condition (10) is only approximately satisfied; compare the discussion of eq. (12)].

The system possesses four degrees of freedom. We could introduce apart from the Cartesian coordinates x, y, z and the momenta p_x, p_y, p_z, a fourth pair of variables p_s, q_s by using equation (17). The

* L. H. Thomas, *Nature*, **117**, 514, 1926; *Phil. Mag.*, **3**, 1, 1927.

** "Solid" is used here to indicate that the body retains it shape in the inertial system in which it is at rest.

* One such term was neglected when $\boldsymbol{v}$ was replaced by $\boldsymbol{p}/m$ instead of by $[\boldsymbol{p}-(e/c)\mathfrak{A}]/m$.

equations of motion take on a simple and more symmetrical form if we use the Poisson bracket formalism of classical mechanics. From the canonical equations of motion and the fact that $\boldsymbol{S}$ depends on p_s and q_s only we have

$$S_x=\{S_x,H\}=(\partial H/\partial S_x)\{S_x,S_x\}+(\partial H/\partial S_y)\{S_x,S_y\}+(\partial H/\partial S_z)\{S_x,S_z\}, \tag{24}$$

$$\{S_k, S_l\} = (\partial S_k/\partial q_s)(\partial S_l/\partial p_s)-(\partial S_k/\partial p_s)(\partial S_l/\partial q_s). \tag{25}$$

Using equations (17) we find for the Poisson brackets of the components of $\boldsymbol{S}$,

$$\{S_x,S_y\}=S_z,\{S_y,S_z\}=S_x,\{S_z,S_x\}=S_y,\{S_x,S_x\}=\{S_y,S_y\}=\{S_z,S_z\}=0 \tag{26}$$

If we denote the vector with components $\partial/\partial S_x$, $\partial/\partial S_y$, $\partial/\partial S_z$ by $\boldsymbol{\nabla}_S$ we get by combining equations (26) and (24)

$$\boldsymbol{S} = [\boldsymbol{\nabla}_S H \wedge \boldsymbol{S}], \tag{27}$$

which is equivalent to equation (18). Equations (24) and (27) have the advantage of being closely connected with the general theory of canonical equations and also they do not need the explicit introduction of p_s and q_s.

If we describe the motion of the electron formally by means of equation (23), this equation will automatically lead to a reaction of the spin on the orbit, since the last term of expression (23) contains $\boldsymbol{x}$ and $\boldsymbol{p}$. Without entering into a discussion of the rather complicated equations of motion * we can easily investigate the consequences of these equations on the *rate of change of the total angular momentum* $\boldsymbol{J}$ of the system. This total angular momentum $\boldsymbol{J}$ is the vector sum of the orbital angular momentum $\boldsymbol{M}$ and spin angular momentum $\boldsymbol{S}$, $\boldsymbol{J} = \boldsymbol{M}+\boldsymbol{S} = [\boldsymbol{x}\wedge\boldsymbol{p}]+\boldsymbol{S}$. If we denote by $\boldsymbol{\nabla}_p$ the vector with components $\partial/\partial p_x$, $\partial/\partial p_y$, $\partial/\partial p_z$, we have

$$\dot{\boldsymbol{J}}=[\dot{\boldsymbol{x}}\wedge\boldsymbol{p}]+[\boldsymbol{x}\wedge\dot{\boldsymbol{p}}]+\dot{\boldsymbol{S}}=[\boldsymbol{\nabla}_p H\wedge\boldsymbol{p}]-[\boldsymbol{x}\wedge\boldsymbol{\nabla}H]+[\boldsymbol{\nabla}_S H\wedge\boldsymbol{S}], \tag{28}$$

* It is of some interest to note that in the inertial system moving with the electron ($\boldsymbol{p}=0$) the expression for mass times acceleration which follows from expression (23) in the case where there are no currents (curl $\mathfrak{H}=0$) is of the form $m\ddot{x} = -\nabla_x H = -e\mathfrak{E}_x-(e/mc)\{S_x\nabla_x\mathfrak{H}_x+\frac{1}{2}S_y[\nabla_x\mathfrak{H}_y+\nabla_y\mathfrak{H}_x]+\frac{1}{2}S_z[\nabla_x\mathfrak{H}_z+\nabla_z\mathfrak{H}_x]\}$. [The term in braces comes from the term $\nabla_x(\boldsymbol{S}\cdot\mathfrak{H})$ by using curl $\mathfrak{H}=0$ and the term with $\mathfrak{A}^2$ in H is neglected]. This expression is not the same as expression (12). The difference which depends on the spin is half the sum of $-(e/mc)(\boldsymbol{S}\,.\,\boldsymbol{\nabla})\mathfrak{H}$ and $-(e/mc)\boldsymbol{\nabla}(\boldsymbol{S}\cdot\mathfrak{H})$.

where we have used for $\dot{\boldsymbol{x}}$ and $\boldsymbol{p}$ the canonical equations of motion.

The expression on the right hand side of equation (28) has a simple physical meaning. Let us investigate what happens to expression (23) when we introduce a new orthogonal system of coordinates which is rotated over a small angle ω_x around the x-axis, ω_y around the y-axis, and ω_z around the z-axis (vector of rotation $\boldsymbol{\omega}$). The result is that $\boldsymbol{p}$ goes over into $\boldsymbol{p}+[\boldsymbol{p}\wedge\boldsymbol{\omega}]$, $\boldsymbol{x}$ into $\boldsymbol{x}+[\boldsymbol{x}\wedge\boldsymbol{\omega}]$, $\boldsymbol{S}$ into $\boldsymbol{S}+[\boldsymbol{S}\wedge\boldsymbol{\omega}]$, and H into $H+\delta H$ where δH is given by the equation

$$\delta H=(\nabla_p H\cdot[\boldsymbol{p}\wedge\boldsymbol{\omega}])+(\nabla H\cdot[\boldsymbol{x}\wedge\boldsymbol{\omega}])+(\nabla_S H\cdot[\boldsymbol{S}\wedge\boldsymbol{\omega}])=(\dot{\boldsymbol{J}}\cdot\boldsymbol{\omega}) \quad (29)$$

If the form of H as a function of $\boldsymbol{p}$, $\boldsymbol{x}$, $\boldsymbol{S}$ is independent of the orientation of the system of coordinates — which is, for instance, the case for an electron in a radially symmetric electrical field, since $\mathfrak{A}$ and $\mathfrak{H}$ are zero and Φ and $\mathfrak{E}$ are invariant against rotations — δH will be equal to zero and we see from equation (29) that *the total angular momentum is constant in time.*

We can easily generalise expression (23) to find the Hamilton function of a *system of spinning electrons* in an external field of force. The rate of change of the spin vector $\boldsymbol{S}_i$ of the i-th electron is given by an equation analogous to equation (20),

$$\dot{\boldsymbol{S}}_i=-(e/mc)[\boldsymbol{S}_i\wedge\{\mathfrak{H}_i'+(2mc)^{-1}[\mathfrak{E}_i'\wedge\boldsymbol{p}_i]\}], \quad (30)$$

where $\mathfrak{H}_i'$ and $\mathfrak{E}_i'$ are the electromagnetic field vectors at the position of the i-th electron. They are built up out of two terms. The first one is the external field ($\mathfrak{H}_i$ and $\mathfrak{E}_i$), the second part is due to the presence of the other electrons. If $\boldsymbol{x}_{ij}=\boldsymbol{x}_i-\boldsymbol{x}_j$, we get neglecting terms of order greater than c^{-1},

$$\mathfrak{E}_i'=\mathfrak{E}_i+e\sum_j\frac{\boldsymbol{x}_{ij}}{r_{ij}^3},\quad \mathfrak{H}_i'=\mathfrak{H}_i+\frac{e}{c}\sum_j\frac{[\boldsymbol{v}_j\wedge\boldsymbol{x}_{ij}]}{r_{ij}^3}+\frac{e}{mc}\sum_j\frac{r_{ij}^2\boldsymbol{S}_j-3\boldsymbol{x}_{ij}(\boldsymbol{S}_j\cdot\boldsymbol{x}_{ij})}{r_{ij}^5} \quad (31)$$

The last term derives from the magnetic field which comes directly from the spins.

We must now add to expression (3.76) for the energy H_0 of the electrons without spin a spin contribution H_s,

$$H=H_0+H_s,\quad H_s=(e/mc)\,\Sigma_i(\boldsymbol{S}_i\cdot\{\mathfrak{H}_i''+(2mc)^{-1}[\mathfrak{E}_i'\wedge\boldsymbol{p}_i]\}), \quad (32)$$

where $\mathfrak{H}_i''$ differs from $\mathfrak{H}_i'$ of equations (31) by an extra factor $\frac{1}{2}$ for the terms containing the $\boldsymbol{S}_j$, as otherwise the canonical equations

$\dot{\boldsymbol{S}}_i = [\boldsymbol{\nabla}_{Si} H \wedge \boldsymbol{S}_i]$ will not have the form (30). We thus get for H_s the expression *

$$\left.\begin{aligned} H_s &= \Sigma_i (e/mc)(\boldsymbol{S}_i \cdot \{\mathfrak{H}_i + (2mc)^{-1}[\mathfrak{E}_i \wedge \boldsymbol{p}_i]\}) \\ &+ \Sigma_{i \neq j} (e^2/m^2 c^2 r_{ij}^3) \,.\, (\boldsymbol{S}_i \cdot [\{\boldsymbol{p}_j - \tfrac{1}{2}\boldsymbol{p}_i\} \wedge \boldsymbol{x}_{ij}]) \\ &+ \Sigma_{i \neq j} (e^2/2m^2 c^2 r_{ij}^5)\{r_{ij}^2 (\boldsymbol{S}_i \cdot \boldsymbol{S}_j) - 3(\boldsymbol{x}_{ij} \cdot \boldsymbol{S}_i)(\boldsymbol{x}_{ij} \cdot \boldsymbol{S}_j)\}. \end{aligned}\right\} \quad (33)$$

The last sum corresponds to the spin-spin interaction; each pair of spins occurs twice in the summation.

The discussion of the rate of change of the total angular momentum $\boldsymbol{J}$ $(= \Sigma_i [\boldsymbol{x}_i \wedge \boldsymbol{p}_i] + \Sigma_i \boldsymbol{S}_i)$ is completely analogous to the discussion a moment ago of the rate of change of the total angular momentum of one electron. Once again $\dot{\boldsymbol{J}}$ will be equal to zero, if H as a function of the $\boldsymbol{x}_i$, $\boldsymbol{p}_i$, and $\boldsymbol{S}_i$ does not alter its form when the system of coordinates is rotated. The simplest example of this theorem of the conservation of angular momentum (including spin) is the case of a free atom where a number of electrons are bound in the field of the nucleus.

Equations (32) and (33) can easily be generalised to the case where nuclei are present.

§ 58. The non-relativistic quantum mechanical treatment of spin

As we mentioned at the beginning of this chapter, Pauli** has shown how one can include the spin of the electron into the quantum mechanical description of a system. We are led in a natural way to Pauli's formulation, if we use the classical formalism of the preceding section, in which we considered in particular the spin angular momentum of the electron, and follow closely the quantum mechanical treatment of the properties of a system with an angular momentum. We showed in § 46 how it is possible to describe the physical situation of a particle by means of an infinite series of functions $a_{lm}(r, t)$ where l and m refer to the eigenvalues of the total angular momentum $M^2 = l(l+1)\hbar^2$ and the z-component of the angular momentum $M_z = m\hbar$, while the distance r from the origin and the time t still occur as continuous variables. These functions were the coefficients in the expansion of the wave function $\psi(\boldsymbol{x}, t)$ in terms of the normalised polar spherical harmonics $\mathscr{P}_l^m$,

* See W. Heisenberg, *Z. Phys.*, **39**, 499, 1926.

** W. Pauli, *Z. Phys.*, **43**, 601, 1927.

$$\psi(\mathbf{x}, t) = \Sigma_{l,m} a_{lm}(r, t)\mathscr{P}_l^m, \quad m = l, l-1, \ldots, -l, \quad l = 0, 1, 2, \ldots \tag{34}$$

If ψ is normalised, $|a_{lm}|^2 r^2 dr$ will be the probability that at the time t r lies within the interval $(r, r+dr)$, and l and m have definite values.

The rate of change of the $a_{lm}(r, t)$ was given by the transformed Schrödinger equation,

$$i\hbar \dot{a}_{lm} = \Sigma_{l'm'} H_{lm,\, l'm'}\, a_{l'm'}, \tag{35}$$

where the matrix elements $H_{lm,\, l'm'}$ of the energy are in general operators as far as r is concerned.

We can use equations (35) to describe the electron spin. We need only to introduce two simplifications and one new feature. The first simplification consists in the fact that in the case of the spin the absolute value of the angular momentum has always the same value. This corresponds to the case where all a_{lm} which do not belong to one particular value of l ($= l_0$ say) are and remain equal to zero. This means that all matrix elements $H_{lm,\, l_0m'}$ are zero, unless $l = l_0$ *. Secondly we have in the classical model of the spinning electron nothing which corresponds to the r-coordinate. In the case of the orbital angular momentum we would get a similar situation if we assume that the matrix elements $H_{l_0m,\, l_0m'}$ are simply numbers and the ratio of one a_{l_0m} to another is independent of r. To obtain the description of the spin we drop the r completely; a_{l_0m} depends on t only, and $|a_{l_0m}|^2$ is the probability that the z-component of the spin angular momentum is equal to $m\hbar$.

A new feature enters when we take into account the empirical evidence which as Uhlenbeck and Goudsmit showed leads to a value $\frac{1}{2}\hbar$ for the spin angular momentum. This means that l_0, *the quantum number of the total angular momentum, is equal to* $\frac{1}{2}$, and the quantum number m which should take on all values from $-l$ to $+l$ can take on only the values $+\frac{1}{2}$ and $-\frac{1}{2}$. This last result means that *there are only two eigenvalues,* $\frac{1}{2}\hbar$ *and* $-\frac{1}{2}\hbar$, *for the spin angular momentum in a given direction.* The situation in which an electron finds itself is, as far as the spin properties are concerned, *described by two complex numbers* a_s ($s = \pm\frac{1}{2}$) which satisfy the two simultaneous equations **

* This will, for instance, occur in the case where we are dealing with the orbital angular momentum, if the energy operator commutes with M_{op}^2, that is, if M_{op}^2 is an integral of the equations of motion.

** In indices we shall for the sake of simplicity write now and henceforth $+$ or $-$ instead of $+\frac{1}{2}$ or $-\frac{1}{2}$.

$$H_{++}a_{+}+H_{+-}a_{-} = i\hbar\dot{a}_{+}, \qquad H_{-+}a_{+}+H_{--}a_{-} = i\hbar\dot{a}_{-},$$

$$\text{or,} \quad \Sigma_{s'} H_{ss'} a_{s'} = i\hbar\dot{a}_{s}, \qquad s, s' = \pm\tfrac{1}{2}. \tag{36}$$

The $H_{ss'}$ are elements of a two-by-two Hermitean matrix,

$$H_{ss'} = H^{*}_{s's}. \tag{37}$$

The solution of equation (36) can always be normalised such that the condition

$$|a_{+}|^2+|a_{-}|^2 = 1 \tag{38}$$

is satisfied. If that is done, $|a_{+}|^2$ and $|a_{-}|^2$ are the probabilities that the z-component of the spin angular momentum is equal to $+\frac{1}{2}\hbar$ or $-\frac{1}{2}\hbar$.

This formal introduction of so-called *half-integer quantum numbers* l and m is something essentially new, since beforehand l and m were always integers. The fact that up to now l and m were integers was due to the circumstance that the quantum mechanical description of particles involved only uniquely defined wave functions $\psi(\boldsymbol{x}, t)$. Similarly it will now be impossible to go over, say, by means of a transformation similar to the one given by equation (34) to a description of the physical situation including the spin by one-valued, continuous wave functions with only continuous arguments.

The energy operator $H_{mm'}$ which occurred previously could be obtained from the energy of the system, that is, from a mechanical quantity which classically was defined as a function of the fundamental mechanical variables q_k and p_k (space coordinates and corresponding momenta). Even though the situation is very different now, we can still interpret the matrix $H_{ss'}$ in equation (36) in a similar way. The most general matrix of this kind can be considered to be a linear combination of the following four fundamental matrices,

$$\begin{gathered} \mathbf{1} = \left\Vert \begin{matrix} 1 & 0 \\ 0 & 1 \end{matrix} \right\Vert, \; S_x = \tfrac{1}{2}\hbar \left\Vert \begin{matrix} 0 & 1 \\ 1 & 0 \end{matrix} \right\Vert, \\ S_y = \tfrac{1}{2}\hbar \left\Vert \begin{matrix} 0 & -i \\ i & 0 \end{matrix} \right\Vert, \; S_z = \tfrac{1}{2}\hbar \left\Vert \begin{matrix} 1 & 0 \\ 0 & -1 \end{matrix} \right\Vert, \end{gathered} \tag{39}$$

where $\mathbf{1}$ is the unit matrix. Clearly we have

$$H_{\text{op}} \equiv \left\Vert \begin{matrix} H_{++} & H_{+-} \\ H_{-+} & H_{--} \end{matrix} \right\Vert = c_0\mathbf{1}+c_x S_x+c_y S_y+c_z S_z, \tag{40}$$

where the c's depend on the $H_{ss'}$ as follows,

$$\left.\begin{aligned} c_0 &= \tfrac{1}{2}(H_{++}+H_{--}), \quad \hbar c_x = H_{+-}+H_{-+}, \\ \hbar c_y &= i(H_{+-}-H_{-+}), \quad \hbar c_z = H_{++}-H_{--}; \\ H_{++} &= c_0+\tfrac{1}{2}\hbar c_z, \quad H_{--} = c_0-\tfrac{1}{2}\hbar c_z, \\ H_{+-} &= \tfrac{1}{2}\hbar(c_x-ic_y), \quad H_{-+} = \tfrac{1}{2}\hbar(c_x+ic_y). \end{aligned}\right\} \tag{41}$$

The matrices $\left\|\begin{matrix} 0 & 1 \\ 1 & 0 \end{matrix}\right\|$, $\left\|\begin{matrix} 0 & -i \\ i & 0 \end{matrix}\right\|$, $\left\|\begin{matrix} 1 & 0 \\ 0 & -1 \end{matrix}\right\|$ are often called the *Pauli matrices.*

If we compare expressions (39) for S_x, S_y, and S_z with the general expressions in § 45 for the matrices of the components M_x, M_y, M_z of the angular momentum, we see that these latter matrices reduce to the matrices (39) when we take formally l to be $\frac{1}{2}$. *We can thus consider S_x, S_y, S_z to be the operators corresponding to the components of the spin angular momentum. The most general observable which refers to the electron spin can always be contructed by adding a constant (the term $c_0\mathbf{1}$) to a linear combination of the components of the spin angular momentum.*

The matrices S_x, S_y, S_z are Hermitean so that the condition that $H_{ss'}$ be Hermitean (corresponding to a real energy function) entails that the c's are all real. Consider the special case of an electron in a magnetic field $\mathfrak{H}$. If we consider only the contribution due to the spin, we get for the energy from equation (16), replacing $\boldsymbol{A}$ by $\boldsymbol{S}$,

$$H_{\mathrm{op}} = (e/mc)(\boldsymbol{S}\cdot\mathfrak{H}), \tag{42}$$

and using equations (40) and (41) we have

$$H_{\mathrm{op}} = \frac{e\hbar}{2mc}\left\|\begin{matrix} \mathfrak{H}_z & \mathfrak{H}_x-i\mathfrak{H}_y \\ \mathfrak{H}_x+i\mathfrak{H}_y & -\mathfrak{H}_z \end{matrix}\right\|. \tag{43}$$

The two numbers $a_\pm$ which occur in the Schrödinger equation (36) can be considered to constitute a function a_s of an argument s which can take on only two values $+\frac{1}{2}$ and $-\frac{1}{2}$. One often uses the notation

$$a_s \to \begin{pmatrix} a_+ \\ a_- \end{pmatrix}, \tag{44}$$

and the action of an arbitrary operator Ω (matrix $\Omega_{ss'}$) on a_s which is given by the equation [compare eq. (3.57)] $(\Omega_{\mathrm{op}}a)_s = \Sigma_{s'}\Omega_{ss'}a_{s'}$ is written in the form

$$\begin{Vmatrix} \Omega_{++} & \Omega_{+-} \\ \Omega_{-+} & \Omega_{--} \end{Vmatrix} \begin{pmatrix} a_+ \\ a_- \end{pmatrix} = \begin{pmatrix} \Omega_{++}\, a_+ + \Omega_{+-}\, a_- \\ \Omega_{-+}\, a_+ + \Omega_{--}\, a_- \end{pmatrix}. \tag{45}$$

If c and c' are arbitrary complex numbers, we have

$$c \begin{pmatrix} a_+ \\ a_- \end{pmatrix} + c' \begin{pmatrix} a'_+ \\ a'_- \end{pmatrix} = \begin{pmatrix} ca_+ + c'\, a'_+ \\ ca_- + c'\, a'_- \end{pmatrix},$$

and we can in general consider any "wave function" a_s to be constructed as a linear combination of the two normalised basis functions

$$\xi = \begin{pmatrix} 1 \\ 0 \end{pmatrix}, \quad \eta = \begin{pmatrix} 0 \\ 1 \end{pmatrix} \tag{46}$$

as follows

$$\begin{pmatrix} a_+ \\ a_- \end{pmatrix} = a_+ \begin{pmatrix} 1 \\ 0 \end{pmatrix} + a_- \begin{pmatrix} 0 \\ 1 \end{pmatrix} \to a_s = a_+ \xi + a_- \eta. \tag{47}$$

The functions ξ and η correspond respectively to situations where the z-component of the spin angular momentum has the well defined value $+\frac{1}{2}\hbar$ or $-\frac{1}{2}\hbar$. Of course, the same would be true of the functions $e^{i\alpha}\xi$ and $e^{i\alpha}\eta$, if α is real. As we might have expected from our considerations in § 45 [see eq. (4.158)], ξ and η are the normalised eigenfunctions of the operator S_z, since we have

$$S_{z\,\mathrm{op}}\, \xi = \tfrac{1}{2}\hbar \begin{Vmatrix} 1 & 0 \\ 0 & -1 \end{Vmatrix} \begin{pmatrix} 1 \\ 0 \end{pmatrix} = \tfrac{1}{2}\hbar \begin{pmatrix} 1 \\ 0 \end{pmatrix} = \tfrac{1}{2}\hbar\, \xi,$$

$$S_{z\,\mathrm{op}}\, \eta = \tfrac{1}{2}\hbar \begin{Vmatrix} 1 & 0 \\ 0 & -1 \end{Vmatrix} \begin{pmatrix} 0 \\ 1 \end{pmatrix} = \tfrac{1}{2}\hbar \begin{pmatrix} 0 \\ -1 \end{pmatrix} = -\tfrac{1}{2}\hbar\, \eta.$$

The eigenvalues are, indeed, $+\frac{1}{2}\hbar$ and $-\frac{1}{2}\hbar$.

One verifies easily that the eigenvalues of S_x and S_y are also $+\frac{1}{2}\hbar$ and $-\frac{1}{2}\hbar$, as we might have expected. The corresponding normalised eigenfunctions are, apart from an arbitrary phase factor, given by the scheme

$$\begin{array}{lll|l} \text{Eigenvalues:} & & +\tfrac{1}{2}\hbar & -\tfrac{1}{2}\hbar \\ \text{Eigenfunctions of} & S_{x\,\mathrm{op}}: & 2^{-\frac{1}{2}}\begin{pmatrix} 1 \\ 1 \end{pmatrix} = 2^{-\frac{1}{2}}(\xi+\eta) & 2^{-\frac{1}{2}}\begin{pmatrix} 1 \\ -1 \end{pmatrix} = 2^{-\frac{1}{2}}(\xi-\eta) \\ & S_{y\,\mathrm{op}}: & 2^{-\frac{1}{2}}\begin{pmatrix} 1 \\ i \end{pmatrix} = 2^{-\frac{1}{2}}(\xi+i\eta) & 2^{-\frac{1}{2}}\begin{pmatrix} 1 \\ -i \end{pmatrix} = 2^{-\frac{1}{2}}(\xi-i\eta) \\ & S_{z\,\mathrm{op}}: & \xi & \eta \end{array} \tag{48}$$

If the H_{op} in the Schrödinger equation (36) does not contain the time explicitly, we can use the substitution

$$a_s = \varphi_s \exp(-iEt/\hbar) \tag{49}$$

to go over to the time-independent Schrödinger equation $\Sigma_{s'} H_{ss'} \varphi_{s'} = E\varphi_s$. This equation corresponds to the quadratic equation $\begin{vmatrix} H_{++}-E & H_{+-} \\ H_{-+} & H_{--}-E \end{vmatrix} = 0$, that is, the problem of the principal axes of $H_{ss'}$.

For instance, in the case of an electron in a constant magnetic field which led to the energy matrix (43) we find for the eigenvalues the equations *,

$$E_1 = \mu_B \mathfrak{H}, \qquad E_2 = -\mu_B \mathfrak{H} \tag{50}$$

where $\mathfrak{H}$ is the absolute magnitude of the magnetic field and where μ_B is given by the equation

$$\mu_B = e\hbar/2mc = eh/4\pi mc = 0.93 \times 10^{-20} \text{ e.s.u.} \tag{51}$$

This quantity, the so-called *Bohr magneton* is the natural quantum theoretical measure for magnetic moments of atomic systems. In the old quantum theory it appeared very naturally, since in that theory any constant angular momentum had a value $l\hbar$, with l integral. This means, in connexion with equation (14), that the magnetic moment of a system of electrons would be an integral multiple of $e\hbar/2mc$, that is, of μ_B, if its angular momentum were constant. The magnetic moment of a spinning electron should according to Uhlenbeck and Goudsmit (see § 56) also be equal to μ_B. Equation (50) thus corresponds exactly to the "classical" model where the spin vector of the electron would be either parallel ($E = \mu_B\mathfrak{H}$) or antiparallel ($E = -\mu_B\mathfrak{H}$) to the external magnetic field.

We must finally draw attention to the fact that the operators (39) corresponding to the components of $\boldsymbol{S}$ satisfy the commutation relations (4.132),

$$S_x S_y - S_y S_x = [S_x, S_y]_- = i\hbar S_z, \ldots \tag{52}$$

The classical counterpart of these equations are the expressions (26) for the Poisson brackets. Moreover, we have the relations

* The corresponding eigenfunctions are $\varphi_1 = \cos \frac{1}{2}\vartheta \exp(-\frac{1}{2}i\chi)\xi + \sin \frac{1}{2}\vartheta \exp(\frac{1}{2}i\chi)\eta$ and $\varphi_2 = \sin\frac{1}{2}\vartheta \exp(-\frac{1}{2}i\chi)\xi - \cos \frac{1}{2}\vartheta \exp(\frac{1}{2}i\chi)\eta$, where ϑ and χ are the polar angles determining the direction of the magnetic field $\boldsymbol{\mathfrak{H}}$ [$\mathfrak{H}_z = \mathfrak{H} \cos \vartheta$, $\mathfrak{H}_x + i\mathfrak{H}_y = \mathfrak{H} \sin \vartheta \exp(i\chi)$].

$$S_x^2=S_y^2=S_z^2=\tfrac{1}{4}\hbar^2,\quad S_xS_y+S_yS_x\equiv[S_x,S_y]_+=[S_y,S_z]_+=[S_z,S_x]_+=0 \tag{53}$$

The last equations express the fact that the components of the spin form a set of *anticommutative quantities*. This is characteristic for the value $\frac{1}{2}$ of the total angular momentum quantum number.

The value of the square of the total angular momentum is given by the equation

$$S^2 = S_x^2+S_y^2+S_2^2 = \tfrac{3}{4}\hbar^2, \tag{54}$$

corresponding to the value $l = \frac{1}{2}$ substituted into the general expression $l(l+1)\hbar^2$ [see eq. (4.153)].

The determination of the two numbers a_s needed to describe the spin is connected in an essential way with the existence of a fixed orthogonal coordinate system x, y, z in space. This was also the case when we were considering the transformation (34). Just as in that case we should expect that a new description of the spin by a'_s which refer to a new system of coordinates x', y', z' is completely equivalent to the description by a_s. We shall prove in § 61 the so-called invariance of the spin description against rotations of the system of coordinates.

We have up to now neglected the orbital motion of the electron. To get the complete quantum mechanical description of the behaviour of a spinning electron in an external electromagnetic field it is easiest to follow the classical treatment of the same problem. The physical situation is now described by a wave function ψ which depends not only on the time and the spin variables but also on three more variables, for instance the three spatial coordinates x, y, z of the electron,

$$\psi = \psi_s(\boldsymbol{x}, t) = \begin{pmatrix}\psi_+(\boldsymbol{x}, t)\\ \psi_-(\boldsymbol{x}, t)\end{pmatrix} = \psi_+(\boldsymbol{x}, t)\,\xi+\psi_-(\boldsymbol{x}, t)\,\eta. \tag{55}$$

We have used both the notation (44) and (47) in writing down equation (55). One often calls ψ a *two-component wave function* with components ψ_+ and ψ_-. It was first of all introduced by Pauli *. One can interpret $|\psi_+|^2 d\boldsymbol{x}$ and $|\psi_-|^2 d\boldsymbol{x}$ to be the probabilities that the electron is within a volume element in space $d\boldsymbol{x}$ while its spin is parallel or antiparallel to the positive z-axis. In order that this interpretation be justified, ψ must be normalised, or

* *Z. Phys.*, **43**, 601, 1927.

$$\int\psi^*\psi \equiv \int |\psi_+|^2 d\mathbf{x} + \int |\psi_-|^2 d\mathbf{x} = 1. \tag{56}$$

The rate of change of ψ is governed by a Schrödinger equation, $H_{\text{op}}\psi = i\hbar\partial\psi/\partial t$, where the energy operator acts on the spin argument s as well as on $\mathbf{x}$ and t. As in equation (40) we can write

$$H_{\text{op}} = H^{(0)}\,\mathbf{1} + H^{(x)}\,S_x + H^{(y)}\,S_y + H^{(z)}\,S_z, \tag{57}$$

where the Hermitean operators $\mathbf{1}$, S_x, S_y, and S_z are defined by equations (39).

The Hermitean operators $H^{(0)}$, $H^{(x)}$, . . . operate on $\mathbf{x}$ and t only. One obtains these operators from the classical expression (23) by using first of all the decomposition (57) and then replacing $\boldsymbol{p}$ by $-i\hbar\boldsymbol{\nabla}$.

We have now obtained *the formalism of Pauli's non-relativistic treatment of the electron spin*. It differs from our earlier treatment of the one-particle system in principle only by the introduction of a fourth (discrete) spin variable. This means that we can immediately apply the formalism of transformation theory (Chapter IV) and the formalism of perturbation theory (Chapter V). For these applications it does not matter that there does not exist an observable which is canonically conjugate to the spin variable, since every possible operator can be interpreted as a linear combination of the components of the spin. We must, however, mention that a more detailed analysis of possible transformations shows a peculiar property of the wave functions which did not show up in the earlier treatment. This property is often called the "two-valuedness" of the spin functions. To avoid misunderstandings we must stress that the two components ψ_+ are both one-valued functions of $\mathbf{x}$ and t just like the "one-component" wave functions studied earlier. We shall return to this point in § 61.

§ 59. The spinning electron in a central field of force

An important application of the theory is that to the problem of the stationary state of *a spinning electron in a spherically symmetric electric field*. The energy function for this case is obtained by putting $\mathfrak{H}$ and $\mathfrak{A}$ in equation (23) equal to zero, and by putting

$$\Phi = -eF(r), \qquad \mathfrak{E} = eF'\mathbf{x}/r, \tag{58}$$

where F has the dimensions of a reciprocal length. The energy operator is given by the equation

$$\left.\begin{aligned} H &= (H_0+H_r)\,\mathbf{1}+H_s, \qquad H_0 = -(\hbar^2/2m)\nabla^2+e^2F(r), \\ H_r &= -(\hbar^4/8m^3c^2)\nabla^4 = -(2mc^2)^{-1}[-H_0+e^2F(r)]^2, \\ H_s &= (e^2F'/2m^2c^2r)(\boldsymbol{S}\cdot[\boldsymbol{x}\wedge\boldsymbol{p}]) = (e^2/2m^2c^2)(F'/r)(\boldsymbol{S}\cdot\boldsymbol{M}), \end{aligned}\right\} \tag{59}$$

where $\boldsymbol{M}$ is, as before, the orbital angular momentum operator.

We treat the problem by perturbation theory. The non-relativistic problem with energy $H_0 \cdot \mathbf{1}$ which does not contain any spin components is perturbed by the relativistic term $H_r \cdot \mathbf{1}$ which takes into account the relativistic change of mass, and by the spin-orbit interaction H_s. The energy eigenvalues $E_{nl}^{(0)}$ of $H_0 \cdot \mathbf{1}$ are those of H_0 and are characterised by two quantum numbers, the principal quantum number n and the orbital angular momentum quantum number l *. The corresponding eigenfunctions φ_0 differ, however, from the eigenfunctions encountered in § 46 by being multiplied by an arbitrary "spin function", since $H_0 \cdot \mathbf{1}$ contains as far as the spin variable is concerned only the unit operator. We write therefore [see eqq. (4.175) and (4.182)]

$$\varphi_0 = [R_{nl}(r)/r]\{Y_l^{(+)}(\boldsymbol{\omega})\,\xi+Y_l^{(-)}(\boldsymbol{\omega})\eta\}, \tag{60}$$

where $Y_l^{(+)}$ and $Y_l^{(-)}$ are two arbitrary spherical harmonics of the l-th degree.

The stationary state characterised by expression (60) is $2(2l+1)$-fold degenerate since both $Y_l^{(+)}$ and $Y_l^{(-)}$ contain $2l+1$ arbitrary coefficients. This degeneracy will in general be only partially lifted by the perturbation. The term $H_r \cdot \mathbf{1}$ will not play a role in this respect, since it refers to the problem without spin and it will thus only slightly shift the eigenvalues $E_{nl}^{(0)}$ and also change the $R_{nl}(r)$ in expression (60) only slightly. This change in the $R_{nl}(r)$ we do not want to consider here. The shift of the energy eigenvalues is obtained

* In § 46 we discussed the non-relativistic problem of a particle in a central field of force in detail only for the case of an attractive Coulomb force. In the discussion of atoms with several electrons it is of advantage to discuss more general attractive central forces (see § 70). Each electron is in a central field of force which is produced partly by the nucleus and the other electrons. This field has a potential $U(r)$ which behaves both as $r \to \infty$ and as $r \to 0$ as r^{-1}. The qualitative discussion of the eigenfunctions and eigenvalues in such a potential leads to the same conclusions as in the case of a Coulomb field (see § 46). In particular we can distinguish for a given orbital angular momentum quantum number l the different discrete eigenfunctions $b_l r$ of equation (4.177) by the number $n-l-1$ of its zeros ($n = l+1, l+2, \ldots$). We write again $b_l r = R_{nl}$ [see, eq. (4.196)] and call n the *principal quantum number*. The corresponding energy eigenvalues ($E_{nl}{}^{(0)}$) depend in general both on n and on l, although in the special case of the Coulomb field, the energy depends on n only [see eq. (4.192)].

from the normal formalism of perturbation theory and we have

$$\delta_r E_{nl}^{(0)} = \overline{H_r} = \int \varphi_0^* H_r \varphi_0 \, d\boldsymbol{x} = -(2mc^2)^{-1} \int \varphi_0^* [-H_0 + e^2 F(r)]^2 \varphi_0 \, d\boldsymbol{x}$$
$$= -(2m c^2)^{-1} [E_0^2 - 2 e^2 E_0 \overline{F} + e^4 \overline{F^2}], \tag{61}$$

where the bars indicate as before [for instance, eq. (3.86)] the expectation value in the unperturbed state.

The perturbation H_s, however, will lead to a splitting up of the unperturbed energy level $E_{nl}^{(0)}$ and in accordance with the general prescription we must solve for each n, l pair a secular problem of the $2(2l+1)$-st degree. In the present case the problem is considerably simplified if we use the fact that the Hamiltonian is invariant against a rotation of the system of coordinates. As in the case of the analogous classical problem, the total angular momentum $\boldsymbol{J} = \boldsymbol{M} + \boldsymbol{S}$ is an integral of the equations of motion. It is easily verified that J_x, J_y, J_z commute with the energy operator (compare also § 61), if one takes into account that the components of $\boldsymbol{S}$ are operators which commute with any function of the coordinates and momenta of the electron.

That $\boldsymbol{M}$ and $\boldsymbol{S}$, and thus $\boldsymbol{J}$, commute with $H_0 \cdot \mathbf{1}$ and $H_r \cdot \mathbf{1}$ is trivial. To discuss the term H_s we note that [compare eqq. (59), (52), and (4.132)] $\boldsymbol{M}(\boldsymbol{S} \cdot \boldsymbol{M}) - (\boldsymbol{S} \cdot \boldsymbol{M})\boldsymbol{M} = -i\hbar[\boldsymbol{S} \wedge \boldsymbol{M}]$, $\boldsymbol{S}(\boldsymbol{S} \cdot \boldsymbol{M}) - (\boldsymbol{S} \cdot \boldsymbol{M})\boldsymbol{S} = -i\hbar[\boldsymbol{M} \wedge \boldsymbol{S}]$. Adding these two equations we get $[\boldsymbol{J}, (\boldsymbol{S} \cdot \boldsymbol{M})]_- = 0$. The term $(\boldsymbol{S} \cdot \boldsymbol{M})$ differs from H_s only by a factor which is a function of r and which thus commutes with $\boldsymbol{S}$, $\boldsymbol{M}$, and $\boldsymbol{J}$.

It is also easily verified that the commutators of the components of $\boldsymbol{J}$ are similar to those of the components of $\boldsymbol{M}$. We shall return to the significance of this result [see eq. (110) in § 61]. At this moment we just calculate them directly,

$$[J_x, J_y]_- = i\hbar J_z, \ldots; \qquad [J^2, J_x]_- = 0, \ldots \tag{62}$$

We note that H, J^2, and J_z commute one with another. This means that it is possible to find wave functions which are simultaneously eigenfunctions of these three operators (compare § 38). We look especially for simultaneous eigenfunctions of J^2 and J_z which are also eigenfunctions of M^2 (M^2 commutes with J^2) and which are thus of the form (60).

If we introduce once again the normalised polar spherical harmonics [eq. (4.161)], we find that the eigenfunctions of $J_z = M_z + S_z$ and the corresponding eigenvalues are of the form

$$\text{Eigenfunction: } Q^{(m+\frac{1}{2})} = c_+ \mathscr{P}_l^m \xi + c_- \mathscr{P}_l^{m+1} \eta; \quad \text{eigenvalue: } (m+\tfrac{1}{2})\hbar, \tag{63}$$

where c_+ and c_- are arbitrary constants. Indeed, M_z operating on the first term of equation (63) multiplies it by $m\hbar$ and the second term by $(m+1)\hbar$ while the multiplication factors in the case of S_z are $\frac{1}{2}\hbar$ and $-\frac{1}{2}\hbar$, leading for both terms to a factor $(m+\frac{1}{2})\hbar$ for J_z.

The condition that expression (63) must be an eigenfunction of $J^2 = M^2+S^2+2(\boldsymbol{M}\cdot\boldsymbol{S})$ gives us the ratio of c_+ to c_-. The calculations are rather tedious, if we use the mathematical tools which we have introduced up to now. For each m value between $-l-1$ and l we find a quadratic secular equation for the eigenvalues of J^2 with two roots which are different and independent of m. The result is

Eigenvalues of J^2		Normalised eigenfunctions
$\hbar^2(l+\frac{1}{2})(l+\frac{3}{2})$	;	$Q^{m+\frac{1}{2}}_{l+\frac{1}{2}} = [(l+m+1)/(2l+1)]^{\frac{1}{2}}\mathscr{P}^m_l\xi + [(l-m)/(2l+1)]^{\frac{1}{2}}\mathscr{P}^{m+1}_l\eta;$ (64a)
$\hbar^2(l-\frac{1}{2})(l+\frac{1}{2})$	;	$Q^{m+\frac{1}{2}}_{l-\frac{1}{2}} = [(l-m)/(2l+1)]^{\frac{1}{2}}\mathscr{P}^m_l\xi - [(l+m+1)/(2l+1)]^{\frac{1}{2}}\mathscr{P}^{m+1}_l\eta.$ (64b)

These formulae are also valid for $m = l$ and $m = -l-1$, if we put $\mathscr{P}^m_l = 0$ for $|m| > l$. In these cases the function (64b) vanishes and the function (64a) reduces to $\mathscr{P}^l_l\xi$ or $\mathscr{P}^{-l}_l\eta$ with the eigenvalue $\hbar^2(l+\frac{1}{2})(l+\frac{3}{2})$.

We must evaluate $J^2{}_{\text{op}}\,Q^{(m+\frac{1}{2})}$ and we can use the relation $J^2 = M^2+S^2+2(M_xS_x+M_yS_y+M_zS_z)$. From equations (4.162) and (4.158) we have

$$\left.\begin{aligned} M_x\mathscr{P}_l{}^m &= \tfrac{1}{2}\hbar\{[(l+m)(l-m+1)]^{\frac{1}{2}}\mathscr{P}_l{}^{m-1}+[(l-m)(l+m+1)]^{\frac{1}{2}}\mathscr{P}_l{}^{m+1}\},\ M_z\mathscr{P}_l{}^m = m\hbar\mathscr{P}_l{}^m,\\ M_y\mathscr{P}_l{}^m &= \tfrac{1}{2}i\hbar\{[(l+m)(l-m+1)]^{\frac{1}{2}}\mathscr{P}_l{}^{m-1}+[(l-m)(l+m+1)]^{\frac{1}{2}}\mathscr{P}_l{}^{m+1}\},\ M^2\mathscr{P}_l{}^m = l(l+1)\hbar^2\mathscr{P}_l{}^m \end{aligned}\right\} \quad (65)$$

From equations (39) and (45) we get

$$\left.\begin{aligned} S_x\xi &= \tfrac{1}{2}\hbar\,\eta, & S_y\xi &= \tfrac{1}{2}i\hbar\eta, & S_z\xi &= \tfrac{1}{2}\hbar\xi; & S^2\xi &= \tfrac{3}{4}\hbar^2\xi;\\ S_x\eta &= \tfrac{1}{2}\hbar\,\xi, & S_y\eta &= -\tfrac{1}{2}i\hbar\,\xi, & S_z\eta &= -\tfrac{1}{2}\hbar\,\eta; & S^2\eta &= \tfrac{3}{4}\hbar^2\,\eta. \end{aligned}\right\} \quad (66)$$

From equations (65) and (66) we now get

$$2(\boldsymbol{M}\cdot\boldsymbol{S})[c_+\mathscr{P}_l{}^m\xi+c_-\mathscr{P}_l{}^{m+1}\eta] = \hbar^2\{(c_+m+c_-[(l-m)(l+m+1)]^{\frac{1}{2}})\mathscr{P}_l{}^m\xi+$$
$$(c_+[(l-m)(l+m+1)]^{\frac{1}{2}}-c_-(m+1))\mathscr{P}_l{}^{m=1}\eta\} = \lambda\hbar^2\{c_+\mathscr{P}_l{}^m\xi+c_-\mathscr{P}_l{}^{m+1}\eta\},$$

where λ is the eigenvalue of $2(\boldsymbol{M}\cdot\boldsymbol{S})/\hbar^2 = [J^2-M^2-S^2]/\hbar^2 = (J^2/\hbar^2)-l(l+1)-\frac{3}{4}$. Equating the coefficients of $\mathscr{P}_l{}^m\xi$ and $\mathscr{P}_l{}^{m+1}\eta$ gives us the following equations for c_+, c_-, and λ: $\lambda(\lambda+1) = l(l+1)$, $(m-\lambda)c_++c_-[(l-m)(l+m+1)]^{\frac{1}{2}} = 0$, $c_+[(l-m)(l+m+1)]^{\frac{1}{2}}-c_-(m+1+\lambda) = 0$. This leads to $\lambda = l$, $c_+/c_- = [(l+m+1)/(l-m)]^{\frac{1}{2}}$, or $\lambda = -l-1$, $c_+/c_- = -[(l-m)/(l+m+1)]^{\frac{1}{2}}$. The eigenvalues of $J^2 = \hbar^2(l^2+l+\frac{3}{4}+\lambda)$ are thus equal to $\hbar^2(l^2+2l+\frac{3}{4}) = \hbar^2(l+\frac{1}{2})(l+\frac{3}{2})$ and $\hbar^2(l^2-\frac{1}{4}) = \hbar^2(l-\frac{1}{2})(l+\frac{1}{2})$ and we have equations (64).

We have obtained here by a tedious straightforward method a

result which we shall understand later on from a much more general point of view. We are dealing with *the combination of two angular momentum vectors into a resultant angular momentum.* In our special case the two vectors are the orbital angular momentum $\boldsymbol{M}$ [quantum number l, eigenvalue $M^2 = \hbar^2 l(l+1)$] and the spin angular momentum $\boldsymbol{S}$ [quantum number $\frac{1}{2}$, eigenvalue $S^2 = \hbar^2 \frac{1}{2} \cdot \frac{3}{2}$] and the resultant vector is $\boldsymbol{J}$ with eigenvalues $J^2 = \hbar^2(l+\frac{1}{2})(l+\frac{3}{2})$ and $J^2 = \hbar^2(l-\frac{1}{2})(l+\frac{1}{2})$. The number of eigenfunctions of J_z of the form (64a) is $2l+2$ corresponding to eigenvalues $J_z = (l+\frac{1}{2})\hbar$, $(l-\frac{1}{2})\hbar$, $\ldots$, $(-l-\frac{1}{2})\hbar$. The number of eigenfunctions of J_z of the form (64b) is $2l$ corresponding to eigenvalues $J_z = (l-\frac{1}{2})\hbar, \ldots, (-l+\frac{1}{2})\hbar$. Altogether there are $4l+2$ eigenfunctions corresponding to the degree of degeneracy of functions of the form (60).

The situation is thus very similar to that in the case of the eigenfunctions and eigenvalues of M^2 and M_z and we are led to introduce a new quantum number, the *quantum number j of the total angular momentum of the system.* In the case (a) $j = l+\frac{1}{2}$ and in case (b) $j = l-\frac{1}{2}$. In both cases we have

$$J^2 = j(j+1)\hbar^2, \tag{67}$$

while J_z takes on the values

$$J_z = m_j\hbar, \quad m_j = -j, -j+1, \ldots, j-1, j \quad (2j+1 \text{ values}). \tag{68}$$

The eigenfunctions Q of equations (64) can be compared with the $\mathscr{P}_l^m$. This all corresponds to a classical model where two vectors of length $l\hbar$ and $\frac{1}{2}\hbar$ are combined to form a vector of length $j\hbar$. The quantity j can take on several values, each differing from its neighbours by 1 while the maximum value of j corresponds to parallel vectors ($j = l+\frac{1}{2}$). For $l \neq 0$ there are two possibilities ($j = l\pm\frac{1}{2}$), but there is only one for $l = 0$ ($j = \frac{1}{2}$). We shall see later on that the same procedure and the same equations (67) and (68) apply to the case of two vectors with arbitrary integral or half-integral quantum numbers (§ 79).

Let us now return to the perturbation H_s. We know from the above analysis that the eigenfunctions which "correspond" to this perturbation (see § 49) are of the form

$$\varphi_0 = (R_{nl}(r)/r)\, Q_{l\pm 1/2}^{m+1/2}, \qquad -(l\pm\tfrac{1}{2}) \leq m+\tfrac{1}{2} \leq l\pm\tfrac{1}{2}. \tag{69}$$

The influence of H_s on the energy levels is thus obtained by evaluating the expectation value of H_s in the states corresponding to these functions.

We have just seen that $(\boldsymbol{S}\cdot\boldsymbol{M})$ has the value $\frac{1}{2}l\hbar^2$ or $-\frac{1}{2}(l+1)\hbar^2$ for $j=l+\frac{1}{2}$ or $j=l-\frac{1}{2}$ and we find thus

$$\delta_s E_{nl}^{(0)} = \overline{H}_s = \frac{e^2\hbar^2}{4m^2c^2}\frac{\int (F'R^2/r)dr}{\int R_{nl}^2\,dr}(k-1)$$

$$= \mu_{\rm B}^2\overline{\left(\frac{F'}{r}\right)}(k-1)\begin{Bmatrix} k = l+1, \text{ if } j = l+\frac{1}{2} \\ k = -l, \text{ if } j = l-\frac{1}{2}\end{Bmatrix}. \qquad (70)$$

We shall meet with the quantum number $k=\pm(j+\frac{1}{2})$ later on in the discussion of the Dirac theory (§ 66).

We see that each level of the unperturbed system is split into two, corresponding to the two possible values of the total angular momentum. Only for $l=0$ do we have no splitting since j has only the value $\frac{1}{2}$. Since eF' is the electrical field strength in the direction of the radius vector [see eq. (58)], F'/r will always be positive in an attractive central field of force and the level corresponding to $j=l-\frac{1}{2}$ lies lower than the one corresponding to $j=l+\frac{1}{2}$.

These results can immediately be applied to the discussion of the alkali spectra (one valence electron in the central field of force produced by the nucleus and the closed shells). If applied to the Coulomb field they lead to the energy levels of the hydrogen atom including (approximately) the relativistic change of mass and the spin. From equations (61) and (70) we can evaluate the shifts in the energy levels. We use for $E_{nl}^{(0)}$ the values $-Z^2me^4/2n^2\hbar^2$ ($=-W_1/n^2$) of equation (4.192) * and evaluate the expectation values of $F=-Z/r$, $F^2=Z^2/r^2$, and $F'/r=Z/r^3$ by using for the R_{nl} equation (4.196). The result is

$$\delta_{\rm r}E_{nl} = W_1\left(\frac{Ze^2}{\hbar c}\right)^2\left[\frac{3}{4n^2}-\frac{1}{(l+\frac{1}{2})n}\right],\ \delta_s E_{nl} = W_1\left(\frac{Ze^2}{\hbar c}\right)^2\frac{k-1}{2n^3l(l+\frac{1}{2})(l+1)}. \qquad (71)$$

It is remarkable that the energy — at least to the approximation used here — depends on n and j, but not on l,

$$E_{nlj} = E_{nl}^{(0)} + \delta_{\rm r}E_{nl} + \delta_s E_{nl} = \frac{Z^2me^4}{2\hbar^2}\left[-\frac{1}{n^2}+\left(\frac{Ze^2}{\hbar c}\right)^2\left(\frac{3}{4n^4}-\frac{1}{n^3(j+\frac{1}{2})}\right)\right]. \qquad (72)$$

* We include the nuclear charge Z in our equations.

This means that a peculiar degeneracy occurs in the case of the hydrogen atom *. Every energy level corresponding to a given value of n splits up into n levels corresponding to $j=\frac{1}{2},\frac{3}{2},\ldots,n-\frac{1}{2}$. Each of these levels is $2(2j+1)$-fold degenerate apart from the last one which corresponds only to one l-value ($l=n-1$) and which is $2j+1$-fold degenerate ($g=2j+1=2(n-\frac{1}{2})+1=2n$). The sum total of all the degrees of degeneracy g is $2n^2$. We have the following scheme

$$\left.\begin{array}{llllllllllll}
l: & 0 & & 1 & & 2 & & \ldots & & n-1 & & \\
j: & & \frac{1}{2} & & \frac{3}{2} & & \frac{5}{2} & \ldots & n-\frac{3}{2} & & & n-\frac{1}{2} \\
g: & & 2.2 & & 2.4 & & 2.6 & & 2(2n-2) & & & 2n
\end{array}\right\} \quad (73)$$

The calculation of the *expectation values of the powers of r in the hydrogen atom* which was necessary to derive equations (71) proceeds more easily from equation (4.185) than from equation (4.196).**

Multiplication of equation (4.185) by $\varrho^{\lambda+1}(dR/d\varrho)-\frac{1}{2}(\lambda+1)\varrho^{\lambda}R$ and integration over r leads after a few partial integrations to the recursion formula

$$[(\lambda+1)/n^2]\overline{\varrho^{\lambda}}-(2\lambda+1)\overline{\varrho^{\lambda-1}}+\lambda(l+\tfrac{1}{2}+\tfrac{1}{2}\lambda)(l+\tfrac{1}{2}-\tfrac{1}{2}\lambda)\overline{\varrho^{\lambda-2}}=0. \quad (74)$$

$\lambda=0$ leads to $n^{-2}-\overline{\varrho^{-1}}=0$, or, $\overline{r^{-1}}=(a_1n^2)^{-1}$ [comp. eq. (4.184)]; (α)

$\lambda=1$ leads to $(2\overline{\varrho}/n^2)-3+l(l+1)\overline{\varrho^{-1}}=0$, or, $\overline{r}=\frac{1}{2}a_1[3n^2-l(l+1)]$;

$\lambda=2,3,\ldots$ lead to values for $\overline{\varrho^2},\overline{\varrho^3},\ldots$

$\lambda=-1$ leads to $\overline{\varrho^{-2}}-l(l+1)\overline{\varrho^{-3}}=0$, or, $\overline{r^{-3}}=\overline{r^{-2}}/a_1l(l+1)$. (β)

$\lambda=-2,-3,\ldots$ leads to $\overline{\varrho^{-4}},\overline{\varrho^{-5}},\ldots$ expressed in terms of $\overline{\varrho^{-2}}$. If $\lambda\leqq-2l-3$, $\overline{\varrho^{\lambda}}$ diverges. We see that we do not obtain $\overline{\varrho^{-2}}$ from equation (74). This expectation value is obtained by considering l in equation (4.185) to be a continuous variable. From the considerations of § 46 it follows that we can consider n to be a function of l by fixing $n-l$ to be a definite integer. The eigenfunction R_{nl} is then a continuous, differentiable function of l and we find

$$\partial(R''/R)/\partial l=-(\partial/\partial l)[-n^{-2}+2\varrho^{-1}-l(l+1)\,\varrho^{-2}]=-2n^{-3}+(2l+1)\,\varrho^{-2}.$$

Multiplying this equation by R^2 and integrating over r the left hand side vanishes and we have

$$-2n^{-3}+(2l+1)\overline{\varrho^{-2}}=0,\text{ or, }\overline{r^{-2}}=a_1{}^{-2}/n^3(l+\tfrac{1}{2}). \quad (\gamma)$$

Equations (α), (β) and (γ) were used to obtain equations (71).

Expression (71) for $\delta_s E_{nl}$ is not defined for $k=l+1$, $l=0$. This is connected with the fact that $\overline{r^{-3}}$ diverges for $l=0$. One assumes that one may put in this case also $(k-1)/l=1$, since equation (72) is then universally valid. This difficulty does not occur in the rigorous relativistic electron theory of Dirac which leads exactly

* We do not want to discuss here the complications due to the Lamb-shift.

** See also J. H. Van Vleck, *Proc. Roy. Soc.*, **A143**, 679, 1934.

to equation (72) if one neglects higher order terms in $Ze^2/\hbar c$ (see § 66).

Sommerfeld had obtained in 1915 a formula which was equivalent to equation (72) (he wrote $k-\frac{1}{2}$ for j, where k was the azimuthal quantum number of the old quantum theory) by applying his method of phase integrals. The correctness of this formula was shown most impressively by Paschen's measurements of the H and He^+-spectra. If one takes only the relativistic change of mass into account in quantum mechanics, one obtains a formula which does not agree with experimental data [compare the expression for $\delta_r E_{nl}$ in equations (71) and (72)], but if one takes the spin as well into account, one obtains again exactly Sommerfeld's formula *.

If we are dealing with heavy nuclei, $Ze^2/\hbar c$ in equation (72) is no longer small compared to unity, as the dimensionless number $e^2/\hbar c$, the so-called *fine structure constant* is approximately equal to 1/137. We see that the influence of the spin on the energy levels may be considerable for elements towards the end of the periodic system. This is seen, for instance, in the so-called "relativity doublets" in the X-ray spectra. These were discovered by Sommerfeld and analysed by him using equation (72), or rather using a closed equation of which equation (72) is a first approximation [see eq. (212)] **.

If the system consisting of a spinning electron in a central field of force is placed in a homogeneous magnetic field, an application of perturbation theory to this problem leads immediately to a quantitative explanation of the anomalous Zeeman effect (weak fields) and of the Paschen-Back effect (strong fields) of the alkali spectra***. These effects have played an important role in the development of the concept of the spinning electron as we mentioned in § 56.

§ 60. Many electron systems

We can now easily give the *quantum mechanical treatment of many electron systems in a field of force*. Classically we include the spin by introducing a fourth pair of canonically conjugate variables for each

* The quantum mechanical formulae (71) were first derived by Heisenberg and Jordan (*Z. Phys.*, **37**, 263, 1926). However, the situation indicated by the scheme (73) was known earlier to exist (See, for instance, G. E. Uhlenbeck and S. Goudsmit, *Physica* (old series), **5**, 266, 1925).

** See A. Sommerfeld, *Atomic structure and spectral lines*, Methuen, London, 1934.

*** Heisenberg and Jordan (*loc. cit.*) gave the quantum mechanical explanation of these effects.

electron. As in the case of a one-electron system, we describe the physical situation by means of a wave function which depends not only on the time and on the spatial coordinates of all the particles, but also on as many spin variables $s_1, s_2, \ldots$ as there are electrons. Each spin variable can take on only two values, $s_i = \pm\frac{1}{2}$. We express this as follows (N: number of electrons)

$$\psi = \psi_{s_1, s_2, \ldots, s_N}(\boldsymbol{x}_1, \boldsymbol{x}_2, \ldots, \boldsymbol{x}_N, t), \tag{75}$$

where as in equation (55) the spin variable is used as an index.

As there are 2^N combinations of values for the N spin variables, ψ is completely determined, if we give 2^N continuous complex functions of the time and of the spatial coordinates. This means that ψ is a 2^N-component wave function. To write such a wave function in the form of a one-column matrix, as was done in equation (55), is usually not convenient. However, the introduction of the functions ξ and η for each spin is often advantageous. Let $\xi_i(\eta_i)$ be again a function of the spin variable s_i which is 1(0) or 0(1) according to whether s_i is $+\frac{1}{2}$ or $-\frac{1}{2}$. A product such as $\xi_1\xi_2\eta_3\xi_4$ is then a function of the four spin variables s_1, s_2, s_3, and s_4 which is different from zero only if $s_1 = s_2 = -s_3 = s_4 = \frac{1}{2}$ and in that case it is equal to 1. If one is dealing with many electrons it is often convenient to use the notation $\xi_i \to \xi_i^{(+)}$, $\eta_i \to \xi_i^{(-)}$.

Any function of the s_i can be written as a linear combination of the 2^N possible products

$$\xi_1^{(s_1)}\,\xi_2^{(s_2)}\,\xi_3^{(s_3)} \ldots \xi_N^{(s_N)} = \Pi_i\,\xi_i^{(s_i)}, \tag{76}$$

and instead of equation (75) we can write

$$\begin{aligned} \psi &= \Sigma_{s_1} \ldots \Sigma_{s_N}\, \psi_{s_1 \ldots s_N}(\boldsymbol{x}_1, \ldots, \boldsymbol{x}_N, t)\, \Pi_i \xi_i^{(s_i)}, \\ \text{or,} \quad \psi &= \Sigma_{s_i} \psi_{s_i}(\boldsymbol{x}_i, t) \Pi_i \xi_i^{(s_i)}. \end{aligned} \tag{77}$$

In these equations the s_i are now really indices and not arguments as in equation (75) *.

The rate of change of ψ is governed by a Schrödinger equation, $H_{\text{op}}\psi = i\hbar\partial\psi/\partial t$, with an H_{op} which can act on the spin variables as well as on the $\boldsymbol{x}_i$ and t. The most general operator which acts on s_i is a linear function of the components of the spin angular momentum of the i-th electron, S_{xi}, S_{yi}, and S_{zi} [see eq. (57)]. The action

* Similarly one can denote an ordinary vector with components A_x, A_y, A_z by A_k ($k = x, y, z$) or describe it by means of unit vectors: $A_x\boldsymbol{i}+A_x\boldsymbol{j}+A_z\boldsymbol{k}$. The $\Pi\xi$ can be considered to be unit vectors in a 2^N-dimensional space.

of these three operators on the wave function follows from the form of the Pauli matrices. Equation (66) shows how they act on the functions ξ_i and η_i, or, $\xi_i^{(+)}$ and $\xi_i^{(-)}$,

$$S_x \xi^{(s)} = \tfrac{1}{2}\hbar \xi^{(-s)}; \quad S_y \xi^{(s)} = \tfrac{1}{2} i\hbar(-1)^{s-\frac{1}{2}} \xi^{(-s)} = i\hbar s\, \xi^{(-s)}; \qquad S_z \xi^{(s)} = \tfrac{1}{2}\hbar(-1)^{s-\frac{1}{2}} \xi^{(s)} = \hbar s \xi^{(s)}. \tag{78}$$

The most general H_{op} can thus be written as a "multilinear" polynomial in terms of the components of all the $\boldsymbol{S}_i$ — as far as its action on the s_i is concerned. This polynomial is linear in the components of each of the $\boldsymbol{S}_i$ *. If we drop all matrices which are unit matrices with respect to the spin variables we get the following expansion, which is similar to equation (57)

$$\left.\begin{aligned} &H = H^{(0)} + H^{(1)} + H^{(2)} + \ldots; \; H^{(1)} = \Sigma_i (H^{x_i} S_{xi} + H^{y_i} S_{yi} + H^{z_i} S_{zi}), \\ &H^{(2)} = \Sigma_{i \neq j} (H^{x_i x_j} S_{xi} S_{xj} + H^{x_i y_j} S_{xi} S_{yj} + H^{y_i x_j} S_{yi} S_{xj} + \ldots), \ldots \end{aligned}\right\} \tag{79}$$

In these equations $H^{(0)}$, H^{x_i}, $H^{x_i x_j}$, ... are all operators acting on the $\boldsymbol{x}_i$ and possibly containing the time explicitly.

If we are considering a concrete problem, we construct the energy operator by starting from a classical energy function which depends on the spatial coordinates and the momenta and which expresses the influence of the spin through terms which are polynomial in the spin components. This corresponds exactly to the general form (32) and (33) of the energy function which we wrote down for a many electron problem in a first approximation. This expression contained a spin independent term H_0 as well as terms of the first and second degree in the S_{xi}, S_{yi}, and S_{zi} which terms correspond to the magnetic spin interaction. Although terms of higher degree in the S would not have made any difficulties, they did not occur. To find now the explicit expression for the energy operator, one replaces in equations (32) and (33) $\boldsymbol{p}_i$ by $-i\hbar\nabla$ and the S_{xi}, ... by the corresponding Pauli matrices.

It is not yet the place to illustrate the influence of the spin for the case of a concrete many electron problem. For such problems one finds realised in nature only a very restricted selection from all possible solutions of the corresponding Schrödinger equation because of the intervention of *Pauli's exclusion principle*. This principle will be discussed in Chapter VII. There is, however, another

* Powers and products of S_{xi}, S_{yi}, and S_{zi} reduce by means of equations (52) and (53) to a constant or to a linear term in these quantities.

point which needs further elucidation and which we shall discuss here.

The atomic systems occurring in nature contain apart from electrons also one or several nuclei. Up to now we considered the nuclei to be charged mass points — like the electrons — and there were no difficulties in including them into the quantum mechanical treatment. We know, however, that atomic nuclei are in fact practically always compound, and usually very complicated, structures. To consider them to be charged mass points is only a first approximation. This approximation is quite a good one as long as we are dealing with systems and processes where the nucleus does not enter into the discussion apart from providing a centre of electrical field. This means that if we are discussing the constitution of atoms and molecules, their spectra *, chemical reactions and so on, we can treat the nuclei as being in first instance characterised only by their charge and mass.

At the same time we are faced with the problem of whether nuclei, even if we may neglect their structure, do not possess spin properties like those of the electron. Indeed, we know that many nuclei possess an intrinsic angular momentum, the so-called *nuclear spin* and that their interaction with other systems is due not only to their charge, but also to the existence of a magnetic dipole moment. These are also many nuclei which possess electric quadrupole moments. These nuclear interactions are shown up, for instance, in the *hyperfine structure* of atomic spectra, or in *nuclear magnetic resonance* experiments **. In Chapter VII we shall discuss how the presence of nuclear spin can influence physical phenomena, even though we may neglect the terms in the Hamiltonian corresponding to the nuclear spin (ortho- and para-hydrogen, intensities in band spectra). It is thus in order to discuss the general point of view from which we may construct a wave function and a Schrödinger equation if we want explicitly to take into account the influence of nuclei and their spin when discussing atomic systems containing spinning electrons.

If we are dealing with nuclei of spin zero, such as the ^{4}He nucleus, the wave functions will depend on the spatial coordinates of these nuclei as well as on the spatial and spin coordinates of the electrons.

* Except if we want to include such details as isotope shifts which depend on the electronic wave functions at the nuclear surface.

** See, for instance, E. R. Andrew, *Nuclear Magnetic Resonance*, Cambridge University Press, 1955.

As far as the nuclei are concerned we replace in the classical expression for the energy their momenta $\boldsymbol{p}$ in the usual manner by $-i\hbar\boldsymbol{\nabla}$.

If, however, we are dealing with the ^{1}H-nucleus, the proton, we must take into account the fact that we know from the properties of the hydrogen molecule, that the proton possesses a spin which corresponds to an angular momentum quantum number $\frac{1}{2}$ as in the case of the electron. This means that the wave function will depend on the three spatial and one spin coordinate of each proton. This latter is a discrete variable which can take on two values. We expect then in the energy operator terms proportional to the components of the proton spin. The exact form of these terms is not known at the present time and hence must be deduced from experimental evidence. According to experimental data the proton possesses a magnetic moment of absolute magnitude

$$\mu \doteq 2.85\mu_{\mathrm{N}} = 2.85(e\hbar/2m_{\mathrm{p}}c) = 2.85(m/m_{\mathrm{p}})\mu_{\mathrm{B}} = (2.85/1836)\mu_{\mathrm{B}}, \qquad (80)$$

where m_{p} is the proton mass and μ_{N} the nuclear magneton which is equal to the Bohr magneton multiplied by the ratio of the electron to the proton mass. If the situation were as simple in the case of the proton as it is in the case of the electron, we would expect $\mu = \mu_{\mathrm{N}}$. Up to now there does not exist a completely satisfactory explanation of the fact that $\mu \doteq 2.85\mu_{\mathrm{N}}$ *. All the same as a first approximation one can take the proton spin into account by including in the energy operator a term which is equal to the scalar product of the proton magnetic moment times the magnetic field at the proton.

A new feature enters when the nuclear spin i is larger than $\frac{1}{2}$ ($i = 1$ for the deuteron, $i = \frac{3}{2}$ for ^{23}Na, . . .). In these cases one introduces a nuclear spin variable s which can take on more than 2 values; it takes on 3 values for $i = 1$ and in general $2i+1$ values ($s = -i, -i+1, \ldots, i-1, i$). If such a nucleus possesses a magnetic moment which is α times its spin angular momentum, the energy operator will include as a first approximation a term $H_s = -\alpha(\boldsymbol{S}\cdot\mathfrak{H})$ where $\mathfrak{H}$ is the magnetic field at the nucleus and $\boldsymbol{S}$ the nuclear spin angular momentum. In this case S_x, S_y, S_z will be matrices of $2i+1$ rows and columns and they will have the same form as the M_x, M_y, and M_z matrices given by figure 11 of § 45 with l replaced by i. We can not express any matrix operating on a function of s as a linear

* The extra magnetic moment is interpreted to be due to the meson fields which accompany each nucleon (proton or neutron). Similarly the magnetic moment of the electron is not exactly equal to μ_{B}; there is a slight difference which is due to the electromagnetic self-field of the electron (compare Chapter VIII).

combination of S_x, S_y, S_z, and the unit matrix, because equation (57) is no longer valid. This means that we must expect that the influence of the nuclear spin on the energy of the system is not only given by linear terms in the components of $\boldsymbol{S}$. An example of such more complicated interactions is the quadrupole interaction of a nucleus. This interaction is described by terms which are quadratic in the S_x, S_y, and S_z *. In general any matrix of $2i+1$ rows and columns can be expressed as a polynomial in S_x, S_y, S_z of a degree which is at most equal to $2i$. We can thus expect terms of this kind in the energy operator.

From the equations (4.162) for M_x, M_y, and M_z it follows that a matrix $a_{mm'}$ all elements of which are zero, except $a_{m_0m_0'}$ can be written as a product $\mathfrak{P}_{2l-m_0-m_0'}(M_z)(M_x+iM_y)^{m_0-m_0'}(m_0 > m_0')$ or $(M_x-iM_y)^{m_0'-m_0}\mathfrak{P}_{2l-m_0'-m_0}(M_z)$ $(m_0<m_0')$. Here, $\mathfrak{P}_k(M_z)$ denotes a polynomial of degree k in M_z of such a kind that the elements b_{mm} of the corresponding diagonal matrix are different from zero only if $m = m_0$, $m_0 < l-k$ $(m_0>m_0')$ or $m=m_0'$, $m_0' > l-k$ $(m_0' > m_0)$. *Any matrix* $a_{mm'}$ *can thus always be written as a polynomial of the 2l-th degree in the* M_x, M_y, M_z.

§ 61. Spinors and rotations in space

In this section we will construct a foundation for Pauli's treatment of the spin by connecting the components of the wave functions with certain mathematical entities, the so-called *spinors*.

We start by considering a vector in three-dimensional space with complex components a, b, and c,

$$\boldsymbol{V} = a\boldsymbol{i}+b\boldsymbol{j}+c\boldsymbol{k} = \boldsymbol{V}_1+i\boldsymbol{V}_2. \tag{81}$$

This vector is completely defined by two real vectors $\boldsymbol{V}_1 = \text{Re}(\boldsymbol{V})$ and $\boldsymbol{V}_2 = \text{Re}(-i\boldsymbol{V})$. The time dependent real vector $\boldsymbol{V}_t = \text{Re}\,[\boldsymbol{V}\exp(-i\omega t)]$ represents clearly a harmonic elliptical oscillation in the plane of $\boldsymbol{V}_1$ and $\boldsymbol{V}_2$ with $\boldsymbol{V}_1$ and $\boldsymbol{V}_2$ conjugate diameters of the ellipse. Multiplying $\boldsymbol{V}$ by a phase factor we obtain a new vector $\boldsymbol{W} = \boldsymbol{W}_1+i\boldsymbol{W}_2$ where $\boldsymbol{W}_1$ and $\boldsymbol{W}_2$ are a second pair of conjugate diameters of the same ellipse. If we perform a transformation from the original system of coordinates x, y, z to a new orthogonal system x', y', z' (the two systems being either both right-handed or both left-handed) a, b, c will transform linearly and the determinant of the transformation will be equal to unity,

* There are five quadratic terms which together with S_x, S_y, S_z and **1** form a linearly independent set of nine quantities. We have the four relations $[S_x, S_y]_- = i\hbar S_z, \ldots, S_x^2+S_y^2+S_z^2 = i(i+1)\hbar^2$ and for the five quadratic terms we can, for instance, choose $[S_x, S_y]_+$, $[S_y, S_z]_+$, $[S_z, S_x]_+$, $S_x^2-S_y^2$, and $S_y^2-S_z^2$.

$$\left.\begin{array}{c} a'=\alpha_{11}a+\alpha_{12}b+\alpha_{13}c, \quad b'=\alpha_{21}a+\alpha_{22}b+\alpha_{23}c, \quad c'=\alpha_{31}a+\alpha_{32}b+\alpha_{33}c; \\ |\alpha_{kl}| = 1. \end{array}\right\} \quad (82$$

The sum of the squares of the components will not change its value,

$$a^2+b^2+c^2 = a'^2+b'^2+c'^2. \quad (83)$$

Consider now the case where $\boldsymbol{V}$ is a null-vector $\boldsymbol{V}^{(0)}$, that is, where

$$V^{(0)2} = a^2+b^2+c^2 = 0. \quad (84)$$

We then have $(\boldsymbol{V}_1^{(0)}+i\boldsymbol{V}_2^{(0)})^2 = V_1^{(0)2}-V_2^{(0)2}+2i(\boldsymbol{V}_1^{(0)}\cdot\boldsymbol{V}_2^{(0)}) = 0$, which means that $\boldsymbol{V}_1^{(0)}$ and $\boldsymbol{V}_2^{(0)}$ have the same length and are orthogonal to each other. We use them to construct an *orthogonal coordinate system* with X-axis parallel to $\boldsymbol{V}_1^{(0)}$ and Y-axis parallel to $\boldsymbol{V}_2^{(0)}$. The Z-axis is chosen in such a way that the XYZ system is right (left) handed, if the xyz system was right (left) handed. The null-vector is completely determined by this coordinate system, provided we also give a positive number which characterises the absolute magnitude of $\boldsymbol{V}_1^{(0)}$ (or $\boldsymbol{V}_2^{(0)}$). We choose for this number

$$G = 2V_1^{(0)} = 2V_2^{(0)} = +[2(a^*a+b^*b+c^*c)]^{1/2}. \quad (85)$$

We now introduce a vector $\boldsymbol{Z}$ of length G in the direction of the positive Z-axis by the equation $G\boldsymbol{Z}=4[\boldsymbol{V}_1^{(0)}\wedge\boldsymbol{V}_2^{(0)}]=-2i[\boldsymbol{V}^{(0)*}\wedge\boldsymbol{V}^{(0)}]$, where $\boldsymbol{V}^{(0)*}$ has the components a^*, b^*, and c^*. The components of $\boldsymbol{Z}$ are

$$Z_x=(2i/G)(c^*b-b^*c), \; Z_y=(2i/G)(a^*c-c^*a), \; Z_z=(2i/G)(b^*a-a^*b). \quad (8$$

If a second null vector $\boldsymbol{W}^{(0)}$ is obtained by multiplying $\boldsymbol{V}^{(0)}$ by $e^{-i\omega}$ the corresponding new coordinate system is rotated over an angle ω around the Z-axis with respect to the old system.

We now express the three complex numbers a, b, c which satisfy equation (84) in terms of two independent complex numbers u and v *,

$$u=(a-ib)^{1/2}, \; v=(-a-ib)^{1/2}; \; a=\tfrac{1}{2}(u^2-v^2), \; b=\tfrac{1}{2}i(u^2+v^2), \; c=-uv. \quad (8$$

These equations fix u and v apart from a common factor ± 1. Substituting equations (87) into equations (85) and (86) we get

$$G=u^*u+v^*v, \; Z_x=u^*v+v^*u, \; Z_y=-i(u^*v-v^*u), \; Z_z=u^*u-v^*v. \quad (8$$

* We have chosen the signs so as to obtain the usual form of the Pauli matrices (see later on in this §).

A transformation of the system of coordinates $xyz \to x'y'z'$ corresponds to a linear transformation of u and v, that is, if a, b, c transform into a', b', c' through equations (82), u' $[= (a'-ib')^{1/2}]$ and v' $[= (-a'-ib')^{1/2}]$ depend on u and v in the following way,

$$u' = \alpha u + \beta v, \qquad v' = \gamma u + \delta v. \tag{89}$$

Proof: The expressions u'^2, $u'v'$, and v'^2 are linear expressions in a', b', and c', and thus [see eqq. (82) and (87)] in u^2, uv, and v^2:

$$\left.\begin{array}{l} u'^2 = \beta_{11}u^2+\beta_{12}v^2+\beta_{13}uv, \ [1]; \ v'^2 = \beta_{21}u^2+\beta_{22}v^2+\beta_{23}uv, \ [2]; \\ \qquad u'v' = \beta_{31}u^2+\beta_{32}v^2+\beta_{33}uv. \ [3] \end{array}\right\} \tag{90}$$

The product of the right hand sides of [1] and [2] is equal to the square of the right hand side of [3]. If the right hand sides of [1] and [2] were not squares themselves, the ratio of u'^2 to v'^2 would be independent of u and v, and one could not express u^2, v^2, and uv in terms of u'^2, v'^2, and $u'v'$. Since that is always possible, equations (89) follow.

From the assumption that the x', y', z' and the x, y, z systems are either both right-handed or both left-handed it follows that

$$\begin{vmatrix} \alpha & \beta \\ \gamma & \delta \end{vmatrix} = \alpha\delta - \beta\gamma = 1, \tag{91}$$

and we get from equation (89)

$$u = \delta u' - \beta v', \qquad v = -\gamma u' + \alpha v'. \tag{92}$$

Equation (91) follows from the fact that the determinant of the transformation (90) is equal to $+1$, as can be seen by comparing equations (87) and (82), but it is also equal to $(\alpha\delta-\beta\gamma)^3$ as can be seen by using equation (89).

Since G is invariant, because of its definition, we find

$$G = u^*u + v^*v = u'^*u' + v'^*v'. \tag{93}$$

This invariance entails that *the linear transformation of u and v is unitary* (see § 37) and we have

$$\delta = \alpha^*, \quad \gamma = -\beta^*, \quad \alpha^*\alpha + \beta^*\beta = |\alpha|^2 + |\beta|^2 = 1. \tag{94}$$

Proof: It follows from equation (93) that u^* and v^* transform contravariant to u and v. Substituting equations (89) into equation (93) and comparing coefficients we find that $u^* = \alpha u'^* + \gamma v'^*$, $v^* = \beta u'^* + \delta v'^*$. Comparing this with equations (92) leads to equations (94).

Equations (89) and (92) can be simplified to read

$$u' = \alpha u + \beta v, \qquad v' = -\beta^* u + \alpha^* v; \qquad u = \alpha^* u' - \beta v', \qquad v = \beta^* u' + \alpha v'. \tag{95}$$

If ϑ, φ, and ψ are the Eulerian angles defining the $x'y'z'$ system with respect to the xyz system (see fig. 13), the following relations hold,

$$\alpha = \cos \tfrac{1}{2}\vartheta \exp [\tfrac{1}{2}i(\varphi+\psi)], \quad \beta = i \sin \tfrac{1}{2}\vartheta \exp [\tfrac{1}{2}i(\varphi-\psi)] \quad (96)$$

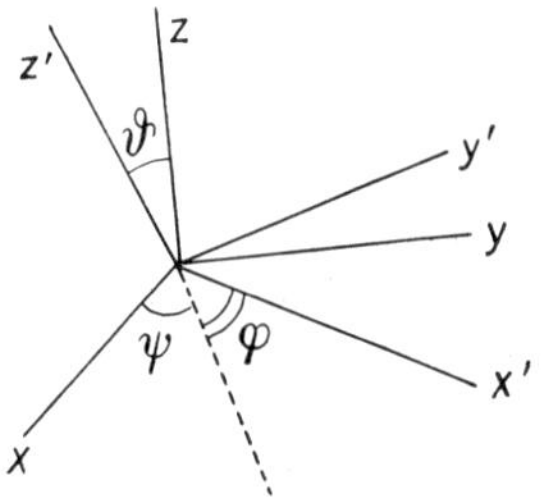

Fig. 13

To prove this we assume that the XYZ system coincides with the $x'y'z'$ system. From the usual spherical geometry formulae for direction cosines we find $a = \frac{1}{2}G(\cos\psi - i\cos\vartheta\sin\psi)e^{-i\varphi}$, $b = \frac{1}{2}G(\sin\psi + i\cos\vartheta\cos\psi)e^{-i\varphi}$, $c = \frac{1}{2}iG\sin\vartheta e^{-i\varphi}$, or,

$$u = G^{\frac{1}{2}} \cos \tfrac{1}{2}\vartheta \exp [-\tfrac{1}{2}i(\varphi-\psi)], \quad v = -iG^{\frac{1}{2}} \sin \tfrac{1}{2}\vartheta \exp [-\tfrac{1}{2}i(\varphi-\psi)], \quad (97)$$

while $u' = G^{\frac{1}{2}}$ and $v' = 0$, because of our choice of coordinate system. From equations (97) and (95) we now deduce equation (96).

Two complex numbers u and v which define in this manner a system of coordinates and a real number G are called the components of a *spinor*. We use the symbol $\binom{u}{v}$ for a spinor.

Multiplication of a spinor by a complex number and addition of spinors is defined by the equation $c_1\begin{pmatrix}u_1\\v_1\end{pmatrix} + c_2\begin{pmatrix}u_2\\v_2\end{pmatrix} = \begin{pmatrix}c_1u_1+c_2u_2\\c_1v_1+c_2v_2\end{pmatrix}$.

If we introduce in particular $\begin{pmatrix}1\\0\end{pmatrix} \equiv \xi$, $\begin{pmatrix}0\\1\end{pmatrix} \equiv \eta$, we can write for the spinor $\binom{u}{v}$, $\psi = u\xi + v\eta$. Geometrically speaking a spinor corresponds to a (complex) vector in a plane, the so-called "spinor plane". Addition of spinors corresponds to vector addition in this plane and ξ and η are the "unit vectors" with respect to which the components are measured. *The real orthogonal transformations in three-dimensional space correspond to unitary transformations in the spinor plane.*

To each spinor which is not identically zero there corresponds one null vector or a system of coordinates in three-dimensional space according to equations (87). However, to each null vector there correspond always two different spinors, since $u\xi+v\eta$ and $-u\xi-v\eta$ correspond to the same null vector. It follows that addition of spinors can not correspond to a simple combination of the corresponding null vectors.

If we change from xyz to $x'y'z'$ we have $abc \rightarrow a'b'c'$, and $uv \rightarrow u'v'$. We can interpret this as follows in three-dimensional space. We keep the same entity, the null vector $\boldsymbol{V}^{(0)}$, but describe it in a different system of coordinates with unit vectors $\boldsymbol{i}'$, $\boldsymbol{j}'$, and $\boldsymbol{k}'$, $\boldsymbol{V}^{(0)} = a\boldsymbol{i}+b\boldsymbol{j}+c\boldsymbol{k} = a'\boldsymbol{i}'+b'\boldsymbol{j}'+c'\boldsymbol{k}'$. We can do the same in the spinor plane,

$$\psi = u\xi+v\eta = u'\xi'+v'\eta'. \tag{98}$$

The spinor is considered to be an invariant entity ψ. A change of the unit vectors in three-dimensional space corresponds to a change of the unit vectors in the spinor plane $\xi\eta \rightarrow \xi'\eta'$ in such a way that $\xi\eta$ transform contravariant to uv, that is, as u^*v^*,

$$\xi' = \alpha^*\xi+\beta^*\eta, \quad \eta' = -\beta\xi+\alpha\eta; \quad \xi = \alpha\xi'-\beta^*\eta', \quad \eta = \beta\xi'+\alpha^*\eta'. \tag{99}$$

The square of the absolute value of the spinor, $|u|^2+|v|^2$ is according to equations (85) and (93) equal to the absolute value, $[|a|^2+|b|^2+|c|^2]^{\frac{1}{2}}$, of the null vector multiplied by $2^{\frac{1}{2}}$.

To each null vector $\boldsymbol{V}^{(0)}$ there corresponds a negative complex conjugate null vector $\boldsymbol{V}^{(0)\dagger}$ with components $-a^*$, $-b^*$, $-c^*$. The system of coordinates corresponding to $\boldsymbol{V}^{(0)\dagger}$ is obtained from that corresponding to $\boldsymbol{V}^{(0)}$ by a rotation over π around the y-axis*. From equation (87) we see that the spinor $u^\dagger$, $v^\dagger$ which corresponds to $\boldsymbol{V}^{(0)\dagger}$ has the components

$$u^\dagger=[-a^*+ib^*]^{\frac{1}{2}}=v^*, \quad v^\dagger=[a^*+ib^*]^{\frac{1}{2}}=-u^*, \quad [u^\dagger v^\dagger=-(-c^*)]. \tag{100}$$

This spinor must be distinguished from the complex conjugate spinor u^*, v^*. We call

$$\psi^\dagger = u^\dagger\,\xi+v^\dagger\eta = v^*\,\xi-u^*\,\eta \tag{101}$$

the *spin conjugate* spinor of ψ. The spin conjugate $\psi^{\dagger\dagger}$ of $\psi^\dagger$ has the components $u^{\dagger\dagger} = v^{\dagger *} = -u$, $v^{\dagger\dagger} = -u^{\dagger *} = -v$. Both compon-

* We have $V_1^{(0)\dagger} = -V_1^{(0)}$, $V_2^{(0)\dagger} = V_2^{(0)}$.

ents have changed their sign. The corresponding null vector is the same as the one corresponding to ψ. A special case of equation (100) is

$$\xi^\dagger = -\eta, \qquad \eta^\dagger = +\xi. \tag{102}$$

Since a^*, b^*, c^* transform as a, b, c v^* and $-u^*$ *will transform as* u *and* v. Indeed, from equation (95) we see that $v^{*\prime} = \alpha v^* + \beta(-u^*)$, $(-u^*)' = -\beta^* v^* + \alpha^*(-u^*)$.

From the unitary character of the transformation (95) the invariance of the *inner product* (see § 37) of two spinors ψ_1 and ψ_2 follows,

$$\int \psi_1^* \psi_2 = u_1^* u_2 + v_1^* v_2 = u_1^{\prime *} u_2' + v_1^{\prime *} v_2'. \tag{103}$$

Equation (93) is a special case of equation (103).

If two spinors are orthogonal to each other, their inner product is equal to zero and the one spinor is, apart from a factor, equal to the spin conjugate of the other, $\int \psi_1^* \psi_2 = 0 \rightarrow \psi_1^\dagger = c\psi_2$. A special case are the unit vectors which are orthogonal to one another and have the absolute value 1.

The cross product of ordinary vectors correspond to the *alternating product of two spinors,*

$$[\psi_1 \wedge \psi_2] \equiv u_1 v_2 - v_1 u_2 = u_1' v_2' - v_1' u_2' = -[\psi_2 \wedge \psi_1]. \tag{104}$$

Its invariance is a consequence of the equations (89) and (91) and is thus independent of the unitary character of the transformation. Equation (104) follows also immediately from the fact that v_2 and $-u_2$ transforms as u_2^* and v_2^*, that is, contravariant to u_1 and v_1 so that $v_2 u_1 + (-u_2) v_1$ is invariant. The alternating product of a spinor with itself is zero.

Let us now consider in somewhat more detail the *ambiguity in the definition of a spinor corresponding to a given null vector.* The collective of all possible unitary transformations in the spinor plane corresponds to the collective of all those pairs of numbers α, β for which $|\alpha|^2 + |\beta|^2 = 1$. If we write $\alpha = x_1 + ix_2$, $\beta = x_3 + ix_4$, we see that this corresponds to all points on the four-dimensional sphere,

$$x_1^2 + x_2^2 + x_3^2 + x_4^2 = 1. \tag{105}$$

To each point of this sphere there corresponds a rotation in space, that is, a real orthogonal transformation of the form (82). Two points at the end of a diameter (x_1, x_2, x_3, x_4 and $-x_1$, $-x_2$, $-x_3$,

$-x_4$) correspond, however, to the same position of the $x'y'z'$-system relative to the xyz system. This is immediately clear from equation (96). The angles φ and ψ are defined modulo 2π, that is, apart from additive terms $2n\pi$ (n: integer), and α and β are thus defined apart from a factor ± 1. If we consider a continuous sequence of positions of the $x'y'z'$-system such that its final position is the same as its initial position, this corresponds to a continuous curve on the sphere (105). If the initial position corresponds to a point P on the sphere, the final position would correspond either to P or to the point Q at the other end of the diameter through P. The components $u'v'$ of a spinor will have returned to their original values in the first case, but will have taken on the same values with opposite sign in the second case. The ambiguity of the spinors expresses the fact that the continuous mapping of the (three-dimensional) collective of all real orthogonal rotations on a closed unbounded three-dimensional point set is a one to two correspondence. As an example we note that α and β of equation (96) change into $-\alpha$ and $-\beta$ if ϑ and φ (or ϑ and ψ) are kept constant, but ψ (or φ) increases continuously from 0 to 2π.

Pauli's two-component wave functions are spinors of the type (98) *whose components* $u = \psi_+$, $v = \psi_-$ *are functions of* x, y, z, t [compare eq. (55)]. We can thus illustrate these wave functions geometrically by assigning to them for each set of values x, y, z, t a null vector whose absolute magnitude is $[|a|^2+|b|^2+|c|^2]^{1/2} = 2^{-1/2}[|\psi_+|^2+|\psi_-|^2]$. The position of the corresponding system of coordinates follows from equation (97). It changes continuously with the values of x, y, z, t. If the wave function is multiplied by an arbitrary phase factor exp $(\frac{1}{2}i\omega)$ each system of coordinates rotates around its own Z-axis over an angle $-\omega$. The whole situation is very similar to the vectorial description of an electromagnetic field. The mathematical structure corresponding to the physical situation does not depend on the particular choice of the xyz coordinate system — as one would require. It is not difficult to find the components of the wave function in the $x'y'z'$ system, if they were known in the xyz system. We could even choose different coordinate systems to which the spinor components at different space-time points refer. The direction of the positive Z-axis at a point P is that direction for which the spin angular momentum component is certainly equal to $+\frac{1}{2}\hbar$. If we consider a direction which makes an angle ϑ with the Z-axis the probabilities of finding the values $\frac{1}{2}\hbar$ and $-\frac{1}{2}\hbar$ for the spin angular momentum component in that direction are in the ratio $\cos^2\frac{1}{2}\vartheta$ to $\sin^2\frac{1}{2}\vartheta$, that is, in the ratio $|u|^2 : |v|^2$ [see eq. (97)].

We must now prove that Pauli's wave functions are, indeed, spinors. Let

$$H_{\text{op}}\psi = i\hbar\partial\psi/\partial t \tag{106}$$

be the Pauli-Schrödinger equation of a spinning electron is the xyz-system. We use equation (99) to transform $\psi = \psi_+\xi + \psi_-\eta$ into a new spinor representation $\psi' = \psi_-\xi' + \psi_+\eta'$ corresponding to an $x'y'z'$-system. The equation $H'_{\text{op}}\psi' = i\hbar\partial\psi'/\partial t$ obtained by transforming equation (106) is identical with the Pauli-Schrödinger equation in the primed coordinate system (see § 58). In other words, if H_{op} were of the form (57), $H_{\text{op}} = H^{(0)} + H^{(x)}S_x + H^{(y)}S_y + H^{(z)}S_z$, H'_{op} will be of the form $H'_{\text{op}} = H'^{(0)} + H'^{(x')}S'_{x'} + H'^{(y')}S'_{y'} + H'^{(z')}S'_{z'}$, where $S'_{x'}$, $S'_{y'}$, $S'_{z'}$ have again the form of the spin matrices (39) and represent the components of the spin angular momentum in the $x'y'z'$-system, while $H'^{(x')}$, $H'^{(y')}$, $H'^{(z')}$ are those linear combinations of the $H^{(x)}$, $H^{(y)}$, $H^{(z)}$ which correspond to the transformation (82) of a vector. The operator $H'^{(0)}$ is the same operator acting on the space-time coordinates as $H^{(0)}$; the prime indicates as in the case of $H'^{(x')}$ that it now operates on functions of $x'y'z't$ and has therefore in general a different form (point transformation; see § 34) *.

Proof: We want to find the transformed form of the operators $\mathbf{1}$, S_x, S_y, S_z corresponding to the transformation $\psi_+\xi + \psi_-\eta = \psi_+'\xi' + \psi_-'\eta'$. The unit matrix remains, of course, the unit operator.

Let uv be the components of an arbitrary spinor and consider the expression $A = (u\psi_- - v\psi_+)(u\xi + v\eta)$.

We can write this expression as follows, using equation (66),

$$\begin{aligned} A &= \psi_+(-uv\xi - v^2\eta) + \psi_-(u^2\xi + uv\eta) = \\ &= (2/\hbar)[\psi_+\{-uvS_z\xi - \tfrac{1}{2}v^2(S_x - iS_y)\xi\} + \psi_-\{-uvS_z\eta + \tfrac{1}{2}u^2(S_x + iS_y)\eta\}] \\ &= (2/\hbar)[\tfrac{1}{2}(u^2 - v^2)S_x + \tfrac{1}{2}i(u^2 + v^2)S_y - uvS_z](\psi_+\xi + \psi_-\eta) \\ &= (2/\hbar)(aS_x + bS_y + cS_z)(\psi_+\xi + \psi_-\eta), \end{aligned}$$

where a, b, c are the components of the null vector corresponding to uv. The expression A is invariant being the product of an alternating and an inner product. This means that $aS_x + bS_y + cS_z = a'S'_{x'} + b'S'_{y'} + c'S'_{z'}$ is also invariant. It follows that the operators S_x, S_y, S_z transform under a rotation exactly as the components of a vector in complete accordance with their meaning as spin angular momentum components. We could, of course, have obtained the same result by explicitly using the transformation scheme of spinors and its relation with the transformation (82) of vectors.

* It would be possible to retain the old xyz as arguments of the functions, so that only the decomposition into components $H^{(x')}$, ... refers to the rotated position of the $x'y'z'$-system.

A deep insight into the nature of the operators S_x, S_y, S_z is obtained by considering them to be *the operators which correspond to infinitesimal rotations of the system of coordinates.* Consider again a spinor described once in the xyz-system and once in the $x'y'z'$-system: $\psi = u\xi + v\eta = u'\xi' + v'\eta'$. Let now the $x'y'z'$-system be obtained from the xyz-system by a very small rotation characterised by the rotation vector $\boldsymbol{\omega}$, that is, a rotation composed out of a rotation ω_x around the x-axis, a rotation ω_y around the y-axis, and a rotation ω_z around the z-axis *. To find the corresponding transformations of uv and $\xi\eta$ it is simplest to start from the transformation of the null vector ($\boldsymbol{V}' = \boldsymbol{V} + [\boldsymbol{V} \wedge \boldsymbol{\omega}] + \ldots$) and we find

$$a' = a + (b\omega_z - c\omega_y) + \ldots, \quad b' = b + (c\omega_x - a\omega_z) + \ldots, \qquad c' = c + (a\omega_y - b\omega_x) + \ldots, \tag{107}$$

where we have neglected quantities which are quadratic in the components of $\boldsymbol{\omega}$. Using equation (87) we have

$$u'^2 = u^2 + iuv\omega_x + uv\omega_y + iu^2\omega_z + \ldots, \quad v'^2 = v^2 + iuv\omega_x - uv\omega_y - iv^2\omega_z + \ldots$$

Hence we have

$$\left.\begin{aligned} \delta u &= u' - u = \tfrac{1}{2}iv(\omega_x - i\omega_y) + \tfrac{1}{2}iu\omega_z + \ldots, \\ \delta v &= v' - v = \tfrac{1}{2}iu(\omega_x + i\omega_y) - \tfrac{1}{2}iv\omega_z + \ldots \end{aligned}\right\} \tag{108}$$

We can also consider $a'b'c'$ to be the components of a null vector in the xyz-system which is rotated over $-\boldsymbol{\omega}$ with respect to the original null vector. Similarly we can consider $u'v'$ to be the components of the corresponding spinor, which differs only slightly from uv and which refers to the same unit spinors $\xi\eta$. It follows that δu and δv are also the components of a spinor.

According to equation (108) δu and δv can be expressed as linear homogeneous expressions in u and v. This means that the spinor $\begin{pmatrix}\delta u\\ \delta v\end{pmatrix}$ is obtained by the action of a linear operator on $\begin{pmatrix}u\\ v\end{pmatrix}$. Using equations (39) this linear operator can be expressed in terms of S_x, S_y, S_z. We have the following relations [compare eq. (45)]

$$S_x\begin{pmatrix}u\\ v\end{pmatrix} = \frac{\hbar}{2}\begin{Vmatrix}0 & 1\\ 1 & 0\end{Vmatrix}\begin{pmatrix}u\\ v\end{pmatrix} = \frac{\hbar}{2}\begin{pmatrix}v\\ u\end{pmatrix}, \quad S_y\begin{pmatrix}u\\ v\end{pmatrix} = i\frac{\hbar}{2}\begin{pmatrix}-v\\ u\end{pmatrix}, S_z\begin{pmatrix}u\\ v\end{pmatrix} = \frac{\hbar}{2}\begin{pmatrix}u\\ -v\end{pmatrix},$$

and equations (108) can be written in the form

* We take $\omega_x(\omega_y, \omega_z)$ positive, if the rotation around the x-axis (y-axis, z-axis) is from the positive y-axis (z-axis, x-axis) to the positive z-axis (x-axis, y-axis).

$$\begin{pmatrix}\delta u\\ \delta v\end{pmatrix} = \frac{i}{2}\frac{2}{\hbar}(\omega_x S_x + \omega_y S_y + \omega_z S_z)\begin{pmatrix}u\\ v\end{pmatrix} = \left(\frac{i}{\hbar}\right)(\boldsymbol{\omega}\cdot\mathbf{S})\begin{pmatrix}u\\ v\end{pmatrix}. \tag{109}$$

This equation means the following. *If we interpret the component in a given direction of the spin angular momentum, multiplied by $i\omega/\hbar$, as an operator, it is equal to the operator which corresponds to an infinitesimal rotation over an angle ω of the spatial coordinate system around an axis along the given direction.*

Conversely we could have *defined* the operator of the angular momentum around an axis as the operator of the infinitesimal rotation around this axis, multiplied by $\hbar/i\omega$. From equation (108) we would thus have obtained immediately the expression (39) that is, the Pauli matrices, for its components. The vectorial character of S_x, S_y, S_z under transformations which we proved a moment ago by explicit evaluations would then have been self-evident right from the beginning.

The commutation relations (52) of the spin components correspond to the fact that in general the order of two successive rotations cannot be interchanged (compare the Euler-Hamilton theorem in the kinematics of rigid bodies).

In particular one sees geometrically the following. Let the transition $xyz \to x'y'z'$ correspond to a rotation ω around the y-axis and $x'y'z' \to x''y''z''$ to a rotation ω around the x'-axis, while $xyz \to \bar{x}'\bar{y}'\bar{z}'$ corresponds to a rotation ω around the x-axis and $\bar{x}'\bar{y}'\bar{z}' \to \bar{x}''\bar{y}''\bar{z}''$ to a rotation ω around the y'-axis. The transition $x''y''z'' \to \bar{x}''\bar{y}''\bar{z}''$ corresponds then (apart from higher order terms) to a rotation ω^2 around the z''-axis, or,

$$(1+\omega^2 D_z)(1+\omega D_x+\omega^2 E_x+\ldots)(1+\omega D_y+\omega^2 E_y+\ldots) = (1+\omega D_y+\omega^2 E_y+\ldots)(1+\omega D_x+\omega^2 E_x+\ldots),$$

where D_x, E_x, D_y, E_y, ... are operators such that the series $1+\omega D_x+\omega^2 E_x+\ldots$ corresponds uniquely to a rotation ω around the x-axis, ... From this operator equation we get by comparing the coefficients of ω^2 on the left and on the right hand side $D_z+D_xD_y = D_yD_x$, or, $D_xD_y-D_yD_x = -D_z$.

Putting $S_x = -i\hbar D_x, \ldots$, we get $S_xS_y-S_yS_x = i\hbar S_z$.

In general, *the operator J of the total angular momentum in a given direction is for any atomic system equal to the product of $-i\hbar/\omega$ and the operator of an infinitesimal rotation* (over an angle ω) *around this direction.* If we denote this operator, divided by ω, by D we have

$$J_x = -i\hbar D_x,\ J_y = -i\hbar D_y,\ J_z = -i\hbar D_z,\ [J_x, J_y]_- = i\hbar J_z, \ldots \tag{110}$$

We prove this by considering a system of particles. The physical situation may be described by a function ψ of all the spatial coordinates x_i, y_i, z_i, and the time. An infinitesimal rotation around the

x-axis [compare eq. (107)], $x = x'$, $y = y' - z'\omega$, $z = z' + y'\omega$, corresponds to the transformation *

$$\psi(\mathbf{x}_i, t) = \psi'(\mathbf{x}_i', t) = (1+\omega D_x)\,\psi(\mathbf{x}_i', t)$$

$$= \psi(\mathbf{x}_i', t) + \sum_i (\nabla_i\psi \cdot \mathbf{x}_i - \mathbf{x}_i') = \psi(\mathbf{x}_i', t) + \omega \sum_i \left[-z_i' \frac{\partial \psi}{\partial y_i'} + y_i' \frac{\partial \psi}{\partial z_i'} \right]$$

$$\text{or,}\quad D_x = \sum_i \left[-z_i \frac{\partial}{\partial y_i} + y_i \frac{\partial}{\partial z_i} \right] = \frac{i}{\hbar} \sum_i (y_i p_{zi} - z_i p_{yi}) = \frac{i}{\hbar} \sum_i M_{xi}. \quad (111)$$

This equation, is indeed, the same as equation (110) since M_{xi} $(= y_i p_{zi} - z_i p_{yi})$ is the x-component of the angular momentum of the i-th particle.

If we are dealing with a system of spinning electrons, the operator of the infinitesimal rotation will be the sum of two terms. The first term refers to the change in the spatial coordinates, the second term to the transformation of the spin coordinates. One finds now

$$\psi_{s_i}(\mathbf{x}_i, t) \to \psi_{s_i}(\mathbf{x}_i, t) + (i\omega/\hbar)\,[\Sigma_i M_{xi} + \Sigma_i S_{xi}]_{\text{op}}\, \psi_{s_i}(\mathbf{x}_i, t),$$

$$\text{or,}\quad J_x = \Sigma_i M_{xi} + \Sigma_i S_{xi}, \quad (112)$$

a result which we had assumed to be true in the case of a single electron in § 59. This equation is also valid in the case when nuclei with spin $\frac{1}{2}$ are present.

If the energy operator of such a system is the same function of coordinates, momenta and spins in all orthogonal systems of coordinates (free system), it will commute with the operators of all infinitesimal rotations and the angular momentum is an integral of the equations of motion.

Proof: Let q denote all the arguments of the wave function, including the spin coordinates, when we are describing the system in the xyz-system of coordinates and let q' denote them in the $x'y'z'$-system.

If $g(q) = H_{\text{op}}(q) f(q) \ldots$ (A), and if $f(q) \to f'(q')$, $g(q) \to g'(q')$ under the transformation $q \to q'$ we get $g'(q') = H'_{\text{op}}(q') f'(q') \ldots$ (B) as the definition of the transformed energy operator (see § 41).

If we are dealing with an infinitesimal rotation, and if ωD_x is the corresponding operator, we have $g'(q') = (1+\omega D_x) g(q')$, $f'(q') = (1+\omega D_x) f(q')$ and from equations (A) and (B) we get

$$H_{\text{op}}'(q')(1+\omega D_x) f(q') = (1+\omega D_x) H_{\text{op}}(q') f(q'),$$

$$\text{or,}\quad H_{\text{op}}'(q') = (1+\omega D_x) H_{\text{op}}(q')(1-\omega D_x)$$

$$= H_{\text{op}}(q') + \omega[D_x H_{\text{op}}(q') - H_{\text{op}}(q') D_x].$$

* To avoid the many primes one often writes these equations as follows: $\psi(\mathbf{x}_i, t) \to \psi(\mathbf{x}_i, t) + \omega\Sigma_i[-z_i(\partial\psi/\partial y_i) + y_i(\partial\psi/\partial z_i)]$, and reads this as "$\psi$ goes over into $\psi + \delta\psi$ under a rotation".

Invariance under rotations means $H'_{\text{op}}(q') = H_{\text{op}}(q')$ and we find thus $[D_x, H_{\text{op}}]_- = 0$.

We may remark in passing that the total momentum operator of a system is similarly proportional to the operator corresponding to an infinitesimal *translation* of the coordinate system. If the energy operator is invariant under such a translation, the law of conservation of momentum will hold (compare the discussion in § 31).

§ 62. Gauge transformations

It is important to note that our general definition of momentum and angular momentum still depends on the special definition of the scalar potential Φ and the vector potential $\mathfrak{A}$ of the external electromagnetic field, that is, on the *gauge* of these potentials. From the classical electron theory we know that if we change the potentials according to the equations

$$\Phi' = \Phi - c^{-1}\partial\zeta/\partial t, \qquad \mathfrak{A}' = \mathfrak{A} + \nabla\zeta, \tag{113}$$

the electromagnetic field derived by using equations (2.99) remains the same, and that thus the equations of motion and the physical phenomena also remain unchanged. In equations (113) ζ is arbitrary, but because of equation (2.100) it is often restricted to those functions which satisfy the equation

$$\nabla^2\zeta - c^{-2}\partial^2\zeta/\partial t^2 = 0. \tag{114}$$

This property is usually called *gauge invariance* and equations (113) describe a *gauge transformation*. Under such a transformation the meaning of the energy and the momentum change. For a particle of charge $-e$ we have $E' = E + (e/c)(\partial\zeta/\partial t)$, $\boldsymbol{p}' = \boldsymbol{p} - (e/c)\nabla\zeta$. This follows from the fact that the kinetic energy E_{kin} and the kinetic momentum $\boldsymbol{\pi}$ remain invariant, where $E_{\text{kin}} = E + e\Phi = E' + e\Phi'$, $\boldsymbol{\pi} = \boldsymbol{p} + (e/c)\mathfrak{A} = \boldsymbol{p}' + (e/c)\mathfrak{A}'$.

This invariance follows from the fact that the kinetic energy and momentum, which depend in the well known way on the mass and velocity of the particle, must remain the same if the canonical equations of motion should lead to the same curvature of the world line of the particle. The angular momentum $[\boldsymbol{x} \wedge \boldsymbol{p}]$ is not gauge invariant, but the kinetic angular momentum $[\boldsymbol{x} \wedge \boldsymbol{\pi}]$ will have an invariant meaning. The rate of change of the latter and not of the former will be equal to the torque of the external forces.

Gauge invariance means for the case of a many particle system (charge of the i-th particle: $-e_i$) treated non-relativistically that the Hamilton function contains the potentials and the momenta in the form $H = -\Sigma_i e_i \Phi_i + f(\boldsymbol{p}_i + [e/c]\mathfrak{A}_{ij}; \boldsymbol{x}_i)$, where f is a function of the arguments $p_{xi} + (e_i/c)\mathfrak{A}_{xi}, \ldots$ which may contain the coordinates $\boldsymbol{x}_i$ as long as its dependence on the $\boldsymbol{x}_i$ is independent of the gauge. The general expression (3.76) did not satisfy this condition, and in all correction terms with c^{-2} we should really have replaced each $\boldsymbol{p}_i$ by $\boldsymbol{p}_i + (e_i/c)\mathfrak{A}_i$. For many applications, however, the difference would not be noticeable. The same also applies to equations (23) and (33). To ensure gauge invariance we should, for instance, have written $[\boldsymbol{p} + (e/c)\mathfrak{A}]^4/8m^3c^2$ instead of $p^4/8m^3c^2$ in equation (23).

We see that in the non-relativistic quantum mechanics of these systems the Schrödinger equation is not uniquely defined if we allow a freedom of choice of Φ and $\mathfrak{A}$ corresponding to equations (113). If we compare the two Schrödinger equations,

$$H_{\text{op}}\,\psi = i\hbar\partial\psi/\partial t, \qquad (115\text{a}); \qquad H'_{\text{op}}\psi' = i\hbar\partial\psi'/\partial t, \qquad (115\text{b})$$

where H_{op} is obtained in the usual way, while H'_{op} differs from H_{op} by replacing Φ by Φ', and $\mathfrak{A}$ by $\mathfrak{A}'$, we see that to each solution ψ of equation (115a) there corresponds a solution ψ' of equation (115b) given by the equation *

$$\psi' = \psi \exp\,[-i\Sigma_i e_i \zeta_i/\hbar c], \qquad \zeta_i = \zeta(\boldsymbol{x}_i, t). \qquad (116)$$

This equation shows us the gauge transformation of the wave functions. The same equation holds if we include the spins in our considerations in the way done by Pauli.

We might hope that the answer to any physical question would be independent of whether we use ψ or ψ'. This hope is fulfilled only if we are considering observables which are defined in a gauge invariant manner. We see therefore that the problem of the angular momentum of a system (see the end of § 60) and even the extremely important problem of the energy have not a gauge invariant meaning. The fact that those problems all the same have a physical meaning is connected with the fact that we are really considering problems which refer to the possible behaviour of certain measuring apparatus (see, for instance, § 52) which is described in classical terms. It happens that in those cases only one particular choice of

* One can see this in principle from the identities $\{-i\hbar\nabla_i + (e_i/c)\mathfrak{A}_i'\}\psi' = \exp[-i\Sigma_i e_i \zeta_i/\hbar c]\{-i\hbar\nabla_i + (e_i/c)\,\mathfrak{A}_i\}\psi$, $\{i\hbar(\partial/\partial t) + \Sigma_i e_i \Phi_i'\}\psi' = \exp[-i\Sigma_i e_i \zeta_i/\hbar c]\{i\hbar(\partial/\partial t) + \Sigma_i e_i \Phi_i\}\psi$.

Φ and $\mathfrak{A}$ is suitable for the usual description of the experimental data ("energy production", "transfer of momentum"). If, for instance, an atom is present in a constant external field, we shall use a time-independent Φ and $\mathfrak{A}$.

If the system can freely rotate in space, we choose $\mathfrak{A} = 0$, $\Phi = \Phi(r)$ and the angular momentum, which we can now easily prove to be a constant of motion, is the same as the gauge invariant kinetic angular momentum.

B. Relativistic Spin Theory

§ 63. Relativistic spinor calculus

We have up to now considered spinors to be entities which were determined by the orientation of an orthogonal system of coordinates in three-dimensional space, but the properties of which were invariant under changes of this orientation. We shall now show how we can extend their definition in such a way that they are relativistically invariant. This makes it possible to proceed to a relativistic quantum mechanical treatment of the spinning electron.

We start again from the definition of two spinor components u and v which are connected to a null vector $\boldsymbol{V}^{(0)}(a, b, c)$ [see eq. (81)]. We saw that if we change the system of coordinates according to equation (82), u, v and the new spinor components u', v' are connected to each other by means of a linear transformation (89) with determinant unity ("unimodular transformation"). We used nowhere in § 61 the fact that the transformation coefficients in equation (82) were real. The only essential point was that $a^2+b^2+c^2$ was invariant. Hence, also any complex orthogonal transformation of a, b, c will correspond to a linear transformation of u and v.

It can be shown that to any complex orthogonal transformation of a, b, c there corresponds a Lorentz transformation, that is, a transformation under which the space-time coordinates $\boldsymbol{x}$, t are changed linearly, homogeneously, and (in agreement with the physical meaning of the transformation) with real coefficients into the coordinates $\boldsymbol{x}'$, t' of a system of coordinates which is moving uniformly with respect to the old system of coordinates. A general Lorentz transformation is characterised by the invariance property,

$$x^2+y^2+z^2-c^2t^2 = x'^2+y'^2+z'^2-c^2t'^2. \tag{117}$$

Such a transformation can always be considered to be a combination of a *special Lorentz transformation*, that is, a transformation, where the spatial coordinates perpendicular to the direction of the relative motion of the two systems do not change, and a pure rotation in three-dimensional space, that is, a real orthogonal transformation of the spatial coordinates. If we restrict ourselves for the time being to those Lorentz transformations which can be obtained from the identical transformation by a continuous change of the coefficients (this means no time reversal or inversion in space), we find that these transformations are completely determined by six real numbers (for instance, three velocity components and three Eulerian angles).

The complex orthogonal transformations of three numbers a, b, c, that is, the linear, homogeneous transformations satisfying the invariance condition

$$a^2+b^2+c^2 = a'^2+b'^2+c'^2, \tag{118}$$

are also determined by six real numbers (for instance, three complex Eulerian angles), if we restrict the transformations again to those which follow continuously from the identity. The connexion between equations (117) and (118) is now obtained by requiring that the components of the real and the imaginary part of the "vector" $\boldsymbol{V}$ (components a, b, c) should transform as the components of a relativistic six-vector.

If we chose as the best known example of such a six-vector the two vectors $\mathfrak{E}$ and $\mathfrak{H}$ of the electromagnetic field, we want the connexion to be such that $\boldsymbol{V}(a, b, c)$ transform as $\mathfrak{H}-i\mathfrak{E}$ or as

$$\mathfrak{E}+i\mathfrak{H}. \tag{119}$$

The fact that the components of the complex vector $\mathfrak{F} = \mathfrak{E}+i\mathfrak{H}$ undergo an orthogonal transformation [$\mathfrak{F}^2 = \mathfrak{E}^2-\mathfrak{H}^2+2i(\mathfrak{E}\cdot\mathfrak{H})$ is invariant] under a Lorentz transformation follows from the relativistic invariance $\mathfrak{H}^2-\mathfrak{E}^2$ and $(\mathfrak{E}\cdot\mathfrak{H})$ and was proved explicitly in § 57 when we discussed the relativistic invariance of the equations of motion of the electron.

When we used equation (87) to define a spinor u, v we connected this spinor with a null-vector $\boldsymbol{V}^{(0)}$ whose components satisfied the relation $a^2+b^2+c^2 = 0$. We retain this definition, but we can no longer connect $\boldsymbol{V}^{(0)}$ with an XYZ-system fixed in (three-dimensional) space. We consider $\boldsymbol{V}^{(0)} = \boldsymbol{V}_1^{(0)}+i\boldsymbol{V}_2^{(0)}$ now to be a *null-six-vector*, that is, a six-vector for which $\boldsymbol{V}_1^{(0)}$ and $\boldsymbol{V}_2^{(0)}$ have the same length and are orthogonal to one another in each frame of reference or for which $\boldsymbol{V}_1^{(0)}$ and $\boldsymbol{V}_2^{(0)}$ transform as the electrical and magnetic field strengths of a plane light wave. We could thus have called such a vector a *light-six-vector*. A spinor u, v is now connected with a null- or light-six-vector.

A Lorentz transformation corresponds to a transformation of the components of a null-six-vector $\boldsymbol{V}^{(0)}(a, b, c)$ and this corresponds, according to equations (89) and (91) to a unimodular transformation of the corresponding spinor,

$$u' = \alpha u + \beta v, \quad v' = \gamma u + \delta v; \quad \alpha\delta - \beta\gamma = 1, \tag{120}$$

where the coefficients $\alpha, \beta, \gamma, \delta$ are uniquely determined, but for a common factor ± 1. Four complex numbers $\alpha, \beta, \gamma, \delta$ which satisfy condition (120) are just determined by six real numbers!

The pure rotations correspond to $\delta = \alpha^*$, $\beta = -\gamma^*$ [see eq. (94)]. The special transformation

$$u' = \tau^{-1} u, \qquad v' = \tau v, \qquad (\tau\text{: real}) \tag{121}$$

corresponds to [compare eq. (87)],

$$a' = \tfrac{1}{2}(\tau^2+\tau^{-2})a + \tfrac{1}{2}i(\tau^2-\tau^{-2})b, \quad b' = -\tfrac{1}{2}i(\tau^2-\tau^{-2})a + \tfrac{1}{2}(\tau^2+\tau^{-2})b,$$
$$c' = c.$$

If we compare this with equation (9) we see that we are dealing with a special Lorentz transformation corresponding to a motion of the $x'y'z'$-system with respect to the xyz-system with a velocity in the direction of the positive z-axis and that τ and v are connected to one another by the equation

$$\tfrac{1}{2}(\tau^2+\tau^{-2}) = \beta, \quad \tfrac{1}{2}(\tau^2-\tau^{-2}) = \beta v/c, \quad \beta = [1-(v^2/c^2)]^{-1/2},$$
$$\text{or,} \quad \tau = \pm\{[1+(v/c)]/[1-(v/c)]\}^{1/4}. \tag{122}$$

An arbitrary Lorentz transformation can always be written in matrix formalism as follows

$$\begin{pmatrix} u' \\ v' \end{pmatrix} = \begin{Vmatrix} \bar{p} & \bar{q} \\ -\bar{q}^* & \bar{p}^* \end{Vmatrix} \begin{Vmatrix} \tau^{-1} & 0 \\ 0 & \tau \end{Vmatrix} \begin{Vmatrix} p & q \\ -q^* & p^* \end{Vmatrix} \begin{pmatrix} u \\ v \end{pmatrix}; |p|^2+|q|^2=|\bar{p}|^2+|\bar{q}|^2=1.$$

A special Lorentz transformation is obtained by putting $\bar{p} = q^*$, $\bar{q} = -q^*$ since in that case the p, q rotation and the $\bar{p}, \bar{q}$ rotation cancel each other [see eq. (95)]. In that case α and δ in equation (120) will be real and $\beta = \gamma^*$.

Under a given Lorentz transformation not only the components of $\mathfrak{E}+i\mathfrak{H}$ but also those of $\mathfrak{E}-i\mathfrak{H}$ transform orthogonally. This means that one can assign to each Lorentz transformation two complex orthogonal transformations. If the first one transforms a, b, c according to equation (82) into a', b', c', the second will transform a^*, b^*, c^* into a'^*, b'^*, c'^* and its coefficients will be the complex conjugates of those of the first transformation.

The components u^*, v^* of the *complex conjugate spinor* transform according to a scheme which is the complex conjugate of equations (120),

$$u^{*\prime} = \alpha^* u^* + \beta^* v^*, \quad v^{*\prime} = \gamma^* u^* + \delta^* v^*; \quad \alpha^*\delta^* - \beta^*\gamma^* = 1. \tag{123}$$

In the case of pure rotations u^*, v^* transform contravariant to u, v (invariance of u^*u+v^*v), but now there exists no analytical relation between the coefficients of equation (120) and those of equations (123).

Since the choice of the sign of i in expression (119) was arbitrary, we see the following. *To each Lorentz transformation there correspond two unimodular transformations of two variables, the two transformations are each others complex conjugates, and to each null-six-vector there correspond two spinors,* which are related in such a way that for a given system of space-time coordinates the components of the one are the complex conjugates of those of the other.

The transformation properties of a spinor are very simply related to those of the space-time coordinates themselves. One finds that x, y, z, t transform as u^*v+v^*u, $-i(u^*v-v^*u)$, u^*u-v^*v, $(u^*u+v^*v)/c$ or schematically,

$$x \longleftrightarrow u^*v+v^*u, \; y \longleftrightarrow -i(u^*v-v^*u), \; z \longleftrightarrow u^*u-v^*v, \; ct \longleftrightarrow u^*u+v^*v; \tag{124}$$

$$x+iy \longleftrightarrow u^*v, \quad x-iy \longleftrightarrow v^*u, \quad z+ct \longleftrightarrow u^*u, \quad -z+ct \longleftrightarrow v^*v. \tag{125}$$

We prove this as follows. First of all we note from equations (120) and (123) that u^*u, u^*v, v^*u, v^*v undergo a linear transformation,

$$\begin{aligned}(u^*u') &= u^{*\prime}u' = (\alpha^*u^*+\beta^*v^*)(\alpha u+\beta v)\\ &= \alpha^*\alpha(u^*u)+\alpha^*\beta(u^*v)+\beta^*\alpha(v^*u)+\beta^*\beta(v^*v), \dots\end{aligned}$$

If four quantities X_1, X_2, X_3, X_4 transform in the same way, that is, if

$$X_1' = \alpha^*\alpha X_1+\alpha^*\beta X_2+\beta^*\alpha X_3+\beta^*\beta X_4, \dots$$

the following identity will hold,

$$X_1'X_4'-X_2'X_3' = X_1X_4-X_2X_3. \tag{126}$$

This relation is immediately seen to hold for the case where $X_1 = u^*u$, $X_2 = u^*v, \dots$ since in that case both sides of the equation are identically equal to zero. It must, however, be a property of the coefficients of the transformation as $X_1X_4-X_2X_3$ is the only quadratic polynomial in $X_1, X_2, \ldots$ which vanishes identically. We might have expected a slightly more general relation, $X_1'X_4'-X_2'X_3' = D(X_1X_4-X_2X_3)$, but it is not difficult to verify that $D = (\alpha^*\delta^*-\beta^*\gamma^*)(\alpha\delta-\beta\gamma) = 1$.

If we put

$$X_1 = z+ct, \qquad X_2 = x+iy, \qquad X_3 = x-iy, \qquad X_4 = -z+ct, \tag{127}$$

x, y, z, t undergo a linear transformation and $(z+ct)(-z+ct)-(x+iy)(x-iy) = c^2t^2-x^2-y^2-z^2$ remains invariant. This invariance means that we are dealing with a Lorentz transformation.

The relations (127), (125) and (124) between $X_1, \ldots, X_4$ and x, y, z, t correspond to a very definite choice from among several possibilities (of signs), but it is the only one agreeing with our sign conventions. We can see this as follows. First of all, the first three right hand sides of equations (124) correspond according to equations (88) to the x-, y-, and z-components of a vector in ordinary space. Secondly, the special transformation (121) corresponds to the following transformations of $X_1, \ldots, X_4$ and $xyzt$,

$$X_1' = \tau^{-2}X_1, \qquad X_2' = X_2, \qquad X_3' = X_3, \qquad X_4' = \tau^2 X_4;$$
$$z'+ct' = \tau^{-2}(z+ct), \qquad x'\pm iy' = x\pm iy, \qquad -z'+ct' = \tau^2(-z+ct).$$

By using equations (122) we can write the equations connecting z' and ct' with z and ct in the form $z' = \beta(z-vt)$, $t' = \beta[t-(vz/c^2)]$ and they are thus equivalent to the Lorentz transformation which leads to equation (121).

From equation (124) we see that unitary transformations of u and v leave the time invariant and correspond thus to pure rotations.

We can assign to each spinor a four vector W^0 with components

$$\left.\begin{aligned} W^0_x = u^*v+v^*u,\ \ W^0_y = -i(u^*v-v^*u),\\ W^0_z = u^*u-v^*v,\ \ cW^0_t = u^*u+v^*v, \end{aligned}\right\} \tag{128}$$

and we see that $W^{02}_x+W^{02}_y+W^{02}_z-c^2W^{02}_t = 0$, that is, W^0 is a null-four-vector or a light-(four-)vector with positive time component. The spatial part of this four-vector has always the direction of the Z-axis of the system of coordinates which has its X- and Y-axis in the directions of the two vectors which make up the corresponding null-six-vector. In the case where this null-six-vector corresponds to the electromagnetic field of a plane light wave $[|\mathfrak{E}| = |\mathfrak{H}|$, $(\mathfrak{E}\cdot\mathfrak{H}) = 0]$, W^0 is the light vector corresponding to this wave [the components of $\mathfrak{E}$ and $\mathfrak{H}$ depend on $xyzt$ as they are functions of the argument $(xW^0_x + yW^0_y+zW^0_z-c^2tW^0_t)$].

If we multiply the spinor components u and v by a phase factor $e^{i\omega}$ the corresponding null-six-vector changes (the system of coordinates rotates around the Z-axis over an angle -2ω), but the corresponding null-four-vector does not.

A four-vector W which is not a null-vector can always be represented by two spinors, or rather, by two pairs of complex conjugate spinors,

$$\left.\begin{aligned} &W_x+iW_y = -u_1^* v_1 \pm u_2^* v_2,\; W_x-iW_y = v_1^* u_1 \pm v_2^* u_2,\\ &W_z+cW_t = u_1^* u_1 \pm u_2^* u_2,\; -W_z+cW_t = v_1^* v_1 \pm v_2^* v_2;\\ &W_x^2+W_y^2+W_z^2-c^2W_t^2 = \mp(u_1^* v_2^* - v_1^* u_2^*)(u_1 v_2 - v_2 u_1). \end{aligned}\right\} \tag{129}$$

The vector W is of spatial or of temporal character according to whether we choose the upper or lower sign.

Multiplication of a spinor by a complex number and addition of two spinors is defined as in § 61. We use again unit vectors ξ and η to construct a Lorentz-invariant quantity representing the spinor,

$$\psi = \begin{pmatrix} u \\ v \end{pmatrix} = u \begin{pmatrix} 1 \\ 0 \end{pmatrix} + v \begin{pmatrix} 0 \\ 1 \end{pmatrix} = u\xi + v\eta. \tag{130}$$

We shall denote the unit spinors which are used to represent the complex conjugate spinor ψ^* simply by ξ^* and η^*,

$$\psi^* = u^*\xi^* + v^*\eta^*. \tag{131}$$

According to their definition the unit spinors transform contravariant to the spinor components themselves, since $u\xi+v\eta = u'\xi'+v'\eta'$; $u^*\xi^*+v^*\eta^* = u^{*\prime}\xi^{*\prime}+v^{*\prime}\eta^{*\prime}$, or [compare eq. (120)]

$$\xi' = \delta\xi - \gamma\eta,\;\; \eta' = -\beta\xi + \alpha\eta;\;\; \xi^{*\prime} = \delta^*\xi^* - \gamma^*\eta^*,\;\; \eta^{*\prime} = -\beta^*\xi^* + \alpha^*\eta^*.$$

The ξ^* and η^* transform in the same way as the components $u^\dagger$ and $v^\dagger$ of the *spin conjugate spinor* which we introduced earlier [see eq. (100)] by the equations

$$u^\dagger = [-a^*+ib^*]^{1/2} = v^*, \quad v^\dagger = [a^*+ib^*]^{1/2} = -u^*. \tag{132}$$

Clearly $u^\dagger u^* + v^\dagger v^*$ is invariant which means that, indeed, $u^\dagger$ and $v^\dagger$ transform as ξ^* and η^* (and $u^{\dagger *}$, $v^{\dagger *}$ transform as ξ, η).

We represent the spin conjugate spinor by the following invariant expression

$$\psi^\dagger = \begin{pmatrix} u^\dagger \\ v^\dagger \end{pmatrix} = u^\dagger \begin{pmatrix} 1 \\ 0 \end{pmatrix} + v^\dagger \begin{pmatrix} 0 \\ 1 \end{pmatrix} = u^\dagger \xi^\dagger + v^\dagger \eta^\dagger. \tag{133}$$

Since in relativistic spinor theory we are no longer dealing exclusively with unitary transformations, we can no longer introduce an invariant inner product, but the alternating product of two spinors remains invariant, $u_1v_2 - v_1u_2 = u_1'v_2' - v_1'u_2'$.

If we make explicit use of spin conjugate spinors we can introduce invariant inner products of the following kind, $\int \psi_1^* \psi_2^\dagger = u_1^* u_2^\dagger + v_1^* v_2^\dagger \quad (= u_1^* v_2^* - v_1^* u_2^*)$.

We obtain the transformation properties of the operators S_x, S_y, S_z which are defined by equations (39) by considering the Lorentz invariant expression $A = (pv - qu)(p\xi + q\eta)$, where p, q and u, v are both spinors. We denote the Pauli matrices by $\sigma_x, \sigma_y, \sigma_z$,

$$\sigma_x = \begin{Vmatrix} 0 & 1 \\ 1 & 0 \end{Vmatrix}, \quad \sigma_y = \begin{Vmatrix} 0 & -i \\ i & 0 \end{Vmatrix}, \quad \sigma_z = \begin{Vmatrix} 1 & 0 \\ 0 & -1 \end{Vmatrix}. \tag{134}$$

Each of them has the eigenvalues ± 1. The symbolical vector with components $\sigma_x, \sigma_y, \sigma_z$ is denoted by $\boldsymbol{\sigma}$ and we have $\boldsymbol{S} = \frac{1}{2}\hbar\boldsymbol{\sigma}$. The $\sigma_x, \sigma_y, \sigma_z$ satisfy the relations [see eq. (66)]

$$\left.\begin{aligned} &\sigma_x \xi = \eta, \quad \sigma_y \xi = i\eta, \quad \sigma_z \xi = \xi; \quad \sigma_x \eta = \xi, \quad \sigma_y \eta = -i\xi, \quad \sigma_z \eta = -\eta; \\ &\sigma_x^2 = \sigma_y^2 = \sigma_z^2 = \begin{Vmatrix} 1 & 0 \\ 0 & 1 \end{Vmatrix} = \mathbf{1}, \quad \sigma_x \sigma_y = -\sigma_y \sigma_x = i\sigma_z, \ldots \end{aligned}\right\} \tag{135}$$

As in § 61 we find that $A = [\frac{1}{2}(p^2 - q^2)\sigma_x + \frac{1}{2}i(p^2 + q^2)\sigma_y - pq\sigma_z]$ $(u\xi + v\eta)$ and we conclude that under a Lorentz transformation the components of $\boldsymbol{\sigma}$ undergo an orthogonal transformation and that they transform as the components of the complex vector $\mathfrak{E} + i\mathfrak{H}$.

It must be noted that the components of $\boldsymbol{\sigma}$ are Hermitean matrices in all space-time coordinate systems and that they thus always possess real eigenvalues, and from this point of view may be considered to have always a zero imaginary part. This is, of course, not true of the components of $\mathfrak{E} + i\mathfrak{H}$.

The components of the vector operator $\boldsymbol{\sigma}^\dagger$ which act on a spin conjugate spinor,

$$\sigma_x^\dagger = \begin{Vmatrix} 0 & 1 \\ 1 & 0 \end{Vmatrix}, \quad \sigma_y^\dagger = \begin{Vmatrix} 0 & -i \\ i & 0 \end{Vmatrix}, \quad \sigma_z^\dagger = \begin{Vmatrix} 1 & 0 \\ 0 & -1 \end{Vmatrix},$$

transform clearly as

$$\tfrac{1}{2}(p^{\dagger 2} - q^{\dagger 2}) = -\tfrac{1}{2}(p^{*2} - q^{*2}), \quad \tfrac{1}{2}i(p^{\dagger 2} + q^{\dagger 2}) = -\tfrac{1}{2}(-i)(p^{*2} + q^{*2}),$$
$$-p^\dagger q^\dagger = p^* q^*,$$

that is as the components of the complex vector $-\mathfrak{E} + i\mathfrak{H}$ (or $\mathfrak{E} - i\mathfrak{H}$).

Weyl and van der Waerden * were the first to consider the connexion between spinors and Lorentz transformations. The latter

* H. Weyl, *Theory of groups and quantum mechanics*, Methuen, London, 1931; B. L. v. d. Waerden, *Gött. Nachr.*, **1929**, 100. See also G. Uhlenbeck and O. Laporte, *Phys. Rev.*, **37**, 1380, 1931.

has shown on the instigation of Ehrenfest how one can build up a general spinor calculus, similar to vector calculus. His main interest was the consideration of quantities the components of which transformed as products of an arbitrary number of spinor components of the four kinds present $(uv;\, u^*v^*;\, u^\dagger v^\dagger;\, u^{\dagger *}v^{\dagger *})$.

We shall show now how *one can construct a spin conjugate spinor from a four-vector and a spinor*. Consider the expression $B^\dagger = (pv-qu)(-q^*\xi^\dagger+p^*\eta^\dagger)$ which is invariant since q^* and $-p^*$ form a spin conjugate spinor. Considering the two components of the spinor, and not the unit spinors, we have

$$B^\dagger = \begin{pmatrix} q^*qu-q^*pv \\ -p^*qu+p^*pv \end{pmatrix}$$

$$= \tfrac{1}{2}(p^*p+q^*q)\begin{pmatrix} u \\ v \end{pmatrix} - \tfrac{1}{2}(p^*q+q^*p)\begin{pmatrix} v \\ u \end{pmatrix}$$

$$+\tfrac{1}{2}i(p^*q-q^*p)\begin{pmatrix} -iv \\ iu \end{pmatrix} - \tfrac{1}{2}(p^*p-q^*q)\begin{pmatrix} u \\ -v \end{pmatrix}$$

$$= \Big\{ \tfrac{1}{2}(p^*p+q^*q) - \tfrac{1}{2}(p^*q+q^*p)\,\sigma_x + \tfrac{1}{2}i(p^*q-q^*p)\,\sigma_y - \tfrac{1}{2}(p^*p-q^*q)\,\sigma_z \Big\}\begin{pmatrix} u \\ v \end{pmatrix}.$$

The coefficients transform as the components of a four vector W [see eq. (124)] and we see thus that

$$\Big\{cW_t - W_x\sigma_x - W_y\sigma_y - W_z\sigma_z\Big\}\begin{pmatrix} u_1 \\ v_1 \end{pmatrix} = \begin{pmatrix} u_2^\dagger \\ v_2^\dagger \end{pmatrix}, \text{ or, } \{cW_t - (\mathbf{W}\cdot\boldsymbol{\sigma})\}\psi_1 = \psi_2^\dagger, \tag{136}$$

that is, if $W(W_t, \mathbf{W})$ is an arbitrary four-vector we can construct according to equation (136) a spin conjugate spinor $\psi_2^\dagger(u_2^\dagger, v_2^\dagger)$ from a spinor $\psi_1(u_1, v_1)$.

Similarly one can show that one can combine a four vector W and a spin conjugate spinor $\psi_1^\dagger$ to form an ordinary spinor ψ_2 by the following relation

$$\Big\{cW_t + W_x\sigma_x^\dagger + W_y\sigma_y^\dagger + W_z\sigma_z^\dagger\Big\}\begin{pmatrix} u_1^\dagger \\ v_1^\dagger \end{pmatrix} = -\begin{pmatrix} u_2 \\ v_2 \end{pmatrix}, \tag{137}$$

$$\text{or, } \{cW_t + (\mathbf{W}\cdot\boldsymbol{\sigma})\}\psi_1^\dagger = -\psi_2.$$

The signs are chosen in such a way that, if $\psi_1^\dagger$ in equation (137) is the spin conjugate of ψ_1 in equation (136), $\psi_2^\dagger$ in equation (136) will be the spin conjugate of ψ_2 in equation (137). In the last of equations (137) we have written simply $\boldsymbol{\sigma}$ for $\boldsymbol{\sigma}^\dagger$.

§ 64. Derivation of the Dirac equations

The strictly relativistic form of the classical equations of motion played an important role in the consideration of § 57. In the present section we shall show how these equations lead to a relativistic quantum mechanics of the spinning electron in a way very similar to the one which led in § 58 to the approximate treatment. We obtain a natural generalisation of Pauli's spin theory and this is the previously mentioned Dirac theory.

We obtained the non-relativistic treatment by considering the Hamiltonian (23). This Hamiltonian differed from the one of a point particle by the addition of a linear term in $\boldsymbol{S}$ which was considered to be small. The corresponding energy expression for a point particle (mass m, charge $-e$) in an electromagnetic field (scalar potential Φ, vector potential $\mathfrak{A}$) had been derived from a relativistically invariant Hamilton equation [compare eq. (2.101)],

$$\mathscr{H}_0 \equiv c^{-2}(E+e\Phi)^2-(\boldsymbol{p}+[e/c]\mathfrak{A}\cdot\boldsymbol{p}+[e/c]\mathfrak{A})-m^2c^2 = 0. \tag{138}$$

Under a Lorentz transformation Φ/c and $\mathfrak{A}$, as well as E/c^2 and $\boldsymbol{p}$ behave as a four-vector, that is, they transform as t and $\boldsymbol{x}$ *. If $d\tau$ is an element of eigentime, the equation of motion can be written in canonical form

$$\left.\begin{aligned}(-2m)(dE/d\tau) &= \partial\mathscr{H}_0/\partial t, \quad (-2m)(dt/d\tau) = -\partial\mathscr{H}_0/\partial E,\\ (-2m)(dp_x/d\tau) &= -\partial\mathscr{H}_0/\partial x, \quad (-2m)(dx/d\tau) = \partial\mathscr{H}_0/\partial p_x, \ldots\end{aligned}\right\} \tag{139}$$

The spin term in equation (23) derived from the expression (21) for the spin energy and the latter expression was based on the consideration that it should be possible to write the equations of motion (15) in the canonical form starting from the energy function (16).

We now want to complete expression (138) for $\mathscr{H}_0$ by adding a small spin term. This spin term we want to derive from the relativistically invariant equations of motion (4),

$$d\boldsymbol{S}/d\tau = -(e/mc)[\boldsymbol{S}\wedge\mathfrak{G}], \qquad \mathfrak{G} = \mathfrak{H}-i\mathfrak{E}, \tag{140}$$

by using the following formal expression for the spin energy,

$$H_s = (e/mc)(\boldsymbol{S}\cdot\mathfrak{G}) = (e/mc)(\boldsymbol{S}\cdot\mathfrak{H}-i\mathfrak{E}), \tag{141}$$

from which equations (140) can be derived as the canonical equations of motion [compare eq. (24)],

* For instance, in the case of a free particle we have (see § 1) $E = mc^2dt/d\tau$, $p = md\boldsymbol{x}/d\tau$, where $d\tau$ is an element of eigentime.

$$\frac{dS_x}{d\tau} = \frac{\partial H_s}{\partial S_x}\{S_x, S_x\} + \frac{\partial H_s}{\partial S_y}\{S_x, S_y\} + \frac{\partial H_s}{\partial S_z}\{S_x, S_z\}, \ldots$$

As in equation (26) we have for the Poisson brackets the relations

$$\{S_x, S_y\} = -\{S_y, S_x\} = S_z, \ldots \tag{142}$$

There is a difference between the present case and the usual form of canonical equation of motion, in that S_x, S_y, S_z, and H_s are complex quantities *. If both the electron velocity and the electrical field strength are equal to zero, equation (141) goes over into equation (16). If, however, $\mathfrak{E} \neq 0$, H_s contains an imaginary part $-(ie/mc)(\boldsymbol{S} \cdot \mathfrak{E})$, a term which suggests an imaginary electrical moment $(ie/mc)\boldsymbol{S}$ for the electron.

If we now wish to combine equations (139) and (140) into one set of canonical equations of motion, we must clearly add to $\mathscr{H}_0$ expression (141), multiplied by $-2m$,

$$\begin{aligned} \mathscr{H} &\equiv \mathscr{H}_0 - 2mH_s \\ &\equiv c^{-2}(E+e\Phi)^2 - (\boldsymbol{p}+[e/c]\mathfrak{A} \cdot \boldsymbol{p}+[e/c]\mathfrak{A}) - (2e/c)(\boldsymbol{S} \cdot \mathfrak{H} - i\mathfrak{E}) - m^2c^2 = 0. \end{aligned} \tag{143}$$

We must now replace $\mathscr{H}_0$ by $\mathscr{H}$ in equations (139) and by doing this we introduce spin-orbit interaction, since $\mathfrak{H} - i\mathfrak{E}$ may depend on $\boldsymbol{x}, t$.

Although one can voice certain objections to expression (143) as a classical Hamiltonian equation**, its quantum mechanical application leads to the very useful Dirac spin theory. This application starts by using a spinor,

$$\psi = \begin{pmatrix} \psi_+ \\ \psi_- \end{pmatrix} = \psi_+(\boldsymbol{x}, t)\,\xi + \psi_-(\boldsymbol{x}, t)\,\eta, \tag{144}$$

to describe the quantum mechanical situation. This spinor must satisfy the Schrödinger equation [compare eq. (3.23)]

* Instead of considering equation (140) to be a system with one complex degree of freedom, we could consider it to be a canonical system with two real degrees of freedom. If we want to use also a real expression, $H_s = \frac{1}{2}(e/mc)[(\boldsymbol{S} \cdot \mathfrak{G}) + (\boldsymbol{S}^* \cdot \mathfrak{G}^*)]$, for the energy, we have instead of equations (142) the relations $\{S_x, S_y\} = 2S_z, \ldots$, and we do not obtain a generalisation of Pauli's spin theory. Compare H. A. Kramers *Zeeman commemorative volume*, Nyhoff, The Hague, 1935, p. 403.

** The fact that $\mathscr{H}$ is not real means that E, $\boldsymbol{p}$, and $\boldsymbol{x}$ can take on complex values, even if we start them off with real values, and $d\tau$ cannot be exactly an element of eigentime. These difficulties are only partly met by the observation that $\boldsymbol{S}$ must be very small since otherwise equation (140) will not hold.

$$\mathscr{H}_{\text{op}}\psi = 0, \tag{145}$$

where E, $\boldsymbol{p}$, and $\boldsymbol{S}$ in equation (143) are the following operators,

$$E_{\text{op}} = i\hbar\partial/\partial t, \quad \boldsymbol{p}_{\text{op}} = -i\hbar\boldsymbol{\nabla}, \quad \boldsymbol{S}_{\text{op}} = \tfrac{1}{2}\hbar\boldsymbol{\sigma}, \tag{146}$$

where $\boldsymbol{\sigma}$ is given by equation (134), that is, where S_x, S_y, S_z are given by equations (39). We have thus for $\mathscr{H}_{\text{op}}$ the equation,

$$\mathscr{H}_{\text{op}} = c^{-2}(E_{\text{op}}+e\Phi)^2-(\boldsymbol{p}_{\text{op}}+[e/c]\mathfrak{A}\cdot\boldsymbol{p}_{\text{op}}+[e/c]\mathfrak{A}) \\ -(e\hbar/c)(\boldsymbol{\sigma}\,.\,\mathfrak{H}-i\mathfrak{E})-m^2c^2. \tag{147}$$

Since $\mathscr{H}_0$ and H_s are both Lorentz invariant [$-i\hbar\boldsymbol{\nabla}$, $(i\hbar/c^2)\partial/\partial t$ transform as the components of a four-vector], equations (145) to (147) form a Lorentz invariant set of equations. In the limit of small velocities, we replace $(E+e\Phi)^2/c^2$ by $m^2c^2+2m(E-mc^2+e\Phi)$ [compare § 25]. Neglecting terms proportional to c^{-2} we obtain again Pauli's equation, provided we may drop the non-Hermitean term $(ie\hbar/c)(\boldsymbol{\sigma}\cdot\mathfrak{E})$ in expression (147). We shall discuss in § 65 the transition to Pauli's treatment.

How can we use solutions of equation (145) to make quantitative predictions about the physical properties of the system in the corresponding physical situation?

Equation (145) is equivalent to two partial differential equations which are of the second order in the time. This means that if at a given time four functions, for instance, ψ_+, ψ_-, $\partial\psi_+/\partial t$, and $\partial\psi_-/\partial t$ are given, ψ will be determined at all times. It is advantageous to introduce instead of the two time derivatives (which do not transform according to simple laws in a relativistic theory) two other functions, namely, the components of a spin conjugate spinor. We obtain this spinor by noting that *the operator* $\mathscr{H}_{\text{op}}+m^2c^2$ *can be written as the product of two linear expressions in* E *and* $\boldsymbol{p}$,

$$\mathscr{H}_{\text{op}} \equiv \{(E_{\text{op}}+e\Phi)/c+(\boldsymbol{p}_{\text{op}}+[e/c]\mathfrak{A}\cdot\boldsymbol{\sigma})\} \\ \{(E_{\text{op}}+e\Phi)/c-(\boldsymbol{p}_{\text{op}}+[e/c]\mathfrak{A}\cdot\boldsymbol{\sigma})\}-m^2c^2. \tag{148}$$

If the components a_t, $\boldsymbol{a}(a_x, a_y, a_z)$ commute with one another, equations (135) will lead to $(\boldsymbol{a}\cdot\boldsymbol{\sigma})^2 = a^2$, $[ca_t+(\boldsymbol{a}\cdot\boldsymbol{\sigma})][ca_t-(\boldsymbol{a}\cdot\boldsymbol{\sigma})] = ca_t^2-a^2$.

If we could neglect the fact that E_{op} does not commute with t, $p_{x\text{op}}$ not with $x, \ldots$, equation (148) would correspond to expression (138) for $\mathscr{H}_0$. Because these quantities do not commute, when we write out the product in expression (148) there will be extra terms which are just the spin term $-(e\hbar/c)(\boldsymbol{\sigma}\cdot\mathfrak{H}-i\,\mathfrak{E})$. For instance, we find for the term in σ_x,

$$(\sigma_x/c)\{-(E_{\text{op}}+e\Phi)(p_{x\text{op}}+[e/c]\mathfrak{A}_x)+(p_{x\text{op}}+[e/c]\mathfrak{A}_x)(E_{\text{op}}+e\Phi)\} \\ -\sigma_y\sigma_z(p_{y\text{op}}+[e/c]\mathfrak{A}_y)(p_{z\text{op}}+[e/c]\mathfrak{A}_z)-\sigma_z\sigma_y(p_{z\text{op}}+[e/c]\mathfrak{A}_z)(p_{y\text{op}}+[e/c]\mathfrak{A}_y)$$

$$= \frac{\sigma_x}{c}\left\{\frac{e\hbar}{ic}\frac{\partial\mathfrak{A}_x}{\partial t}+\frac{e\hbar}{i}\frac{\partial\Phi}{\partial x}-i\left[\frac{e\hbar}{i}\frac{\partial\mathfrak{A}_z}{\partial y}-\frac{e\hbar}{i}\frac{\partial\mathfrak{A}_y}{\partial z}\right]\right\} = -\frac{e\hbar}{c}\sigma_x(\mathfrak{H}_x-i\mathfrak{E}_x),$$

where we have used equations (146) and (2.99).

We now introduce two quantities $\chi_+^\dagger$ and $\chi_-^\dagger$ by the equation

$$\left\{\frac{E_{\text{op}}+e\Phi}{c}-\left(\boldsymbol{p}_{\text{op}}+\frac{e}{c}\,\mathfrak{A}\cdot\boldsymbol{\sigma}\right)\right\}\begin{pmatrix}\psi_+\\ \psi_-\end{pmatrix}=mc\begin{pmatrix}\chi_+^\dagger\\ \chi_-^\dagger\end{pmatrix}, \tag{149}$$

and combining equations (148), (145) and (149) we get

$$\left\{\frac{E_{\text{op}}+e\Phi}{c}+\left(\boldsymbol{p}_{\text{op}}+\frac{e}{c}\,\mathfrak{A}\cdot\boldsymbol{\sigma}\right)\right\}\begin{pmatrix}\chi_+^\dagger\\ \chi_-^\dagger\end{pmatrix}=mc\begin{pmatrix}\psi_+\\ \psi_-\end{pmatrix}. \tag{150}$$

From our earlier considerations [see eq. (136)] it follows from equation (149) that $\chi_+^\dagger$ and $\chi_-^\dagger$ are the components of a spin conjugate spinor, which means that we know their transformation properties.

Equations (149) and (150) form a set of four simultaneous homogeneous partial differential equations of the first order which are Lorentz invariant because of the way they were derived. They are the *Dirac equations of the electron in an external electromagnetic field.* Using equations (146) and (134) for the various operators in these equations we can write them explicitly as follows,

$$\left.\begin{aligned}
&\left(\frac{1}{c}\frac{\partial}{\partial t}+\frac{\partial}{\partial z}\right)\psi_1-\frac{ie}{\hbar c}(\Phi-\mathfrak{A}_z)\,\psi_1\\
&\qquad+\left(\frac{\partial}{\partial x}-i\frac{\partial}{\partial y}\right)\psi_2+\frac{ie}{\hbar c}(\mathfrak{A}_x-i\mathfrak{A}_y)\psi_2=-\frac{imc}{\hbar}\psi_3,\\
&\left(\frac{1}{c}\frac{\partial}{\partial t}-\frac{\partial}{\partial z}\right)\psi_2-\frac{ie}{\hbar c}(\Phi+\mathfrak{A}_z)\,\psi_2\\
&\qquad+\left(\frac{\partial}{\partial x}+i\frac{\partial}{\partial y}\right)\psi_1+\frac{ie}{\hbar c}(\mathfrak{A}_x+i\mathfrak{A}_y)\psi_1=-\frac{imc}{\hbar}\psi_4,\\
&\left(\frac{1}{c}\frac{\partial}{\partial t}-\frac{\partial}{\partial z}\right)\psi_3-\frac{ie}{\hbar c}(\Phi+\mathfrak{A}_z)\,\psi_3\\
&\qquad-\left(\frac{\partial}{\partial x}-i\frac{\partial}{\partial y}\right)\psi_4-\frac{ie}{\hbar c}(\mathfrak{A}_x-i\mathfrak{A}_y)\psi_4=-\frac{imc}{\hbar}\psi_1,\\
&\left(\frac{1}{c}\frac{\partial}{\partial t}+\frac{\partial}{\partial z}\right)\psi_4-\frac{ie}{\hbar c}(\Phi-\mathfrak{A}_z)\,\psi_4\\
&\qquad-\left(\frac{\partial}{\partial x}+i\frac{\partial}{\partial y}\right)\psi_3-\frac{ie}{\hbar c}(\mathfrak{A}_x+i\mathfrak{A}_y)\,\psi_3=-\frac{imc}{\hbar}\psi_2,\\
&\psi_+=\psi_1,\ \psi_-=\psi_2,\ \chi_+^\dagger=\psi_3,\ \chi_-^\dagger=\psi_4.
\end{aligned}\right\} \tag{151}$$

The physical situation is described by the four functions $\psi_1, \ldots, \psi_4$. We can prescribe the values of the ψ at one time, and their values at another time then follow from equations (151).

Under a Lorentz transformation ψ_1 and ψ_2 (and also ψ_3 and ψ_4) transform among themselves according to the rules [see eqq. (120) and (132)]

$$\begin{pmatrix}\psi_1'\\ \psi_2'\end{pmatrix} = \begin{Vmatrix}\alpha & \beta\\ \gamma & \delta\end{Vmatrix}\begin{pmatrix}\psi_1\\ \psi_2\end{pmatrix}, \qquad \begin{pmatrix}\psi_3'\\ \psi_4'\end{pmatrix} = \begin{Vmatrix}\delta^* & -\gamma^*\\ -\beta^* & \alpha^*\end{Vmatrix}\begin{pmatrix}\psi_3\\ \psi_4\end{pmatrix}.$$

If we are dealing with pure rotations the two transformations are the same, since in that case $\delta = \alpha^*$, $\gamma = -\beta^*$.

To *discuss the physical interpretation* of the four component wave function we shall treat the four ψ as one function $\psi_{s,r}(\boldsymbol{x}, t)$ which depends not only on $\boldsymbol{x}$, t but also on two discrete arguments s and r both of which can take on two values $+\frac{1}{2}$ and $-\frac{1}{2}$. The variable s is the spin variable which we have already introduced when writing * $\psi_\pm$, $\chi_\pm^\dagger$; $r = +\frac{1}{2}$ $(-\frac{1}{2})$ indicates that we are dealing with $\psi_\pm$ $(\chi_\pm^\dagger)$,

$$\psi_1 = \psi_{++}, \quad \psi_2 = \psi_{-+}, \quad \psi_3 = \psi_{+-}, \quad \psi^4 = \psi_{--}. \tag{152}$$

We introduce an operator $\boldsymbol{\rho}(\varrho_x, \varrho_y, \varrho_z)$ which acts on the r-argument in the same way as $\boldsymbol{\sigma}(\sigma_x, \sigma_y, \sigma_z)$ acts on the s-argument,

$$\varrho_x = \begin{Vmatrix}0 & 1\\ 1 & 0\end{Vmatrix}, \quad \varrho_y = \begin{Vmatrix}0 & -i\\ i & 0\end{Vmatrix}, \quad \varrho_z = \begin{Vmatrix}1 & 0\\ 0 & -1\end{Vmatrix}, \quad \varrho_x^2 = \varrho_y^2 = \varrho_z^2 = 1. \tag{153}$$

It must be borne in mind that the indices x, y, z of ϱ_x, ϱ_y, ϱ_z have no connexion whatever with the spatial coordinates. The action of $\boldsymbol{\rho}$ (and $\boldsymbol{\sigma}$) on the wave function is given by the equations

$$\left.\begin{aligned} \varrho_x\psi_{s,r} &= \psi_{s,-r}, \ \varrho_y\psi_{s,r} = -i(-1)^{r-\frac{1}{2}}\,\psi_{s,-r}, \ \varrho_z\psi_{s,r} = (-1)^{r-\frac{1}{2}}\,\psi_{s,r};\\ \sigma_x\psi_{s,r} &= \psi_{-s,r}, \ \sigma_y\psi_{s,r} = -i(-1)^{s-\frac{1}{2}}\,\psi_{-s,r}, \ \sigma_z\psi_{s,r} = (-1)^{s-\frac{1}{2}}\,\psi_{s,r}, \end{aligned}\right\} \tag{154}$$

or, using once again ψ and $\chi^\dagger$ we have

$$\varrho_x\begin{pmatrix}\psi\\ \chi^\dagger\end{pmatrix} = \begin{pmatrix}\chi^\dagger\\ \psi\end{pmatrix}, \quad \varrho_y\begin{pmatrix}\psi\\ \chi^\dagger\end{pmatrix} = i\begin{pmatrix}-\chi^\dagger\\ \psi\end{pmatrix}, \quad \varrho_z\begin{pmatrix}\psi\\ \chi^\dagger\end{pmatrix} = \begin{pmatrix}\psi\\ -\chi^\dagger\end{pmatrix}. \tag{155}$$

We see thus that equations (150) and (149) can be written in the form

$$[(E_{\rm op}+e\Phi)/c - (\boldsymbol{p}_{\rm op}+[e/c]\mathfrak{A}\cdot\boldsymbol{\sigma})\,\varrho_z]\begin{pmatrix}\psi\\ \chi^\dagger\end{pmatrix} = mc\,\varrho_x\begin{pmatrix}\psi\\ \chi^\dagger\end{pmatrix}.$$

If we introduce the abbreviations

$$\varepsilon = (E_{\rm op}+e\Phi)/c, \qquad \boldsymbol{\pi} = \boldsymbol{p}_{\rm op}+(e/c)\,\mathfrak{A}, \tag{156}$$

* As before, we shall drop the $\frac{1}{2}$ from the indices and write $+$ $(-)$ instead of $+\frac{1}{2}$ $(-\frac{1}{2})$.

where ε is the kinetic energy divided by c, and $\boldsymbol{\pi}$ the kinetic momentum, we can write this equation in the form

$$[\varepsilon - \varrho_z(\boldsymbol{\pi} \cdot \boldsymbol{\sigma}) - \varrho_x mc]\,\psi_{s,r} = 0. \tag{157}$$

Writing

$$\varrho_z\sigma_x = \alpha_1\ (=\alpha_x),\ \varrho_z\sigma_y = \alpha_2\ (=\alpha_y),\ \varrho_z\sigma_z = \alpha_3\ (=\alpha_z),\ \varrho_x = \alpha_4, \tag{158}$$

we have the following commutation relations for the α,

$$\alpha_k\alpha_l + \alpha_l\alpha_k = [\alpha_k, \alpha_l]_+ = 2\,\delta_{kl}, \qquad k = 1, 2, 3, 4, \tag{159}$$

and we can write equation (157) in the form

$$(\varepsilon - \alpha_1\pi_x - \alpha_2\pi_y - \alpha_3\pi_z - \alpha_4 mc)\,\psi = 0. \tag{160}$$

We have now arrived at what was the starting point of Dirac's considerations *. In non-relativistic quantum mechanics a unique physical interpretation of the wave function and especially the introduction of a positive probability density was possible (see § 28) because the Schrödinger equation could be written in the form

$$H_{\rm op}\psi = i\hbar\partial\psi/\partial t, \tag{161}$$

where $H_{\rm op}$ was a Hermitean operator acting on the arguments of ψ. This equation was derived from the classical formula which gave the energy as a polynomial in the momenta, with coefficients which were functions of the coordinates. Such a procedure did not seem to be possible if we wanted to use the *relativistic* relation between the energy and the other quantities, since the expression for the energy contained a square root. Using the notation (156) we had [compare eq. (138)],

$$\varepsilon^2 - (\pi_x^2 + \pi_y^2 + \pi_z^2 + m^2c^2) = 0 \rightarrow \varepsilon - [\pi_x^2 + \pi_y^2 + \pi_z^2 + m^2c^2]^{1/2} = 0. \tag{162}$$

Dirac succeeded in obtaining a polynomial instead of the square root expression by introducing four symbolical quantities $\alpha_1, \ldots, \alpha_4$, which had to satisfy the anticommuting rules (159), and wrote

$$\varepsilon - (\alpha_1\pi_x + \alpha_2\pi_y + \alpha_3\pi_z + \alpha_4 mc) = 0. \tag{163}$$

The α were considered to be Hermitean operators and Dirac constructed a representation of these operators by means of 4×4 matrices introducing two sets of anticommuting quantities $\boldsymbol{\sigma}$ and $\boldsymbol{\rho}$ similar to the relations (158). [The rows and columns of these ma-

* P. A. M. Dirac, *Proc. Roy. Soc.*, **A117**, 610, 1928; **A118**, 351, 1928.

trices refer to the four combinations s, r which are possible, or, to the indices of the $\psi_1, \psi_2, \psi_3, \psi_4$.] It turned out that multiplying equation (163) by $\varepsilon + (\alpha_1\pi_x + \alpha_2\pi_y + \alpha_3\pi_z + \alpha_4 mc)$ led to the first of equations (162) only, if the ε, $\boldsymbol{\pi}$ commuted with one another, that is, if there is no external field. If there is an external field, equation (162) must be completed by adding a term, corresponding to the "spin term" in equation (147). It thus turned out that *the linear expression* (163) *leads automatically to the existence of an electron spin with an angular momentum* $\frac{1}{2}\hbar$ *and a magnetic moment* μ_B.

The method we used to obtain equation (160) is clearly not according to such clear principles, but our method has the advantage that the relativistic invariance of equation (163) is immediately clear, while Dirac had to prove this invariance by explicitly using operator calculus *.

If we start, as Dirac did, from equation (163) and the wave equation (160), it is not yet clear what will be the arguments of ψ and how the α will operate on these arguments. Van der Waerden has shown, however, that every finite representation of the α is essentially equivalent to the particular representation (158)** Connected to this is the fact that it is possible—as was shown especially by Dirac—to obtain many results from equation (163) without discussing the Schrödinger equation.

§ 65. Discussion of the Dirac equations

The Dirac equations (160) have the same form as the Schrödinger equation (161) of non-relativistic quantum mechanics with a Hermitean energy operator,

$$H_{\text{op}} = -e\Phi + c\alpha_1(p_x + [e/c]\mathfrak{A}_x) + c\alpha_2(p_y + [e/c]\mathfrak{A}_y) + c\alpha_3(p_z + [e/c]\mathfrak{A}_z) + \alpha_4 mc^2. \qquad (164)$$

We can therefore immediately apply all considerations referring to the physical interpretation of the wave functions, in particular transformation theory to the solutions of the Dirac equations. The fact that there are now two variables s, r of a new kind, does not make a difference since the operators $\alpha_1, \ldots, \alpha_4$ are Hermitean [see eq. (158); each has a two-fold eigenvalue $+1$ and a two-fold eigenvalue -1]. The question will arise whether it is possible to interpret s and r as

* Pauli's proof (*Handb. Phys.*, **24₁**, 330, 1933) is slightly different and more systematic than Dirac's.

** More exactly we should have said any "irreducible representation". The most general irreducible representation of the α is of the form $\alpha_1 = \omega^{-1}\varrho_z\sigma_x\omega$, $\alpha_2 = \omega^{-1}\varrho_z\sigma_y\omega, \ldots$, where ω is an arbitrary operator. Compare B. L. v. d. Waerden, *Die gruppentheoretische Methode in der Quantenmechanik*, Springer, Berlin, 1932, p. 55.

„observables". A partial answer to this question is obtained by considering the *probability density I* and the *probability current density* $\mathfrak{S}$.

We define $|\psi_{s,r}(\boldsymbol{x}, t)|^2 d\boldsymbol{x}$ as the probability that at t the electron is within the volume element $d\boldsymbol{x}$ and that the two new variables — whatever their meaning — have the values s and r. The probability $I d\boldsymbol{x}$ that the electron is in the volume elemnt $d\boldsymbol{x}$, independent of the values of s and r, is then the sum of four terms,

$$\left.\begin{aligned} I = \Sigma_{s,r}\psi^*_{s,r}\psi_{s,r} &= \psi_1^*\psi_1+\psi_2^*\psi_2+\psi_3^*\psi_3+\psi_4^*\psi_4 \\ &= \psi_+^*\psi_++\psi_-^*\psi_-+\chi_+^{\dagger\,*}\chi_+^{\dagger}+\chi_-^{\dagger *}\chi_-^{\dagger} = (\psi_+^*\psi_++\psi_-^*\psi_-) \\ &\qquad + (\chi_+^*\chi_++\chi_-^*\chi_-), \end{aligned}\right\} \tag{165}$$

where $\chi_\pm$ is the spinor of which $\chi_\pm^\dagger$ is the spin conjugate [compare eq. (132)].

Equation (165) gives a positive definite expression for the probability density I. The integral of I over the whole of space should be time independent and equal to unity, $\Sigma\int|\psi|^2 d\boldsymbol{x} = 1$.

The fact that the normalising integral is time independent follows as before [compare § 28], that is, by multiplying equation (161) by ψ^* and integrating the imaginary part of the ensuing equation over all coordinates (that means here, integration over $\boldsymbol{x}$ and summation over s and r).

We now define the *expectation value of a quantity* F by the equation,

$$\overline{F} = \int\psi^* F\psi = \Sigma_{s,r}\int\psi^*_{s,r} F_{\text{op}}\psi_{s,r} d\boldsymbol{x}, \tag{166}$$

where F_{op} is the operator corresponding to the quantity F which may act on the variables $\boldsymbol{x}$, s, and r.

If $G(\boldsymbol{x})$ is a function which is zero everywhere but in a small region u of space where $G = 1$, we find that $\int\psi^* G\psi$ is the probability of finding the electron within u [see eq. (165)]. Moreover, according to the considerations of § 44 the expression

$$\tfrac{1}{2}\int\psi^*(G\dot{x}+\dot{x}G)\psi = \int\mathfrak{S}_x G\,d\boldsymbol{x} \tag{167}$$

will be the integral of the probability current density $\mathfrak{S}_x$ over u. Since H_{op} is linear in the momenta, the equation of continuity, $(\partial I/\partial t)+(\boldsymbol{\nabla}\cdot\mathfrak{S}) = 0$, will automatically be satisfied. To find the operators corresponding to the components $\dot{x}$, $\dot{y}$, $\dot{z}$ of the velocity we use equation (164) and obtain

$$\dot{x} = (i/\hbar)(H_{\text{op}}x - xH_{\text{op}}) = (i/\hbar)c\alpha_1(p_x x - xp_x) = c\alpha_1, \dot{y} = c\alpha_2, \dot{z} = c\alpha_3. \tag{168}$$

The first three Dirac α are thus, apart from a factor c, the components of the velocity operator; they are therefore often written as a vector $\boldsymbol{\alpha}(\alpha_x, \alpha_y, \alpha_z)$. In the Dirac theory the velocity is not proportional to the kinetic momentum $\boldsymbol{\pi} = \boldsymbol{p} + (e/c)\mathfrak{A}$, a fact which we shall discuss later on.

Combining equations (167) and (168) leads to the following equations for the *components of the probability current density* [compare eq. (154)],

$$\left.\begin{aligned}
\mathfrak{S}_x/c = \Sigma\psi^*\alpha_1\psi &= \Sigma_{s,r}\psi^*_{s,r}\varrho_z\sigma_x\psi_{s,r}\\
&= \psi_1^*\psi_2+\psi_2^*\psi_1-\psi_3^*\psi_4-\psi_4^*\psi_3\\
&= (\psi_+^*\psi_-+\psi_-^*\psi_+)-(\chi_+^{\dagger *}\chi_-^{\dagger}+\chi_-^{\dagger *}\chi_+^{\dagger})\\
&= (\psi_+^*\psi_-+\psi_-^*\psi_+)+(\chi_+^*\chi_-+\chi_-^*\chi_+),\\
\mathfrak{S}_y/c = \Sigma\psi^*\alpha_2\psi &= \Sigma_{s,r}\psi^*_{s,r}\varrho_z\sigma_y\psi_{s,r}\\
&= -i(\psi_1^*\psi_2-\psi_2^*\psi_1)+i(\psi_3^*\psi_4-\psi_4^*\psi_3)\\
&= -i(\psi_+^*\psi_--\psi_-^*\psi_+)-i(\chi_+^*\chi_--\chi_-^*\chi_+),\\
\mathfrak{S}_z/c = \Sigma\psi^*\alpha_3\psi &= \Sigma_{s,r}\psi^*_{s,r}\varrho_z\sigma_z\psi_{s,r}\\
&= (\psi_1^*\psi_1-\psi_2^*\psi_2)-(\psi_3^*\psi_3-\psi_4^*\psi_4)\\
&= (\psi_+^*\psi_+-\psi_-^*\psi_-)+(\chi_+^*\chi_+-\chi_-^*\chi_-).
\end{aligned}\right\} \quad (169)$$

As in equation (164) we have introduced the spinor $\chi_\pm$ of which $\chi_\pm^\dagger$ is the spin conjugate. Comparing equations (165) and (169) with equation (169) we see that *I*, $\mathfrak{S}$ *are the components of a relativistic time-like four vector, the four-vector of the probability density.* The equation of continuity is also Lorentz invariant, as it ought to be. Finally, if $\psi_{s,r}$ is normalised, $\psi'_{s',r'}$ obtained from $\psi_{s,r}$ by a Lorentz transformation will also be normalised.

Consider in $\boldsymbol{x}$, t-space two three-dimensional hypersurfaces which are rotated one with respect to the other and of which one corresponds to t=constant, and the other to t' = constant. Connect these two hypersurfaces so that we obtain a closed three-dimensional hypersurface, and let the connecting parts be in those regions where ψ is practically equal to zero. From Gauss' theorem we get immediately $\int I'd\boldsymbol{x}' = \int Id\boldsymbol{x} = 1$.

We can construct two Lorentz invariant quantities from two spinors ψ and $\chi^\dagger$, namely, $\psi_+{}^*\chi_+{}^\dagger+\psi_-{}^*\chi_-{}^\dagger$, and its complex conjugate $\chi_+{}^{\dagger *}\psi_++\chi_-{}^{\dagger *}\psi_-$. One verifies easily that $\Sigma\psi^*\alpha_4\psi = \Sigma\psi^*\varrho_x\psi$ is equal to their sum, that is, equal to twice their real part, and that $\Sigma\psi^*\alpha_1\alpha_2\alpha_3\alpha_4\psi = -\Sigma\psi^*\varrho_y\psi$ is equal to twice their imaginary part multiplied by i. Their product is equal to $I^2-\mathfrak{S}^2/c^2$, and gives thus a measure for the "intensity" of the wave phenomenon in a space-time point.

According to our considerations in § 61 we can find the *angular momentum* operator of the electron by considering the change in the $\psi_{s,r}$ when the coordinate system undergoes an infinitesimal rotation. As in the case of Pauli's theory we find that the rotation operator consists of two parts. One of them corresponds to the change in the space coordinates [compare eq. (111)], the other one to the transformation of the ψ-components. The two pairs ψ_{s+} and ψ_{s-} behave in exactly the same manner (see the end of § 64) and we have from equation (109),

$$\begin{pmatrix}\delta\psi_{+r}\\ \delta\psi_{-r}\end{pmatrix} = \frac{i}{\hbar}(\boldsymbol{\omega}\cdot\boldsymbol{S})\begin{pmatrix}\psi_{+r}\\ \psi_{-r}\end{pmatrix} = \tfrac{1}{2}i(\boldsymbol{\omega}\cdot\boldsymbol{\sigma})\begin{pmatrix}\psi_{+r}\\ \psi_{-r}\end{pmatrix}.$$

The total rotation operator is thus simply given by the equation $\boldsymbol{D}_{\text{op}} = (i/\hbar)[\boldsymbol{x}\wedge\boldsymbol{p}_{\text{op}}]+\frac{1}{2}i\boldsymbol{\sigma}$ and we find for the angular momentum operator [compare eq. (110)],

$$\boldsymbol{J}_{\text{op}} = -i\hbar\,\boldsymbol{D}_{\text{op}} = [\boldsymbol{x}\wedge\boldsymbol{p}_{\text{op}}]+\tfrac{1}{2}\hbar\,\boldsymbol{\sigma} = [\boldsymbol{x}\wedge\boldsymbol{p}_{\text{op}}]-\tfrac{1}{4}i[\boldsymbol{\alpha}\wedge\boldsymbol{\alpha}], \tag{170}$$

where we have used equation (158) to express $\boldsymbol{\sigma}(\sigma_x, \sigma_y, \sigma_z)$ in terms of $\boldsymbol{\alpha}(\alpha_x, \alpha_y, \alpha_z)$ *. If expression (164) does not change under a rotation, that is, if Φ, $\mathfrak{A}$ are the same functions of $\boldsymbol{x}$, t as Φ', $\mathfrak{A}'$ of $\boldsymbol{x}'$, t', H_{op} will commute with $\boldsymbol{J}_{\text{op}}$ [compare the discussion in § 61] and the angular momentum is constant in time.

As a matter of fact the situation with regard to the angular momentum is the same as in the non-relativistic case, since expression (170) for $\boldsymbol{J}$ is equal to $\boldsymbol{M}+\boldsymbol{S}$ as in § 61. The commutation relations (62) are again valid, and the eigenvalues of J^2 and J_z are given again by equations (67) and (68). We need only remember that the corresponding eigenvalues contain the argument r (see § 66).

The Dirac equations are gauge invariant (see § 62) since they contain the energy, momenta and potentials only in the combinations of the kinetic energy ($=\varepsilon c$) and kinetic momenta ($=\boldsymbol{\pi}$). If we change the gauge according to equation (113) the wave function changes according to equation (116) and the considerations of § 62 remain valid in the Dirac theory. The angular momentum is not

* The most general operator which can act on a four-component function is a matrix with sixteen arbitrary coefficients, which means that there exist sixteen linearly independent operators of this kind. This corresponds to the fact that we can construct from the four operators $\alpha_1, \ldots, \alpha_4$ which satisfy the relations (159) eleven other linearly independent operators which with the α_k and the unit operator form the complete set. Those eleven operators are the six products $\alpha_1\alpha_2$, $\alpha_1\alpha_3, \ldots$, the four products $\alpha_1\alpha_2\alpha_3$, $\alpha_1\alpha_2\alpha_4, \ldots$, and the product $\alpha_1\alpha_2\alpha_3\alpha_4$. If we use $\boldsymbol{\sigma}$ and $\boldsymbol{\varrho}$ instead of the α_k, these sixteen operators correspond to the sixteen products of $\mathbf{1}$, σ_x, σ_y, σ_z with $\mathbf{1}$, ϱ_x, ϱ_y, ϱ_z. A moment ago we saw that, indeed, $\alpha_1\alpha_2\alpha_3\alpha_4 = -\varrho_y$.

gauge invariant, but the kinetic angular momentum is. The rate of change of the latter and not of the former is equal to the resultant torque of the external forces *. If these forces derive from a radially symmetric electric field, their total torque is zero and if we use the "natural" gauge $\Phi = \Phi(r)$, $\mathfrak{A} = 0$, the angular momentum is equal to the kinetic angular momentum.

Let us first of all discuss the solution of the Dirac equations for the case of a *free particle* ($\Phi = 0$, $\mathfrak{A} = 0$). A general solution is obtained most simply by taking for ψ_+ and ψ_- two arbitrary solutions of the relativistic wave equation (1.63), $\nabla^2\psi_\pm - c^{-2}(\partial^2\psi_\pm/\partial t^2) - (mc/\hbar^2)\psi_\pm = 0$, and then determining $\chi^\dagger_\pm$ from equation (149), $mc\chi^\dagger = (i\hbar/c)(\partial\psi/\partial t) + i\hbar(\nabla\cdot\boldsymbol{\sigma})\psi$.

If in particular we are trying to find solutions corresponding to well defined values E and $\boldsymbol{p}$ of the energy and momentum, we obtain the expressions for the *de Broglie waves in the Dirac theory*,

$$\left.\begin{aligned}
&\psi_1=\psi_+=a_1\exp\{(i/\hbar)[(\boldsymbol{p}\cdot\boldsymbol{x})-Et]\},\ \psi_2=\psi_-=a_2\exp\{(i/\hbar)[(\boldsymbol{p}\cdot\boldsymbol{x})-Et]\},\\
&\psi_3=\chi^\dagger_+=a_3\exp\{(i/\hbar)[(\boldsymbol{p}\cdot\boldsymbol{x})-Et]\},\ \psi_4=\chi^\dagger_-=a_4\exp\{(i/\hbar)[(\boldsymbol{p}.\boldsymbol{x})-Et]\};\\
&mca_3=[(E/c)-p_z]a_1-[p_x-ip_y]a_2,\ mca_4=[(E/c)+p_z]a_2-[p_x+ip_y]a_1;
\end{aligned}\right\} \quad (1$$

$$E = \pm c[m^2c^2+p^2]^{1/2}, \tag{172}$$

where p is the absolute value of the momentum. Characteristic is the ambiguity in sign in equation (172). If we choose the positive sign, we are in agreement with classical theory where the energy, $E = c(m^2c^2+p^2)^{1/2} = mc^2[1-(v^2/c^2)]^{-1/2}$, is always positive, corresponding to a positive rest mass. The negative sign is, however, also possible and corresponds to a negative rest mass,

$$E = -c[m^2c^2+p^2]^{1/2} = -mc^2[1-(v^2/c^2)]^{-1/2}. \tag{173}$$

* Pauli (*Handb. Phys.*, **24₁**, 235, 1933) has discussed this theorem and similar ones using the energy momentum tensor $T_{\mu\nu}$ which was introduced by Tetrode (*Z. Phys.*, **49**, 853, 1928) into the Dirac theory. This tensor is obtained in a way completely analogous to the way we introduced the probability density and the probability current density, namely by quantising the sixteen classical expressions

$$T_{\mu\nu}^{\text{class}} = \left\|\begin{matrix}
\pi_x\dot{x} & \pi_x\dot{y} & \pi_x\dot{z} & \pi_x\\
\pi_y\dot{x} & \pi_y\dot{y} & \pi_y\dot{z} & \pi_y\\
\pi_z\dot{x} & \pi_z\dot{y} & \pi_z\dot{z} & \pi_z\\
\varepsilon\dot{x} & \varepsilon\dot{y} & \varepsilon\dot{z} & \varepsilon
\end{matrix}\right\|$$

as follows

$$T_{\mu\nu}^{\text{qu}} = \lim_{G\to 0}\,(4G)^{-1}\int\psi^*[G(T_{\mu\nu}+T_{\nu\mu})+(T_{\mu\nu}+T_{\nu\mu})G]\psi.$$

It is no longer possible, as it was in the classical theory, to exclude from our considerations of physical processes solutions corresponding to equation (173). We meet here the famous *appearance of negative masses in the Dirac theory* to which Dirac drew attention in his paper.

The most general solution of the Dirac equations for a free particle is obtained by a superposition of solutions of the kind (171). It consists of two sums (or integrals), one corresponding to positive and one to negative energies,

$$\psi = \Sigma a^{(+)} \exp \{(i/\hbar)[(\boldsymbol{p} \cdot \boldsymbol{x}) - c(p^2+m^2c^2)^{\frac{1}{2}}t]\}$$
$$+ \Sigma a^{(-)} \exp \{(i/\hbar)[(\boldsymbol{p} \cdot \boldsymbol{x}) + c(p^2+m^2c^2)^{\frac{1}{2}}t]\},$$

where the summations (integrals) extend over all possible values of the momentum $\boldsymbol{p}$.

As in the discussion in Chapter I we can construct wave packets, and especially "special" wave packets or "wave groups" the propagation of which illustrates the classical motion of a particle obeying Newton's first law. Such a wave group is constructed from de Broglie waves the momenta and energies of which differ very little from a certain average momentum and average energy, and the amplitudes of which are chosen in such a way that at a more or less well defined point in space, the approximate "position" of the particle, the constituent waves are in phase. The group moves with the group velocity [compare eqq. (1.15), (1.24)] $\boldsymbol{v} = (\boldsymbol{p}/p)(dE/dp) = \boldsymbol{p}c^2/E$. Expressing $\boldsymbol{v}$ in $\boldsymbol{p}$, or $\boldsymbol{p}$ in $\boldsymbol{v}$, we have

$$\boldsymbol{v} = \pm c\boldsymbol{p}(m^2c^2+p^2)^{-\frac{1}{2}} = \pm \frac{\boldsymbol{p}}{m}\left(1+\frac{p^2}{m^2c^2}\right)^{-\frac{1}{2}},$$
$$\boldsymbol{p} = \pm m\boldsymbol{v}\left(1-\frac{v^2}{c^2}\right)^{-\frac{1}{2}}. \tag{174}$$

The group velocity is parallel or antiparallel to the momentum according to whether we are dealing with a positive or a negative rest mass and we can understand how it came about that in the general Dirac theory the kinetic momentum and the velocity corresponded to different operators.

We shall now consider a system of reference $(x'y'z')$ in which the electron is at rest and the z'-axis of which falls along the direction of the electron momentum in the xyz-system, and we shall use the following notation

The question arises whether an experimental set-up which is meant to deal with such a situation is able to detect any spin effect, that is, whether electron experiments which can be approximately interpreted by means of the classical particle concept can lead us to the realisation of the presence of spin. We might, for instance, consider the deflexion of electrons in a magnetic field, — corresponding to the Stern-Gerlach experiment for neutral atoms. Bohr * has always denied this possibility. He has shown by considering special cases that the unavoidable smearing out of the wave packet always leads to an uncertainty which is at least of the same order of magnitude as the influence of the spin. Pauli** has discussed the situation in more detail using an approximate solution of the Dirac equation of the kind discussed in § 27. The fact that our classical Hamilton equation (147) has really a meaning only in the limit of vanishing $\boldsymbol{S}$ indicates also that we should drop the terms containing the spin if we construct approximate solutions of the Dirac equations which can be interpreted classically. We do not want to discuss this question in any more detail.

We now wish to show that the *solution of the Dirac equations in the non-relativistic limit* leads to Pauli's non-relativistic treatment which started from the Hamiltonian (23). We mentioned this limiting process briefly in § 64. It is advantageous to introduce four wave functions $\chi_1, \chi_2, \chi_3, \chi_4$ instead of the $\psi_1, \psi_2, \psi_3, \psi_4$ by means of the equations

$$\begin{aligned} \chi_1 &= \psi_1+\psi_3 = \psi_+ + \chi_+^\dagger, \quad & \chi_2 &= \psi_2+\psi_4 = \psi_- + \chi_-^\dagger, \\ \chi_3 &= \psi_1-\psi_3 = \psi_+ - \chi_+^\dagger, \quad & \chi_4 &= \psi_2-\psi_4 = \psi_- - \chi_-^\dagger. \end{aligned} \tag{182}$$

These χ_k possess the property that in a situation where the electron velocity is small compared to velocity of light and where the energy of the electron is nearly equal to mc^2, χ_3 and χ_4 are very small compared to χ_1 and χ_2 [of the order of $v/c \doteqdot p/mc$; compare eq. (179)]. It is of some advantage to introduce these functions for the discussions of certain problems [see §§66 and 82].

Using the abbreviations (156) we get from equations (149) and (150)

$$\varepsilon\begin{pmatrix}\chi_1\\ \chi_2\end{pmatrix} - (\boldsymbol{\pi}\cdot\boldsymbol{\sigma})\begin{pmatrix}\chi_3\\ \chi_4\end{pmatrix} = mc\begin{pmatrix}\chi_1\\ \chi_2\end{pmatrix}, \quad \varepsilon\begin{pmatrix}\chi_3\\ \chi_4\end{pmatrix} - (\boldsymbol{\pi}\cdot\boldsymbol{\sigma})\begin{pmatrix}\chi_1\\ \chi_2\end{pmatrix} = -mc\begin{pmatrix}\chi_3\\ \chi_4\end{pmatrix}, \text{ or,}$$

$$(\varepsilon - mc)\underset{2}{\chi_1} = (\boldsymbol{\pi}\cdot\boldsymbol{\sigma})\underset{4}{\chi_3}, \qquad (\varepsilon + mc)\underset{4}{\chi_3} = (\boldsymbol{\pi}\cdot\boldsymbol{\sigma})\underset{2}{\chi_1}. \tag{183}$$

* N. Bohr, *Proceedings* 1930 *Solvay Congress,* Gauthier-Villars, Paris, 1931.

** W. Pauli, *Helv. Phys. Acta,* **5**, 179, 1932.

These equations can be written in the form

$$[\varepsilon - (\boldsymbol{\pi} \cdot \boldsymbol{\sigma})\varrho_x - mc\varrho_z]\chi = 0 \tag{184}$$

where ϱ_x and ϱ_z operate on the χ as they did on the ψ [see eqq. (153) and (155)]. Equation (184) gives us an example of a choice of the α-operators in Dirac's equation different to the one made in writing down equations (157) and (158),

$$\varrho_x\sigma_x = \alpha_1, \quad \varrho_x\sigma_y = \alpha_2, \quad \varrho_x\sigma_z = \alpha_3, \quad \varrho_z = \alpha_4. \tag{185}$$

It is immediately clear that equations (157) and (184) are equivalent. The explicit expressions for the probability current density, . . . in terms of the χ are, however, different. Equation (185) was the choice of Dirac in his original paper *.

Eliminating $\chi_{\substack{3\\4}}$ from equation (183) we get

$$\chi_{\substack{3\\4}} = (\varepsilon + mc)^{-1}(\boldsymbol{\pi} \cdot \boldsymbol{\sigma})\chi_{\substack{1\\2}}, (\varepsilon - mc)\ \chi_{\substack{1\\2}} = (\boldsymbol{\pi} \cdot \boldsymbol{\sigma})(\varepsilon + mc)^{-1}(\boldsymbol{\pi} \cdot \boldsymbol{\sigma})\chi_{\substack{1\\2}}, \text{ or,}$$

$$[\varepsilon - mc - (\boldsymbol{\pi} \cdot \boldsymbol{\sigma})(\varepsilon + mc)^{-1}(\boldsymbol{\pi} \cdot \boldsymbol{\sigma})]\chi_{\substack{1\\2}} = 0. \tag{186}$$

Assuming that the operator $(\varepsilon + mc)^{-1}$ is well defined, we have now a wave equation (186) with a Hermitean operator — we note that the operator (147) was not Hermitean — operating on $\chi_{\substack{1\\2}}$. We now introduce the non-relativistic energy E_{nr}, $\varepsilon = mc + (E_{\text{nr}} + e\Phi)/c$ and expand, $(\varepsilon + mc)^{-1} = (2mc)^{-1} - (E_{\text{nr}} + e\Phi)/4m^2 c^3 + \ldots$.

Multiplying by c and using only the first two terms of the expression we get from equation (186)

$$[E_{\text{nr}} + e\Phi - (\boldsymbol{\pi} \cdot \boldsymbol{\sigma})^2/2m + (\boldsymbol{\pi} \cdot \boldsymbol{\sigma})(E_{\text{nr}} + e\Phi)(\boldsymbol{\pi} \cdot \boldsymbol{\sigma})/4m^2c^2]\chi_{\substack{1\\2}} = 0. \tag{187}$$

Using the usual expressions for $\boldsymbol{p}_{\text{op}}$ and $E_{\text{nr op}}$ [compare eq. (146)] we find after a straightforward calculation that equation (187) leads to the following gauge invariant equation (terms of order c^{-3} or smaller have been neglected),

$$[E_{\text{nr}} + e\Phi - |\boldsymbol{\pi}|^2/2m + |\boldsymbol{\pi}|^4/8m^3 c^2 - (e\hbar/2\,mc)(\boldsymbol{\sigma} \cdot \mathfrak{H} + \{[\mathfrak{E} \wedge \boldsymbol{\pi}]/2\,mc\})]\chi_{\substack{1\\2}} = 0 \tag{188}$$

which is just the Schrödinger equation (23) of the Pauli theory.

The term $\pi^2/2m$ derives from the term $(\boldsymbol{\pi} \cdot \boldsymbol{\sigma})^2/2m$; however, since the components of $\boldsymbol{\pi} = \boldsymbol{p} + (e/c)\mathfrak{A}$ do not commute, this term also gives rise to a term proportional

* P. A. M. Dirac, *Proc. Roy. Soc.* **A117**, 610, 1928.

$\mathfrak{S}''/c = |A^{(+)}|^2\cos^2\chi \cos\chi'/\cos^2\frac{1}{2}(\chi-\chi')$, and one easily verifies the conservation law $\mathfrak{S}+\mathfrak{S}' = \mathfrak{S}''$.

The probabilities for reflexion, w_r, and for transmission, w_t, are given by the expressions $w_r = |B^{(+)}|^2/|A^{(+)}|^2 = [1-\cos(\chi+\chi')]/[1+\cos(\chi+\chi')]$, $w_t = 1-w_r$.

In the limit where $e\Phi_0 = T+mc^2$, so that it is just possible for the particle to slip through into II, we have $\chi' = \frac{1}{2}\pi$, $w_r = 1$, $w_t = 0$. In the limit $e\Phi_0 \rightarrow \infty$ we have $\chi' = 0$, $w_r = (1-\cos\chi)/(1+\cos\chi)$, $w_t = 2\cos\chi/(1+\cos\chi) = \{\frac{1}{2}+\frac{1}{2}[1-(mc^2/T)^2]^{-1/2}\}^{-1}$.

If $e\Phi_0$ lies between $T-mc^2$ and $T+mc^2$, the wave motion in II will decrease exponentially with increasing z (p' is negative imaginary) and it is not contained in the solutions of the Dirac equations given by equations (176).

Klein's considerations show, indeed, that it is impossible to avoid in the Dirac theory states of negative mass. His argument would be even more convincing, if it were possible to construct a situation where the particle which was originally in I would occur at a later time with a negative mass in I (and not in II). Because of the conservation of energy one must in this case either consider the possibility of the collision of the particle in II with another particle during which it loses energy at least to the amount mc^2+T or if one does not want to introduce a second particle one must consider external forces which change in time. The second alternative is not a very promising one, as an exchange of energy between an electron and an external electromagnetic field without quantisation of the field ("quanta of radiation") would in any case lead to results in disagreement with experiment. If suffices here to remark that the difficulty of the occurrence of negative masses also occurs in the quantum theory of radiation discussed in Chapter VIII. This theory shows that there is no foundation for the hope that the forces occurring in nature will be such that no negative masses occur and that we do not need to consider potentials such as the one considered by Klein, or similar ones.

In 1930 Dirac * showed a way out of this difficulty of the negative masses in his "hole theory". In this theory Pauli's exclusion principle, which we have not yet discussed, plays a fundamental role. An especially characteristic fact is that as well as electrons, positively charged particles with the same mass and spin properties occur, and that the total number of positive and negative electrons does not remain constant, while their total charge does.

Even if the results of this theory must contain a certain grain of truth, vide experimental data about the creation and annihilation of pairs, it is still far from a closed definitive theory which is physically

* P. A. M. Dirac, *Proc. Roy. Soc.*, **A126**, 360, 1930. See also § 72.

satisfactory *. Its shortcomings are connected with the unsatisfactory present-day theory of the interaction between particles and radiation (compare § 90), but a discussion of these difficulties is outside the scope of this book.

The Dirac theory of the electron also leads to a better theory of the *interaction between electrons*. Each electron is described by its spatial coordinates $\boldsymbol{x}_i$ as well as its two spin coordinates s_i, r_i (see § 64) and instead of a 2^N-component wave function (75) we are dealing with a 4^N-component wave function $\Psi_{s_1 r_1, s_2 r_2, \ldots s_N r_N}(\boldsymbol{x}_1, \ldots, \boldsymbol{x}_N; t)$. In the limit of free electrons the energy operator would be $H^{(0)}_{\text{op}} = c\Sigma_i[(\boldsymbol{\alpha}^{(i)}\boldsymbol{p}_i)+\alpha_4^{(i)}mc]$ (compare eq. (164)], where $\boldsymbol{\alpha}^{(i)}(\alpha_1^{(i)}, \alpha_2^{(i)}, \alpha_3^{(i)})$, $\alpha_4^{(i)}$ operate on the s_i, r_i in the way indicated by equation (154). The eigenfunctions can be written as the product of de Broglie waves of type (171) corresponding to the different electrons.

We might try to add terms to $H^{(0)}_{\text{op}}$ in order to arrive at a description of the interaction in the frame work of point particles. This procedure can never lead to a general relativistic theory (compare § 81)**, but the following treatment should lead to results which are correct to a first approximation.

Consider two electrons 1 and 2. The action of 2 on 1 can be described according to equation (164) by adding to the Hamiltonian operator a term

$$-e\Phi_1+e(\boldsymbol{\alpha}^{(1)}\cdot\mathfrak{A}_1), \tag{188A}$$

where Φ_1 and $\mathfrak{A}_1$ are the potentials of the electromagnetic field produced by 2 at the position 1. It is essential that the electromagnetic field can be derived from these potentials by using equations (2.99), but the potentials need not necessarily satisfy the equations (2.100). We shall choose $(\boldsymbol{\nabla}\cdot\mathfrak{A}_1) = 0$, so that $\Phi_1 = -e/r_{12}$ [compare eq. (8.143)], where r_{12} is the distance between the two electrons. The magnetic field $\mathfrak{H}_1$ at the position of 1 is to a first approximation in classical theory given by the Biot-Savart law, $\mathfrak{H}_1 = -(e/c)[\boldsymbol{x}_{12}\wedge\boldsymbol{v}_2]/r_{12}^3$, where $\boldsymbol{x}_{12}$ is the vector from 1 to 2, and where $\boldsymbol{v}_2$ is the velocity of 2. The vector potential of this field satisfying $(\boldsymbol{\nabla}\cdot\mathfrak{A}) = 0$ is given by the expression ***

* See P. A. M. Dirac, *Proc. Camb. Phil. Soc.*, **30**, 150, 1934; W. Heisenberg, *Z. Phys.*, **90**, 209, 1934; V. Weisskopf, *Proc. Dan. Acad. Sc.*, **24**, Nr 6, 1936. Compare also the discussion in Chapter VIII.

** Møller (*Z. Phys.*, **70**, 786, 1931) has shown how one can give a relativistic treatment of the case of *weakly* interacting fast electrons.

*** In the case where spin is neglected, this expression leads to the Darwin terms fo equation (3.76).

Proof. Introduce for a moment new coordinates $x'y'z'$ by the equations $x' = -x$, $y' = -y$, $z' = -z$ (inversion with respect to the origin). The radius $r = [x^2+y^2+z^2]^{1/2}$ is invariant, $r = [x'^2+y'^2+z'^2]^{1/2}$.* The operator $\boldsymbol{p}_{\text{op}}(= -i\hbar\nabla)$ goes over into $-\boldsymbol{p}_{\text{op}}'(= i\hbar\nabla')$ so that the transformed equations (189) will be of the form

$$\left.\begin{aligned} \{[E-U(r')]/c-mc\}\varphi_{\substack{1\\2}}' &= (\boldsymbol{p}_{\text{op}}'\cdot\boldsymbol{\sigma})\varphi_{\substack{3\\4}}', \quad \varphi_{\substack{1\\2}}'(\boldsymbol{x}') = \varphi_{\substack{1\\2}}(-\boldsymbol{x}), \\ \{[E-U(r')]/c+mc\}\varphi_{\substack{3\\4}}' &= (\boldsymbol{p}_{\text{op}}'\cdot\boldsymbol{\sigma})\varphi_{\substack{1\\2}}', \quad \varphi_{\substack{3\\4}}'(\boldsymbol{x}') = -\varphi_{\substack{3\\4}}(-\boldsymbol{x}). \end{aligned}\right\} \tag{193}$$

Since the $\mathscr{P}_l^m$ are derived from the harmonic polynomials, $H_l(\boldsymbol{x})$, of the l-th degree we have $\mathscr{P}_l^m(\boldsymbol{x}) = \mathscr{P}_l^m(-\boldsymbol{x}') = (-1)^l\,\mathscr{P}_l^m(\boldsymbol{x}')$, or, using a vector $\boldsymbol{\omega}$ to indicate a direction in space, $\mathscr{P}_l^m(\boldsymbol{\omega}) = (-1)^l\,\mathscr{P}_l^m(\boldsymbol{\omega}')$. It then follows from equation (190) that $Q_{j\,\mu}{}^{(\pm)}(\boldsymbol{\omega}, s) = (-1)^{j\pm 1/2}\,Q_{j\mu}{}^{(\pm)}(\boldsymbol{\omega}', s)$. From equations (191) and (193) we then get $\varphi_{\substack{1\\2}} = (-1)^{j-1/2}(AQ^{(+)}-BQ^{(-1)})$, $\varphi_{\substack{3\\4}} = (-1)^{j-1/2}(-CQ^{(+)}+DQ^{(-)})$.

Since it follows from equations (193) and (189) that the φ' satisfy the same equations as the φ, we see that if the φ given by equations (191) are eigenfunctions, we obtain another eigenfunction φ' corresponding to the same eigenvalue by changing the sign of B and C. This means that $\varphi+\varphi'$ and $\varphi-\varphi'$ are also eigenfunctions corresponding to the same eigenvalue and this corresponds to the simplification (192).** Only, if the eigenvalue happens to be degenerate, will both $\varphi+\varphi'$ and $\varphi-\varphi'$ be different from zero.

We shall first consider $B = C = 0$ and put

$$\varphi_{\substack{1\\2}} = [F^{(+)}(r)/r]Q_{j\,\mu}^{(+)}, \qquad \varphi_{\substack{3\\4}} = [G^{(+)}(r)/r]Q_{j\,\mu}^{(-)}, \tag{194}$$

and equations (189) are of the form

$$\left.\begin{aligned} [(E-U)/c-mc](F^{(+)}/r)\,Q_{j\,\mu}^{(+)} &= (\boldsymbol{p}_{\text{op}}\cdot\boldsymbol{\sigma})(G^{(+)}/r)\,Q_{j\,\mu}^{(-)}, \\ [(E-U)/c+mc](G^{(+)}/r)\,Q_{j\,\mu}^{(-)} &= (\boldsymbol{p}_{\text{op}}\cdot\boldsymbol{\sigma})(F^{(+)}/r)\,Q_{j\,\mu}^{(+)}. \end{aligned}\right\} \tag{195}$$

The action of the operator $(\boldsymbol{p}_{\text{op}}\cdot\boldsymbol{\sigma})$ on the $Q^{(\pm)}$ is such that it produces $Q^{(\mp)}$ multiplied by a function of r. Equations (195) lead thus to two equations for F and G and from the solution of these equations the eigenvalues of E follow. The equations for F and G are

$$\left.\begin{aligned} [(E-U)/c-mc]F^{(+)} &= -i\hbar[G^{(+)\prime}+(j+\tfrac{1}{2})(G^{(+)}/r)], \\ & \qquad G^{(+)\prime} = dG^{(+)}/dr, \\ [(E-U)/c+mc]G^{(+)} &= -i\hbar[F^{(+)\prime}-(j+\tfrac{1}{2})(F^{(+)}/r)], \\ & \qquad F^{(+)\prime} = dF^{(+)}/dr. \end{aligned}\right\} \tag{196}$$

If we take $A = D = 0$, we put

$$\varphi_{\substack{1\\2}} = [F^{(-)}(r)/r]Q_{j\,\mu}^{(-)}, \quad \varphi_{\substack{3\\4}} = [G^{(-)}(r)/r]Q_{j\,\mu}^{(+)}, \tag{197}$$

* It is better not to describe the spinors in the new system of coordinates, but to use for them the old xyz-system (compare the discussion in § 61). Van der Waerden (*Die gruppentheoretische Methode in der Quantenmechanik*, Springer, Berlin 1932) has shown how the spinors transform under an inversion.

** In the language of group theory we are performing the reduction of (191) with respect to the inversion group.

and we get the following equations for $F^{(-)}$ and $G^{(-)}$,

$$\left.\begin{aligned}[(E-U)/c-mc]F^{(-)} &= -i\hbar[G^{(-)\prime}-(j+\tfrac{1}{2})(G^{(-)}/r)],\\ [(E-U)/c+mc]G^{(-)} &= -i\hbar[F^{(-)\prime}+(j+\tfrac{1}{2})(F^{(-)}/r)].\end{aligned}\right\} \quad (198)$$

To derive equations (196) and (198) we use the identity

$$(\boldsymbol{\nabla}\cdot\boldsymbol{\sigma})\{U(r)\,V_s(\boldsymbol{x})\} = (\boldsymbol{\sigma}\cdot V_s\boldsymbol{\nabla}U)+U(\boldsymbol{\nabla}\cdot\boldsymbol{\sigma})V_s$$
$$= (dU/dr)(\boldsymbol{\sigma}\cdot\boldsymbol{x})(V_s/r)+U(\boldsymbol{\nabla}\cdot\boldsymbol{\sigma})V_s, \quad (199)$$

where U is an arbitrary function of r, while $V_s(\boldsymbol{x})$ is a two component spin function of $\boldsymbol{x}$ and s.

If we put $(\boldsymbol{\sigma}\cdot\boldsymbol{x})Q^{(+)} = \alpha r Q^{(-)}$, $(\boldsymbol{\sigma}\cdot\boldsymbol{x})Q^{(-)} = \beta r Q^{(+)}$, it follows that $\alpha\beta = 1$, since $(\boldsymbol{\sigma}\cdot\boldsymbol{x})^2 = r^2$. Since the Q of equations (190) are normalised $\begin{Vmatrix} 0 & \alpha \\ \beta & 0 \end{Vmatrix}$ is a Hermitean matrix, and α and β are complex conjugate phase factors. We have defined the polar spherical harmonics which occur in equations (190) in such a way [see eqq. (4.148) (4.161)] that for all values of j and μ we have $\alpha = \beta = 1$. A different choice of the Q would lead only to an additional constant factor in one of the equations (194).

We now note that the components of the spin functions $r^{\pm j-\frac{1}{2}}Q^{(+)}$ and $r^{\pm(j+1)-\frac{1}{2}}$ $Q^{(-)}$ are harmonic functions of x, y, z (they are of the form H_l or $r^{-2l-1}H_l$, where H_l is a harmonic polynomial of the l-th degree). This means that $\boldsymbol{\nabla}^2 = (\boldsymbol{\nabla}\cdot\boldsymbol{\sigma})^2$ acting on them will give zero. Since apart from a constant factor $(\boldsymbol{\nabla}\cdot\boldsymbol{\sigma})\ r^{\pm(j+1)-\frac{1}{2}}$ $Q^{(-)}$ is equal to $r^{\pm(j+1)-\frac{3}{2}}\,Q^{(+)}$, and similarly $(\boldsymbol{\nabla}\cdot\boldsymbol{\sigma})\ r^{\pm j-\frac{1}{2}}\,Q^{(+)}$ equal to $r^{\pm j-\frac{3}{2}}$ $Q^{(-)}$, it follows that at least one of the pair $(\boldsymbol{\nabla}\cdot\boldsymbol{\sigma})r^a\,Q^{(+)}$, $(\boldsymbol{\nabla}\cdot\boldsymbol{\sigma})\ r^b\,Q^{(-)}$ must be zero when a and b are one of the four pairs $(j-\frac{1}{2}, j+\frac{1}{2})$, $(-j-\frac{5}{2}, -j-\frac{3}{2})$, $(j-\frac{1}{2}, j-\frac{3}{2})$, $(-j-\frac{1}{2}, -j-\frac{3}{2})$. Hence we have $(\boldsymbol{\nabla}\cdot\boldsymbol{\sigma})\ r^{j-\frac{1}{2}}\ Q_{j\mu}{}^{(+)} = 0$ and $(\boldsymbol{\nabla}\cdot\boldsymbol{\sigma})r^{-j-\frac{3}{2}}$ $Q_{j\mu}{}^{(-)} = 0$.

If we now combine equations (199) and (195) we get $(\boldsymbol{p}_{\text{op}} = -i\hbar\boldsymbol{\nabla})$

$$(\boldsymbol{p}_{\text{op}}\cdot\boldsymbol{\sigma})(G^{(+)}/r)Q^{(-)} = -i\hbar(\boldsymbol{\nabla}\cdot\boldsymbol{\sigma})\{(G^{(+)}r^{j+\frac{1}{2}})(r^{-j-\frac{3}{2}}\,Q^{(-)})\}$$
$$= -i\hbar r^{-j-\frac{3}{2}}\,Q^{(+)}d/dr)(G^{(+)}\,r^{j+\frac{1}{2}}) = -i\hbar[G^{(+)\prime}+(j+\tfrac{1}{2})(G^{(+)}/r)]Q^{(+)},$$
$$(\boldsymbol{p}_{\text{op}}\cdot\boldsymbol{\sigma})(F^{(+)}/r)\,Q^{(+)} = -i\hbar(\boldsymbol{\nabla}\cdot\boldsymbol{\sigma})\{(F^{(+)}r^{-j-\frac{1}{2}})(r^{j-\frac{1}{2}}\,Q^{(+)})\}$$
$$= -i\hbar r^{j-\frac{1}{2}}\,Q^{(-)}(d/dr)(F^{(+)}r^{-j-\frac{1}{2}}) = -i\hbar[F^{(+)\prime}-(j+\tfrac{1}{2})(F^{(+)}/r)]Q^{(-)}.$$

Inserting these formulae into equations (195) leads to equations (196). The derivation of equations (198) is similar. Of course, one could have derived these equations directly from equations (4.163) and (4.164).

In the non-relativistic approximation * E is equal to mc^2 up to terms of the order c^{-2} and we see from equations (196) [and (198)] that $G^{(\pm)}$ is small compared to $F^{(\pm)}$ of the order of c^{-1}. This we should have expected from the considerations of the previous section. If we introduce again E_{nr} by the equation $E = mc^2+E_{\text{nr}}$ and if we expand $F^{(+)}$ and $cG^{(+)}$ in powers of c^{-1} we find to a first approximation from equations (196),

$$(E_{\text{nr}}-U)F^{(+)} = -i\hbar[cG^{(+)\prime}+(j+\tfrac{1}{2})(cG^{(+)}/r)],$$
$$2mcG^{(+)} = -i\hbar[F^{(+)\prime}-(j+\tfrac{1}{2})(F^{(+)}/r)].$$

* This approximation has a meaning if it is justifiable to consider the potential energy to be small compared to mc^2 in the stationary states under consideration.

Up to now no nuclei have been produced with Z as large as 137 so that β is always smaller than one [see eq. (204)]. This means that even for the smallest value of $k(|k| = 1)$ λ is real. It is interesting to note that according to equations (194) and (197) as $r \to 0$ the wave function φ in the case $|k| = 1$ goes to infinity as $\exp\{(-1+[1-\beta^2]^{1/2}) \ln r\}$ while in the non-relativistic case the wave functions always remain finite. This does, however, not impair the physical usefulness of the formulae.

From the requirement that the wave functions must be normalisable it follows that only the positive sign can be used in equation (207). Substituting equations (206) into equations (205) we get recurrence relations between the a_s and b_s. These relations can be written in the form

$$\left.\begin{aligned} &a_s/(s+P) = \pm b_s/(s-Q) = c_s, \\ &\qquad P = k \mp \beta(A/B)^{1/2},\ Q = k \mp \beta(B/A)^{1/2}, \\ &c_s/c_{s-1} = \mp 2(AB)^{1/2}\{[s-1-\tfrac{1}{2}(Q-P)]/(s+\lambda)(s-\lambda)\}. \end{aligned}\right\} \quad (208)$$

If we put $c_\lambda = 1$ we get *

$$\begin{aligned} \Sigma_\lambda^\infty c_s r^s &= r^\lambda {}_1F_1(\lambda - \tfrac{1}{2}[Q-P],\ 2\lambda+1;\ \mp 2r[AB]^{1/2}) \\ &= r^\lambda {}_1F_1(\lambda \pm \beta[B-A]/2[AB]^{1/2},\ 2\lambda+1;\ \mp 2r[AB]^{1/2}), \end{aligned} \quad (209)$$

$$\left.\begin{aligned} \Sigma_\lambda^\infty a_s r^s &= \Sigma_\lambda^\infty (s+P) c_s r^s = r^{-P+1}(d/dr)\,\Sigma c_s r^{s+P} \\ &= r^{-P+1}(d/dr)(r^{\lambda+P}\,{}_1F_1), \\ \Sigma_\lambda^\infty b_s r_s &= \mp r^{Q+1}(d/dr)(r^{\lambda-Q}\,{}_1F_1). \end{aligned}\right\} \quad (210)$$

The discrete stationary states of the hydrogen atom correspond to $0 < E < mc^2$. In that case both A and B are positive and the function ${}_1F_1$ behaves asymptotically as $\exp[2r(AB)^{1/2}]$ if we take the lower sign in equations (206) and (209). The only exception occurs when $\lambda - \beta[(B-A)/2(AB)^{1/2}]$ is equal to zero or to a negative integer, since then the series (209) stops after a finite number of terms and thus is a finite polynomial. Both F and G in equation (206) behave asymptotically as $\exp[-r(AB)^{1/2}]$ and we have obtained a normalisable eigenfunction (compare the analogous discussion of the non-relativistic case in § 46).

We thus find for the discrete energy levels

$$\lambda - \beta[(B-A)/2(AB)^{1/2}] = -n', \qquad n' = 0, 1, 2, \ldots, \quad (211)$$

* Gordon (*Z. Phys.*, **48**, 11, 1928) was the first to give the explicit expression for the eigenfunctions in terms of the confluent hypergeometric series. See also Bethe, *Handb. Phys.*, **24**$_1$, 311, 1933.

where $n'+1$ is the number of terms of the polynomial (210).

From equation (204) we get now

$$\beta[(B-A)/2(AB)^{1/2}] = \beta E(m^2c^4-E^2)^{-1/2} = n'+\lambda = n'+(k^2-\beta^2)^{1/2},$$

$$\text{or,} \quad E = mc^2[1+\{\beta/[n'+(k^2-\beta^2)^{1/2}]\}^2]^{-1/2}. \tag{212}$$

We have thus obtained an expression for the energy levels of the discrete stationary states of the hydrogen atom. This expression is identically the same as the one derived in 1915 by Sommerfeld using the old quantum theory (see § 59).

An expansion in powers of β leads to

$$E = mc^2-[mc^2\beta^2/2(n'+|k|)^2] + \tfrac{1}{2}mc^2\beta^4[\tfrac{3}{4}(n'+|k|)^{-4}-(n'+|k|)^{-3}|k|^{-1}]+\ldots$$

$$= mc^2+(Z^2e^4m/2\hbar^2)\{-n^{-2}+(Ze^2/\hbar c)^2[\tfrac{3}{4}n^{-4}-n^{-3}(j+\tfrac{1}{2})^{-1}]+\ldots\}$$

This formula is the same as the one derived in the non-relativistic theory [compare eq. (72)] apart from the rest mass energy mc^2. The integer $n'+|k|$ plays the role of the principal quantum number in Bohr's theory. The difficulties which occurred for the case $l = 0$, $j = \frac{1}{2}$, $k = 1$ in the non-relativistic theory (see §§ 46 and 59) do not occur in the Dirac theory. The energy (212) is well defined for these values of the quantum numbers.

If k is positive, it is equal to $l+1$ and $j = l+\frac{1}{2}$, while for negative k, $k = -l$, $j = l-\frac{1}{2}$, where l is the orbital angular momentum quantum number of the corresponding solution of the non-relativistic theory (see § 59).

Since the principal quantum number n must be at least $l+1$ [compare eq. (4.191)] $n' = 0$ cannot correspond to a solution, if k is negative. We see that this is, indeed, the case as ${}_1F_1$ reduces to 1 in this case while $\lambda+P = \lambda-Q = 0$.

If $n' = 0$ we see from equation (212) that $E = mc^2[1-(\beta^2/k^2)]^{1/2}$ and from equations (204) we see that $\beta(A/B)^{1/2} = |k|-\lambda$, $\beta(B/A)^{1/2} = |k|+\lambda$, so that $P = k+|k|-\lambda$, $Q = k+|k|+\lambda$, since we must take the lower sign in equation (208). For negative k these expressions reduce to $P = -\lambda$, $Q = \lambda$.

We can introduce two quantities n_1 and n_2, both smaller than the classical principal quantum number $n = n'+|k|$ which both converge to n in the limit $\beta \to 0$,

$$\left.\begin{aligned} n=n'+|k|,\ \ n_1=n'+(k^2-\beta^2)^{1/2}=n'+\lambda,\ \ n_2=(n_1^2+\beta^2)^{1/2}, \\ k=l+1, \text{ or } -l. \end{aligned}\right\} \tag{213}$$

We now find for P and Q, using the lower sign in equations (208)

$$P = k+n_2-n_1,\ Q = k+n_2+n_1,\ AB = (a_1 n_2)^{-2},\ a_1 = \hbar^2/Zme^2,$$

where a_1 is the Bohr orbit for the case of a nuclear charge Ze. Using equations (206) and (210) we get for F and G the expressions

$$\left.\begin{aligned} F &= C(n_2+n_1)^{1/2}\exp(-r/a_1n_2)r^{-k-n_2-n_1+1}(d/dr)[r^{\lambda+k+n_2-n_1}\,{}_1F_1], \\ G &= -iC(n_2-n_1)^{1/2}\exp(-r/a_1n_2)r^{k+n_2+n_1+1}(d/dr)[r^{\lambda-k-n_2-n_1}\,{}_1F_1], \\ {}_1F_1 &= {}_1F_1(\lambda-n_1,\ 2\lambda+1;\ 2r/a_1n_2). \end{aligned}\right\} \quad (214)$$

From the definition of ${}_1F_1$ it follows that the expression (214) for F goes over into the non-relativistic expression [compare eqq. (4.186) and (4.190)] both for $k = -l$ and for $k = l+1$, if we put $\beta = 0$,

$$F = \text{constant}\ r^{l+1} \exp(-r/a_1 n)\,{}_1F_1(l-n+1,\ 2l+2;\ 2r/a_1 n).$$

One can write ${}_1F_1$ in equation (214) in a form similar to expression (4.196),

$${}_1F_1(\lambda-n_1,\ 2\lambda+1;\ 2r/a_1n_2) = \text{const.}\ r^{-2\lambda}\exp(2r/a_1n_2) \\ (d/dr)^{n_1-\lambda}[r^{n_1+\lambda}\exp(-2r/a_1n_2)], \quad (215)$$

from which one can find the number of zeros of $F(r)$ and $G(r)$.

An eigenfunction φ_k is normalised when $\Sigma_k\int|\varphi_k|^2 d\boldsymbol{x} = 1$, or, if we use equations (194) [or (197)] and the fact that the $Q^{(\pm)}$ are normalised on the unit sphere [compare eqq. (64) and (4.161)], when $\int_0^\infty(|F(r)|^2+|G(r)|^2)dr = 1$.

From equations (214) it follows that the normalisation constant C is given by the equation

$$C = (2/a_1n_2)^{\lambda+1/2}[\Gamma(2\lambda+n'+1)/(k+n_2)n'!]^{1/2}[2n_2\Gamma(2\lambda+1)]^{-1}. \quad (216)$$

Bechert * has shown how one can normalise the hydrogen eigenfunctions in the Dirac theory. His calculations are straightforward but lengthy and tedious. A simpler method is the following one ** which can also be applied to other normalisation problems. We want to calculate the integral $\int|f|^2 = \int f^*f$, f being an eigenfunction of the equation $(H_{\text{op}}-E)f = 0$, where H_{op} is Hermitean. Let $f(E)$ be a solution of this equation which is a continuous function of E and which goes over into the eigenfunction if E is one of the (discrete) eigenvalues. We have then $(H_{\text{op}}-E)f_E = f$, $f_E = \partial f/\partial E$ and $I = \int f^*f = \int f^*(H_{\text{op}}-E)f_E = \int (f^*H_{\text{op}}f_E - f_E H_{\text{op}}f^*)$. If both f and f_E satisfied the boundary conditions which characterise the eigenvalue problem, the last integral would be equal to zero. If E is one of the discrete eigenvalues (and only in this case is I convergent), f satisfies these boundary condition, but f_E will not satisfy them. The Hermitean character of H_{op} means, however, that we can express this integral in terms of quantities which describe the

* K. Bechert, *Ann. Phys.*, **6**, 700, 1930.

** Compare, H. B. G. Casimir, *Arch. Teyler*, **8**, 201, 1936.

behaviour of f and f_E at the "boundaries". If, for instance, H_{op} is of the form $A(d^2/dx^2)+BU(x)$, we have $\int_a^b |f|^2 dx = |f^*(df_E/dx) - f_E(df^*/dx)|_a^b$.

In the present case equations (205) for F and G can be written in the form

$$(H_{op}-E)\binom{F}{G} = 0, \; H_{op} = \varrho_z mc^2 - i\hbar c\, \varrho_x \frac{d}{dr} + \frac{\hbar c}{r} k\, \varrho_y - \frac{Ze^2}{r}, \tag{217}$$

where F and G can be considered to be the two components of a wave function $f_w(r)$ $(w = \pm\frac{1}{2})$ and where ϱ_x, ϱ_y, and ϱ_z have the form of the Pauli matrices (153). For the normalisation integral we have

$$\left.\begin{aligned} \int_0^\infty (|F|^2+|G|^2)dr &= \lim_{r\to\infty} \Sigma_w \int_0^r f_w{}^* f_w\, dr \\ &= \lim_{r\to\infty} \Sigma_w \int_0^r (f^* H_{op}{}^* f_E - f_E H_{op} f^*)dr \\ = \lim_{r\to\infty} -i\hbar c \int_0^r (F^* G_E{}' + G^* F_E{}' &+ F_E G^{*\prime} + G_E F^{*\prime})\,dr \\ &= -i\hbar c \lim_{r\to\infty} (F^* G_E + G^* F_E). \end{aligned}\right\} \tag{218}$$

We can obtain F, G, and their derivatives with respect to E from equations (206) and (210). The asymptotic behaviour of those functions follows from the asymptotic behaviour of ${}_1F_1$ for which we have *

$$\lim_{x\to\infty} {}_1F_1(a, c; x) = [\Gamma(c)/\Gamma(a)]x^{a-c}e_y, \quad (a \neq 0, -1, -2, \ldots);$$
$$\lim_{x\to\infty} {}_1F_1(-n', c; x) = (-1)^{n'}[\Gamma(c)/\Gamma(c+n')]x^{n'}, \quad (a = 0, -1, -2, \ldots).$$

If a is nearly equal to $-n'$ $(a = -n'+\delta)$, $[\Gamma(a)]^{-1}$ is nearly equal to zero and can be expanded in powers of δ. From a well-known formula for Γ-functions we have $[\Gamma(a)]^{-1} = (\sin \pi a/\pi)\Gamma(1-a) \doteqdot (-1)^{n'} n'!\,\delta$.

For the values of a and c which interest us (see eqq. (214)] we get

$$\lim_{r\to\infty} {}_1F_1(-n', 2\lambda+1; 2r[AB]^{1/2}) \doteqdot (-1)^{n'}\Gamma(2\lambda+1)[2r(AB)^{1/2}]^{n'}/\Gamma(2\lambda+n'+1),$$
$$\lim_{r\to\infty} (\partial/\partial E)\, {}_1F_1(-n', 2\lambda+1; 2r[AB]^{1/2}) \doteqdot (-1)^{n'}\Gamma(2\lambda+1)[2r(AB)^{1/2}]^{-2\lambda-n'-1}.$$
$$\exp\,[2r(AB)^{1/2}](d/dE)[-\beta(B-A)/2(AB)^{1/2}].$$

From these equations equation (216) follows. We may note that the two terms in expression (218) are equal.

The discrete eigenvalues (212) are the only eigenvalues of the hydrogen atom with E between mc^2 and $-mc^2$. If $E > mc^2$, or $E < -mc^2$, A or B in equations (204) becomes negative and F and G behave oscillatorily for large values of r (compare the discussion in § 46). Each value of E in those regions corresponds to an improper eigenfunction. The case $E > mc^2$ corresponds especially to the continuous eigenvalue spectrum of the non-relativistic theory. The states of negative kinetic energy which are characteristic for the Dirac theory occur for $E < -mc^2$. As we mentioned earlier, the fact that those solutions exist is a difficulty especially since the Dirac theory must contain a great deal of truth since it gives results in agreement with experimental data. We mention especially the excellent agreement between the energy levels predicted by equation (212) and those found from atomic spectra.

* See E. T. Whittaker and G. N. Watson, *Modern Analysis*, Cambridge University Press, 1927, Ch. XVI.

CHAPTER VII

THE EXCLUSION PRINCIPLE

§ 67. THE PAULI PRINCIPLE FOR ELECTRONS

Pauli's exclusion principle expresses an extremely important physical law and it must rigorously be taken into account when quantum mechanics is applied in atomic physics. We can formulate it as follows. *The physical situation of an atomic system which contains two or more electrons is always such that the corresponding wave function is antisymmetric* (or *alternating*) *in the coordinates of the electrons.*

This statement means the following. The wave function ψ depends, *firstly* on the spatial and spin coordinates $\boldsymbol{x}_i$, s_i ($i = 1, 2, \ldots, n$) of the n electrons in the system (these coordinates refer to the same (orthogonal) spatial coordinate system for all the electrons), *secondly*, on the spatial [and possibly spin (see § 60)] coordinates $\boldsymbol{X}_k$ ($k = 1, 2, \ldots$) of the nuclei in the system, and, *thirdly*, on the time t. It satisfies the Schrödinger equation

$$H_{\mathrm{op}}\psi = i\hbar\partial\psi/\partial t, \tag{1}$$

where H_{op} acts in general on all coordinates of the first and second group and may contain the time explicitly. According to the exclusion principle the only solutions of equation (1) to be realized in nature are those which satisfy the relation

$$\left.\begin{aligned}&\psi(\boldsymbol{x}_1, s_1; \ldots; \boldsymbol{x}_i, s_i; \ldots; \boldsymbol{x}_j, s_j; \ldots; \boldsymbol{x}_n, s_n; \boldsymbol{X}_k; t) = \\ -&\psi(\boldsymbol{x}_1, s_1; \ldots; \boldsymbol{x}_j, s_j; \ldots; \boldsymbol{x}_i, s_i; \ldots; \boldsymbol{x}_n, s_n; \boldsymbol{X}_k; t),\end{aligned}\right\} \tag{2}$$

where i and j represent two arbitrarily chosen electrons.

This means that if we consider two sets of values of the arguments of ψ which differ only in that two electrons have interchanged their positions and the directions of their spin, the two values of ψ will differ only in sign. There may be solutions of equation (1) which do not satisfy this condition, but physical situations corresponding to such wave functions do not occur in nature *.

* One often uses the legal phrase "such solutions are *forbidden*".

We have formulated the exclusion principle in such a way that it fits in with our development of the mathematical foundations of quantum mechanics. That the exclusion principle is logically compatible with these foundations is due to the essential circumstance that *electrons are physically indistinguishable, equivalent entities.* All electrons have exactly the same charge, mass and spin properties. The formulation of general laws governing the motion of the electrons in an external field and the interaction of electrons with nuclei or with one another remains unchanged, when we change arbitrarily the originally chosen numbering of the electrons. This is the case both in classical and in quantum theory.

In order that this "exchangeability" of the electrons can be expressed in a simple way it is, of course, advisable to use the same coordinate systems to describe all the electrons. We made this choice explicitly before expressing the exclusion principle in the form (2). It is immaterial for our present discussion that it is sometimes advantageous to use coordinates which do not satisfy this condition.

The equations of motion in quantum mechanics are determined by the form of H_{op} in equation (1). If we use equivalent coordinates the exchangeability of the electrons is expressed by the fact that H_{op} *is symmetric in the coordinates of the electrons.* This means that the "form" of the energy operator does not change if we interchange the indices which refer to the numbering of a pair of electrons. Consider, for instance, electrons 1 and 2 and take the general expression for the energy derived in the previous chapter [eqq. (6.32) and (6.33)] *. If in this expression for the energy we replace $\boldsymbol{x}_1$ by $\boldsymbol{x}_2$, $\boldsymbol{x}_2$ by $\boldsymbol{x}_1$, ∇_1 by ∇_2, ∇_2 by ∇_1, $\boldsymbol{S}_1$ by $\boldsymbol{S}_2$, and $\boldsymbol{S}_2$ by $\boldsymbol{S}_1$, the operator remains unchanged, since $e_1 = e_2$ and $m_1 = m_2$ **. This symmetry is, of course, based on the fact that also in the classical Hamiltonian the electronic arguments occurred symmetrically, even when spin was taken into account [compare eqq. (3.76) and (6.33)].

* If we neglect spin and relativistic details, the energy operator of a system of n electrons in an external elective field with a potential $\Phi(\boldsymbol{x})$ is given by the equation [compare, for instance, eq. (3.26)], $H_{op} = -(\hbar^2/2m)\Sigma_i \nabla_i^2 - e\Sigma_i \Phi(\boldsymbol{x}_i) + \Sigma_{i<j}(e^2/r_{ij})$. The symmetry of H_{op}, that is, its invariance under a change in the numbering of the electrons, is immediately clear. This symmetry is also retained in the more general case which includes spin.

** We have not yet taken into account the possible presence of nuclei. It is, however, clear that the interaction of an electron with a nucleus will correspond to additional terms in the energy operator which are the same for all electrons, because all electrons have the same charge and mass.

We can define the property of a linear operator acting on functions of two variables q_1 and q_2 to be symmetric in these variables in the following way. If $g(q_1, q_2) = \Omega f(q_1, q_2)$, and if we denote by a prime the function of q_1 and q_2 which is obtained from the original function by exchanging q_1 and q_2, that is, $f'(q_1, q_2) = [f(q_1, q_2)]' = f(q_2, q_1)$, the following relation will hold, $g' = \Omega' f'$, where Ω' is the transformed operator which also acts on functions of q_1 and q_2 (see § 34). The operator Ω is "symmetric" in q_1 and q_2, that is, invariant under their exchange, if Ω and Ω' are identical,

$$\Omega = \Omega', \quad \text{or,} \quad (\Omega f)' = \Omega f'. \tag{3}$$

If we are dealing with two electrons, we have the variables $\boldsymbol{x}_1$, s_1 and $\boldsymbol{x}_2$, s_2 instead of q_1 and q_2. The exchange is now the one where $\boldsymbol{x}_1 \rightleftarrows \boldsymbol{x}_2$, $s_1 \rightleftarrows s_2$. One sees immediately that the energy operator of a many electron problem satisfies also the conditions of symmetry defined in this way.

The Schrödinger equation (1) enables us, at any rate in principle, to calculate ψ at any time, once ψ is given at one particular moment. If we choose at $t = 0$ ψ to be equal to a function ψ_0 which is antisymmetric with respect to the exchange of the arguments of two electrons, $(\partial\psi/\partial t)_{t=0}$ will also be antisymmetric, because H_{op} is symmetric, since $i\hbar(\partial\psi/\partial t)' = (H_{\text{op}}\,\psi)' = H_{\text{op}}\,\psi' = H_{\text{op}}(-\psi) = -i\hbar(\partial\psi/\partial t)$, where the primes now indicate the exchange of the arguments of an arbitrarily chosen pair of electrons. It follows that ψ will remain antisymmetric during a small interval of time $(0, \Delta t)$. By repeating the argument we find that ψ will remain antisymmetric, if it were antisymmetric once. We have thus shown the existence of antisymmetric solutions of equation (1). They all correspond to the possibility of constructing at a given time an antisymmetric function ψ.

The logical *possibility* of introducing the exclusion principle has thus been proven. The ideas of the (non-relativistic) wave mechanics do not give us any a priori arguments for the *necessity* of this principle. For the time being we shall accept it as correct because it enables us to interpret physical phenomena.

In particular it seems impossible to derive from classical physics, using the correspondence principle, arguments in favour of the preference in nature of antisymmetric wave functions. Consider a situation which approximately corresponds to a classical state, that is, a situation where the particles possess as well defined positions and momenta as is compatible with the Heisenberg relations. If such a situation is described by a wave function ψ of the kind discussed in § 27, the way ψ changes will correspond approximately to the classical laws of motion of the particles. This exhausts the range of applicability of the correspondence principle. Any function ψ' which is obtained from ψ by interchanging the coordinates of two, or more, electrons will illustrate just as well the correspondence with the classical motion, and so will any linear combination of those functions ψ'. There exists one linear combination which satisfies the exclusion principle, but nothing distinguishes it from the other ones. This state of affairs is underlined by the fact that one cannot use correspondence arguments with respect to the electron spin (see § 65), while the spin coordinates have to be taken into account on the same basis as the spatial coordinates for the formulation of the exclusion principle.

One consequence of the exclusion principle is the fact that the wave function is zero for those values of its arguments where $\boldsymbol{x}_i = \boldsymbol{x}_j$, $s_i = s_j$, since ψ should be equal to $-\psi'$ because of the exclusion principle and is equal to $+\psi'$ because of the relations between the arguments. If we remember that ψ plays the role of a probability amplitude, we see that this means that one can never find two electrons with their spins in the same direction at the same point in space.

It is tempting to see here a connexion with the fact that the Coulomb force between two electrons becomes infinite when they approach each other. One might imagine that the character of the interaction between two electrons, paying due regard to their spin properties, is such that among the solutions of the exact Schrödinger equation the antisymmetric solutions are the only possible ones, or at any rate the only ones which remain regular for all values of t. Investigations of the relativistic many particle problem give us, however, no reasons to believe that this interpretation of the exclusion principle is correct.*

§ 68. Exclusion principles for other equivalent particles

The mathematical counterpart of the antisymmetric solutions of a Schrödinger equation with a symmetric energy operator is the set of solutions symmetric in the arguments of all the electrons. These solutions of equation (1) are the ones for which ψ does not

* It is possible to construct an ad hoc energy operator, which is not relativistically invariant, but which for not too small distances apart of the electrons corresponds to the interaction found from the theory of electrons and which for vanishing distances apart becomes singular in such a way that the only solutions possible are the antisymmetric ones (see G. Jaffé, *Z. Phys.*, **66**, 748, 1930).

change its value if the arguments corresponding to two electrons are interchanged, that is, for which

$$\psi(\ldots; \boldsymbol{x}_i, s_i; \ldots; \boldsymbol{x}_j, s_j; \ldots) = \psi(\ldots; \boldsymbol{x}_j, s_j; \ldots; \boldsymbol{x}_i, s_i; \ldots) \qquad (4)$$

That such solutions exist is proven in the same way as we proved the existence of antisymmetric solutions. If ψ is symmetric at $t = 0$, it will be symmetric at all times.

If there are only two electrons, any solution ψ of equation (1) can be written as the sum of a symmetric and an antisymmetric solution, since the function ψ' which is obtained from ψ by interchanging the arguments corresponding to the two electrons satisfies the Schrödinger equation, $i\hbar\partial\psi'/\partial t = i\hbar(\partial\psi/\partial t)' = (H_{\text{op}}\psi)' = H_{\text{op}}\psi'$ and ψ can be written as the sum of $\frac{1}{2}(\psi+\psi')$ (symmetric) and $\frac{1}{2}(\psi-\psi')$ (antisymmetric). If there are more than two electrons, there are other solutions than the symmetric and the antisymmetric ones. These solutions possess different "symmetry properties" and from them and the symmetric and anti-symmetric solutions one can construct the most general solutions. We do not wish to discuss these other solutions *.

In the case of electrons the symmetric solutions are never realised, but they play an important role in the case of *other physically indistinguishable particles* occurring in nature. In general such particles need not be elementary particles such as the electron, positron, proton and neutron. Nuclei with atomic number 2 or larger are composite structures; indeed, the idea that they are built up out of protons and neutrons is essential in nuclear theory. As we mentioned earlier (§ 60) their detailed structure is often unimportant and nuclei behave in their interaction with other particles as point particles characterised, as the electrons are, by their mass, charge and spin properties. In particular we can describe their behaviour by means of a wave function which depends only on their spatial coordinates $\boldsymbol{X}$ and their spin coordinate s. The latter may take on $2i+1$ different values, $s = i, i-1, \ldots, -i$, corresponding to the $2i+1$ different possible values of the component of its intrinsic angular momentum in a given direction. The *angular momentum quantum number* or *spin quantum number* i is either a (non-negative) integer or a half-integer.

Nuclei of the same kind also behave as equivalent particles and the energy operator of the system is symmetric in the coordinates of

* See E. P. Wigner, *Z. Phys.*, **40**, 492, 883, 1927.

those particles. It has now been found empirically that for any kind of such equivalent particles there exists an exclusion principle. *If the spin quantum number is half integral* ($i = \frac{1}{2}, \frac{3}{2}, \ldots$) *the particles satisfy Pauli's exclusion principle,* that is, the wave functions are antisymmetric with respect to the corresponding coordinates. Examples of this kind of particles are electrons, positrons, neutrons, and protons ($i = \frac{1}{2}$ in all those cases). *If, however, the spin quantum number is an integer* ($i = 0, 1, 2, \ldots$) *the physical situations occurring in nature correspond to wave functions which are symmetric with respect to the corresponding coordinates.*

The empirical arguments for the validity of these rules * are manifold and convincing in the case of electrons and protons (for instance, the properties of molecular hydrogen **). There are no experimental reasons to doubt their correctness in other cases — intensities of molecular spectra and scattering experiments support them — while there are strong theoretical reasons that they are valid***.

For certain problems — for instance, the equation of state of gases — the interaction between atoms or molecules, that is, between entities of much larger dimensions, is such that their internal structure does not play a role. In this case also it is possible to work formally with point particles each being described by their spatial and spin coordinates and to use the above rules for the wave functions.

There is another way to state these rules. *Particles with half integer angular momentum quantum numbers are said to obey Fermi-Dirac statistics, and those with integer angular momentum quantum number Bose-Einstein statistics.* This nomenclature refers to the thermodynamic behaviour of "perfect gases" of equivalent particles ****. We do not wish to enter into statistical considerations, but may mention here that they plan an essential part in the analysis of intensities of molecular spectra.

As far as the better established parts of quantum mechanics are concerned, the exclusion rules are logically consistent. If we assume that a composite particle is always built up out of elementary particles with spin quantum number $\frac{1}{2}$—for instance, nuclei out of protons and neutrons—one would expect that the spin quantum number will be integer or half integer according to whether the number n of constituent elementary particles is even or odd. This is a consequence of the theory

* W. Heisenberg, *Z. Phys.*, **41**, 239, 1927; F. Hund, *Z. Phys.*, **42**, 93, 1927.

** D. M. Dennison, *Proc. Roy. Soc.*, **A115**, 483, 1927.

*** See, for instance, W. Pauli and F. J. Belinfante, *Physica*, **7**, 177, 1940.

**** See, for instance, D. ter Haar, *Elements of statistical mechanics*, Rinehart, New York, 1954.

which has been well borne out by experimental data in the case of the electron cloud surrounding the nucleus, but all experimental data about nuclei also agree with this idea. If now for $i = \frac{1}{2}$ Fermi-Dirac statistics holds, we see that an exchange of two equivalent particles leads to the simultaneous exchange of n pairs of equivalent elementary particles and thus, according to the Pauli principle, to the multiplication of the wave function by $(-1)^n$. The wave function would thus be symmetric or antisymmetric according to whether n is even or odd, or according to whether i is integral or half-integral in accordance with the exclusion rules. This argument is not rigorous * but contains some grains of truth.**

If a system contains several kinds of equivalent particles, the corresponding wave function must satisfy the exclusion rules for each of the kinds. The simplest example is the hydrogen molecule containing two electrons and two protons. The complete wave function depends on $2 \cdot 4+2 \cdot 4 = 16$ coordinates and must change its sign if either the arguments corresponding to the two protons or those corresponding to the two electrons are interchanged.

§ 69. Permutations

By successively interchanging two numbers out of a sequence, one can obtain every *permutation* of these numbers. The sequence

$$1, 2, \ldots, i, \ldots, N \tag{I}$$

goes over into

$$n_1, n_2, \ldots, n_i, \ldots, n_N, \tag{II}$$

where in the last sequence every number from the first sequence occurs once, but only once. The total number of different permutations is $N!$. A permutation is called even or odd according to whether the number of pair interchanges which lead to it is even or odd. If we denote a permutation by P we can introduce a quantity δ_{P} which is $+1$ or -1 according to whether P is even or odd. If a permutation is characterised by its cycles, and if there are N' cycles, δ_{P} is given by the equation $\delta_{\mathrm{P}} = (-1)^{N+N'}$.

A cycle $(a_1a_2 \ldots a_M)$ means that where in the sequence (I) the number a_1 (or a_2, $a_3, \ldots, a_{M-1}, a_M$) occurs the number a_2 (or $a_3, a_4, \ldots, a_M, a_1$) occurs in the sequence (II). Consider, for instance, the sequence 1, 2, 3, 4, 5, 6, 7. The permutation corresponding to the cycles (1532)(47)(6)—or (47)(1532)(6), since the order of the cycles is immaterial—is the permutation 5127364. In this case $N = 7$, $N' = 3$ so that $\delta_{\mathrm{P}} = +1$. This permutation can be obtained by the consecutive interchange of four pairs.

The cycles are usually arranged in such a way that the lengths of the cycles form a non-increasing sequence. A permutation is then characterised by the lengths of

* Especially one must investigate the connexion between the wave function as a function of the coordinates of the elementary particles and the wave function as a function of the coordinates of the composite particles treated as separate entities.

** Compare, J. R. Oppenheimer and P. Ehrenfest, *Phys. Rev.*, **37**, 333, 1931.

the consecutive cycles, $M_1, M_2, \ldots, (M_{j+1} \leq M_j)$, or, by a *partitio* of the number N, since $N = M_1+M_2+\ldots$ $(M_{j+1} \leq M_j)$. Permutations with the same partitio form a *class* of permutations.

If in a function f of N arguments $q_1, \ldots, q_N$ the argument q_1 is replaced by q_{n_1}, q_2 by $q_{n_2}, \ldots$ according to a permutation P of the numbers 1 to N, we get a function f' of the same arguments. We express this by the following equation

$$f'(q_1, q_2, \ldots, q_N) = f(q_{n_1}, q_{n_2}, \ldots, q_{nN}) = \mathrm{P}f(q_1, q_2, \ldots, q_N). \quad (5)$$

We can consider P to be a *real linear operator*. The corresponding operator function [see eq. (3.65)] can be expressed in terms of Dirac's δ-function as follows, $\delta(q_{n_1}-q_1')\delta(q_{n_2}-q_2')\ldots\delta(q_{n'N}-q_N')$. In the special case where the sequences (I) and (II) are the same, so that P is the identical permutation, P is simply the unit operator.

The product of two permutations $\mathrm{P}_2\mathrm{P}_1$ is clearly again a permutation P_3 and the following relation holds,

$$\delta_{\mathrm{P}_2\mathrm{P}_1} = \delta_{\mathrm{P}_2} \cdot \delta_{\mathrm{P}_1}. \quad (6)$$

In general $\mathrm{P}_2\mathrm{P}_1$ will be different from $\mathrm{P}_1\mathrm{P}_2$.

A P-operator is in general not Hermitean since the adjoint operator $\mathrm{P}^\dagger$, defined by the equation $\int f\mathrm{P}g d\tau = \int g\mathrm{P}^\dagger f d\tau$, is just that permutation which applied after P restores the original sequence, that is, $\mathrm{P}^\dagger\mathrm{P} = 1$, or, $\mathrm{P}^\dagger = \mathrm{P}^{-1}$. The permutation P^{-1} is obtained from P by reversing in each cycle the order of the numbers. That is, if P corresponds to the cycles $(a_1a_2\ldots a_{M_1})(a_{M_1+1}a_{M_1+2}\ldots a_{M_1+M_2})$ $\ldots$, P^{-1} will correspond to $(a_{M_1}\ldots a_2a_1)(a_{M_1+M_2}\ldots a_{M_1+2}a_{M_1+1})\ldots$ We see that P is only Hermitean if all cycles have the length 2 or 1, since in that case $\mathrm{P}^{-1} = \mathrm{P}$. One can, of course, construct two Hermitean operators $\mathrm{P}^{(+)} = \frac{1}{2}(\mathrm{P}+\mathrm{P}^{-1})$ and $\mathrm{P}^{(-)} = \frac{1}{2}(\mathrm{P}-\mathrm{P}^{-1})$ from P and P^{-1} (see § 29). One sees easily that $\mathrm{P}^{(+)}$ and $\mathrm{P}^{(-)}$ commute with one another.

Consider now an atomic system containing N equivalent particles. If in equation (5) q_i denotes the coordinates $\boldsymbol{x}_i, s_i$ of the i-th particle and if we understand by the replacement of q_i by q_{n_i} the replacement of $\boldsymbol{x}_i, s_i$ by $\boldsymbol{x}_{n_i}, s_{n_i}$, then we can use equation (5) to discuss this N-particle problem.

The symmetry op H_{op} in equation (1) means that [compare eq. (3)]

$$\mathrm{P}(H_{\mathrm{op}}f) = H_{\mathrm{op}}(\mathrm{P}f), \quad \text{or,} \quad \mathrm{P}H_{\mathrm{op}} - H_{\mathrm{op}}\mathrm{P} = 0, \quad (7)$$

which means that the energy operator commutes with all permutation operators. The permutations are thus integrals of the quantum mechanical problem, when we use the definition given in § **31**. There are, however, no observables in the ordinary sense which correspond to them, as they cannot be written as functions of the coordinates, momenta and spins of the particles *.

From equations (7) and (1) it follows that $i\hbar\partial \mathrm{P}\psi/\partial t = i\hbar \mathrm{P}\partial\psi/\partial t = \mathrm{P}H_{\mathrm{op}}\psi = H_{\mathrm{op}}(\mathrm{P}\psi)$, and we see that if ψ is a solution of equation (1) so are all the $\mathrm{P}\psi$ solutions. Every linear combination of the $\mathrm{P}\psi$ with arbitrary coefficients c_{P} will thus also be a solution,

$$\psi' = \Sigma_{\mathrm{P}}\, c_{\mathrm{P}}\, \mathrm{P}\psi, \tag{8}$$

where the summation extends over all $N!$ permutations.

The antisymmetric solutions (2) are clearly characterised by the relation

$$\mathrm{P}\psi = \delta_{\mathrm{P}}\psi, \tag{9}$$

and the symmetric solutions (4) by

$$\mathrm{P}\psi = \psi, \tag{10}$$

where in both equation (9) and equation (10) P is any of the $N!$ permutations.

We can always construct an antisymmetric and a symmetric solution from an arbitrary solution ψ of equation (1) as follows,

$$\psi_{\mathrm{antis}} = \text{constant.}\ \Sigma_{\mathrm{P}}\, \delta_{\mathrm{P}}\, \mathrm{P}\psi, \tag{11}$$

$$\psi_{\mathrm{symm}} = \text{constant.}\ \Sigma_{\mathrm{P}}\, \mathrm{P}\psi. \tag{12}$$

Apart from a possible multiplying factor, expressions (11) and (12) are, indeed, the only antisymmetric and symmetric functions contained among the functions (8).

If we fix P′ and let P go through all $N!$ permutations, $\mathrm{P}'' = \mathrm{P}'\mathrm{P}$ will also go through all $N!$ permutations, albeit in a different order. The antisymmetry of the function (11) is then proved as follows [compare eq. (6)], $\mathrm{P}'\Sigma_{\mathrm{P}}\,\delta_{\mathrm{P}}\,\mathrm{P}\psi = \Sigma_{\mathrm{P}}\,\delta_{\mathrm{P}}\,\mathrm{P}'\mathrm{P}\psi = \Sigma_{\mathrm{P}''}\,\delta_{\mathrm{P}'}\,\delta_{\mathrm{P}''}\,\mathrm{P}''\psi = \delta_{\mathrm{P}'}\Sigma_{\mathrm{P}}\,\delta_{\mathrm{P}}\,\mathrm{P}\psi$.

If $\psi' = \Sigma_{\mathrm{P}}\, b_{\mathrm{P}}\, \delta_{\mathrm{P}}\, \mathrm{P}\psi$ were also antisymmetric, we have

$$\begin{aligned}\psi' &= \delta_{\mathrm{P}'}\,\Sigma_{\mathrm{P}}\, b_{\mathrm{P}}\, \delta_{\mathrm{P}}\, \mathrm{P}'\mathrm{P}\psi = \Sigma_{\mathrm{P}''}\, b_{\mathrm{P}'^{-1}\mathrm{P}''}\, \delta_{\mathrm{P}''}\, \mathrm{P}''\psi \\ &= \Sigma_{\mathrm{P}}\, b_{\mathrm{P}'^{-1}}\, \delta_{\mathrm{P}}\, \mathrm{P}\psi = \Sigma_{\mathrm{P}}\, [(\Sigma_{\mathrm{P}'}\, b_{\mathrm{P}'^{-1}\mathrm{P}})/N!]\delta_{\mathrm{P}}\, \mathrm{P}\psi \\ &= (\Sigma_{\mathrm{P}}\, b_{\mathrm{P}}/N!)\, \Sigma_{\mathrm{P}}\, \delta_{\mathrm{P}}\, \mathrm{P}\psi.\end{aligned}$$

Either $\Sigma_{\mathrm{P}}\, b_{\mathrm{P}} \neq 0$ and ψ' is essentially the same as function (11), or $\Sigma_{\mathrm{P}}\, b_{\mathrm{P}} = 0$, that means that $\psi' = 0$. There is thus only one ψ_{antis}, but expression (11) can have different forms, when the $\mathrm{P}\psi$ are linearly dependent.

The proof in the symmetric case is completely similar.

* In this connexion we may remind ourselves of the close connexion between integrals of motion and infinitesimal transformations. There is, of course, no infinitesimal transformation corresponding to the invariance of the Hamiltonian under permutations.

It is, of course, possible that either expression (11) or expression (12) (or both) is identically equal to zero. Expression (11) will, for instance, be zero, if φ itself is symmetric.

§ 70. Stationary states of several independent electrons in a common field of force; the shell structure of the atom

We discussed in § 32 the problem of the stationary states of several independent particles. We shall now discuss the particular case of N electrons in a common, constant field of force. The coordinates $\boldsymbol{x}_i$, s_i of the i-th electron are together denoted by q_i and the q's are defined in the same way for all the electrons (see the discussion in § 67). The energy operator of the system is of the form [compare eq. (3.106)],

$$H_{\text{op}} = \Sigma_{i=1}^{N} G_{\text{op}}(q_i), \tag{13}$$

where $G_{\text{op}}\,(q_i)$ is a time independent energy operator acting on the coordinates of the i-th electron only and having the same form for all i. For the sake of simplification we shall assume that $G_{\text{op}}(q)$ possesses a discrete eigenvalue spectrum. Let $\varphi_k(q)$ be a complete set of normalised eigenfunctions of G_{op} with E_k the corresponding eigenvalues. Here k stands for one or for a group of quantum numbers. A complete system of eigenvalues and eigenfunctions of the time independent Schrödinger equation of the complete system,

$$(H_{\text{op}} - E)\,\varphi = 0 \tag{14}$$

is given by

$$\varphi = \Pi_i\, \varphi_{ki}(q_i), \qquad E = \Sigma_i\, E_{ki}, \tag{15}$$

where each k_i can take on all possible values of the quantum number(s) of the one particle problem. Equations (15) correspond to a situation where the first electron is in the quantum state k_1, the second electron in the state k_2, . . .

This energy of a stationary state (15) depends clearly only on *which* quantum numbers $k_1, k_2, \ldots, k_N$ are represented, but not in the order in which they are assigned to the various electrons. This means that $\mathrm{P}\varphi$—P a permutation of the electrons — or in general any linear combination of the $\mathrm{P}\varphi$ [compare eq. (8)] will be a solution of equation (14). If we wish to interpret the physical situation corresponding to such a linear combination, it has clearly no longer a meaning to say that the first electron is in this, the second electron in that, state of the one-electron problem, and so on. This would be true for a φ-function of the form (15) or for any of the $P\varphi$, but now we are dealing with a superposition, a kind of mixture of all these situations. It still remains true to say that there are N electrons which are distributed over the stationary states $k_1, \ldots, k_N$.

Among the linear combinations of the $P\varphi$ there is according to the considerations of the previous section at most one which satisfies the Pauli principle, namely the one given by expression (11),

$$\Phi_{\text{antis}} = \text{constant.}\ \Sigma_P\, \delta_P\, P\varphi. \tag{16}$$

One sees immediately that expression (16) is identically equal to zero as soon as two or more (groups of) quantum numbers from the set $k_1, \ldots, k_N$ are the same. If we are dealing with two electrons, expression (16) is of the form $\Phi = \varphi_{k1}(q_1)\, \varphi_{k2}(q_2) - \varphi_{k1}(q_2)\varphi_{k2}(q_1)$ and this expression is equal to zero as soon as k_1 and k_2 are the same, that is, when φ_{k1} and φ_{k2} are the same function. In the case of more than two particles one sees easily that the terms in expression (16) cancel each other in pairs, when $k_i = k_j$. We can put this as follows. If we perform a permutation P' of the k_i, expression (16) is multiplied by $\delta_{P'}$. If k_i and k_j are interchanged, expression (15) for φ, and thus also expression (16) remains unchanged, if $k_i = k_j$. Since $\delta_{P'} = -1$ in this case, we see that $\Phi = -\Phi = 0$. *According to the Pauli principle situations where two or more electrons are in the same stationary state of the one electron problem will never occur.*

We have now reached the historical starting point out of which the present-day ideas about the exclusion principle have gradually developed. By using the models of the old quantum theory Bohr * in particular developed the idea that the state of each electron in an atom could approximately be described as a stationary state in a central field of force. This central field described the total action of both the nucleus and the other electrons on the electron in question. It will not be too far removed from reality to assume this field to be the same for all the electrons.

The energies of the stationary states are determined by the values of the principal quantum number n ($n = 1, 2, 3, \ldots$) [see the discussion in § 59] and the azimuthal quantum number l ($l = 0, 1, 2, \ldots, n-1$)**. The stationary state of an atom with N electrons is to a first approximation characterised by the numbers of electrons

* N. Bohr, *Z. Phys.*, **9**, 1, 1922.

* Bohr uses the quantum number k which is equal to $l+1$, ($k = 1, 2, \ldots, n$).

which are in the various states of given n and l. From chemical and spectroscopic data it followed that there could be at most $2n^2$ electrons in a state of principal quantum number n. Electrons with the same principal quantum number move within a spherical region around the nucleus and form the so-called n-shell. In general the radius of the n-th shell increases with increasing n corresponding to the increase of the orbital dimensions in a Coulomb field (see § 46), while the energy needed to remove an electron from the n-th shell decreases in general with increasing n similar to expression (4.192). Because of this the shells with the lowest principal quantum numbers are in general *fully occupied* in the ground state of the atom, which means two electrons in the first or K-shell, eight in the second or L-shell, eighteen in the third or M-shell, . . . Exceptions occur only for the two or three largest of the principal quantum numbers which occur. These exceptions occur for the transition elements in the periodic system and are explained by Bohr as being due to "penetrating orbits" *.

The final answer to the problem of distribution of the electrons in a completely filled shell over the possible values of l ($l = 0, 1, 2, \ldots, n-1$) was given by Main Smith and Stoner **: *In a completely filled shell there are $2(2l+1)$ electrons with quantum numbers n, l.* From an analysis of the spectroscopic data this statement was confirmed and even strengthened since it appeared that there could never be more than $2(2l+1)$ electrons with the same values of n and l in one atom. An important role in Stoner's considerations was played by the formal analogy between the electronic states in a completely filled shell and the states, obtained from spectroscopic data, of the most loosely bound electron in an alkali atom in a magnetic field. Pauli*** went a step further in considering this analogy. He made the assumption that the stationary state of an electron in a central field of force is characterised by four quantum numbers n, l, j, m_j**** ($j = l \pm \frac{1}{2}$, except for $l = 0$, when $j = \frac{1}{2}$; $m_j = j, j-1, j-2, \ldots, -j$) and introduced the principle that *two independent electrons in a common*

* We refer to H. Kuhn, *Hand. u. Jahrb. Chem. Phys.*, Vol. 9, part I, Akademische Verlag, Leipzig, 1934; L. Pauling and S. Goudsmit, *The structure of line spectra*, McGraw-Hill, New York, 1930.

** J. D. Main Smith, *Chemistry and atomic structure*, Benn, London, 1924; E. C. Stoner, *Phil. Mag.*, **48**, 719, 1924.

*** W. Pauli, *Z. Phys.*, **31**, 765, 1925.

**** These quantum numbers entered into the discussion in § 59 of the spinning electron in a central field of force.

central field of force can never occur in states with the same values of all four quantum numbers. In this way he was able to explain the maximum number $2(2l+1)$ of electrons in a state of given values of n and l *. It is true that the physical meaning of the quantum number m_j and even of the quantum number j is not very clear because of the interaction between the electrons which is in fact present. Pauli, however, was able to show in how far one could still interpret physically the number of electrons in the various states.

The interpretation of Pauli's four quantum numbers in the vector model using Uhlenbeck and Goudsmit's hypothesis of the spinning electron (see §§ 56 and 59) rounded off this line of development in the framework of the old quantum theory. Dirac and Heisenberg ** were the first to use the wave mechanical formulation of the exclusion principle which we discussed in § 67.

If the permutation P changes the number i into n_i, the explicit expression for $\mathrm{P}\varphi$ is [compare eq. (15)] $\mathrm{P}\varphi = \Pi_i \varphi_{ki}(q_{n_i})$ and we see that expression (16) can be written in the form of a determinant,

$$\Phi_{k1,\ldots,kN} = (N!)^{-\frac{1}{2}} \begin{vmatrix} \varphi_{k1}(q_1) & \varphi_{k2}(q_1) & \cdots & \varphi_{kN}(q_1) \\ \varphi_{k1}(q_2) & \varphi_{k2}(q_2) & \cdots & \varphi_{kN}(q_2) \\ \cdots & \cdots & \cdots & \cdots \\ \varphi_{k1}(q_N) & \varphi_{k2}(q_N) & \cdots & \varphi_{kN}(q_N) \end{vmatrix} = (N!)^{-\frac{1}{2}} |\varphi_{kj}(q_i)|. \quad (17)$$

We have chosen the constant in equation (16) equal to $(N!)^{-\frac{1}{2}}$. If the φ_{ki} are normalised, Φ is now also normalised and Φ can be used as a probability amplitude,

$$\int \Phi^* \Phi = 1. \quad (18)$$

The integral sign indicates here integration over all Cartesian coordinates $\boldsymbol{x}_1, \ldots \boldsymbol{x}_N$ (density function $= 1$) as well as summation over all possible 2^N values of the spin coordinates $s_1, \ldots, s_N$. The correctness of equation (18) is based on the orthogonality of the φ_{k_i}. In Φ there are $N!$ terms $\mathrm{P}\varphi$ and hence in $\Phi^*\Phi$ $(N!)^2$ terms $(\mathrm{P}'\varphi)^*(\mathrm{P}\varphi)$. Because of the orthogonality of the φ_{k_i} $\int (\mathrm{P}'\varphi)^*(\mathrm{P}\varphi)$ is different from zero only when P′ and P are the same permutation and then this expression is equal to unity, if the φ_{k_i} are normalised.

From the form of equation (17) we see immediately that Φ is identically zero as soon as two functions φ_{k_i} and φ_{k_j} are the same.

* There are $2l+2$ states corresponding to $j = l+\frac{1}{2}$ and $2l$ corresponding to $j = l-\frac{1}{2}$, giving together $2(2l+1)$ states.

** P. A. M. Dirac, *Proc. Roy. Soc.*, **A112**, 661, 1926; W. Heisenberg, *Z. Phys.*, **40**, 501, 1926.

Slater * was one of the first authors to use expression (17) for many problems and this expression is therefore often called a *Slater determinant.*

§ 71. Quantum theory of N-electron systems

Let the $\varphi_k(q)$ be a complete orthonormal set of functions. Any antisymmetric function of the coordinates of the N electrons can now be expanded in terms of Slater determinants,

$$\Psi_{\text{antis}} = \Sigma_{[k_1, \ldots, k_N]} A_{[k_1, \ldots, k_N]} \Phi_{[k_1, \ldots, k_N]} = \Sigma_{[k_i]} A_{[k_i]} \Phi_{[k_i]} (i = 1, 2, \ldots, N),$$

where the summation is over all possible combinations of N quantum numbers (or groups of quantum numbers) in such a way that of all combinations which differ only by a permutation of the k_i only one representative is chosen. This prescription is indicated by the square brackets around the indices. In many cases $[k_i]$ is a useful abbreviation for $[k_1, \ldots, k_N]$. Of course, those combinations for which two (or more) k_i are the same are automatically excluded. This expansion is completely similar to the expansions of arbitrary functions in terms of a set of normalised functions (see § 36). The coefficients A are similarly determined,

$$A_{[k_i]} = \int \Phi^*_{[k_i]} \Psi. \tag{20}$$

The proof of equation (19) is simple if we assume that the general expansion theorem is proved (see § 32). We have then $\Psi = \Sigma A'_{k_1, k_2, \ldots, k_N} \varphi_{k_1}(q_1) \varphi_{k_2}(q_2) \ldots \varphi_{k_N}(q_N)$. In this expression all k_i can go through all possible values independently. The different products $\Pi \varphi_{k_i}(q_i)$ are linearly independent. If Ψ is antisymmetric, we have

$$\Psi = (N!)^{-1} \Sigma_{\mathrm{P}} \delta_{\mathrm{P}} \mathrm{P} \Psi = (N!)^{-1} \Sigma_{k_i} A_{k_i}' \Sigma_{\mathrm{P}} \delta_{\mathrm{P}} \mathrm{P} \Pi_i \varphi_{k_i}(q_i) = (N!)^{-1/2} \Sigma A'_{k_i} \Phi'_{k_i}.$$

Those terms which only differ by a permutation of the k_i can be combined into a representative $\Phi_{[k_i]}$ and we obtain equation (19). Equation (20) then follows from the orthonormality of the $\Phi_{[k_i]}$.

We can easily generalise equations (19) and (20) to include the case where the eigenfunctions $\varphi(q)$ which occur in the Slater determinants belong wholly or partly to a continuous spectrum. In addition to (or instead of) the discrete argument k of $\varphi_k(q)$ we must introduce a continuous parameter k or several such parameters [compare eq. (4.74)] and the summation in equation (19) is replaced by a process in which we sum over some arguments and integrate over the others. It is also possible that the process would be integration over a many dimensional region which, for instance, is

* J. C. Slater, *Phys. Rev.*, **34**, 1293, 1929.

bounded by the relations $k_1 \leqq k_2 \leqq k_3 \ldots \leqq k_N$. For the sake of simplicity we shall assume in the following that the k_i are always discrete arguments.

The coefficients $A_{[k_i]}$ can be considered to be a *transformed representation* of the physical situation of the N-electron problem characterised by Ψ. If Ψ satisfies the wave equation (1) so that the $A_{[k_i]}$ are time dependent, the transformed Schrödinger equation has the form (see § 42)

$$H'_{\text{op}} A_{[k_i]}(t) = i\hbar\, \partial A_{[k_i]}(t)/\partial t, \tag{21}$$

where H'_{op} is a matrix with the matrix elements [compare eq. (4.109)]

$$(H'_{\text{op}} A)_{[k_i]} = \Sigma_{[k'_i]} H_{[k_i][k'_i]} A_{[k'_i]};\ H_{[k_i][k'_i]} = \int \Phi^*_{[k_i]} H_{\text{op}} \Phi_{[k'_i]}. \tag{22}$$

Any other representation $\bar{A}$ of the same situation by means of Slater determinants $\overline{\Phi}_{[\bar{k}_i]}$, which are constructed by using a different orthonormal set $\bar{\varphi}_{\bar{k}}$, corresponds to a Schrödinger equation of the type (21) which could have been obtained by transforming equation (21), $\overline{H}'_{\text{op}} \bar{A}_{[\bar{k}_i]}(t) = i\hbar \partial \bar{A}_{[\bar{k}_i]}(t)/\partial t$. The transformation function which would lead to this transformation is a unitary matrix which can be obtained from the considerations of § 41 [see eqq. (4.80), (4.81), (4.83), (4.96) and (4.97)],

$$\bar{A}_{[\bar{k}_i]} = \Sigma_{[\bar{k}_i]} U^{[\bar{k}_i]}_{[k_i]} A_{[\bar{k}_i]};\ \ H'_{[\bar{k}_i][\bar{k}'_i]} = \Sigma_{[\bar{k}_i],[\bar{k}'_i]} U^{[\bar{k}_i]}_{[k_i]} H_{[k_i][k'_i]} U^{[\bar{k}'_i]}_{[k'_i]}; \tag{23}$$

$$U^{[\bar{k}_i]}_{[k_i]} = \int \overline{\Phi}^*_{[\bar{k}_i]} \Phi_{[k_i]};\ \overline{\Phi}_{[\bar{k}_i]} = \Sigma_{[k_i]} \Phi_{[k_i]} U^{[\bar{k}_i]*}_{[k_i]}. \tag{24}$$

Under a transformation of the type (23) the symmetry properties of the wave functions will remain invariant, which means that the transformed function will also satisfy the Pauli principle. One could even choose for $U^{[\bar{k}_i]}_{[k_i]}$ an *arbitrary* unitary matrix in which case the $\overline{\Phi}$ which correspond to the new $\bar{A}$-representation are no longer Slater determinants, (although the $\overline{\Phi}$ are still antisymmetric!), and $[\bar{k}_i]$ is a symbol for some group of quantum numbers (whose permutations are no longer considered).

In the discussion of physical experiments, that is, in applications of the Schrödinger equation, one is in principle always dealing with probability statements about the values of observables which are measured by the apparatus under consideration (see §§ 35, 39, 43, and 52). Because the electrons are indistinguishable the operations corresponding to these observables will always be symmetric with respect to all electrons in the system.

It is true that the apparatus itself will contain electrons and it looks as if we should have to take these into account. However, the idea of measurement is such that we study the interaction between a given system and a piece of apparatus and assume that to describe this interaction we need only a few uniquely measureable parameters of the apparatus which, moreover, are defined in classical terms (see § 52). This means that a certain amount of "idealisation" enters in any planning or interpreting of an experiment. Such an idealisation leads us to neglect the atomistic structure of the apparatus. Of course, we must require from such an idealisation that it will be compatible with considerations where the structure of the apparatus is explicitly introduced. It is, in fact, not difficult to show that the idealisation to a system with only N electrons, namely those of the system under consideration, is completely justified in this respect.

The probability statements referred to a moment ago are obtained by using the probability amplitudes of the given situation with respect to the observable F under consideration (§ 39). These amplitudes are the coefficients in the expansion of the wave function Ψ in terms of the eigenfunctions S_k of F_{op}

$$\Psi = \Sigma_l b_l S_l(q_1, \ldots, q_N), \quad (F_{\text{op}} - F_l) S_l(q_1, \ldots, q_N) = 0. \tag{25}$$

If F_{op} is symmetric and if S is one of its eigenfunctions, all PS will be eigenfunctions belonging to the same eigenvalue, since we have from equation (25) that $F_l \mathrm{P} S_l = \mathrm{P} F_{\text{op}} S_l = F_{\text{op}} \mathrm{P} S_l$.

If Ψ is antisymmetric, it follows from equation (25) that

$$\Psi = (N!)^{-1} \Sigma \delta_{\mathrm{P}} \mathrm{P} \Psi = (N!)^{-1} \Sigma_l b_l \Sigma_{\mathrm{P}} \delta_{\mathrm{P}} \mathrm{P} S_l. \tag{26}$$

The expression $\Sigma_{\mathrm{P}} \delta_{\mathrm{P}} \mathrm{P} S_l$ is an antisymmetric eigenfunction. *The antisymmetric wave function can thus always be expanded in terms of the antisymmetric eigenfunctions of a symmetric* F_{op}; *these eigenfunctions form in this respect a complete set which can always be chosen to be orthonormal.* The expansion in terms of the Slater determinants was a particular case of this theorem.

We may thus assume that the S_l in equation (25) are orthonormal antisymmetric eigenfunctions of a symmetric F_{op}. The coefficients in the expansion follow from $b_l = \int S_l^* \Psi$, provided the S_l are normalised. The $|b_l|^2$ are connected with the occurrence of the eigenvalues F_l The expectation value of F_{op} (see § 39) is given by the expression $\overline{F} = \Sigma_l |b_l|^2 F_l = \int \Psi^* F_{\text{op}} \Psi$.

Let us consider the special case where we try to find out whether or not at the time t an electron is present within a given region V of space (compare § 7). Let $V(\boldsymbol{x})$ be a function of the spatial coordinates which is equal to unity within V and equal to zero outside. Our experiment then determines the "value" of V. If there are N electrons, the experiment cannot determine "which" of the electrons was found in V, but only gives the result that within V no electrons were present, or one, or several. The number present is the value of $F = \sum_{i=1}^{N} V(\underset{\sim}{x}_i)$ Quantum mechanically speaking, F is a symmetric operator with eigenvalues $0, 1, \ldots, N$ and its expectation value gives us the probability of finding an electron within V, or rather gives us the average number of electrons within V*. This expectation value is given by the equation

$$\overline{F} = \int \Psi^* F_{\text{op}} \Psi = \Sigma_i \int \Psi^* V(\boldsymbol{x}_i) \Psi = N \int V(\boldsymbol{x}_1) |\Psi|^2. \quad (27)$$

In this sum, we have thus N terms $\int V(\boldsymbol{x}_i)|\Psi|^2$ which are the probabilities that the first, second, . . . electron is within V in accordance with the general interpretation of $|\Psi|^2$ as a probability density. As is indicated in the last equality of equation (27) these terms are all equal to each other. This is because a change in the numbering of the electrons will not change the value of the integral. If we renumber the electrons, Ψ goes over into $\delta_{\text{P}} \Psi$ and $|\Psi|^2$ remains the same, since $\delta_{\text{P}}^2 = 1$. The last integral in equation (27) goes therefore over into $\int V(\boldsymbol{x}_i)|\Psi|^2$ where now i can be any number from 1 to N. Every electron has the same chance of being found inside V. If V is the total space available to the electrons, $\overline{F}$ will be equal to N as is to be expected. A similar result would have been obtained, if Ψ has been a symmetric function.

This example clearly illustrates the indistinguishability of the electrons. We may point out a difference from the classical electron theory. In that theory electrons are also equivalent entities, but in the description of their motions it was in principle always possible to follow each electron separately along its own orbit. If we had had them "in our hands" once and had labelled them, the question of which electron was found inside V, say, at a later time would have a physical meaning. However, such a question can have a meaning, that is, it will be possible to give a unique experimental answer to it, only in such situations where the exclusion of all solutions of equation (1) which do not satisfy the Pauli principle does not play a role.

One often uses the expression *"the electrons exchange continually their position (and the direction of their spin)* in a manner which cannot be described classically". By this we mean the following. We may have observed the presence of one electron. This electron interacts afterwards with another electron and after this we observe again the presence of an electron. There is now a certain probability that this "is"

* The probability of finding at least one electron within V is clearly given by the expectation value of the symmetric operator $F' = 1 - \Pi_i[1 - V(\boldsymbol{x}_i)]$. In the limit where $V \to 0$ F and F' become the same. The case where exactly *one* electron is within V can also easily be treated.

still the same electron, but there is also a probability that we are actually observing the other electron. The stronger the interaction, the larger will be the probability per unit time that the exchange takes place. We do not wish to enter here into a more detailed discussion. We may mention that Bohr in his early considerations about the shell structure of the atom at times expressed the idea that it was necessary for the stability of the atom that there occur discontinuous processes, where two electrons exchange their roles (that is, for instance, their quantum numbers).

All problems concerning symmetric operators can be described in terms of the representation by means of coefficients $A_{[k_i]}$. These coefficients were particularly suited to represent situations corresponding to an antisymmetric Ψ and could not be used in the case of an arbitrary operator which would change the antisymmetrical character of Ψ. However, the symmetric operators conserve the antisymmetry.

The operator equation $\chi_{\text{antis}} = F_{\text{op}}^{(\text{symm})} \zeta_{\text{antis}}$ corresponds to the equation

$$B_{[k_i]} = \int \Phi^*_{[k_i]} F_{\text{op}} \Sigma_{[k'_i]} C_{[k'_i]} \Phi_{[k'_i]} = \Sigma_{[k'_i]} F_{[k_i][k'_i]} C_{[k'_i]},$$

if we have performed the transformation $C_{[k_i]} = \int \Phi^*_{[k_i]} \zeta$, $B_{[k_i]} = \int \Phi^*_{[k_i]} \chi$. The F-matrix is given by the equation [compare eq. (22)]

$$F_{[k_i][k'_i]} = \int \Phi^*_{[k_i]} F_{\text{op}} \Phi_{[k'_i]}. \tag{28}$$

The eigenfunctions $S^l_{[k_i]}$ of this matrix follow from the antisymmetric S_l of equation (25), $S_l = \Sigma \Phi_{[k_i]} S^l_{[k_i]}$, or, $S^l_{[k_i]} = \int \Phi^*_{[k_i]} S_l(q_1, \ldots, q_N)$. They satisfy the eigenvalue equation $\Sigma_{[k'_i]} F_{[k_i][k'_i]} S^l_{[k'_i]} = F_l S^l_{[k_i]}$. The expansion (25) corresponds to the equations $A_{[k_i]} = \Sigma_l b_l S^l_{[k_i]}$ and $b_l = \Sigma_{[k_i]} S^{l*}_{[k_i]} A_{[k_i]}$.

The mathematical formalism of quantum mechanics can thus in full force be applied to a restricted class of problems in which we deal only with symmetric operators and with solutions which satisfy the Pauli principle. One works with a $A_{[k_i]}$-representation or with a representation obtained from the $A_{[k_i]}$ by an unitary transformation and one satisfies automatically the requirements of the exclusion principle. (One sees easily that the same would be true if one considered only symmetric wave functions; compare the considerations of the next section). There is one difference from the case of arbitrary operators and solutions. In the case of the general quantum mechanical problem, one can always construct the operators corresponding to observables from a small number of basic operators, $\boldsymbol{x}_i$, $\boldsymbol{P}_i$, $\boldsymbol{S}_i$. One might ask now that all operators which are symmetric in these $9N$ variables could be simply constructed in a similar fashion from a finite number of symmetric basic operators. This is, however, not possible in a way which is simple, practical, and useful.

§ 72. FORMULATION OF THE MANY PARTICLE PROBLEM INDEPENDENT OF THE NUMBER OF PARTICLES

We consider again the coefficients $A_{[k_i]}$ defined by equation (19) which gives us a representation of the physical situation of the N-electron problem. This representation can be formulated as follows. The A form a function of infinitely many arguments $n_1, n_2, \ldots, n_k \ldots$ each of which can take on only two values, 0 and 1,

$$A_{[k_i]} \to A(n_1, \ldots, n_k, \ldots). \tag{29}$$

This means that the different possible values of the number (or group of numbers) k which characterise the different members of the basic orthonormal set φ_k (see the beginning of § 71) may be put in some sequence which is fixed once and for all. From now on k will indicate the number in this sequence. Furthermore, we will arrange it in such a way that $\Phi_{[k_i]}$ in equation (19) is always that Slater determinant (17) for which $k_1 < k_2 < \ldots < k_N$. This fixes the sign of the $A_{[k_i]}$. The arguments in equation (29) are defined in such a way that those n_k which correspond to a k contained in the set of numbers $k_1, \ldots, k_i, \ldots, k_N$ are equal to one, while the other n_k are equal to zero. A physical situation for which A is equal to zero except for one definite combination of the n_k, corresponds to the presence of a number of electrons distributed over those situation φ_k of the one-electron problem for which the corresponding n_k are equal to one.

This formulation of the situation is independent of the number of electrons present. We are used to having a fixed number N of electrons present and we are therefore inclined to assume that only those A-functions have a physical meaning for which A is different from zero only for those values of its arguments for which $\Sigma_1^\infty n_k = N$.

It is, however, easy to give examples where the number of electrons during an experiment is undetermined so that the functions $A(n_1, n_2, \ldots)$ are the natural means to describe the situation. This is, for instance, the case if a region of space containing a given number of electrons is divided by a fixed screen into two parts and if we consider the situation in one of the two parts (see, for instance, the discussion in § 7). Moreover, we know nowadays that even in a closed system the number of particles of a given kind is not necessarily constant (pair-annihilation or pair-creation; see the discussion at the end of § 65). The representation (29) gives us a possibility to apply the theory to annihilation and creation processes *.

If $(p)_k$ denotes that function of n_k which is equal to 1 for $n_k = p$ and is equal to zero for $n_k \neq p$, we can write equation (29) in the form

$$A(n_1, \ldots, n_k, \ldots) = \Sigma_{[k_i]} A_{[k_i]} \Pi_k (\Sigma_i \delta_{k k_i})_k, \tag{30}$$

where δ_{kk_i} is again the Weierstrass δ-function.

The discussion of the physical properties of a many electron system is connected with those operators which correspond to observables. These can easily be expressed in the $(n'_1, \ldots, n'_k, \ldots)$ representation. We write for such an operator [compare eq. (28)]

$$F_{[k_i][k'_i]} \rightarrow F_{n_1, \ldots, n_k, \ldots; n'_1, \ldots, n'_k, \ldots}. \tag{31}$$

Each term of the last matrix corresponds to a transition from the situation $n_1, \ldots, n_k, \ldots$ to the situation $n'_1, \ldots, n'_k, \ldots$ Up to now we have only met with such operators where Σn_k is the same for the initial and the final situation. More general operators can easily be conceived, but they have as yet no physical meaning.

An arbitrary operator of the type (31) can always be constructed from certain simple basic operators. As far as the mathematics is concerned we are dealing with a situation similar to the one where we dealt with operators acting on functions F of a number of spin variables $s_1, \ldots, s_k$ ($s_i = \pm\frac{1}{2}$). Each variable can here also take on two values and the most general operator can be written as a multilinear expression in the Pauli matrices corresponding to each of the variables (see § 60). Our F-operators can thus be expressed as a multilinear expression in basic operators, of which there are three operators corresponding to each value of k. We do not use here the Pauli-matrices, but rather the following operators,

$$U_k \rightarrow \begin{array}{c|cc} n_k \backslash n'_k & 1 & 0 \\ \hline 1 & 0 & 0 \\ 0 & 1 & 0 \end{array}, \quad V_k \rightarrow \begin{array}{c|cc} n_k \backslash n'_k & 1 & 0 \\ \hline 1 & 0 & 1 \\ 0 & 0 & 0 \end{array}, \quad T_k \rightarrow \begin{array}{c|cc} n_k \backslash n'_k & 1 & 0 \\ \hline 1 & -1 & 0 \\ 0 & 0 & 1 \end{array} \tag{32}$$

* Dirac (*Proc. Roy. Soc.*, **A114**, 243, 1927) introduced the $A(n_1, \ldots)$ to discuss light quanta. If these are considered to be "particles" one must of necessity consider their creation (emission) and annihilation (absorption) [see § 88].

The operators U_k, V_k, and T_k are unit matrices as far as the other $n_{k'}$ variables are concerned ($k' \neq k$). The U_k and V_k are not Hermitian; they correspond to $\frac{1}{2}(\sigma_x - i\sigma_y)$ and $\frac{1}{2}(\sigma_x + i\sigma_y)$ if we express them in terms of the Pauli-matrices [compare eq. (6.134)], and U_k and V_k can thus be considered to be each other's complex conjugates *. The following relations hold

$$UT+TU=VT+TV=0,\ UV=\begin{Vmatrix}0 & 0\\ 0 & 1\end{Vmatrix},\ VU=\begin{Vmatrix}1 & 0\\ 0 & 0\end{Vmatrix},\ U^2=V^2=0,\ \ T^2=1. \tag{33}$$

We define the following operators

$$\mathbf{a}_k = T_1 T_2 \ldots T_{k-1} U_k, \qquad \mathbf{a}_k^* = T_1 T_2 \ldots T_{k-1} V_k. \tag{34}$$

The $\mathbf{a}_k$ and $\mathbf{a}_k^*$ are unit matrices with respect to the $n_{k'}$-variables with $k' > k$.

From equations (33) and (34) it follows that

$$\mathbf{a}_k^* \mathbf{a}_l + \mathbf{a}_l \mathbf{a}_k^* = \mathbf{1}\,\delta_{kl}, \quad \mathbf{a}_k \mathbf{a}_l + \mathbf{a}_l \mathbf{a}_k = \mathbf{a}_k^* \mathbf{a}_l^* + \mathbf{a}_l^* \mathbf{a}_k^* = 0. \tag{35}$$

All $\mathbf{a}$ and $\mathbf{a}^*$ are thus anticommuting except $\mathbf{a}$ and $\mathbf{a}^*$ of the same index. The $\mathbf{a}$ and $\mathbf{a}^*$ of equation (34) are called the *Jordan-Wigner matrices.*

We now want to show how one can obtain the F-matrix (31) which corresponds to an arbitrary symmetric operator F_{op}. We first of all introduce the following matrix, the coefficients of which are functions of the four coordinates $q(=\boldsymbol{x}, s)$ of an electron,

$$\boldsymbol{\varphi}\,(q) = \Sigma_1^\infty \mathbf{a}_k \varphi_k\,(q), \tag{36}$$

where the φ_k are again the complete orthonormal set of functions which we introduced earlier **.

An operator which refers to an N-electron problem can be built up linearly from operators $G^{(1)}(q)$ which act on the coordinates of only one electron, operators $G^{(2)}(q, q')$ which act on two electrons ***, . . .,

* Considered as operators (matrices) they are complex conjugate *and* adjoint (see § 29) or, Hermitean conjugate. The same is true for the $\mathbf{a}$ and $\mathbf{a}^*$ introduced by equation (34).

** Formally equation (36) corresponds to the expansion of a function $f(q)$ in terms of the $\varphi_k(q)$. Because of equations (35) and the completeness of the set of the $\varphi_k(q)$ the $\varphi(q)$ obey the relations $\boldsymbol{\varphi}^*(q)\,\boldsymbol{\varphi}(q') + \boldsymbol{\varphi}(q')\,\boldsymbol{\varphi}^*(q) = \delta(q'-q)$ where $\delta(q'-q)$ is the Dirac δ-function as far as $\boldsymbol{x}$ is concerned and the unit matrix as far as s is concerned.

*** We could have included in equation (37) a constant $G^{(0)}$ which did not depend on the number of particles.

$$F_{\text{op}} = \Sigma_i G^{(1)}(q_i) + \Sigma_{ij} G^{(2)}(q_i, q_j) + \ldots \tag{37}$$

The first sum is over all electrons, the second one over all combinations of two electrons, . . . The F-matrix we are looking for is a multilinear expression in the $\mathbf{a}$ and $\mathbf{a}^*$ which is obtained as follows

$$F_{n_1, \ldots, n_k \ldots; n'_1, \ldots, n'_k, \ldots} = \mathbf{F} = \tag{38}$$

$$\int \boldsymbol{\varphi}^*(q) G^{(1)}(q) \boldsymbol{\varphi}(q) + (2!)^{-1} \iint \boldsymbol{\varphi}^*(q) \boldsymbol{\varphi}^*(q') G^{(2)}(q, q') \boldsymbol{\varphi}(q') \boldsymbol{\varphi}(q) + (3!)^{-1} \iiint \ldots + \ldots$$

As before $\int$ means integration over $\boldsymbol{x}$ and summation over the two values of s.

Equation (38) can be proved as follows. One substitutes for the $\boldsymbol{\varphi}(q)$ expression (36) and notes that only those matrix elements of $\mathbf{a}_k^* \mathbf{a}_{k'}$ are different from zero for which n_k and $n_{k'}$ both change, the one from 0 to 1, the other from 1 to 0. Using equation (31) and (28) one then finds equation (38) by a process of carefully sorting out of the non-zero terms.

If we put in particular $G^{(1)} = V(\boldsymbol{x})$ [$V(\boldsymbol{x}) = 1$ within V and $V(\boldsymbol{x}) = 0$ outside V], $G^{(2)} = G^{(3)} = \ldots = 0$, expression (37) will be the operator corresponding to the number of electrons within V, or,

$$\mathbf{N}_{V\text{op}} = \int_V \boldsymbol{\varphi}^*(q)\, \boldsymbol{\varphi}(q), \tag{39}$$

where the integral extends only over the region V. We see that the operator $\boldsymbol{\varphi}^*(q)\, \boldsymbol{\varphi}(q)$ is the operator corresponding to the probability density. If V is the whole of space, we have obtained the operator corresponding to the total number of electrons. From equations (35) and (36) it follows that

$$\mathbf{N}_{\text{op}} = \int \boldsymbol{\varphi}^* \boldsymbol{\varphi} = \Sigma_k \mathbf{a}_k^* \mathbf{a}_k \; (= \Sigma_k V_k U_k). \tag{40}$$

The eigenvalues of $V_k U_k$ are 1 and 0, and the V and U corresponding to different k commute. It follows that the eigenvalues of $\mathbf{N}_{\text{op}}$ are all non-negative integers. All operators (38) commute with $\mathbf{N}_{\text{op}}$, since it follows from equations (35) that

$$(\Sigma_k \mathbf{a}_k^* \mathbf{a}_k)\{\mathbf{a}_l^* \mathbf{a}_m^* \ldots \mathbf{a}_{l'} \mathbf{a}_{m'} \ldots\} = \{\mathbf{a}_l^* \mathbf{a}_m^* \ldots \mathbf{a}_{l'} \mathbf{a}_{m'} \ldots\}(\Sigma_k \mathbf{a}_k^* \mathbf{a}_k),$$

and expression (38) in a linear combination of operators of the type given in braces. This commutability means that all elements of $\mathbf{F}$ corresponding to a change in the total number of particles will be equal to zero.

Equations (34) to (40) give us *a description of many-electron systems independent of the number of particles.* The physical situation is described by a function $A(n_1, \ldots, n_k, \ldots, t)$ which is time-dependent and which may, moreover, depend on the coordinates of other kinds of particles. It satisfies the Schrödinger equation

$$\mathbf{H} A(n_1, \ldots, n_k, \ldots, t) = i\hbar \partial A/\partial t, \qquad (41)$$

where $\mathbf{H}$ is an operator of the form (38). At the moment only those solutions have a physical meaning where A is different from zero only for a given number of electrons. According to our considerations up to now H_{op} will contain only terms $G^{(1)}$ which refer to one electron and terms $G^{(2)}$ which refer to pairs of electrons. The more important one of the $G^{(2)}$ is, in the non-relativistic approximation, the Coulomb interaction between two electrons *. This means that only two terms will occur in the series on the right hand side of equation (38). If there is only a term $G^{(1)}$ we can write

$$H_{n_1, \ldots n_k, \ldots; n'_1, \ldots, n'_k, \ldots} = \mathbf{H} = \Sigma \mathbf{a}_k^* H_{kl} \mathbf{a}_l, \quad H_{kl} = \int \varphi_k^* G^{(1)} \varphi_l. \quad (42)$$

If we are dealing with a system of independent particles this linear form in the $\mathbf{a}^*$ and $\mathbf{a}$ is sufficient. It would also be sufficient if the electrons were not interacting directly, but each interacted indirectly through other parts of the system under consideration. This is just the basic idea of quantum electrodynamics where the electrons — as in the electromagnetic theory of electrons — interact with one another through the intermediary of another part of the system, namely the electromagnetic field. In those considerations equation (42) is very useful.

The formalism given here is due to Jordan and Wigner**. It is called the *method of quantisation of the wave function*, or the *method of second quantisation* since the function appearing in equation (36) is certainly not a *c*-number, which it would be if the $\mathbf{a}$ were just numbers. However, the quantum of action enters nowhere so that these names are slightly misleading.

Jordan and Wigner's theory was the completion, or counterpart, of the earlier work of Jordan and Klein*** where an analogous formalism was developed for systems consisting of particles obeying

* If $G(q, q')$ in equation (38) is the Coulomb energy corresponding to two electrons at P and Q, the energy term would be analogous to the classical expression U of the potential energy of a charge distribution $\varrho = e\Sigma \boldsymbol{\varphi}^* \boldsymbol{\varphi}$ (the summation is over the spin coordinate), $U = \frac{1}{2}\iint (\varrho_P \varrho_Q / r_{PQ})\, d\boldsymbol{x}_P\, d\boldsymbol{x}_Q$.

** P. Jordan and E. Wigner, *Z. Phys.*, **47**, 631, 1928.

*** P. Jordan and O. Klein, *Z. Phys.*, **45**, 751, 1927.

Bose-Einstein statistics. We shall give here the formulae for this case.

Instead of equation (16), that is, instead of the expression (19) of an antisymmetric function in terms of Slater determinants, we now will have the expansion of a symmetric function in terms of normalised symmetric wave functions [compare eq. (12)],

$$\Phi_{k_1,\ldots,kN} = K\,\Sigma_{\mathrm{P}}\,\mathrm{P}\,\Pi_{i=1}^{N}\,\varphi_{k_i}(q_i). \tag{43}$$

where the φ_k are again a discrete complete orthonormal set of functions of the four coordinates $\boldsymbol{x}$, s of one particle, and where the summation is over all $N!$ permutations. Since now several k_i may be the same, the normalisation integral $\int|\Sigma_{\mathrm{P}}\,\mathrm{P}\,\Pi\,\varphi_{k_i}|^2$ will be larger than $N!$. It is multiplied by the product $n_1!n_2!\ldots$ if there are groups of $n_1, n_2, \ldots$ equal k-values. Thus the normalisation constant K in equation (43) satisfies the equation

$$K^{-2} = N!\,n_1!\,n_2!\ldots \tag{44}$$

The expansions are now

$$\left.\begin{aligned} \Psi_{\mathrm{symm}} &= \Sigma_{\{k_1,\ldots,k_N\}}\,A_{\{k_1,\ldots,k_N\}}\,\Phi_{\{k_1,\ldots,k_N\}} = \Sigma_{\{k_i\}}\,A_{\{k_i\}}\,\Phi_{\{k_i\}}, \\ A_{\{k_i\}} &= \int \Phi^*_{\{k_i\}}\,\Psi,\quad 1 = \int|\Psi|^2 = \Sigma_{\{k_i\}}\,|A_{\{k_i\}}|^2. \end{aligned}\right\} \tag{45}$$

Because of its definition Φ is symmetric in its indices $k_1, \ldots, k_N$. These indices are written in braces in equation (45) to indicate that of all combinations of the N numbers k which differ only by a permutation of the k_i only one representative combination is chosen. If only *one* coefficient A is different from zero, we are dealing with a situation where one particle is in the state φ_{k_1} of the one-particle problem, a second one in the state $\varphi_{k_2}, \ldots$ If the particles are independent, this might characterise a stationary state. Since $\Phi_{\{k_i\}}$ is symmetric, several of the k_i might be equal, that is, several particles could be in the same states of the one-particle problem.

The $A_{\{k_i\}}$ are the transformed representation of the Ψ-situation. Any symmetric operator F_{op} acting on a symmetric function will give another symmetric function. In the frame work of the symmetric exclusion principle F_{op} can be transformed into a matrix in the $\{k_i\}$-representation, which is defined by analogy with equation (28),

$$F_{\{k_i\}\{k'_i\}} = \int \Phi^*_{\{k_i\}}\,F_{\mathrm{op}}\,\Phi_{\{k'_i\}}. \tag{46}$$

The Schrödinger equation is now in analogy to equations (21) and (22),

$$\Sigma_{\{k'_i\}} H_{\{k_i\}\,\{k'_i\}} A_{\{k'_i\}}(t) = i\hbar(\partial/\partial t) A_{\{k_i\}}(t). \tag{47}$$

We now introduce the notation similar to equations (29) and (30),

$$A_{\{k_i\}} \to A(n_1, n_2, \ldots, n_k, \ldots);\; F_{\{k_i\}\,\{k'_i\}} \to F_{n_1,\ldots,\, n_k \ldots,\, n'_1 \ldots,\, n'_k,\ldots}. \tag{48}$$

The only difference is now that *each* n_k *can take on all non-negative values,* and not only 0 and 1. We can again use equation (30). Once again it is possible to express the F-matrices in basic matrices such that we can find a simple formalism to find the F-matrix from F_{op}.

The basic matrices will again be denoted by $\mathbf{a}_k$ and $\mathbf{a}_k^*$, but they are now no longer given by equations (32) and (34). We do not need any auxiliary matrices $U, V, T,$ but write immediately

$$\mathbf{a}_k \to \begin{array}{c|ccccc} n_k \backslash n'_k & 0 & 1 & 2 & 3 & \ldots \\ \hline 0 & 0 & 1^{1/2} & 0 & 0 & \ldots \\ 1 & 0 & 0 & 2^{1/2} & 0 & \ldots \\ 2 & 0 & 0 & 0 & 3^{1/2} & \ldots \\ 3 & 0 & 0 & 0 & 0 & \ldots \\ \cdot & \cdot & \cdot & \cdot & \cdot & \ldots \\ \cdot & \cdot & \cdot & \cdot & \cdot & \ldots \end{array}, \qquad \mathbf{a}_k^* \to \begin{array}{c|ccccc} n_k \backslash n'_k & 0 & 1 & 2 & 3 & \ldots \\ \hline 0 & 0 & 0 & 0 & 0 & \ldots \\ 1 & 1^{1/2} & 0 & 0 & 0 & \ldots \\ 2 & 0 & 2^{1/2} & 0 & 0 & \ldots \\ 3 & 0 & 0 & 3^{1/2} & 0 & \ldots \\ \cdot & \cdot & \cdot & \cdot & \cdot & \ldots \\ \cdot & \cdot & \cdot & \cdot & \cdot & \ldots \end{array} \tag{49}$$

As far as all $n_{k'}$ for which $k' \neq k$ are concerned, $\mathbf{a}_k$ and $\mathbf{a}_k^*$ are unit matrices. The $\mathbf{a}_k$ are not Hermitean, and neither are the $\mathbf{a}_k^*$. Both $\mathbf{a}_k + \mathbf{a}_k^*$ and $i(\mathbf{a}_k - \mathbf{a}_k^*)$ are Hermitean which means that $\mathbf{a}_k$ and $\mathbf{a}_{k'}^*$ are again Hermitean conjugate [see the discussion of eqq. (32)]. The following relations hold

$$\left.\begin{array}{c} -\mathbf{a}_k^*\mathbf{a}_l + \mathbf{a}_l\mathbf{a}_k^* = \mathbf{1}\delta_{kl}, \quad \mathbf{a}_k\mathbf{a}_l - \mathbf{a}_l\mathbf{a}_k = 0, \quad \mathbf{a}_k^*\mathbf{a}_l^* - \mathbf{a}_l^*\mathbf{a}_k^* = 0, \\[1ex] \mathbf{a}_k^*\mathbf{a}_k \to \begin{array}{c|ccccc} n_k \backslash n'_k & 0 & 1 & 2 & 3 & \ldots \\ \hline 0 & 0 & 0 & 0 & 0 & \ldots \\ 1 & 0 & 1 & 0 & 0 & \ldots \\ 2 & 0 & 0 & 2 & 0 & \ldots \\ 3 & 0 & 0 & 0 & 3 & \ldots \\ \cdot & \cdot & \cdot & \cdot & \cdot & \ldots \\ \cdot & \cdot & \cdot & \cdot & \cdot & \ldots \end{array}, \qquad \mathbf{a}_k\mathbf{a}_k^* \to \begin{array}{c|ccccc} n_k \backslash n'_k & 0 & 1 & 2 & 3 & \ldots \\ \hline 0 & 1 & 0 & 0 & 0 & \ldots \\ 1 & 0 & 2 & 0 & 0 & \ldots \\ 2 & 0 & 0 & 3 & 0 & \ldots \\ 3 & 0 & 0 & 0 & 4 & \ldots \\ \cdot & \cdot & \cdot & \cdot & \cdot & \ldots \\ \cdot & \cdot & \cdot & \cdot & \cdot & \ldots \end{array} \end{array}\right\} \tag{50}$$

The further development is the same as in the antisymmetric case. Equation (36) introduces a matrix $\boldsymbol{\varphi}(q)$ and $\mathbf{F}$ is given by equation (38) and the number operator by equation (40).

The different terms in expression (40) commute again and each has the eigenvalues 0, 1, 2, . . . [compare eq. (50)]. Once again the eigenvalues of $\mathbf{N}_{\text{op}}$ are all non-negative numbers and the fact that $\mathbf{N}_{\text{op}}$ commutes again with expression (38) for **F** confirms that there are no terms in **F** which correspond to "transitions" with changing number of particles.

The proof that equations (38) and (46) are equivalent is different from the proof of the equivalence of equations (38) and (28). It is once again a tedious proof and is left to the reader *.

If a system contains apart from particles of one kind also particles of another kind (for instance, electrons as well as nuclei), the function A which is used to characterise the situation will depend not only on $n_1, n_2, \ldots$ and t, but also on the coordinates of the second kind of particle** and the matrix elements of **F** are themselves operators acting on those coordinates. If there are more than one particle of the second kind, we can generalise our formalism. For each kind we introduce a matrix function of the kind (36).

$$\boldsymbol{\varphi}_{\text{I}} = \Sigma_k \mathbf{a}_{\text{I}k} \varphi_{\text{I}k}, \qquad \boldsymbol{\varphi}_{\text{II}} = \Sigma_l \mathbf{a}_{\text{II}l} \varphi_{\text{II}l}, \tag{51}$$

where $\mathbf{a}_{\text{I}}$ and $\mathbf{a}_{\text{II}}$ are chosen in accordance with the exclusion principle obeyed by the particles in question, and where each of the $\mathbf{a}_{\text{I}}$, $\mathbf{a}^*_{\text{I}}$ commutes with each of the $\mathbf{a}_{\text{II}}$, $\mathbf{a}^*_{\text{II}}$. A situation is described by a function $A(n_{\text{I}1}, \ldots, n_{\text{I}k}, \ldots; n_{\text{II}1}, \ldots, n_{\text{II}l}, \ldots; t)$ and the transformed matrix of an operator F_{op} which is symmetric in both kinds of particle is given by the equation

$$\left.\begin{aligned} \mathbf{F} = \int \boldsymbol{\varphi}^*_{\text{I}}(q_{\text{I}})\, G^{(1)}_{\text{I}}\, \boldsymbol{\varphi}_{\text{I}}(q_{\text{I}}) + \int \boldsymbol{\varphi}^*_{\text{II}}(q_{\text{II}})\, G^{(1)}_{\text{II}}\, \boldsymbol{\varphi}_{\text{II}}(q_{\text{II}}) + \\ \iint \boldsymbol{\varphi}^*_{\text{I}}(q_{\text{I}})\, \boldsymbol{\varphi}^*_{\text{II}}(q_{\text{II}})\, G^{(2)}_{\text{I, II}}(q_{\text{I}}, q_{\text{II}})\, \boldsymbol{\varphi}_{\text{I}}(q_{\text{I}})\, \boldsymbol{\varphi}_{\text{II}}(q_{\text{II}}) + \ldots, \end{aligned}\right\} \tag{52}$$

if F_{op} is written by analogy with equation (37) in the form

$$F_{\text{op}} = \Sigma_i G^{(1)}_{\text{I}}(q_{\text{I}i}) + \Sigma_j G^{(1)}_{\text{II}}(q_{\text{II}j}) + \Sigma_{ij} G^{(2)}_{\text{I,II}}(q_{\text{I}i}, q_{\text{II}j}) + \ldots \tag{53}$$

The $G^{(2)}_{\text{I, II}}$ refer to terms in the operator which act on *one* particle of the first kind and on *one* particle of the second kind. If $G^{(2)}_{\text{I, II}}$ were a product,

$$G^{(2)}_{\text{I, II}}(q_{\text{I}i}, q_{\text{II}j}) = G^{(1)}_{\text{I, II}}(q_{\text{I}i})\, G^{(1)}_{\text{I, II}}(q_{\text{II}j}), \tag{54}$$

the last term on the right hand side of equation (52) would be the product of two commuting matrices.

* See, for instance, V. Fock, *Z. Phys.*, **75**, 622, 1934.

** We assume that we have not yet performed a unitary transformation on these coordinates.

If I and II refer to electrons and positrons, both $\mathbf{a}_I$ and $\mathbf{a}_{II}$ would be Jordan-Wigner matrices. Processes involving the creation or annihilation of a pair are described by adding to the energy matrix terms of the kind

$$\mathbf{a}_I^* \mathbf{a}_{II}^* \Omega^* + \mathbf{a}_I \mathbf{a}_{II} \Omega. \tag{55}$$

The first term corresponds to the annihilation, the second one to the creation of a pair [compare eqq. (32) and (34)]. Together they are Hermitean. In expression (55) Ω can be an operator acting on that part of the total system (for instance, the electromagnetic field) under whose "influence" these processes take place.

These remarks are meant only to give an example of the formal possibility of describing annihilation and creation processes. This description goes over into Dirac's hole theory (see § 65) only when we assume that every $\mathbf{a}_I$, $\mathbf{a}_I^*$ *anticommutes* with every $\mathbf{a}_{II}$, $\mathbf{a}_{II}^*$. This is so because the hole theory operates with only one kind of electrons [eqq. (36 to (42)], and $G^{(1)}$ is the operator (6.164). All $\mathbf{a}_k^*$ (or $\mathbf{a}_k$) corresponding to states of negative energy are replaced by $\mathbf{a}_k$ (or $\mathbf{a}_k^*$), that is, the *absence* of an electron in a state of negative energy is interpreted as the *presence* of a particle. Particles of this second kind are supposed to be the *positrons* and the commutability of $\mathbf{N}_{op}$ with expression (42) does no longer mean conservation of the total number of particles, but conservation of charge.

§ 73. Systems with two electrons without spin forces

Up to now we have kept our discussions of the Pauli principle very general. If it is applied to particular problems one notices especially the peculiar features due to the special character of the spin coordinates. Their character enables one especially to classify the possible "antisymmetric situations" in terms of "multiplet situations". The fact that the energetic influence of the spin forces is often very small, or even negligible, makes it possible to survey the consequences of the theory from a rather more simple point of view. We may especially mention in this respect the Russell-Saunders coupling scheme in which spectral terms are classified in (spin) multiplet terms.

In the present and in the next section we shall illustrate this by discussing the two-electron problem while in §§ 75 to 80 the discussion is extended to many-electron systems.

If we resolve a function ψ of the coordinates of two electrons into its four spin components, we have (see § 60)

$$\psi(\boldsymbol{x}_1, s_1, \boldsymbol{x}_2, s_2) = f_{++}(1, 2)\,\xi_1\xi_2 + f_{+-}(1, 2)\,\xi_1\eta_2 + f_{-+}(1, 2)\,\eta_1\xi_2 + f_{--}(1, 2)\,\eta_1\eta_2, \tag{56}$$

where 1, 2 is a shorthand notation for $\boldsymbol{x}_1$, $\boldsymbol{x}_2$. If P denotes the exchange permutation of the two electrons, we have

$$\mathrm{P}\,\psi = f_{++}(2,1)\,\xi_2\xi_1 + f_{+-}(2,1)\,\xi_2\eta_1 + f_{-+}(2,1)\,\eta_2\xi_1 + f_{--}(2,1)\,\eta_2\eta_1.$$

If ψ is antisymmetric in the two electrons, it follows from $\psi = -\mathrm{P}\psi$ that the following relations hold,

$$\begin{aligned} f_{++}(1,2) &= -f_{++}(2,1), \quad & f_{+-}(1,2) &= -f_{-+}(2,1),\\ f_{-+}(1,2) &= -f_{+-}(2,1), \quad & f_{--}(1,2) &= -f_{--}(2,1). \end{aligned} \tag{57}$$

We now introduce the following notation,

$$\begin{aligned} f_{++}(1,2) &= g_1^{\mathrm{a}}(1,2), \quad & 2^{-\frac{1}{2}}[f_{+-}(1,2)+f_{-+}(1,2)] &= g_0^{\mathrm{a}}(1,2),\\ f_{--}(1,2) &= g_{-1}^{\mathrm{a}}(1,2), \quad & 2^{-\frac{1}{2}}[f_{+-}(1,2)-f_{-+}(1,2)] &= g_0^{\mathrm{s}}(1,2). \end{aligned}$$

From equations (57) we see that the three functions g^{a} are antisymmetric and that g^{s} is symmetric. The index of the g is the sum of the spin-indices of the corresponding f-functions. We see that we can write any antisymmetric function in the form

$$\psi = \underbrace{\{g_1^{\mathrm{a}}\xi_1\xi_2 + g_0^{\mathrm{a}}[\xi_1\eta_2+\eta_1\xi_2]2^{-\frac{1}{2}} + g_{-1}^{\mathrm{a}}\eta_1\eta_2\}}_{\text{triplet terms}} + \underbrace{g_0^{\mathrm{s}}[\xi_1\eta_2-\eta_1\xi_2]2^{-\frac{1}{2}}}_{\text{singlet term}} = {}^3\psi + {}^1\psi, \tag{58}$$

where the g are arbitrary functions — apart from their symmetry properties. The first three terms are called triplet terms and the last term a singlet term. We do not wish to discuss here the historical reasons for these names (see § 79). For the moment we may mention in this connexion that the triplet terms are constructed from three different linearly independent "spin functions" while there is only *one* spin function involved in the singlet term. A spin function is a function ζ depending only on the spin coordinates.

In the case of N electrons there are 2^N linearly independent spin functions, since there are 2^N possible combinations of the spin arguments. Hence there are four in the case of two electrons. These can always be chosen in accordance with equation (58) in such a way that they fall into two groups, the triplet and the singlet spin functions. We use the following notation *

* The superscripts to the left of the ζ in equations (59) and to the left of the ψ in equation (58) are chosen to agree with the spectroscopic Russell-Saunders designation of triplet and singlet terms.

$$\left.\begin{aligned}&\text{Triplet:}\quad {}^3\zeta_1 = \xi_1\xi_2,\ {}^3\zeta_0 = 2^{-\frac{1}{2}}[\xi_1\eta_2+\eta_1\xi_2],\ {}^3\zeta_{-1} = \eta_1\eta_2;\\ &\text{Singlet:}\qquad\qquad\qquad {}^1\zeta_0 = 2^{-\frac{1}{2}}[\xi_1\eta_2-\eta_1\xi_2].\end{aligned}\right\}\tag{59}$$

The ${}^3\zeta$ are *symmetric* in the spin coordinates and ${}^1\zeta_0$ *antisymmetric.* The antisymmetry of ψ follows immediately from equation (58). The spatial coordinates produce the antisymmetry in the case of triplet terms and the spin coordinates in the case of the singlet term.

The spin functions in equations (59) are chosen in such a way that they form an orthonormal set. This can be seen by noting that the set (59) can be obtained by a unitary transformation of the set $\xi_1\xi_2$, $\xi_1\eta_2$, $\eta_1\xi_2$, $\eta_1\eta_2$ which is obviously orthonormal. The four terms in equation (58) are also orthogonal to one another, but not necessarily normalised.

If ψ of equation (58) is a wave function corresponding to a physical situation so that ψ contains the time, the g will also contain the time. The extreme right hand side of equation (58) expresses the fact that the situation is a superposition, or a *mixture of a triplet and a singlet situation.* We can speak of the triplet and the singlet contribution to the situation corresponding to the two integrals in the normalisation equation

$$1=\int|\psi|^2=\int\{|g_1^{\mathrm{a}}|^2+|g_0^{\mathrm{a}}|^2+|g_{-1}^{\mathrm{a}}|^2\}d\boldsymbol{x}_1d\boldsymbol{x}_2+\int|g_0^{\mathrm{s}}|^2d\boldsymbol{x}_1d\boldsymbol{x}_2=\int|{}^3\psi|^2+|\int{}^1\psi|^2.$$

If ψ corresponds to a stationary state, ${}^3\psi$ and ${}^1\psi$, and also all the g, contain the time in a common exponential factor and the "mixing ratio" of triplet and singlet is constant in time.

Consider now a system of two electrons in an external field of force and let the energetic influence of the spin forces be very small. To a first approximation the energy operator will act on the spatial coordinates of the electrons only*. Let this operator be denoted by H^0_{op}; it will be symmetric in the two electrons. If we substitute expression (58) into the Schrödinger equation

$$H^0_{\mathrm{op}}\psi = i\hbar\partial\psi/\partial t,\tag{60}$$

we see that, since H^0_{op} does not change the spin functions, each of the functions g and in particular ${}^3\psi$ and ${}^1\psi$ will be a solution of this equation. This can be expressed as follows. Each solution of the Schrödinger equation $H^0_{\mathrm{op}}g = i\hbar\partial g/\partial t$ (g: a function of the spatial coordinates only) can be written as a sum of an antisymmetric solution g^{a} and a symmetric solution g^{s} according to $g = \frac{1}{2}(g-\mathrm{P}g)$

* This means that the energy does not depend on the relative directions of the spin and the orbital angular momenta.

$+\frac{1}{2}(g+Pg) = g^a + g^s$, since Pg is a solution if g is (see § 69). A solution g^a multiplied by one of the ${}^3\zeta$ leads to a triplet solution of equation (60) and a solution g^s multiplied by ${}^1\zeta_0$ to a singlet solution. Any solution is thus a superposition of those two kinds of solution.

If we want to find a *stationary state*, $\psi = \varphi \exp(-iEt/\hbar)$, $(H^0_{op} - E)\varphi = 0$, and if we write again $\varphi = (g_1^{a\,3}\zeta_1 + g_0^{a3}\zeta_0 + g^a_{-1}{}^3\zeta_{-1}) + g_0^{s1}\zeta_0$, each of the terms will be a solution, and each of the g-functions satisfies the following time-independent Schrödinger equation which contains only spatial coordinates,

$$(H^0_{op} - E)g = 0, \tag{61}$$

with the same value of E for all four g-functions. This E-value is thus at least fourfold degenerate. The fact that eigenvalues belonging to different eigenfunctions would be the same, would be fortuitous * and in general one can reach the conclusion that the four g-functions will differ only by a multiplying factor (which may be zero). The g_0^s and g^a must as a matter of principle be different, and we see thus that *the stationary states can be divided into two groups, the triplet states and the singlet states*. The corresponding eigenfunctions are of the form

Triplet: ${}^3\varphi = g^a(c_1\,{}^3\zeta_1 + c_0{}^3\zeta_0 + c_{-1}\,{}^3\zeta_{-1})$, eigenvalue ${}^3E$; (62a)

or, Singlet: ${}^1\varphi = g^{s\,1}\zeta_0$, eigenvalue ${}^1E$, (62b)

where the c's are arbitrary numbers. In equations (62) g^a is an antisymmetric and g^s a symmetric solution of equation (61). We may assume in both cases that the corresponding eigenvalue is non-degenerate, as long as we may neglect the spin coordinates. The complete triplet eigenfunction, which contains the spin coordinates, is always threefold degenerate, but the complete singlet eigenfunction remains nondegenerate. It will be only by accident if a ${}^3E$ or ${}^1E$ or several ${}^3E$, ${}^1E$ are equal to one another.

The part played by the spins in the solutions (62) can be interpreted as follows. *The two spin vectors of the electrons corresponding both to a spin quantum number* $\frac{1}{2}$ *appear in the triplet states to be combined into a resultant spin vector with "spin quantum number"* $\frac{1}{2}+\frac{1}{2} = 1$ (*spins parallel*); *in the singlet states they appear to be combined into a resultant spin vector with "spin quantum number"* $\frac{1}{2}-\frac{1}{2} = 0$ (*spins antiparallel*). This terminology refers to the com-

* In the case of more than two electrons we shall meet with cases where eigenvalues of the time independent Schrödinger equation for functions of spatial coordinates only show an essential degeneracy due to the symmetry of H_{op}.

bination of ordinary vectors (that is, vectors the components of which are c-numbers) into a resultant vector. One compares the situation with the combination of two vectors of length $\frac{1}{2}\hbar$ into a new vector. The length of this vector could be between $\hbar$ (parallel position) and 0 (anti-parallel position), but only those two extreme values will occur.

An analysis of this interpretation requires the investigation of the vector $\boldsymbol{S}$ which is the ordinary vector sum of the two spin vectors $\boldsymbol{S}_1$ and $\boldsymbol{S}_2$,

$$\boldsymbol{S} = \boldsymbol{S}_1 + \boldsymbol{S}_2, \quad \boldsymbol{S}^2 = (\boldsymbol{S}\cdot\boldsymbol{S}) = S_1^2 + S_2^2 + 2(\boldsymbol{S}_1\cdot\boldsymbol{S}_2). \tag{63}$$

The components of $\boldsymbol{S}_1$ and $\boldsymbol{S}_2$ are operators of the kind discussed in § 58. They act on all functions of s_1 and s_2, that is, on $\xi_1\xi_2$, $\xi_1\eta_2$, $\eta_1\xi_2$, $\eta_1\eta_2$, and linear combinations of them, according to the rules (6.78). Since the components of $\boldsymbol{S}_1$ commute with the components of $\boldsymbol{S}_2$, the commutation rules (6.52) which held for $\boldsymbol{S}_1$ and $\boldsymbol{S}_2$ will also hold for $\boldsymbol{S}$, that is, $S_xS_y - S_yS_x = i\hbar S_z, \ldots$ The situation is, indeed, very similar to the one discussed in § 59 where the orbital angular momentum vector $\boldsymbol{M}$ and the spin vector $\boldsymbol{S}$ were combined. For instance, we have also the relations [compare eq. (6.62)] $S^2S_x - S_xS^2 = 0, \ldots$, and we can find a set of functions which are eigenfunctions of both S^2 and S_z and can determine the corresponding eigenvalues.

We might proceed as in § 59 but we prefer the following method. Both S^2 and S_z are symmetric in the two electrons, which means that they commute with the exchange permutation operator P (see § 69),

$$\mathrm{P}S^2 - S^2\mathrm{P} = 0, \qquad \mathrm{P}S_z - S_z\mathrm{P} = 0 \tag{64}$$

The eigenfunctions we are looking for can thus in principle be arranged so as to be eigenfunctions of P also. The eigenfunctions of P are the symmetric spin functions ${}^3\zeta$ and the antisymmetric spin function ${}^1\zeta_0$ [see eq. (59)] and they correspond to the eigenvalues 1 and -1 of P, $\mathrm{P}\,{}^3\zeta = {}^3\zeta$, $\mathrm{P}\,{}^1\zeta_0 = -{}^1\zeta_0$. Since there is only *one* ${}^1\zeta$, this function must automatically be an eigenfunction of S^2 and S_z. Indeed, it corresponds to the eigenvalue 0 of both of them since [see eq. (6.66)]

$$\left.\begin{aligned}
S_z\,{}^1\zeta_0 &= (S_{z1}+S_{z2})2^{-1/2}(\xi_1\eta_2-\eta_1\xi_2) = 2^{-3/2}\hbar\{(\xi_1\eta_2+\eta_1\xi_2)+(-\xi_1\eta_2-\eta_1\xi_2)\}=0,\\
(\boldsymbol{S}_1\cdot\boldsymbol{S}_2)\,{}^1\zeta_0 &= (S_{x1}S_{x2}+S_{y1}S_{y2}+S_{z1}S_{z2})2^{-1/2}(\xi_1\eta_2-\eta_1\xi_2)\\
&= 2^{-3/2}\hbar[S_{x1}(\xi_1\xi_2-\eta_1\eta_2)-iS_{y1}(\xi_1\xi_2+\eta_1\eta_2)-S_{z1}(\xi_1\eta_2+\eta_1\xi_2)]\\
&= 2^{-5/2}\hbar[(\eta_1\xi_2-\xi_1\eta_2)+(\eta_1\xi_2-\xi_1\eta_2)-(\xi_1\eta_2-\eta_1\xi_2) = -\tfrac{3}{4}\hbar^2\,{}^1\zeta_0,\\
S^2\,{}^1\zeta_0 &= [S_1^2+S_2^2+2(\boldsymbol{S}_1\cdot\boldsymbol{S}_2)]\,{}^1\zeta_0 = \tfrac{3}{4}\hbar^2(1+1-2)\,{}^1\zeta_0 = 0.
\end{aligned}\right\}$$

The term $2(\boldsymbol{S}_1 \cdot \boldsymbol{S}_2)$ compensates thus the terms S_1^2 and S_2^2 [see eq. (6.66)] as would be expected from the geometrical picture of an antiparallel alignment of $\boldsymbol{S}_1$ and $\boldsymbol{S}_2$. The function ${}^1\zeta_0$ is also an an eigenfunction of S_x and S_y with eigenvalue 0 [compare the remark at the end of § 39].

We must now investigate the behaviour of the ${}^3\zeta_s$ with respect to S^2 and S_z. The eigenvalue of S^2 is the same for all three, namely $2\hbar^2$. The eigenvalue of S_z corresponding to ${}^3\zeta_s$ is $s\hbar$,

$$S^2\,{}^3\zeta_s = 2\hbar^2, \qquad S_z\,{}^3\zeta_s = s\hbar \qquad (s = -1, 0, 1). \tag{66}$$

This shows at the same time the physical meaning of the subscript s which we introduced in equation (59). It is the quantum number of the z-component of the resultant spin angular momentum. The three triplet functions correspond to the same value of the absolute square of the resultant spin vector $\boldsymbol{S}$. The "spin quantum number" of this vector is defined as the maximum value $\bar{s}$ of S_z [compare eq. (6.68)] *,

$$S_z = s\hbar, \qquad s = \bar{s}, s-1, \ldots, -\bar{s}, \tag{67}$$

and the eigenvalue of S^2 can be expressed in a form similar to equation (6.67)

$$S^2 = \bar{s}(\bar{s}+1)\hbar^2 \tag{68}$$

Equations (67) and (68) hold both for triplet ($\bar{s}=1$) and for singlet ($\bar{s} = 0$) states [compare eq. (6.66)].

In terms of vector models the situation can be described as follows. In a system of two electrons when spin forces can be neglected the two spins will combine to form a resultant spin vector which has either the "length" $\hbar$ (that is, $\bar{s} = 1$: triplet) or the length 0 ($\bar{s} = 0$: singlet). In the latter case it has no sense to talk about the direction of the spin vector, while in the former case the direction is arbitrary, but constant in time. The fact that the direction is constant follows from the fact that the c_s in equation (62a) are constants and is due to the neglect of all forces which tend to change the direction of the spin.

In quantum mechanics this "unchanging direction" of the spin has its own meaning. It is, for instance, possible to say that a certain situation characterised by equation (62a) corresponds to the result-

* In the literature the spin quantum number is usually denoted by s and the quantum number of the z-component of the spin angular momentum often by m_s by analogy with the notation m for the magnetic quantum number. Since we have used s instead of m_s we shall denote the spin quantum number by $\bar{s}$.

ing spin being along the positive (or negative) z-axis, namely, when only c_1 (or c_{-1}) is different from zero. There are also other combinations of c_1, c_0, and c_{-1} (namely those for which $(c_1+c_{-1})/c_0$ is real)* which correspond to well-defined directions of the resulting spin. In general, however, one cannot interpret a triplet situation in this way. If c_1, c_0, and c_{-1} are arbitrary numbers we are dealing with a superposition of three independent situations corresponding to the usual case of a threefold degeneracy.

The analysis of the situation can be very much extended by an explicit investigation of the influence of spatial rotations on the spin functions, that is, by an investigation of the changes occurring when the system of coordinates with respect to which ξ and η are defined is rotated (compare § 61). Those considerations are, however, special cases of the general discussion in §§ 75 to 77.

A special case of the spinless two electron problem occurs when the interaction between the electrons is such that to a first approximation they can be considered to be independent, at any rate as far as the description of one particularly considered stationary state Z (or a group of such states) is concerned. This means mathematically (see § 70) that the wave function corresponding to Z is to a first approximation a product of two functions,

$$g = \varphi_1(1)\varphi_2(2)+g'(1, 2), \tag{69}$$

where g' is "small of the first order" compared to, and orthogonal to, $\varphi_1\varphi_2$. The functions φ_1 and φ_2 will be assumed to be normalised, but they need not be orthogonal to each other. They can even be identical. An approximate value of the eigenvalue is given by the equation

$$E_0 = \iint \varphi_1^*(1)\,\varphi_2^*(2)\,H_{\text{op}}\,\varphi_1(1)\,\varphi_2(2)\,d\boldsymbol{x}_1\,d\boldsymbol{x}_2. \tag{70}$$

A special case occurs when the energy operator is of the form

$$H^0{}_{\text{op}} = G_{\text{op}}(1)+G_{\text{op}}(2)+W_{\text{op}}(1, 2), \tag{71}$$

where the interaction energy $W_{\text{op}}(1, 2)$ is small compared to $G_{\text{op}}(1)$ and $G_{\text{op}}(2)$. This case corresponds (approximately) to two independent electrons in a common field of force. In many applications $H^0{}_{\text{op}}$ will be of this form and $W_{\text{op}}(1, 2)$ will be the Coulomb energy of the two electrons. This energy can certainly not be assumed to be always small. All the same, expression (69) will often be a good approximation, but with functions φ which do not necessarily belong to a set of eigenfunctions of a given one-particle operator which is the same for the two electrons. The case of the helium atom is an example of this. In the stationary states corresponding to the normal spectrum, the inner electron "moves" to a first approximation in a "1-

* In this case only one can find a solution of the equation $(aS_x+bS_y+cS_z)(c_1{}^3\zeta_1+c_0{}^3\zeta_0+c_{-1}{}^3\zeta_{-1}) = A(c_1{}^3\zeta_1+c_0{}^3\zeta_0+c_{-1}{}^3\zeta_{-1})$ with real values of a, b, c, and A.

quantum orbit" in a Coulomb field with $Z = 2$, the outer electron in an n-quantum orbit in a Coulomb field with $Z = 1$. Expression (69) should be a fair first approximation, but φ_1 and φ_2 are not always orthogonal to each other.

Such applications were the reason for the general development of perturbation theory in § 51. It is easy to adapt that theory to the case where W in equation (71) is small.

Because H^0_{op} is symmetric $\text{P}g = \varphi_1(2)\varphi_2(1)+g'(2, 1)$ will be a solution, if expression (69) is one, and the approximate energy eigenvalue (70) is thus two-fold degenerate. If we remember that all solutions of equation (61) can be split into symmetric and antisymmetric ones, we can say that in the zeroth approximation there exists an eigenvalue E_0 which belongs to both an approximate symmetric solution g^{s} and an approximate antisymmetric solution g^{a},

$$g^{\text{s}} = C^{\text{s}}[\varphi_1(1)\varphi_2(2)+\varphi_1(2)\varphi_2(1)]+g'^{\text{s}}, \tag{72}$$

$$g^{\text{a}} = C^{\text{a}}[\varphi_1(1)\varphi_2(2)-\varphi_1(2)\varphi_2(1)]+g'^{\text{a}}. \tag{73}$$

The C^{s} and C^{a} are normalisation constants. If φ_1 and φ_2 are orthogonal to each other, $C^{\text{s}} = C^{\text{a}} = 2^{-\frac{1}{2}}$. We have up to terms of the second order

$$\left.\begin{aligned} 1=(C^{\text{s}})^2\int[\varphi_1^*(1)\varphi_2^*(2)+\varphi_1^*(2)\varphi_2^*(1)][\varphi_1(1)\varphi_2(2) \\ +\varphi_1(2)\varphi_2(1)]d\boldsymbol{x}_1 d\boldsymbol{x}_2=2(C^{\text{s}})^2(1+C), \\ C=|\int\varphi_1^*(1)\varphi_2(1)d\boldsymbol{x}_1|^2,\ C^{\text{s}}=[2(1+C)]^{-\frac{1}{2}},\ C^{\text{a}}=[2(1-C)]^{-\frac{1}{2}}. \end{aligned}\right\} \tag{74}$$

If φ_1 and φ_2 are the same, only the symmetric solution remains *. Even if they are nearly the same, expression (73) is still not a very good approximation. We shall assume that φ_1 and φ_2 are very different. If we remind ourselves of the results of the perturbation theory of degenerate systems (§ 51) we must solve a secular equation of the second degree [see eq. (5.59)] to find the first approximation to the correct energy. In the present case, we know, however, that g^{s} and g^{a} are solutions which are adapted to the present problem. It is easily seen that if F_{op} is a symmetric operator all matrix elements of the type $\int g^{\text{a}*}F_{\text{op}}g^{\text{s}}\,d\boldsymbol{x}_1\,d\boldsymbol{x}_2$ will be equal to zero, since the integral should remain the same, if the two electrons are exchanged, but it will also change its sign because of the symmetry properties of the factors in the integrand. We can thus immediately use equation (5.53) to find the second approximation to the energy by

* Example: The ground state of the helium atom. In this case one chooses for φ the wave function of the 1-quantum state in a Coulomb field due to a nuclear charge Ze and uses the Ritz method to find the "best" Z-value, which should lie between 1 and 2.

putting first $\varphi_0 = g^{\text{s}}$ and then $\varphi_0 = g^{\text{a}}$. We get in that way

$$E_0^{\text{s}} = (C^{\text{s}})^2 \iint [\varphi_1^*(1)\varphi_2^*(2)+\varphi_1^*(2)\varphi_2^*(1)] H_{\text{op}} [\varphi_1(1)\varphi_2(2) + \varphi_1(2)\varphi_2(1)] d\boldsymbol{x}_1 d\boldsymbol{x}_2 \tag{75a}$$

$$= (E_0 + A)/(1+C),$$

$$E_0^{\text{a}} = (C^{\text{a}})^2 \iint [\varphi_1^*(1)\varphi_2^*(2)-\varphi_1^*(2)\varphi_2^*(1)] H_{\text{op}} [\varphi_1(1)\varphi_2(2) - \varphi_1(2)\varphi_2(1) d\boldsymbol{x}_1 d\boldsymbol{x}_2 \tag{75b}$$

$$= (E_0 - A)/(1-C),$$

$$A = \iint \varphi_1^*(1)\varphi_2^*(2) H_{\text{op}} \varphi_1(2)\varphi_2(1) d\boldsymbol{x}_1 d\boldsymbol{x}_2. \tag{76}$$

The integral A is called the *exchange integral* *. It is always real since an exchange of the electrons will not change A though it will lead to its complex conjugate [see eq. (3.47)].

The energy value E_0 is split into two. The first level belongs to a symmetric state and the other one to an antisymmetric one. If φ_1 and φ_2 are orthogonal to each other, $C = 0$ and the splitting is symmetric with respect to E_0. If we use expression (71) and if $W_{\text{op}}(1, 2)$ is the Coulomb energy, we get the following often used expression for the exchange integral

$$A = \iint \varphi_1^*(1)\varphi_2^*(2)(e^2/r_{12})\varphi_1(2)\varphi_2(1) d\boldsymbol{x}_1 d\boldsymbol{x}_2. \tag{77}$$

We are now able to discuss the question of the theoretically "best" choice of φ_1 and φ_2. There are really two questions, namely, which choice gives us the best representation of g^{s} according to equation (72), and which gives us the best representation of g^{a} according to equation (73). The answers to those two questions will in general be different. Using the Ritz method (5.54) one investigates for which φ-functions the integral (75a), or the integral (75b) is an extremum. These conditions lead to a set of rather complicated simultaneous differential equations** which is not easily solved exactly and which is different for the two cases. In many cases there is a "natural" choice for the φ (see, for instance, the case of the helium atom mentioned a little while ago) which will then usually lead to the same φ-functions for equations (72) and (73). Often one obtains good results by using for φ_1 and φ_2 functions of a form to be expected containing a few adjustable parameters. The best values of the parameters are then again determined by using for instance, the Ritz method.***

In order that the approximation (75a) and (75b) can be considered to be a good one, the exchange integral must be "small", that is, small compared to the distances between E_0 and the other energy

* W. Heisenberg, *Z. Phys.*, **38**, 411, 1926; **41**, 239, 1297.
** V. Fock, *Z. Phys.*, **61**, 126, 1930.
*** E. A. Hylleraas, *Z. Phys.*, **48**, 469, 1928; **54**, 347, 1929; J. C. Slater, *Phys. Rev.*, **32**, 349, 1928.

levels of the system. This is seen, for instance, from a consideration of the higher approximations according to the scheme given in § 51. In most cases it will be difficult to give a numerical estimate of the accuracy obtained.

The peculiar character of the splitting of the energy level E_0 in a symmetric and an antisymmetric state is further illustrated by the following discussion. Let us forget for the moment the spin completely and let us consider the following solution of the general Schrödinger equation of the two-electron problem,

$$\psi = [\varphi_1(1)\varphi_2(2)+\varphi_1(2)\varphi_2(1)] \exp [-i(E+A)t/\hbar] \\ +[\varphi_1(1)\varphi_2(2)-\varphi_1(2)\varphi_2(1)]\exp [-i(E-A)t/\hbar], \qquad (78)$$

which is indeed to a fair approximation a solution, provided the electrons are independent to a first approximation. If at $t = t_0 = 0$ $\psi_{t0} = 2\varphi_1(1)\varphi_2(2)$, the first electron will definitely be in the state 1 and the second electron in the state 2. At $t = t_1 = \pi\hbar/2A$ we have, however, $\psi_{t1} = 2 \exp [-i(E+A)t_1/\hbar]\varphi_1(2)\varphi_2(1)$, and now the first electron is in the state 2 and the second electron in state 1. At $t = t_2 = 2t_1$ we have $\psi_{t_2} = 2 \exp [-i(E+A)t_2/\hbar]\varphi_1(1)\varphi_2(2)$. The electrons "oscillate" with the frequency $\nu = A/\pi\hbar$ between the two one-electron states. One often talks about the frequency with which exchange takes place. If one of the symmetry types is forbidden, the possibility of such processes, which enable us to think in terms of a classical picture, will be lost. To describe physical phenomena one must introduce the spin and thus the Pauli principle. Although both g^a and g^s occur, equation (78) has strictly speaking no longer a physical meaning. If one wants all the same to talk about the exchange of electrons (see § 71), it means that one wants to consider those aspects of the problem for the understanding of which the Pauli principle, and usually also the spin can be neglected to a first approximation.

§ 74. Systems with two electrons with spin forces

We return now to the general two-electron problem and investigate the influences of the spin forces, which we neglected up to now, on the stationary states described by the wave functions (62a) and (62b). We assume for the time being that these states are nondegenerate as far as the spinless problem is concerned.

The perturbation energy operator is of the form (compare eq. (6.79))

$$H^{(1)}_{\text{op}} = (H^{x1} S_{x1}+H^{y1} S_{y1}+H^{z1} S_{z1})+(H^{x2} S_{x2}+\ldots) \\ +(H^{x1x2} S_{x1} S_{x2}+H^{x1y2} S_{x1} S_{y2}+\ldots). \qquad (79)$$

The operators $H^{x1}, \ldots$ act on the spatial coordinates of the two electrons which can be indicated by writing $H^{x1}(1, 2), \ldots$ Since $H^{(1)}_{\text{op}}$ will be symmetric, we have

$$H^{x1}(1, 2) = H^{x2}(2, 1), \ldots; \\ H^{x1\,x2}(1, 2) = H^{x1\,x2}(2, 1),\ H^{x1\,y2}(1, 2) = H^{y1\,x2}(2, 1), \ldots \qquad (80)$$

The influence of $H^{(1)}_{\text{op}}$ on the energy depends to a first approximation on its expectation value in the unperturbed state; if this state is g-fold degenerate, we must, however, solve a secular problem of the g-th degree (§§ 48, 49).

As far as the singlet state (62b) is concerned, the situation is simple. The expectation value of expression (79) is equal to zero [see eqq. (6.78) and (80) and remember that $H^{x1x2}+H^{y1y2}+H^{z1z2}=0$ according to eq. (6.33)]. The eigenvalue ${}^1E$ is not changed in first approximation.

The situation is different in the triplet state, which in general will split into three. We have to evaluate the matrix elements of $H^{(1)}_{\text{op}}$ with respect to the three states $g^{\text{a}\,3}\zeta_1$, $g^{\text{a}\,3}\zeta_0$, $g^{\text{a}\,3}\zeta_{-1}$ (g^{a} normalised). Integration over the spatial coordinates of the two electrons means averaging over the state g^{a} of the spinless problem, for instance, $\overline{H^{x1}} = \int g^{\text{a}*} H^{x1} g^{\text{a}}\, d\boldsymbol{x}_1\, d\boldsymbol{x}_2$, $\overline{H^{x2}} = \int g^{\text{a}*} H^{x2} g^{\text{a}}\, d\boldsymbol{x}_1\, d\boldsymbol{x}_2$, . . . These integrals do not change their value if we change the numbering of the electrons. Since g^{a} is antisymmetric it follows by using equation (80) that

$$\left.\begin{aligned}&\overline{H^{x1}} = \overline{H^{x2}} = A_x, \quad \overline{H^{y1}} = \overline{H^{y2}} = A_y, \quad \overline{H^{z1}} = \overline{H^{z2}} = A_z;\\ &\overline{H^{x1x2}} = A_{xx}, \quad \overline{H^{x1y2}} = \overline{H^{y1x2}} = A_{xy}\ (= A_{yx}), \ldots,\end{aligned}\right\} \tag{81}$$

where $\boldsymbol{A}$ is a vector fixed in space and A_{kl} a fixed symmetric second rank tensor. From equation (6.33) it follows that $A_{xx}+A_{yy}+A_{zz}=0$.

As far as the perturbation of the triplet term is concerned, the operator (79) can be replaced by the following operator which is symmetric in the spin coordinates,

$$\overline{H^{(1)}_{\text{op}}} = (\boldsymbol{A}\cdot\boldsymbol{S}_1+\boldsymbol{S}_2)+A_{xx}S_{x1}S_{x2}+A_{xy}(S_{x1}S_{y2}+S_{y1}S_{x2})+\ldots \tag{82}$$

We must now evaluate the matrix elements of $H^{(1)}_{\text{op}}$ with respect to ${}^3\zeta_{-1}$, ${}^3\zeta_0$, and ${}^3\zeta_1$. These correspond to a matrix which (considered as operator) acts on the discrete argument s ($s = 1, 0, -1$). As we saw in § 60 such a matrix can always be expressed in terms of the three basic matrices S_x, S_y, S_z [see eq. (63)] which corresponds to the components of the total spin angular momentum $\boldsymbol{S}$ and which are constructed according to the scheme of § 45 (see fig. 11),

$$\frac{S_x}{\hbar}\rightarrow\left|\begin{matrix}0 & 2^{-\frac12} & 0\\ 2^{-\frac12} & 0 & 2^{-\frac12}\\ 0 & 2^{-\frac12} & 0\end{matrix}\right|,\quad \frac{S_y}{\hbar}\rightarrow\left|\begin{matrix}0 & -i2^{-\frac12} & 0\\ i2^{-\frac12} & 0 & -i2^{-\frac12}\\ 0 & i2^{-\frac12} & 0\end{matrix}\right|,\quad \frac{S_z}{\hbar}\rightarrow\left|\begin{matrix}1 & 0 & 0\\ 0 & 0 & 0\\ 0 & 0 & -1\end{matrix}\right|. \tag{83}$$

It is not difficult to verify that the operators S_x, . . . defined by equation (63) possesses this form with respect to the ${}^3\zeta_s$. The spin operators acting on the ${}^3\zeta_s$ which occur in equation (82) can be expressed as follows in terms of S_x, S_y, and S_z,

$$\boldsymbol{S}_1+\boldsymbol{S}_2=\boldsymbol{S};\ S_{x1}S_{x2}=\tfrac{1}{2}S_x^2-\tfrac{1}{4}\hbar^2,\ S_{x1}S_{x2}+S_{y1}S_{x2}=\tfrac{1}{2}(S_xS_y+S_yS_x),\ldots \tag{84}$$

Equation (84) can be obtained by a straightforward evaluation. However, the calculation can be shortened considerably if we take the way $\boldsymbol{S}$ behaves under a rotation explicitly into account *.

The perturbation of the triplet state is thus finally determined by the operator

$$\overline{H_{\mathrm{op}}^{(1)}}=(\boldsymbol{A}\cdot\boldsymbol{S})+\tfrac{1}{2}[A_{xx}S_x^2+A_{xy}(S_xS_y+S_yS_x)+\ldots] \\ -\tfrac{1}{4}\hbar^2(A_{xx}+A_{yy}+A_{zz}). \tag{85}$$

This result can be interpreted as follows, using correspondence language. The action of the spin forces on a triplet state is such that the resultant spin vector seems to be no longer free but to be situated in a field influencing its direction. The energy in this field contains only linear and quadratic terms in the spin components.

The linear terms correspond exactly to the influence of a homogeneous magnetic field on the spin vector. If there is, indeed, an external homogeneous magnetic field $\mathfrak{H}$ [corresponding to the terms with $\mathfrak{H}_i$ in eq. (6.33)] and if in the averaging process (81) everything drops out except the terms containing the magnetic field (and this often happens in applications) the linear term in equation (85) will be

$$\overline{H_{\mathrm{op}}^{(1)}} = (e/mc)(\mathfrak{H}\cdot\boldsymbol{S}), \tag{86}$$

just as in equation (6.42), the only difference being that the spin quantum number $\bar{s}$ is now 1 and not $\frac{1}{2}$.

The splitting up of the energy of the triplet state follows from the calculation of the three eigenvalues of expression (85). The corresponding spin eigenfunctions show which linear combinations of the ${}^3\zeta_s$ must be used in equation (62a) to obtain the eigenfunctions which in the zeroth approximation are adapted to the perturbation. In general this means a complicated secular problem of the third degree. In the special case where only linear terms occur, for instance, when we are dealing with the operator (86) the result is very simple.

* One verifies, for instance, the relation $S_{z1}S_{z2}=\frac{1}{2}S^2{}_z-\frac{1}{4}\hbar^2$, which is easily done since only diagonal terms occur. The other relations follow then immediately from the fact that they form together a second rank tensor.

Let the magnetic field $\mathfrak{H}$ be along the z-axis. In that case the three ${}^3\zeta_s$ are eigenfunctions of $\overline{H^{(1)}_{\text{op}}}$ and the corresponding eigenvalues are

$$ {}^3E' = {}^3E + s(e\hbar/mc)\mathfrak{H}, \qquad s = 1, 0, -1, \tag{87} $$

where ${}^3E$ is the eigenvalue of g^{a} of equation (62a) and where ${}^3E'$ denotes the three energy values of the perturbed system. In terms of the vector model: the resulting spin vector is parallel, perpendicular, or antiparallel to the magnetic field.

If, however, only the quadratic terms in expression (85) are different from zero, $\overline{H^{(1)}_{\text{op}}}$ can be written in the form

$$ \overline{H^{(1)}_{\text{op}}} = \tfrac{1}{2}(A_{xx}S_z^2 + A_{yy}S_y^2 + A_{zz}S_z^2) - \tfrac{1}{4}\hbar^2(A_{xx} + A_{yy} + A_{zz}) \tag{88} $$

by a suitable choice of the system of coordinates. The last term means only a general shift of the energy, which is zero if the energy is of the form (6.33) as we mentioned a moment ago. It is easily verified that ${}^3\zeta_0$ is an eigenfunction of expression (88) with eigenvalue $\frac{1}{2}\hbar^2(A_{xx}+A_{yy}) - \frac{1}{4}\hbar^2(A_{xx}+A_{yy}+A_{zz})$. The other eigenvalues are obtained by cyclic permutations *. The shift of the centre of gravity of the three components is $(\hbar^2/12)(A_{xx}+A_{yy}+A_{zz})$.

Up to now we have discussed the influence of the spin forces only on the energy. If we want to investigate their influence on the wave functions, we find by using the methods of perturbation theory instead of expressions (62a) and (62b) the following ones

$$ {}^3\varphi_i = g^{\text{a}}(c_{1i}\,{}^3\zeta_1 + c_{0i}\,{}^3\zeta_0 + c_{-1i}\,{}^3\zeta_{-1}) + {}^3\varphi_i', \qquad i = 1, 2, 3, \tag{89} $$

$$ {}^1\varphi = g^{\text{s}}\,{}^3\zeta_0 + \qquad {}^1\varphi', \tag{90} $$

where the φ' are small additional terms which may be expanded in terms of the eigenfunctions of the problem without spin forces. The index i refers to the three states into which the triplet state is split.

The determination of ${}^1\varphi'$ is simple, since g^{s} is non-degenerate and we have from equation (5.20b)

$$ {}^1\varphi' = -\Sigma_l[H^{(1)}_{lk}/(E_l - E_k)]\varphi_l^{(0)}; \quad H^{(1)}_{lk} = \int \varphi_l^{(0)*} H^{(1)}_{\text{op}} \varphi_k^{(0)}, \tag{91} $$

where the state (90) is denoted by the index k and where l refers to all other states of the problem without the spin perturbation. There are two kinds of $\varphi_l^{(0)}$, namely, the singlet functions of the type (62b) and the triplet functions of the type (62a). It is important to note

* These formulae can, for instance, be applied to the case of a triplet state without "orbital degeneracy" of a polyatomic molecule the nuclei of which are considered to be fixed centres.

that in general both kinds of function occur in ${}^1\varphi'$, that is, that the spin perturbation will "mix" into the original singlet state a small fraction of triplet functions. To prove this, we show that the matrix elements $H^{(1)}_{lk}$ in equation (91) where l refers to a triplet and k to a singlet state are in general different from zero. The expansion (79) shows that $H^{(1)}_{lk}$ is the sum of a number of terms each of which is a product of an integral over the spatial coordinates and a sum over the spin coordinates. For instance, the third and sixth term in expression (79) will lead to contributions of the form

$$\Big(\int (g^{a*} H^{z1}_{op} g^{s}\, d\boldsymbol{x}_1\, d\boldsymbol{x}_2\Big)\big(\Sigma_{s_1 s_2}\, {}^3\zeta_0{}^*\, S_{z1}\, {}^1\zeta_0\big)$$
$$+\Big(\int g^{a*} H^{z2}_{op} g^{s}\, d\boldsymbol{x}_1\, d\boldsymbol{x}_2\Big)\big(\Sigma_{s_1 s_2}\, {}^3\zeta_0{}^*\, S_{z2}\, {}^1\zeta_0\big). \qquad (92)$$

The sums and integrals in equation (92) would be equal to zero, if all operators H^{z1}_{op}, H^{z2}_{op}, S_{z1}, S_{z2} were symmetric in the two electrons. They are, however, not symmetric, but H^{z1} and H^{z2} satisfy the relation (80). The sums are just numbers which can be evaluated by using equations (59) and (6.66),

$$\Sigma_{s_1 s_2}\, {}^3\zeta_0^*\, S_{z1}\, {}^1\zeta_0 = -\Sigma_{s_1 s_2}\, {}^3\zeta_0^*\, S_{z2}\, {}^1\zeta_0 = \tfrac{1}{2}\hbar.$$

The two integrals are also the same, but with opposite sign; they will only be equal to zero, if H^{z1}_{op} happened to be symmetric.

We can use the same procedure for the triplet states. *In general, the stationary states will be superpositions of a triplet (singlet) state with a small admixture of singlet (triplet) states.* All the same the corresponding spectral terms are called in spectroscopy triplet (or singlet) terms. It may happen that triplet and singlet states are thoroughly mixed. This is, for instance, often the case when a triplet and a singlet term are lying very near to each other [for instance, because the exchange integral (76) happens to be very small]. Because the denominator in equation (91) is very small, the admixture is large. If $E_l - E_k$ is of the same order of magnitude as the spin perturbation of the triplet energy, one must obtain the influence of $H^{(1)}_{op}$ on the two unperturbed states by means of a secular equation of the fourth degree which takes both the triplet and the singlet terms into account (see the end of § 50).

If the singlet state is non-degenerate, the perturbation will not change the energy in the first approximation. This is clearly still true if the singlet state is degenerate, since all matrix elements of the perturbation matrix are equal to zero due to the properties of ${}^1\zeta_0$. The situation is different when we consider the perturbation of a *degenerate triplet state* by the spin forces. In the important applications to the case of a nucleus with two electrons (He I, Li II, ... spectra) degeneracy is the rule for the "spinless" states. We can see this in the vector model where we notice the arbitrariness of the "orientation" of the resulting orbital angular momentum which is obtained by adding the orbital angular momenta of the electrons. All atomic problems where all electrons bar two are in completed n,l-shells can to a fair approximation be treated as two-electron problems and in exactly the same manner we have in the zeroth approximation degenerate triplet states.

Let a degenerate triplet state be characterised in the zeroth approximation by the wave function [compare eq. (62a)]

$$^3\varphi = (\Sigma_{m=-l}^{+l} b_m g_m^{\mathrm{a}}) \cdot (\Sigma_{s=-1}^{+1} c_s {}^3\zeta_s), \qquad \text{eigenvalue } {}^3E, \qquad (93)$$

where the g_m^{a} are an orthonormal set of antisymmetric eigenfunctions of the time-independent Schrödinger equation (61) all corresponding to the eigenvalue ^3E. They are characterised by a quantum number m which can take on one of the $2l+1$ values $m = l, l-1, \ldots, -l$. We have chosen the notation in such a way that it is adapted to the application to atomic problems of the kind mentioned, but we need not yet attach any physical meaning to the quantum numbers m and l. The $2l+1$ coefficients b_m are arbitrary numbers as are the three coefficients c_s; the degree of degeneracy of ^3E is thus $3(2l+1)$. We assume explicitly that there are no perturbing states with energies near ^3E.

We should now construct according to the rules of perturbation theory all matrix elements of the perturbing energy (79) with respect to the states $g_m^{\mathrm{a}} {}^3\zeta_s$ and then determine the eigenvalues and eigenfunctions, which would be functions of the discrete arguments m and s, by solving the corresponding secular problem. The matrix elements of each of the terms of expression (79) is the product of an integral over the spatial coordinates of the electrons and a sum over the spin coordinates. The results of the summation are contained in equations (84) and (85). We shall use the following notation for the matrix elements obtained by performing the integrations,

$$(A_x)_{mm'} = \int g_m^* H^{x1} g_{m'} d\boldsymbol{x}_1 d\boldsymbol{x}_2 = \int g_m^* H^{x2} g_{m'} d\boldsymbol{x}_1 d\boldsymbol{x}_2,$$

$$(A_{xy})_{mm'} = \int g_m^* H^{x1y2} g_{m'} d\boldsymbol{x}_1 d\boldsymbol{x}_2 = \int g_m^* H^{y1x2} g_{m'} d\boldsymbol{x}_1 d\boldsymbol{x}_2 = (A_{yx})_{mm'}.$$

The perturbation matrix is now of the form [compare eq. (85)],

$$\overline{H_{\text{op}}^{(1)}} = (A_x)_{mm'} S_x + \ldots + \tfrac{1}{2}(A_{xx})_{mm'} S_x^2 + \tfrac{1}{2}(A_{xy})_{mm'}(S_x S_y + S_y S_x) + \ldots$$
$$- \tfrac{1}{4}\hbar^2 [(A_{xx})_{mm'} + (A_{yy})_{mm'} + (A_{zz})_{mm'}]. \qquad (94)$$

Formally we can express the $(A_x)_{mm'}, \ldots$ in terms of three basic matrices $(M_x)_{mm'}$, $(M_y)_{mm'}$, $(M_z)_{mm'}$, which would have the same form as the M-matrices of figure 11 in § 45. We would get polynomials of at most $2l$-th degree (see the end of § 60),

$$(A_x)_{mm'} = \alpha_x + \alpha_x^x M_x + \alpha_x^y M_y + \alpha_x^z M_z + \alpha_x^{xx} M_x^2 + \alpha_x^{xy}(M_x M_y + M_y M_x) + \ldots$$

We have written this expression in such a way that all α are real, if A_x is Hermitean. The energy operator is thus finally a polynomial of the $2l$-th degree in the M_x, M_y, M_z and of the 2-nd degree in the S_x, S_y, S_z,

$$\overline{H_{\text{op}}^{(1)}} = \alpha_x S_x + \ldots \quad + \alpha_x^x M_x S_x + \alpha_y^x M_x S_y + \ \ldots \quad + \alpha_{xx}^x M_x S_x^2 + \ldots$$
$$+ \alpha_x^{xx} M_x^2 S_x + \ldots \quad + \alpha_{xx}^{xx..} M_x^{2l} S_x^2 + \ldots \qquad (95)$$

The representation (95) has as yet no advantages over (94), since the indices x, y, z of the M do not refer to a spatial system of coordinates in the way that the indices of the S do. As soon, however, as we apply this perturbation operator to the case of triplet spectra, that is, to the case of two electrons in a central field of force, it is possible to use a representation such that M_x, M_y, M_z are the components of a vector just as S_x, S_y, S_z are. This is connected to the general theorems on the angular momentum discussed in § 61. In the problem where we neglected the spin perturbation and where ^3E is degenerate the energy operator $H_{\text{op}}^{(0)}$ is invariant under rotations which means that $H_{\text{op}}^{(0)}$ commutes with the components $M_{x\text{op}}$, $M_{y\text{op}}$, $M_{z\text{op}}$ of the total angular momentum and that the wave functions of the stationary states can be expanded in terms of eigenfunctions of M_{op}^2 and $M_{z\text{op}}$ and can be classified according to the eigenvalues of these operators. We shall show in § 79 that the eigenvalues of M^2 are the same as in the one-electron problem: $M^2 = l(l+1)\hbar^2$ [see eq. (4.151)] where l is an integer, the orbital angular momentum quantum number. A stationary state characterised by a certain value of l is $2l+1$-fold degenerate and the corresponding eigenfunctions can be chosen to be eigenfunctions of M_z corresponding to the eigenvalues $m\hbar$ ($|m| \leq l$). Finally the matrices of M_x, M_y, M_z with respect to the φ_m are of the form given by figure 11.

If M_x, M_y, and M_z have this meaning in equation (95), this equation can be simplified considerably because, like $H_{\text{op}}^{(0)}$, also $H_{\text{op}}^{(0)} + H_{\text{op}}^{(1)}$

[and thus also expression (95)] is invariant under rotation of the x, y, z-coordinate system. That means that expression (95) must be invariant under any simultaneous orthogonal transformation of the components of $\boldsymbol{M}$ and $\boldsymbol{S}$. As a result expression (95) can contain only three arbitrary constants and must be of the form

$$\overline{H_{\text{op}}^{(1)}} = \alpha_0+\alpha_1(\boldsymbol{M}\cdot\boldsymbol{S})+\alpha_2(\boldsymbol{M}\cdot\boldsymbol{S})^2. \tag{96}$$

In terms of the vector model this means that the orbital momentum vector and the spin angular momentum vector are coupled by a "force" the energy of which can be expanded in powers of the „cosine" of the angle between the two vectors. The term with α_1 is called cosine-coupling and it would not be difficult to discuss its physical meaning using the explicit expressions for $H^{x1}, \ldots$ [compare eqq. (79) and (6.33)]. The term with α_2, the "cosine squared coupling", is due to spin-spin interaction [see again eq. (6.33)].

The problem of the helium spectrum * and of triplet spectra in general leads, as we see from equation (96), first of all to the problem of the eigenvalues and eigenfunctions of the matrix $(\boldsymbol{M}\cdot\boldsymbol{S})$. This is a special case of the general problem of the addition of vectors, to be discussed in § 79.

§ 75. Analysis of multiplet situations in the N-electron problem

We shall now generalise the analysis of § 73 to the case of N electrons.

We write a function ψ of the $\boldsymbol{x}_i$, s_i $(i = 1, 2, \ldots, N)$ in the following form [compare eq. (6.77)]

$$\psi = \Sigma_{s_1,s_2,\ldots,s_N} f_{s_1,s_2,\ldots,s_N}(1, 2, \ldots, N)\, \xi_1^{(s_1)} \xi_2^{(s_2)} \ldots \xi_N^{(s_N)}$$
$$= \Sigma_{s_i} f_{s_i}(\boldsymbol{x}_i) \Pi_i \xi_i^{(s_i)}. \tag{97}$$

which for $N = 2$ reduces the expression (56). The $f(1, 2, \ldots)$ are functions of the spatial coordinates of the electrons, where 1, 2, ... is a shorthand notation for $\boldsymbol{x}_1$, $\boldsymbol{x}_2$, ... Instead of ξ and η we use the notation ξ^+ and ξ^- (or $\xi^{(+\frac{1}{2})}$ and $\xi^{(-\frac{1}{2})}$, in general $\xi^{(s)}$). There are 2^N normalised and mutually orthogonal spin functions $\Pi_i \xi^{(s_i)}$. An arbitrary *spin function* ζ can be expressed as a linear combination of them,

$$\zeta = \Sigma a_{s_1,\ldots,s_N} \Pi_i \xi_i^{(s_i)}, \tag{98}$$

* See especially H. Bethe, *Handb. Phys.*. **24₁**, 324, 1933.

where the $a_{s_1, \ldots, s_N}$ form a set of 2^N numerical coefficients. If we have a set of 2^N of such ζ-functions, $\zeta_1, \ldots, \zeta_k, \ldots$, ψ can clearly be "expanded" in terms of the ζ,

$$\psi = \Sigma_k f_k(1, 2, \ldots, N)\, \zeta_k, \tag{99}$$

as long as the ζ_k are linearly independent. It is not necessary that the ζ_k are orthogonal to each other.

For the time being we use expression (97) and we introduce the condition that ψ be antisymmetric in the electrons,

$$\mathrm{P}\psi = \delta_{\mathrm{P}}\psi. \tag{100}$$

For each permutation P we can find by comparing terms with the same spin functions a number of relations between the functions f_{s_i} which reduce to equations (57) for the case $N = 2$. Each s_i can be either $+\frac{1}{2}$ or $-\frac{1}{2}$. Since under a permutation the s_i-values are interchanged, we see immediately that from a function $f_{s_i}(\boldsymbol{x}_i)$ corresponding to a given set of s_i-values we can derive the form of all those functions $f_{s_i'}(\boldsymbol{x}_i)$ for which the number of s_i which are equal to $+\frac{1}{2}$ is the same as for f_{s_i}. If for a moment we denote those values of i for which $s_i = +\frac{1}{2}$ by p (there may be N_+ of them) and those values of i for which $s_i = \frac{1}{2}$ by q (there may be N_- of them), $f_{s_i}(\boldsymbol{x}_i)$ must also be antisymmetric in all $\boldsymbol{x}_p$ and at the same time be antisymmetric in all $\boldsymbol{x}_q$ *. The most general antisymmetric ψ is constructed by giving for every splitting up of N into two non-negative integers

$$N = N_+ + N_- \quad (N_+ = N, N-1, \ldots, 0) \tag{101}$$

a function of $\boldsymbol{x}_1, \ldots, \boldsymbol{x}_N$ which is antisymmetric in a group of N_+ of its arguments and also antisymmetric in the other N_- arguments. Altogether ψ is determined by giving $N+1$ such functions **.

We must now find an expansion of expression (97) which is similar to equation (58),

* This is so because a permutation of the N_+ (N_-) electrons p (q) does not change the spin function $\Pi_i \xi_i^{(s_i)}$.

** Example for $N=3$: $\psi = \underbrace{f_{+++}\xi_1\xi_2\xi_3}_{N_+=3} + \underbrace{(f_{++-}\xi_1\xi_2\eta_3 + f_{+-+}\xi_1\eta_2\xi_3 + f_{-++}\eta_1\xi_2\xi_3)}_{N_+=2}$

$+ \underbrace{(f_{+--}\xi_1\eta_2\eta_3 + f_{-+-}\eta_1\xi_2\eta_3 + f_{--+}\eta_1\eta_2\xi_3)}_{N_+=1} + \underbrace{f_{---}\eta_1\eta_2\eta_3}_{N_+=0}.$

In this case $f_{+++}(1, 2, 3) = -f_{+++}(2, 1, 3) = \ldots$; similarly for f_{---}; $f_{++-}(1, 2, 3) = -f_{++-}(2, 1, 3) = -f_{+-+}(1, 3, 2) = -f_{-++}(3, 2, 1) = \ldots$; similarly for f_{+--}.

$$\psi = {}^{N+1}\psi + {}^{N-1}\psi + {}^{N-3}\psi + \ldots + ({}^{1}\psi \text{ or } {}^{2}\psi), \tag{102}$$

where the last superscript is 2 or 1 according to whether N is odd or even. Each term is a product of a space function and a spin function or a sum of such products,

$$^{M}\psi = \Sigma_{k_M} {}^{M}g_{k_M}(1, 2, \ldots, N)\, {}^{M}\zeta_{k_M}, \quad M = N+1, N-1, \ldots \tag{103}$$

Equation (102) corresponds to an expansion (99) in terms of a new set of spin functions ζ_k which fall into subgroups in a way to be specified.

This division of the wave function will be according to *two physical viewpoints*. *First of all* we require that *each of the terms* $^{M}\psi$ *is antisymmetric,* which means that each of them corresponds to a situation which satisfies the Pauli principle,

$$\mathrm{P}^{M}\psi = \delta_{\mathrm{P}}{}^{M}\psi \tag{104}$$

Secondly, the totality of all situations corresponding to one of the terms should be independent of the orientation of the spatial coordinate system. The description of the spin functions is connected with a definite choice of the x, y, z-system. If we choose three new axes, x', y', z', we get for each electron two new unit spin functions ξ_i', η_i' which are connected with the previous ξ_i, η_i according to equation (6.99). If ψ is now expressed in the ξ_i', η_i' (transformed ψ'), $\psi' = \Sigma_{s_i} f'_{s_i} \Pi_i \xi_i'^{(s_i)}$, and if we expand ψ' in terms of spin functions $^{M}\zeta'_{k_M}$ which are obtained from the $\Pi_i \xi_i'^{(s_i)}$ in the same way as the ζ from the $\Pi\xi$, $\psi' = {}^{N+1}\psi' + {}^{N-1}\psi' + \ldots$, our second requirement means that for each value of M, $^{M}\psi'$ is just that function which is obtained by a "rotation" from the original $^{M}\psi$. We emphasise that we are at the moment not at all interested in a possible transformation of the spatial coordinates. These coordinates may, for instance, be referred to a curvilinear system of coordinates which we have chosen once and for all (compare the discussion in § 61).

In writing down equation (102) we have tacitly assumed that an expansion of this kind is, indeed, possible and will contain at most $\frac{1}{2}(N+2)$ or $\frac{1}{2}(N+1)$ terms according to whether N is even or odd. We may also mention at this moment that two different $^{M}\psi$ are always orthogonal to one another.

The physical meaning of the second requirement can be seen from the following considerations. Classically the instantaneous situation of an N-electron system is characterised by the positions $\boldsymbol{x}_i$ and velocities $\boldsymbol{v}_i$ of all the electrons and by the directions of their spin vectors $\boldsymbol{S}_i$. A rotation of the system of coordinates corresponds from the point of view of this system to an opposite rotation of all these vectors. We now want to find a collection of states which are as far as possible restricted with respect to the combinations of the spin directions occurring and which go over into the same collection when a rotation of the kind just mentioned occurs. As far as the $\boldsymbol{x}_i$ and $\boldsymbol{v}_i$ are concerned, this is immediately the case, as all possible combinations of these vectors occur anyhow. We can thus neglect them as far as the present discussion is concerned. A collection of this sort is obtained by first of all considering one combination of $\boldsymbol{S}_i$-directions and then adding to this combination all those systems of $\boldsymbol{S}_i'$-directions which are obtained from the first one by rotations. Since the electrons are equivalent, the order in which the $\boldsymbol{S}_i$ or $\boldsymbol{S}_i'$ are taken will be immaterial. The classical classification of the states corresponding to the expansion of ψ is thus the one where *the states are distinguished by the way in which the various spin directions are orientated with respect to each other.* Classically this would clearly lead to a division into infinitely many collections of states. In the case of two electrons these would, for instance, be characterised by the angle between the two spin vectors or by the length of the resulting spin vector, at any rate by a continuous variable. In the case of N electrons there would be $2N-3$ variables (2 for the direction of each spin; the -3 occurs because the whole system of spins can be arbitrarily orientated in space).

These considerations help us to understand the method employed to find the expansion (102). It is obtained from the following theorem. *The expansion* (102) *corresponds exactly to the expansion in terms of the eigenfunction of the square of the resultant spin angular momentum* S^2, *that is, the different terms correspond to the different eigenvalues of* S^2. In the classical case we were looking for the values of symmetrical functions of the spin-directions which depend only on the relative orientations of the various spins, but not on the orientation in space. The "collections" of states were then to be classified according to these "values". In quantum mechanics these functions will be observables or operators symmetric in the S_{xi}, S_{yi}, S_{zi} ($i = 1, \ldots N$) which are invariant under rotations, that is, under a simultaneous real orthogonal transformation of all triples S_{xi}, S_{yi}, S_{zi} (see § 61). From the fact that the most general operator can always be expressed as a linear combination of the spin components of each of the electrons (since the spin quantum number of the electron is $\frac{1}{2}$) it follows that there is *essentially only one operator of the kind we are looking for*, namely, the absolute square of the resultant spin angular momentum,

$$S^2_{\text{op}} = (\Sigma_i \boldsymbol{S}_i \cdot \Sigma_i \boldsymbol{S}_i) = \Sigma_i S_i^2 + \Sigma_{ij} (\boldsymbol{S}_i \cdot \boldsymbol{S}_j) = N \cdot \tfrac{3}{4}\hbar^2 + \Sigma_{ij} (\boldsymbol{S}_i \cdot \boldsymbol{S}_j). \quad (105)$$

Any polynomial in S^2 would also be sufficient, but it would lead to exactly the same classification of the eigenfunctions, and the latter is the only point of interest.

Terms of n-th degree in an operator symmetric in the $\boldsymbol{S}_i$ can always be written in the form $\mathscr{P}(\boldsymbol{S}_i) = \Sigma_k c_k \Sigma \text{P} \text{P} \Pi_{i=1}{}^n (\boldsymbol{A}_k \cdot \boldsymbol{S}_i)$, where the $\boldsymbol{A}_k$ is a sufficiently large set of fixed vectors. This expression satisfies the symmetry requirements (the P refer to the electrons i). One can first expand the right-hand side in terms of products of the components of the $\boldsymbol{A}_k$. If $\mathscr{P}$ is invariant under rotations one may average these products over all possible orientations of the coordinate system. This leads to the result that each term of the sum over k will be the same polynomial of the $\boldsymbol{S}_i$, provided n is even, while each term will be equal to zero, if n is odd.

There is thus only *one* polynomial of this kind for each even value of n. This must just be the $\frac{1}{2}n$-th power of the expression (105), as can also be shown directly.

It is interesting to see how only one quantity, the total angular momentum, remains of the $2N-3$ quantities which were needed for the classical classification of the various states. In the language of the vector model one explains this by saying that each spin can only be parallel or antiparallel to $\boldsymbol{S}$. If, moreover, the order in which the spins are combined is immaterial, the relative orientations are completely determined by the length of $\boldsymbol{S}$. If N_+ electron spins are parallel to $\boldsymbol{S}$ and N_- antiparallel, the total length is $S = \frac{1}{2}\hbar(N_+ - N_-)$, and there are exacty $N+1$ possibilities, $N_+ = N, N-1, \ldots, 0$ corresponding to a total spin $S = N \cdot \frac{1}{2}\hbar$, $(N-2) \cdot \frac{1}{2}\hbar, \ldots$

We have even obtained the number of possible expansions (102) but the proof we just gave is not rigorous. A rigorous procedure is the following one. Let there be given 2^N linearly independent functions ζ_k which are divided in some way in groups ${}^M\zeta_{k_M}$. One can construct an operator Ω which satisfies the relation $\Omega^M \zeta_{k_M} = a_M {}^M\zeta_{k_M}$, where the a_M are arbitrary numbers which are once and for all chosen for each group characterised by M. In a representation where the rows and columns of the matrix Ω refer to the k_M this matrix will be a diagonal matrix of the form

$$
\begin{array}{c|cc|cc|c}
 & \multicolumn{2}{c}{{}^{M'}\zeta_{k_{M'}}} & \multicolumn{2}{c}{{}^{M''}\zeta_{k_{M''}}} & \\
\hline
{}^{M'}\zeta_{k_M} & \begin{matrix} a_{M'} & 0 \\ 0 & a_{M'} \\ & & \ddots \end{matrix} & & 0 & & \\
\hline
{}^{M''}\zeta_{k_M} & 0 & & \begin{matrix} a_{M''} & 0 \\ 0 & a_{M''} \\ & & \ddots \end{matrix} & & \\
\hline
\end{array}
$$

If we know that the action of another operator Ω' on the ζ_{k_M} leads only to linear combinations of the ζ_{k_M} belonging to the same M-value, Ω' will commute with Ω, since Ω acts as a unit matrix as far as each M-group is concerned. In our case all permutations (first requirement) and all rotations (second requirement) are the only operators Ω' which satisfy this requirement. This means that Ω must commute with every P, that is, Ω must be symmetric in the electrons, and Ω must be invariant under rotations, that is, Ω must commute with the components of $\Sigma \boldsymbol{S}_i$ (see § 61). We showed a moment ago that there is essentially only one operator of this type, namely S^2_{op}. As this operator is Hermitean it follows at the same time that two functions ${}^M\zeta$ which correspond to different values of M — and thus also two of the terms ${}^M\psi$ — will be orthogonal to each other. There is no danger that the use of S^2_{op} will not lead to the most general expansion (102). If there exist subdivisions, there must exist another operator of the type Ω *.

§ 76. Rotations and angular momentum operators

We shall now consider the further characterisation of the spin functions occurring in the term ${}^M\psi$ of the expansion (102), that is, we shall consider the eigenfunctions and eigenvalues of the operator S^2. In view of later applications we shall first of all consider the following more general problem. Consider the functions $\varphi(q) \equiv \varphi(q_1, q_2, \ldots)$ of the variables $q_1, q_2, \ldots$ which are defined with respect to a spatial orthogonal system of coordinates x, y, z. The q_i may be continuous or they may be discrete variables. The coordinates which are defined in the same way with respect to another

* One is dealing here with the simultaneous reduction of the spin functions with respect to the permutation group and the rotational group. See B. L. v. d. Waerden, *Die gruppentheoretische Methode in der Quantenmechanik*, Springer, Berlin, 1932.

system of coordinates x', y', z' which follows from x, y, z by rotation are denoted by $q'(q_1', q_2', \ldots)$. Let it be known how one can obtain from a function of the q the corresponding function of the q' and vice versa (transformation of functions). A linear combination of two functions will go over into the same linear combination of the transformed functions *. Let φ be a given function. We can now transform $\varphi(q')$ into a function $\varphi'(q)$ of the original coordinates, $\varphi(q') \to \varphi'(q)$. The function $\varphi'(q)$ originates, so to say, by letting „the function φ rotate with the coordinate system".

The linear operator relation which produces φ' from φ depends on the rotation producing x', y', z' from x, y, z. We shall denote the operator, which is not necessarily Hermitean, by R, $\varphi' = R\varphi$.

The operators corresponding to an infinitesimal rotation around the x-, y-, and z-axis, divided by the angle of rotation, are denoted as in § 61, by** $-(i/\hbar)J_x$, $-(i/\hbar)J_y$, $-(i/\hbar)J_z$, and we speak of the angular momentum vector operator $\boldsymbol{J}$ with components J_x, J_y and J_z.

The J_x, J_y, J_z satisfy automatically the commutation relations $[J_x, J_y]_- = i\hbar J_z, \ldots$; $[J^2, J_x]_- = 0, \ldots$, where $J^2 = (\boldsymbol{J} \cdot \boldsymbol{J}) = J_x^2 + J_y^2 + J_z^2$.

For the sake of simplicity we assume J_x, J_y, J_z to be *Hermitean operators* ***. If φ is an eigenfunction of J^2, $\varphi' = R\varphi$ will also be an

* Examples: (1) A scalar field is determined by a given function of x, y, z. The transformed function is obtained by expressing this function in terms of x', y', z'.

(2) A vector $\boldsymbol{V}$ is considered to be a function of an argument k which can take on only three values, $k = 1, 2, 3$; for instance, $f_1 = V_x$, $f_2 = V_y$, $f_3 = V_z$ (or $f_1 = V_x + iV_y$, $f_2 = V_x - iV_y$, $f_3 = V_z$). The introduction of the new system of coordinates leads to the new function $f_1' = V_{x'}$, $f_2' = V_{y'}$, $f_3' = V_{z'}$ (or $f_1' = V_x' + iV_y', \ldots$) the values of which follow from the values of f according to a linear operator law.

(3) A spinor, the components of which are functions of x, y, z (Pauli's two-component wave function).

Only one condition is characteristic, namely, that two successive rotations must lead to the same result as the resultant rotation applied directly. (This may lead to an uncertainty in sign). In group theory one would say that one is dealing with a representation of the rotational group.

** The difference in sign [see, for instance, eq. (6.110)] is due to the fact that in Chapter VI—where we considered the invariance of the description of physical systems under rotations of the system of coordinates—we always considered infinitesimal rotations of the system of coordinates [$\varphi(q) \to \varphi'(q')$] while we now consider infinitesimal rotations of the function, while the system of coordinates stays fixed.

*** If this were not the case, one could find a non-unitary transformation $q \to \bar{q}$ which would transform $\varphi(q)$ into $\bar{\varphi}(\bar{q})$ in such a way that J_x, J_y, J_z become Hermitean. This point is not important for our discussions. See H. Casimir and B. v. d. Waerden, *Math. Ann.*, **111**, 1, 1935.

eigenfunction, since from $J^2\varphi = A\varphi$ it follows that $J_x J^2 \varphi = J^2 J_x \varphi = A J_x \varphi, \ldots$, so that each function $[1-(i/\hbar)(\boldsymbol{\omega}\cdot\boldsymbol{J})]\varphi$ which is obtained from φ by an infinitesimal rotation ω is also an eigenfunction of J^2. A finite rotation can always be constructed out of infinitesimal rotations. One can appeal to the fact that the action of J^2_{op} is invariant under rotation, because $\boldsymbol{J}$ behaves as a vector.

Since there are ∞^3 operators R, we can obtain ∞^3 functions φ'. We assume that there are among those only a *finite* number of linearly independent functions. Let $\varphi_1, \varphi_2, \ldots, \varphi_k$ be a set of such linearly independent functions. We can construct from them linear combinations which are eigenfunctions of J_z. This is always possible since each $J_z\varphi_k$ will be a linear combination of all the φ_k, since it is obtained from φ_k by an infinitesimal rotation. Let $j\hbar$ be the largest eigenvalue of J_z with corresponding eigenfunction φ_j. Let $J_+ = J_x + iJ_y$, $J_- = J_x - iJ_y$. We then see that

$$J_+J_- = J_x^2+J_y^2-i(J_xJ_y-J_yJ_x) = J^2-J_z^2+\hbar J_z, \tag{106}$$

$$\begin{aligned} J_-J_+ &= \qquad\qquad\qquad\qquad = J^2-J_z^2-\hbar J_z, \\ [J_z, J_+]_- = \hbar J_+, \quad &[J_z, J_-]_- = -\hbar J_-, \quad [J_+, J_-]_- = 2\hbar J_z. \end{aligned} \tag{107}$$

From

$$J_z\varphi_j = j\hbar\varphi_j \tag{108}$$

it follows that

$$\begin{aligned} J_z(J_+\varphi_j) &= J_+(J_z+\hbar)\varphi_j = \hbar(j+1)(J_+\varphi_j), \\ J_z(J_-\varphi_j) &= J_-(J_z-\hbar)\varphi_j = \hbar(j-1)(J_-\varphi_j), \end{aligned}$$

or, $J_+\varphi_j$ and $J_-\varphi_j$ are eigenfunctions of J_z corresponding to eigenvalues $(j+1)\hbar$ and $(j-1)\hbar$. As $j\hbar$ was the largest eigenvalue of J_z, $J_+\varphi_j$ must be equal to zero, and thus also $J_-J_+\varphi_j$. From equation (107) it follows that

$$0 = (J^2-J_z^2-\hbar J_z)\varphi_j = (J^2-j^2\hbar^2-j\hbar^2)\varphi_j, \text{ or, } J^2 = j(j+1)\hbar^2. \tag{109}$$

The eigenvalue of J^2 is thus expressed in terms of the largest eigenvalue of J_z.

Consider the eigenfunctions $J_-\varphi_j, J_-^2\varphi_j, \ldots$ corresponding to eigenvalues $(j-1)\hbar$, $(j-2)\hbar$, ... This sequence must terminate at an eigenfunction $\varphi_{\bar{j}}$ corresponding to the smallest eigenvalue $\bar{j}\hbar$ of J_z, and for that eigenfunction we have $J_-\varphi_{\bar{j}} = 0$, and thus from equation (106), $0 = J_+J_-\varphi_{\bar{j}} = (J^2J_z^2+\hbar J_z)\varphi_{\bar{j}} = (J^2_-\bar{j}^2\hbar^2+\bar{j}\hbar)\varphi_{\bar{j}}$. or, $J^2 = \bar{j}(\bar{j}-1)\hbar^2$, which is in agreement with equation (109), only if

$\bar{j} = -j$. Since $j-\bar{j}$ is an integer, j must be an *integer* or a *half-integer.* One calls j the *angular momentum quantum number.*

We have now found a set of eigenfunctions and eigenvalues,

$$\left.\begin{array}{ll}\text{Eigenvalues:} & j\hbar,\ (j-1)\hbar,\ \ldots,\ m\hbar,\ \ldots,\ -j\hbar; \\ \text{Eigenfunctions:} & \varphi_j,\ \varphi_{j-1},\ \ldots,\ \varphi_m,\ \ldots,\ \varphi_{-j},\end{array}\right\} \tag{110}$$

where m is usually called the *magnetic quantum number.*

The φ_m which can still be multiplied by an arbitrary constant will be chosen from now on to satisfy the relations

$$J_+\varphi_m = (j-m)\hbar\varphi_{m+1}, \qquad J_-\varphi_m = (j+m)\hbar\varphi_{m-1}, \tag{111}$$

where the second relation follows from the first one and equations (106) and (107).

If the set of functions (110) does not form a complete set of eigenfunctions of J^2, one can find in addition one or more different sets, each of $2j+1$ functions φ_m. It is clearly possible to arrange these sets in such a way that functions belonging to different sets are orthogonal to each other.

The φ_m belonging to one set are orthogonal to each other since they correspond to different eigenvalues of J_z, but they are not normalised. From equation (111) it follows that

$$\int|\varphi_{m+1}|^2 = \frac{\hbar^{-2}}{(j-m)^2}\int(J_+\varphi_m)^*(J+\varphi_m) = \frac{\hbar^{-2}}{(j-m)^2}\int\varphi_m J_-^* J_+^* \varphi_m^*$$

$$= \frac{\hbar^{-2}}{(j-m)^2}\int\varphi_m^* J_- J_+\varphi_m = \frac{j(j+1)-m(m+1)}{(j-m)^2}\int\varphi_m^*\varphi_m = \frac{j+m+1}{j-m}\int|\varphi_m|^2.$$

Introducing the normalised functions $\underline{\varphi}_m$ by writing

$$\underline{\varphi}_m = C_m\varphi_m, \quad C_m^2/C_{m+1}^2 = (j+m+1)/(j-m), \int|\underline{\varphi}_m|^2 = 1,$$

the matrix elements of J_x and J_y which are different from zero follow from the equation

$$\begin{aligned}(J_+)_{m,\,m-1} &= \int\underline{\varphi}_m^* J_+\underline{\varphi}_{m-1} = \hbar C_m C_{m-1}(j-m+1)|\varphi_m|^2 \\ &= \hbar[(j+m)(j-m+1)]^{1/2}.\end{aligned} \tag{112}$$

We thus finally obtain for the matrices of J_x, J_y, J_z exactly the same expressions as we found in § 45 for the components of the angular momentum of a single electron. The only difference is that what was called l there, is now called j and that j can be both integral and half integral.

We have here derived in a very general way the simple properties which any angular momentum operator possesses and which we had met several times before (§§ 45, 58, 59, 66).

The functions φ_m refer to a given x, y, z system of coordinates. If we use again the rotational operator R which corresponds to a rotation from x, y, z to x', y', z', each of the φ_m goes over into a φ'_m according to the relation $\varphi'_m = R\varphi_m$. These φ'_m play the same role in the x', y', z' system as the φ_m in th x, y, z system. Each of the φ'_m must, however, be a linear combination of the φ_m,

$$\varphi'_m = \Sigma_{m'=-j}^{+j} \alpha^R_{m'\,m} \varphi_{m'}. \tag{113}$$

One can easily show that the matrices $\alpha^R_{m'\,m}$ which correspond to the rotations R satisfy the rules of matrix multiplication. That means that if R'' is the result of first a rotation R and then a rotation R', $R''\varphi = R'R\varphi$, we have $\Sigma_{m'} \alpha^{R'}_{m''\,m'} \alpha^R_{m'\,m} = \alpha^{R''}_{m''\,m}$.

A simple and complete characterisation of the rotational transformation (113) can be obtained by noting that the following expression is invariant,

$$\psi = \Sigma_{m=-j}^{+j} u^{j+m} v^{j-m} \binom{2j}{j+m} \varphi_m = \Sigma_m u'^{j+m} v'^{j-m} \binom{2j}{j+m} \varphi'_m. \tag{114}$$

In this expression u, v (u', v') are the components of a spinor referred to the x, y, z- $(x', y', z'$-) system of coordinates. This equation is completely analogous to the equation of § 61, where we had

$$u\xi + v\eta = u'\xi' + v'\eta', \tag{115}$$

which expressed the fact that the description of a spinor by means of the expression $u\xi + v\eta$ is invariant. The analogy follows from the fact that the unit spin functions ξ, η (ξ', η') are eigenfunctions of S_z (S'_z) just as φ_m (φ'_m) are eigenfunctions of $J_z (J'_z)$.

To prove that expression (114) is invariant, we show that it is invariant under an infinitesimal rotation. A small change in u, v, and φ_m corresponds to the following change in ψ

$$\begin{aligned} \delta\psi &= \delta\Big[\Sigma_m u^{j+m} v^{j-m} \binom{2j}{j+m} \varphi_m\Big] \\ &= \Sigma_m \binom{2j}{j+m} [(j+m)\, u^{j+m-1} v^{j-m} \varphi_m \delta u + (j-m)\, u^{j+m} v^{j-m-1} \varphi_m \delta v + u^{j+m} v^{j-m} \delta\varphi_m]. \end{aligned} \tag{116}$$

An infinitesimal rotation $\boldsymbol{\omega}$ corresponds to the following changes in u, v, and φ_m [compare eq. (6.109)]

$$\begin{pmatrix} u'-u \\ v'-v \end{pmatrix} = \begin{pmatrix} \delta u \\ \delta v \end{pmatrix} = \frac{i}{\hbar}(\boldsymbol{\omega} \cdot \mathbf{S}) \begin{pmatrix} u \\ v \end{pmatrix}, \quad \varphi_m' - \varphi_m = \delta\varphi_m = -\frac{i}{\hbar}(\boldsymbol{\omega} \cdot \boldsymbol{J})\, \varphi_m,$$

where we have used a self-explanatory notation. The components S_z, J_z correspond to the changes (we omit the common factor $i\omega_z$; compare eq. (110)] $\delta u \to \frac{1}{2}u$ $\delta v \to -\frac{1}{2}v$, $\delta\varphi_m \to -m\varphi_m$, while we have corresponding to the „components" $S_+(=S_x+iS_y)$, $J_+(=J_x+iJ_y)$ the changes (compare eq. (111)] $\delta u \to v$, $\delta v \to 0$, $\delta\varphi_m \to -(j-m)\varphi_{m+1}$, and corresponding to the „components" $S_-(=S_x-iS_y)$, $J_-(=J_x-iJ_y)$ the changes [see again eq. (111)] $\delta u \to 0$, $\delta v \to u$, $\delta\varphi_m \to -(j+m)$ φ_{m-1}. Each of these three possibilities leads to $\delta\psi = 0$ when substituted into equation (116).

We know how u, v transform under a rotation [see eq. (6.95), (6.96)]

$$u' = \alpha u+\beta v,\ v'= -\beta^* u+\alpha^* v;\ u=\alpha^* u'-\beta v',\ v = \beta^* u'+\alpha v'. \quad (117)$$

If we now substitute the equations for u, v into equation (114) we find by comparing the coefficients of $u'^{j+m}v'^{j-m}$ how φ_m can be expressed in terms of the φ_m. A simpler way to obtain the transformation of the φ_m is by taking the $2j$-th power of equation (115),

$$(u\xi+v\eta)^{2j} = (u'\,\xi'+v'\,\eta')^{2j},$$

$$\Sigma_m\, u^{j+m}\, v^{j-m}\binom{2j}{j+m}\xi^{j+m}\eta^{j-m} = \Sigma_m\, u'^{j+m}\, v'^{j-m}\binom{2j}{j+m}\xi'^{j+m}\,\eta'^{j-m}. \quad (118)$$

Comparing equations (114) and (118) we see that *the "monomials" $\xi^{j+m}\eta^{j-m}$ transform in the same way as the φ_m.*

The (unnormalised) monomials have themselves no physical meaning. We are only interested in their transformation properties which follow immediately from the unitary transformation of the ξ and η [see eq. (6.99)]. In the cases $j = \frac{1}{2}$ and $j = 1$ we have

$$\left.\begin{aligned}
&j=\tfrac{1}{2}\begin{cases}\xi' = \alpha^*\xi+\beta^*\eta \\ \eta' = -\beta\xi+\alpha\eta\end{cases} \to \begin{aligned}\varphi'_{1/2} &= \alpha^*\varphi_{1/2}+\beta^*\varphi_{-1/2}\\ \varphi'_{-1/2} &= -\beta\varphi_{1/2}+\alpha\varphi_{-1/2}\end{aligned}\\
&j=1\begin{cases}\xi'^2 = (\alpha^*\xi+\beta^*\eta)^2 = \alpha^{*2}\xi^2+2\alpha^*\beta^*\xi\eta+\beta^{*2}\eta^2\\
\xi'\eta' = (\alpha^*\xi+\beta^*\eta)(-\beta\xi+\alpha\eta) = -\alpha^*\beta\xi^2+(\alpha^*\alpha-\beta^*\beta)\xi\eta+\beta^*\alpha\eta^2\\
\eta'^2 = (-\beta\xi+\alpha\eta)^2 = \beta^2\xi^2-2\beta\alpha\,\xi\eta+\alpha^2\eta^2\\
\qquad\qquad\downarrow\\
\varphi'_1 = \alpha^{*2}\varphi_1+2\alpha^*\beta^*\varphi_0+\beta^{*2}\varphi_{-1}\\
\varphi'_0 = -\alpha^*\beta\varphi_1+(\alpha^*\alpha-\beta^*\beta)\varphi_0+\beta^*\alpha\varphi_{-1}\\
\varphi'_{-1} = \beta^2\varphi_1-2\beta\alpha\varphi_0+\alpha^2\varphi_{-1}\end{cases}
\end{aligned}\right\} \quad (119)$$

The $\alpha^R_{m'\,m}$ of equation (113) can thus for every rotation and every value of j be expressed without any difficulty in terms of the α, α^*, β, β^*. If $2j$ is odd, we have the same ambiguity in the sign of the φ_m as we had in the case of the spinors. If $2j$ is even, the coefficients

$\alpha^R_{m'm}$ are uniquely determined by the relative orientation of x, y, z with respect to x', y', z'. The spherical harmonics P^m_{lt} of chapter 4 are the simplest example of a set of functions of the type φ_m for integral j $(= l)$ *.

§ 77. Multiplet situations (continued)

The expansion (102) needed the determination of those spin-functions which belonged to the different eigenvalues of the operator S^2 [see eq. (105)]. According to our considerations these eigenvalues will certainly be of the form $S^2 = \bar{s}(\bar{s}+1)\hbar^2$. The terms ${}^M\psi$ in equation (102) correspond to the different possible values of the spin quantum number $\bar{s}$. For each value $\bar{s}$ there will be a set (or several sets) of $M = 2\bar{s}+1$ functions of the type (110). We call M the *multiplicity* of such functions. One can determine as follows which values of $\bar{s}$ and how many linearly independent sets of φ_m will occur. The functions φ_m are always eigenfunctions of $S_{z\text{op}}$. The original 2^N unit spin functions $\Pi_i\xi_i^{(s_i)}$ are themselves eigenfunctions of $S_{z\text{op}}$, and we have [see eq. (6.78)]

$$S_{z\text{op}}\Pi_i\xi_i^{(s_i)} = \Sigma_{i'}S_{zi'\text{op}}\Pi_i\xi_i^{(s_i)} = \hbar(\Sigma_i s_i)(\Pi_i\xi_i^{(s_i)}) = \tfrac{1}{2}\hbar(N_+-N_-)\Pi_i\xi_i^{(s_i)},$$

where, as before, $N_+(N_-)$ is the number of positive (negative) s_i According to equation (101) N_++N_- is equal to the number of electrons. The number of unit spin functions corresponding to the same eigenvalue $\frac{1}{2}\hbar(N_+-N_-)$ is equal to the number of possibilities of finding among N numbers s_i exactly N_+ positive values $s_i = +\frac{1}{2}$, and is thus equal to the binomial coefficient $\binom{N}{N_+}$. We have thus the following scheme

N_+	Eigenvalue of S_z $[= \frac{1}{2}\hbar(N_+-N_-) = m\hbar]$	Number of eigenfunctions of S_z $\left[= \binom{N}{N_+}\right]$
N	$\frac{1}{2}N\hbar$	1

* We can immediately compare the definition (4.145) of the spherical functions with equation (114). The ξ and η occurring in that equation transform in exactly the same way as the ξ and η here, and it follows from equation (4.145) that the Q and thus also the $P_l{}^m$ transform as $(-1)^m\binom{2l}{l+m}\xi^{l-m}\eta^{l+m}$.

$$\begin{array}{lll}
N-1 & (\tfrac{1}{2}N-1)\hbar & N \\
N-2 & (\tfrac{1}{2}N-2)\hbar & \dfrac{N(N-1)}{1\cdot 2} \\
\cdot & \cdot & \cdot \\
\cdot & \cdot & \cdot \\
\cdot & \cdot & \cdot \\
N-n & (\tfrac{1}{2}N-n)\hbar & \dfrac{N!}{n!(N-n)!}=\dfrac{N(N-1)\ldots(N-n+1)}{1\cdot 2\ldots n} \\
\cdot & \cdot & \cdot \\
\cdot & \cdot & \cdot \\
\cdot & \cdot & \cdot \\
0 & -\tfrac{1}{2}N\hbar & 1
\end{array} \qquad (120)$$

Among the sets of φ_m which together form a complete set of eigenfunctions of S^2_{op} there is thus one for which the spin quantum number $\bar{s}$ is equal to $\frac{1}{2}N$, since this set will contain both that eigenfunction of S_z for which $m = \frac{1}{2}N$, that is, for which all s_i are qeual to $+\frac{1}{2}(\varphi_{\frac{1}{2}N} = \Pi_i \xi_i^{(+)} = \Pi_i \xi_i)$ and that eigenfunction $\varphi_-{}^{\frac{1}{2}N} = \Pi_i \xi_i^{(-)} = \Pi \eta_i$ for which all s_i are equal to $-\frac{1}{2}$. In this set of φ_m each of the eigenvalues of S_z between $\frac{1}{2}N\hbar$ and $-\frac{1}{2}N\hbar$ must be represented once; the φ_m corresponding to each m are certain linear combinations of the eigenfunctions of S_z corresponding to the eigenvalue $m\hbar$. The eigenvalue $\bar{s} = \frac{1}{2}N$ of S^2_{op} is thus exactly $2(\frac{1}{2}N)+1 = N+1$-fold degenerate and corresponds — in the vector model — to all spins being parallel (see § 75)

There are also sets of eigenfunctions of S^2 corresponding to the spin quantum number $\bar{s} = \frac{1}{2}N - 1$. Of those there will be $N-1$ sets. This can be seen as follows. If one removes from the scheme (12) the $N+1$ eigenfunctions corresponding to $\bar{s} = \frac{1}{2}N$ and then considers all those spin functions orthogonal to them, there remain $N-1$ linearly independent eigenfunctions of S^2 corresponding to the largest remaining eigenvalue $(\frac{1}{2}N-1)\hbar$. There must thus be exactly the number of sets of corresponding to $\bar{s} = \frac{1}{2}N-1$. Proceeding in this way one arrives at the following classification of the eigenfunctions of S^2_{op}.

Spin quantum number $\bar{s}$	Multiplicity $M(=2\bar{s}+1)$	Number $A_{\bar{s}}$ of sets of φ_m	Degree of degeneracy $M \cdot A_{\bar{s}}$
$\frac{1}{2}N$	$N+1$	1	$N+1$
$\frac{1}{2}N-1$	$N-1$	$\binom{N}{N-1}-\binom{N}{N}=N-1$	$(N-1)^2$
$\frac{1}{2}N-2$	$N-3$	$\binom{N}{N-2}-\binom{N}{N-1}=\frac{1}{2}N^2-\frac{3}{2}N$	$\frac{1}{2}N(N-3)^2$
.	.	.	.
.	.	.	.
.	.	.	.
$\frac{1}{2}N-n$	$N-2n+1$	$\binom{N}{N-n}-\binom{N}{N-n+1}$	$\dfrac{N!(N-2n+1)^2}{(N-n+1)!n!}$
.	.	.	.
.	.	.	.
.	.	.	.
0 or $\frac{1}{2}$	1 or 2		

(121)

The nature of the splitting up into "multiplet situations" and the meaning of the number M is now completely clarified. We see that M is the *multiplicity* of the spin functions occurring in ${}^M\psi$ and, as was shown in equation (102), it can take on the values $N+1, N-1, \ldots,$ 2 (or 1). The function ${}^M\psi$ corresponds to a situation where N_+ $[=\frac{1}{2}(N+M-1)]$ of the N spins are parallel and $N_-[=\frac{1}{2}(N-M+1)]$ of them are antiparallel to the resultant spin vector which is exactly the same result as the one obtained in § 75 by using the vector model. In the case of two electrons only $M = 3$ and $M = 1$ can occur corresponding to triplet and singlet terms. In § 73 we discussed this case in detail. For $N = 3$ there are quartet ($M = 4$) and doublet ($M = 2$) states; for $N = 4$ quintet ($M = 5$), triplet ($M = 3$), and singlet ($M = 1$) states, . . . In general one speaks of M-tuplet states. If N is even all M are odd, and vice versa.

We must now obtain explicit expressions for the spin eigenfunctions ${}^M\zeta$ corresponding to different multiplicities. The easiest way to obtain these is by using invariants of the type (114). We shall first of all consider the expression

$$ {}^{N+1}Z = (u\xi_1+v\eta_1)(u\xi_2+v\eta_2)\ldots(u\xi_N+v\eta_N) = \Pi_i(u\xi_i+v\eta_i), \qquad (122) $$

where u, v is a fixed spinor. If we write equation (122) out in detail we have

$$^{N+1}Z = u^N\xi_1\xi_2 \ldots \xi_N + u^{N-1}v(\eta_1\xi_2\xi_3 \ldots \xi_N + \xi_1\eta_2\xi_3 \ldots \xi_N + \ldots + \xi_1\xi_2 \ldots \eta_N) + \ldots + v^N\eta_1 \ldots \eta_N .$$

Equation (122) is invariant, since each factor is an invariant. Considered as a polynomial in u and v, Z is homogeneous of the N-th degree. The coefficients of the monomials $u^{\bar{s}+m}v^{\bar{s}-m}$ ($\bar{s} = \frac{1}{2}N$, $m = \frac{1}{2}N, \frac{1}{2}N-1, \ldots$) are thus — apart from a numerical factor $\binom{2\bar{s}}{\bar{s}-m}$ — just the φ_m set corresponding to the spin quantum number $\bar{s} = \frac{1}{2}N$. The φ_m are simply the symmetric linear combinations of the unit spin functions. Equation (122) is a simultaneous representation of all spin functions corresponding to the largest multiplicity.

In the case $\bar{s} = \frac{1}{2}N-1$ we consider

$$^{N-1}Z = \left(\xi_1 \frac{\partial}{\partial v} - \eta_1 \frac{\partial}{\partial u}\right)(u\xi_2+v\eta_2) \ldots (u\xi_N+v\eta_N). \tag{123}$$

The first factor containing the differential operators is invariant since it follows from equation (117) that $\partial/\partial u' = \alpha^*(\partial/\partial u) + \beta^*(\partial/\partial v)$, $\partial/\partial v' = -\beta(\partial/\partial u)+\alpha(\partial/\partial v)$, which means that $\partial/\partial u$, $\partial/\partial v$ transform as ξ, η [see eq. (119)] and $[\xi_1(\partial/\partial v)-\eta_1(\partial/\partial u)]$ is invariant just as $\xi_1\eta-\eta_1\xi$ is invariant. The expression (123) is a homogeneous polynomial of degree $N-2$ in u, v. The coefficients of the monomials u^{N-2}, $u^{N-3}v$, ... give us a φ_m-set of spin functions corresponding to the spin quantum number $\bar{s} = \frac{1}{2}N-1$. If we write equation (123) out in detail there will be $N-1$ terms,

$$^{N-1}Z = (\xi_1\eta_2-\eta_1\xi_2)(u\xi_3+v\eta_3) \ldots (u\xi_N+v\eta_N) + (\xi_1\eta_3-\eta_1\xi_3)(u\xi_2+v\eta_2) \ldots (u\xi_N+v\eta)_N + \ldots \tag{124}$$

In particular for the case of two electrons equation (123) leads simply to

$$^1Z = \left(\xi_1 \frac{\partial}{\partial v} - \eta_1 \frac{\partial}{\partial u}\right)(u\xi_2+v\eta_2) = \xi_1\eta_2-\eta_1\xi_2,$$

corresponding to the expression (59) for the singlet spin function $^1\zeta_0$.

Apart from expression (123) there are $N-1$ similar expressions. If we use the abbreviations

$$D_i = \xi_i \frac{\partial}{\partial v} - \eta_i \frac{\partial}{\partial u}, \qquad A_i = \xi_i u + \eta_i v, \tag{125}$$

we get in general a φ_m-set corresponding to $\bar{s} = \frac{1}{2}N - 1$ by writing

$$^{N-1}Z_j = D_j \Pi_{i(\neq j)} A_i. \tag{126}$$

In equation (123) we had $j = 1$.

We have thus N sets of each $N-1$ functions φ_m. The different φ corresponding to the same m are not orthogonal to each other. They are not even linearly independent, since there exists exactly one linear relation between the $^{N-1}Z_j$,

$$\Sigma_j {}^{N-1}Z_j = 0. \tag{127}$$

That equation (127) is correct follows, for instance, by writing each term in the form (124). One thus sees that the term containing the factor $\xi_i \eta_j - \eta_i \xi_j$ from Z_j cancels the term containing the factor $\xi_j \eta_i - \eta_j \xi_i$ from Z_i. Linear combinations other than the symmetric one (127) cannot be equal to zero. If such a linear combination existed, one could find at least $N-1$ different ones by permutations and all Z would be equal to zero.

From equation (127) it follows that of the N functions (126) there are only $N-1$ linearly independent ones, in accordance with the scheme (121). In principle one can construct in many different ways a set of $N-1$ linear combinations of the $^{N-1}Z_j$ which are orthogonal to each other. The symmetry in the electron numbers will, however, disappear during this orthogonalising process.

In general one can construct the φ_m sets, that is, the spin functions $^M\zeta$ of equation (103), which correspond to $\bar{s} = \frac{1}{2}N - n$, $M = N+1-2n$ by means of the prescription

$$^{N+1-2n}\zeta \rightarrow {}^{N+1-2n}Z_{j_1, j_2, \ldots, j_n} = D_{j_1} D_{j_2} \ldots D_{j_n} \Pi_{i(\neq j_1, \neq j_2, \ldots, \neq j_n)} A_i, \; 2n \leqq N, \tag{128}$$

where the j_k are chosen arbitrarily from among the N electron numbers i. There are clearly $\binom{N}{n}$ possibilities of choosing the j_k and therefore $\binom{N}{n}$ functions of the type (128) (the order of the j_k is immaterial). The Z are neither orthogonal to each other, nor linearly independent, since there are $\binom{N}{N-n}$ essentially different relations between them. One such relation is clearly

$$\Sigma_{j(\geqq n)} {}^{N+n-1}Z_{1,2,\ldots,n-1,j} = D_1 D_2 \ldots D_{n-1} \Sigma_{j(\geqq n)} \{D_j \Pi^N_{i=n(i\neq j)} A_i\} = 0. \tag{129}$$

The other relations are obtained by permutations of the electron numbers.

Expression (129) is equal to zero since the sum over j itself is equal to zero for the same reasons as in the discussion of equation (127). If there existed relations between the Z other than those deriving from equation (129) all Z would be equal to zero.

We want to draw attention to one particular property of the representation (128) of the ${}^{M}\zeta$. We required that the ${}^{M}\zeta$ would transform among themselves both under rotations and under permutations. The first requirement is met by the invariance of Z under rotations: the $N+1-2n$ functions transform under a rotation according to the law applying to the transformation of the monomials $\xi^{\bar{s}-m}\eta^{\bar{s}-m} \to \xi'^{\bar{s}+m}\eta'^{\bar{s}-m}$. In the case of a permutation a given $Z_{j_1,\ldots,j_n}$ will go over into another $Z_{j'_1,\ldots,j'_n}$. We only need to know into which numbers $j'_1, \ldots, j'_n$ the $j_1, \ldots, j_n$ will go over under the given permutation. The spin functions of given m go thus over into each other, just like the Z.

A simpler representation of the ${}^{M}\zeta$ would be the following one

$$\bar{s} = \tfrac{1}{2}N, \qquad {}^{N+1}X = (u\xi_1+v\eta_1)(u\xi_2+v\eta_2)\ldots(u\xi_N+v\eta_N)[= {}^{N+1}Z];$$

$$\bar{s} = \tfrac{1}{2}N-1, \qquad {}^{N-1}X = (\xi_1\eta_2-\eta_1\xi_2)(u\xi_3+v\eta_3)\ldots(u\xi_N+v\eta_N) \text{ and all } \mathrm{P}\,{}^{N-1}X;$$

$$\bar{s} = \tfrac{1}{2}N-2, \qquad {}^{N-3}X = (\xi_1\eta_2-\eta_1\xi_2)(\xi_3\eta_4-\eta_3\xi_4)(u\xi_5+v\eta_5)\ldots(u\xi_N+v\eta_N)$$

$$\text{and all } \mathrm{P}\,{}^{N-3}X; \ldots$$

A factor such as $\xi_1\eta_2-\eta_1\xi_2$ in ${}^{N-1}X$ is interpreted as an antiparallel position of the first and the second electron, while the other electrons are all parallel. A disadvantage of this description is that the number of the ${}^{M}X$ is very large, because the number of possible $\mathrm{P}\,{}^{M}X$ is so large, and the occurrence of the correct number of vanishing independent linear combinations is not so easily verified.

We can now start to investigate in more detail the special properties of the *functions of the spatial coordinates* ${}^{M}g(1, 2, \ldots, N)$ occurring in the terms ${}^{M}\psi$. These properties follow from the Pauli principle [see eq. (104)]. If we choose, for instance, the largest multiplicity $M = N+1$, all spin functions — which are summarised in ${}^{N+1}Z$ — are symmetric in all electrons. As a result the g in ${}^{N+1}\psi$ must all be antisymmetric in all the electrons. The most general expression for ${}^{N+1}\psi$ is thus

$${}^{N+1}\psi = g^{\mathrm{a}}_{\frac{1}{2}N}({}^{N+1}Z)^{(\frac{1}{2}N)} + g^{\mathrm{a}}_{\frac{1}{2}N-1}({}^{N+1}Z)^{(\frac{1}{2}N-1)} + \ldots + g^{\mathrm{a}}_{m}({}^{N+1}Z)^{(m)} + \ldots, \tag{130}$$

where $({}^{N+1}Z)^{(m)}$ is the factor of $u^{\bar{s}+m}v^{\bar{s}-m}$ in ${}^{N+1}Z$ and where all g^{a}_{m} are antisymmetric. For the case $N = 2$ we get the triplet terms of equation (58).

In the case $M = N+1-2n$ ($n = 1, 2, \ldots$) the situation is more complicated. We write

$$^{M}\psi = \Sigma_{j_1,\ldots,j_n} g_{j_1,\ldots,j_n} {}^{M}Z_{j_1,\ldots,j_n}. \tag{131}$$

We should also have taken into account the fact that the magnetic quantum number m can take on M different values and we should have summed over them, as we did in equation (130). Since the m value corresponding to a spin function does not change under a permutation, we can consider expression (131) to be representative of each value of m occurring in the $Z_{j_1,\ldots}$.

Since the $Z_{j_1,\ldots}$ are not linearly independent [see eq. (129)] the expansion (131) is not unique. This is, however, rather an advantage when we wish to characterise the g in a simple manner. If $^{M}\psi$ of equation (131) is antisymmetric, we can write

$$^{M}\psi = (N!)^{-1}\Sigma_{\mathrm{P}}\delta_{\mathrm{P}}\mathrm{P}^{M}\psi = (N!)^{-1}\Sigma_{j_1',\ldots,j_n'}\Sigma_{\mathrm{P}}\,\delta_{\mathrm{P}}\,\mathrm{P}g_{(j_1',\ldots,j_n')}{}^{M}Z_{j_1',\ldots,j_n'},$$

where we sum over all possible permutations of the electron numbers and where we have put the subscripts of the g within brackets to indicate that the action of P on the g means a permutation of the position of the electrons, but not a change in the subscripts. Since a permutation will produce from $^{M}Z_{j_1',\ldots,j_n'}$ a $^{M}Z_{j_1'',\ldots,j_n''}$ the result of the double summation is really a simple summation over all $^{M}Z_{j_1,\ldots,j_n}$ which we write as follows

$$^{M}\psi = \Sigma_{j_1,\ldots,j_n} {}^{M}g_{(j_1,\ldots,j_n)} {}^{M}Z_{j_1,\ldots,j_n}, \tag{132}$$

$$^{M}g_{(j_1,\ldots,j_n)} = (N!)^{-1}\Sigma_{j_1',\ldots,j_n'}[\Sigma_{jj'}\delta_{\mathrm{P}_{jj'}}\mathrm{P}_{jj'}g_{(j')}]. \tag{133}$$

The last equation is obtained by dividing the permutations into groups, each group referring to n definite electron numbers $j_1', \ldots, j_n'$ and containing all those permutations $\mathrm{P}_{jj'}$ where $j_1', \ldots, j_n'$ goes over into $j_1, \ldots, j_n$ or any permutation of $j_1, \ldots, j_n$. If we have found *one* permutation of the type $\mathrm{P}_{jj'}$, we get all other permutations $\mathrm{P}_{jj'}$ by permuting both the $j_1, \ldots, j_n$ among each other and all the numbers which are not equal to $j_1, \ldots, j_n$ among each other. This means that there are always $n!(N-n)!$ permutations $\mathrm{P}_{jj'}$. We shall denote by P_j a permutation which permutes only the $j_1, \ldots, j_n$ among each other and by $\mathrm{P}_{\tilde{j}}$ one which permutes only the other numbers. If $\mathrm{P}^0_{jj'}$ is one permutation of the type $\mathrm{P}_{jj'}$, we have

$$^{M}g_{(j_1,\ldots,j_n)} = (N!)^{-1}\Sigma_{j_1',\ldots,j_n'}[\Sigma_{\mathrm{P}_j}\Sigma_{\mathrm{P}_{\tilde{j}}}\delta_{\mathrm{P}_j}\delta_{\mathrm{P}_{\tilde{j}}}\{\delta_{\mathrm{P}^0_{jj'}}\mathrm{P}^0_{jj'}g_{(j')}\}].$$

From this way of expressing the ^{M}g it follows immediately that *the functions* ${}^{M}g_{(j_1,\ldots,j_n)}$ *are antisymmetric in the positions of the electrons* $j_1, \ldots, j_n$, *and also antisymmetric in the positions of all the other electrons*. It then follows from equation (133) that *one can obtain from one function* ${}^{M}g_{(j_1,\ldots,j_n)}$ *a different* ${}^{M}g_{(\bar{j}_1,\ldots,\bar{j}_n)}$ *simply by permutation*. If $\mathrm{P}_{\bar{j}j}$ is a permutation which changes $j_1, \ldots, j_n$ to $\bar{j}_1, \ldots, \bar{j}_n$, $\mathrm{P}_{\bar{j}j}$ operating on the right hand side of equation (133) means clearly that one is no longer summing over all permutations which change j' into j, but over all those which change j' into $\bar{j}$, or $\mathrm{P}_{\bar{j}j}\,{}^{M}g_{(j)} = \delta_{\mathrm{P}_{\bar{j}j}}\,{}^{M}g_{(\bar{j})}$. If this equation is substituted into equation (132) we get for ${}^{M}\psi$ the expression

$$ {}^{M}\psi = [n!(N-n)!]^{-1} \Sigma_{\mathrm{P}} \delta_{\mathrm{P}}\, {}^{M}g_{(1,2,\ldots,n)}\, \mathrm{P}\, {}^{M}Z_{1,2,\ldots,n}, \quad (2n \leqq N) \qquad (134) $$

where ${}^{M}g_{(1,2,\ldots,n)}$ is antisymmetric both in $1, 2, \ldots, n$ and in $n+1, \ldots, N$. From equation (134) it is clear that ${}^{M}\psi$ is antisymmetric. The $n!(N-n)!$ permutations in the sum which change $1, 2, \ldots, n$ in the same set of numbers $j_1, \ldots, j_n$ lead to $n!(N-n)!$ equal terms corresponding to one term in equation (132).

As we mentioned before, because of the relations (129) between the Z, neither (132) nor (134) is a unique way of describing ${}^{M}\psi$. Only $\binom{N}{n} - \binom{N}{n-1}$ of the Z are linearly independent and that number of linear combinations of the ${}^{M}g_{(j)}$ would be sufficient to characterise ${}^{M}\psi$ uniquely. The most general linear combination which can occur in such a characterisation can clearly always be written in the form

$$ G(1, 2, \ldots, N) = \Sigma\, {}^{M}g_{(j_1,\ldots,j_n)}(1, 2, \ldots, N)\, a_{j_1,\ldots,j_n}, \qquad (135) $$

where the coefficients a — n different numbers in an arbitrary order — are arbitrary, but for the requirement that they satisfy the following $\binom{N}{n-1}$ linear relations [compare eq. (129)]

$$ \Sigma_{j_n} a_{j_1,\ldots,j_n} = 0, \qquad (136) $$

where $j_1, \ldots, j_{n-1}$ is one of the $\binom{N}{n-1}$ possible combinations of $n-1$ numbers and where the summation extends over all possible values of j_n. *Because of equation* (136) *it is impossible to find a* G *which is antisymmetric in more than* $N-n$ *electron numbers.*

Proof. If, for instance, G of equation (135) were antisymmetric in the numbers $n, n+1, \ldots, N$ (that is, in $N-n+1$ numbers) and if P' denotes a permutation which interchanges *only these* numbers, we would have

$$G = [(N-n+1)!]^{-1}\Sigma_{P'}\delta_{P'}P'G = [(N-n+1)!]^{-1}\Sigma_j[\Sigma_{P'}\delta_{P'}P'\,{}^M g_{(j)}]a_j. \quad (137)$$

Since ${}^M g_{(j)}$ is antisymmetric in the numbers j, and also in all the other numbers, the expression within square brackets depends only on which of the numbers 1, 2, ... $n-1$ are among the $j_1, \ldots, j_n$. If these numbers are $k_1, \ldots, k_\nu$ $(\nu \leqq n-1)$ we can write

$$G = \Sigma_{k_1,\ldots,k_\nu} G_{k_1,\ldots,k_\nu}[\Sigma_{k_{\nu+1},\ldots,k_N} a_{k_1,\ldots,k_\nu;\,k_{\nu+1},\ldots,k_N}], \quad \nu \leqq n-1,$$

where the second sum extends over all a_j for which exactly $n-\nu$ of the j are larger than $n-1$ while $k_1, \ldots, k_\nu$ are fixed. From equation (136) it follows, however, that such sums are equal to zero. In the case where $\nu = n-1$ it follows immediately from equation (136) itself, while for $\nu < n-1$ it follows by considering a suitable linear combination of all those equations (136) for which $k_1, \ldots, k_\nu$ are among the $j_1, \ldots, j_{n-1}$.

This property of the G enables us to *choose the* ${}^M g_{(j)}$ *in equation* (132) *in a unique way,* namely in such a way that there are altogether *no* linear combinations of them which are antisymmetric in more than $N-n$ electrons. This means that all the sums over expressions in square brackets in equation (137) must be identically equal to zero,

$$\Sigma_{P'}\delta_{P'}P'\,{}^M g_{(j_1,\ldots,j_n)} = 0. \quad (138)$$

This identity should be valid for any choice of $N-n+1$ numbers. The summation refers to all the permutations of these numbers.

If the original g in equation (132) do not satisfy this requirement, one can always replace them by linear combinations $g'_{(j)}$ which (*i*) leave the expression (132) unchanged, (*ii*) are still antisymmetric in the j and in the other numbers, and (*iii*) satisfy the relations (138).

I am indebted to Dr. F. J. Belinfante for the following prescription for the "reduction" of the g-functions. If ${}^{N+1-2}Z_{j'_1,\ldots,j'_n}(j_1, \ldots, j_n)$ [abbreviated to ${}^M Z_{j'}(j)$] is the value of the spin function (128), if the spin variables $s_{j_1}, \ldots, s_{j_n}$ are all equal to $+\frac{1}{2}$, and all other spin variables are equal to $-\frac{1}{2}$, we find from expressions (125) for D and A,

$${}^M Z_{j'}(j) = \left(\frac{\partial}{\partial v}\right)^{n-\lambda}\left(-\frac{\partial}{\partial u}\right)^{\lambda} u^\lambda v^{N-n+\lambda} = (-1)^\lambda \binom{N-n}{\lambda}^{-1} v^{N-2n}\frac{(N-n)!}{(N-2n)!},$$

where λ is the number of $j_1, \ldots, j_n$ which are not contained in the $j'_1, \ldots, j'_n$. We define the new functions ${}^M g'_{(j)}$ by the equation

$$\begin{aligned} {}^M g'_{(j)} &= v^{-N+2n}\frac{(N-2n+1)!}{(N-n+1)!}\sum_{(j')} {}^M g_{(j')}\,{}^M Z_{j'}(j) \\ &= \frac{N-2n+1}{N-n+1}\sum_{\lambda=0}^{n}\left\{(-1)^\lambda\binom{N-n}{\lambda}^{-1}[\Sigma'\,{}^M g_{(j')}]\right\}, \end{aligned} \quad (139)$$

where the first summation is over all possible $j'_1, \ldots, j'_n$ combinations, while the Σ' summation is over all those combinations $j'_1, \ldots, j'_n$ for which λ has a given value.

We see that the second member, apart from a multiplying factor is equal to the function ${}^M\psi$ for the special case where the spin coordinates are chosen in the way mentioned a moment ago. This expression is therefore antisymmetric both in the positions of the electrons $j_1, \ldots, j_n$ and in the positions of the other electrons. The

${}^{M}g'$ satisfy, however, also the conditions (138), since they are a special case of functions of the type G [see eqq. (135), (136)].

Finally it follows from the right hand side of equation (139) that

$$\Sigma_j\, {}^{M}g'_{(j)}\, Z_j = \frac{N-2n+1}{N-n+1} \Sigma_{\lambda=0}^{n}\, (-1)^{\lambda} \binom{N-n}{\lambda}^{-1} \{\Sigma_{j'}\, {}^{M}g_{(j')}[\Sigma'' Z_j]\}$$

where the Σ'' extends over all j combinations for which λ numbers are not among $j'_1, \ldots, j'_n$. Because of equation (129) one finds for each sum $\Sigma'' Z_j$ the value $(-1)^{\lambda}\binom{n}{\lambda} Z_{j'}$. We get therefore ${}^{M}\psi = \Sigma_j\, {}^{M}g_{(j)}\, Z_j = \Sigma_j\, {}^{M}g'_{(j)}\, Z_j$. We leave to the reader the (straightforward) proof that there are no other functions ${}^{M}g'$ with the same properties.

From now on we shall assume that the g in equation (132) are always such "reduced" functions satisfying equation (138). Two reduced g-functions corresponding to different multiplicities are always orthogonal to each other. We prove as an example that each ${}^{M}g_{(j_1,\ldots,j_n)}$ is orthogonal to each ${}^{M'}g_{(j_1,\ldots,j_{n-1})}$. Let P′ denote the permutations of the $N-n+1$ numbers $j_n, \ldots, j_N$.

$$\begin{aligned}\int {}^{M}g^{*}_{(j_1,\ldots,j_{n-1})}\, {}^{M'}g_{(j'_1,\ldots,j'_n)}\, d\tau \\ &= [(N-n+1)!]^{-1} \int \Sigma_{\mathrm{P}'}\, \delta_{\mathrm{P}'}\, \mathrm{P}'\, {}^{M}g^{*}_{(j_1,\ldots,j_{n-1})}\, \delta_{\mathrm{P}'}\, \mathrm{P}'\, {}^{M'}g_{(j'_1,\ldots,j'_n)}\, d\tau \\ &= [(N-n+1)!]^{-1} \int {}^{M}g^{*}_{(j_1,\ldots,j_{n+1})}\, \Sigma_{\mathrm{P}'}\, \mathrm{P}'\, \delta_{\mathrm{P}'}\, {}^{M'}g_{(j'_1,\ldots,j'_n)}\, d\tau = 0.\end{aligned}$$

A function of N arguments, antisymmetric in $N-n$ arguments and in the other n arguments $(N-n \geqq n)$ which cannot be changed by permutations and linear combinations of permutations into a function antisymmetric in more than $N-n$ arguments is in group theoretical nomenclature called a function of antisymmetry type $[N-n]+[n]$. This name refers not only to the function itself, but also to all its permutations and linear combinations of those. Together they form a reduced set of functions of the given (anti)symmetry type. This set contains exactly $\binom{N}{n} - \binom{N}{n-1}$ linearly independent functions.

In similar manner one defines a set of function of symmetry type $\{N-n\}+\{n\}$. The functions ${}^{M}Z_j$ are an example of such a set. The multiplet term ${}^{M}\psi$ of an arbitrary antisymmetric function is thus characterised by the symmetry (antisymmetry) type of the spin (position) function contained in it.

From a set of functions of (anti)symmetry type $(N-n)+(n)$ one can choose those functions which are as (anti)symmetric in as many coordinates as possible. It is possible to prove that there is at most one function which is (anti)symmetric in n

groups of two arguments; this function will certainly be antisymmetric (symmetric) in the other $N-2n$ arguments (compare the X earlier on in this section). One calls this a type

$$\overbrace{(2)+(2)+\ldots+(2)}^{n} + \overbrace{(1)+(1)+\ldots+(1)}^{N-2n}$$

The situation with regard to the position functions is very similar to the one with regard to the spin functions. In the latter case we had a classification by means of eigenfunctions of S^2. The ${}^M g$ corresponding to different values of M correspond to an analysis of all possible functions in terms of eigenfunctions of a number of commuting operators. These operators form the complete set of those operators χ which are constructed linearly from the P operators and which all commute with P.* If one considers the classification from this group theoretical point of view the orthogonality which we proved a moment ago is immediately obvious.

The simplest example of a set of functions of antisymmetry type $[N-n]+[n]$ is obtained as follows. Let $\Pi_{j_1,\ldots,j_m}$ be the alternating product of m variables $x_{j_1}\, x_{j_2}\, \ldots, x_{j_m}$,

$$\Pi_{j,\ldots,j_m} = (x_{j_2}-x_{j_1})(x_{j_3}-x_{j_1})\ldots(x_{j_m}-x_{j_1})(x_{j_3}-x_{j_2})\ldots (x_{j_m}-x_{j_2})(x_{j_4}-x_{j_3})\ldots(x_{j_m}-x_{j_{m-1}})$$

$$= \begin{vmatrix} 1 & x_{j_1} & x_{j_1}^2 & \ldots & x_{j_1}^{m-1} \\ 1 & x_{j_2} & x_{j_2}^2 & \ldots & x_{j_2}^{m-1} \\ \ldots & \ldots & \ldots & \ldots & \ldots \\ 1 & x_{j_m} & x_{j_m}^2 & \ldots & x_{j_m}^{m-1} \end{vmatrix}, \qquad (140)$$

the antisymmetry of which is clear from this equation. The set of functions we are looking for is "generated" by the function

$$g(\underbrace{1,2,\ldots,n}_{\text{antis}};\underbrace{n+1,\ldots,N}_{\text{antis}}) = \Pi_{1,\ldots,n}\Pi_{n+1,\ldots,N} = (x_2-x_1)\ldots(x_n-x_1)\ldots$$
$$(x_n-x_{n-1})(x_{n+2}-x_{n+1})\ldots(x_N-x_{n+1})\ldots(x_N-x_{N-1}). \qquad (141)$$

The impossibility of constructing from this g a function antisymmetric in $N-n+1$ variables is most easily proved by using the determinant expression for the Π.

A function G antisymmetric in the numbers $n, \ldots, N$ is obtained by writing $G = \Sigma_{\mathrm{P}'}\delta_{\mathrm{P}'}\mathrm{P}'g(1, \ldots, n;\, n+1, \ldots, N)$ where P′ denotes the permutations of the numbers n to N. Apart from a factor this G is equal to the determinant

* See P. A. M. Dirac, *The principles of quantum mechanics*, Oxford University Press, 1930, Ch. XI.

$$\begin{vmatrix} 1 & x_1 & x_1^2 & \dots x_1^{n-1} & & & & \\ \dots & \dots & \dots & \dots & & & 0 & \\ 1 & x_{n-1} & x_{n-1}^2 & \dots x_{n-1}^{n-1} & & & & \\ 1 & x_n & x_n^2 & \dots x_n^{n-1} & 1 & x_n & x_n^2 & \dots x_n^{N-n-1} \\ 1 & x_{n+1} & x_{n+1}^2 & \dots x_{n+1}^{n-1} & 1 & x_{n+1} & x_{n+1}^2 & \dots x_{n+1}^{N-n-1} \\ \dots & \dots & \dots & \dots & \dots & \dots & \dots & \dots \\ 1 & x_N & x_N^2 & \dots x_N^{n-1} & 1 & x_N & x_N^2 & \dots x_N^{N-n-1} \end{vmatrix}$$

If this determinant is expanded in terms of products of subdeterminants to the left and the right of the vertical dividing line, one obtains the above expression for G. If, however, we expand in terms of products of subdeterminants above and below the horizontal dividing line, we get zero, since all subdeterminants below the line are zero as they contain always at least two equal columns.

§ 78. Stationary states of N-electron systems without spin forces

If we may neglect the action of the spin in writing down the energy, the energy operator H^0_{op} will act only on the spatial coordinates of the electrons and will leave all spin functions unchanged. If we introduce into the Schrödinger equation $H^0_{\text{op}}\psi = i\hbar\partial\psi/\partial t$ the expansion (102) of ψ in multiplet situations, $\psi = \Sigma_M {}^M\psi$ ($M = N+1$, $N-1, \dots$) the action of H^0_{op} on ${}^M\psi$ will lead to another M-tuplet function, and each term ${}^M\psi$ satisfies itself the Schrödinger equation. In particular it will be possible to divide all stationary states in groups of *multiplet states* with wave functions corresponding to one definite multiplicity,

$$^M\psi = {}^M\varphi \exp[-i\,{}^MEt/\hbar]; \qquad (H^0_{\text{op}} - {}^ME)\,{}^M\varphi = 0. \tag{142}$$

It is only by chance that two eigenvalues corresponding to different multiplicities will be the same.

From the previous section we know that the most general form of the ${}^M\varphi$ is given by the equation

$$\left.\begin{aligned} &{}^M\varphi = \Sigma_m[\Sigma_{j_1,\dots,j_n}\,{}^Mg^{(m)}_{(j_1,\dots,j_n)}\,{}^MZ^{(m)}_{j_1,\dots,j_n}], \\ &M = N-2n+1;\; m = \tfrac{1}{2}N-n,\; \tfrac{1}{2}N-n-1, \dots, \; -\tfrac{1}{2}N+n, \end{aligned}\right\} \tag{143}$$

where in contradistinction to equation (132) we have taken the magnetic quantum number explicitly into account [compare eq. (130)]. If expression (143) satisfies the Schrödinger equation (142) each of the terms ${}^Mg$ will satisfy the Schrödinger equation

$$(H^0_{\text{op}} - E)g = 0, \tag{144}$$

where there is now no vestige left of the spin coordinates. The validity of equation (144) follows from the fact that the totality of all functions of the spatial coordinates of the electrons represented by equations (143) [by taking all possible combinations of the spin coordinates], including any linear combination of these functions, will satisfy equation (144). These functions form, however, a set of reduced functions of which the g are particular examples [see, for instance, eq. (139)].

If one function ${}^{M}g^{(m)}_{(j)}$ satisfies equation (144), it follows automatically that also all other functions ${}^{M}g$ with the same value of m will be solutions. From the symmetry of H^0_{op} it follows that $\mathrm{P}H^0_{\text{op}} - H^0_{\text{op}}\mathrm{P} = 0$, and $\delta_{\mathrm{P}}\,\mathrm{P}\,{}^{M}g^{(m)}_{(j)}\ (= {}^{M}g^{(m)}_{(j')})$ is a solution, if ${}^{M}g^{(m)}_{(j)}$ is one. *An eigenvalue of equation* (144) *corresponding to an eigenfunction of antisymmetry type* $[N-n]+[n]$ *is automatically* $\binom{N}{n}-\binom{N}{n-1}$*-fold degenerate* *. As a representative of the eigenfunctions we can, for instance, choose the function ${}^{M}g^{(m)}_{1,2,\ldots,n}$ which is antisymmetric in both $1,\ldots,n$ and $n+1,\ldots,N$.

According to equation (143) there will be such a representative for each value of m. If for different values of m the corresponding $g^{(m)}_{(j)}$ were linearly independent, we would be dealing with a chance degeneracy of ${}^{M}E$; "chance" because the degeneracy is not a consequence of the equivalence of the electrons. It is probably better to write instead of equation (143)

$$\left.\begin{aligned} {}^{M}\varphi &= \Sigma_j\, {}^{M}g_{(j)}\left[\Sigma_m c_m\, {}^{M}Z^{(m)}_j\right] \\ &= [n!(N-n)!]^{-1}\,\Sigma_{\mathrm{P}}\{\delta_{\mathrm{P}}\,\mathrm{P}\,{}^{M}g_{(1,\ldots,n)}\left[\Sigma_m c_m \mathrm{P}\,{}^{M}Z^{(m)}_{1,\ldots n}\right]\}, \end{aligned}\right\} \quad (145)$$

and we have complete analogy with the expression (62a) for the triplet states of the two electron problem. Since the c_m are arbitrary the degree of degeneracy of a multiplet state is equal to M. The much higher degree of degeneracy of the spinless problem has completely disappeared, since the function ${}^{M}\varphi$ considered as a function of both spatial and spin coordinates is simply antisymmetric in all electrons (symmetry type $[N]$).

* From the point of view of group theory one discusses the solutions of equation (144) by using the arguments sketched in the small type section at the end of the preceding section. Since H^0_{op} commutes with all permutations, one can classify its eigenfunctions in terms of eigenfunctions of the χ-operators mentioned there. Among them some are of the antisymmetry type $[N-n]+[n]$. One "completes" them by means of spin functions Z into antisymmetric solutions (143) of the Schrödinger equation of the system containing the spin ("composition" of two representations of the permutation group).

Using a model picture we can say that a multiplet state is characterised by a certain state of motion of the electrons while the spins are dragged along without changing their directions. The freedom in the choice of the spin directions corresponds in quantum mechanics to a freedom in the choice of the resultant total spin angular momentum S^2 (see § 75). A consequence of the Pauli principle is that for a given value fo S^2 only a certain type — as far as permutations are concerned — of electronic orbits are realised. It is, however, not possible to give a physical interpretation of the peculiar properties of this type of orbit using a classical picture *.

If the N electrons of a system without spin forces are independent, we have [see eq. (13)]

$$H^0_{\text{op}} = \Sigma_{i=1}^N G^0_{\text{op}}(\boldsymbol{x}_i), \tag{146}$$

where G^0_{op} operates only on the spatial coordinates of one electron. Let the functions $f_k(\boldsymbol{x})$ $(k = 1, 2, \ldots)$ form an orthonormal set of eigenfunctions of G^0_{op} (for the sake of simplicity we have assumed the eigenvalue spectrum to be discrete; the eigenvalues will be denoted by E_k). The index k distinguishes the different values of a quantum number (or a group of quantum numbers). The eigenfunctions and eigenvalues of the spinless Schrödinger equation are then given by the equations

$$f^{k_1 \ldots k_N} = \Pi_{i=1}^N f_{k_i}(i), \qquad E = \Sigma_{i=1}^N E_{k_i}, \tag{147}$$

where i is a shorthand notation for $\boldsymbol{x}_i$.

The eigenvalue (147) is non-degenerate only when all k_i $(i = 1, 2, \ldots, N)$ are the same; otherwise degeneracy occurs. If $N_1, N_2, \ldots$ are the numbers of electrons in the groups of equal k_i $(N_1+N_2+\ldots = N)$, the degree of degeneracy will be equal to $N!/N_1!N_2!\ldots$

The functions $f^{k_1\cdots}$ have in general not yet the properties of the reduced ${}^M g$-functions of the general representation (145) of the stationary states. If f is a solution of the Schrödinger equation, all $\mathrm{P}f$ will also be solutions, and we must construct linear combinations g of the $\mathrm{P}f$ which are of the antisymmetry type $[N-n]+[n]$.

We see immediately that this is only possible if among the quantum numbers $k_1, \ldots, k_n$ a given number occurs at most twice,

* The fact that there are problems in classical physics with equivalent coordinates, where the mathematical formalism of symmetry characters is the most suitable means to describe the system (for instance, the case of symmetrically coupled oscillators: W. Heisenberg, *Z. Phys.*, **40**, 501, 1926) is something completely different.

that is, if there are never more than two electrons in the same state of the spinless one-electron problem.

If ${}^{N+1-2n}g_{(1,2,\ldots,n)} = \Sigma_{\mathrm{P}}\, a_{\mathrm{P}}\, \mathrm{P}\, f^{k_1 \cdots k_N}$ is a linear combination which is antisymmetric both in the numbers $1, \ldots, n$ and in the numbers $n+1, \ldots, N$, and if P′ (P″) denotes the permutations of $1, 2, \ldots, n$ $(n+1, n+2, \ldots, N)$ we have the following relations

$$\left.\begin{aligned} {}^{M}g_{1,2,\ldots,n} &= [n!(N-n)!]^{-1}\, \Sigma_{\mathrm{P}'} \Sigma_{\mathrm{P}''}\, \delta_{\mathrm{P}'} \delta_{\mathrm{P}''}\, \mathrm{P}'\mathrm{P}''\, {}^{M}g_{(1,2,\ldots,n)} \\ &= [n!(N-n)!]^{-1}\, \Sigma_{\mathrm{P}'} \Sigma_{\mathrm{P}''} \Sigma_{\mathrm{P}}\, a_{\mathrm{P}}\, \delta_{\mathrm{P}'} \delta_{\mathrm{P}''}\, \mathrm{P}'\mathrm{P}''\mathrm{P}\, f^{k_1 \cdots k_N} \end{aligned}\right\} \tag{148}$$

In each Pf a given electron number corresponds to one of the numbers $k_1, \ldots, k_N$. If two or more electron numbers which are less than $n+1$ corresponded to the same quantum number in Pf, Pf would be symmetric in those numbers and the summation $\Sigma_{\mathrm{P}'}\, \delta_{\mathrm{P}'}\, \mathrm{P}'$ would lead to zero. The same would apply, if two or more numbers larger than n in Pf corresponded to the same quantum number, since in that case $\Sigma_{\mathrm{P}''}\, \delta_{\mathrm{P}''}\, \mathrm{P}''$ would lead to zero.

It is easy to construct as the product of two Slater determinants linear combinations of the Pf which are antisymmetric both in electrons 1 to n and in electrons $n+1$ to N,

$$\left.\begin{aligned} f^{(l_1 \cdots l_n)(l_{n+1} \cdots l_N)}_{1,2,\ldots,n} &= \begin{vmatrix} f_{l_1}(1) & f_{l_2}(1) & \cdots & f_{l_n}(1) \\ f_{l_1}(2) & \cdots & \cdots & \cdots \\ \cdots & \cdots & \cdots & \cdots \\ f_{l_1}(n) & \cdots & \cdots & f_{l_n}(n) \end{vmatrix} \begin{vmatrix} f_{l_{n+1}}(n+1) & \cdots & f_{l_N}(n+1) \\ f_{l_{n+1}}(n+2) & \cdots & \cdots \\ \cdots & \cdots & \cdots \\ f_{l_{n+1}}(N) & \cdots & f_{l_N}(N) \end{vmatrix} \\ &= |f_{l_1}(i) \ldots f_{l_n}(i)| \cdot |f_{l_{n+1}}(j) \ldots f_{l_N}(j)| \quad i = 1, 2, \ldots, n;\, j = n+1, \ldots, N, \end{aligned}\right\} \tag{149}$$

where the $l_1, \ldots, l_N$ are a permutation of the $k_1, \ldots, k_N$ occurring in equation (147). One must take care, of course, that there are no two l's which are equal in the same determinant. If necessary, one can arrange the l always as follows, $l_1 < l_2 < \ldots < l_n$, $l_{n+1} < l_{n+2} < \ldots < l_N$.

The f functions (149) and the corresponding Pf are in general not reduced ${}^{M}g$ functions; * this is the case only if the numbers $l_1, \ldots, l_n$ are all included among the numbers $l_{n+1}, \ldots, l_N$ (the proof of this statement runs along the lines of the proof in the last paragraph of § 77). From their antisymmetry character it follows, however, that substitution of $f_{1,2,\ldots n}$ for ${}^{M}g_{(1,\ldots,n)}$ into equation (145) will lead to a correct multiplet function ${}^{M}\varphi$; because of the property (129) of

* One can, for instance, construct out of $f_1(1) \begin{vmatrix} f_2(2) & f_3(2) \\ f_2(3) & f_3(3) \end{vmatrix}$ and its permutations the determinant $\begin{vmatrix} f_1(1) & f_2(1) & f_3(1) \\ \cdots & \cdots & \cdots \\ \cdots & \cdots & f_3(3) \end{vmatrix}$ which is antisymmetric in all these particles.

the spin functions Z, the functions f are invisibly reduced. If we now substitute f from equation (149) into the right hand side of equation (145) and if we write for the sake of simplicity ${}^{M}Z_{j_1\dots j_n}$ instead of $\Sigma c_m {}^{M}Z^{(m)}_{j_1\dots j_n}$ — which means that later on the monomials $u^{\bar{s}+m}v^{\bar{s}-m}$ in ${}^{M}Z$ [see eqq. (125) and (128)] must be replaced by arbitrary numbers with an index m — we get the following expression for ${}^{M}\varphi$,

$$
\begin{aligned}
&{}^{M}\varphi^{(l_1\dots l_n)(l_{n+1}\dots l_N)}\\
&= [n!(N-n)!]^{-1}\Sigma_{\mathrm{P}}\,\delta_{\mathrm{P}}\, f^{(l_1\dots l_n)(l_{n+1}\dots l_N)}_{1,\dots,n}\,\mathrm{P}\,D_1D_2\dots D_nA_{n+1}\dots A_N\\
&= \left|\begin{array}{cccc:cccc}
f_{l_1}(1)\,D_1 & f_{l_2}(1)\,D_1 & \dots & f_{l_n}(1)\,D_1 & f_{l_{n+1}}(1)\,A_1 & \dots & & f_{l_N}(1)A_1\\
f_{l_1}(2)\,D_2 & \dots & \dots & f_{l_n}(2)\,D_2 & f_{l_{n+1}}(2)\,A_2 & \dots & & \dots\\
\dots & \dots & \dots & \dots & \dots & \dots & & \dots\\
f_{l_1}(N)\,D_N & \dots & \dots & f_{l_n}(N)\,D_N & f_{l_{n+1}}(N)A_N & \dots & & f_{l_N}(N)\,A_N
\end{array}\right| \qquad (150)\\
&\equiv |\,f_{l_1}(i)\,D_i f_{l_2}(i)\,D_i\dots f_{l_n}(i)\,D_i f_{l_{n+1}}(i)\,A_i\dots f_{l_N}(i)A_i|,\\
&D_i = \xi_i(\partial/\partial v)-\eta_i(\partial/\partial u),\quad A_i = u\xi_i+v\eta_i.
\end{aligned}
$$

If one expands the determinant in terms of subdeterminants to the left and the right of the broken line, one gets, indeed, the second member of equation (150). The determinant has the form of a single Slater determinant [see eq. (177)]. Its elements are functions of both spin and spatial coordinates.

Without considering an expansion in terms of multiplet states one could also immediately write down a Slater determinant as wave function of a stationary state by taking into account the fact that one can construct from the f_k a complete set of eigenfunctions of the one-electron problems including spin by taking $f_k\xi^{(+)}$ and $f_k\xi^{(-)}$. The solution of the time-independent Schrödinger equation can thus be written in the form

$$\varphi = |f_{k_1}(i)\,\xi_i^{(\pm)}\dots f_{k_n}(i)\,\xi_i^{(\pm)} f_{k_{n+1}}(i)\,\xi_i^{(\pm)}\dots f_{k_N}(i)\,\xi_i^{(\pm)}|. \qquad (151)$$

If all k_i are different, the spin function in any of the columns of this determinant can be either $\xi^{(+)}$ or $\xi^{(-)}$ and there will be 2^N different eigenfunctions where the electrons are distributed (as far as their spatial coordinates are concerned) over N different states. If there are, however, ν pairs of equal numbers among the k_i, the spin functions in the two columns corresponding to such a pair must be different, that is $\xi^{(+)}$ in the one and $\xi^{(-)}$ in the other column, since otherwise the determinant is equal to zero. In this case there are only $2^{N-2\nu}$ different eigenfunctions (151).

In principle each Slater determinant (151) can be expanded in terms of its multiplet contributions. These have the form of linear combinations of the $2^{N-2\nu}$ functions just mentioned, which are eigenfunctions of the total spin angular momentum S^2. One can determine the number of combinations which give rise to a given spin quantum number $\bar{s}$, that is, which correspond to a given multiplicity, by the same methods which were used in § 77 to expand a pure spin function in terms of multiplet contributions. The z-component $m\hbar$ of the total spin angular momentum can take on the values $(\frac{1}{2}N-\nu)\hbar$, $(\frac{1}{2}N-\nu-1)\hbar, \ldots, -(\frac{1}{2}N-\nu)\hbar$, and a given value of m occurs $\binom{N-2\nu}{\frac{1}{2}N-\nu-m}$ times. The spin quantum number $\bar{s}$ occurs therefore $\left\{\binom{N-2\nu}{\frac{1}{2}N-\nu-\bar{s}} - \binom{N-2\nu}{\frac{1}{2}N-\nu-\bar{s}-1}\right\}$ times. This number is thus the number of independent φ_m sets each M-fold degenerate corresponding to a multiplicity $M = 2\bar{s}+1$. The maximum multiplicity which occurs is $M = N+1-2\nu$.

The multiplet contributions of expression (151) *can be expressed explicitly by means of the multiplet functions* (150). For a given value of n the number of different functions (150) is equal to the number of possibilities of choosing the numbers $l_1, \ldots, l_n$. The numbers l are a permutation of the $k_1, \ldots, k_N$ of such a nature that each k-value which occurs twice (there are ν of them) will be both among $l_1, \ldots, l_n$ and among $l_{n+1}, \ldots, l_N$. There are then left $\binom{N-2\nu}{n-\nu} = \binom{N-2\nu}{\frac{1}{2}N-\nu-\bar{s}}$ possibilities of distributing the other k-values.

Comparing this number with the number found a moment ago we see that the functions considered are linearly dependent and that there exist $\binom{N-2\nu}{\frac{1}{2}N-\nu-\bar{s}-1}$ independent linear relations between them. They can also not be orthogonal to each other. Indeed, it is possible to show by using equations (149) and (150) that there are exactly that number of relations (we shall not prove this here). They express the fact that it is impossible to construct from a linear combination of expression (150) and its permutations a function which is antisymmetric in more than $N-n-\nu$ quantum numbers [expression (150) is antisymmetric in exactly $N-n-\nu$ quantum numbers].

If the particles are no longer strictly independent, that is, if the operator H^0_{op} of equation (146) must be completed by terms containing the coordinates of several electrons (especially Coulomb terms!), the energy value E of equation (147) which was originally $2^{N-2\nu}$ fold degenerate will in general split up into a number of terms each with its own multiplicity. The splitting will be governed by the matrix elements $H^0_{kk'}$ of the total energy operator with respect to the unperturbed functions. As in the two-electron problem these matrix elements will be different from zero only if k and k' refer to unperturbed eigenfunctions of the same multiplicity. The perturbation problem is thus much simplified by expanding the $2^{N-2\nu}$ Slater determinants (151) in its multiplet contributions. It is now only necessary to calculate the matrix elements $H^0_{kk'}$, referring to one multiplicity and the magnetic quantum number can be neglected completely since also in the perturbed systems the spins are still free. If we consider especially the case of a small splitting, we see that for each multiplicity M (if ν is the number of k-pairs, the largest M is equal to $N-2\nu+1$) we must solve a secular problem of a degree equal to the number of the linearly independent functions ${}^M\varphi$ of the type (150), that is, of degree equal to $\binom{N-2\nu}{n-\nu} - \binom{N-2\nu}{n-\nu-1}$, where $M = N-2n+1$ ($n \geqq \nu$). If the complete energy operator corresponding to the electron interactions contains only terms referring to pairs of electrons — and this is the case in all applications — it follows easily that the matrix elements $H^0_{kk'}$ (possibly we need only the average values H^0_{kk}, for instance, if we are dealing with a two-electron problem) are *linear combinations of the expectation value integrals of the type* (70) *and exchange integrals of the type* (76) [or (77) if we are dealing with Coulomb potentials]. This is the case because each matrix element can be expressed as a sum of integrals of the kind [see eq. (47)] $\int \{\mathrm{P}_1 \Pi_{i=1}^N f_{k_i}(i)\}^* H^0_{\text{op}} \{\mathrm{P}_2 \Pi f_{k_i}(i)\}$, where P_1 and P_2 are two permutations. Because of the orthogonality of the f_k, integration of those terms in the energy operator H^0_{op} which refer to pairs of electrons gives a non-zero contribution only when all but two of the factors $f_{k_i}(i)$ in the products in front and behind are identical. The integral reduces thus to the integrals (70) or (76).

If we want to know the stationary states of an N-electron system without spin forces where the interaction is not particularly small, we can always expand the wave functions in terms of Slater determinants (151) where the f are an orthogonal set of functions of the spatial coordinates. It will always turn out that such a representation can be written as an expansion in terms of determinants of the type (150) where the multiplicity M of the state is related to n by the equation $M = N-2n+1$. The terms in this expansion fall into groups, each group referring to a particular choice of the quantum numbers $k_1, \ldots, k_N$.

Even in problems where the electron interaction is strong it is possible to consider successfully the electrons to be *approximately independent*. This question was discussed when we considered two-electron systems. Its significance here is that in order to describe a situation it is possible to choose the functions $f_{k_1}, \ldots, f_{k_N}$ (which all depend on the spatial coordinates of one electron only) in such a way that the determinant (150) or — more generally — a linear combination of determinants of the type (150) which correspond to different permutations of the quantum numbers $k_1, \ldots, k_N$ is a reasonable approximation to the required wave function. For instance, by using the Ritz method we can find the "best" form of the functions f, the "best" values of the numerical coefficients in the linear combination considered and the "best" value of the energy.

For many problems of this kind we can consider the interaction between the electrons to be replaced by the action of a common field of force which is characteristic for the stationary state and which acts on each electron. This field of force combines with the nuclear field to give a resultant field U. The $f_{k_1} \ldots f_{k_N}$ are eigenfunctions of the corresponding one-electron problem. The resultant wave equation will be of the form $[-(\hbar^2/2m)\nabla^2+U-E_{k_i}]f_{k_i} = 0$, where the nuclei are supposed to be fixed in space. Indeed, one can often obtain in this way a reasonable form for the f. It is, however, not essential that the different f all satisfy an equation of this kind with a universal U. An example is Hartree's method * of obtaining a reasonable approximation to the f_{k_i} in atoms. He looks for such f_{k_i} that each of them is an eigenfunction in a field of force corresponding to the nuclear field and the spherically symmetric "smeared out" charge distribution of all the other electrons. For each electron in a "n_l-orbit" there is thus a field of force characteristic of the n_l.

In many investigations of atomic and molecular structure it is important to construct wavefunctions from a small number of func-

* D. R. Hartree, *Proc. Camb. Phil. Soc.*, **24**, 89, 1928.

tions referring to the spinless one-particle problem *. One of the simplest cases is the description of an atom with only closed shells (see § 70). Another typical example is Bloch's treatment of the conduction electrons in a metal **. In the case of closed shells the situation is simplified because each state is doubly occupied, so that we are always dealing with a singlet state. Using equation (151) we try the following wave function,

$$\left.\begin{aligned}
\varphi &= \begin{vmatrix} f_1(1)\eta_1 \ldots . f_n(1)\eta_1 & f_1(1)\xi_1 \ldots . f_n(1)\xi_1 \\ \ldots\ldots\ldots\ldots & \ldots\ldots\ldots\ldots \\ f_1(2n)\eta_{2n} \ldots f_n(2n)\eta_{2n} & f_1(2n)\xi_{2n} \ldots f_n(2n)\xi_{2n} \end{vmatrix} \\
&= \frac{1}{(n!)^2}\sum_{\mathrm{P}}\delta_{\mathrm{P}}\left\{\mathrm{P}\begin{vmatrix} f_1(1) \ldots f_n(1) \\ \ldots\ldots\ldots \\ f_1(n) \ldots f_n(n)\end{vmatrix}\cdot\begin{vmatrix} f_1(n+1) \ldots f_n(n+1) \\ \ldots\ldots\ldots \\ f_1(2n) \ldots\ldots f_n(2n)\end{vmatrix}\right\}. \\
&\quad \mathrm{P}\eta_1 \ldots \eta_n \xi_{n+1} \ldots \xi_{2n} \\
&= (n!)^{-2}\Sigma_{\mathrm{P}}\,\delta_{\mathrm{P}}\,[\mathrm{P}f(1, 2, \ldots, n)f(n+1, \ldots, 2n)]\,P\eta_1 \ldots \xi_{2n}.
\end{aligned}\right\} \quad (153)$$

The number of electrons is $2n$, and the wave functions $f_1, \ldots, f_n$ are assigned to the doubly occupied states. The product of two f-functions (and each product obtained by permutations from it) is supposed to be an approximate solution of the spinless problem. By using the variational principle (5.54) one tries to determine the f as accurately as possible. In the case of the electron shells in the atom one uses equation (152) for the f ***.

If we had used for φ expression (150) we would have obtained instead of equation (153) the more complicated one,

$$\varphi = (n!)^{-2}\Sigma_{\mathrm{P}}\delta_{\mathrm{P}}\{\mathrm{P}f(1,2,\ldots,n)f(n+1,\ldots,2n)\}\mathrm{P}\,D_1 D_2 \ldots D_n A_{n+1} \ldots A_{2n} \quad (154)$$

In expression (154) we have as spin functions the correct singlet eigenfunctions of S^2 which are rather complicated when written out explicitly because of the fact that the D are differentiation symbols. For expression (153), however, the simple product $\eta_1 \ldots \eta_n \xi_{n+1} \ldots \xi_{2n}$ and its permutations occur as spin functions. All the same, expressions (153) and (154) are identical. This arises because there are linear relations between the product of $f(1, \ldots, n)f(n+1, \ldots, 2n)$

* R. S. Mulliken (for instance, *Rev. Mod. Phys.*, **2**, 60, 506, 1930) calls such functions *orbitals*.

** F. Bloch, *Z. Phys.*, **52**, 555, 1928.

*** See, for instance, V. Fock, *Z. Phys.*, **61**, 126, 1930.

and its permutations. These relations express the fact that these products form automatically a reduced set of functions of the antisymmetry type $[n]+[n]$. This is proved simply by replacing in the considerations at the end of § 77 the functions $x^0, x^1, \ldots, x^{n-1}$ by $f_1, f_2, \ldots, f_n$.

This is a special case of a general theorem according to which we can simplify expression (132) for an arbitrary multiplet function by using simple expressions for the ${}^MZ_{j_1 \ldots j_n}$ if the ${}^Mg$ belong to a reduced set of functions. We still write ${}^MZ_{1,2,\ldots n} = \overline{D}_1 \overline{D}_2 \ldots \overline{D}_n A_{n+1} \ldots A_N$ and obtain the other ${}^MZ$ from permutations. The $\overline{D}$, however, are now given by

$$\overline{D}_i = \xi_i v^* - \eta_i u^* \qquad (A_i = \xi_i u + \eta_i v) \tag{155}$$

The Z are still rotational invariants and also still of the symmetry type $\{n\}+\{N-n\}$, but they no longer satisfy the linear relations (129). However, now the g-functions produce a "reduction" of these simpler spin functions because of the relations existing between them. The situation is parallel to the earlier one where the g might not be reduced, but where the relations between the Z produced the reduction.

This theorem is based on the fact that the multiplet function ${}^M\psi$ is uniquely defined. It follows that when expression (132) is written out in detail using the new Z, it must contain automatically the rotationally invariant factor $(u^*u+v^*v)^n$, and this factor must then necessarily be multiplied by a homogeneous polynomial in u and v of degree $N-2n = M-1$, the same polynomial which occurred when the old Z were used.

In connexion with this we can introduce the following *symbolic representation of an arbitrary multiplet function* (compare the end of § 77),

$$\begin{aligned} {}^M\psi &= |\overline{D}_i, x_i \overline{D}_i, \ldots, x_i^{n-1} \overline{D}_i; A_i, x_i A_i, \ldots, x_i^{N-n-1} A_i| \\ &= [n!(N-n!)]^{-1} \Sigma_P \delta_P \{P\Pi_{1\ldots n} \Pi_{n+1\ldots N}\} P(\overline{D}_1 \ldots \overline{D}_n A_{n+1} \ldots A_N). \end{aligned} \tag{156}$$

The coefficients of the separate $\overline{D} \ldots A \ldots$ products are reduced functions of the antisymmetry type $[N-n]+[n]$. They transform under permutations exactly like arbitrary reduced ${}^Mg$ functions and they possess the same linear properties. This was the reason why this representation could be used. We can write even more simply

$$\begin{aligned} \psi &= A_1 \ldots A_N \Pi_{1\ldots N} = A_1 \ldots A_N (x_2-x_1)(x_3-x_1) \ldots (x_N-x_1)(x_3-x_2) \ldots (x_N - x_{N-1}); \\ x_i^m &= (\overline{D}_i/A_i) x_i^{m-1}, \quad m \geqq N-n. \end{aligned} \tag{157}$$

This equation is to be interpreted as follows: The alternating product is written out explicitly and all powers of any x_i higher than x_i^{N-n-1} are reduced according to the second equation (157).

§ 79. N-ELECTRON SYSTEMS WITH SPIN FORCES; RUSSELL-SAUNDERS COUPLING

For very many atomic problems the presence of spin forces is only a small perturbation. We have extensively discussed their action in the case of the two-electron problem (§ 74). All characteristics which we saw then occur slightly generalised in the N-electron problem. We restrict our discussion (as in § 74) to the influence of spin forces on the stationary states.

The perturbing energy operator is [see eq. (6.79)]

$$H^{(1)}_{\text{op}} = \Sigma_i [H^{x_i} S_{x_i} + H^{y_i} S_{y_i} + H^{z_i} S_{z_i}] + \Sigma_{i\,j} [H^{x_i x_j} S_{x_i} S_{x_j} + H^{x_i y_j} S_{x_i} S_{y_j} + \ldots], \quad (158)$$

where the second sum extends over all electron pairs. The operators $H^{x_i}, \ldots$ which act on the spatial coordinates of the electrons need not be symmetric in them. The symmetry of $H^{(1)}_{\text{op}}$ requires only that H^{x_i}, H^{y_i}, and H^{z_i} be symmetric in all electrons, but the i-th one, and that $H^{x_i x_j}, \ldots$ are symmetric in all electrons but the i-th and j-th. If, however, in H^{x_i} the i-th electron is interchanged with the i'-th $(\mathrm{P}_{i\longleftrightarrow i'})$ we must have

$$\mathrm{P}_{i\longleftrightarrow i'} H^{x_i}(1\ldots i\ldots i'\ldots N) = H^{x_i}(1\ldots i'\ldots i\ldots N) = H^{x_{i'}}(1\ldots i\ldots i'\ldots N). \quad (159)$$

Similarly we must have for the $H^{x_i y_j} \ldots$ the relations

$$\mathrm{P}_{i\longleftrightarrow i',\, j\longleftrightarrow j'} H^{x_i y_j}(\ldots i \ldots j \ldots i' \ldots j' \ldots) = H^{x_i y_j}(\ldots i' \ldots j' \ldots i \ldots j \ldots) = H^{x_{i'} y_{j'}}(\ldots i \ldots j \ldots i' \ldots j' \ldots)$$

Let an unperturbed multiplet state be characterised by the following M eigenfunctions [see eq. (145)]

$$^{M}\varphi^{(m)} = \Sigma_{j_1 \ldots j_n} {}^{M}g_{(j_1 \ldots j_n)} Z^{(m)}_{j_1 \ldots j_n}, \qquad m = \bar{s}, \bar{s}-1, \ldots, -\bar{s} \quad (160)$$

Let $^{M}g_{(j_1 \ldots j_n)}$ be a reduced eigenfunction of the spinless problem belonging to the eigenvalue ^{M}E. For the time being we assume that the only degeneracy of this eigenvalue is the one following automatically from its antisymmetry character $[N-n]+[n]$ (see § 78). The eigenvalue ^{M}E is M-fold degenerate corresponding to the M possible orientations of the spin vector. This degeneracy expresses itself in the M different possible values of the magnetic quantum number m. We assume that no eigenvalues of different multiplicity are lying near ^{M}E (or possibly even coincide with ^{M}E).

According to perturbation theory the energetic influence of $H^{(1)}_{\text{op}}$ on ^{M}E is found from the eigenvalues of the perturbation matrix. In computing this matrix we might expect complications because of the intricate form of the $^{M}\varphi^{(m)}$. However, we know for certain that the matrix elements will be of the form

$$H^{(1)}_{mm'} \to A + (A_x S_x + A_y S_y + A_z S_z) + \tfrac{1}{2}[A_{xx} S_x^2 + A_{xy}(S_x S_y + S_y S_x) + \ldots] \quad (161)$$

where S_x, S_y, S_z are the components of the resulting spin angular momentum expressed as matrices of the form given in figure 11, while $A, A_x \ldots$ are constants obtained by integration over all spatial coordinates. The term $A_x S_x$ in expression (161) derives from the terms $\Sigma_i H^{x_i} S_{x_i}$ in equation (158), the terms $A_{xx} S_x^2$, $A_{xy}(S_{xy}+S_{yx}), \ldots$ from the terms involving $S_{x_i} S_{x_j}$, $S_{x_i} S_{y_j}$, ... in equation (158), and the constant A from the terms $S_{x_i} S_{x_j}$, $S_{y_i} S_{y_j}$, and $S_{z_i} S_{z_j}$ in equation (158) all in the same way as in the case of the perturbation of a triplet state [compare eq. (85)].

The general form of expression (158) is clear (see the end of § 60). The correspondence between the terms of expression (161) and those of equation (158) follows from the vector character of $S_{x_i}, \ldots$

To prove expression (161) we consider, for instance, the three mm'-matrices corresponding to the operators $F_{\text{op}} S_{x_i}$, $F_{\text{op}} S_{y_i}$ and $F_{\text{op}} S_{z_i}$, where F_{op} is an arbitrary operator acting on the spatial coordinates. These operators transform as the components of a vector into $F_{\text{op}} S_{x'_i}$, $F_{\text{op}} S_{y'_i}$, and $F_{\text{op}} S_{z'_i}$ where the primes refer to a new $x'y'z'$-system of coordinates. The three corresponding mm'-matrices, which—as we know—can be expressed in terms of S_x, S_y, and S_z, must now have the same form if they are expressed in terms of the S'_x, S'_z, and S'_y referring to the $x'y'z'$. However, the components S_x, S_y, S_z themselves are the only polynomials in S_x, S_y, S_z which transform as a vector. Hence the matrix corresponding to $F_{\text{op}} S_{x_i}$ must be equal to S_x, apart from a factor. Similarly, we can consider the nine quantities $F_{\text{op}} S_{x_i} S_{x_j}$, $F_{\text{op}} S_{x_i} S_{y_j}$, ... Out of these we must construct those groups of less than nine linear combinations which transform into one another under rotations, since only of those it is true to say that their transformation properties will be followed by the corresponding matrices—which are functions of S_x, S_y, and S_z. We do not wish to consider this point in any more detail, but only mention that the constant A in expression (161) derives from the invariance of $S_{x_i} S_{x_j} + S_{y_i} S_{y_j} + S_{z_i} S_{z_j}$ under rotations.

We could also have obtained expression (161) directly, using the relations (159).

The determination of the coefficients A in expression (161) can sometimes be very difficult. One uses the operators $H^{x_i}, \ldots$ of equation (158) and the ${}^M g$-functions of equation (145). The general character of the splitting up of ${}^M E$ follows, however, from expression (161); it involves solving a secular problem of the M-th degree. If the terms involving $S_x^2, \ldots$ (and the term A) are negligibly small — they derive from the small spin-spin interaction — the solution is simple. The perturbation $A_x S_x + A_y S_y + A_z S_z$ corresponds to the perturbation by a homogeneous magnetic field $\mathfrak{H}$ whose perturbation operator was given by equation (86). We choose our system of coordinates in such a way that the z-axis is parallel to the magnetic field. This brings the perturbation matrix into diagonal form. Its diagonal elements, that is, the energy splittings, are equal to [compare eq. (87)]

$$ {}^M E' - {}^M E = m(e\hbar/m_0 c)\mathfrak{H} = 2m\mu_{\text{B}}\mathfrak{H}. \tag{162} $$

The energy level is split into a number of equidistant levels, and the number of these levels is equal to the multiplicity of the state considered.

If in expression (161) only the terms quadratic in S_x, S_y, S_z are present, the perturbation matrix can be brought into the form $A_{xx}S_x^2+A_{yy}S_z^2+A_{zz}S_z^2$ by a rotation. The corresponding secular problem can immediately be solved only when two of the A's are equal in which case ${}^ME' - {}^ME = C_1+C_2m^2$. If all three A are different the problem is closely connected with the theory of the Lamé-functions *. We mention here only that if the number of electrons is odd (M even) the eigenvalue will be two-fold degenerate, but not when the number of electrons is even (except when two of the A's are equal).

This two-fold degeneracy is a special case of a very general theorem** about the eigenfunctions of a system containing several spinning electrons. The theorem is independent of the Pauli principle. This theorem states that if $\varphi_{s_1, s_2 ..}(\boldsymbol{x}_1, .., \boldsymbol{x}_N)$ [see eq. (6.75)] is an eigenfunction of the energy operator $H_{\text{op}} = H_{\text{op}}(\boldsymbol{x}_1, ..., \boldsymbol{x}_N, \boldsymbol{p}_1, ..., \boldsymbol{p}_N, \boldsymbol{S}_1, ..., \boldsymbol{S}_N)$ corresponding to an eigenvalue E, the spin conjugate function [see eq. (6.100)]

$$\varphi^\dagger{}_{s_1 .. s_N}(\boldsymbol{x}_1, ..., \boldsymbol{x}_N) = (-1)^{\frac{1}{2}N - \Sigma s_k} \varphi^*{}_{-s_1, -s_2, .., -s_N}(\boldsymbol{x}_1, ..., \boldsymbol{x})$$

will be an eigenfunction of the operator $H'_{\text{op}} = H_{\text{op}}(\boldsymbol{x}_1, ..., \boldsymbol{x}_N, -\boldsymbol{p}_1, ..., -\boldsymbol{p}_N, -\boldsymbol{S}_1, ..., -\boldsymbol{S}_N)$ corresponding to the same eigenvalue. If H'_{op} and H_{op} are identical, E will not be degenerate provided $\varphi^\dagger = a\varphi$. From this last equation it follows that $(\varphi^\dagger)^\dagger = a^*\varphi^\dagger = a^*a\varphi$. Since $(\varphi^\dagger)^\dagger$ is also equal to $(-1)^N\varphi$ we see that *every eigenvalue is degenerate with an even degree of degeneracy when there is an odd number of electrons, provided the energy is invariant under a reversal of sign of the momenta and spins*. Wigner*** has drawn attention to the close connexion between this theorem and the problem of time reversal in quantum mechanics.

Let us now consider the question of the influence of the spin forces on the wave function of a multiplet state. We write [see eqq. (89) and (90)] ${}^M\varphi = {}^M\overline{\varphi} + {}^M\varphi'$, where ${}^M\overline{\varphi}$ is a linear combination of the zeroth approximation multiplet functions given by equation (160) which is appropriate to the spin perturbation. The extra term can be expanded in terms of all other zeroth approximation multiplet functions, ${}^M\varphi' = \Sigma_{M'}\, {}^{M'}\chi$ which expresses the fact that in general ${}^M\varphi'$ contains terms possessing a multiplicity different from M, or, that *spin perturbation will lead to an admixture of other multiplicities in a given multiplet state*. Whether a given multiplicity

* H. A. Kramers and G. P. Ittmann, *Z. Phys.*, **60**, 663, 1930.
** H. A. Kramers, *Proc. Acad. Sc., Amsterdam*, **33**, 959, 1930.
*** E. Wigner, *Göttinger Nachr.*, **1932**, 546.

M' (in this approximation) occurs in ${}^M\varphi'$ depends on whether there are non-vanishing integrals of the kind [see eq. (91)]

$$\int ({}^{M'}\overline{\varphi})^* H^{(1)}_{\text{op}} ({}^{M}\overline{\varphi}). \tag{163}$$

It is possible to prove that ${}^M\varphi'$ *can contain terms of multiplicities* $M+2$, M, $M-2$ *due to the terms in expression* (158) *linear in* $S_{x_i}, \ldots$, *while the terms quadratic in* $S_{x_i}, \ldots$ *can lead to multiplicities* $M+4$, $M+2$, M, $M-2$, $M-4$. In order that $M-2$, or $M-4$, occurs, M must be larger than 2, or 4.

We shall not give the complete proof of this statement. We only draw attention to the fact that expression (163) after substitution of expressions (158) and (160) leads among other things to sums over all spin coordinates of the form

$$\Sigma_{s_1,\ldots,s_N}({}^{M'}Z)^* S_{x_i}({}^{M}Z), \quad \Sigma_{s_1,\ldots,s_N}({}^{M'}Z)^* S_{x_i} S_{y_j}({}^{M}Z), \tag{164}$$

where ${}^{M'}Z$ and ${}^{M}Z$ are eigenfunctions of S^2_{op}. All sums of this kind are contained in the expressions

$$\Sigma_{s_1,\ldots,s_N}({}^{M'}Z)^*[\tfrac{1}{2}(\overline{u}^2-\overline{v}^2)\, S_{x_i}+\tfrac{1}{2}i(\overline{u}^2+\overline{v}^2)S_{y_i}-\overline{u}\,\overline{v}\, S_{z_i}]^p\, ({}^{M}Z), \tag{165}$$

where now ${}^{M}Z$ and ${}^{M'}Z$ are the rotationally invariant expressions (128) which contain a spinor u, v, where $\overline{u}$, $\overline{v}$ is another spinor, and where the exponent p is either 1 or 2. The separate expressions (164) are found by considering in expression (165) definite powers of u, v, u^*, v^*, $\overline{u}$, $\overline{v}$. The expression (165) is rotationally invariant; the result of summation must thus also be rotationally invariant in the six spin components. It is possible to prove that invariants in those components must be polynomials in the three expressions $A = u^*u+v^*v$, $B = u^*\overline{u}+v^*\overline{v}$, $C = u\overline{v}-v\overline{u}$. If $p=1$ there are only three possibilities: $A^{M-1}B^2$, $A^{M-2}BC$, $A^{M-3}C^2$, since otherwise the powers of u, v and $\overline{u}$, $\overline{v}$ would not match. These three possibilities correspond to $M' = M+2$, M, $M-2$. In the case $p = 2$ there are five possibilities: $A^{M-1}B^4$, $A^{M-2}B^3C, \ldots, A^{M-5}C^4$, corresponding to $M' = M+4, M+2, \ldots, M-4$.

To complete the proof we must still show that due to the symmetry properties of the g and H integrals over the spatial coordinates of the type $\int ({}^{M'}g)^* H^{x_i}_{\text{op}}({}^{M}g)d\tau$, $\int ({}^{M'}g)^* H^{x_j y_j}_{\text{op}}({}^{M}g)d\tau$ are not identically equal to zero, provided $M'-M$ is equal to one of the possibilities found a moment ago.

Finally we want to consider the case where the unperturbed multiplet state shows not only spin degeneracy, but also *orbital degeneracy*. This means that there are several sets of reduced functions ${}^{M}g^{(m)}_{(1,\ldots,j_n)}$ which satisfy the spinless Schrödinger equation (114) for the same ${}^{M}E$-value. These sets of functions are distinguished by a quantum number m, which is chosen in such a way that it can take on the values $l, l-1, \ldots, -l$ ($2l+1$ values) (compare § 74). To avoid misunderstandings we shall denote here the magnetic spin quantum number by $\overline{m}$.

The energy perturbation due to the spin forces follows from a secular problem of the degree $(2l+1)(2\overline{s}+1)$. The perturbation matrix is of the form [see eqq. (94) and (158)]

$$H^{(1)}_{m\overline{m}, m'\overline{m}'} \rightarrow A_{mm'} + (A_x)_{mm'}(S_x)_{\overline{m}\overline{m}'} + \ldots + \tfrac{1}{2}(A_{xx})_{mm'}(S_x)^2_{\overline{m}\overline{m}'} + \ldots \quad (166)$$

Formally we can again construct the matrices $A_{mm'}, \ldots$ using three basic matrices $(M_x)_{mm'}$, $(M_y)_{mm'}$, and $(M_z)_{mm'}$ which will be of the form illustrated in Fig. 11. This means that the perturbation matrix is again of the form (95).

Of great physical importance is the case of *an atom with N electrons not under the influence of any external forces.* In this case the stationary states can always be classified according to the eigenvalues of the total angular momentum of the sytem. It is true both for the spinless problem and for the case where spin forces are included. The orbital degeneracy mentioned a moment ago can now be the $2l+1$-fold degeneracy corresponding to an orbital angular momentum quantum number l; its occurrence and properties follow from the discussion of § 76. The matrices $M_x, \ldots$ in expression (95) can now be identified — as in the two-electron problem — with the components of the orbital angular momentum. We know then immediately that l is integral, since M_z possesses only integral eigenvalues. From the rotational invariance of the total system we conclude that expression (96) is again the most general form of the perturbation matrix which governs in the first approximation the splitting up of the energy of the multiplet state. This matrix is the sum of a cosine coupling and a cosine-squared coupling,

$$H^{(1)}_{m\overline{m}, m'\overline{m}'} \rightarrow \alpha_0 + \alpha_1(\boldsymbol{M}\cdot\boldsymbol{S}) + \alpha_2(\boldsymbol{M}\cdot\boldsymbol{S})^2. \quad (167)$$

The considerations of § 76 enable us to find immediately the solutions of this eigenvalue problem. We introduce the total angular momentum $\boldsymbol{J}$ of the atom,

$$\boldsymbol{J} = \boldsymbol{M} + \boldsymbol{S}. \quad (168)$$

We know now that the new stationary states which are perturbed by the spin forces can be classified according to the eigenvalues and eigenfunctions of J^2 and J_z. A simple enumeration as in § 77 shows us immediately what values the quantum number j of the total angular momentum can take on. We consider all possible eigenvalues $m_j\hbar$ of J_z; these are the sum of an eigenvalue $m\hbar$ of M_z and an eigenvalue $\overline{m}\hbar$ of S_z, and there are thus $(2l+1)(2\overline{s}+1)$ possible combinations $m_j = m+\overline{m}$. The maximum (minimum) value of m_j is clearly $l+\overline{s}$ $(-l-\overline{s})$ and occurs only once. The values $m_j = \pm(l+\overline{s}-1)$ occur twice, $m_j = \pm(l+\overline{s}-2)$ thrice, . . ., finally $\pm|l-\overline{s}|$ $2\overline{s}+1$-times $(\overline{s} < l)$ or $2l+1$ times $(\overline{s} > l)$. All $|m_j|$ values smaller than $|l-\bar{s}|$ occur exactly as often as $m_j = |l-\bar{s}|$ itself. From this we conclude that the j-values

$$j=l+\bar{s}, l+\bar{s}-1, \ldots, |l-\bar{s}| \, [2\bar{s}+1 \; j\text{-values } (\bar{s}<l);\; 2l+1 \; j\text{-values } (\bar{s}>l)] \tag{169}$$

occur each exactly once when $\boldsymbol{M}$ and $\boldsymbol{S}$ are combined to give $\boldsymbol{J}$ (compare § 59).

The corresponding eigenfunctions are linear combinations of the different ${}^M\varphi^{(m,\bar{m})}$ for which $m+\bar{m} = m_j$. The function ${}^M\varphi^{(m,\bar{m})}$ $[= \Sigma_{j_1 \ldots j_n} {}^M g^{(m)}_{(j_1 \ldots j_n)} Z^{(\bar{m})}_{(j_1 \ldots j_n)}]$ is the unperturbed wave function [see eq. (160)] corresponding to $M_z = m\hbar$, $S_z = \bar{m}\hbar$. The $2l+1$ functions ${}^M g^{(m)}_{(j_1 \ldots j_n)}$ and the $2\bar{s}+1$ functions $Z^{(\bar{m})}_{j_1 \ldots j_n}$ are for giving values of $j_1 \ldots j_n$ defined with respect to a given spatial xyz-system. They are normalised in accordance with equation (111). Under rotations the $g^{(m)}$ and $Z^{(\bar{m})}$ transform linearly among each other like the monomials $\xi^{l+m}\eta^{l-m}$ and $\bar{\xi}^{\bar{s}+\bar{m}}\bar{\eta}^{\bar{s}-\bar{m}}$, where ξ, η and $\bar{\xi}$, $\bar{\eta}$ transform both as the unit spinors [see eqq. (114) and (118)]. From the *symbolical representation*

$${}^M\varphi^{(m,\bar{m})} = \xi^{l+m}\,\eta^{l-m}\,\bar{\xi}^{\bar{s}+\bar{m}}\,\bar{\eta}^{\bar{s}-\bar{m}} \tag{169A}$$

it can now be deduced how the ${}^M\varphi^{(m,\bar{m})}$ transform under rotations. If u, v is a constant spinor in the expression

$$\begin{aligned} \Phi_j &= (\xi\bar{\eta}-\eta\bar{\xi})^{-j+l+\bar{s}}(u\xi+v\eta)^{j+l-\bar{s}}(u\bar{\xi}+v\bar{\eta})^{j-l+\bar{s}} \\ &= \Sigma^{-j}_{m_j=j}\, u^{j+m_j} v^{j-m_j} \binom{2j}{j+m_j} \Phi^{(m_j)}, \end{aligned} \tag{169B}$$

this expression will be rotationally invariant and the last sum can be compared with expression (114). By comparing coefficients we can express $\Phi_j^{(m_j)}$ explicitly as a linear combination of the $\xi^{l+m}\eta^{l-m}\bar{\xi}^{\bar{s}+\bar{m}}\bar{\eta}^{\bar{s}-\bar{m}}$. If we then replace these expressions by the ${}^M\varphi^{(m,\bar{m})}$ we have obtained an explicit representation of the wave function corresponding to given values of the quantum numbers j, m' of the total system. These wave functions are symbolically represented in expression (169B). It can be seen that j in expression (169B) can take on only the values (169).

Many properties of systems of coupled vectors can immediately be calculated using expression (169B);* especially for calculating quantum mechanical *intensities* (intensity ratios in the Zeeman effect and within a multiplet) this equation is very useful.

* H. A. Kramers, *Proc. Acad. Sc., Amsterdam*, **33**, 953, 1930; **34**, 965, 1931; H. C. Brinkman, *Thesis*, Utrecht, 1932; *Z. Phys.*, **79**, 753, 1932; *Applications of spinor invariants in atomic physics*, North Holland Publ. Cy. Amsterdam 1956.

The multiplet state is thus split up into $2\bar{s}+1$ (or $2l+1$) states by the spin forces ($\bar{s} < l$ or $\bar{s} > l$); each state corresponds to one of the values, j, of the total angular momentum and is $2j+1$-fold degenerate. The magnitude of the splitting follows from equation (167). We have

$$(\boldsymbol{M} \cdot \boldsymbol{S}) = \tfrac{1}{2}[J^2 - M^2 - S^2] = \tfrac{1}{2}\hbar^2[j(j+1) - l(l+1) - \bar{s}(\bar{s}+1)]$$

and hence

$$\left.\begin{aligned} {}^{M}E' - {}^{M}E = \alpha_0 + \tfrac{1}{2}\alpha_1 \hbar^2\{j(j+1) - l(l+1) - \bar{s}(\bar{s}+1)\} \\ + \tfrac{1}{4}\alpha_2 \hbar^4\{j(j+1) - l(l+1) - \bar{s}(\bar{s}+1)\}^2. \end{aligned}\right\} \quad (170)$$

The first term is an unimportant constant, the second one is the Landé term for the *splitting up of multiplets in atomic spectra*, and the third term is an extra term which for most applications is very small compared to the second one, but which plays an important role in the He I and Li II spectra *.

In many cases the multiplicity of spectral terms derives immediately from the simple scheme just discussed: a stationary state without spin forces is $2l+1$-fold degenerate with respect to the orbital quantum number l, and possesses also a $2\bar{s}+1$-fold spin degeneracy. There are no spectral terms of a different multiplicity in the neighbourhood. The spin forces lead to a splitting up in the manner described in this section into terms with wave functions into which other spin multiplicities are only slightly mixed. In this case we are dealing with Russell-Saunders coupling and one used the notation ${}^{M}S_j$, ${}^{M}P_j$, ${}^{M}D_j$, ${}^{M}F_j$, ${}^{M}G_j$, ... The superscript refers to the spin multiplicity, the subscript to the value of the quantum number j. The symbols S, P, D, F, G, ... are retained for historical reasons and refer to the values 0, 1, 2, 3, 4, ... of the quantum number l. More systematically one could have used the symbol ${}^{M}l_j$.

The different spectral terms of an atom all corresponding to the same spin multiplicity $M = 2\bar{s}+1$ will split up into $2\bar{s}+1$ terms if $l \geqq \bar{s}$, or into $2l+1$ terms, if $l < \bar{s}$. *The maximum multiplicity which occurs in a system of connected spectral terms is thus equal to the spin multiplicity.* Whether or not spectral terms are connected is empirically bound up with the spectra themselves, and theoretically is connected with spontaneous radiative transitions (see § 84). The tate of affairs of atomic spectra which we have just discussed was

* See, for instance, H. Bethe, *Handb. Phys.*, **24$_1$**, 385, 1933.

the historical starting point for the development of the theory of many electron systems. Hence the terminology "multiplet" which we have used as a basis for our general theoretical treatment of many electron systems, using the Pauli principle.

§ 80. Coupling of many electron systems; homopolar chemical bonds

Consider two systems with N_1 and N_2 electrons at different positions in space, so that to a first approximation they may be independent. We neglect spin forces for the sake of simplicity, and we assume that the first system is in a degenerate multiplet state of multiplicity $M_1 = 2\bar{s}_1+1$, and the second system in a state of multiplicity $M_2 = 2\bar{s}_2+1$. Let N equal N_1+N_2 and let all electrons be numbered consecutively. The eigenfunctions (eigenvalues) of the two states are denoted by

$$^{M_1}\varphi_1{}^{m_1}(1, \ldots, N_1) \text{ and } {}^{M_2}\varphi_2{}^{m_2}(N_1+1, \ldots, N_1+N_2) \quad (E_1 \text{ and } E_2),$$

where m_1 and m_2 are the magnetic spin quantum numbers. Their product $\varphi_1\varphi_2$ is an approximate solution of the Schrödinger equation of the total system. In this solution all interactions between the first N_1 electrons and the other N_2 electrons are neglected. If we permute in this product the numbers $1, \ldots, N$ we have a different distribution of the electrons over the two systems, but the product is still a solution. We can now construct from all permuted products that approximate solution which satisfies the exclusion principle (P: permutation of $1, \ldots, N$),

$$\Phi^{(m_1,\, m_2)} = \Sigma_{\mathrm{P}}\, \delta_{\mathrm{P}}\, \mathrm{P}\, {}^{M_1}\varphi_1^{(m_1)}\, (1, \ldots, N_1)\, {}^{M_2}\varphi_2{}^{(m_2)}\, (N_1+1, \ldots, N_1+N_2). \tag{171}$$

In many applications it is important to take into account the fact that expression (171) can still be a good approximation when, to be sure, the interaction between the two systems is large, but of such a nature that the action of system 1 on system 2 can be replaced by a fixed external field of force, and vice versa. These fields of force must have been taken into account when φ_1 and φ_2 are constructed.

The $(2\bar{s}_1+1)(2\bar{s}_2+1)$ functions $\Phi^{(m_1,\, m_2)}$ do not form a pure multiplet state, but they can be expanded in their multiplet contributions. This means that one can look for such linear combinations

$$^{M}\Phi = \Sigma^{M} a_{m_1 m_2}\, \Phi^{(m_1\, m_2)}, \tag{172}$$

that the ${}^{M}\Phi$ are multiplet functions of the total system. As can be seen, this problem is mathematically exactly the same as the problem of the coupling of two vectors treated in the previous section. These vectors can justifiably be called the total spin vectors $\boldsymbol{S}_1$ and $\boldsymbol{S}_2$ of the two systems although they are not linear combinations of the N_1 (or N_2) electrons *. Even without discussing the question of the coefficients ${}^{M}a_{m_1 m_2}$ in equation (172) we can thus say that when the two systems are formally coupled the following spin quantum numbers (and corresponding multiplicities) of the total system are possible

$$\bar{s} = \bar{s}_1+\bar{s}_2,\ \bar{s} = \bar{s}_1+\bar{s}_2-1, \ldots\ \bar{s} = |\bar{s}_1-\bar{s}_2|,\ M = 2\bar{s}+1 \qquad (173)$$

There corresponds *one* system of M states ${}^{M}\Phi^{(m)}$ to each value of $\bar{s}$, corresponding to the M possible values of the magnetic quantum number m ($|m| \leqq \bar{s}$).

If now the two systems approach each other so closely that we can no longer neglect the interaction between the electrons, the energy of the system will split up in as many multiplet states as is indicated by equation (173) (we still neglect spin forces), that is into $2\bar{s}_1+1$ or $2\bar{s}_2+1$ states according to whether $\bar{s}_1 < \bar{s}_2$ or $\bar{s}_1 > \bar{s}_2$. If we take the average of the exact energy operator over the zeroth approximation states, we obtain a matrix, the elements of which are numbered by the pairs m_1, m_2, and m_1', m_2', $H_{m_1 m_2, m'_1 m'_2} = \int \Phi^{(m_1 m_2)*} H_{\text{op}}\, \Phi^{(m'_1 m'_2)}$. This matrix must be rotationally invariant, since the spins are free, and it must thus be of the form

$$H_{m_1 m_2, m'_1 m'_2} \to C_0+C_1(\boldsymbol{S}_1 \cdot \boldsymbol{S}_2)+C_2(\boldsymbol{S}_1 \cdot \boldsymbol{S}_2)^2+\ldots, \qquad (174)$$

where the maximum power is the smaller of the two numbers $2\bar{s}_1$ and $2\bar{s}_2$ (see § 60). In practice only the first two terms occur, and the energies are of the form

$${}^{M}E = {}^{M_1}E_1+{}^{M_2}E_2+C_0+C_1\{\bar{s}(\bar{s}+1)-\bar{s}_1(\bar{s}_1+1)-\bar{s}_2(\bar{s}_2+1)\}. \qquad (175)$$

We encountered an example of this situation in § 74 when we discussed the case of two electrons which were independent in a first approximation. The splitting up into a triplet term of energy $(E_0-A)/(1-C)$ and a singlet term of energy $(E_0+A)/(1+C)$ [see

* Under a rotation $\Phi^{(m_1 m_2)}$ will transform into new $\Phi'^{(m'_1 m'_2)}$ which can be expressed in terms of the old $\Phi^{(m_1 m_2)}$ by using the invariance property

$$\Sigma_{m_1 m_2}\, u_1^{\bar{s}_1+m_1}\, v_1^{\bar{s}_1-m_1}\, u_2^{\bar{s}_2+m_2}\, v_2^{\bar{s}_2-m_2}\, \Phi^{(m_1 m_2)} = \Sigma_{m_1 m_2}\, u'^{\bar{s}_1+m_1}_1 \ldots \Phi'^{(m_1 m_2)},$$

where the Φ's are suitably normalised, in the same way as if the $\Phi^{(m_1 m_2)}$ were products of eigenfunctions of two given angular momentum vectors.

eq. (75)] corresponds, indeed, to a perturbation energy $C_0 + C_1(\boldsymbol{S}_1 \cdot \boldsymbol{S}_2)$; the constants C_0 and C_1 are simply connected with the expectation value integral (70) and the exchange integral (76). The coefficients $C_1, C_2, \ldots$ in expression (174) are also connected with exchange integrals of a more general nature; they are obtained from integrals of the kind $\int \varphi'^* W \varphi'' \, d\tau$ where W contains the interaction energy of the electrons, and where φ' and φ'' are two different approximate solutions of the spinless Schrödinger equation of the total system, with the electrons distributed over the two separate systems in different ways.

The *schematic theory of the homopolar valence,* which Heitler and London * developed starting from their calculations on the hydrogen molecule, is based on the considerations just given. In this theory one investigates the energy of a system of two atoms, both in their ground state, as a function of their distance apart, a. The coefficients $C_0, C_1, \ldots$ are functions of a. One assumes: (*i*) $C_0(a)$ is always positive and increases monotonically from zero as a decreases from ∞; (*ii*) $C_1(a)$ is also positive for large values of a. (We shall not consider other coefficients, or the complications entering because φ' and φ'' are not exactly orthogonal to each other [compare the occurrence of C in eq. (74)].) A consequence of these assumptions is that the lowest energy state of the total system corresponds to a spin quantum number $|\bar{s}_1 - \bar{s}_2|$. If, for instance, $\bar{s}_1 < \bar{s}_2$ in this ground state, the $\bar{s}_1$ electron spin of the first system is antiparallel to the electron spin of the second system: *the spins have the "tendency" to compensate each other.* Finally one assumes: (*iii*) $C_1(a)$ is so large that the energy of the ground state is smaller than the energy for $a = \infty$. This means that the atoms will form a molecule.

These assumptions are correct in the cases of a few simple systems which can be treated quantitatively. Two hydrogen atoms in their ground state (doublet state) lead to a triplet term and a singlet term when they approach each other. The singlet term is the lowest and corresponds to a bound state (H_2-molecule). An He-atom (in its singlet ground state) repels both another He-atom and a hydrogen atom. In this case only $C_0(a)$ enters.

These considerations — as well as the more general considerations given above — can easily be generalised to more than two atoms. In general the number of "valence bonds" of an atom would be equal to the number of the free spins (that is $2\bar{s}$, twice the spin quan-

* W. Heitler and F. London, *Z. Phys.*, **44**, 455, 1927.

tum number) in its ground state. Further investigations * have shown that this valence theory is not able to account for all phenomena. One of the greatest difficulties of developing a simple physical valence theory is the fact that the interaction between two atoms in a molecule is so strong that in the molecule the peculiarities characterising the states of the loosely bound electrons in the separate atoms have practically all disappeared. One can thus not expect good results from a perturbation theory — and this is essentially what all schematic valence theorems are.

* See, for instance, W. Heitler, *Handb. Radiologie*, **6**$_2$, 485, 1934; F. Hund, *Handb. Phys.*, **24**$_1$, 561, 1933; L. Pauling, *J. Am. Chem. Soc.*, **53**, 1367, 1931.

CHAPTER VIII

ELECTROMAGNETIC RADIATION

§ 81. Quantum theory of radiation and quantum electrodynamics

The quantum mechanics of atomic systems treated in the preceding chapters was a generalisation of classical mechanics of point particles. We used the concepts "position" and "momentum" of a particle; the influence of external fields on particles is taken into account by including in the Hamiltonian terms which are functions of spatial and momentum coordinates. The introduction of the ,,spin coordinate" and the "spin angular momentum" of a particle in chapter 6 did not lead to an essential deviation from the general framework of classical mechanics. We could still obtain an interpretation of physical phenomena using wave functions, the arguments of which are the particle coordinates.

Even in classical physics one was aware of the fact that the framework of the classical mechanics of point particles was insufficient. The simplest way to see this is by considering the requirement of the invariance of the laws of nature under Lorentz transformations; this leads to a finite velocity of propagation of all actions. The causal interaction between spatially separated particles can only be transmitted by a "field' which itself carries energy and momentum. Such a field theory reminds us of the classical theory of electrons with its linear field equations for the electromagnetic $\mathfrak{E}$, $\mathfrak{H}$-field: the existence of charged elementary particles enters into this theory as a strange element, needed because of experimental evidence, and the difficulties which occurred in classical physics because of the field-particle dualism have not yet been solved (problem of the extension and structure of electrons; see § 1 *). Although it was an approxi-

* Several people have tried to change the classical theory of electrons in such a way that the existence of point electrons appears to be a natural consequence of the theory. In the framework of the general theory of relativity, in which there are anyhow no strictly linear field equations, Einstein (*Proc. Acad. Sc., Berlin,* **1919**, 349) has considered such a possibility. Born and Infeld (*Proc Roy. Soc.,* **A143**, 410, 1933, **A144**, 425, 1934) have developed in the framework of the special theory of relativity a non-linear field theory with point singularities.

mate theory, the classical theory of electrons was able to give a description of the behaviour of elementary particles which could be applied to a large variety of problems. To a first approximation this description in the non-relativistic theory is a classical mechanics one [see eq. (3.76)]; but also it was able to take into account "radiation phenomena", that is, all those interaction effects between particles for which one must explicitly introduce energy and momentum conditions in the field in order that those quantities are conserved.

Radiative phenomena played a deciding role in the historical development of quantum theory (Planck 1900, Einstein 1905, Bohr 1913). The later development of a systematic quantum theory in the period 1924—1926 was mainly concerned with "quantising" the classical mechanics of point particles. Soon afterwards it proved possible — especially because of a paper by Dirac in 1927 * — to write down a quantum theory of radiation which was derived in a similar way by quantising the classical theory of radiation. Einstein's light quanta and their consequences, Bohr's frequency condition, the quantum theoretical dispersion formula, . . . appeared now in a natural and necessary connexion with the purely mechanical properties of atomic systems consisting of charged particles.

Dirac's radiation theory was based on a separation of the electromagnetic field into a conservative part, which played a role even in the purely mechanical description, and the real radiation field of which Dirac showed how it could be treated as a field of light quanta. Certain simple properties of the classical theory of electrons were lost; for instance, this separation is not a Lorentz-invariant process. Also there are problems — for instance, when very fast particles collide — where it is not at all clear how one can effect this separation. Several people** have therefore tried to construct a consistent "quantum electrodynamics" which pays due respect to all aspects of the classical theory of electrons while at the same time being able to solve those problems where the more restricted theory of Dirac's failed. It is indeed possible to construct such a theory, and in many respects it has achieved what was expected of it. However, the difficulties of the classical theory of electrons connected with the particle-field dualism which we mentioned a moment ago, also crop up in quantum electrodynamics. Quantum electrodynamics is, as the classical theory of electrons was, an approximate theory

* P. A. M. Dirac, *Proc. Roy. Soc.*, **A114**, 243, 710, 1927.

** See especially W. Heisenberg and W. Pauli, *Z. Phys.*, **56**, 1, 1929; **59**, 168, 1929.

and the expectation that the known methods of quantum theory would because of their very nature be able to give more information about the nature of elementary particles has by and large not been fulfilled.

Because of the preliminary character of present-day quantum electrodynamics we shall restrict ourselves in this chapter mainly to Dirac's theory of radiation and its established applications, and we shall not consider the particular problems of quantum electrodynamics *.

In §§ 82 to 85 we show first how far one can proceed without quantising the radiation field. We see then clearly the necessity for the quantisation of the radiation field, and at the same time we are able to derive some important formulae for the absorption and emission of radiation.

§ 82. THE UNQUANTISED RADIATION FIELD; ABSORPTION OF RADIATION

The principles of quantum mechanics discussed earlier enable us to take into account the action of a given external electromagnetic field on the state of an atomic system. Such a field is characterised by a scalar potential Φ and a vector potential $\mathfrak{A}$ [see eq. (2.99)], and its influence is described by certain terms in the energy operator. For a non-relativistic treatment those terms are given by equations (2.102) (one particle, no spin), (3.76) (many particles, no spin), (6.23) (one particle, with spin), and (6.32) and (6.33) (many particles, with spin). If spin is included, the terms include the field quantities $\mathfrak{E}$ and $\mathfrak{H}$ as well as Φ and $\mathfrak{A}$. In Dirac's relativistic theory of one electron [eq. (6.164)] only Φ and $\mathfrak{A}$ occur.

It is possible that the external field derives wholly or partly from electromagnetic radiation. The equations quoted a moment ago enable us thus to calculate the action of radiation on matter. Since a radiation field always varies in time, the energy of the system will not be constant in such a calculation, that is, the probability distribution of the energy will be time dependent. In most applications the influence of the radiation may be considered to be a small perturbation and we can discuss the situation as regards the energy

* We refer for this to W. Pauli, *Handb. Phys.*, **24₁**, 83, 1933; W. Heitler, *Quantum theory of radiation*, Oxford University Press, 1954; G. Wentzel, *Quantum theory of fields*, Interscience Publ., New York, 1949; W. Thirring, *Einführung in die Quantenelektrodynamik*, Deuticke, Vienna, 1955.

most simply by considering the energy function of the unperturbed problem. We shall find the following situation. The distribution of the unperturbed energy over its possible values (we assume the unperturbed system to be conservative) shows not only a fast, periodic "oscillation" proportional to the amplitude of the radiation field (this would also occur, if the perturbing field were constant) but also a secular change, steadily increasing with time, which can be interpreted as *positive or negative absorption of radiative energy.*

We shall pursue this idea in some detail. A radiation field in classical theory is a solution of the Maxwell equations in vacuo without singularities; it can always be considered to be a superposition of monochromatic plane waves,

$$\mathfrak{E} = -c^{-1}\partial\mathfrak{A}/\partial t, \quad \mathfrak{H} = \operatorname{curl}\mathfrak{A}, \quad (\nabla \cdot \mathfrak{A}) = 0, \tag{1}$$

$$\left.\begin{aligned}\mathfrak{A} &= \Sigma_\lambda \mathfrak{A}_\lambda = \operatorname{Re}\, \Sigma_\lambda \mathfrak{A}^0_\lambda \exp[2\pi i\{(\boldsymbol{\sigma}_\lambda \cdot \boldsymbol{x}) - \nu_\lambda t\}],\, \nu_\lambda = c\sigma_\lambda,\, (\mathfrak{A}^0_\lambda \cdot \boldsymbol{\sigma}_\lambda) = 0 \\ &= \tfrac{1}{2}\Sigma_\lambda[\mathfrak{A}^0_\lambda \exp\{2\pi i[(\boldsymbol{\sigma}_\lambda \cdot \boldsymbol{x}) - \nu_\lambda t]\} + \mathfrak{A}^{0*}_\lambda \exp\{-2\pi i[(\boldsymbol{\sigma}_\lambda \cdot \boldsymbol{x}) - \nu_\lambda t]\}].\end{aligned}\right\} \tag{2}$$

The different plane waves are distinguished by an index λ. Each of them is described by a divergence free vector potential $\mathfrak{A}_\lambda$ which is completely characterised by a wave vector $\boldsymbol{\sigma}_\lambda$ and a complex amplitude vector $\mathfrak{A}^0_\lambda$ perpendicular to $\boldsymbol{\sigma}_\lambda$. Both the real and the imaginary part of $\mathfrak{A}^0_\lambda$ are perpendicular to $\boldsymbol{\sigma}_\lambda$; if they are parallel we have linearly polarised light, and if they are perpendicular to each other, we have circularly polarised light. The frequencies ν_λ are all positive.

The perturbed system may contain an arbitrary number of particles; for the time being we neglect spins. Let $H^{(0)}_{\text{op}}$ be the unperturbed energy function. The wave functions of the stationary states form the orthonormal set φ_k (eigenvalues E_k). Let $\mathfrak{A}^H$ be that part of the vector potential which does not correspond to the perturbing radiation field — such a potential is present when the system is in a constant external magnetic field. According to equations (2.102) and (3.76) $\mathfrak{A}^H$ and the radiation vector potential $\mathfrak{A}$ occur in the terms $-\Sigma_i(e_i/mc)(\boldsymbol{p}_i \cdot \mathfrak{A}^H_i + \mathfrak{A}_i) + \Sigma_i(e_i^2/2m_ic^2)(\mathfrak{A}^H_i + \mathfrak{A}_i \cdot \mathfrak{A}^H_i + \mathfrak{A}_i)$ in the energy operator. In those terms e_i and m_i are the charge and mass of the i-th particle and the index i of $\mathfrak{A}^H_i$ and $\mathfrak{A}_i$ indicate that we must take the values of these vectors at the position of the i-th particle. The perturbation energy of the radiation is thus

$$\left.\begin{aligned}H^{(1)}_{\text{op}} &= -\Sigma_i(e_i/m_ic)(\boldsymbol{p}_i - [e_i/c]\mathfrak{A}^H_i \cdot \mathfrak{A}_i) + \Sigma_i(e_i^2/2m_ic^2)\mathfrak{A}_i^2 \\ &= -c^{-1}\Sigma_i(e_i\boldsymbol{v}_i \cdot \mathfrak{A}_i) + \Sigma_i(e_i^2/2m_ic^2)\mathfrak{A}_i^2,\end{aligned}\right\} \tag{3}$$

where $\boldsymbol{v}_i$ is the operator corresponding to the velocity $\dot{\boldsymbol{x}}_i$ of the i-th particle * (see § 43),

$$\boldsymbol{v}_i = \dot{\boldsymbol{x}}_i = (i/\hbar)[H^{(0)}\boldsymbol{x}_i - \boldsymbol{x}_i H^{(0)}] = (\boldsymbol{p}_i - [e_i/c]\mathfrak{A}_i^H)/m_i. \tag{4}$$

As $(\nabla \cdot \mathfrak{A}) = 0$, the $\boldsymbol{v}_i$ commute with the $\mathfrak{A}_i$ so that the order of the factors in equation (3) is immaterial.

Using equations (5.64) and (5.67) and writing for the wave function $\psi = \Sigma_k \gamma_k \varphi_k \exp(-iE_k t/\hbar)$, $\Sigma_k |\gamma_k|^2 = 1$, we have according to equation (5.68) (method of the variation of constants)

$$i\hbar\dot{\gamma}_k = \Sigma_l H_{kl}^{(1)} \exp(2\pi i \nu_{kl} t)\gamma_l, \qquad \nu_{kl} = (E_k - E_l)/h, \tag{5}$$

$$H_{kl}^{(1)} = \int \varphi_k^* H_{\text{op}}^{(1)} \varphi_l \quad \text{(integration over all particle coordinates).} \tag{6}$$

From equations (2) and (3) it follows that the matrix elements $H_{kl}^{(1)}$ are sums of a number of components which vary harmonically in time. We write $H_{kl}^{(1)} = \Sigma_\varrho (kl)_\varrho \exp(-2\pi i \nu_\varrho t)$, where the frequencies ν_ϱ can be both positive and negative. Restricting ourselves to first order perturbations, we may neglect the second term on the right hand side of equation (3), and using equations (6) and (2) we have

$$H_{kl}^{(1)} = \Sigma_\lambda \{(kl)_\lambda \exp(-2\pi i \nu_\lambda t) + (kl)_{-\lambda} \exp(2\pi i \nu_\lambda t)\}, \tag{7}$$

$$\left.\begin{aligned} (kl)_\lambda &= -(2c)^{-1} \left(\mathfrak{A}_\lambda^0 \cdot \int \varphi_k^* \; [\exp\{2\pi i(\boldsymbol{\sigma}_\lambda \cdot \boldsymbol{x}_i)\} e_i \boldsymbol{v}_i] \varphi_l\right), \\ &= -(2c)^{-1} \left(\mathfrak{A}_\lambda^0 \cdot \int \exp[2\pi i(\boldsymbol{\sigma}_\lambda \cdot \boldsymbol{x}_i)] (\varrho \boldsymbol{v})_{kl} \, d\boldsymbol{x}\right), \\ (kl)_{-\lambda} &= -(2c)^{-1} \left(\mathfrak{A}_\lambda^{0*} \cdot \int \exp[-2\pi i(\boldsymbol{\sigma}_\lambda \cdot \boldsymbol{x}_i)] (\varrho \boldsymbol{v})_{kl} \, d\boldsymbol{x}\right), \end{aligned}\right\} \tag{8}$$

$$(\varrho \boldsymbol{v})_{kl} = \Sigma_i \left[\int \tfrac{1}{2}(\varphi_k^* e_i \boldsymbol{v}_i \varphi_l + \varphi_l e_i \boldsymbol{v}_i^* \varphi_k^*) \, \Pi_{j(\neq i)} d\boldsymbol{x}_j\right]_{\boldsymbol{x}_i = \boldsymbol{x}}. \tag{9}$$

In equation (8) we have introduced a vector field in space $(\varrho \boldsymbol{v})_{kl}$ which is defined by equation (9) and which is usually called *the kl-element of the electrical current density of the unperturbed system* **. Equation (9) must be interpreted as follows. We integrate the expression $\frac{1}{2}(\varphi_k^* e_i \boldsymbol{v}_i \varphi_l + \varphi_l e_i \boldsymbol{v}_i^* \varphi_k^*)$ which is constructed from the "current operator" $e_i \boldsymbol{v}_i$ of the i-th particle over the coordinates of all particles bar the i-th one, consider then the value of this integral at $\boldsymbol{x}_i = \boldsymbol{x}$ and sum the result over all particles.

* We have neglected here the terms of the magnetic interaction between the electrons [Darwin's terms in eq. (3.76)].

** Equation (9) is analogous to the expression for the current density in the classical theory of electrons.

The matrix $(\varrho\boldsymbol{v})_{kl}$ is Hermitean, and $(\varrho\boldsymbol{v})_{kl}d\boldsymbol{x}$ can be interpreted as the matrix element of an operator corresponding to the current due to all electrons within $d\boldsymbol{x}$. This can be seen by considering the function $F(\boldsymbol{x})$ which is equal to unity within $d\boldsymbol{x}$ and equal to zero outwith $d\boldsymbol{x}$. If we write $F_i = F(\boldsymbol{x}_i)$, the operator just mentioned will be equal to (we have made the operator Hermitean; compare similar considerations in § 44)

$$(\varrho\boldsymbol{v})_{\text{op}}\, d\boldsymbol{x} = \tfrac{1}{2}\Sigma_i e_i(F_i \boldsymbol{v}_i + \boldsymbol{v}_i F_i). \tag{10}$$

The kl-element of this operator divided by $d\boldsymbol{x}$ is just expression (9) *. If there is only one particle this expression reduces to [compare eq. (2.104)]

$$(\varrho\boldsymbol{v})_{kl} = \frac{e}{2}(\varphi_k^* \boldsymbol{v}\varphi_l + \varphi_l \boldsymbol{v}^* \varphi_k^*) = \frac{e\hbar}{2mi}(\varphi_k^* \boldsymbol{\nabla}\varphi_l - \varphi_l \boldsymbol{\nabla}\varphi_k^*) - \frac{e^2}{2mc}\mathfrak{A}^H \varphi_k^* \varphi_l. \tag{11}$$

The counterparts of the current density operator and matrix are the operator and the *matrix of the charge density*,

$$(\varrho)_{\text{op}}\, d\boldsymbol{x} = \Sigma_i e_i F_i, \tag{12}$$

$$(\varrho)_{kl} = (d\boldsymbol{x})^{-1}\int \varphi_k^*(\Sigma_i e_i F_i)\varphi_l = \Sigma_i e_i\Big[\int \varphi_k^* \varphi_l \Pi_{j(\neq i)}\, d\boldsymbol{x}_j\Big]_{\boldsymbol{x}_i = \boldsymbol{x}}. \tag{13}$$

If $(\varrho\boldsymbol{v})_{\text{op}}$ and $(\varrho)_{\text{op}}$ of equations (10) and (12) are generalised into time independent operators (see § 43) one can prove that $\partial[\varrho(t)]_{\text{op}}/\partial t + (\boldsymbol{\nabla}\cdot[\varrho\boldsymbol{v}(t)]_{\text{op}}) = 0$ (compare a similar proof in § 44). If the φ_k in equations (9) and (13) are eigenfunctions of the energy (this is immaterial for the definition of these matrices) we have ** [see eq. (4.122)] $2\pi i\nu_{kl}(\varrho)_{kl} + (\boldsymbol{\nabla}\cdot[\varrho\boldsymbol{v}]_{kl}) = 0$.

Let us return to the radiative perturbation. Substitution of expression (7) into expression (5) shows that secular perturbation

* To prove the first of equations (8) we note that

$$\int \varphi_k^*[\Sigma_i \exp\{2\pi i(\boldsymbol{\sigma}\cdot\boldsymbol{x}_i)\}\boldsymbol{v}_i]\varphi_l = \int \varphi_l[\Sigma_i \boldsymbol{v}_i^* \exp\{2\pi i(\boldsymbol{\sigma}\cdot\boldsymbol{x}_i)\}]\varphi_k^*$$
$$= \tfrac{1}{2}\int\{\varphi_k^*[\Sigma_i \exp\{2\pi i(\boldsymbol{\sigma}\cdot\boldsymbol{x}_i)\}\boldsymbol{v}_i]\varphi_l + \varphi_l[\Sigma_i \exp\{2\pi i(\boldsymbol{\sigma}\cdot\boldsymbol{x}_i)\ \boldsymbol{v}_i^*]\varphi_k^*\}$$
$$-\tfrac{1}{2}(h\boldsymbol{\sigma}/m_i)\int \varphi_l[\exp 2\pi i(\boldsymbol{\sigma}\cdot\boldsymbol{x}_i)\}]\varphi_k^*.$$ The last term when multiplied by $\mathfrak{A}_\lambda^0$ vanishes.

** If we are dealing with a "free" atomic system, the stationary states form a continuum set, corresponding to the continuous variability of the total momentum, and the position of the system in space is completely undetermined. If it is necessary to take this freedom of the system into account (see § 91), it is often advisable to modify the current density matrix and charge density matrix in such a way that they refer to coordinates relative to the centre of gravity.

effects can occur as soon as the radiation field contains frequencies ν_λ which are equal to, or nearly equal to, a Bohr frequency $|\nu_{kl}|$. In this case there will be on the right hand side of equation (5) terms of the kind $\exp[2\pi i(\pm\nu_\lambda+\nu_{kl})t]$ which do not (or only slowly) change with time, and certain γ_k-values will show considerable changes in time. As $|\gamma_k|^2$ is the probability of finding the system in a state φ_k, we see that *the incoming radiation can cause transitions from one stationary state of an atomic system to another, if it contains components with frequencies near a Bohr frequency.*

We shall apply our equations to the important case where a parallel beam of polarised white light falls on the system. Let the system at $t = 0$ be in the state φ_l, $(\gamma_l)_{t=0} = 1$. We consider the transition to a state φ_k. Let $I(\nu)d\nu$ be the light energy in a frequency interval $(\nu, \nu+d\nu)$ which passes per unit time through a unit area, and let $I(\nu)$ be a "smooth" function in the neighbourhood of $|\nu_{kl}|$. Equation (2) can now be interpreted to mean that there are many ν_λ-values which are lying closely together. Their number within an interval $d\nu$ may be $N_\lambda d\nu$. The corresponding $\boldsymbol{\sigma}_\lambda$ have all the same, or practically the same, direction, while the $\mathfrak{A}_\lambda^0$ only differ by complex numerical factors, so that we can write

$$\mathfrak{A}_\lambda^0 = q_\lambda \mathfrak{a}, \tag{14}$$

where $\mathfrak{a}$ is a complex unit vector perpendicular to $\boldsymbol{\sigma}$,

$$(\mathfrak{a}\cdot\boldsymbol{\sigma}) = 0, \qquad |\mathfrak{a}|^2 = 1. \tag{15}$$

We assume that we can talk about a well defined mean value of $|q_\lambda|^2$ for each ν_λ in the neighbourhood of $|\nu_{kl}|$ and that the phase differences of the q_λ may be assumed to be randomly distributed. From Poynting's expression for the energy current density in a radiation field we then get the following expression for $I(\nu)$,

$$I(\nu_\lambda) = (\pi\nu_\lambda^2/2c)\overline{|q_\lambda|^2}N_\lambda, \tag{16}$$

as $(2\pi i\nu_\lambda/c)q_\lambda\mathfrak{a}_\lambda$ is the (complex) amplitude of the electric field for the λ-th component [see eq. (1)], and the time average of the Poynting vector is its absolute square multiplied by $c/8\pi$.

During a period T within which γ_l is still always practically equal to unity, we have from equations (5) and (7)

$$i\hbar\dot{\gamma}_k = \Sigma_\lambda (kl)_\lambda \exp[2\pi i(\nu_{kl}-\nu_\lambda)t]+\Sigma_\lambda(kl)_{-\lambda}\exp[2\pi i(\nu_{kl}+\nu_\lambda)t]. \tag{17}$$

If $\nu_{kl} > 0$, that is, if we are considering a transition to a state of higher energy (normal or *positive absorption*) only terms of the first sum need be considered, and only those for which $|(\nu_{kl}-\nu_\lambda)T|$ remains less than a small integer. We assume that the perturbation is weak and that we thus can choose T so large that only a small frequency region satisfies the above requirement. Integrating equation (17) we obtain

$$\gamma_k = \hbar^{-1}\Sigma_\lambda(kl)_\lambda\{[1-\exp(i\Delta_\lambda t)]/\Delta_\lambda\},\ t < T,\ \Delta_\lambda = 2\pi(\nu_{kl}-\nu_\lambda). \quad (18)$$

For the absolute square of γ_k we have

$$|\gamma_k|^2 = \hbar^{-2}\overline{|(kl)_\lambda|^2}_{\nu_\lambda \doteq \nu_{kl}} \cdot N_{\nu_{kl}}\int_{-\infty}^{+\infty}|1-e^{i\Delta t}|^2(d\Delta/2\pi\,\Delta^2) = \hbar^{-2}\overline{|(kl)_\lambda|^2}N_{\nu_{kl}}t. \quad (19)$$

This expression derives from the absolute squares of the separate terms of the series (18). The summation could be replaced by an integration over Δ after multiplication by $N_{\nu_{kl}}/2\pi$. The other products in the complete expression for $\gamma_k^*\gamma_k$ vanish on the average because of the randomness of the phases of the q_λ.

The average value of the absolute square of $(kl)_\lambda$ occurring in equation (19) refers to those components of the radiation field the frequencies of which lie very closely to ν_{kl}. From equations (8) and (14) it follows that

$$\overline{|(kl)_\lambda|^2} = (4c^2)^{-1}\overline{|q_\lambda|^2}\,|(\mathfrak{a}\cdot\int\exp[2\pi i(\boldsymbol{\sigma}_\lambda\cdot\boldsymbol{x})](\varrho\boldsymbol{v})_{kl}d\boldsymbol{x})|^2.$$

Using equation (16) we can now write equation (19) in the following form

$$|\gamma_k|^2 = B_{l\to k}(\boldsymbol{\sigma}, \mathfrak{a})I(|\nu_{kl}|)t/c, \quad (20)$$

$$B_{l\to k}(\boldsymbol{\sigma}, \mathfrak{a}) = (2\pi\hbar^2\nu_{kl}^2)^{-1}|(\mathfrak{a}\cdot\exp[2\pi i(\boldsymbol{\sigma}\cdot\boldsymbol{x})](\varrho\boldsymbol{v})_{kl}d\boldsymbol{x})|^2;$$
$$\nu_{kl} > 0,\quad |\boldsymbol{\sigma}| = \nu_{kl}/c. \quad (21)$$

From a similar calculation and using the fact that $(\varrho\boldsymbol{v})_{kl}$ is Hermitean we could have obtained the following expression for a transition to a state of lower energy (*negative* absorption, sometimes called *induced emission*),

$$B_{l\to k}(\boldsymbol{\sigma}, \mathfrak{a}) = (2\pi\hbar^2\nu_{kl})^{-1}|(\mathfrak{a}\cdot\int\exp[2\pi i(\boldsymbol{\sigma}\cdot\boldsymbol{x})](\varrho\boldsymbol{v})_{lk}d\boldsymbol{x})|^2;$$
$$\nu_{kl} < 0,\qquad |\boldsymbol{\sigma}| = |\nu_{kl}|/c. \quad (22)$$

From equation (20) we see that *the radiation produces a probability for a quantum jump which is proportional both to the time and to the intensity of the radiation.* Such transition probabilities were introduced by Einstein in 1917 in his quantum theory of black body radiation *. *The coefficients B introduced by equation* (20) *are a natural generalisation of Einstein's B's* which referred to the action of isotropically incoming unpolarised radiation. If $\varrho(\nu)d\nu$ denotes the density per unit volume of this radiation in the frequency interval $d\nu$ at the position of the atomic system, Einstein put the induced transition probability per unit time equal to $B_{l\to k}\varrho(|\nu_{kl}|)$. In agreement with this definition we introduced $B(\boldsymbol{\sigma}, \mathfrak{a})$ in equation (20), since $(I/c)d\nu$ is the energy density of the incoming beam in the frequency interval $d\nu$. The quantity $B(\boldsymbol{\sigma}, \mathfrak{a})$ refers to a given direction $\boldsymbol{\sigma}$ and a given state of polarisation $\mathfrak{a}$ of the incoming radiation. Also we have always

$$B_{l\to k}(\boldsymbol{\sigma}\cdot\mathfrak{a}) = B_{k\to l}(\boldsymbol{\sigma}\cdot\mathfrak{a}). \tag{23}$$

Einstein's B's are clearly obtained by averaging $B(\boldsymbol{\sigma}, \mathfrak{a})$ over all values of $\boldsymbol{\sigma}$ and $\mathfrak{a}$. This is done most easily by summing first over two mutually orthogonal directions of polarisation. If necessary, the result of averaging can be written in the form of two integrals over space.

If the extension of the irradiated atomic system is small compared to the wave length of the radiation, the exponentials in equations (21) and (22) may be put equal to unity. It is advantageous to introduce *the operator* $\boldsymbol{P}_{\text{op}}$ *of the electrical polarisation,* or *of the dipole moment*

$$\boldsymbol{P}_{\text{op}} = \Sigma_i e_i \boldsymbol{x}_i, \qquad \dot{\boldsymbol{P}}_{\text{op}} = \Sigma_i e_i \boldsymbol{v}_i. \tag{24}$$

The integral in equation (21) is then equal to [compare eq. (9)]

$$\int (\varrho\boldsymbol{v})_{kl}\, d\boldsymbol{x} = \dot{\boldsymbol{P}}_{kl} = 2\pi i\,\nu_{kl}\boldsymbol{P}_{kl} = 2\pi i\,\nu_{kl}\int \varphi_k^* \boldsymbol{P}\varphi_l,$$

and B now only depends on the polarisation vector. We have

$$B_{l\to k}(\mathfrak{a}) = (2\pi/\hbar^2)|(\mathfrak{a}\cdot\boldsymbol{P}_{kl})|^2, \qquad \nu_{kl} > 0, \tag{25}$$

and also,

$$B_{l\to k}(\mathfrak{a}) = (2\pi/\hbar^2)|(\mathfrak{a}\cdot\boldsymbol{P}_{lk})|^2, \qquad \nu_{kl} < 0. \tag{26}$$

Averaging over all possible directions of $\mathfrak{a}$ gives us for B the expression

$$B_{l\to k} = B_{k\to l} = (2\pi/3\hbar^2)|\boldsymbol{P}_{kl}|^2. \tag{27}$$

Equations (21) and (22) can easily be generalised to cover the case where we take the spin of the electrons present into account. We complete equation (3) by including the term [compare eq. (6.23)]

* A. Einstein, *Phys. Z.*, **18**, 121, 1917.

$$H_{\rm op}^{(1)\rm spin} = \Sigma_{i(\rm electrons)}(e/mc)(\boldsymbol{S}_i \cdot \mathfrak{H}_i + (2mc)^{-1}[\mathfrak{E}_i \wedge \boldsymbol{p}_i]).$$

From equations (1) and (2) it follows that

$$\mathfrak{E} = \mathrm{Re}\,\Sigma_\lambda\, \mathfrak{A}_\lambda^0\, 2\pi i\, \sigma_\lambda \exp\{2\pi i[(\boldsymbol{\sigma}_\lambda \cdot \boldsymbol{x}) - \nu_\lambda t]\},$$

$$\mathfrak{H} = \mathrm{Re}\,\Sigma_\lambda\, 2\pi i[\boldsymbol{\sigma}_\lambda \wedge \mathfrak{A}_\lambda^0] \exp\{2\pi i[(\boldsymbol{\sigma}_\lambda \cdot \boldsymbol{x}) - \nu_\lambda t]\},$$

and we get the following additional term in equation (8)

$$(kl)_\lambda^{\rm spin} = -\frac{e}{2mc}\left(\mathfrak{A}_\lambda^0 \cdot \int \varphi_k^* \sum_{\substack{i\\(\rm el)}} [2\pi i(\boldsymbol{\sigma}_\lambda - \frac{\sigma_\lambda}{2c}\boldsymbol{v}_i) \wedge \boldsymbol{S}_i] \exp[2\pi i(\boldsymbol{\sigma}_\lambda \cdot \boldsymbol{x}_i)]\varphi_l\right). \tag{28}$$

The first part of the integral in equation (28) including the factor $\boldsymbol{\sigma}_\lambda$ in the integrand derives from the magnetic field. In the non-relativistic approximation — and equation (28) is only valid in this approximation — this part is more important than the second part (with factor $\sigma_\lambda \boldsymbol{v}_i/2c$) deriving from the electric field. The first part can be changed * into a form analogous to equation (8)

$$(kl)_\lambda^{\rm spin} = -(2c)^{-1}\left(\mathfrak{A}_\lambda^0 \cdot \int \exp[2\pi i(\boldsymbol{\sigma}_\lambda \cdot \boldsymbol{x})](\varrho \boldsymbol{v})_{kl}^{\rm spin}\, d\boldsymbol{x}\right), \tag{29}$$

where the vector field $(\varrho \boldsymbol{v})_{kl}^{\rm spin}$, the matrix element of the so-called *spin current* density, whose divergence is zero and which is independent of $\boldsymbol{\sigma}_\lambda$, is given by

$$(\varrho \boldsymbol{v})_{kl}^{\rm spin} = \frac{i}{\hbar}\sum_{i(\rm el)}\left[\int (\varphi_k^*[\boldsymbol{S}_i \wedge e\boldsymbol{v}_i]\varphi_l - \varphi_l[\boldsymbol{S}_i^* \wedge e\boldsymbol{v}_i^*]\varphi_k^*) \prod_{j(\neq i)} d\boldsymbol{x}_j\right]_{\boldsymbol{x}_i = \boldsymbol{x}}. \tag{30}$$

The integral sign indicates not only the indicated integration over spatial coordinates but also summation over all spin coordinates. The matrix (30) is Hermitean, and when integrated over the whole of space it vanishes. In the case of a one-electron problem it is simply equal to $(\varrho \boldsymbol{v})_{kl}^{\rm spin} = (e/m)\Sigma\{(\varphi_k^*[\boldsymbol{S} \wedge \nabla\varphi_l] + \varphi_l[\boldsymbol{S}^* \wedge \nabla\varphi_k^*]\}$, where Σ indicates summation over the spin coordinate. The second part of the integral (28) cannot be represented by a vector field independent of $\boldsymbol{\sigma}$. We shall neglect it in the following.

Even apart from the problem of the influence of a perturbing radiation field, the spin current is a natural way to describe the magnetic action of the electron spin. This can be seen by writing the expression for the spin energy in the following way.

$$H_{\rm op}^{(1)\rm spin} = \frac{e}{mc}\Sigma_i(\boldsymbol{S}_i \cdot \mathfrak{H}_i) = \frac{e}{mc}\Sigma_i(\boldsymbol{S}_i \cdot \mathrm{curl}\,\mathfrak{A}_i) = \frac{ie}{\hbar mc}\Sigma(\boldsymbol{S}_i \cdot [\boldsymbol{p}_i \wedge \mathfrak{A}_i] + [\mathfrak{A}_i \wedge \boldsymbol{p}_i])$$

$$= \frac{ie}{\hbar mc}\Sigma\{([\boldsymbol{S}_i \wedge \boldsymbol{p}_i] \cdot \mathfrak{A}_i) - (\mathfrak{A}_i \cdot [\boldsymbol{S}_i \wedge \boldsymbol{p}_i]) = \frac{i}{\hbar c}\Sigma\{([\boldsymbol{S}_i \wedge e\boldsymbol{v}_i] \cdot \mathfrak{A}_i) - (\mathfrak{A}_i \cdot [\boldsymbol{S}_i \wedge e\boldsymbol{v}_i])\}.$$

* Note that $2\pi i\boldsymbol{\sigma}_\lambda \exp[2\pi i(\boldsymbol{\sigma}_\lambda \cdot \boldsymbol{x}_i)]\varphi_l = \nabla_i\{\exp[2\pi i(\boldsymbol{\sigma}_\lambda \cdot \boldsymbol{x}_i)]\varphi_l\} - \exp[2\pi i(\boldsymbol{\sigma}_\lambda \cdot \boldsymbol{x}_i)]\nabla_i\varphi_l$.

The determination of the magnetic action of a system is equivalent to the calculation of the interaction energy in an external magnetic field. If $\mathfrak{A}$ is the vector potential of this field and if we want to know the matrix elements of this energy with respect to an orthonormal set φ_k of eigenfunctions of the unperturbed system we get for the part deriving from the electron spin

$$H_{kl}^{(1)\text{spin}} = \int \varphi_k^* \Sigma_j \frac{i}{\hbar c} \{([\boldsymbol{S}_j \wedge e\boldsymbol{v}_j] \cdot \mathfrak{A}_j) - (\mathfrak{A}_j \cdot [\boldsymbol{S}_j \wedge e\boldsymbol{v}_j])\}\varphi_l = -\frac{1}{c}\int (\mathfrak{A} \cdot [\varrho\boldsymbol{v}]_{kl}^{\text{spin}})\, d\boldsymbol{x},$$

with $(\varrho\boldsymbol{v})_{kl}^{\text{spin}}$ given by equation (30).

The part not deriving from the spin is according to equation (3) given by the expression

$$H_{kl}^{(1)} = -\frac{1}{2c}\int \varphi_k^* \Sigma_i\{(e_i\, \boldsymbol{v}_i \cdot \mathfrak{A}_i) + (\mathfrak{A}_i \cdot e_i\, \boldsymbol{v}_i)\}\varphi_l = -\frac{1}{c}\int (\mathfrak{A} \cdot [\varrho\boldsymbol{v}]_{kl})d\boldsymbol{x},$$

where $(\varrho\boldsymbol{v})_{kl}$—the current density if spin is neglected—is defined by equation (9). Comparing the two expressions shows the reason for the nomenclature "spin current".

To illustrate all this we shall calculate the spin current matrix elements for a one-particle system referred to the orthonormal functions $\varphi_{k(+)} = \varphi_k(\boldsymbol{x})\xi$ and $\varphi_{k(-)} = \varphi_k(\boldsymbol{x})\eta$ (ξ and η are again the unit spin vectors). Equation (30) now becomes simply (Σ indicates summation over the spin quantum number),

$$\left.\begin{aligned}
(\varrho\boldsymbol{v})_{k(+),\,l(+)}^{\text{spin}} &= -(\varrho\boldsymbol{v})_{k(-)\,l(-)}^{\text{spin}} = \frac{ie}{m\hbar}\Sigma\{\varphi_k^*\, \xi^*[\boldsymbol{S} \wedge \boldsymbol{p}]\,\varphi_l\xi - \varphi_l\xi[\boldsymbol{S}^* \wedge \boldsymbol{p}^*]\,\varphi_k^*\,\xi^*\}\\
&= \frac{e\hbar}{2m}\left\{-\boldsymbol{e}_x\frac{\partial}{\partial y} + \boldsymbol{e}_y\frac{\partial}{\partial x}\right\}(\varphi_k^*\,\varphi_l),\\
(\varrho\boldsymbol{v})_{k(-),\,l+)}^{\text{spin}} &= \frac{ie}{m\hbar}\Sigma\{\varphi_k^*\,\eta^*[\boldsymbol{S} \wedge \boldsymbol{p}]\varphi_l\xi - \varphi_l\xi[\boldsymbol{S}^* \wedge \boldsymbol{p}^*]\,\varphi_k^*\,\eta^*\}\\
&= \frac{e\hbar}{2m}\left\{(i\boldsymbol{e}_x - \boldsymbol{e}_y)\frac{\partial}{\partial z} + \boldsymbol{e}_z\Big(\frac{\partial}{\partial y} - i\frac{\partial}{\partial x}\Big)\right\}(\varphi_k^*\,\varphi_l),
\end{aligned}\right\} \tag{31}$$

where $\boldsymbol{e}_x$, $\boldsymbol{e}_y$, $\boldsymbol{e}_z$ are unit vectors along the x-, y-, and z-axes. Integration of expressions (31) over the whole of space clearly gives zero. The diagonal element $(\varrho\boldsymbol{v})_{k(+),\,k(+)}^{\text{spin}}$ would describe the spin current in a stationary state $\varphi_k\xi$. If we put $\varphi_k^*\varphi_k = I(\boldsymbol{x})$, we have

$$(\varrho\boldsymbol{v})_{k(+),\,k(+)}^{\text{spin}} = \frac{e\hbar}{2m}\left\{-\boldsymbol{e}_x\frac{\partial I}{\partial y} + \boldsymbol{e}_y\frac{\partial I}{\partial x}.\right\} \tag{32}$$

In the case of a spherically symmetrical I-distribution—for instance, the ground state of the hydrogen atom—it corresponds to a spherically symmetrical charge distribution of density $\varrho = eI$, where every spherical shell (radius r) rotates uniformly with angular velocity $(\hbar/m)d\ln I/dr^2$.

According to the methods of the classical theory of electrons the resulting magnetic moment of the spin current is given by the matrix $(\boldsymbol{\mu})_{k(\pm),\,l(\pm)} = (2c)^{-1}$ $\int [\boldsymbol{x} \wedge (\varrho\boldsymbol{v})_{k(\pm),\,l(\pm)}^{\text{spin}}]\, d\boldsymbol{x}$ which leads, as one would expect, to $(\boldsymbol{\mu})_{k(\pm),\,l(\pm)} \to -(e/mc)(\boldsymbol{e}_x S_x + \boldsymbol{e}_y S_y + \boldsymbol{e}_z S_z)\,\delta_{kl}$ when expressions (31) are used.

If in expressions (21) and (22) for the transition probabilities we wish to take the electron spin into account, we need only add the spin current $(\varrho\boldsymbol{v})_{kl}^{\text{spin}}$ to the vector field $(\varrho\boldsymbol{v})_{kl}$. It has to be borne in mind that we must now sum over the spin coordinates of all particles not only in expression (30), but also in expression (9) for $(\varrho\boldsymbol{v})_{kl}$.

In Dirac's rigorously relativistic theory of the one-electron system equations (21) and (22) can also be used immediately. Formally the calculation of $(\varrho \boldsymbol{v})_{kl}$ is much simpler here than in the non-relativistic case as the perturbation energy of a radiation field with vector potential $\mathfrak{A}$ is according to equation (6.164) given by

$$H_{\mathrm{op}}^{(1)} = e(\boldsymbol{\alpha} \cdot \mathfrak{A}) = (e/c)(\boldsymbol{v} \cdot \mathfrak{A}), \tag{33}$$

where now $\boldsymbol{v}$ is the velocity operator of the electron. The term with $\mathfrak{A}^2$ which occurred in equation (3) is no longer present. Equation (8) for the matrix elements $(kl)_\lambda$ is still valid, and the current density matrix $(\varrho \boldsymbol{v})_{kl}$ is simply given by the equation

$$(\varrho \boldsymbol{v})_{kl}^{\mathrm{Dirac}} = -ec\,\Sigma\,\boldsymbol{\alpha}_{kl} = -ec\,\Sigma_{s,r}\,(\varphi_{s,r})_k^*\,\boldsymbol{\alpha}(\varphi_{s,r})_l, \tag{34}$$

where the summation is over the discrete arguments s, r of the Dirac theory so that there will be four terms [see, for instance, eq. (6.169)].

If we go over to the not rigorously relativistic approximation corresponding to Pauli's representation with two components, $(\varrho \boldsymbol{v})_{kl}^{\mathrm{Dirac}}$ will separate into two parts, the first one being the "spinless" current (9), and the second one corresponding to the spin current (30) — in both cases in the simple form of the one-electron problem. If we choose for the $\varphi_{s,r}$ the representation which was the basis of equation (6.184) this separation comes about automatically.

In this representation $\varphi_{s,+}$ $(s = \pm\frac{1}{2})$ are the two "large" components of the wave function, corresponding to Pauli's components φ_s $(s = \pm\frac{1}{2})$, while $\varphi_{s,-}$ are the small components which satisfy the relation (see § 65) $\varphi_{s,-} = [\varepsilon + mc]^{-1}(\boldsymbol{\pi} \cdot \boldsymbol{\sigma})\varphi_{s,+}$. As $\boldsymbol{\alpha} = \varrho_x \boldsymbol{\sigma}$, we have $-ec\,\Sigma_{s,r}(\varphi_{s,r})_k^*\,\boldsymbol{\alpha}(\varphi_{s,r})_l = -ec\,\Sigma_s\{(\varphi_s)_k^*\,\boldsymbol{\sigma}[\varepsilon+mc]^{-1}(\boldsymbol{\pi} \cdot \boldsymbol{\sigma})(\varphi_s)_l + [\varepsilon^*+mc]^{-1}(\boldsymbol{\pi}^* \cdot \boldsymbol{\sigma}^*)(\varphi_s)_k^*\,\boldsymbol{\sigma}(\varphi_s)_l\} \doteqdot -(e/2m)\Sigma_s\{(\varphi_s)_k^*\,\boldsymbol{\sigma}(\boldsymbol{\pi} \cdot \boldsymbol{\sigma})(\varphi_s)_l + (\varphi_s)_l \boldsymbol{\sigma}^*(\boldsymbol{\pi}^* \cdot \boldsymbol{\sigma}^*)(\varphi_s)_k^*\}$, where we put $\varepsilon = mc$, that is, where we neglected terms with $c, \ldots$ in the denominator.

From the properties of $\boldsymbol{\sigma}$ [eq. (6.135)] it follows immediately that $\boldsymbol{\sigma}(\boldsymbol{\pi} \cdot \boldsymbol{\sigma}) = \boldsymbol{\pi} - i[\boldsymbol{\sigma} \wedge \boldsymbol{\pi}]$, $\boldsymbol{\sigma}^*(\boldsymbol{\pi}^* \cdot \boldsymbol{\sigma}^*) = \boldsymbol{\pi}^* + i[\boldsymbol{\sigma}^* \wedge \boldsymbol{\pi}^*]$. These equations are the basis of the separation under discussion. Dropping the index s and taking into account the meaning of $\boldsymbol{\sigma}(\boldsymbol{\sigma} = 2\boldsymbol{S}/\hbar)$ and $\boldsymbol{\pi}$ [eq. (6.156)] we have

$$(\varrho \boldsymbol{v})_{kl}^{\mathrm{Dirac}} \doteqdot -\tfrac{1}{2}e\Sigma(\varphi_k^*\,\boldsymbol{v}\varphi_l + \varphi_l \boldsymbol{v}^*\varphi_k^*)$$
$$+ (ie/\hbar)\,\Sigma(\varphi_k^*[\boldsymbol{S} \wedge \boldsymbol{v}]\varphi_l - \varphi_l[\boldsymbol{S}^* \wedge \boldsymbol{v}^*]\varphi_k^*) = (\varrho \boldsymbol{v})_{kl} + (\varrho \boldsymbol{v})_{kl}^{\mathrm{spin}}.$$

In this equation $\boldsymbol{v}$ is again the operator defined by equation (4), $\boldsymbol{v} = [\boldsymbol{p} + (e/c)\mathfrak{A}]/m = \boldsymbol{\pi}/m$.

§ 83. The Insufficiency of an Unquantised Radiation Theory; Classical Theory of the Emission of Radiation

Even if the results of the previous sections were correct, they are certainly incomplete, and the correspondence with the classical theory of electrons lacks an essential feature. If in the classical theory a system of charged particles absorbs radiation energy, this energy disappears from the radiation field — in accordance with the principle of conservation of energy. In detail this compensation will be obtained by the absorbing system becoming a source of secondary radiation. This secondary radiation is superimposed upon the primary radiation and shows such phase relations ("coherent scattering") that the energy is conserved. However, in the theory of the previous section there was included no secondary radiation.

According to the classical theory a system will "emit" radiation as soon as the charged particles in the system are accelerated, and the secondary radiation derives from that part of the acceleration which is due to the perturbation of the radiation. The shortcoming of the theory of the previous sections is indeed the lack of emission of radiation, or, *in this theory the radiation field influences the particles, but the reaction of the particles on the radiation field is lacking.*

In the classical theory of electrons the inhomogeneous differential equations (2.100) for the electromagnetic potentials Φ and $\mathfrak{A}$ contain the mathematical expression for this reaction. Integration of these equations leads to

$$\Phi_P = \int \{\varrho\}_Q r_{PQ}^{-1} d\mathbf{x}_Q + \Phi'_P, \qquad \mathfrak{A}_P = \int (\varrho \mathbf{v})_Q c^{-1} r_{PQ}^{-1} d\mathbf{x}_Q + \mathfrak{A}'_P, \quad (35)$$

where the braces indicate that the quantity within must be taken at time $t-(r_{PQ}/c)$ (retardation) and where Φ'_P and $\mathfrak{A}'_P$ are solutions of the homogeneous differential equations. In the case of discrete particles, equations (35) give us the well-known expressions for the Wiechert-Liénard potentials. Using equation (2.99) we can now calculate the electromagnetic field produced by the charges and currents. It contains the normal electric (Coulomb) and magnetic forces, but also a radiation part which is most simply characterised by expanding, for large values of the distance R from the origin, $\mathfrak{E}$ and $\mathfrak{H}$ in powers of R^{-1}. The terms with R^{-1} are then the radiation terms.

These terms are most simply found by expanding ϱ and $\varrho \mathbf{v}$ in Fourier series,

$$\left.\begin{aligned} \varrho &= \Sigma_\lambda \{(\varrho)_\lambda \exp(2\pi i \nu_\lambda t) + (\varrho)^*_\lambda \exp(-2\pi i \nu_\lambda t)\}, \\ \varrho \mathbf{v} &= \Sigma_\lambda \{(\varrho \mathbf{v})_\lambda \exp(2\pi i \nu_\lambda t) + (\varrho \mathbf{v})^*_\lambda \exp(-2\pi i \nu_\lambda t), \\ 2\pi i \nu_\lambda (\varrho)_\lambda &+ (\nabla \cdot [\varrho \mathbf{v}]_\lambda) = 0, \qquad \nu_\lambda > 0 \end{aligned}\right\} \quad (36)$$

Let $\boldsymbol{\sigma}_\lambda$ be a wave vector of absolute value $\sigma_\lambda = \nu_\lambda/c$ parallel to the radius vector $\boldsymbol{X}$ of the point where we consider the electromagnetic field while the origin is taken somewhere in the system of charged particles. We have ($R = |\boldsymbol{X}|$; $\boldsymbol{x}$ is a point within the system)

$$\{\varrho\} = 2\,\mathrm{Re}\,\Sigma_\lambda(\varrho)_\lambda \exp[2\pi i(\boldsymbol{\sigma}_\lambda\cdot\boldsymbol{x})]\exp[2\pi i\nu_\lambda(t-R/c)],$$
$$\{\varrho\boldsymbol{v}\} = 2\,\mathrm{Re}\,\Sigma_\lambda(\varrho\boldsymbol{v})_\lambda \exp[2\pi i(\boldsymbol{\sigma}_\lambda\cdot\boldsymbol{x})]\exp[2\pi i\nu_\lambda(t-R/c)].$$

The radiation terms in Φ and $\mathfrak{A}$ are now

$$\Phi_{\mathrm{rad}} = 2\,\mathrm{Re}\{\exp[2\pi i\nu_\lambda(t-R/c)]/R\}\int(\varrho)_\lambda \exp[2\pi i(\boldsymbol{\sigma}_\lambda\cdot\boldsymbol{x})]\,d\boldsymbol{x}, \tag{37}$$

$$\mathfrak{A}_{\mathrm{rad}} = 2\,\mathrm{Re}\{\exp[2\pi i\nu_\lambda(t-R/c)]/R\}\int(\varrho\boldsymbol{v})_\lambda \exp[2\pi i(\boldsymbol{\sigma}_\lambda\cdot\boldsymbol{x})]\,d\boldsymbol{x}. \tag{38}$$

If we want to use equation (2.99) to get from these equations the $\mathfrak{E}$, $\mathfrak{H}$-field, it is convenient to remember that if curl $(\varrho\boldsymbol{v})_\lambda$ were equal to zero, $\mathfrak{E} = \mathfrak{H} = 0$, so that if we decompose $(\varrho\boldsymbol{v})_\lambda$ in an irrotational and a source-free part according to the equation

$$(\varrho\boldsymbol{v})_\lambda = (\varrho\boldsymbol{v})_{\lambda\mathrm{I}}+(\varrho\boldsymbol{v})_{\lambda\mathrm{II}},\quad [\boldsymbol{\nabla}\wedge(\varrho\boldsymbol{v})]_{\lambda\mathrm{I}} = 0,\quad (\boldsymbol{\nabla}\cdot[\varrho\boldsymbol{v}]_{\lambda\mathrm{II}}) = 0, \tag{39}$$

only $(\varrho\boldsymbol{v})_{\lambda\mathrm{II}}$ will contribute to the radiation field, and we get the following vector potential

$$\left.\begin{aligned}
&\mathfrak{A}'_{\mathrm{rad}} = \tfrac{1}{2}\Sigma_\lambda(\mathfrak{A}_\lambda+\mathfrak{A}^*_\lambda),\\
&\mathfrak{A}_\lambda = (2/cR)\exp[2\pi i\nu_\lambda(t-R/c)]\int(\varrho\boldsymbol{v})_{\lambda\mathrm{II}}\exp[2\pi i(\boldsymbol{\sigma}_\lambda\cdot\boldsymbol{x})]\,d\boldsymbol{x}\\
&= (2/cR)\exp[2\pi i\nu_\lambda(t-R/c)]\\
&\qquad\int\{(\varrho\boldsymbol{v})_\lambda-(\boldsymbol{\sigma}_\lambda/\sigma_\lambda^2)(\boldsymbol{\sigma}_\lambda\cdot[\varrho\boldsymbol{v}]_\lambda)\}\exp[2\pi i(\boldsymbol{\sigma}_\lambda\cdot\boldsymbol{x})]\,d\boldsymbol{x},
\end{aligned}\right\} \tag{40}$$

which is different from expression (38). This potential is perpendicular to $\boldsymbol{\sigma}$ and thus to $\boldsymbol{X}$. The radiation field itself is given by the equations

$$\left.\begin{aligned}
&\mathfrak{E} = \tfrac{1}{2}\Sigma_\lambda(\mathfrak{E}_\lambda+\mathfrak{E}^*_\lambda),\quad \mathfrak{E}_\lambda = -2\pi i\sigma_\lambda\mathfrak{A}_\lambda;\quad \mathfrak{H} = \tfrac{1}{2}\Sigma_\lambda(\mathfrak{H}_\lambda+\mathfrak{H}^*_\lambda),\\
&\mathfrak{H}_\lambda = -2\pi i[\boldsymbol{\sigma}_\lambda\wedge\mathfrak{A}_\lambda].
\end{aligned}\right\} \tag{41}$$

Proof. The integral in equation (38) can be transformed as follows

$$\int(\varrho\boldsymbol{v})_\lambda\exp[2\pi i(\boldsymbol{\sigma}_\lambda\cdot\boldsymbol{x})] = \sigma_\lambda^{-2}\int\{[\boldsymbol{\sigma}_\lambda\wedge[(\varrho\boldsymbol{v})_\lambda\wedge\boldsymbol{\sigma}_\lambda]]+\boldsymbol{\sigma}_\lambda(\boldsymbol{\sigma}_\lambda\cdot[\varrho\boldsymbol{v}]_\lambda\}\exp[2\pi i(\boldsymbol{\sigma}_\lambda\cdot\boldsymbol{x})]\,d\boldsymbol{x}$$
$$= (2\pi i\sigma_\lambda^2)^{-1}[\boldsymbol{\sigma}\wedge\int[\boldsymbol{\nabla}\wedge(\varrho\boldsymbol{v})_\lambda]\exp\{2\pi i(\boldsymbol{\sigma}_\lambda\cdot\boldsymbol{x})\}d\boldsymbol{x}]-(\boldsymbol{\sigma}_\lambda/2\pi i\sigma_\lambda^2)\int(\boldsymbol{\nabla}\cdot[\varrho\boldsymbol{v}]_\lambda)\exp[2\pi i(\boldsymbol{\sigma}_\lambda\boldsymbol{x})]\,d\boldsymbol{x}$$
$$= \int(\varrho\boldsymbol{v})_{\lambda\mathrm{II}}\exp[2\pi i(\boldsymbol{\sigma}_\lambda\cdot\boldsymbol{x})]\,d\boldsymbol{x}+\int(\varrho\boldsymbol{v})_{\lambda\mathrm{I}}\exp[2\pi i(\boldsymbol{\sigma}_\lambda\cdot\boldsymbol{x})]\,d\boldsymbol{x} = \int(\varrho\boldsymbol{v})_{\lambda\mathrm{II}}\exp[2\pi i(\boldsymbol{\sigma}_\lambda\cdot\boldsymbol{x})]\,d\boldsymbol{x}$$
$$+(c\boldsymbol{\sigma}_\lambda/\sigma_\lambda)\int(\varrho)_\lambda\exp[2\pi i(\boldsymbol{\sigma}_\lambda\cdot\boldsymbol{x})]\,d\boldsymbol{x}.$$

In the corresponding separation of $\mathfrak{A}$ in equation (38) we get first of all a contribution [given by expression (40)] perpendicular to $\boldsymbol{\sigma}$, and secondly a contribution parallel to $\boldsymbol{\sigma}$ which is equal to Φ of equation (37), apart from a factor $\boldsymbol{\sigma}/\sigma$. Calculating now $\mathfrak{E}$ and $\mathfrak{H}$ from equation (2.99) we need only consider differentiation of the exponential $\exp\,[2\pi i\nu_\lambda(t-R/c)]$, that is, if we are considering the radiation terms we can write

$$c^{-1}(\partial/\partial t)\{\exp[2\pi i\nu_\lambda(t-R/c)]\ldots\} = 2\pi i\sigma_\lambda\{\exp\,[2\pi i\nu_\lambda(t-R/c)]\ldots\},$$
$$\text{curl}\,\{\exp[2\pi i\nu_\lambda(t-R/c)]\ldots\} = -2\pi i[\boldsymbol{\sigma}_\lambda \wedge \{\exp\,[2\pi i\nu_\lambda(t-R/c)]\ldots\}].$$

It turns out that the fields due to Φ_{rad} and that part of $\mathfrak{A}_{\text{rad}}$ which is parallel to $\boldsymbol{\sigma}$ vanish, while that part of $\mathfrak{A}_{\text{rad}}$ which is perpendicular to $\boldsymbol{\sigma}$ leads to the fields (41).

We can now attempt to extend equations (40) and (41) to cover the case of the emission of radiation by quantised atoms. Immediately we are confronted with the difficulty that $\varrho\boldsymbol{v}$ is an operator in quantum theory; this means that the electro-magnetic radiation field will be an operator according to equation (40). This indicates that *in a correct quantum theory we must consider the field quantities from the beginning to be operators, that is, q-numbers, or, we must quantise the radiation field.*

Intermezzo on multipole radiation *. Equation (40) defines a vector field which on each sphere R = constant is perpendicular to the radius and which is the same on each sphere, apart from a factor. Such a unique *vector field on a sphere* can be expanded systematically in terms of its elementary parts in complete analogy to the expansion of a scalar function on a sphere in terms of spherical harmonics.

We first of all define a divergence ($\underset{\circ}{\text{div}}$) and curl ($\underset{\circ}{\text{curl}}$) of a vector field on the sphere by the equations

$$\underset{\circ}{\text{div}}\,\mathfrak{E} = \lim_{A\to 0} A^{-1}\oint \mathfrak{E}_n\,ds, \qquad \underset{\circ}{\text{curl}}\,\mathfrak{E} = \lim_{A\to 0} A^{-1}\oint \mathfrak{E}_s\,ds, \tag{42}$$

where ds is the line element of a closed curve enclosing the (infinitesimal) area A, where $\mathfrak{E}_n$ and $\mathfrak{E}_s$ are the components of $\mathfrak{E}$ in the directions of the outward normal $\boldsymbol{n}$ and of $\boldsymbol{ds}$ (the sense in which the integrals are to be taken is determined once and for all). The vector $\mathfrak{E}$ is completely fixed by $\underset{\circ}{\text{div}}\,\mathfrak{E}$ and $\underset{\circ}{\text{curl}}\,\mathfrak{E}$, since the difference $\mathfrak{E}'-\mathfrak{E}$ between $\mathfrak{E}$ and a second vector field $\mathfrak{E}'$ on the sphere with the same $\underset{\circ}{\text{div}}$ and $\underset{\circ}{\text{curl}}$ must be irrotational and thus be derivable from a potential Φ and secondly must have a divergence equal to zero so that Φ must be a constant. We can separate $\mathfrak{E}$ into two parts, as follows

$$\mathfrak{E} = \mathfrak{E}_{\text{I}} + \mathfrak{E}_{\text{II}}, \qquad \underset{\circ}{\text{curl}}\,\mathfrak{E}_{\text{I}} = 0, \qquad \underset{\circ}{\text{div}}\,\mathfrak{E}_{\text{II}} = 0. \tag{43}$$

* See also, H. A. Kramers, *Physica*, **10**, 261, 1943.

Let $\mathfrak{H}$ be the field obtained from $\mathfrak{E}$ by rotating the vector $\mathfrak{E}$ over an angle $\frac{1}{2}\pi$. If $\mathfrak{H}_{\mathrm{I}}(\mathfrak{H}_{\mathrm{II}})$ corresponds to $\mathfrak{E}_{\mathrm{I}}\,(\mathfrak{E}_{\mathrm{II}})$ it follows from equations (42) and (43) that

$$\underset{\circ}{\operatorname{div}}\, \mathfrak{H}_{\mathrm{I}} = \underset{\circ}{\operatorname{curl}}\, \mathfrak{H}_{\mathrm{II}} = 0, \quad \underset{\circ}{\operatorname{curl}}\, \mathfrak{H}_{\mathrm{I}} = \underset{\circ}{\operatorname{div}}\, \mathfrak{E}_{\mathrm{I}}, \quad \underset{\circ}{\operatorname{div}}\, \mathfrak{H}_{\mathrm{II}} = -\underset{\circ}{\operatorname{curl}}\, \mathfrak{E}_{\mathrm{II}} \tag{44}$$

If a point on the sphere is characterised by the unit vector $\boldsymbol{\omega}$, we have for the expansion of $\underset{\circ}{\operatorname{div}}\, \mathfrak{E}_{\mathrm{I}}$ in terms of Laplacian spherical harmonics,

$$\underset{\circ}{\operatorname{div}}\, \mathfrak{E}_{\mathrm{I}} = \Sigma_{l=0}^{\infty} Y_l^{\mathrm{I}}(\boldsymbol{\omega}) = \underset{\circ}{\operatorname{curl}}\, \mathfrak{H}_{\mathrm{I}}. \tag{45}$$

It can be shown that these equations are solved by the expressions

$$\mathfrak{E}_{\mathrm{I}} = -R^2 \Sigma_l [l(l+1)]^{-1} \underset{\circ}{\operatorname{grad}}\, Y_l^{\mathrm{I}}(\boldsymbol{\omega}),$$
$$\mathfrak{H}_{\mathrm{I}} = -R^2 \Sigma_l [l(l+1)]^{-1} [\boldsymbol{\omega} \wedge \underset{\circ}{\operatorname{grad}}\, Y_l^{\mathrm{I}}(\boldsymbol{\omega})], \tag{46}$$

where $\underset{\circ}{\operatorname{grad}}\, \Phi$ (Φ a scalar) is defined in a natural way as the vector of absolute magnitude $\partial\Phi/\partial n$ in the direction $\boldsymbol{n}$ perpendicular to the curve Φ = constant.

We now write for the expansion of $\underset{\circ}{\operatorname{curl}}\, \mathfrak{E}_{\mathrm{II}}$ in terms of spherical harmonics,

$$-\underset{\circ}{\operatorname{curl}}\, \mathfrak{E}_{\mathrm{II}} = \Sigma_{l=0}^{\infty} Y_l^{\mathrm{II}}(\boldsymbol{\omega}) = \underset{\circ}{\operatorname{div}}\, \mathfrak{H}_{\mathrm{II}}. \tag{47}$$

This equation has the following solution

$$\mathfrak{E}_{\mathrm{II}} = R^2 \Sigma_l [l(l+1)]^{-1} [\boldsymbol{\omega} \wedge \underset{\circ}{\operatorname{grad}}\, Y_l^{\mathrm{II}}(\boldsymbol{\omega})],$$
$$\mathfrak{H}_{\mathrm{II}} = -R^2 \Sigma_l [l(l+1)]^{-1} \underset{\circ}{\operatorname{grad}}\, Y_l^{\mathrm{II}}(\boldsymbol{\omega}). \tag{48}$$

If the vector field on the sphere is rotated as a whole, each Y_l goes over into a Y_l' which means that the separate terms in equations (46) and (48) have a rotationally invariant meaning.

If we now identify the $\mathfrak{E}$ and $\mathfrak{H}$ of our formulae with the $\mathfrak{E}_\lambda$, $\mathfrak{H}_\lambda$-field of the radiation field in equations (41), we can write for the expansions (45) and (47),

$$\begin{aligned} \Sigma_l Y_l^{\mathrm{I}}(\boldsymbol{\omega}) &= \underset{\circ}{\operatorname{div}} \int (\varrho \boldsymbol{v})_{\lambda \mathrm{II}} \exp[2\pi i \sigma_\lambda (\boldsymbol{\omega} \cdot \boldsymbol{x})]\, d\boldsymbol{x} \\ &= R^{-1} \int \{2\pi i \sigma_\lambda [(\varrho \boldsymbol{v}_\lambda \cdot \boldsymbol{x}) - (\varrho \boldsymbol{v}_\lambda \cdot \boldsymbol{\omega})(\boldsymbol{x} \cdot \boldsymbol{\omega})] - 2(\varrho \boldsymbol{v}_\lambda \cdot \boldsymbol{\omega})\} \\ &\qquad \exp[2\pi i \sigma_\lambda (\boldsymbol{\omega} \cdot \boldsymbol{x})]\, d\boldsymbol{x} \qquad (49) \\ &= R^{-1} \int (\varrho \boldsymbol{v}_\lambda \cdot 2\pi i \sigma_\lambda [\boldsymbol{\omega} \wedge [\boldsymbol{x} \wedge \boldsymbol{\omega}]] - 2\boldsymbol{\omega}) \exp[2\pi i \sigma_\lambda (\boldsymbol{\omega} \cdot \boldsymbol{x})]\, d\boldsymbol{x}, \end{aligned}$$

$$\Sigma_l Y_l^{\text{II}}(\boldsymbol{\omega}) = -\underset{\circ}{\text{curl}}\int (\varrho \boldsymbol{v})_{\lambda\text{II}} \exp[2\pi i \sigma_\lambda (\boldsymbol{\omega} \cdot \boldsymbol{x})] d\boldsymbol{x}$$

$$= -R^{-1}\int 2\pi i \sigma_\lambda (\boldsymbol{\omega} \cdot [\boldsymbol{x} \wedge \varrho \boldsymbol{v}_\lambda]) \exp [2\pi i \sigma_\lambda (\boldsymbol{\omega} \cdot \boldsymbol{x})] d\boldsymbol{x}. \quad (50)$$

Since in electrostatics $Y_l(\boldsymbol{\omega})$ is the general expression for the dependence of the potential of a multipole (2^l-pole) on the direction $\boldsymbol{\omega}$, the separate terms in the expression (46) of the $\mathfrak{E}_\text{I}$, $\mathfrak{H}_\text{I}$ field are called *electrical multipole radiations,* and the terms in the expansion (48) of the $\mathfrak{E}_{\text{II}}$, $\mathfrak{H}_{\text{II}}$-field *magnetic multipole radiations* ($l = 1$: dipole radiation; $l = 2$: quadrupole radiation; $l = 3$: octupole radiation; . . .). This division of the radiation in its multipole components is not related to the static multipole fields of the system which may be present; these depend on the charge distribution and behave like R^{-l-1}, while the radiations depend on the source free part of the current distribution and behave like R^{-1}.

In writing down equations (37), (38), (40) and (41) we have neglected systematically higher powers of R^{-1}; the field (41) was thus not a rigorous solution of the Maxwell equations in vacuo. The same is true of the multipole radiation fields just discussed; they are the largest terms, in a finite expansion in powers of R^{-1}, of the fields which satisfy the Maxwell equations rigorously everywhere except at $R = 0$. (These terms are proportional to R^{-1}). These fields are the ones usually called multipole fields in the literature. (Compare the well-known Hertzian dipole radiation from which one can obtain the higher multipole radiations by differentiation).

If the dimensions of the radiating $(\varrho \boldsymbol{v})_\lambda$-field are small compared to the wavelength σ_λ^{-1} of the emitted light, we can use an expansion in powers of σ_λ. The zeroth power of σ_λ given us the *electrical dipole radiation,* $Y_1^{\text{I}}(\boldsymbol{\omega}) = -2R^{-1}(\boldsymbol{\omega} \cdot \int (\varrho \boldsymbol{v})_\lambda d\boldsymbol{x})$. The first power of σ_λ gives us the *electrical quadrupole radiation,* $Y_2^{\text{I}}(\boldsymbol{\omega}) = 2\pi i \sigma_\lambda R^{-1} \int\{(\boldsymbol{x} \cdot [\varrho \boldsymbol{v}]_\lambda) - 3(\boldsymbol{x} \cdot \boldsymbol{\omega})([\varrho \boldsymbol{v}]_\lambda \cdot \boldsymbol{\omega})\} d\boldsymbol{x}$, and the *magnetic dipole radiation,* $Y_1^{\text{II}}(\boldsymbol{\omega}) = -2\pi i \sigma_\lambda R^{-1}(\boldsymbol{\omega} \cdot \int [\boldsymbol{x} \wedge (\varrho \boldsymbol{v})_\lambda] d\boldsymbol{x})$.

If it is not possible to expand in powers of σ_λ, Y_l^{I} and Y_l^{II} can be represented by Bessel functions. We do not wish to discuss this case.

§ 84. THE "SEMI-CLASSICAL" THEORY OF SPONTANEOUS TRANSITIONS

Without using a consistent quantum theory of radiation we can derive from equation (40) formulae which probably enable us to represent the most important aspects of the emission of radiation by quantised systems. This derivation which does not use explicitly the q-number character of the radiation field, is often called the *correspondence treatment,* or the *semiclassical theory of the emission of radiation* *.

* See, O. Klein, *Z. Phys.*, **41**, 407, 1927.

We consider an atomic system which is not perturbed by time-dependent external forces. In that case, the time-dependent operator corresponding to the current density is particularly simple, if we use a representation by means of the energy eigenfunctions φ_k (eigenvalues E_k), that is, if we look for the Heisenberg matrices of $(\varrho\mathbf{v})_{\text{op}}$ (see § 43). The result is [see eq. (4.122)],

$$[\varrho\mathbf{v}(t)]_{kl} = \varrho(\mathbf{v})_{kl} \exp(2\pi i\nu_{kl}t), \qquad h\nu_{kl} = E_k - E_l, \tag{51}$$

with $(\varrho\mathbf{v})_{kl}$ given by equation (9). In the classical theory of radiation we are interested in the harmonic components $(\varrho\mathbf{v})_\lambda$ in which the current density field can be expanded. In the *kl*-representation each harmonic component of $\varrho\mathbf{v}(t)$ possesses only one element different from zero according to equation (51), unless fortuitously several Bohr frequencies are the same. We expect thus that *the frequencies which can occur in the emitted radiation will be the Bohr frequencies* $|\nu_{kl}|$. For each pair of stationary states k, l there are two harmonic current density fields, which are each other's complex conjugates, $(\varrho\mathbf{v})_{kl} \exp(2\pi i\nu_{kl}t)$, and $(\varrho\mathbf{v})_{lk} \exp(2\pi i\nu_{lk}t) = (\varrho\mathbf{v})^*_{kl}\exp(-2\pi i\nu_{kl}t)$. They are related to each other in the same way as $(\varrho\mathbf{v})_\lambda \exp(2\pi i\nu_\lambda t)$ and $(\varrho\mathbf{v})^*_\lambda \exp(-2\pi i\nu_\lambda t)$ in the classical expansion (36).

To each transition from a stationary state of higher energy (k) to one of lower energy (l) there corresponds a real radiation field with the vector potential,

$$\begin{aligned} \mathfrak{A}'_{\text{emission}} &= \tfrac{1}{2}(\mathfrak{A}_{kl} + \mathfrak{A}^*_{kl}), \\ \mathfrak{A}_{kl} &= 2(cR)^{-1} \exp[2\pi i\nu_{kl}(t - R/c)] \int (\varrho\mathbf{v})_{kl\text{II}} \exp[2\pi i(\boldsymbol{\sigma}\cdot\mathbf{x})]d\mathbf{x}; \\ &\qquad \sigma = \nu_{kl}/c. \end{aligned} \tag{52}$$

The semi-classical theory of the emission of radiation now gives the following *prescription*. If we are dealing with a large number of atomic systems of the same kind, randomly distributed in space, and if N_k is the number of atoms in the k-th state, we shall find at a large distance from the radiating system the radiation field (52) but with an N_k-times larger intensity.

This means that each atom in the state k acts as a classical source of radiation with a current density field

$$\left.\begin{aligned} (\varrho\mathbf{v})_k &= \Sigma_l\, 2\,\text{Re}\,(\varrho\mathbf{v})_{kl} \exp(2\pi i\nu_{kl}t) \\ &= \Sigma_l\{(\varrho\mathbf{v})_{kl} \exp(2\pi i\nu_{kl}t) + (\varrho\mathbf{v})_{lk} \exp(-2\pi i\nu_{kl}t)\}, \\ &\qquad E_l < E_k, \end{aligned}\right\} \tag{53}$$

where the summation is over all transitions to states of lower energy

E_l. Expression (53) is sometimes called the *virtual source of radiation* in the state k.

The radiation field we have described produces radiation energy in empty space; to compensate this the atoms show *spontaneous Bohr transitions* to the lower E_l-levels, and each transition corresponds to a loss of an amount of energy $h\nu_{kl}$ by the atom. The number of transitions $k \to l$ per unit time is put equal to $N_k A_{k\to l}$, as was done by Einstein, and $A_{k\to l}$ is called the coefficient of the *spontaneous radiation transitions.*

The picture of such quantum jumps belongs to the old quantum theory; if we wish to adapt it to the wave-mechanical description of the state of an atomic system, we are led to the quantum theory of the radiation field which we do not want to consider as yet. We may mention that the semi-classical prescription seems fully justifiable in the framework of the quantum theory of radiation (see § 91). We shall discuss its plausibility, using correspondence arguments, in the next section.

We shall now determine the intensity of the radiation field (52) and calculate the value of $A_{k\to l}$ required by the energy compensation. The (complex) amplitude of $\mathfrak{E}$ in the field (52) is equal to $-2\pi i(\nu_{kl}/c)\mathfrak{A}_{kl}$, its absolute square multiplied by $c/8\pi$ gives us the time average of the Poynting vector. From equation (52) we find thus for the energy $Id\boldsymbol{\omega}dt$ emitted within the solid angle $d\boldsymbol{\omega}$ during a time interval dt,

$$\begin{aligned} I_{\nu_{kl}} d\boldsymbol{\omega}dt &= R^2 d\boldsymbol{\omega}dt(c/8\pi)(2\pi\nu_{kl}/c)^2|\mathfrak{A}_{kl}|^2 \\ &= (2\pi\nu_{kl}^2/c^3)|\int(\varrho\boldsymbol{v})_{k,\mathrm{II}}\exp\,[2\pi i(\boldsymbol{\sigma}\cdot\boldsymbol{x})]d\boldsymbol{x}|^2\,d\boldsymbol{\omega}dt. \end{aligned} \tag{54}$$

The energy balance requires that $\oint I_{\nu_{kl}} d\boldsymbol{\omega} = 4\pi\overline{I_{\nu_{kl}}} = h\nu_{kl}A_{k\to l}$, where the bar indicates averaging over all directions $\boldsymbol{\sigma}$ of emission. We get thus

$$A_{k\to l} = (4\pi\nu_{kl}/\hbar c^3)\overline{|\int(\varrho\boldsymbol{v})_{kl\mathrm{II}}\exp[2\pi i(\boldsymbol{\sigma}\cdot\boldsymbol{x})]d\boldsymbol{x}|^2}. \tag{55}$$

As the vector which we get from the integral is the component of the integral $\int(\varrho\boldsymbol{v})_{kl}\exp\,[2\pi i(\boldsymbol{\sigma}\cdot\boldsymbol{x})]d\boldsymbol{x}$ which is perpendicular to $\boldsymbol{\sigma}$ [see eq. (40)], the average value of its absolute square will be equal to twice the average over all directions of $\mathfrak{a}$ of the absolute square occurring in equation (21), so that we have

$$A_{k\to l} = (4\pi\nu_{kl}/\hbar c^3)\cdot 4\pi\hbar^2\nu_{kl}^2\,\overline{B_{l\to k}(\boldsymbol{\sigma},\mathfrak{a})} = (8\pi h\nu_{kl}^3/c^3)\,B_{l\to k}, \tag{56}$$

where $B_{l\to k}$ is Einstein's coefficient B for uniformly incoming radiation. This relation between A and B was derived by Einstein in 1917 from Planck's radiation formula.

The state of polarisation of the emitted radiation in a given direction is completely described by the complex components of $\mathfrak{A}_{kl}$, that is, by the projection of the complex vector $\int (\varrho\boldsymbol{v})_{kl}\exp[2\pi i(\boldsymbol{\sigma}\cdot\boldsymbol{x})]d\boldsymbol{x}$ on the plane perpendicular to $\boldsymbol{\sigma}$. If the wave length σ^{-1} of the emitted radiation is large compared with the dimensions of the radiating system, $\exp[2\pi i(\boldsymbol{\sigma}\cdot\boldsymbol{x})]$ may be put equal to unity, and the total action of the virtual source of radiation corresponds to the action of a number of virtual oscillators $\dot{\boldsymbol{P}}_{kl}(t)$, where $\boldsymbol{P}(t)$ is the time-dependent operator of the total electrical polarisation of the system [see eq. (24)] *. Equation (52) simplifies as follows

$$\mathfrak{A}_{kl} = 2(cR)^{-1}\exp[2\pi i\nu_{kl}(t-R/c)]\dot{\boldsymbol{P}}_{kl\mathrm{II}} = (4\pi i\nu_{kl}/cR)\exp[2\pi i\nu_{kl}(t-R/c)]\boldsymbol{P}_{kl\mathrm{II}}. \quad (57)$$

The radiation from a virtual oscillator is the same as that of a charged particle (charge e) the position of which changes with time according to the equation

$$e\boldsymbol{x} = \boldsymbol{P}_{kl}\exp(2\pi i\nu_{kl}t) + \boldsymbol{P}_{lk}\exp(-2\pi i\nu_{kl}t) = 2\,\mathrm{Re}\,\boldsymbol{P}_{kl}\exp(2\pi i\nu_{kl}t) \quad (58)$$

We have assumed that the dimensions of the (in general elliptic) orbit of the particle is negligibly small compared to the wave length c/ν_{kl}. Einstein's A is in this case of the so-called *electrical dipole radiation* given by the equation [see eqq. (56), 27)]

$$A_{k\to l} = (32\pi^3\nu_{kl}^3/3\hbar c^3)|\boldsymbol{P}_{kl}|^2. \quad (59)$$

The state of polarisation of the radiation in a given direction corresponds exactly to the shape of the elliptic orbit (58) projected on a plane perpendicular to this direction.

If in the radiation field we wish to take into account the action of the electron spin, equation (54) remains valid, provided we add to $(\varrho\boldsymbol{v})_{kl}$ the spin current $(\varrho\boldsymbol{v})_{kl}^{\mathrm{spin}}$ of equation (30). As $\int(\varrho\boldsymbol{v})_{kl}^{\mathrm{spin}}d\boldsymbol{x} = 0$, the spin will never contribute to the dipole radiation, discussed a moment ago, but it may lead to a non-vanishing contribution when we cannot put the exponential in equation (54) equal to unity.

If we may approximate $\exp(2\pi i(\boldsymbol{\sigma}\cdot\boldsymbol{x})]$ by $1+2\pi i(\boldsymbol{\sigma}\cdot\boldsymbol{x})$ we get in the integral in equation (54) a term proportional to σ, which leads to the *contribution of the spin*

* It is of historical interest to note that Heisenberg's first paper on matrix calculus, in 1925, started from the idea of virtual oscillators, an idea which had been developed earlier. Heisenberg compared these oscillators with the coefficients of the Fourier expansion of the particle coordinates in their classical motion.

to the magnetic dipole radiation in $I_{\nu_{kl}}$. According to the classical theory of radiation, a system with a variable magnetic moment emits a radiation, just like a system with a variable electrical moment. One would thus expect classically that a system in which the electrons change the directions of their spins would contribute to this radiation; this contribution is the correspondence counterpart of the spin radiation just considered (see § 83).

Equations (52), (53), *and* (54) *enable us in principle to calculate the intensity and polarisation of the radiation emitted during the Bohr transitions.* For most applications equations (57) and (59), which refer to dipole radiation, are sufficient; everything depends thus in this case on the discussion of the matrix

$$\boldsymbol{P}_{kl} = \int \varphi_k^* (\Sigma_i e_i \boldsymbol{x}_i) \varphi_l. \tag{60}$$

If the dimensions of the $(\varrho \boldsymbol{v})_{kl}$-field are not negligibly small compared to the wavelength, we can analyse the radiation in terms of multipole radiations (see the end of § 83). If we put $\exp[2\pi i(\boldsymbol{\sigma} \cdot \boldsymbol{x})] = 1+2\pi i(\boldsymbol{\sigma} \cdot \boldsymbol{x})+\ldots$, the second term can give rise to electrical quadrupole- * and magnetic dipole radiation. The evaluation of these weaker radiations has a meaning especially when the dipole radiation, which is in general stronger, is not present. A *kl*-transition may take place through a higher multipole radiation, when the corresponding matrix element $\boldsymbol{P}_{kl}$ of the dipole moment is equal to zero. Such a case often occurs in the case of atomic or molecular spectra because of the so-called *selection rules.*

We mention here only the following important selection rule. *If a transition between two states of a free atomic system* (with j' and j'' as the values of the total angular momentum quantum number) *is accompanied by the emission of (electrical or magnetic*)2^l*-pole radiation, we have* $j'+j'' \geqq l$ *and* $|j'-j''| \leqq l$, *that is, there exists a triangle with sides* j', j'', *and* l. In particular in the case of (electrical or magnetic) dipole radiation we have $j'-j'' = \pm 1$, or 0 (If $j' = 0$, $j' = j''$ is forbidden), and in the case of quadrupole radiation $j'-j'' = \pm 2, \pm 1$, or 0 (If $j' = 0$, or $\frac{1}{2}$, $j' = j''$ is forbidden).

The proof of this rule follows by considering equations (49) and (50) [$(\varrho \boldsymbol{v})_\lambda$ must be replaced by $(\varrho \boldsymbol{v})_{kl}$ in these equations] according to which the intensity of a multipole radiation is determined by the matrix element of operators Ω_l^m which occur as coefficients in the expansion of a certain scalar operator Ω, which depends on a direction $\boldsymbol{\omega}$, in terms of spherical harmonics of $\boldsymbol{\omega}$, $\Omega = \Sigma_{l,m} \Omega_l^{-m} P_l^m(\boldsymbol{\omega})$. A rotation of the system of coordinates with respect to which the P_l^m are defined corresponds to a linear transformation of the $2l+1$ operators Ω_l^m (l fixed) among each other. These Ω_l^m transform as the monomials $\bar{\xi}^{l+m}\bar{\eta}^{l-m}$, since each sum $\Sigma_m \Omega_l^{-m} P_l^m$ is rotationally invariant [see eq. (4.147) and the footnote at the end of § 76]. Symbol-

* A. Rubinowicz, *Z. Phys.*, **53**, 267, 1929.

ically we can thus represent the matrix elements of the Ω_l^m with respect to two states characterised by j' and j'' by [see eqq. (7.114) and (7.118)],

$$I = \int (u^* \xi^* + v^* \eta^*)^{2j'} (\bar{u}\bar{\xi} + \bar{v}\bar{\eta})^{2l} (u\xi + v\eta)^{2j''}, \tag{60A}$$

where u, v, and $\bar{u}$, $\bar{v}$ are two fixed spinors, and where ξ, η, and $\bar{\xi}$, $\bar{\eta}$ transform as the unit spinors. If we expand in terms of powers of u^*, v^*, $\bar{u}$, $\bar{v}$, u, and v, the coefficient of $u^{*j'+m'} v^{*j'-m'} \bar{u}^{l+m} \bar{v}^{l-m} u^{j''+m''} v^{j''-m''}$ will be the symbolical expression for the matrix element

$$\binom{2l}{l+m}\binom{2j'}{j'+m'}\binom{2j''}{j'+m''}\int \varphi^*_{j'm'}\, \Omega_l^m\, \varphi_{j''m''}. \tag{60B}$$

As expression (60A) is rotationally invariant, the value of the integral must be a polynomial of the invariants $A = u^*u + v^*v$, $B = u^*\bar{u} + v^*\bar{v}$, and $C = u\bar{v} - v\bar{u}$ *. Clearly the only possibility is to write $I = \text{constant} \cdot A^{j'+j''-l} B^{j'+l-j''} C^{-j'+j''+l}$, and we see that j', j'', and l must form a triangle, as the exponents of A, B, and C can never be negative. From this expression, by comparing coefficients, it is simple to calculate the relative values of the various matrix elements (60B). Their squares give us the *relative intensities of the components of a "multipole line" in the Zeeman effect.*

In this section we have applied the semi-classical theory of the emission of radiation only to spontaneous transitions. It is possible to generalise in a natural way the prescriptions of this theory in such a way that they can also be applied to evaluate the secondary radiation emitted by an atomic system when it is perturbed by an external radiation field. In this way we can develop the semi-classical theory of coherent and non-coherent scattering, which includes the discussion of dispersion, Raman-scattering and Compton-effect. We prefer, however, to discuss all this using the quantum theory of radiation (see §§ 92 to 95).

§ 85. Emission of radiation and correspondence principle

In the case of many atomic systems there are solutions of the time-dependent Schrödinger equation which — at any rate during a finite time interval — can to a fair approximation be interpreted as the propagation of wave packets according to the laws of classical mechanics (see §§ 27 and 30). From the theory of electrons we can evaluate the classical radiation field corresponding to such a classical orbital motion. If we consider such a wave packet solution as the superposition of stationary states, $\psi = \Sigma_k c_k \varphi_k \exp(-iE_k t/\hbar)$ [see eq. (3.99)], the main contributions to ψ will come from a relatively small group of states, since only the c_k corresponding to those states are appreciably different from zero. These states belong to a narrow energy range, corresponding to the approximately defined

* For a proof see H. C. Brinkman, *Applications of spinor invariants in atomic physics*, North Holland Publ. Cy., 1956.

energy of the classical motion. If we now use the semi-classical prescription to discuss the radiative transitions from one of these states to another one of nearly the same energy, we can show — at any rate in simple cases * — that the accompanying radiation, as far as its frequency, intensity and polarisation are concerned is the same as one of the harmonic components of the classical radiation just mentioned. This is thus a correspondence justification of the semi-classical prescription. One often talks about the *correspondence between classical and Bohr radiation.*

This correspondence with regard to radiation was the particular requirement which Bohr wanted to cover when in 1918 he introduced the correspondence principle. It enabled him to draw quantitative as well as qualitative conclusions regarding the intensity and polarisation of the radiation emitted during quantum "jumps" **. The basis of his discussion was the illustration of stationary states by classical orbits satisfying certain "quantum conditions" (see § 54).

We shall discuss this correspondence in the case of a simple example. Consider a charged particle (mass m) moving along a circle of unit radius (two-dimensional rotator). Let its position be determined by the angle x and let p be its momentum. The Hamiltonian may be given by the equation $H(p, x) = (p^2/2m)+U(x)$, but we are not interested in its exact form. For the sake of simplicity let us consider a motion where x increases monotonically with time; let its energy be E and its frequency ω,

$$E = H(p, x). \tag{61}$$

Each single-valued function G of p and x (periodic in x, with period 2π) can be expanded in a Fourier series of the time,

$$G(p, x)=\Sigma_{\tau=-\infty}^{+\infty} G_\tau \exp(2\pi i\tau\omega t), \quad G_\tau=\omega\int_0^{1/\omega} G(p, x)\exp(-2\pi i\tau\omega t)dt.$$

We consider p to be a function of E and x and transform G_τ into an integral over x. Differentiating expression (61) with respect to E we have

* The cases where this is possible are those where the classical motion can be expanded in harmonic components; these are the so-called multiply periodic systems, with well defined periodicity moduli of the action integrals (see § 54).

** See also H. A. Kramers, *Trans. Roy. Danish Acad. Sci.*, **3**, 284, 1919.

$$\left.\begin{aligned} & 1=\frac{\partial H}{\partial p}\frac{\partial p}{\partial E}, \quad dt=\frac{\partial p}{\partial E}dx, \quad t=\int\frac{\partial p}{\partial E}dx=\frac{\partial S}{\partial E}, \\ & S(E,x)=\int_0^x p(E,x)dx, \quad p=\frac{\partial S}{\partial x}, \\ & \qquad\qquad G_\tau=\omega\int_0^{2\pi}G(p,x)\exp\left(-2\pi i\tau\omega\frac{\partial S}{\partial E}\right)\frac{\partial p}{\partial E}dx. \end{aligned}\right\} \tag{62}$$

We now introduce I, the periodicity modulus of the action S (usually called the *phase integral*),

$$I=S(E,2\pi)=\int_0^{2\pi}p\,dx, \qquad \frac{dI}{dE}=\int_0^{2\pi}\frac{\partial p}{\partial E}dx=\frac{1}{\omega}.$$

We can now replace E by I $[S(E,x)\to S(I,x),\ldots]$,

$$G_\tau=\int_0^{2\pi}G(p,x)\exp\left(-2\pi i\tau\frac{\partial S}{\partial I}\right)\frac{\partial p}{\partial I}dx, \qquad \frac{\partial p}{\partial I}=\omega\frac{\partial p}{\partial E}=\omega\frac{dt}{dx}.$$

By using the condition

$$I_n=nh, \qquad n \text{ integral}, \tag{63}$$

we introduce a discrete set of classical orbital motions characterised by an integer n. Let us assume that the value of I which we were considering is approximately equal to $n'h$, where n' is a very large integer. Let τ be $\ll n'$ and let n'' be $n'-\tau$, so that we have the relations $2\pi\tau(\partial S/\partial I)=(S_{n'}-S_{n''})/\hbar$ with $S_n=S(nh,x)$. It then follows that

$$\left.\begin{aligned} & G_\tau \doteqdot \int_0^{2\pi}\varphi_{n'}^{*}G\varphi_{n''}dx, \quad \varphi_n=\left[\left(\frac{\partial p}{\partial I}\right)_n\right]^{1/2}\exp\left(iS_n/\hbar\right), \\ & \int_0^{2\pi}|\varphi_n|^2dx=\int_0^{2\pi}\omega_n\left(\frac{dt}{dx}\right)_n dx=1 \end{aligned}\right\} \tag{64}$$

This way of expressing G_τ corresponds formally to the writing down of a matrix element with respect to two wave functions $\varphi_{n'}$, $\varphi_{n''}$ corresponding to two stationary states,

$$G_\tau \doteqdot G_{n';n''}, \quad n''-n'=\tau. \tag{65}$$

Indeed, in the region of large n-values the expressions

$$\varphi_n=\left[\left(\frac{\partial p}{\partial I}\right)_n\right]^{1/2}\exp\left(\frac{i}{\hbar}S_n\right)=\left[\left(\frac{\partial p}{\partial I}\right)_n\right]^{1/2}\exp\left[\frac{i}{\hbar}\int^x p(x,I)\,dx\right] \tag{66}$$

are the approximate normalised solutions of the time-independent

Schrödinger equation which according to quantum mechanics corresponds to our system

$$H_{\rm op}\varphi \equiv H(-i\hbar[\partial/\partial x], x)\varphi = E\varphi. \tag{67}$$

This is an example of the so-called W.K.B.-Solution *. To see this we write the Hamiltonian in the following general form

$$H_{\rm op}=U_0(x)+\tfrac{1}{2}[U_1(x)p_{\rm op}+p_{\rm op}U_1(x)]+\tfrac{1}{2}[U_2(x)p^2_{\rm op}+p^2_{\rm op}U_2(x)]+\ldots, p_{\rm op}=-i\hbar\partial/\partial x, \tag{68}$$

and we look for a solution of equation (67) in the form

$$\varphi = C(x)\exp[i\chi(x)]. \tag{69}$$

In equation (69) χ and C are real functions as in equation (3.29). As in § 27 we also assume that C and $\partial\chi/\partial x$ change so little in an interval in which χ increases by 2π, that inequalities (3.32) and (3.33) are valid. We calculate $H_{\rm op}\varphi$. The main difference between this and the calculation of § 27 is that now we have written H explicitly in Hermitean form. Neglecting all second and higher derivatives of $C(\chi)$ with respect to x and also terms involving $(\partial C/\partial x)\cdot(\partial^2\chi/\partial x^2), \ldots$ we get

$$\left.\begin{aligned} H_{\rm op}\varphi \doteqdot \Sigma_{n=0,1,\ldots}&\left\{CU_n\left(\frac{\partial\hbar\chi}{\partial x}\right)^n + \frac{\hbar}{i}\frac{\partial C}{\partial x}U_n n\left(\frac{\partial\hbar\chi}{\partial x}\right)^{n-1} + \frac{\hbar}{i}C\frac{dU_n}{dx}\frac{n}{2}\left(\frac{\partial\hbar\chi}{\partial x}\right)^{n-1}\right. \\ &\left.+\frac{\hbar}{i}CU_n\frac{n(n-1)}{2}\frac{\partial^2\hbar\chi}{\partial x^2}\left(\frac{\partial\hbar\chi}{\partial x}\right)^{n-2}\right\}e^{i\chi} \\ =&\left\{CH(p,x)+\frac{\hbar}{i}\left[\frac{\partial C}{\partial x}\frac{\partial H}{\partial p}+\tfrac{1}{2}C\frac{\partial^2 H}{\partial p\partial x}+\tfrac{1}{2}C\frac{\partial p}{\partial x}\frac{\partial^2 H}{\partial p^2}\right]\right\}e^{i\chi};\ p=\hbar\frac{\partial\chi}{\partial x}. \end{aligned}\right\} \tag{70}$$

The term involving $\partial^2 H/\partial p\partial x$ derives from the Hermitean form of expression (68); the other terms also occur in equation (3.35). Substitution into equation (67) yields the equations

$$H(p,x)=E, \qquad \frac{\partial}{\partial x}\ln C = -\tfrac{1}{2}\left[\frac{\partial}{\partial x}\ln\frac{\partial H}{\partial p}+\frac{\partial p}{\partial x}\frac{\partial}{\partial p}\left(\ln\frac{\partial H}{\partial p}\right)\right]. \tag{71}$$

The first of these equations is simply the Hamilton-Jacobi equation of classical mechanics and is equivalent to $\hbar\chi = \int^x p(E,x)dx = S$.

From the requirement that φ of equation (69) is single-valued, it follows immediately that the periodicity modulus of χ is an integral multiple of 2π, in agreement whith the phase integral condition (63). We have here the quantum condition which was used in the model theory to determine the energies of the stationary states. Only in our particular, simple case where x always increases (or decreases) monotonically in the classical motion, can this condition be taken over immediately in quantum mechanics.

The second equation (71) can immediately be integrated; it is clearly essential in this case that expression (68) is Hermitean. We find

$$\ln C = -\tfrac{1}{2}\ln\frac{\partial H}{\partial p}+\text{constant},$$

$$C = \text{constant}\cdot\left(\frac{\partial H}{\partial p}\right)^{-1/2} = \text{constant}\cdot\dot{x}^{-1/2} = \text{constant}\cdot\left(\frac{\partial p}{\partial I}\right)^{1/2},$$

in accordance with equation (66).

In order that the integral (64) for G_τ corresponds exactly to the quantum me-

* G. Wentzel, *Z. Phys.*, **38**, 518, 1926; H. A. Kramers, *Z. Phys.*, **39**, 828, 1926; L. Brillouin, C.R., **183,** 24, 1926.

chanical matrix element, $G(-i\hbar[\partial/\partial x], x)$ should have appeared in the integrand instead of $G(p, x)$. By using equations (70), with G instead of H, one sees easily that the difference is of the relative order of magnitude n'^{-1} and thus negligible.

Equation (65) *expresses the general correspondence between the properties of the classical motion and the quantum mechanical description.* The separate Fourier components $G_\tau \exp(2\pi i\tau\omega t)$ of a classical motion of energy E are *in the region of large quantum numbers* equal to the Heisenberg matrix elements, $G_{n'\to n''} \exp(2\pi i\nu_{n'n''} t)$, of the operator G_{op}, if the energies of the stationary states characterised by n' and n'' are relatively little different and both approximately equal to E. If in particular we are interested in the correspondence with respect to the emission of radiation, we choose for G_{op} the operator of the current density.

§ 86. The radiation field in vacuo as a canonical system

The Maxwell equations in empty space,

$$[\nabla \wedge \mathfrak{H}] = c^{-1}\partial\mathfrak{E}/\partial t, \quad [\nabla \wedge \mathfrak{E}] = -c^{-1}\partial\mathfrak{H}/\partial t, \quad (\nabla \cdot \mathfrak{E}) = (\nabla \cdot \mathfrak{H}) = 0,$$

can be written in the following form

$$[\nabla \wedge \mathfrak{F}] = ic^{-1}\partial\mathfrak{F}/\partial t, \qquad (72); \qquad\qquad (\nabla \cdot \mathfrak{F}) = 0, \qquad (73)$$

if we introduce the complex vector, $\mathfrak{F}$, by the equations $\mathfrak{F} = \mathfrak{E}+i\mathfrak{H}$, $\mathfrak{F}^* = \mathfrak{E}-i\mathfrak{H}$. From equation (72) we see that $(\nabla \cdot \mathfrak{F})$ is independent of time, and we can thus consider equation (73) as a special integral of equation (72).

The energy density of the electromagnetic field is $(\mathfrak{E}^2+\mathfrak{H}^2)/8\pi = (\mathfrak{F}^* \cdot \mathfrak{F})/8\pi$; the total energy is thus given by an integral over the whole of space,

$$H = (8\pi)^{-1}\int(\mathfrak{F}^* \cdot \mathfrak{F})\, d\boldsymbol{x}. \qquad (74)$$

Equations (72) can be considered to be the Hamiltonian, canonical equations of motion of a system of infinitely many degrees of freedom, the energy of which is given by equation (74). To prove this we shall expand the complex vector field $\mathfrak{F}$ in plane waves, that is, to consider $\mathfrak{F}$ as a superposition of vector fields of the form

$$\boldsymbol{A}_\lambda \exp[2\pi i(\boldsymbol{\sigma}_\lambda \cdot \boldsymbol{x})], \qquad (75)$$

where the components of the wave vectors $\boldsymbol{\sigma}_\lambda$ can take on all possible real values. The complex amplitude $\boldsymbol{A}_\lambda$ can be resolved into a component parallel to $\boldsymbol{\sigma}_\lambda$ (*longitudinal waves*) and a component perpendicular to $\boldsymbol{\sigma}_\lambda$ (*transverse waves*). The latter can again be resolved into two mutually perpendicular components (*linearly polarised waves*). For our purpose it is advantageous to consider this second component to be a sum of two complex vectors of zero absolute length (compare § 61) the real and imaginary parts of which are both perpendicular to $\boldsymbol{\sigma}_\lambda$ (*right hand and left hand circulārly polarised waves*).

In order to obtain the resolution of an arbitrary vector field $\mathfrak{F}(x, y, z, t)$ in a convenient way, it is advisable to introduce the formal assumption that the field is enclosed within a very large cubical box (edge length L) on the surface of which the so-called *periodicity conditions* are satisfied. If the axes of coordinates are chosen parallel to the edges of the cube and if we choose the origin of the system of coordinates in the centre of the cube, we require that

$$\mathfrak{F}(\tfrac{1}{2}L, y, z) = \mathfrak{F}(-\tfrac{1}{2}L, y, z), \quad \mathfrak{F}(x, \tfrac{1}{2}L, z) = \mathfrak{F}(x, -\tfrac{1}{2}L, z), \ldots \tag{76}$$

It is assumed that the physical laws in this "cubical" space are practically the same as those in infinite space and go over into these in the limit as $L \to \infty$.

Because of the periodicity conditions (76) only such wave vectors $\boldsymbol{\sigma}_\lambda$ are allowable in expression (75) for which $L\sigma_x$, $L\sigma_y$, and $L\sigma_z$ are integers, or, $\sigma_x = k_x/L$, $\sigma_y = k_y/L$, $\sigma_z = k_z/L$ and the expansion of the vector field becomes a triple Fourier series where each wave number is characterised by the three numbers k_x, k_y, and k_z. In the following we shall indicate such a triple of numbers by the index λ,

$$(\boldsymbol{\sigma}_\lambda \cdot \boldsymbol{x}) = (k_x x + k_y y + k_z z)/L, \qquad \sigma_\lambda = [k_x^2 + k_y^2 + k_z^2]^{1/2}/L. \tag{77}$$

The wave vector with the same length as $\boldsymbol{\sigma}_\lambda$ but in the opposite direction shall be denoted by $\boldsymbol{\sigma}_{-\lambda}$,

$$\boldsymbol{\sigma}_\lambda = -\boldsymbol{\sigma}_{-\lambda}, \qquad \sigma_\lambda = \sigma_{-\lambda}. \tag{78}$$

The number n_λ of all wave vectors $\boldsymbol{\sigma}_\lambda$ with directions within the volume element $d\boldsymbol{\sigma}$ in $\boldsymbol{\sigma}$-space is given by the equation $n_\lambda = \Omega d\boldsymbol{\sigma}$, where Ω ($= L^3$) is the volume of the cube. The summation of a quantity f_λ over all possible wave vectors can thus be changed into an integration as follows,

$$\Sigma_\lambda f_\lambda = \Sigma n_\lambda f(\boldsymbol{\sigma}) = \Omega \int f(\boldsymbol{\sigma}) d\boldsymbol{\sigma}. \tag{79}$$

We now introduce the following decomposition of $\mathfrak{F}$,

$$\left.\begin{aligned}\mathfrak{F} = \mathfrak{F}_{\text{I}}+\mathfrak{F}_{\text{II}}+\mathfrak{F}_{\text{III}}, \quad \mathfrak{F}_{\text{I}} &= \Omega^{-\frac{1}{2}} \Sigma_\lambda a_\lambda \mathfrak{c}_\lambda \exp\,[2\pi i(\boldsymbol{\sigma}_\lambda \cdot \boldsymbol{x})],\\ \mathfrak{F}_{\text{II}} &= \Omega^{-\frac{1}{2}} \Sigma_\lambda b_\lambda \mathfrak{c}_\lambda \exp\,[-2\pi i(\boldsymbol{\sigma}_\lambda \cdot \boldsymbol{x})],\\ \mathfrak{F}_{\text{III}} &= \Omega^{-\frac{1}{2}} \Sigma_\lambda c_\lambda \boldsymbol{\sigma}_\lambda \exp\,[2\pi i(\boldsymbol{\sigma}_\lambda \cdot \boldsymbol{x})].\end{aligned}\right\} \tag{80}$$

The a_λ, b_λ, and c_λ are complex coefficients, and $\mathfrak{c}_\lambda$ is a complex vector perpendicular to $\boldsymbol{\sigma}_\lambda$, the square of which is zero and the absolute square of which is σ_λ^2,

$$(\mathfrak{c}_\lambda \cdot \boldsymbol{\sigma}_\lambda) = 0, \qquad (\mathfrak{c}_\lambda \cdot \mathfrak{c}_\lambda) = 0, \qquad (\mathfrak{c}_\lambda^* \cdot \mathfrak{c}_\lambda) = \sigma_\lambda^2. \tag{81}$$

If we write for the moment $\mathfrak{c}_\lambda = \mathfrak{r}_1 + i\mathfrak{r}_2$, where $\mathfrak{r}_1$ and $\mathfrak{r}_2$ are two real vectors of length $2^{-\frac{1}{2}}\sigma_\lambda$, perpendicular to one another and to $\boldsymbol{\sigma}_\lambda$, we shall assume that $\mathfrak{r}_1$, $\mathfrak{r}_2$, and $\boldsymbol{\sigma}_\lambda$ will be in the same relative position as the x-, y-, and z-axes. It thus follows that

$$[\mathfrak{c}_\lambda \wedge \boldsymbol{\sigma}_\lambda] = i\sigma_\lambda \mathfrak{c}_\lambda, \qquad [\mathfrak{c}_\lambda^* \wedge \boldsymbol{\sigma}] = -i\sigma_\lambda \mathfrak{c}_\lambda^*. \tag{82}$$

The plane waves $\mathfrak{c}_\lambda \exp\,[2\pi i(\boldsymbol{\sigma}_\lambda \cdot \boldsymbol{x})]$ which together form $\mathfrak{F}_{\text{I}}$ are called *right hand circularly polarised.* The vector $\mathfrak{c}_\lambda$ is defined completely but for a complex factor of absolute magnitude unity; we choose it for a given $\boldsymbol{\sigma}_\lambda$ in such a way that

$$\mathfrak{c}_\lambda^* = \mathfrak{c}_{-\lambda}. \tag{83}$$

The complex conjugate of a right hand circularly polarised wave is again right hand circularly polarised,

$$\{\mathfrak{c}_\lambda \exp\,[2\pi i(\boldsymbol{\sigma}_\lambda \cdot \boldsymbol{x})]\}^* = \mathfrak{c}_\lambda^* \exp\,[-2\pi i(\boldsymbol{\sigma}_\lambda \cdot \boldsymbol{x})] = \mathfrak{c}_{-\lambda} \exp\,[2\pi i(\boldsymbol{\sigma}_{-\lambda} \cdot \boldsymbol{x})];$$

one might say that the wave vector has changed sign. This remark is of importance if we now consider the *left hand circularly polarised* waves. These could be represented either by $\mathfrak{c}_\lambda^* \exp\,[2\pi i(\boldsymbol{\sigma}_\lambda \cdot \boldsymbol{x})]$ or by $\mathfrak{c}_\lambda \exp\,[-2\pi i(\boldsymbol{\sigma}_\lambda \cdot \boldsymbol{x})]$. Once we have characterised the right hand circularly polarised part of $\mathfrak{F}$ by the terms $\mathfrak{c}_\lambda \exp\,[2\pi i(\boldsymbol{\sigma}_\lambda \cdot \boldsymbol{x})]$ we have thus a twofold choice for the terms representing the left hand circularly polarised part. It will become clear later on why we have made the particular choice (80).

There are also two possible ways of choosing the *longitudinal waves* $\mathfrak{F}_{\text{III}}$ of $\mathfrak{F}$; we have arbitrarily chosen to have $+2\pi i(\boldsymbol{\sigma}_\lambda \cdot \boldsymbol{x})$ in the exponent.

Any vector field satisfying the periodicity conditions can be expanded similarly to the expansion (80). As $\exp\,[2\pi i(\boldsymbol{\sigma}_\lambda - \boldsymbol{\sigma}_{\lambda'} \cdot \boldsymbol{x})]$ integrated over Ω is equal to Ω, if $\lambda = \lambda'$, and otherwise is equal to zero, we get the expansion coefficients from the equations

$$\Omega^{-\frac{1}{2}} \int (\mathfrak{F} \cdot \mathfrak{c}_\lambda^*) \exp [-2\pi i(\boldsymbol{\sigma}_\lambda \cdot \boldsymbol{x})] d\boldsymbol{x} = a_\lambda \sigma_\lambda^2,$$

$$\Omega^{-\frac{1}{2}} \int (\mathfrak{F} \cdot \mathfrak{c}_\lambda^*) \exp [2\pi i(\boldsymbol{\sigma}_\lambda \cdot \boldsymbol{x})] d\boldsymbol{x} = b_\lambda \sigma_\lambda^2,$$

$$\Omega^{-\frac{1}{2}} \int (\mathfrak{F} \cdot \boldsymbol{\sigma}_\lambda) \exp [-2\pi i(\boldsymbol{\sigma}_\lambda \cdot \boldsymbol{x})] d\boldsymbol{x} = c_\lambda \sigma_\lambda^2.$$

The energy (74) expressed in terms of the a_λ and b_λ is of the form

$$H = (8\pi)^{-1} \Sigma_\lambda \sigma_\lambda^2 (a_\lambda^* a_\lambda + b_\lambda^* b_\lambda) + (8\pi)^{-1} \int (\mathfrak{F}_{\mathrm{III}}^* \cdot \mathfrak{F}_{\mathrm{III}}) d\boldsymbol{x}. \qquad (84)$$

Using equation (82) one can show that the following relations hold,

$$\begin{aligned}
&\operatorname{curl} \{\mathfrak{c}_\lambda \exp [2\pi i(\boldsymbol{\sigma}_\lambda \cdot \boldsymbol{x})]\} \\
&= -2\pi i [\mathfrak{c}_\lambda \wedge \boldsymbol{\sigma}_\lambda] \exp [2\pi i(\boldsymbol{\sigma}_\lambda \cdot \boldsymbol{x})] = 2\pi \sigma_\lambda \mathfrak{c}_\lambda \exp [2\pi i(\boldsymbol{\sigma}_\lambda \cdot \boldsymbol{x})], \\
&\operatorname{curl} \{\mathfrak{c}_\lambda \exp [-2\pi i(\boldsymbol{\sigma}_\lambda \cdot \boldsymbol{x})]\} = -2\pi \sigma_\lambda \mathfrak{c}_\lambda \exp [-2\pi i(\boldsymbol{\sigma}_\lambda \cdot \boldsymbol{x})], \qquad (85) \\
&\operatorname{curl} \{\boldsymbol{\sigma}_\lambda \exp [2\pi i(\boldsymbol{\sigma}_\lambda \cdot \boldsymbol{x})] = 0, \\
&\operatorname{div} \{\boldsymbol{\sigma}_\lambda \exp [2\pi i(\boldsymbol{\sigma}_\lambda \cdot \boldsymbol{x})]\} = 2\pi i\, \sigma_\lambda^2 \exp [2\pi i(\boldsymbol{\sigma}_\lambda \cdot \boldsymbol{x})].
\end{aligned}$$

The first two equations show the advantages of introducing circularly polarised waves. Substituting equations (80) into equations (72) leads now to the equations

$$2\pi \sigma_\lambda a_\lambda = (i/c) \dot{a}_\lambda, \qquad -2\pi \sigma_\lambda b_\lambda = (i/c) \dot{b}_\lambda, \qquad 0 = (i/c) \dot{c}_\lambda.$$

The third of these equations shows that c_λ is constant. From equations (73) and (85) it follows that $c_\lambda = 0$, or $\mathfrak{F}_{\mathrm{III}} = 0$, so that we get for the energy

$$H = (8\pi)^{-1} \Sigma_\lambda \sigma_\lambda^2 (a_\lambda^* a_\lambda + b_\lambda^* b_\lambda),$$

while we get for the equations of motion

$$\begin{aligned}
\dot{a}_\lambda &= -2\pi i \nu_\lambda a_\lambda, \quad \dot{a}_\lambda^* = 2\pi i \nu_\lambda a_\lambda^*, \\
\dot{b}_\lambda &= 2\pi i \nu_\lambda b_\lambda, \quad \dot{b}_\lambda^* = -2\pi i \nu_\lambda b_\lambda^*, \quad \nu_\lambda = c\sigma_\lambda.
\end{aligned} \qquad (87)$$

Equations (87) are solved by the expressions $a_\lambda = a_\lambda^{(0)} \exp(-2\pi i \nu_\lambda t)$ $b_\lambda = b_\lambda^{(0)} \exp(2\pi i \nu_\lambda t)$, and the electromagnetic field appears, as we would expect, to be a superposition of right hand and left hand circularly polarised plane light waves,

$$\begin{aligned}
\mathfrak{E} + i\mathfrak{H} = \mathfrak{F} = \Omega^{-\frac{1}{2}} \Sigma_\lambda \{ & a_\lambda^{(0)} \mathfrak{c}_\lambda \exp [2\pi i\{(\boldsymbol{\sigma}_\lambda \cdot \boldsymbol{x}) - \nu_\lambda t\}] \\
& + b_\lambda^{(0)} \mathfrak{c}_\lambda \exp [-2\pi i\{(\boldsymbol{\sigma}_\lambda \cdot \boldsymbol{x}) - \nu_\lambda t\}] \qquad (88)
\end{aligned}$$

It is now clear why we have chosen the particular expansion (80); $\boldsymbol{\sigma}_\lambda$ is everywhere the wave vector in the direction of the propagation of the light. The nomenclature right hand and left hand circularly polarised is also in agreement with the usual convention that light propagated in the positive z-direction while the $\mathfrak{E}$-vector rotates over $\frac{1}{2}\pi$ from the positive x-axis to the positive y-axis is called right hand circularly polarised.

Equations (87) can be considered to be the Hamiltonian canonical equations of motion corresponding to the Hamiltonian energy (86). Let us introduce for the moment new variables p_λ, q_λ, $\bar{p}_\lambda$, and $\bar{q}_\lambda$ through the equations,

$$a_\lambda = (4\pi)^{1/2}\sigma_\lambda^{-1}(p_\lambda - 2\pi i\nu_\lambda q_\lambda), \quad b_\lambda = (4\pi)^{1/2}\sigma_\lambda^{-1}(\bar{p}_\lambda + 2\pi i\nu_\lambda \bar{q}_\lambda). \tag{89}$$

Equations (86) and (87) now become

$$\left.\begin{aligned} H &= \tfrac{1}{2}\Sigma_\lambda\{p_\lambda^2 + (2\pi\nu_\lambda)^2 q_\lambda^2 + \bar{p}_\lambda^2 + (2\pi\nu_\lambda)^2\bar{q}_\lambda^2\}, \\ \dot{p}_\lambda &= -(2\pi\nu_\lambda)^2 q_\lambda = -\frac{\partial H}{\partial q_\lambda}, \; \dot{q}_\lambda = p_\lambda = \frac{\partial H}{\partial p_\lambda}, \; \dot{\bar{p}}_\lambda = -\frac{\partial H}{\partial \bar{q}_\lambda}, \; \dot{\bar{q}}_\lambda = \frac{\partial H}{\partial \bar{p}_\lambda}. \end{aligned}\right\} \tag{90}$$

In this way *the field equations of empty space appear as the canonical equations of a mechanical system of infinitely many degrees of freedom. There are two infinitely large groups of "coordinates", the q_λ and $\bar{q}_\lambda$, and their canonical conjugates, the p_λ and $\bar{p}_\lambda$.* The energy function contains each pair of conjugate variables p_λ, q_λ (or $\bar{p}_\lambda$, $\bar{q}_\lambda$) in the form $\frac{1}{2}[p_\lambda^2 + (2\pi\nu_\lambda)^2 q_\lambda^2]$ (or $\frac{1}{2}[\bar{p}_\lambda^2 + (2\pi\nu_\lambda)^2\bar{q}_\lambda^2]$), *corresponding to a linear harmonic oscillator of frequency ν_λ* (that is, a particle of unit mass in a linear field of force with potential energy $\frac{1}{2}[2\pi\nu_\lambda]^2 q_\lambda^2$). If necessary one can combine the two oscillators belonging to a given λ into one two-dimensional isotropic harmonic oscillator.

The idea that the empty radiation field can be considered to be an infinite number of linear harmonic oscillators, *the eigenvibrations of empty space*, was the basis of Rayleigh and Jeans' discussion of the law of radiation called after them. In the literature this idea is often introduced starting from the equation satisfied by the vector potential $\mathfrak{A}$ of the $\mathfrak{E}$, $\mathfrak{H}$-field, $\nabla^2\mathfrak{A} - c^{-2}\partial^2\mathfrak{A}/\partial t^2 = 0$. In view of the quantisation of the radiation field we have preferred not to introduce potentials and only to consider the $\mathfrak{E}$, $\mathfrak{H}$-field (the vector $\mathfrak{F}$) itself. The decomposition into circularly polarised waves was a natural one in this case. Using a linear canonical transformation we can introduce instead of the p_λ, q_λ, $\bar{p}_\lambda$, $\bar{q}_\lambda$ corresponding to a given wave vector $\boldsymbol{\sigma}_\lambda$ suitable new variables, and we can obtain a description corresponding to a different state of polarisation. The number of eigenvibrations with wave vectors corresponding to a volume element $d\boldsymbol{\sigma}$ in $\boldsymbol{\sigma}$-space is clearly equal to $2\Omega d\boldsymbol{\sigma}$ [see eq. (79)].

§ 87. Quantisation of the Radiation Field; Light Quanta

It is easy to quantise a system of independent linear harmonic oscillators (see §§ 32 and 17), and we can formally extend the corresponding methods of quantum mechanics to the case where the number of oscillators is infinite. Let the physical situation of the radiation field be characterised by a wave function $\psi(\ldots q_\lambda \ldots, \ldots \bar{q}_\lambda \ldots, t)$ depending on all the q_λ, $\bar{q}_\lambda$ and on the time. The function H of equation (90) is the energy operator, with the p's being replaced by $-i\hbar\partial/\partial q$. According to equation (3.108) the eigenfunctions will be products of the eigenfunctions of the separate oscillators, and the eigenvalues [see eq. (3.109)] sums of the separate eigenvalues. If $n_\lambda(\bar{n}_\lambda)$ denotes the quantum number of the oscillator corresponding to $q_\lambda(\bar{q}_\lambda)$, the energy eigenvalues are characterised by an infinite set of integers. We arrange the wave vectors $\boldsymbol{\sigma}_\lambda$ according to some scheme in a sequence: $\boldsymbol{\sigma}_1, \boldsymbol{\sigma}_2, \ldots, \boldsymbol{\sigma}_\lambda, \ldots$ (the equation $\boldsymbol{\sigma}_\lambda = -\boldsymbol{\sigma}_{-\lambda}$ loses its significance) and we have now for the eigenvalues and eigenfunctions of the radiation field [see eq. (2.34)],

$$E_{n_1, n_2, \ldots} = \Sigma_\lambda h\nu_\lambda [(n_\lambda + \tfrac{1}{2}) + (\bar{n}_\lambda + \tfrac{1}{2})], \tag{91}$$

$$\varphi_{n_1, n_2, \ldots} = \Pi_\lambda \varphi_{n\lambda}(q_\lambda)\, \varphi_{\bar{n}\lambda}(\bar{q}_\lambda), \tag{92}$$

where $\varphi_{n\lambda}$ is an eigenfunction (quantum number n_λ) of the harmonic oscillator of frequency ν_λ. The energy of the radiation field is thus equal to the sum of a "zero point energy",

$$E_{0,0,\ldots} = \Sigma_\lambda h\nu_\lambda, \tag{93}$$

and a number of "energy quanta" of the radiation, each of magnitude $h\nu_\lambda$.

We have obtained here the famous formula of Einstein's *light quanta*. At that time there was no mention of the idea of "radiation particles" which is nowadays so closely connected with the concept of light quanta.

We wish to investigate in how far it has a meaning to talk about particles of radiation, and for that purpose we shall describe the radiation field slightly differently. We transform from the "q_λ, $\bar{q}_\lambda$-language" to the "n_λ, $\bar{n}_\lambda$-language", that is, the situation of the radiation field is described by a set of numbers $A(n_1, \ldots, n_\lambda, \ldots; \ldots, \bar{n}_\lambda, \ldots; t)$ [in short, $A(n_\lambda, \bar{n}_\lambda, t)$], which are the coefficients in the expansion of ψ in terms of the eigenfunctions (92), $\psi(q_\lambda, \bar{q}_\lambda, t) = \Sigma_{\ldots n_\lambda \ldots \bar{n}_\lambda \ldots} A(n_\lambda, \bar{n}_\lambda, t)\varphi_{n_1, n_2, \ldots}$. The coefficients A form the transformed wave function which depends on the discrete arguments $\ldots n_\lambda \ldots \bar{n}_\lambda \ldots$ and on t. Operators $F_{\rm op}$ that is, functions of the $p, q, \bar{p}$, and $\bar{q}$, which acted on ψ are now transformed into operators, or matrices, which act on these arguments n, $F_{\rm op} \to F_{n_\lambda \bar{n}_\lambda; n'_\lambda, \bar{n}'_\lambda}$. We are in particular interested in the matrices corresponding to the canonical variables $p, q, \bar{p}$, and $\bar{q}$. It is, however, more useful to investigate immediately the matrices corresponding to the original a_λ, b_λ, a_λ^*, and b_λ^* [see eqq. (89)]. First of all we determine their commutation relations. From the relations $[p_\lambda, q_\lambda]_- = [\bar{p}_\lambda, \bar{q}_\lambda]_- = -i\hbar$ (all other commutators vanish), it follows easily that only a_λ, a_λ^* and b_λ, b_λ^* form non-commuting pairs, and that we have $[a_\lambda^*, a_\lambda]_- = -[b_\lambda^*, b_\lambda]_- = -8\pi hc/\sigma_\lambda$. We now introduce instead of a_λ, b_λ, a_λ^*, and b_λ^* new variables $\boldsymbol{\alpha}_\lambda$, $\boldsymbol{\alpha}_\lambda^*$, $\boldsymbol{\beta}_\lambda$, and $\boldsymbol{\beta}_\lambda^*$ as follows

$$\boldsymbol{\alpha}_\lambda = \tau a_\lambda,\quad \boldsymbol{\alpha}_\lambda^* = \tau a_\lambda^*,\quad \boldsymbol{\beta}_\lambda^* = \tau b_\lambda,\quad \boldsymbol{\beta}_\lambda = \tau b_\lambda^*,\quad \tau = (\sigma_\lambda/8\pi hc)^{1/2}, \tag{94}$$

and we have the following commutation relations,

$$\left.\begin{array}{l} -[\boldsymbol{\alpha}_\lambda^*, \boldsymbol{\alpha}_{\lambda'}]_- = \mathbf{1}\cdot\delta_{\lambda\lambda'} = \delta_{\lambda\lambda'}, \quad -[\boldsymbol{\beta}_\lambda^*, \boldsymbol{\beta}_{\lambda'}]_- = \mathbf{1}\cdot\delta_{\lambda\lambda'} = \delta_{\lambda\lambda'}. \\ \text{(all other pairs of variables commuting)} \end{array}\right\} \tag{95}$$

These are exactly the commutation relations which we met with earlier [see eq. (7.50)] in the description of systems of equivalent Bose-Einstein particles. The characterisation of the situation by a function $A(n_\lambda, \bar{n}_\lambda, t)$ is adapted to this description. By analogy with equation (7.49) we can write down for the $\boldsymbol{\alpha}_\lambda(\boldsymbol{\beta}_\lambda)$ matrices which are unit matrices with respect to all $n_{\lambda'}(\bar{n}_{\lambda'})$ except $n_\lambda(\bar{n}_\lambda)$ and we have

$$\boldsymbol{\alpha}_\lambda \rightarrow \begin{array}{c|cccccc} n_\lambda \backslash n'_\lambda & 0 & 1 & 2 & 3 & \ldots \\ \hline 0 & 0 & \sqrt{1} & 0 & 0 & \ldots \\ 1 & 0 & 0 & \sqrt{2} & 0 & \ldots \\ 2 & 0 & 0 & 0 & \sqrt{3} & \ldots \\ \cdot & \cdot & \cdot & \cdot & \cdot & \ldots \\ \cdot & \cdot & \cdot & \cdot & \cdot & \ldots \\ \cdot & \cdot & \cdot & \cdot & \cdot & \ldots \end{array}, \qquad \boldsymbol{\beta}_\lambda \rightarrow \begin{array}{c|cccccc} \bar{n}_\lambda \backslash \bar{n}'_\lambda & 0 & 1 & 2 & 3 & \ldots \\ \hline 0 & 0 & \sqrt{1} & 0 & 0 & \ldots \\ 1 & 0 & 0 & \sqrt{2} & 0 & \ldots \\ 2 & 0 & 0 & 0 & \sqrt{3} & \ldots \\ \cdot & \cdot & \cdot & \cdot & \cdot & \ldots \\ \cdot & \cdot & \cdot & \cdot & \cdot & \ldots \\ \cdot & \cdot & \cdot & \cdot & \cdot & \ldots \end{array}, \tag{96}$$

while we have for $\boldsymbol{\alpha}_\lambda^*(\boldsymbol{\beta}_\lambda^*)$ the adjoint *and* complex conjugate matrices *,

* The $\boldsymbol{\alpha}_\lambda$ and $\boldsymbol{\alpha}_\lambda^*$ (also the $\boldsymbol{\beta}_\lambda$ and $\boldsymbol{\beta}_\lambda^*$) correspond to two quantities the sum of which and the difference of which divided by i are Hermitean operators—which would correspond to real classical quantities. Therefore, $\boldsymbol{\alpha}_\lambda$ and $\boldsymbol{\alpha}_\lambda^*$ are not complex conjugate operators in the normal sense, but *Hermitean conjugate.* It would be more consistent to write $\widetilde{\boldsymbol{\alpha}}_\lambda^*(\widetilde{\boldsymbol{\beta}}_\lambda^*)$ instead of $\boldsymbol{\alpha}_\lambda^*(\boldsymbol{\beta}_\lambda^*)$ [see eq. (4.91)].

$$
\boldsymbol{\alpha}_\lambda^* \rightarrow \begin{array}{c|cccccc}
n_\lambda \diagdown n'_\lambda & 0 & 1 & 2 & 3 & \ldots \\
\hline
0 & 0 & 0 & 0 & 0 & \ldots \\
1 & \sqrt{1} & 0 & 0 & 0 & \ldots \\
2 & 0 & \sqrt{2} & 0 & 0 & \ldots \\
3 & 0 & 0 & \sqrt{3} & 0 & \ldots \\
. & . & . & . & . & \ldots \\
. & . & . & . & . & \ldots
\end{array}, \qquad
\boldsymbol{\beta}_\lambda^* \rightarrow \begin{array}{c|cccccc}
\bar{n}_\lambda \diagdown \bar{n}' & 0 & 1 & 2 & 3 & \ldots \\
\hline
0 & 0 & 0 & 0 & 0 & \ldots \\
1 & \sqrt{1} & 0 & 0 & 0 & \ldots \\
2 & 0 & \sqrt{2} & 0 & 0 & \ldots \\
3 & 0 & 0 & \sqrt{3} & 0 & \ldots \\
. & . & . & . & . & \ldots \\
. & . & . & . & . & \ldots
\end{array} . \qquad (97)
$$

We would have obtained exactly these matrices if we had first of all obtained the matrices for the p_λ, q_λ, $\bar{p}_\lambda$, and $\bar{q}_\lambda$ from the oscillator problem, and afterwards used equations (89) and (94) to get $\boldsymbol{\alpha}_\lambda$ and $\boldsymbol{\beta}_\lambda$,

$$
\boldsymbol{\alpha}_\lambda = (2h\nu_\lambda)^{-\frac{1}{2}}(p_\lambda - 2\pi i\nu_\lambda q_\lambda), \qquad \boldsymbol{\beta}_\lambda = (2h\nu_\lambda)^{-\frac{1}{2}}(\bar{p}_\lambda - 2\pi i\nu_\lambda \bar{q}_\lambda). \qquad (98)
$$

One can show that in the case of a linear harmonic oscillator with energy function $H = \frac{1}{2}[(p^2/m) + m(2\pi\nu)^2 q^2]$ the matrices of p and q with respect to the energy eigenfunctions φ_n are of the form

$$
q_{nn'} = \int \varphi_n^*(q)\, q\, \varphi_{n'}(q)\, dq \rightarrow (h/2\pi^2\nu m)^{\frac{1}{2}} \cdot \left[\begin{array}{c|cccccc}
n \diagdown n' & 0 & 1 & 2 & 3 & \ldots \\
\hline
0 & 0 & \frac{1}{2}i\sqrt{1} & 0 & 0 & \ldots \\
1 & -\frac{1}{2}i\sqrt{1} & 0 & \frac{1}{2}i\sqrt{2} & 0 & \ldots \\
2 & 0 & -\frac{1}{2}i\sqrt{2} & 0 & \frac{1}{2}i\sqrt{3} & \ldots \\
. & . & . & . & . & \ldots
\end{array}\right];
$$

$$
p_{nn'} = \int \varphi_n^*(q) \frac{\hbar}{i} \frac{d}{dq} \varphi_{n'}(q)\, dq \rightarrow (2h\nu m)^{\frac{1}{2}} \cdot \left[\begin{array}{c|cccccc}
n \diagdown n' & 0 & 1 & 2 & 3 & \ldots \\
\hline
0 & 0 & \frac{1}{2}\sqrt{1} & 0 & 0 & \ldots \\
1 & \frac{1}{2}\sqrt{1} & 0 & \frac{1}{2}\sqrt{2} & 0 & \ldots \\
2 & 0 & \frac{1}{2}\sqrt{2} & 0 & \frac{1}{2}\sqrt{3} & \ldots \\
. & . & . & . & . & \ldots
\end{array}\right]. \qquad (99)
$$

According to equation (2.33) the normalised eigenfunctions φ_n are equal to $\varphi_n = (-i)^n N_n^{-\frac{1}{2}} \exp(\frac{1}{2}\xi^2)\, D^n \exp(-\xi^2)$, where $\xi = (4\pi^2\nu m/h)^{\frac{1}{2}} q$, $D = d/d\xi$, and $N_n = \int_{-\infty}^{+\infty} \exp(\xi^2)[D^n \exp(-\xi^2)]^2 dq = (4\pi^2\nu m/h)\int_{-\infty}^{+\infty}[(-2\xi)^n + \ldots]D^n \exp(-\xi^2)\, d\xi = 2^n n!(4\pi^3\nu m/h)^{\frac{1}{2}}$. From the properties of D it follows that $D^{n+1}\exp(-\xi^2) = D^n[-2\xi \exp(-\xi^2)] = -2\xi D^n \exp(-\xi^2) - 2n D^{n-1}\exp(-\xi^2)$, or, $i\varphi_{n+1} N_{n+1}^{\frac{1}{2}} = 2q(4\pi^2 \nu m/h)^{\frac{1}{2}} \varphi_n N_n^{\frac{1}{2}} + 2in\, \varphi_{n-1} N_{n-1}^{\frac{1}{2}}$, and $q\,\varphi_n = \frac{1}{2}i(h/2\pi^2\nu m)^{\frac{1}{2}}[\varphi_{n+1}(n+1)^{\frac{1}{2}} - \varphi_{n-1} n^{\frac{1}{2}}]$. From this we obtain the matrix $q_{nn'}$ while $p_{nn'}$ follows from the equation $p_{n,\,n\pm1} = (im/\hbar)[Hq - qH]_{n,\,n\pm1} = \mp 2\pi i\nu m q_{n,\,n\pm1}$. Substitution of expressions (99) into equation (98) leads to expressions (96), if m is put equal to unity.

In Heisenberg's first paper on quantum mechanics in 1925 these expressions for $p_{nn'}$ and $q_{nn'}$ were derived from the commutation relations and the requirement that H be a diagonal matrix. In 1926 Schrödinger derived the same expression from wave mechanics, as we have done here.

*The radiation field can thus be considered to be either a system of independent oscillators, or an ensemble of independent Bose-Einstein particles *. There is a complete mathematical equivalence between these two modes of description, and this justifies in the first instance the nomenclature of the corpuscular concept of light.* The stationary state (91), (92) can in a natural way be interpreted as a situation with n_λ right hand circularly polarised and $\bar{n}_\lambda$ left hand circularly polarised light quanta with a wave vector $\boldsymbol{\sigma}_\lambda$ ($\lambda = 1, 2, \ldots$). In the present interpretation it is immaterial whether we consider the radiation field to be *one* Bose-Einstein ensemble of both right hand and left hand light quanta, or whether we consider left hand and right hand light quanta to be two different kinds of Bose-Einstein particles. The number of right hand quanta in the "state" λ (that is, with wave vector $\boldsymbol{\sigma}_\lambda$) is given by the operator $\boldsymbol{\alpha}_\lambda^*\boldsymbol{\alpha}_\lambda$, and the number of left hand quanta by the operator $\boldsymbol{\beta}_\lambda^*\boldsymbol{\beta}_\lambda$ [see eq. (7.50)]. The eigenvalues of these operators are just the integers n_λ and $\bar{n}_\lambda$ which characterise the stationary states of empty space. The total number of light quanta corresponds to the operator [compare eq. (7.40)]

$$N_{\text{op}} = \Sigma_\lambda(\boldsymbol{\alpha}_\lambda^*\boldsymbol{\alpha}_\lambda + \boldsymbol{\beta}_\lambda^*\boldsymbol{\beta}_\lambda). \tag{100}$$

If we use equations (98) to replace the $p, q, \bar{p}$, and $\bar{q}$ by the operators $\boldsymbol{\alpha}, \ldots$, we get for the energy operator [compare eqq. (86) and (90)], $H_{\text{op}} = \frac{1}{2}\Sigma_\lambda h\nu_\lambda[\boldsymbol{\alpha}_\lambda^*\boldsymbol{\alpha}_\lambda + \boldsymbol{\alpha}_\lambda\boldsymbol{\alpha}_\lambda^* + \boldsymbol{\beta}_\lambda^*\boldsymbol{\beta}_\lambda + \boldsymbol{\beta}_\lambda\boldsymbol{\beta}_\lambda^*]$. Because of the relations (95) this expression differs from the simpler one

$$H_{\text{op}} = \Sigma_\lambda h\nu_\lambda(\boldsymbol{\alpha}_\lambda^*\boldsymbol{\alpha}_\lambda + \boldsymbol{\beta}_\lambda^*\boldsymbol{\beta}_\lambda)\,[\text{eigenvalues: } \Sigma h\nu_\lambda(n_\lambda + \bar{n}_\lambda)] \tag{101}$$

by the zero point energy (93). As the sum (93) is infinite (we can write $E_{0,0,\ldots} = hc\Omega\int_0^\infty \sigma d\boldsymbol{\sigma}$) and as there are no compelling reasons as there are in the case of the "real" oscillators of classical mechanics — to prefer the expression $\frac{1}{2}(\boldsymbol{\alpha}^*\boldsymbol{\alpha} + \boldsymbol{\alpha}\boldsymbol{\alpha}^*)$ to the expression $\boldsymbol{\alpha}^*\boldsymbol{\alpha}$ it is conventional *to consider expression* (101) *to be the "correct" expression for the radiation energy.*

Let us return to the electromagnetic field. We use equation (94) to replace the coefficients in the expansion (80) by the $\boldsymbol{\alpha}_\lambda$ and $\boldsymbol{\beta}_\lambda$. We drop again $\mathfrak{F}_{\text{III}}$ and obtain

$$\mathfrak{F}_{\text{op}}(\boldsymbol{x}) = 4\pi(\hbar c/\Omega)^{1/2}\Sigma_\lambda\sigma_\lambda^{-1/2}\{\boldsymbol{\alpha}_\lambda\mathfrak{c}_\lambda \exp[2\pi i(\boldsymbol{\sigma}_\lambda\cdot\boldsymbol{x})] + \boldsymbol{\beta}_\lambda^*\mathfrak{c}_\lambda \exp[-2\pi i(\boldsymbol{\sigma}_\lambda\cdot\boldsymbol{x})]. \tag{102}$$

If the matrix elements of $\boldsymbol{\alpha}_\lambda(\boldsymbol{\beta}_\lambda^*)$ are considered to be time-depend-

* See P. A. M. Dirac, *Proc. Roy. Soc.*, **A114**, 243, 1927.

ent quantities, they contain the time in the form of a factor $\exp(-2\pi i \nu_\lambda t)[\exp(2\pi i \nu_\lambda t)]$ and they still satisfy the relations (95). This can be seen, for instance, by considering the following equations

$$-i\hbar(d\boldsymbol{\alpha}_\lambda/dt)_{t=0} = [H_{\text{op}}, \boldsymbol{\alpha}_\lambda]_- = h\nu_\lambda(\boldsymbol{\alpha}_\lambda^* \boldsymbol{\alpha}_\lambda \boldsymbol{\alpha}_\lambda - \boldsymbol{\alpha}_\lambda \boldsymbol{\alpha}_\lambda^* \boldsymbol{\alpha}_\lambda) = -h\nu_\lambda \boldsymbol{\alpha}_\lambda,$$
$$(d\boldsymbol{\beta}_\lambda^*/dt)_{t=0} = 2\pi i \nu_\lambda \boldsymbol{\beta}_\lambda^*,$$
$$\boldsymbol{\alpha}_\lambda(t) = \boldsymbol{\alpha}_\lambda \exp(-2\pi i \nu_\lambda t),\ \boldsymbol{\beta}_\lambda^*(t) = \boldsymbol{\beta}_\lambda^* \exp(2\pi i \nu_\lambda t).$$

We have thus an equation completely analogous to the classical equation (88)

$$\mathfrak{F}_{\text{op}}(\boldsymbol{x}, t) = 4\pi(\hbar c/\Omega)^{1/2} \Sigma_\lambda \sigma_\lambda^{-1/2}[\boldsymbol{\alpha}_\lambda \mathfrak{c}_\lambda \exp\{2\pi i[(\boldsymbol{\sigma}_\lambda \cdot \boldsymbol{x}) - \nu_\lambda t]\} + \boldsymbol{\beta}_\lambda^* \mathfrak{c}_\lambda \exp\{-2\pi i[(\boldsymbol{\sigma}_\lambda \cdot \boldsymbol{x}) - \nu_\lambda t]\}]. \tag{103}$$

In this last equation $\boldsymbol{\alpha}_\lambda$ and $\boldsymbol{\beta}_\lambda^*$ are, of course, again time-independent.

For physical phenomena where the size of the cube Ω does not play a role, we may assume that we can always replace the sum by an integral over $\boldsymbol{\sigma}$-space. If we divide this space into small volume elements $D\boldsymbol{\sigma}$ which are nevertheless so large that they contain many $\boldsymbol{\sigma}_\lambda$-points, and if we define for each volume element two operators $\mathbf{A}_{\boldsymbol{\sigma}}$ and $\mathbf{B}_{\boldsymbol{\sigma}}^*$ as sums over the $\boldsymbol{\sigma}_\lambda$ within $D\boldsymbol{\sigma}$, as follows,

$$(D\boldsymbol{\sigma}/\sigma)\mathbf{A}_{\boldsymbol{\sigma}} = (\Omega\sigma)^{-1/2} \Sigma_{\lambda(D\boldsymbol{\sigma})} \boldsymbol{\alpha}_\lambda;\ (D\boldsymbol{\sigma}/\sigma)\mathbf{B}_{\boldsymbol{\sigma}} = (\Omega\sigma)^{-1/2} \Sigma_{\lambda(D\boldsymbol{\sigma})} \boldsymbol{\beta}_\lambda, \tag{104}$$

$$[\mathbf{A}_{\boldsymbol{\sigma}}, \mathbf{A}_{\boldsymbol{\sigma}}^*]_- = [\mathbf{B}_{\boldsymbol{\sigma}}, \mathbf{B}_{\boldsymbol{\sigma}}^*]_- = (\sigma/D\boldsymbol{\sigma})^2 (\Omega\sigma)^{-1} \Sigma_{\lambda(D\boldsymbol{\sigma})} [\boldsymbol{\alpha}_\lambda, \boldsymbol{\alpha}_\lambda^*]_- = \sigma/D\boldsymbol{\sigma} \text{ (all other pairs of quantities commute)}, \tag{105}$$

we get instead of expression (103) for $\mathfrak{F}_{\text{op}}$ the equation

$$\mathfrak{F}_{\text{op}}(\boldsymbol{x}, t) = 4\pi(\hbar c)^{1/2} \int [\mathbf{A}_{\boldsymbol{\sigma}} \mathfrak{c}_{\boldsymbol{\sigma}} \exp\{2\pi i[(\boldsymbol{\sigma} \cdot \boldsymbol{x}) - \nu t]\} + \mathbf{B}_{\boldsymbol{\sigma}}^* \mathfrak{c}_{\boldsymbol{\sigma}} \exp\{-2\pi i[(\boldsymbol{\sigma} \cdot \boldsymbol{x}) - \nu t]\}](D\boldsymbol{\sigma}/\sigma). \tag{106}$$

For the energy operator we get

$$H_{\text{op}} = (8\pi)^{-1} \int (\mathfrak{F}_{\text{op}}^* \cdot \mathfrak{F}_{\text{op}}) d\boldsymbol{x} = h \int \nu (\mathbf{A}_{\boldsymbol{\sigma}}^* \mathbf{A}_{\boldsymbol{\sigma}} + \mathbf{B}_{\boldsymbol{\sigma}}^* \mathbf{B}_{\boldsymbol{\sigma}}) D\boldsymbol{\sigma}/\sigma \quad (+ \text{ zero point energy}). \tag{107}$$

The more detailed, even too detailed, description of the single light quanta by means of the $\boldsymbol{\alpha}_\lambda$, $\boldsymbol{\beta}_\lambda^*$ is now lost. However, for all physical applications it is sufficient to consider the expressions $\mathbf{A}^*\mathbf{A}D\boldsymbol{\sigma}/\sigma$ and $\mathbf{B}^*\mathbf{B}D\boldsymbol{\sigma}/\sigma$ to be the number of right hand—and left hand quanta within $D\boldsymbol{\sigma}$, or, $N_{\text{op}} = \int (\mathbf{A}_{\boldsymbol{\sigma}}^* \mathbf{A}_{\boldsymbol{\sigma}} + \mathbf{B}_{\boldsymbol{\sigma}}^* \mathbf{B}_{\boldsymbol{\sigma}}) D\boldsymbol{\sigma}/\sigma$. In agreement with this we can use the following representation of the $\mathbf{A}$, $\mathbf{B}$, $\mathbf{A}^*$, and $\mathbf{B}^*$ matrices [see eq. (96)],

$$\begin{matrix}\mathbf{A}\\ \\ \mathbf{B}\end{matrix} \to \left(\frac{\sigma}{D\boldsymbol{\sigma}}\right)^{1/2} \left|\begin{matrix} 0 & \sqrt{1} & & & \\ & 0 & \sqrt{2} & & \\ & & & \cdot & \\ & & & & \cdot \\ & & & & & \cdot \end{matrix}\right|, \quad \begin{matrix}\mathbf{B}^*\\ \\ \mathbf{A}^*\end{matrix} \to \left(\frac{\sigma}{D\boldsymbol{\sigma}}\right)^{1/2} \left|\begin{matrix} 0 & & & & \\ \sqrt{1} & 0 & & & \\ & \sqrt{2} & & & \\ & & \cdot & & \\ & & & \cdot & \\ & & & & \cdot \end{matrix}\right|,$$

$$\begin{matrix}\mathbf{A}^*\mathbf{A}\dfrac{D\boldsymbol{\sigma}}{\sigma}\\ \\ \mathbf{B}^*\mathbf{B}\dfrac{D\boldsymbol{\sigma}}{\sigma}\end{matrix} \to \left|\begin{matrix} 0 & & & & 0 & \\ & 1 & & & & \\ & & 2 & & & \\ & & & 3 & & \\ 0 & & & & \cdot & \\ & & & & & \cdot \end{matrix}\right|. \tag{108}$$

This representation is in accordance with the commutation relations (105). The physical situation must now be described by a function $\psi(N_1, N_2, \ldots, \overline{N}_1, \overline{N}_2, \ldots, t)$ where $N_1, N_2, \ldots$ $(\overline{N}_1, \overline{N}_2, \ldots)$ are the numbers of right hand (left hand) quanta in the volume elements $(D\boldsymbol{\sigma})_1$, $(D\boldsymbol{\sigma})_2, \ldots$ in $\boldsymbol{\sigma}$-space.

We have now obtained a *description of the radiation field free from the restriction of the cube.* We expect this description to be logically consistent as far as applications are concerned and we expect it to be independent of the particular choice of the volume elements $D\boldsymbol{\sigma}$ in $\boldsymbol{\sigma}$-space. We do not wish to prove this statement * and only draw attention to the fact that it follows from equations (105), (106), and (108) that *this description is Lorentz-invariant.*

* If we wish to drop the restriction to *finite* volume elements $D\boldsymbol{\sigma}$ in $\boldsymbol{\sigma}$-space and want to consider expressions (106) and (107) to be real integrals, we consider the operator

$$\mathbf{A}_\Sigma = \int_\Sigma \mathbf{A}_{\boldsymbol{\sigma}}\, D\boldsymbol{\sigma}/\sigma \;\; (\to \int_\Sigma \mathbf{A}_{\boldsymbol{\sigma}} d\boldsymbol{\sigma}/\sigma), \tag{109}$$

where the summation extends over an arbitrary, finite region Σ of $\boldsymbol{\sigma}$-space containing many volume elements $D\boldsymbol{\sigma}$. From equation (105) it follows that $[\mathbf{A}_\Sigma, \mathbf{A}^*_{\boldsymbol{\sigma}}]_- = \delta(\Sigma, \boldsymbol{\sigma}) = \int_\Sigma \delta(\boldsymbol{\sigma}-\boldsymbol{\sigma}')d\boldsymbol{\sigma}'$, where $\delta(\Sigma, \boldsymbol{\sigma})$ is equal to unity or zero according to whether $\boldsymbol{\sigma}$ lies within or outwith Σ. Formally this function can be considered to be an integral over Σ of an integrand $\delta(\boldsymbol{\sigma}-\boldsymbol{\sigma}')$ which possesses the properties of the three dimensional Dirac delta-function in $\boldsymbol{\sigma}$-space. We can now consider expressions (106) and (107) as real integrals in which the operators $\mathbf{A}_{\boldsymbol{\sigma}}$, which are functions of $\boldsymbol{\sigma}$, satisfy the commutation relations

$$[\mathbf{A}_{\boldsymbol{\sigma}'}, \mathbf{A}^*_{\boldsymbol{\sigma}}]_- = \sigma\delta(\boldsymbol{\sigma}-\boldsymbol{\sigma}') \; (= [\mathbf{B}_{\boldsymbol{\sigma}'}, \mathbf{B}^*_{\boldsymbol{\sigma}}]_-), \tag{110}$$

The description of the physical situation by means of a function $\psi(N_1, \ldots, \overline{N}_1, \ldots, t)$ has no longer a meaning, since the expressions (108) have no longer any mathematical meaning. The representations (103) and (106) by means of sums here proved useful for actual applications; only when interpreting the results can one, if necessary, go over to statements involving real integrals over $\boldsymbol{\sigma}$-space.

Proof of the relativistic invariance. Under a Lorentz-transformation $\boldsymbol{x}, t \to \boldsymbol{x}', t'$, which does not involve an inversion, the null-four-vector $\boldsymbol{\sigma}, \nu/c^2$ goes over into another null-four-vector $\boldsymbol{\sigma}', \nu'/c^2$. The invariance $d\boldsymbol{\sigma}/\sigma = d\boldsymbol{\sigma}'/\sigma'$ follows, since under the particular transformation

$$x' = x, \quad y' = y, \quad z' = \beta(z-vt), \quad t' = \beta[t-(vz/c^2)], \quad \beta = [1-(v^2/c^2)]^{-\frac{1}{2}},$$

we have

$$\sigma_x' = \sigma_x, \ \sigma_y' = \sigma_y, \ \sigma_z' = \beta[\sigma_z-(v\sigma/c)], \ \sigma' = \beta[\sigma-(v\sigma_z/c)],$$

$$\frac{d\boldsymbol{\sigma}'}{\sigma'} = d\sigma_x\, d\sigma_y\, [d\sigma_z - \frac{v}{c}\frac{\partial\sigma}{\partial\sigma_z}\, d\sigma_z]/[\sigma-\frac{v}{c}\sigma_z] = \frac{d\boldsymbol{\sigma}}{\sigma} \quad \left(\frac{\partial\sigma}{\partial\sigma_z} = \frac{\sigma_z}{\sigma}\right).$$

A Lorentz transformation corresponds to a complex orthogonal transformation under which $\mathfrak{F} = \mathfrak{E}+i\mathfrak{H}$ goes over into $\mathfrak{F}' = \mathfrak{E}'+i\mathfrak{H}'$ while the null-vector $\mathfrak{c}$ which is perpendicular to $\boldsymbol{\sigma}$ and has the absolute length σ goes over into a null-vector $\mathfrak{c}'$ perpendicular to $\boldsymbol{\sigma}'$ and of absolute length σ' (see § 63). This means that $\mathbf{A}_{\boldsymbol{\sigma}}$ and $\mathbf{B}^*_{\boldsymbol{\sigma}}$ in expression (106) must be Lorentz-invariant operators in complete accordance with the fact that in equations (105) and (108) only the Lorentz-invariant factor $D\boldsymbol{\sigma}/\sigma$ occurs. The $\mathbf{A}_{\boldsymbol{\sigma}}$ and $\mathbf{B}^*_{\boldsymbol{\sigma}}$ have the dimensions of a reciprocal length.

According to equations (103) or (106) the six components of $\mathfrak{E}$ and $\mathfrak{H}$ are certain Hermitean operators, that is, q-numbers. Each of them is a linear function of the $\boldsymbol{\alpha}$, $\boldsymbol{\alpha}^*$, $\boldsymbol{\beta}$, and $\boldsymbol{\beta}^*$, or of the p, q, $\bar{p}$, and $\bar{q}$ [see eq. (98)]. It follows that—as we would expect—the eigenvalues of each component are all integers from $-\infty$ to $+\infty$. Because of the commutation relations (95) not all pairs of components will commute, but we have that* *a field component at a space-time point* $\mathrm{P}(\boldsymbol{x}, t)$ *is in general non-commuting with a component at another space-time point* $\mathrm{P}'(\boldsymbol{x}', t)$, *only if* P' *lies on the light cone with* P *as centre, that is when the relation*

$$D^2 \equiv (\boldsymbol{x}-\boldsymbol{x}' \cdot \boldsymbol{x}-\boldsymbol{x}')-c^2(t-t')^2 = 0 \tag{111}$$

is satisfied. In all other cases the two components will commute.

The mathematical expression for the commutation relations of the electromagnetic field can be obtained from equation (103). First of all, we see that the components of $\mathfrak{F}$ all commute with one another. If for the moment we denote the x, y, and z-components of a vector $\boldsymbol{V}$ by V_1, V_2, and V_3 (V_k in general), and if V_k is a component at P and V_k' a component at P′ we have

$$\begin{gathered} 0 = [\mathfrak{F}_k, \mathfrak{F}_l']_- = [\mathfrak{E}_k+i\mathfrak{H}_k, \mathfrak{E}_l'+i\mathfrak{H}_l']_-; \\ [\mathfrak{E}_k, \mathfrak{E}_l']_- = [\mathfrak{H}_k, \mathfrak{H}_l']_-; \qquad [\mathfrak{H}_k, \mathfrak{E}_l']_- = -[\mathfrak{E}_k, \mathfrak{H}_l']_-. \end{gathered} \tag{112}$$

Because of the relations (95) we have, however, for the commutation relations of the components of $\mathfrak{F}$ and $\mathfrak{F}^*$,

* P. Jordan and W. Pauli, *Z. Phys.*, **47**, 151, 1927.

$$[\mathfrak{E}_k - i\mathfrak{H}_k, \mathfrak{E}'_l + i\mathfrak{H}'_l]_- = [\mathfrak{F}^*_k, \mathfrak{F}'_l]_- = -(8\pi hc/\Omega)\Sigma_l \sigma_\lambda^{-1} \mathfrak{c}^*_{\lambda k}\mathfrak{c}_{\lambda l} \cdot$$
$$[\exp\{2\pi i[(\boldsymbol{\sigma}_\lambda \cdot \mathbf{x}' - \mathbf{x}) - \nu_\lambda(t'-t)]\} - \exp\{-2\pi i[(\boldsymbol{\sigma}_\lambda \cdot \mathbf{x}' - \mathbf{x}) - \nu_\lambda(t'-t)]\}]$$
$$= -8\pi hc\int (d\boldsymbol{\sigma}/\sigma)[\delta_{kl}\sigma^2 - \sigma_k\sigma_l + i\sigma\sigma_{kl}]\, i\sin\{2\pi[(\boldsymbol{\sigma}\cdot\mathbf{x}' - \mathbf{x}) - c\sigma(t'-t)]\}$$
$$= 2i\hbar c[\delta_{kl}\frac{1}{c^2}\frac{\partial^2}{\partial t^2} - \frac{\partial^2}{\partial x_k\,\partial x_l} - \frac{i}{c}\frac{\partial^2}{\partial t\,\partial x_{kl}}]\Delta, \tag{113}$$
$$\Delta = 2\int \sin\{2\pi[(\boldsymbol{\sigma}\cdot\mathbf{x}' - \mathbf{x}) - c\sigma(t'-t)]\}(d\boldsymbol{\sigma}/\sigma)$$
$$= \lim_{M\to\infty}\frac{1}{|\mathbf{x}'-\mathbf{x}|}\left[\frac{\sin\{2\pi M[|\mathbf{x}'-\mathbf{x}| + c(t'-t)]\}}{\pi[|\mathbf{x}'-\mathbf{x}| + c(t'-t)]} - \frac{\sin\{2\pi M[|\mathbf{x}'-\mathbf{x}| - c(t'-t)]\}}{\pi[|\mathbf{x}'-\mathbf{x}| - c(t'-t)]}\right]$$

Here σ_{kl} and x_{kl} are defined by the equations

$$\left.\begin{array}{l}\sigma_{11} = \sigma_{22} = \sigma_{33} = 0,\ \sigma_{12} = -\sigma_{21} = \sigma_3,\ \sigma_{23} = -\sigma_{32} = \sigma_1,\ \sigma_{31} = -\sigma_{13} = \sigma_2, \\ \dfrac{\partial}{\partial x_{11}} = \dfrac{\partial}{\partial x_{22}} = \dfrac{\partial}{\partial x_{33}} = 0,\ \dfrac{\partial}{\partial x_{12}} = -\dfrac{\partial}{\partial x_{21}} = \dfrac{\partial}{\partial z},\ \dfrac{\partial}{\partial x_{23}} = -\dfrac{\partial}{\partial x_{32}} = \dfrac{\partial}{\partial x}, \\ \qquad\qquad \dfrac{\partial}{\partial x_{31}} = -\dfrac{\partial}{\partial x_{13}} = \dfrac{\partial}{\partial y}.\end{array}\right\} \tag{114}$$

The function Δ is a relativistically invariant function of $|\mathbf{x}'-\mathbf{x}|$ and $t'-t$; it is an improper mathematical function * which can be expressed as follows in terms of Dirac delta-functions,

$$\Delta = \{\delta[|\mathbf{x}'-\mathbf{x}| + c(t'-t)] - \delta[|\mathbf{x}'-\mathbf{x}| - c(t'-t)]\}/|\mathbf{x}'-\mathbf{x}|. \tag{115}$$

The meaning of this function becomes clear by a consideration of the integral

$$J = \int_G f(\mathbf{x}', t)\Delta\, d\mathbf{x}'dt', \tag{116}$$

where f is some function defined in space-time, and where the integral extends over some region G in space-time. The integral J will depend only on the values of the function f in those points of G which lie on the light cone (**111**) of the point P($\mathbf{x}$, t). We see thus that J is a triple integral. If we integrate first of all over t' we get

$$J(\mathbf{x}, t) = \int_G |\mathbf{x}'-\mathbf{x}|^{-1}\{f(\mathbf{x}', t - |\mathbf{x}'-\mathbf{x}|/c) - f(\mathbf{x}', t + |\mathbf{x}'-\mathbf{x}|/c)\}d\mathbf{x}', \tag{117}$$

corresponding to the difference between the retarded and accelerated potentials (which are solutions of the wave equation

* The second expression for Δ of equation (113) is obtained by integrating the first expression over a sphere of radius M in $\boldsymbol{\sigma}$-space. The delta-functions of equation (115) occur because of the relation $\mathrm{Lim}_{M\to\infty}\int_{-\infty}^{+\infty}[\sin(2\pi Ma)/\pi a]f(a)da = f(0)$.

$\nabla^2 J - c^{-2}\partial^2 J/\partial t^2 = -4\pi f$). The meaning of a function such as $\partial^2 \Delta/\partial x \partial y$ [see eq. (113)] can thus be seen from the equation

$$\int_G f(\mathbf{x}', t)(\partial^2 \Delta/\partial x\, \partial y)\, d\mathbf{x}'\, dt' = \partial^2 J/\partial x\, \partial y.$$

From equations (112) and (113) we get *the commutation relations for the* $\mathfrak{E}$, $\mathfrak{H}$*-field* (note that $i[\mathfrak{E}_k, \mathfrak{E}'_l]_-$ and $i[\mathfrak{E}_k, \mathfrak{H}'_l]_-$ are Hermitean operators)

$$[\mathfrak{E}_k, \mathfrak{E}'_l]_- = [\mathfrak{H}_k, \mathfrak{H}'_l]_- = i\hbar c[\delta_{kl}c^{-2}(\partial^2/\partial t^2) - (\partial^2/\partial x_k\, \partial x_l)]\Delta, \tag{118}$$

$$[\mathfrak{E}_k, \mathfrak{H}'_l]_- = -[\mathfrak{H}_k, \mathfrak{E}'_l]_- = -i\hbar\, \partial^2 \Delta/\partial t\, \partial x_{kl}. \tag{119}$$

The components of $\mathfrak{E}$ and $\mathfrak{H}'$ in the same direction always commute according to equation (119), but all other pairs of components are non-commuting. In order to give an example showing how we can obtain from equation (118) formulae which do no longer contain the improper function Δ we introduce the following operator

$$\overline{f\mathfrak{E}_l} = \int_G f\, \mathfrak{E}'_l\, d\mathbf{x}'\, dt', \tag{120}$$

where the integration is the same as in equation (116) and where f here as there is an arbitrary function of $\mathbf{x}'$, t'. We have in particular

$$[\mathfrak{E}_k, \overline{f\mathfrak{E}_l}]_- = \int f[\mathfrak{E}_k, \mathfrak{E}'_l]_-\, d\mathbf{x}'\, dt' = i\hbar c[\delta_{kl}\, c^{-2}\, (\partial^2 J/\partial t^2) - (\partial^2 J/\partial t\, \partial x_{kl})],$$

where J depends on f through equation (117).

The *physical meaning of the commutation relations* is the following. If we are dealing with a physical situation where a component $\mathfrak{E}_k$ has a well defined value at the space-time point P, the values of the other field components—with the possible exception of $\mathfrak{H}_k$—at a point P′ on the light cone (111) of P are certainly not well defined. If we wish to obtain more information about the nature of the quantised electromagnetic field, we are faced with the following question: What are the consequences of equations (118) and (119) when we try to measure with physical instruments as exactly as possible the values of $\mathfrak{E}_k$, $\mathfrak{H}_k$ at different points of space-time? This question was treated in great detail by Bohr and Rosenfeld *. We shall only mention here a few points from their results.

The measurement requires the use of a charged probe which takes up a certain volume of space-time. Any measurement refers thus not to an $\mathfrak{E}_k$ or $\mathfrak{H}_k$ in just one point, but to a space-time region G, that is, it refers to a quantity corresponding to an operator such as the one given by equation (120). Since we are only interested in

* N. Bohr and L. Rosenfeld, *Proc. Dan. Acad. Sci.*, **12**, nr. 8, 1933; also *Phys. Rev.*, **78**, 794, 1950.

the nature of equations (118) and (119), which do not refer to the atomistic structure of matter, we may leave out of consideration the existence of elementary particles with very definite charges and masses. We can thus take for the charges any values and need discuss only the purely mechanical concepts of positions and momenta of parts of the probe, and these quantities must obey quantum mechanics. The main point of interest is now the field produced by the probe during the measurement. This field is subject to an uncertainty because of the Heisenberg relations. First of all one can show that it is still possible to measure an $\mathfrak{E}_k$- of $\mathfrak{H}_k$-component arbitrarily exactly, even though this uncertainty exists. As soon, however, as we are dealing with two measurements in two regions G_1 and G_2 which can "reach each other" by means of "light signals" (that is, some points of G_1 lie on light cones of points of G_2), it is possible that the measurement in G_1 leads to an extra $\mathfrak{E}$, $\mathfrak{H}$-field in G_2, which in principle has a certain uncertainty. As a consequence there will be certain *uncertainty relations of the radiation field* which couple the uncertainties in the measurements in G_1 and G_2 in a manner completely analogous to the Heisenberg relations. In the expressions for the complementary uncertainties the quantum of action occurs as well as the distance between and the dimensions of the two regions. As a result the relative uncertainties in the measurements can in general become appreciable only when we are dealing with "weak" fields. The occurrence of the factor $|\boldsymbol{x}'-\boldsymbol{x}|^{-1}$ in equations (115) and (117) means that the uncertainties decrease with increasing distance between two regions. Bohr and Rosenfeld were able to prove in detail that the Heisenberg relations lead to the consequences of equations (118) and (119) and vice versa.

In all our considerations up to now we assumed that there were no charges or currents; in particular we put $\mathfrak{F}_{\mathrm{III}}$ in equation (80) equal to zero. With the present status of the theory it is often advisable not to change the commutation relations (118) and (119) when charged particles are present (see § 90).

We wish to consider briefly *the properties of the radiation field, when only one light quantum is present.* If we denote by $(\lambda)[(\bar{\lambda})]$ that function of the numbers $(n_1, n_2, \ldots, \bar{n}_1, \bar{n}_2, \ldots)$ which is equal to unity when $n_\lambda = 1$ $(\bar{n}_\lambda = 1)$ and all other $n_{\lambda'} = 0$, and equal to zero for all other combinations of these numbers, the physical situation at $t = 0$ in this case is described by the equations

$$\Phi = \Sigma_\lambda[a_\lambda(\lambda)+b_\lambda(\bar{\lambda})], \tag{121}$$

$$\Sigma_\lambda\{|a_\lambda|^2+|b_\lambda|^2\} = 1, \tag{122}$$

where the a_λ and b_λ are a set of complex numerical coefficients.

The value of an electromagnetic quantity—represented by an operator O which is built up out of the operators $\mathfrak{E}_{\mathrm{op}}(\boldsymbol{x}, t)$, $\mathfrak{H}_{\mathrm{op}}(\boldsymbol{x}, t)$ —will in general not be well defined, but is distributed according to a certain probability law. The mean value of the n-th power is given by the equation $\overline{O^n} = \int \Phi^* O^n \Phi = \Sigma_{\lambda\lambda'}[a_\lambda^*(O^n)_{\lambda\lambda'}a_{\lambda'}+\ldots]$, where $(O^n)_{\lambda\lambda'}$ is the matrix element of O^n referring to the transition $(\lambda) \to (\lambda')$. For the sake of simplicity we have not included the terms involving $\bar{\lambda}$.

The properties of the radiation field in the situation (121) possess a great similarity to those of a classical radiation field [see eq. (103)]

$$\mathfrak{F}_{\text{class}} = 4\pi(\hbar c/\Omega)^{1/2}\Sigma_\lambda \sigma_\lambda^{-1/2}\,[a_\lambda \mathfrak{c}_\lambda \exp\{2\pi i[(\boldsymbol{\sigma}_\lambda \cdot \boldsymbol{x})-\nu_\lambda t]\} + b_\lambda^* \mathfrak{c}_\lambda \exp\{-2\pi i[(\boldsymbol{\sigma}_\lambda \cdot \boldsymbol{x})-\nu_\lambda t]\}. \quad (123)$$

This field possesses clearly the energy $(8\pi)^{-1}\int(\mathfrak{F}^*_{\text{class}} \cdot \mathfrak{F}_{\text{class}})\,d\boldsymbol{x} = hc\Sigma_\lambda \sigma_\lambda [|a_\lambda|^2+|b_\lambda|^2]$; this value is equal to the quantum mechanical average value of the energy (101) in the situation (121). If the a_λ and b_λ are different from zero practically only for wave vectors $\boldsymbol{\sigma}_\lambda$ which lie very near to a given vector $\boldsymbol{\sigma}_0$, this energy would, according to equation (122), just be the energy of one light quantum of frequency $c\sigma_0 = \nu_0$.

We shall consider in particular the electric field (similar considerations hold for the magnetic field); the corresponding operator is given by the equation (see the beginning of § 86)

$$\left.\begin{aligned} \mathfrak{E}(\boldsymbol{x},t) &= 2\pi(\hbar c/\Omega)^{1/2}\Sigma_\lambda\sigma_\lambda^{-1/2}\{(\boldsymbol{\alpha}_\lambda\mathfrak{c}_\lambda+\boldsymbol{\beta}_\lambda\mathfrak{c}_\lambda^*)\exp(i\omega_\lambda) \\ &\qquad + (\boldsymbol{\alpha}_\lambda^*\mathfrak{c}_\lambda^*+\boldsymbol{\beta}_\lambda^*\mathfrak{c}_\lambda)\exp(-i\omega_\lambda)\}, \\ \omega_\lambda &= 2\pi[(\boldsymbol{\sigma}_\lambda\cdot\boldsymbol{x})-\nu_\lambda t]. \end{aligned}\right\} \quad (124)$$

To simplify our equations we write equation (124) in the form

$$\left.\begin{aligned} \mathfrak{E}_{\text{op}} &= \tfrac{1}{2}\Omega^{-1/2}\Sigma_\mu(\boldsymbol{\gamma}_\mu\mathfrak{f}_\mu+\boldsymbol{\gamma}_\mu^*\mathfrak{f}_\mu^*), \text{ with } \boldsymbol{\gamma}_\mu = \boldsymbol{\alpha}_\lambda \text{ or } \boldsymbol{\beta}_\lambda, \text{ and} \\ \mathfrak{f}_\mu &= 4\pi(\hbar c/\sigma_\lambda)^{1/2}\mathfrak{c}_\lambda\exp(i\omega_\lambda) \text{ or } 4\pi(\hbar c/\sigma_\lambda)^{1/2}\mathfrak{c}_\lambda^*\exp(i\omega_\lambda), \end{aligned}\right\} \quad (125)$$

where now Σ_μ refers to both right hand and left hand quanta, and where $\boldsymbol{\gamma}_\lambda$ commutes with all $\boldsymbol{\gamma}$ and all $\boldsymbol{\gamma}^*$ except $\boldsymbol{\gamma}_\lambda^*$,

$$[\boldsymbol{\gamma}_\mu, \boldsymbol{\gamma}_\mu^*]_- = \mathbf{1},\ \boldsymbol{\gamma}_\mu \to \begin{vmatrix} 0 & \sqrt{1} & 0 & . & . \\ 0 & 0 & \sqrt{2} & . & . \\ . & . & . & . & . \\ . & . & . & . & . \end{vmatrix},\ \boldsymbol{\gamma}_\mu^* \to \begin{vmatrix} 0 & 0 & . & . & . \\ \sqrt{1} & 0 & . & . & . \\ 0 & \sqrt{2} & . & . & . \\ . & . & . & . & . \end{vmatrix}.$$

Similarly we write instead of equations (121) and (122)

$$\Phi = \Sigma_\mu c_\mu(\mu), \qquad \Sigma_\mu|c_\mu|^2 = 1. \quad (126)$$

From equation (123) we now get

$$\mathfrak{E}_{\text{class}} = \tfrac{1}{2}\Omega^{-1/2}[\Sigma_\mu c_\mu\mathfrak{f}_\mu+\Sigma_\mu c_\mu^*\mathfrak{f}_\mu^*]. \quad (127)$$

The expectation value of $\mathfrak{E}_{\text{op}}$ is zero in each space-time point. This is often expressed by saying that in a light quantum *the phase of the light vector is completely undetermined*. This is not true for the classical field (127), but we know that the properties of a situation (126) do not change, if all c_μ are multiplied with the same phase factor, so that to that extent the phase of expression (127) is also undetermined.

The expectation value of $(\mathfrak{E}_{\mathrm{op}})^2$ can be shown to be given by the equation

$$\overline{(\mathfrak{E}_{\mathrm{op}})^2} = \tfrac{1}{2}\Omega^{-1}|\Sigma_\mu c_\mu \mathfrak{k}_\mu|^2 + \tfrac{1}{4}\Omega^{-1}\Sigma_\mu \mathfrak{k}_\mu^2. \tag{128}$$

The second term is infinite, but is formally exactly equal to the expectation value of a situation of the radiation field where there is no light quantum present. The occurrence of this term is closely connected with the zero-point energy. In expression (101) for the total energy H_{op} this term did no longer occur, but H_{op} was not equal to the classical expression $(8\pi)^{-1}\int(\mathfrak{E}^2+\mathfrak{H}^2)d\boldsymbol{x}$. The first term of expression (128) is closely connected with $(\mathfrak{E}_{\mathrm{class}})^2$; it corresponds to the cross products when expression (127) is squared. This means that in equation (128) the squares, $\frac{1}{4}\Omega^{-1}[\Sigma c_\mu \mathfrak{k}_\mu]^2$ and $\frac{1}{4}\Omega^{-1}[\Sigma c_\mu^* \mathfrak{k}_\mu^*]^2$, are omitted; they correspond classically to the periodic vibrations of frequencies $2\nu_\mu$ in $(\mathfrak{E}_{\mathrm{class}})^2$ [see eqq. (124) and (125)]. If all frequencies occurring in expression (127) are nearly equal($\doteqdot \nu_0$) and if the c_μ are chosen in such a way that, at least during a certain finite time interval, expression (127) is appreciably different from zero only in a finite region of space, expression (127) gives us a *classical wave packet* with approximately defined frequency and wave number. Our considerations show that the properties of such a packet illustrate, up to a point, the properties of a quantum mechanical light quantum. The distribution function of the operator $(\mathfrak{E}_{\mathrm{op}})^2$ is such that—as we saw—the expectation values of the higher powers of $(\mathfrak{E}_{\mathrm{op}})^2 - \frac{1}{4}\Omega^{-1}\Sigma_\mu|\mathfrak{k}_\mu|^2$ diverge. If we, however, consider the integral of the operators $\mathfrak{E}_{\mathrm{op}}$ or $(\mathfrak{E}_{\mathrm{op}})^2$ over a small region of space-time, instead of considering their values in a well defined space-time point,—and the integral corresponds to an actual measurement—these divergences will in general disappear.

If the wave function (121) is simply the sum of two terms, corresponding to the excitation of a right hand and a left hand quantum with the same wave vector, $\Phi = a_\lambda \cdot (\lambda) + b_\lambda \cdot (\bar{\lambda})$, $|a_\lambda|^2 + |b_\lambda|^2 = 1$, we are dealing with a situation which can be interpreted as being the presence of a (plane) *elliptically* ($|a_\lambda| \neq |b_\lambda|$) or *linearly* ($|a_\lambda| = |b_\lambda|$) polarised light quantum.

A light quantum which is polarised perpendicularly to the one just considered is represented by the function Φ' [$= -b_\lambda^* \cdot (\lambda) + a_\lambda^* \cdot (\bar{\lambda})$] which is orthogonal to Φ. If we want to describe the radiation field by means of this kind of quanta (instead of circularly polarised quanta) it is advisable to introduce instead of the α_λ, β_λ and their Hermitean conjugates the following new variables,

$$\delta_\lambda = a_\lambda^* \alpha_\lambda + b_\lambda^* \beta_\lambda,\ \delta_\lambda^* = a_\lambda \alpha_\lambda^* + b_\lambda \beta_\lambda^*,\ \epsilon_\lambda = -b_\lambda \alpha_\lambda + a_\lambda \beta_\lambda,\ \epsilon_\lambda^* = -b_\lambda^* \alpha_\lambda^* + a_\lambda^* \beta_\lambda^*,$$

$$[\delta_\lambda, \delta_\lambda^*]_- = [\epsilon_\lambda, \epsilon_\lambda^*]_- = 1, [\delta_\lambda, \epsilon_\lambda^*]_- = [\epsilon_\lambda, \delta_\lambda^*]_- = \ldots = 0,\ \delta_\lambda^* \delta_\lambda + \epsilon_\lambda^* \epsilon_\lambda = \alpha_\lambda^* \alpha_\lambda + \beta_\lambda^* \beta_\lambda^*$$

The δ_λ, ϵ_λ, δ_λ^*, and ϵ_λ^* are described by the same matrices as the α_λ, β_λ, . . . This corresponds to a certain unitary transformation of the old wave function $A(\ldots, n_\lambda, \ldots, \bar{n}_\lambda, \ldots, t)$ into a new wave function $A'(n_1, \ldots, m_\lambda, \ldots, \bar{n}_1, \ldots, \overline{m}_\lambda, \ldots, t)$, where the quantum numbers m_λ, $\overline{m}_\lambda$ for the wave vector σ_λ refer to quanta polarised in the two chosen directions. If we replace in expression (103) α_λ and β_λ by the δ_λ, ϵ_λ, δ_λ^*, and $\epsilon_\lambda^*$the new expression for the operator of the field, $\mathfrak{F}_{\text{op}}$, is clearly less simple than in the case of circularly polarised quanta. We could extend this kind of transformation to light quanta of other wave numbers; we could also use transformations to "mix" light quanta with different directions of the wave vector, that is, consider general linear transformations of the α_λ, β_λ. For certain problems (for instance, the case of the spherical spreading out of radiation from a point) such transformed representations may be useful. However, for interpreting spectroscopic measurements on a radiating system, the description using plane waves is usually preferable.

Let us consider as an example of one linearly polarised light quantum the case where $a_\lambda = b_\lambda = 2^{-\frac{1}{2}}$. The corresponding classical radiation field (123) would always have the electrical vector in the direction $\mathfrak{c}_\lambda + \mathfrak{c}_\lambda^*$. The same is true to some extent in the quantised radiation field. One can show, for instance, that the expectation value of the square of the component of $\mathfrak{E}$ in a direction $\boldsymbol{l}[(\boldsymbol{l} \cdot \boldsymbol{l}) = 1]$ is given by the equation

$$\overline{\mathfrak{E}_{\boldsymbol{l}}^2} = (2\pi hc/\Omega\sigma_\lambda)(\mathfrak{c}_\lambda + \mathfrak{c}_\lambda^* \cdot \boldsymbol{l})^2 + (4\pi hc/\Omega)\Sigma_{\lambda'}|(\mathfrak{c}_{\lambda'}^* \cdot \boldsymbol{l})|^2/\sigma_{\lambda'}.$$

The second term is infinite, but it is formally equal to the expectation value of $\mathfrak{E}_{\boldsymbol{l}}^2$ in a radiation field without light quanta (zero point energy!); the first term vanishes for all directions which are perpendicular to the classical direction of the electrical vector.

In fact, all classical properties of the radiation field have their correspondence analogue in the quantum theory of radiation. If there are many quanta present, the particular way in which the theory has been constructed guarantees that for a classical state of the radiation field there is a quantum mechanical counterpart, which for all practical purposes possesses the same properties as far as results of measurements are concerned, and in which the quantum of action plays practically no role. This can be shown explicitly, and it is, for instance, possible to show why in electromagnetic phenomena involving many quanta (such as radiowaves) one can measure the phase without having explicitly to take into account a quantum reaction of the measuring apparatus, so that in all practical applications the phase is thus completely defined. In the case of only one light quantum, which we discussed in this section, we cannot neglect the quantum theoretical aspects of the radiation field. There are certainly aspects which can be interpreted as the state of polarisation, the possibility of interference, the propagation in space, . . .; however, the finer detail of the situation cannot be derived from a classical "surrogate" field such as the field (127).

§ 88. Field theory and corpuscular theory of radiation

How far and how exactly can one consistently compare the radiation field with an ensemble of independent particles?

When in 1924 de Broglie suggested that material particles should show wave phenomena (see § 1) such a comparison was of great heuristic importance. Now that wave mechanics has become a consistent formalism one could ask whether it is possible to consider the Maxwell equations to be a kind of Schrödinger equation for light particles, instead of considering them, as we have done up to now, to be classical equations of motion which formally look like a wave equation, and which are quantised only later on; or are both ideas equivalent?

Let us first of all note the following. According to the latest investigation of the properties of the "vacuum" *the quantum theory of a charge free radiation field must in its present status certainly be considered to be an approximation.* After all, pairs of negatrons and positrons can be created and annihilated, and even when those processes do not take place, the consistent formalism requires deviations from the principle of the possibility of superposition of radiation fields (two "colliding" light quanta scatter each other *). These circumstances shed a different light on the physical meaning of the above questions.

To begin with we shall sketch—without considering at all questions of relativistic invariance—a formal point of view according to which the quantum mechanics of a system of independent Bose-Einstein particles would always be equivalent to a quantised field theory. Such a system is characterised by an operator wave function $\boldsymbol{\varphi}$ [see eq. (7.36)],

* W. Heisenberg, *Z. Phys.*, **90**, 209, 1934; H. Euler and H. Kockel, *Naturwiss.*, **23**, 246, 1935; H. Euler, *Ann. Phys.*, **26**, 398, 1935; W. Heisenberg and H. Euler, *Z. Phys.*, **98**, 714, 1936; V. Weisskopf, *Proc. Dan. Acad. Aci.*, **14**, No. 6, 1936.

$$\boldsymbol{\varphi} = \Sigma_k \boldsymbol{\alpha}_k \varphi_k(\boldsymbol{x}, s) \tag{129}$$

where the φ_k are the orthonormal eigenfunctions of the energy operator G_{op} of the one-particle problem. The variable s is a possibly occurring spin variable; we assume explicitly that the spin quantum is integral so that there will not be an undetermined factor ± 1 in the definition of φ_k. If the spin quantum number is equal to one, φ_k can be considered to be a vector field.

If $E_k = h\nu_k$ are the eigenvalues of the one-particle problem, the energy operator will be given by the equation, $H_{\text{op}} = h\Sigma_k \nu_k \boldsymbol{\alpha}_k^* \boldsymbol{\alpha}_k$. If the $\boldsymbol{\varphi}$ in equation (129) is generalised to be a time-dependent operator we can write

$$\boldsymbol{\psi}(\boldsymbol{x}, s, t) = \Sigma_k \boldsymbol{\alpha}_k \varphi_k(\boldsymbol{x}, s) \exp(-2\pi i \nu_k t) = \Sigma_k \boldsymbol{\alpha}_k \psi_k(\boldsymbol{x}, s, t). \tag{130}$$

This operator satisfies the Schrödinger equation of the one-particle problem. It can be split into two Hermitean parts, $\boldsymbol{\psi} = \boldsymbol{\psi}_1 + i\boldsymbol{\psi}_2$, each of which can be interpreted as observables in an extended sense. They commute neither with each other, nor with the number of particles $\Sigma_k \boldsymbol{\alpha}_k^* \boldsymbol{\alpha}_k$. In order that they can be measured, there must be measuring apparatus whose action is expressed by these operators.

Formally we can assign to the field $\boldsymbol{\psi}$ another field $\mathbf{F}$, whose absolute square $\Sigma_s \mathbf{F}^* \mathbf{F}$ can be considered to be the energy density,

$$\mathbf{F} = h^{1/2} \Sigma_k \nu_k^{1/2} \boldsymbol{\alpha}_k \psi_k, \qquad \int \Sigma_s \mathbf{F}^* \mathbf{F} dx = H_{\text{op}}. \tag{131}$$

This "energy field" $\mathbf{F}$ satisfies, as $\boldsymbol{\psi}$ does, the Schrödinger equation of the one-particle problem and is also constructed out of two observables, $\mathbf{F} = \mathbf{F}_1 + i\mathbf{F}_2$. If we assume that $\mathbf{F}_1$ and $\mathbf{F}_2$ are measurable, a situation involving many particles can to a fair approximation be represented by a purely classical field,

$$F_{\text{class}} = \Sigma_k c_k \psi_k, \qquad H_{\text{class}} = \Sigma_k |c_k|^2, \tag{132}$$

where the c_k are ordinary complex numbers corresponding to the $(h\nu_k)^{1/2}\boldsymbol{\alpha}_k$. The quantum-representation (131) of the field can again be obtained by "quantising" equation (132).

The analogy of this formal theory with the theory of the electromagnetic field is apparent. If we compare equations (131) and (103), we see, however, an essential difference, namely, that *in the electromagnetic field there occur apart from a set of operators* $\boldsymbol{\alpha}$ *the Hermitean conjugates of a second set of operators* $\boldsymbol{\beta}$. This complication is essential. It enables us to obtain *a relativistically invariant theory containing positive eigenvalues only,* and it makes possible the existence of a field which as far as its space-time properties are concerned is characterised by a *differential equation*. However, against these advantages we must put the disadvantage that there is no longer—as there was in the simple mechanical problem (130)—the *possibility* of defining simple operators representing the number and flux of particles. We must also mention here the disadvantage of the infinite zero-point energy.

To discuss these points in more detail it is advisable to distinguish in the radiation field two kinds of particle, the left hand and right hand (circularly polarised) ones. Let $\boldsymbol{\psi}_{\mathrm{I}}$ and $\boldsymbol{\psi}_{\mathrm{II}}$ be assigned to these two kinds,

$$\boldsymbol{\psi}_{\mathrm{I}} = \Sigma_\lambda \boldsymbol{\alpha}_\lambda \psi_{\mathrm{I}\lambda}, \quad \psi_{\mathrm{I}\lambda} = \Omega^{-\frac{1}{2}}(\mathbf{c}_\lambda/\sigma_\lambda)\exp.\{2\pi i[(\boldsymbol{\sigma}_\lambda\cdot\boldsymbol{x})-v_\lambda t]\},$$
$$\boldsymbol{\psi}_{\mathrm{II}} = \Sigma_\lambda \boldsymbol{\beta}_\lambda \psi_{\mathrm{II}\lambda}, \quad \psi_{\mathrm{II}\lambda} = \Omega^{-\frac{1}{2}}(\mathbf{c}^*_\lambda/\sigma_\lambda)\exp\{2\pi i[(\boldsymbol{\sigma}_\lambda\cdot\boldsymbol{x})-v_\lambda t]\}.$$

The $\psi_{\mathrm{I}\lambda}$ and $\psi_{\mathrm{II}\lambda}$ are normalised wave functions corresponding to the stationary states of the corresponding one-particle problem; each of them corresponds to a well defined value $h\boldsymbol{\sigma}_\lambda$ of the momentum vector of the light quantum. The energies of these states are all positive, $E_{\mathrm{I}\lambda} = E_{\mathrm{II}\lambda} = hc|\boldsymbol{\sigma}_\lambda| = hv_\lambda$. Just because of this the $\psi_{\mathrm{I}\lambda}$ and $\psi_{\mathrm{II}\lambda}$ are not uniquely defined as solutions of a differential equation. The simplest differential equation satisfied by them, $(\boldsymbol{\nabla}^2 - c^{-2}\partial^2/\partial t^2)\psi = 0$ has also solutions proportional to $\exp\{2\pi i[(\boldsymbol{\sigma}_\lambda\cdot\boldsymbol{x})+v_\lambda t]\}$. It is true that $\psi_{\mathrm{I}\lambda}$ and $\psi_{\mathrm{II}\lambda}$ multiplied by σ_λ are simply characterised in a relativistically invariant manner: under a Lorentz transformation the left hand (or right hand) ψ retains its sense of rotation. We shall now assume that the six-vector field $\sigma_\lambda\psi_{\mathrm{I}\lambda}$ transforms as $\mathfrak{E}+i\mathfrak{H}$, and the field $\sigma_\lambda\psi_{\mathrm{II}\lambda}$ as $\mathfrak{E}-i\mathfrak{H}$. Note that $\boldsymbol{\psi}_{\mathrm{I}}$ and $\boldsymbol{\psi}_{\mathrm{II}}$ themselves do not transform simply under a Lorentz transformation—they can by no means be compared to an $\mathfrak{E}+i\mathfrak{H}$ or an $\mathfrak{E}-i\mathfrak{H}$-field.

Only that combination of the one-particle solutions where each $\psi_{\mathrm{I}\lambda}$ and $\psi_{\mathrm{II}\lambda}$ is multiplied by $\sigma_\lambda^{\frac{1}{2}}$ (or $v_\lambda^{\frac{1}{2}}$) behaves as an $\mathfrak{E}+i\mathfrak{H}$-field, that is, a six-vector field [see eq. (103) and the relativistically invariant eq. (106)]. This gives us the "energy field"

$$\mathfrak{F}_{\mathrm{op}} = (8\pi)^{\frac{1}{2}}\Sigma_\lambda(hv_\lambda)^{\frac{1}{2}}[\boldsymbol{\alpha}_\lambda\psi_{\mathrm{I}\lambda}+\boldsymbol{\beta}^*_\lambda\psi^*_{\mathrm{II}\lambda}]. \tag{133}$$

This field is uniquely characterised by the differential equations (72) and (73). Their most general solution splits naturally into two parts I and II; this splitting is invariant with respect to the complete group of all spatial and time translations including all Lorentz transformations which do not involve an inversion.

The differential equations (72) and (73) of the energy field are thus not simply the Schrödinger equation of a one-particle problem; this equation does not exist (in the form of a differential equation), its role is taken over by equations (72) and (73) which might be called the *quasi-Schrödinger equations.*

If $I_{op}d\boldsymbol{x}$ denotes the number of particles in a volume element $d\boldsymbol{x}$, and if $(\boldsymbol{S}_{op} \cdot \boldsymbol{df})$ denotes the number of light particles passing per unit time through the area $\boldsymbol{df}$, we could use for I the equation, $I_{op} = \psi_I^* \psi_I + \psi_{II}^* \psi_{II}$, by analogy with the earlier expressions for the probability density. Indeed, $\int I_{op}d\boldsymbol{x}$, integrated over all space, gives us the total number of left hand- and right hand quanta. However, it is not possible to find an expression for $\boldsymbol{S}_{op}$ which satisfies the equation of continuity (1.50) and at the same time depends only on the values of ψ_I and ψ_{II} and their derivatives at the space-time point considered (see § 7). One should really require that the operators which can represent I_{op} and $\boldsymbol{S}_{op}$ in a space-time point can be derived from the measurable energy field (133) at that point; in that case even the above expression for I_{op} does not meet our requirements, because of the occurrence of factors $\nu_\lambda^{1/2}$ in equation (133). This is connected with the fact that in general I_{op} does not behave as the time component of a four-vector. The answer to the question put at the beginning of this section is thus that *one can not speak of particles in a radiation field in the same sense as in the (non-relativistic*) quantum mechanics of systems of point particles.*

We discussed earlier the energy density operator in the radiation field. If we write this operator in the form $(8\pi)^{-1}(\mathfrak{F}_{op}^* \cdot \mathfrak{F}_{op}) = (8\pi)^{-1}(\mathfrak{F}_{op} \cdot \mathfrak{F}_{op}^*) = (8\pi)^{-1}[\mathfrak{E}_{op}^2 + \mathfrak{H}_{op}^2]$ with $\mathfrak{F}_{op}$ given by equation (133), integration over space gives

$$H_{op} = \Sigma_\lambda h\nu_\lambda [\boldsymbol{\alpha}_\lambda^* \boldsymbol{\alpha}_\lambda + \boldsymbol{\beta}_\lambda \boldsymbol{\beta}_\lambda^*] = \Sigma_\lambda h\nu_\lambda [\boldsymbol{\alpha}_\lambda^* \boldsymbol{\alpha}_\lambda + \boldsymbol{\beta}_\lambda^* \boldsymbol{\beta}_\lambda (+1)]. \quad (134)$$

The $(+1)$ corresponds to the infinite zero-point energy which can simply be dropped [as in eq. (101)] without influencing the discussion of physical measurements.

We have, apart from the energy density and total energy, also the operators of the *momentum density* $(4\pi c)^{-1}[\mathfrak{E}_{op} \wedge \mathfrak{H}_{op}] = (-i/8\pi c) \cdot [\mathfrak{F}_{op}^* \wedge \mathfrak{F}_{op}]$, and of the total momentum $\boldsymbol{P}_{op}$,

* This restriction is necessary as Dirac's hole theory has shown that also in the relativistic treatment of electrons one cannot define rigorously a particle density and a particle current density.

$$\boldsymbol{P}_{\text{op}} = (-i/8\pi c)\int[\mathfrak{F}^*_{\text{op}} \wedge \mathfrak{F}_{\text{op}}]d\boldsymbol{x} = \Sigma_\lambda h\boldsymbol{\sigma}_\lambda[\boldsymbol{\alpha}^*_\lambda \boldsymbol{\alpha}_\lambda + \boldsymbol{\beta}_\lambda \boldsymbol{\beta}^*_\lambda] \\ = \Sigma_\lambda h\boldsymbol{\sigma}_\lambda[\boldsymbol{\alpha}^*_\lambda \boldsymbol{\alpha}_\lambda + \boldsymbol{\beta}^*_\lambda \boldsymbol{\beta}_\lambda(+1)]. \tag{135}$$

Here also we can drop the $(+1)$ corresponding to the zero-point momentum. The value of the infinite sum $\Sigma_\lambda h\boldsymbol{\sigma}_\lambda$ is undetermined. By using the notation (106), (107) we have

$$\boldsymbol{P}_{\text{op}} = \int h\boldsymbol{\sigma}[\mathbf{A}^*\mathbf{A} + \mathbf{B}^*\mathbf{B}](D\boldsymbol{\sigma}/\sigma)\{+ \text{ zero-point momentum}\}. \tag{136}$$

It is easily seen that, as it should be, $\boldsymbol{P}_{\text{op}}$ and H_{op}/c^2 [eq. (107)] form together a four-vector.

Apart from a factor $i\hbar$, $\boldsymbol{P}_{\text{op}}$ should be the operator of an infinitesimal translation (see § 61). Using equation (105) we find, indeed, that

$$[P_{x\text{op}}, \mathfrak{F}_{\text{op}}]_- = 4\pi(\hbar c)^{1/2}\int[-h\sigma_x \mathbf{A}\,\mathfrak{c} \exp\{2\pi i[(\boldsymbol{\sigma}\cdot\boldsymbol{x}) - \nu t]\} \\ + h\sigma_x \mathbf{B}^*\,\mathfrak{c}\exp\{-2\pi i[(\boldsymbol{\sigma}\cdot\boldsymbol{x}) - \nu t]\}](D\boldsymbol{\sigma}/\sigma) = i\hbar\partial\mathfrak{F}_{\text{op}}/\partial x.$$

Finally we have the *angular momentum density* operator (angular momentum with respect to the origin) which is equal to $(4\pi c)^{-1}[\boldsymbol{x} \wedge [\mathfrak{E}_{\text{op}} \wedge \mathfrak{H}_{\text{op}}]] = (-i/8\pi c)[\boldsymbol{x} \wedge [\mathfrak{F}^*_{\text{op}} \wedge \mathfrak{F}_{\text{op}}]]$. If we integrate this quantity over the whole of space, we get the *total angular momentum*. The expression for the total angular momentum can be expressed in the following form involving the operators $\mathbf{A}$, $\mathbf{B}$, and their derivatives with respect to σ_x, σ_y, and σ_z *,

$$\boldsymbol{J}_{\text{op}} = \frac{-i}{8\pi c}\int[\boldsymbol{x} \wedge [\mathfrak{F}^*_{\text{op}} \wedge \mathfrak{F}_{\text{op}}]]d\boldsymbol{x} = -i\hbar\int\{\mathbf{A}^*[\boldsymbol{\sigma}\wedge\boldsymbol{\nabla}_\sigma]\mathbf{A} + \mathbf{B}^*[\boldsymbol{\sigma}\wedge\boldsymbol{\nabla}_\sigma]\mathbf{B}\}\frac{D\boldsymbol{\sigma}}{\sigma}, \tag{137}$$

where $\boldsymbol{\nabla}_\sigma$ is the vector operator with components $\partial/\partial\sigma_x$, $\partial/\partial\sigma_y$, $\partial/\partial\sigma_z$.

The connexion between $\boldsymbol{J}_{\text{op}}$ and the operators of infinitesimal rotations is clear when we remember that the action of such a rotation on $\mathfrak{F}_{\text{op}}$ is such that a wave $\exp\{2\pi i[(\boldsymbol{\sigma}\cdot\boldsymbol{x}) - \nu t]\}$ which in expression (106) is multiplied by the operator $\mathbf{A}_{\boldsymbol{\sigma}}$ after the rotation will be multiplied by the operator $\mathbf{A}_{\boldsymbol{\sigma}'}$ where $\boldsymbol{\sigma}'$ is obtained from $\boldsymbol{\sigma}$ by an infinitesimal rotation, $\boldsymbol{\sigma}' = \boldsymbol{\sigma} + [\boldsymbol{\omega} \wedge \boldsymbol{\sigma}]$; we have thus $\mathbf{A}_{\boldsymbol{\sigma}'} = \mathbf{A}_{\boldsymbol{\sigma}} + (\boldsymbol{\omega}\cdot[\boldsymbol{\sigma}\wedge\boldsymbol{\nabla}_\sigma])\,\mathbf{A}_{\boldsymbol{\sigma}}$, and a similar relation for $\mathbf{B}_{\boldsymbol{\sigma}}$. Hence it follows that

$$[J_{x\text{op}}, \mathfrak{F}_{\text{op}}]_- = i\hbar\cdot 4\pi(\hbar c)^{1/2}\int\{[\boldsymbol{\sigma}\wedge\boldsymbol{\nabla}_\sigma]_x \mathbf{A}_{\boldsymbol{\sigma}}\,\mathfrak{c}_{\boldsymbol{\sigma}} \exp\{2\pi i[(\boldsymbol{\sigma}\cdot\boldsymbol{x}) - \nu t]\} \\ + [\boldsymbol{\sigma}\wedge\boldsymbol{\nabla}_\sigma]\,\mathbf{B}^*_{\boldsymbol{\sigma}}\,\mathfrak{c}_{\boldsymbol{\sigma}} \exp\{-2\pi i[(\boldsymbol{\sigma}\cdot\boldsymbol{x}) - \nu t]\}\}(D\boldsymbol{\sigma}/\sigma).$$

* By a formal integration by parts we get $\int\mathbf{A}^*[\boldsymbol{\sigma}\wedge\boldsymbol{\nabla}_\sigma]\mathbf{A}(D\boldsymbol{\sigma}/\sigma) = \int -[\boldsymbol{\sigma}\wedge\boldsymbol{\nabla}_\sigma\,\mathbf{A}^*\mathbf{A}(D\boldsymbol{\sigma}/\sigma)$. We see thus that the operator (137) is Hermitean.

This equation is satisfied—as can be checked—by expression (137). Apparently the $D\boldsymbol{\sigma}$ notation is more suited to the discussion of the angular momentum than is the representation by means of discrete sums Σ_λ. We do not wish to discuss the question of what is meant by operators of the kind $\partial\mathbf{A}/\partial\sigma_x$. This problem is connected with the definition of the wave-function in the $D\boldsymbol{\sigma}$-notation.

The eigenvalues of the angular momentum in a given direction are equal to $m\hbar$, those of the square of the total angular momentum are equal to $l(l+1)\hbar^2$ as is required by the general analysis of § 76. Here both m and l are integral since in the representation of a situation of the radiation field the wave function returns to its initial value when the system of coordinates is rotated continuously over 2π. The eigenfunctions of the angular momentum can be chosen to be also eigenfunctions of the energy and of the number of particles —since $\boldsymbol{J}_{\rm op}$ commutes with $H_{\rm op}$ and $N_{\rm op}$, but not with $\boldsymbol{P}_{\rm op}$. It is even possible to expand every situation characterised by an energy, which is defined by a definite number of light quanta, in terms of eigenfunctions of $J_{z\rm op}$ and $J^2_{\rm op}$. These eigenfunctions correspond to what one would call classically a superposition of incoming and outgoing multipole radiation. We do not wish to write them down explicitly.

§ 89. THE EQUATIONS OF THE CLASSICAL THEORY OF ELECTRONS IN CANONICAL FORM

The equations of the classical theory of electrons,

$$[\nabla \wedge \mathfrak{H}] = \frac{1}{c}\frac{\partial \mathfrak{E}}{\partial t} + \frac{4\pi\varrho \boldsymbol{v}}{c},\ (\nabla\cdot\mathfrak{E}) = 4\pi\varrho,\ [\nabla \wedge \mathfrak{E}] = -\frac{1}{c}\frac{\partial \mathfrak{H}}{\partial t},\ (\nabla\cdot\mathfrak{H}) = 0,$$

can be written in the following form,

$$[\nabla \wedge \mathfrak{F}] - (i/c)(\partial\mathfrak{F}/\partial t) = 4\pi i\varrho\boldsymbol{v}/c, \quad (\nabla\cdot\mathfrak{F}) = 4\pi\varrho, \qquad (138)$$

where the complex vector $\mathfrak{F} = \mathfrak{E} + i\mathfrak{H}$ is again introduced. As in equation (80) we split $\mathfrak{F}$ into three parts, a right hand (I), a left hand (II), and a longitudinal (III) part. We now also do the same with the current vector field $\varrho\boldsymbol{v}$,

$$\mathfrak{F} = \mathfrak{F}_{\rm I} + \mathfrak{F}_{\rm II} + \mathfrak{F}_{\rm III}, \qquad \varrho\boldsymbol{v} = (\varrho\boldsymbol{v})_{\rm I} + (\varrho\boldsymbol{v}_{\rm II}) + (\varrho\boldsymbol{v}_{\rm III}). \quad (139)$$

These equations are defined independently of the orientation of the spatial system of coordinates. If, however, we go over to a moving system of reference by means of a Lorentz transformation the splitting up in the new system of coordinates is in general not simply connected to the splitting up in the original system of coordinates. The *radiation field* $\mathfrak{F}_{\mathrm{I}}+\mathfrak{F}_{\mathrm{II}}$ is thus defined mathematically precisely, but not in a Lorentz-invariant manner.

If we substitute expressions (139) into equations (138) we get

$$\operatorname{curl} \mathfrak{F}_{\mathrm{I}} - \frac{i}{c}\frac{\partial \mathfrak{F}_{\mathrm{I}}}{\partial t} = \frac{4\pi i}{c}(\varrho \boldsymbol{v})_{\mathrm{I}}, \qquad \operatorname{div} \mathfrak{F}_{\mathrm{I}} = 0, \tag{140}$$

$$\operatorname{curl} \mathfrak{F}_{\mathrm{II}} - \frac{i}{c}\frac{\partial \mathfrak{F}_{\mathrm{II}}}{\partial t} = \frac{4\pi i}{c}(\varrho \boldsymbol{v})_{\mathrm{II}}, \qquad \operatorname{div} \mathfrak{F}_{\mathrm{II}} = 0, \tag{141}$$

$$-\frac{i}{c}\frac{\partial \mathfrak{F}_{\mathrm{III}}}{\partial t} = \frac{4\pi i}{c}(\varrho \boldsymbol{v})_{\mathrm{III}}, \quad \operatorname{div} \mathfrak{F}_{\mathrm{III}} = 4\pi\varrho, \quad (\operatorname{curl} \mathfrak{F}_{\mathrm{III}} = 0). \tag{142}$$

According to equations (142) the real field $\mathfrak{F}_{\mathrm{III}}$ is uniquely determined by the real charge distribution ϱ. It is just the electrostatic field which is produced by the charges ϱ and which can be expressed by a Coulomb potential Φ',

$$\mathfrak{F}_{\mathrm{III}} = \mathfrak{E}_{\mathrm{III}} = -\nabla\Phi', \qquad \Phi'_{\mathrm{P}} = \int (\varrho_{\mathrm{Q}}/r_{\mathrm{PQ}})\, d\boldsymbol{x}_{\mathrm{Q}}. \tag{143}$$

The time dependence of $\mathfrak{F}_{\mathrm{III}}$ is thus determined by the time dependence of ϱ. The first of equations (142) is nothing but the law of conservation of charge, $\operatorname{div}(\varrho\boldsymbol{v})_{\mathrm{III}} = \operatorname{div}(\varrho\boldsymbol{v}) = -\partial\varrho/\partial t$, and thus $(\varrho\boldsymbol{v})_{\mathrm{III}}$ is also uniquely determined by $\partial\varrho/\partial t$.

The real equations of motion of the $\mathfrak{E}$, $\mathfrak{H}$ field are thus only equations (140) and (141). To write them in canonical form we expand $\mathfrak{F}_{\mathrm{I}}$ and $\mathfrak{F}_{\mathrm{II}}$ in plane waves as in equation (80),

$$\operatorname{curl} \mathfrak{F}_{\mathrm{I}} - \frac{i}{c}\frac{\partial \mathfrak{F}_{\mathrm{I}}}{\partial t} = \Omega^{-\frac{1}{2}} \sum_{\lambda} [2\pi\sigma_\lambda a_\lambda - \frac{i}{c}\dot{a}_\lambda]\mathfrak{c}_\lambda \exp[2\pi i(\boldsymbol{\sigma}_\lambda \cdot \boldsymbol{x})] = \frac{4\pi i}{c}(\varrho\boldsymbol{v})_{\mathrm{I}},$$

$$\operatorname{curl} \mathfrak{F}_{\mathrm{II}} - \frac{i}{c}\frac{\partial \mathfrak{F}_{\mathrm{II}}}{\partial t} = \Omega^{-\frac{1}{2}} \sum_{\lambda} [-2\pi\sigma_\lambda b_\lambda - \frac{i}{c}\dot{b}_\lambda]\mathfrak{c}_\lambda \exp[-2\pi i(\boldsymbol{\sigma}_\lambda \cdot \boldsymbol{x})] = \frac{4\pi i}{c}(\varrho\boldsymbol{v})_{\mathrm{II}}.$$

Multiplying these equations by $\mathfrak{c}_\lambda^* \exp[-2\pi i(\boldsymbol{\sigma}_\lambda \cdot \boldsymbol{x})]$ and $\mathfrak{c}_\lambda^* \exp[2\pi i(\boldsymbol{\sigma}_\lambda \cdot \boldsymbol{x})]$ respectively, and integrating over the cube of volume Ω we get

$$\left.\begin{aligned} \dot{a}_\lambda &= -2\pi i\nu_\lambda a_\lambda - (4\pi/\sigma_\lambda^2\Omega^{\frac{1}{2}})\int (\varrho\boldsymbol{v}\cdot\mathfrak{c}_\lambda^*) \exp[-2\pi i(\boldsymbol{\sigma}_\lambda \cdot \boldsymbol{x})]d\boldsymbol{x}, \\ \dot{b}_\lambda &= 2\pi i\nu_\lambda b_\lambda - (4\pi/\sigma_\lambda^2\Omega^{\frac{1}{2}})\int (\varrho\boldsymbol{v}\cdot\mathfrak{c}_\lambda^*) \exp[2\pi i(\boldsymbol{\sigma}_\lambda \cdot \boldsymbol{x})]d\boldsymbol{x}. \end{aligned}\right\} \tag{144}$$

In these equations we can drop the index of the $(\varrho\boldsymbol{v})$. These inhomogeneous equations (144) (and their complex conjugates) take the place of the homogeneous equations (87) and express the influence of the matter on the radiation field. If $\varrho\mathbf{v}$ is known as a function of $\mathbf{x}$, t, equations (144) can be solved by quadrature. By substituting $a_\lambda(t) = a_\lambda^0(t) \exp(-2\pi i\nu_\lambda t)$, $b_\lambda(t) = b_\lambda^0(t) \exp(2\pi i\nu_\lambda t)$ we get

$$\dot{a}_\lambda^0 = -(4\pi/\sigma_\lambda^2\Omega^{1/2})\int(\varrho\mathbf{v}\cdot\mathfrak{c}_\lambda^*)\exp\{-2\pi i[(\boldsymbol{\sigma}_\lambda\cdot\mathbf{x})-\nu_\lambda t]\}d\mathbf{x},$$

$$a_\lambda^0(t)-a_\lambda^0(0) = -(4\pi/\sigma_\lambda^2\Omega^{1/2})\int_0^t\left[\int(\varrho\mathbf{v}\cdot\mathfrak{c}_\lambda^*)\exp\{-2\pi i[(\boldsymbol{\sigma}_\lambda\cdot\mathbf{x})-\nu_\lambda t]\}d\mathbf{x}\right]dt,$$

and similar equations for the $b_\lambda^0(t)$.

The last equation expresses the fact that the amplitudes of the eigenvibrations of empty space change gradually under the influence of a current distribution with a non vanishing curl. According to whether the absolute value of a_λ increases or decreases there is emission or absorption of right hand circularly polarised light with a wave vector $\boldsymbol{\sigma}_\lambda$ *. The contents of this equation are mathematically equivalent to the description of the radiation by means of retarded (or accelerated) potentials (compare § 83). This equation is particularly suited to determine the field at t if it were known at t_0, and required knowledge of $\varrho\mathbf{v}$ during the time interval (t_0, t).

The equations (144) must now be identified with canonical equations corresponding to an energy [see eq. (86)],

$$H = (8\pi)^{-1}\Sigma_\lambda\sigma_\lambda^2(a_\lambda^* a_\lambda+b_\lambda^* b_\lambda)+(8\pi)^{-1}\int(\mathfrak{F}_{\text{III}}^*\cdot\mathfrak{F}_{\text{III}})d\mathbf{x}+H'. \quad (145)$$

The first term is the contribution to the energy of the electromagnetic field from the transverse waves, and the second term the contribution of the "longitudinal" or "Coulomb" field (143). The extra term H' must, in a suitable manner, take care of the interaction between the radiation field and that part of the current density whose divergence vanishes.

To obtain the canonical form we first of all introduce again the variables p_λ, q_λ, $\bar{p}_\lambda$, and $\bar{q}_\lambda$ defined by equations (89), and we assume that now also as in the case of the charge- and current free field these quantities will play the role of canonically conjugate variables**. Remembering equations (144) we expect that it will be simpler not

* If the $\varrho\mathbf{v}$-field is constant in time, equations (144) possess the simple solution of constant a_λ and b_λ corresponding to a stationary magnetic field. This belongs, in our present terminology, also to the radiation field.

** It is by no means obvious that this will be possible. In classical mechanics, for instance, in the case of a one-particle system the coordinates and the components of the (kinetic) momentum are no longer canonically conjugate when an external magnetic field is present (see § 25).

to split the a_λ, a_λ^*, b_λ, and b_λ^* into their real and imaginary parts, but to form immediately canonically conjugate pairs out of them by multiplication by constants. The following pairs should behave as p and q, $\{i\varkappa_\lambda a_\lambda^*, \varkappa_\lambda a_\lambda\}$ and $\{i\varkappa_\lambda b_\lambda, \varkappa_\lambda b_\nu^*\}$, where $\varkappa_\lambda = (\sigma_\lambda/16\pi^2 c)^{1/2}$, that is, the following equations should hold

$$\begin{aligned} i\varkappa_\lambda \dot{a}_\lambda^* &= -\frac{\partial H}{\partial(\varkappa_\lambda a_\lambda)}, & \varkappa_\lambda \dot{a}_\lambda &= \frac{\partial H}{\partial(i\varkappa_\lambda a_\lambda^*)}; \\ i\varkappa_\lambda \dot{b}_\lambda &= -\frac{\partial H}{\partial(\varkappa_\lambda b_\lambda^*)}, & \varkappa_\lambda \dot{b}_\lambda^* &= \frac{\partial H}{\partial(i\varkappa_\lambda b_\lambda)}. \end{aligned} \tag{146}$$

If we use for H expression (145), the first term gives, indeed, the equations of motion for the empty field [$\varrho \mathbf{v} = 0$ in eq. (144)].

We assume now, to begin with, that the charge- and current density are given functions of $\mathbf{x}$, t, that is, we treat them as external parameters; we do not want to discuss at this moment the influence of the $\mathfrak{E}$, $\mathfrak{H}$-field on them. In this case the second term in equation (145), the Coulomb energy, will not depend on the a, a^*, b, and b^*. The third term, H', finally will be linear in those variables, in order that equations (146) will lead to the terms in equations (144) containing $\varrho\mathbf{v}$, or, $H' = \Sigma_\lambda(R_\lambda a_\lambda + R_\lambda^* a_\lambda^* + S_\lambda b_\lambda + S_\lambda^* b_\lambda^*)$. Comparing equations (144) and (146) we find that

$$R_\lambda = (i/4\pi c\sigma_\lambda \Omega^{1/2})\int(\varrho\mathbf{v}\cdot\mathfrak{c}_\lambda)\exp[2\pi i(\boldsymbol{\sigma}_\lambda\cdot\mathbf{x})]\,d\mathbf{x},$$

$$S_\lambda = (-i/4\pi c\sigma_\lambda \Omega^{1/2})\int(\varrho\mathbf{v}\cdot\mathfrak{c}_\lambda)\exp[-2\pi i(\boldsymbol{\sigma}_\lambda\cdot\mathbf{x})]\,d\mathbf{x},$$

$$\begin{aligned} H' = \operatorname{Re} i\Omega^{-1/2}c^{-1}\int(\varrho\mathbf{v}\cdot\Sigma_\lambda(2\pi\sigma_\lambda)^{-1}\{a_\lambda\mathfrak{c}_\lambda\exp[2\pi i(\boldsymbol{\sigma}_\lambda\cdot\mathbf{x})] \\ -b_\lambda\mathfrak{c}_\lambda\exp[-2\pi i(\boldsymbol{\sigma}_\lambda\cdot\mathbf{x}]\})\,d\mathbf{x}. \end{aligned} \tag{147}$$

The sum within braces is clearly very closely connected with the sums (80) for $\mathfrak{F}_{\mathrm{I}}$ and $\mathfrak{F}_{\mathrm{II}}$. If we calculate by differentiating each term the curl of the vector defined by the sum within braces we see, using equations (85), that we have, indeed, exactly the sum of $\mathfrak{F}_{\mathrm{I}}$ and $\mathfrak{F}_{\mathrm{II}}$. We have thus

$$H' = (i/2c)\int(\varrho\mathbf{v}\cdot\mathfrak{B}-\mathfrak{B}^*)d\mathbf{x}, \quad \operatorname{curl}\mathfrak{B} = \mathfrak{F}_{\mathrm{I}}+\mathfrak{F}_{\mathrm{II}} \equiv \mathfrak{F}_{\mathrm{I,\,II}}, \quad \operatorname{div}\mathfrak{B} = 0.$$

In view of the meaning of $\mathfrak{F}$ [note that $\mathfrak{H}_{\mathrm{III}} = 0$, because of eq. (143)] we have curl $(\mathfrak{B}-\mathfrak{B}^*) = 2i(\mathfrak{H}_{\mathrm{I}}+\mathfrak{H}_{\mathrm{II}}) = 2i\mathfrak{H}$. We can write thus

$$H' = -c^{-1}\int(\varrho\mathbf{v}\cdot\mathfrak{A}')d\mathbf{x}; \quad \operatorname{curl}\mathfrak{A}' = \mathfrak{H}, \quad \operatorname{div}\mathfrak{A}' = 0, \tag{148}$$

where the real vector field $\mathfrak{A}'$ is the vector potential with vanishing divergence of the magnetic field,

$$\mathfrak{A}' = (i/2\Omega^{1/2})\Sigma_\lambda (2\pi\sigma_\lambda)^{-1}\{-a_\lambda \mathfrak{c}_\lambda \exp[2\pi i(\boldsymbol{\sigma}_\lambda \cdot \boldsymbol{x})] + a_\lambda^* \mathfrak{c}_\lambda^* \exp[-2\pi i(\boldsymbol{\sigma}_\lambda \cdot \boldsymbol{x})] + b_\lambda \mathfrak{c}_\lambda \exp[-2\pi i(\boldsymbol{\sigma}_\lambda \cdot \boldsymbol{x})] - b_\lambda^* \mathfrak{c}_\lambda^* \exp[2\pi i(\boldsymbol{\sigma}_\lambda \cdot \boldsymbol{x})]\}. \quad (149)$$

We have thus quite naturally obtained the equations of the classical theory of electrons in the form of a set of canonical equations of motion with an infinite number of variables. These variables refer exclusively to the "radiation field", that is, to that part of the $\mathfrak{E}$, $\mathfrak{H}$-field whose divergence vanishes. The term (148) in the energy function (145) describes how the material charges influence this field.

There is not much sense in introducing at this point the quantisation of the radiation field, since in view of possible applications in atomic theory we must first express the atomistic structure of electricity. It would be desirable to find a formulation where the change in time of the total system, field and electrically charged particles, was described by means of canonical equations of motion; these equations could then easily be quantised. However, *an exact solution of this problem is impossible.* This is connected with the well-known difficulties which the field-particle dualism introduces in the classical theory of electrons. It is, however, not difficult to obtain in the frame-work of non-relativistic mechanics an approximate solution which treats only the *secular radiation effects.* From this theory one obtains the Dirac radiation theory by quantisation. The rigorously relativistically invariant methods of present-day quantum electrodynamics are also an approximate theory, in the same sense as Dirac's theory.

We consider again equations (144) and note that only inside a charged particle the $\varrho\boldsymbol{v}$-field will be different from zero so that the integrals in equations (144) become sums over the separate charged particles which are present,

$$\dot{a}_\lambda = -2\pi i\nu_\lambda a_\lambda - (4\pi/\sigma_\lambda^2\Omega^{1/2})\Sigma_i e_i(\boldsymbol{v}_i \cdot \mathfrak{c}_\lambda^*)\exp[-2\pi i(\boldsymbol{\sigma}_\lambda \cdot \boldsymbol{x}_i)], \quad (150)$$

and a similar equation for $\dot{b}_\lambda$, where e_i and $\boldsymbol{v}_i$ are the charge and velocity of the i-th particle.

If we assume that the particles have a finite extension of the order of magnitude R, equation (150) will be valid only for wave vectors satisfying the inequality $\sigma_\lambda \ll R^{-1}$. If, for instance, we assume $R \doteqdot 10^{-13}$ cm for an electron, in agreement with the classical theory of an electromagnetic mass, we could not trust results involving wave numbers of the order of magnitude 10^{13} or larger. We shall see, however, that in many applications such large wave numbers do not occur.

According to classical mechanics the velocity of a particle is expressed as a function of the coordinates and the canonically conjugate momenta of the particles by the following equation,

$$v_{xi} = \partial H^{(\mathrm{mat})}/\partial p_{xi}, \ \ldots \qquad (v_i \doteq m_i^{-1}[\boldsymbol{p}_i - (e_i/c)\mathfrak{A}_i^{\mathrm{ext}}]), \tag{151}$$

where $H^{(\mathrm{mat})}$ is the Hamiltonian corresponding to the material particles. The expression $m_i^{-1}[\boldsymbol{p}_i - (e_i/c)\mathfrak{A}_i^{\mathrm{ext}}]$ appears in particular when we neglect relativistic details [see eqq. (2.102), (3.76)]. No matter what is the form of $H^{(\mathrm{mat})}$ we know from the requirement of gauge invariance (see § 62) that $H^{(\mathrm{mat})}$ contains the $\boldsymbol{p}_i$ always as a function of the kinetic momentum $\boldsymbol{\pi}_i = \boldsymbol{p}_i - (e_i/c)\mathfrak{A}_i^{\mathrm{ext}}$, where $\mathfrak{A}_i^{\mathrm{ext}}$ is the vector potential of the *external* electromagnetic field at the position of the i-th particle. (Note that in general div $\mathfrak{A}_i^{\mathrm{ext}}$ will not be equal to zero.) Equation (151) can thus also be written in the form

$$v_{xi} = -(c/e_i)\partial H^{(\mathrm{mat})}/\partial \mathfrak{A}_{xi}^{\mathrm{ext}}. \tag{152}$$

We are now faced with the important question, which part of the *total* electromagnetic field—the vector potential of which was given by equation (149) and had a vanishing divergence—can be ascribed to the external field? We shall define uniquely the *eigenfield* of the i-th particle, that is, the *field which is dragged along* by it, as the field which in a system of reference in which the particle is at rest is simply the electrostatic field of the point charge at rest. This means that at any time it is the electromagnetic field of a particle with the same values of $\boldsymbol{x}_i$ and $\dot{\boldsymbol{x}}_i$ as the particle considered. The *eigenfield of the whole system* is the sum of the eigenfields of all the particles separately. We define now the *external radiation field* $\mathfrak{F}^{\mathrm{ext}}$ by the equation

$$\mathfrak{F}^{\mathrm{ext}} = \mathfrak{F}^{\mathrm{total}} - \mathfrak{F}^{\mathrm{eigen}}, \tag{153}$$

and similarly we write for the (transverse) vector potential, $\mathfrak{A}'^{\mathrm{ext}} = \mathfrak{A}'^{\mathrm{total}} - \mathfrak{A}'^{\mathrm{eigen}}$.

The total field and the eigenfield are both infinite at the positions of the particles, if we consider them simply to be point charges. However, $\mathfrak{H}^{\mathrm{ext}}$ and $\partial\mathfrak{A}'^{\mathrm{ext}}/\partial t$ are both finite, while $\mathfrak{E}^{\mathrm{ext}}$ (which is *not* equal to $-c^{-1}\partial\mathfrak{A}'^{\mathrm{ext}}/\partial t$) contains a term proportional to $\ddot{\boldsymbol{x}}_i$ which behaves as r^{-1} at a distance r from the particle. The longitudinal part of the external field is rigorously equal to zero since div $\mathfrak{F}_{\mathrm{III}}^{\mathrm{total}}$ = div $\mathfrak{F}_{\mathrm{III}}^{\mathrm{eigen}}$. We take the following series expansions for the transverse parts of $\mathfrak{F}^{\mathrm{ext}}$, $\mathfrak{A}'^{\mathrm{ext}}$, $\mathfrak{F}^{\mathrm{eigen}}$, and $\mathfrak{A}'^{\mathrm{eigen}}$ [see eqq. (80), (149)],

$$\left.\begin{aligned}\mathfrak{F}^{\text{ext}}_{\text{I, II}} &= \Omega^{-1/2}\Sigma_\lambda\{a'_\lambda \mathfrak{c}_\lambda \exp[2\pi i(\boldsymbol{\sigma}_\lambda \cdot \boldsymbol{x})] + b'_\lambda \mathfrak{c}_\lambda \exp[-2\pi i(\boldsymbol{\sigma}_\lambda \cdot \boldsymbol{x})]\},\\ \mathfrak{A}'^{\text{ ext}}_{\text{I, II}} &= \tfrac{1}{2} i\Omega^{-1/2}\Sigma_\lambda(2\pi\sigma_\lambda)^{-1}\{-a'_\lambda \mathfrak{c}_\lambda \exp[2\pi i(\boldsymbol{\sigma}_\lambda \cdot \boldsymbol{x})] + \ldots\},\end{aligned}\right\} \quad (154)$$

$$\left.\begin{aligned}\mathfrak{F}^{\text{eigen}}_{\text{I, II}} &= \Omega^{-1/2}\Sigma_\lambda\{a''_\lambda \mathfrak{c}_\lambda \exp[2\pi i(\boldsymbol{\sigma}_\lambda \cdot \boldsymbol{x})] + b''_\lambda \mathfrak{c}_\lambda \exp[-2\pi i(\boldsymbol{\sigma}_\lambda \cdot \boldsymbol{x})]\},\\ \mathfrak{A}'^{\text{ eigen}}_{\text{I, II}} &= \tfrac{1}{2} i\Omega^{-1/2}\Sigma_\lambda(2\pi\sigma_\lambda)^{-1}\{-a''_\lambda \mathfrak{c}_\lambda \exp[2\pi i(\boldsymbol{\sigma}_\lambda \cdot \boldsymbol{x})] + \ldots\}.\end{aligned}\right\} \quad (155)$$

The $a''(t)$ are obtained by solving the following differential equations [see eq. (150)],

$$\left.\begin{aligned}&d\,a''_\lambda(\tau)/d\tau = -2\pi i\,\nu_\lambda a''_\lambda(\tau) - (4\pi/\sigma_\lambda^2\Omega^{1/2})\Sigma_i(e_i\,\boldsymbol{v}_i \cdot \mathfrak{c}_\lambda^*)\exp[-2\pi i(\boldsymbol{\sigma}_\lambda \cdot \boldsymbol{x}_i)],\\ &d\,b''_\lambda(\tau)/d\tau = \ldots;\ \boldsymbol{v}_i = \boldsymbol{v}_i(t) = \text{constant},\ \boldsymbol{x}_i(\tau) = \boldsymbol{x}_i(t) + (\tau - t)\boldsymbol{v}_i.\end{aligned}\right\} \quad (156)$$

These equations can easily be integrated. The eigenfield clearly corresponds to the following solution

$$\left.\begin{aligned}a''_\lambda &= \frac{2i}{c\Omega^{1/2}}\sum_i \frac{e_i(\boldsymbol{v}_i \cdot \mathfrak{c}_\lambda^*)\exp[-2\pi i(\boldsymbol{\sigma}_\lambda \cdot \boldsymbol{x}_i)]}{\sigma_\lambda^2[\sigma_\lambda - c^{-1}(\boldsymbol{v}_i \cdot \boldsymbol{\sigma}_\lambda)]},\\ b''_\lambda &= \frac{-2i}{c\Omega^{1/2}}\sum_i \frac{e_i(\boldsymbol{v}_i \cdot \mathfrak{c}_\lambda^*)\exp[2\pi i(\boldsymbol{\sigma}_\lambda \cdot \boldsymbol{x}_i)]}{\sigma_\lambda^2[\sigma_\lambda - c^{-1}(\boldsymbol{v}_i \cdot \boldsymbol{\sigma}_\lambda)]}.\end{aligned}\right\} \quad (157)$$

The series (155) diverge at the positions of the particles.

From equation (154) it follows that

$$\mathfrak{c}_\lambda^*\exp[-2\pi i(\boldsymbol{\sigma}_\lambda \cdot \boldsymbol{x})] = -4\pi i\sigma_\lambda\Omega^{1/2}\partial\mathfrak{A}'^{\text{ ext}}/\partial a'^*_\lambda = -4\pi i\sigma_\lambda\Omega^{1/2}\partial\mathfrak{A}^{\text{ext}}/\partial a'^*_\lambda,$$
$$\mathfrak{c}_\lambda^*\exp[2\pi i(\boldsymbol{\sigma}_\lambda \cdot \boldsymbol{x})] = 4\pi i\sigma_\lambda\Omega^{1/2}\partial\mathfrak{A}^{\text{ext}}/\partial b'^*_\lambda.$$

We may here replace $\mathfrak{A}'^{\text{ ext}}$ by $\mathfrak{A}^{\text{ext}}$ since the difference between the two is independent of the a'_λ and b'_λ.

This means that we can use equation (152) to write the sums in equation (150) in the form

$$\begin{aligned}\sum_i e_i(\boldsymbol{v}_i \cdot \mathfrak{c}_\lambda^*)\exp[-2\pi i(\boldsymbol{\sigma}_\lambda \cdot \boldsymbol{x}_i)] &= 4\pi i c\sigma_\lambda\Omega^{1/2}\sum_i\left\{\frac{\partial H^{(\text{mat})}}{\partial\mathfrak{A}^{\text{ext}}_{xi}}\frac{\partial\mathfrak{A}^{\text{ext}}_{xi}}{\partial a'^*_\lambda} + (y) + (z)\right\}\\ &= 4\pi i c\sigma_\lambda\Omega^{1/2}\partial H^{(\text{mat})}/\partial a'^*_\lambda,\\ \Sigma_i e_i(\boldsymbol{v}_i \cdot \mathfrak{c}_\lambda^*)\exp[2\pi i(\boldsymbol{\sigma}_\lambda \cdot \boldsymbol{x}_i)] &= -4\pi i c\sigma_\lambda\Omega^{1/2}\partial H^{(\text{mat})}/\partial b'^*_\lambda.\end{aligned}$$

The differential equation (150) for the total $\mathfrak{E}$, $\mathfrak{H}$ field can now be written in the form of a differential equation for the external field,

$$\left.\begin{aligned}\dot{a}'_\lambda &= -2\pi i\,\nu_\lambda a'_\lambda - (16\pi^2 ic/\sigma_\lambda)(\partial H^{(\text{mat})}/\partial a'^*_\lambda) - (\dot{a}''_\lambda + 2\pi i\,\nu_\lambda a''_\lambda),\\ \dot{b}'_\lambda &= 2\pi i\,\nu_\lambda b'_\lambda + (16\pi^2 ic/\sigma_\lambda)(\partial H^{(\text{mat})}/\partial b'^*_\lambda) - (\dot{b}''_\lambda - 2\pi i\,\nu_\lambda b''_\lambda).\end{aligned}\right\} \quad (158)$$

The expression $\dot{a}''_\lambda + 2\pi i\nu_\lambda a''_\lambda$ would be exactly equal to $-(16\pi^2 ic/\sigma_\lambda)$

$(\partial H^{(\mathrm{mat})}/\partial a_\lambda'^*)$ [and a similar relation would hold for b_λ], if the accelerations of the particles vanished. We have thus

$$\ddot{a}_\lambda' + 2\pi i\nu_\lambda \dot{a}_\lambda' = -\sum_i \left[\frac{\partial a_\lambda''}{\partial v_{xi}}\dot{v}_{xi} + \frac{\partial a_\lambda''}{\partial v_{yi}}\dot{v}_{yi} + \frac{\partial a_\lambda''}{\partial v_{zi}}\dot{v}_{zi}\right]$$

$$= \frac{-2i}{c\sigma_\lambda^2 \Omega^{1/2}} \sum_i \left\{\frac{(\dot{\mathbf{v}}_i \cdot \mathfrak{c}_\lambda^*)}{[\sigma_\lambda - c^{-1}(\mathbf{v}_i \cdot \boldsymbol{\sigma}_\lambda)]} + \frac{(\mathbf{v}_i \cdot \mathfrak{c}_\lambda^*)(\dot{\mathbf{v}}_i \cdot \boldsymbol{\sigma}_\lambda)}{c[\sigma_\lambda - c^{-1}(\mathbf{v}_i \cdot \boldsymbol{\sigma}_\lambda)]^2}\right\} \exp[-2\pi i(\boldsymbol{\sigma}_\lambda \cdot \mathbf{x}_i)]$$

$$\doteq -(2i/c\sigma_\lambda^2 \Omega^{1/2}) \Sigma_i e_i (\dot{\mathbf{v}}_i \cdot \mathfrak{c}_\lambda^*) \exp[-2\pi i(\boldsymbol{\sigma}_\lambda \cdot \mathbf{x}_i)] \qquad (159)$$

and a similar equation for b_λ.

Equation (159) expresses the well known classical result that the radiation field is produced by the accelerations of the particles. It is possible to show that the present equations are equivalent to equations (140) and (141) specialised to the case of discrete particles.

Unfortunately equations (159) are not suited to the transition to a set of canonical equations. It would not be difficult to use them to obtain canonical equations for the radiation field itself. These would, however, refer only to a prescribed motion of the particles, and they could not be extended in a useful manner to a set of canonical equations for the total system which contains the action of the radiation field on the particles. To obtain such a set we return to equations (158) and look for the *secular* changes of the radiation field. We write equation (158) in the following form,

$$\frac{d}{dt}(a_\lambda' \exp[2\pi i\nu_\lambda t]) = -i\frac{16\pi^2 c}{\sigma_\lambda}\frac{\partial H^{(\mathrm{mat})}}{\partial a_\lambda'^*}\exp(2\pi i\nu_\lambda t) - \frac{d}{dt}(a_\lambda''\exp[2\pi i\nu_\lambda t]) \qquad (160)$$

If we may assume that the velocities of the particles remain finite, the quantity $a_\lambda'' \exp(2\pi i\nu_\lambda t)$ will remain bounded. If we are interested only in the secular change of a_λ', we may drop the term $(d/dt)[a_\lambda''\exp(2\pi i\nu_\lambda t)]$; its average value may be taken to be zero *. The fast periodic changes in $a_\lambda' \exp(2\pi i\nu_\lambda t)$ are now not correctly described. A consequence of this is that there will occur incorrect terms in the description of the motion of the particles—which depends on $\mathfrak{A}'^{\,\mathrm{ext}}$. These terms in turn react on the evaluation of $a_\lambda(t), \ldots$ Because of the fact that the interaction between radiation field and matter is "weak", we may, however, neglect these effects.

* The transition to the secular equations can up to a point be compared with the classical, non-relativistic spin treatment of § 57. As in that case we should expect that *also here a rigorous, relativistically invariant treatment will only be possible in the framework of a new, purely quantum theoretical formalism.*

In view of the fact that there is often present a constant or slowly changing magnetic field—which as we mentioned earlier will be considered to be part of the eigenfield it is advisable not to drop the last term on the right hand side of equation (160) completely but to replace it by $(d/dt)(\bar{a}''_\lambda \exp[2\pi i \nu_\lambda t])$ where $\bar{a}''_\lambda$ is the time average of a''_λ.

In that case we get the secular radiation equations in the following form,

$$\left.\begin{aligned} \dot{a}'_\lambda &= -2\pi i \nu_\lambda a'_\lambda - (16\pi^2 ic/\sigma_\lambda)(\partial H^{(\text{mat})}/\partial a'^*_\lambda) - 2\pi i \nu_\lambda \bar{a}''_\lambda, \\ \dot{b}'_\lambda &= 2\pi i \nu_\lambda b'_\lambda + (16\pi^2 ic/\sigma_\lambda)(\partial H^{(\text{mat})}/\partial b'^*_\lambda) + 2\pi i \nu_\lambda \bar{b}''_\lambda. \end{aligned}\right\} \quad (161)$$

These equations are incorrect for short time intervals, but they give us the correct total change in $a'_\lambda \exp(2\pi i \nu_\lambda t)$ and $b'_\lambda \exp(-2\pi i \nu_\lambda t)$ over long periods.

Equations (161) can immediately be written in the canonical form (146), if we take for the energy function

$$\left.\begin{aligned} H &= (8\pi)^{-1} \Sigma_\lambda \sigma_\lambda^2 (a'_\lambda a'^*_\lambda + b'^*_\lambda b'_\lambda) \\ &\qquad + (8\pi)^{-1} \Sigma_\lambda \sigma_\lambda^2 (a'^*_\lambda \bar{a}''_\lambda + \bar{a}''^*_\lambda a'_\lambda + b'^*_\lambda \bar{b}''_\lambda + \bar{b}''_\lambda b'_\lambda) + H^{(\text{mat})} \\ &= (8\pi)^{-1} \int (\mathfrak{F}^{*\text{ext}} \cdot \mathfrak{F}^{\text{ext}}) d\boldsymbol{x} + (8\pi)^{-1} \int [(\mathfrak{F}^{*\text{ext}} \cdot \bar{\mathfrak{F}}^{\text{eigen}}) \\ &\qquad + (\bar{\mathfrak{F}}^{*\text{eigen}} \cdot \mathfrak{F}^{\text{ext}})] d\boldsymbol{x} + H^{(\text{mat})}. \end{aligned}\right\} \quad (162)$$

The first term corresponds to the energy of the external radiation field. The second term derives from the interference of the external field with the eigenfield. In practical applications $\bar{\mathfrak{F}}^{\text{eigen}}$ will usually be equal to i times the average magnetic field of the system. During a period which is long enough for an appreciable exchange of energy between radiation and matter to take place, the $\bar{\mathfrak{F}}^{\text{eigen}}$-field may have changed. It is true that we can calculate this change from the canonical equations, but it is clear that these equations do not refer to a conservative system in the usual sense. The occurrence of average values remind us of the fact that we have neglected the details of the interaction between radiation and matter. Usually this term is dropped altogether corresponding to a practically constant shift of the a'_λ, b'_λ-values.

The last term of equation (162) contains through the vector potential of the external field the a'_λ and b'_λ and it also contains the coordinates and momenta of the material particles in such a way that in accordance with classical mechanics the motion of these particles follows from a set of canonical equations derived from

$H^{(\mathrm{mat})}$. The action of the radiation field on the particles is not described in all details by these equations, since the radiation field itself is correctly described only in as far as its secular aspects are concerned *.

We shall base our non-relativistic treatment of the interaction between radiation and matter on equation (162). In this treatment we may put both the longitudinal part of the vector potential of the external radiation field—which was hardly mentioned up to now—and the scalar potential of this field equal to zero. Usually one defines the vector potential $\mathfrak{A}$ which occurs in classical mechanics in such a way that it forms with the scalar potential Φ a transverse four-vector and thus satisfies the equation [see eq. (2.100)] $(\nabla \cdot \mathfrak{A})+c^{-1}\partial\Phi/\partial t = 0$. In the theory of radiation, however, it is advisable to use consistently the transverse vector potential $\mathfrak{A}'$ and the corresponding scalar potential Φ'. Equations (2.100) are thus replaced by the equations

$$\nabla^2 \Phi' = -4\pi\varrho, \quad \nabla^2 \mathfrak{A}' - c^{-2}\,\partial^2 \mathfrak{A}'/\partial t^2 = -4\pi\varrho\mathbf{v}/c + c^{-1}\,\partial\nabla\Phi'/\partial t,$$

and we have

$$\mathfrak{F}_{\mathrm{I,\,II}} = -c^{-1}(\partial\mathfrak{A}'/\partial t)+i[\nabla \wedge \mathfrak{A}'], \quad \mathfrak{F}_{\mathrm{III}} = -\nabla\Phi'.$$

This choice of the potentials is particularly suited to our division into an external and an eigenfield, which is not Lorentz invariant. Neither of these fields satisfy the Maxwell equations, but their longitudinal parts are exactly the same.

In the case of the one-particle problem $H^{(\mathrm{mat})}$ is of the form [see, for instance, eq. (6.23)]

$$H^{(\mathrm{mat})} = m_1 c^2 (+e_1 \Phi'^{\mathrm{ext}})$$
$$+\frac{1}{2m_1}(\boldsymbol{p}_1 - \frac{e_1}{c}\mathfrak{A}_1'^{\mathrm{ext}})^2 - \frac{1}{8m_1^3 c^2}(\boldsymbol{p}_1 - \frac{e_1}{c}\mathfrak{A}_1'^{\mathrm{ext}})^4 + \ldots \quad (163)$$

where we have added, for formal reasons, the rest energy m_1c^2. Because of our definition of the external field there cannot be a contribution to $\mathfrak{A}_1'^{\mathrm{ext}}$ deriving from a constant or slowly changing mag-

* The radiation field evaluated from equations (161) or (162) undergoes, as we mentioned before, fast periodic vibrations which will be different from the real ones which follow from equations (159). It can be shown by considering examples that those vibrations would possess an infinitely large amplitude in the energy of the radiation field, which they would not have if they had been calculated from equations (159). In evaluating the motion of the particles we must formally proceed as if the influence of the radiation on the particles is small—for instance, we must drop in the sums Σ_λ all terms for which σ_λ is larger than a certain maximum value, which must be suitably chosen.

netic field. Of course, $\mathfrak{A}_1'^{\text{ext}}$ could contain such a term, but we prefer to consider such a magnetic field to derive from the other particles and thus to include it in the discussion of the many-particle problem. For the same reasons there will be no scalar potential of the external field in the strictly one-particle problem; we have for this reason put the corresponding term within brackets.

Equation (163) expresses the fact that $H^{(\text{mat})}$ is the sum of the rest energy and the kinetic energy and thus is equal to the total energy of the eigenfield in the formal terminology of a purely electromagnetic theory of the elementary particles. The energy of the eigenfield would be the sum of the energies of the longitudinal and of the transverse part of $\mathfrak{F}^{\text{eigen}}$. The first one derives from the electric field only and corresponds mainly to the term m_1c^2 (it is exactly equal to m_1c^2 if the particle velocity vanishes), and the second one is contained in the kinetic energy and represents the expression

$$(8\pi)^{-1}\,\Sigma\sigma_\lambda^2(a_\lambda''^{*}\,a_\lambda''+b_\lambda''^{*}\,b_\lambda) = (8\pi)^{-1}\int(\mathfrak{F}_{\text{I, II}}^{*\,\text{eigen}}\cdot\mathfrak{F}_{\text{I, II}}^{\text{eigen}})\,d\boldsymbol{x},$$

which is lacking in equation (162). The energy of the eigenfield which would be infinite in the case of point charges is in reality finite (finite "masses"). (Because of this one should according to classical considerations assume the particles to have finite dimensions.)

We see thus that equation (162) corresponds exactly to the total energy of the field in this formal approach. Because of the peculiar character of the second term in expression (162) the total energy of the system will not necessarily be constant—if one wants to retain this term—over long periods; this is, of course, connected with the fact that the theory is only approximative.

To find the exact expression for $H^{(\text{mat})}$ in the case of the *two-particle problems* we must clearly write down a set of canonical equations of motion which describe correctly the motion of each particle under the influence of the eigenfield of the other particle. If we consider only terms up to the order of magnitude v^2/c^2 the corresponding energy function [see eq. (3.76)] will be (for a proof compare the discussion in § 65)

$$H^{(\text{mat})} = \frac{\pi_1^2}{2m_1}+\frac{\pi_2^2}{2m_2}-\frac{\pi_1^4}{8m_1^3c^2}-\frac{\pi_2^4}{8m_2^3c^2} +\frac{e_1e_2}{r_{12}}-\frac{e_1e_2}{2m_1m_2c^2}\,\frac{(\boldsymbol{\pi}_1\cdot\boldsymbol{\pi}_2)r_{12}^2+(\boldsymbol{\pi}_1\cdot\boldsymbol{x}_{12})(\boldsymbol{\pi}_2\cdot\boldsymbol{x}_{12})}{r_{12}^3}, \tag{164}$$

where $\boldsymbol{\pi}_i = \boldsymbol{p}_i-(e_i/c)\mathfrak{A}_i'^{\text{ext}}$.

The first four terms derive mainly from the kinetic energy of the two particles. Only a small part of them derives —as do the next terms—from the "interference" of the eigenfields. The fifth term, the Coulomb energy, is the "interference energy" of the longitudinal part, and the sixth term derives from the transverse part and corresponds exactly to Darwin's terms [see eq. (3.76)]. Its meaning is thus that it represents the magnetic interaction of the eigenfields up to terms of the order v^2/c^2.

The "interference" energy of the longitudinal electric field is clearly given by the equation

$$\tfrac{1}{2}\int(\varrho_1\Phi_2'+\varrho_2\Phi_1')d\boldsymbol{x} = \tfrac{1}{2}[e_1(e_2/r_{12})+e_2(e_1/r_{12})] = e_1 e_2/r_{12},$$

where the ϱ_i are the charge distributions of the particles and where the Φ' [see eq. (143)] are the potentials of the corresponding eigenfields.

The interference energy of the transverse eigenfield is obtained from equation (157); in that equation the denominators simplify to σ_λ. We get thus for the interference part of $(8\pi)^{-1}\Sigma_\lambda\sigma_\lambda^2[|a_\lambda''|^2+|b_\lambda''|^2]$

$$(e_1 e_2/2\pi c^2 \Omega)\Sigma_\lambda\sigma_\lambda^{-4}[(\boldsymbol{v}_1\cdot\mathfrak{c}_\lambda)(\boldsymbol{v}_2\cdot\mathfrak{c}_\lambda^*)+(\boldsymbol{v}_1\cdot\mathfrak{c}_\lambda^*)(\boldsymbol{v}_2\cdot\mathfrak{c}_\lambda)][\exp\{2\pi i(\boldsymbol{\sigma}_\lambda\cdot\boldsymbol{x}_{12})\}+\exp\{-2\pi i(\boldsymbol{\sigma}_\lambda\cdot\boldsymbol{x}_{12})\}]$$

$$= (e_1e_2/\pi c^2)\int[\sigma^2(\boldsymbol{v}_1\cdot\boldsymbol{v}_2)-(\boldsymbol{v}_1\cdot\boldsymbol{\sigma})(\boldsymbol{v}_2\cdot\boldsymbol{\sigma})]\cos[2\pi(\boldsymbol{\sigma}\cdot\boldsymbol{x}_{12})]d\boldsymbol{\sigma}/\sigma^4$$

$$= (e_1e_2/2c^2r_{12}^3)[(\boldsymbol{v}_1\cdot\boldsymbol{v}_2)\,r_{12}^2+(\boldsymbol{v}_1\cdot\boldsymbol{x}_{12})(\boldsymbol{v}_2\cdot\boldsymbol{x}_{12})].$$

In accordance with this, equation (164) becomes

$$H^{(\mathrm{mat})} = \tfrac{1}{2}m_1v_1^2+\tfrac{1}{2}m_2v_2^2+\frac{3m_1v_1^4}{8c^2}+\frac{3m_2v_2^4}{8c^2}+\frac{e_1e_2}{r_{12}}+\frac{e_1e_2}{2c^2}\,\frac{(\boldsymbol{v}_1\cdot\boldsymbol{v}_2)r_{12}^2+(\boldsymbol{v}_1\cdot\boldsymbol{x}_{12})(\boldsymbol{x}_2\cdot\boldsymbol{x}_{12})}{r_{12}^3},$$

where $\boldsymbol{\pi}$ is replaced by $\boldsymbol{v}$ by means of the relation $v_{1x} = \partial H/\partial p_{1x} = \partial H/\partial \pi_{1x}, \ldots$ The Darwin terms are thus, as far as their absolute value is concerned, equal to the $\boldsymbol{v}$-dependent extra term in the interference energy of the eigenfields, but the sign is different.

Equation (164) can easily be generalised to include more particles. The Darwin terms can be used to take the magnetic action of external currents into account. If a large number of particles (index k) produce a practically continuous current field from separate contributions $e_k\boldsymbol{v}_k \doteq e_k\boldsymbol{\pi}_k/m_k$, and if we are not interested in the influence of the system on these particles, the Darwin terms lead to a contribution to the Hamiltonian which as far as the first particle is concerned is given by the equation

$$\left.\begin{aligned} -(e_1/2m_1c^2)(\boldsymbol{\pi}_1\cdot\Sigma_k(e_k/m_k)[\boldsymbol{\pi}_k r_{1k}^2+\boldsymbol{x}_{1k}(\boldsymbol{\pi}_k\cdot\boldsymbol{x}_{1k})]r_{1k}^{-3} \\ = -(e_1/m_1c)(\boldsymbol{\pi}_1\cdot\mathbf{A}_1), \\ \mathbf{A}_1 \doteq \int[\varrho\boldsymbol{v}r_1^2+\boldsymbol{x}_1(\varrho\boldsymbol{v}\cdot\boldsymbol{x}_1)]d\boldsymbol{x}/2cr_1^3 = \int\varrho\boldsymbol{v}d\boldsymbol{x}/cr_1. \end{aligned}\right\} \quad (165)$$

We have now obtained an integral over that region of space which contains the current density; the radius vector of the first particle to an arbitrary point of this region is denoted by $\boldsymbol{x}_1$. The last equation follows as the continuous current field is a transverse field and equation (165) is in agreement with the corresponding term in equation (2.102).

Similarly one can derive the electrical action of external charges from the potential $\Phi = \Sigma_k(e_k/r_{1k}) \doteq \int \varrho\, d\boldsymbol{x}/r$, obtained by summing over the Coulomb interactions of particle 1 with the other particles k, on which the influence of the system does not interest us explicitly.

Kramers * developed in 1938 the theory described in this section. At the Galvani congress in Bologna in 1937 ** he discussed how this theory was constructed to be structure-independent as suggested by the fact that in the classical theory of electrons all physically significant results depend only on the mass m and the charge e of the electron, and do not contain any reference to the structure of the electron. The mass m appearing in this theory can be called the *experimental* mass, it is the sum of the inertial and the electromagnetic mass. The second point of Kramers' theory was to bring the classical equations into canonical form.

In the original form of the theory, as presented here, Kramers used a Hamiltonian which had practically the same form as the usual one, but with $\mathfrak{F}^{\text{ext}}$ taking the place of $\mathfrak{F}^{\text{total}}$ and with the experimental mass instead of the inertial mass. This shows that this theory was the first case of *mass renormalisation*; indeed, as soon as the Lamb-shift had been experimentally observed, Kramers was able to find its value from his theory ***.

As the eigenfield is removed, the details of the structure of the particles are eliminated, and one can then use a simple model, and even a point charge.

Serpe **** used Kramers' theory to remove the displacement of spectral lines found by Weisskopf and Wigner *****. Opechowski ****** in 1941 constructed a Hamiltonian which in dipole approximation and to the first order in e was exact, while Kramers reported at the 1948 Solvay Congress on a Hamiltonian which is correct to any order in e again in dipole approximation. Essentially this Hamiltonian was used by van Kampen in his treatment of the scattering of light.

§ 90. QUANTUM THEORY OF THE INTERACTION BETWEEN RADIATION AND MATTER

Dirac's radiation theory is obtained by quantising the a'_λ, a'^*_λ, b'_λ, and b'^*_λ of the external radiation field occurring in the Hamil-

* The last paragraphs of this section were added in the translation. They are based on Kramers' 1948 Solvay talk (*Collected Scientific Papers*, North Holland Publ. Cy., 1956, p. 845) and on van Kampen's Leiden thesis (*Proc Dan. Acad. Sci.*, **26**, nr. 15, 1951).

** Nuovo Cim., **15**, 108, 1938 (*Collected Scientific Papers*, p. 831).

*** See *Collected Scientific Papers*, p. 858.

**** J. Serpe, *Physica*, **7**, 133, 1940; **8**, 226, 1941.

***** V. Weisskopf and E. P. Wigner, *Z. Phys.*, **63**, 54, 1930; **65**, 18, 1930.

****** W. Opechowski, *Physica*, **8**, 161, 1941.

tonian (162) in exactly the same way as in the vacuum case the a_λ, a_λ^*, b_λ, b_λ^* were quantised (see § 87). That means that we put [see eq. (94)]

$$\boldsymbol{\alpha}_\lambda = \tau a'_\lambda, \quad \boldsymbol{\alpha}_\lambda^* = \tau a'^*_\lambda, \quad \boldsymbol{\beta}_\lambda^* = \tau b'_\lambda, \quad \boldsymbol{\beta}_\lambda = \tau b'^*_\lambda, \quad \tau = (\sigma_\lambda/8\pi hc)^{1/2},$$

and we introduce the commutation relations

$$[\boldsymbol{\alpha}_\lambda, \boldsymbol{\alpha}_\lambda^*]_- = [\boldsymbol{\beta}_\lambda, \boldsymbol{\beta}_\lambda^*]_- = \mathbf{1}, \quad \text{(all other pairs commuting).}$$

At the same time we must replace the coordinates and momenta occurring in $H^{(\mathrm{mat})}$ by operators in the usual way, $H^{(\mathrm{mat})} \to H^{(\mathrm{mat})}_{\mathrm{op}}$. In this way we obtain the following energy operator of the total system of radiation and particles,

$$H_{\mathrm{op}} = h\Sigma_\lambda \nu_\lambda(\boldsymbol{\alpha}_\lambda^* \boldsymbol{\alpha}_\lambda + \boldsymbol{\beta}_\lambda^* \boldsymbol{\beta}_\lambda) + H^{(\mathrm{mat})}_{\mathrm{op}}. \tag{166}$$

The operator $H^{(\mathrm{mat})}_{\mathrm{op}}$ also contains the $\boldsymbol{\alpha}$ and $\boldsymbol{\beta}$, as it depends on the vector potential $\mathfrak{A}'^{\,\mathrm{ext}}$ of the external field [see eqq. (149) and (154)],

$$\begin{aligned}\mathfrak{A}'^{\,\mathrm{ext}} = (\hbar c/\Omega)^{1/2}\Sigma_\lambda \sigma_\lambda^{-3/2}\{&-i\boldsymbol{\alpha}_\lambda \mathfrak{c}_\lambda \exp[2\pi i(\boldsymbol{\sigma}_\lambda\cdot\boldsymbol{x})]\\ +i\boldsymbol{\alpha}_\lambda^* \mathfrak{c}_\lambda^* \exp[-2\pi i(\boldsymbol{\sigma}_\lambda\cdot\boldsymbol{x})] &- i\boldsymbol{\beta}_\lambda \mathfrak{c}_\lambda^* \exp[2\pi i(\boldsymbol{\sigma}_\lambda\cdot\boldsymbol{x})]\\ &+i\boldsymbol{\beta}_\lambda^* \mathfrak{c}_\lambda \exp[-2\pi i(\boldsymbol{\sigma}_\lambda\cdot\boldsymbol{x})]\}.\end{aligned} \tag{167}$$

In equation (166) we dropped for the sake of simplicity the interaction energy between the average external field and the eigenfield. The longitudinal part $\mathfrak{F}^{\mathrm{eigen}}_{\mathrm{III}}$ of this eigenfield is nothing but the Coulomb field of the charges present * and contains thus only the spatial coordinates of the particles. Equations (155) and (157) gave the exact classical expressions for the transverse part. These expressions can be replaced by the corresponding quantum mechanical operators. Apart from their time averages $(\bar{a}''_\lambda)_{\mathrm{op}}$ and $(\bar{b}''_\lambda)_{\mathrm{op}}$, which correspond to the average magnetic field of the particle (see § 89), there is not much point in considering these terms because our canonical description of the radiation field is approximate. We shall therefore only consider the average value $\bar{\mathfrak{F}}^{\mathrm{eigen}} = \bar{\mathfrak{F}}^{\mathrm{eigen}}_{\mathrm{I,II}} + \mathfrak{F}^{\mathrm{eigen}}_{\mathrm{III}}$ to be the eigenfield. All components of this field commute with one another and for the total $\mathfrak{E}$, $\mathfrak{H}$-field which interests us ($\mathfrak{F}^{\mathrm{total}} = \mathfrak{F}^{\mathrm{ext}} + \bar{\mathfrak{F}}^{\mathrm{eigen}}$) we have exactly the same commutation relations as the ones we met with and discussed in the vacuum case (see § 87).

If the system contains particles with spin, especially electrons, it is an obvious choice to include in $H^{(\mathrm{mat})}$ just the spin-involving

* An "external" electrical and magnetic field also belongs in our present classification to the eigenfield (see § 89). In order not to produce any confusion we have here and henceforth called the external field the "external radiation field".

operators in the manner discussed in chapter six. In the non-relativistic treatment this would mean the expression (6.33) for H_s. As far as the one-electron problem is concerned we could include the Dirac theory.

To obtain a discussion of radiation phenomena we must solve the Schrödinger equation corresponding to expression (166). This problem is usually treated by the methods of *perturbation theory*, as the interaction between radiation field and matter is considered to be a weak perturbation. This treatment is all the more indicated by the fact that the canonical energy function in § 89 was introduced under the assumption that this interaction was small. Because of the approximations introduced then we must obviously be very careful when we are discussing the results, especially if we use second order perturbation theory.

The physical situation of the total system of particles and radiation is described by a normalised wave function $\Psi(n_1, \ldots, n_\lambda, \ldots; \bar{n}_1, \ldots, \bar{n}_\lambda, \ldots; q_1, q_2, \ldots; t)\,(\int|\Psi|^2 = 1)$. The integers $n_\lambda(\bar{n}_\lambda)$ refer to the number of right hand (left hand) quanta corresponding to the wave vector $\boldsymbol{\sigma}_\lambda$. The $q_1, q_2, \ldots$ are, for instance, the spatial and spin coordinates of the particles; we can, however, transform to another set of arguments which can describe the mechanical situation. As such arguments we choose a group of parameters $k_1, k_2, \ldots$ which would describe the stationary states of the mechanical system, if there were no external radiation field. We shall call them "quantum numbers", even though some or all of them may be continuous parameters, and we shall denote them collectively by the symbol k. If φ_k is the normalised wave function corresponding to the quantum number k—the φ_k depend on the spatial and spin coordinates of the particles—we can write

$$\left.\begin{aligned} &\Psi(\ldots, n_\lambda, \ldots; \ldots, \bar{n}_\lambda, \ldots; q_1, q_2, \ldots; t) \\ &= \Sigma_k \chi_k(\ldots, n_\lambda, \ldots; \ldots, \bar{n}_\lambda, \ldots; t)\varphi_k(q_1, q_2, \ldots), \\ &\qquad \Sigma_{n_\lambda, \bar{n}_\lambda, k} |\chi_k|^2 = 1. \end{aligned}\right\} \tag{168}$$

The transformed function χ is thus represented by a set of time-dependent coefficients, each of which corresponds to definite values of the quantum numbers n_λ, $\bar{n}_\lambda$ of the radiation field and the quantum numbers k of the particle system.

If the interaction between radiation and particles can be neglected we could characterise a stationary state of the total system by a function χ which is equal to zero for all combinations of values of the

n_λ, $\bar{n}_\lambda$, and k except one particular one ($n_1 = n_1^0; \ldots; n_\lambda = n_\lambda^0; \ldots;$ $\bar{n}_\lambda = \bar{n}_\lambda^0; \ldots; k = k_0$) and for that particular combination we have

$$\chi_{k_0}(n_\lambda^0, \bar{n}_\lambda^0, t) = \exp(-iE_0 t/\hbar), \quad E_0 = \Sigma_\lambda h\nu_\lambda(n_\lambda^0 + \bar{n}_\lambda^0) + E_{k_0}, \tag{169}$$

with E_{k_0} the particle energy corresponding to φ_{k_0}. In this state the expectation value $(\overline{\mathfrak{F}^* \cdot \mathfrak{F}})$ of the energy density of the radiation will be the same at every point of space. The most general situation of the unperturbed system is a superposition of states (169),

$$\chi = \Sigma_{n_\lambda, \bar{n}_\lambda, k}\, \gamma_k(n_\lambda, \bar{n}_\lambda) \exp[-iE_k(n_\lambda, \bar{n}_\lambda)t/\hbar], \; \Sigma_{n_\lambda, \bar{n}_\lambda, k} |\gamma|^2 = 1. \tag{170}$$

The interaction enters by the circumstance that the coefficients $\gamma_k(n_\lambda, \bar{n}_\lambda)$ are no longer constant, but change with time. Their change is described by equation (5.74) of § 53,

$$i\hbar\dot{\gamma}_k(n_\lambda, \bar{n}_\lambda)$$
$$= \Sigma_{k', n'_\lambda, \bar{n}'_\lambda} G(k, n_\lambda, \bar{n}_\lambda; k'_1, n'_\lambda, \bar{n}'_\lambda)\gamma_{k'}(n'_\lambda, \bar{n}'_\lambda) \exp[i(E-E')t/\hbar]. \tag{171}$$

The absolute square of $\gamma_k(n_\lambda, \bar{n}_\lambda)$ at time t is the probability that at t the particle system is in the state k and that at the same time there are present exactly n_λ right handed and $\bar{n}_\lambda$ left handed quanta corresponding to the wave vector $\boldsymbol{\sigma}_\lambda$.

The $G(k, n_\lambda, \bar{n}_\lambda; k', n'_\lambda, \bar{n}'_\lambda)$ are the matrix elements of the perturbation energy G_{op}; this energy is given by those terms in $H^{(\text{mat})}$ which contain the amplitudes of the external radiation field. In general we can write

$$G_{\text{op}} = \Sigma_\lambda(g_\lambda \boldsymbol{\alpha}_\lambda + g_{\bar{\lambda}} \boldsymbol{\beta}_\lambda + g_\lambda^* \boldsymbol{\alpha}_\lambda^* + g_{\bar{\lambda}}^* \boldsymbol{\beta}_\lambda^*) + \Sigma_{\lambda\lambda'}(g_{\lambda\lambda'} \boldsymbol{\alpha}_\lambda \boldsymbol{\alpha}_{\lambda'} + \ldots). \tag{172}$$

The terms linear in the $\boldsymbol{\alpha}_\lambda, \ldots$ are small of the first order, and those quadratic in the $\boldsymbol{\alpha}_\lambda$ small of the second order. There is hardly any point in going up to higher order terms. As the linear terms do not commute with the operator (100) of the total number of light quanta this number will in general change with time (see § 72).

For the non-relativistic spinless one-particle problem we have

$$G_{\text{op}} = -(e_1/m_1 c)(\boldsymbol{p}_1 \cdot \mathfrak{A}_1'^{\text{ext}}) + (e_1^2/2m_1 c^2)\mathfrak{A}_1'^{\text{ext}\,2} + \ldots, \tag{173}$$

where we neglect the terms deriving from the relativistic correction to the kinetic energy. If a constant external magnetic field (vector potential $\mathfrak{A}'_H$) is present we must replace $\boldsymbol{p}_1$ by $\boldsymbol{p}_1 - (e_1/c)\mathfrak{A}'_H$.

In the case of a many-particle system with spin we have (see § 82)

$$\left.\begin{aligned} G_{\rm op} = & \underbrace{-\sum_i \frac{e_i}{m_i c}\left(\boldsymbol{p}_i - \frac{e_i}{c}\mathfrak{A}'_{Hi} \cdot \mathfrak{A}'^{\rm ext}_i\right)}_{-c^{-1}(\Sigma_i e_i (\boldsymbol{v}^0_i \cdot \mathfrak{A}'^{\rm ext}_i)} - i \sum_{i \atop \text{electrons}} \frac{e_i}{\hbar m_i c} \\ & \{([\boldsymbol{S}_i \wedge \boldsymbol{p}_i] \cdot \mathfrak{A}'^{\rm ext}_i) - (\mathfrak{A}'^{\rm ext}_i \cdot [\boldsymbol{S}_i \wedge \boldsymbol{p}_i])\} + \Sigma_i (e_i^2/2m_i c^2)\, \mathfrak{A}'^{\rm ext\,2}_i . \end{aligned}\right\} \quad (174)$$

We have assumed that the Darwin terms in $G_{\rm op}$ [see eq. (164)] are negligible—except in as far as we can consider $\mathfrak{A}'_H$ to derive from them—and we have also neglected the contribution from the interaction between the spin and $\mathfrak{E}^{\rm ext}$. The form of the terms involving spin in equation (174) is in general corerct only for electrons. The first sum can be expressed simply by means of the velocity operators $\boldsymbol{v}^0_i$, as indicated; these operators refer to the unperturbed particle-system.

If we use for the one-electron problem the Dirac theory, we would obtain for $G_{\rm op}$ the following extremely simple expression,

$$G_{\rm op} = e(\boldsymbol{\alpha} \cdot \mathfrak{A}'^{\rm ext}), \quad (175)$$

where $-e$ is the electronic charge and $c\boldsymbol{\alpha}$ Dirac's velocity operator. In this case there are thus no terms quadratic in the $\boldsymbol{\alpha}_\lambda, \ldots$ This expression can immediately be compared with H' of equation (148); this last quantity was the extra term in the Hamiltonian of the radiation field when the current field was assumed to be given. The difference between expressions (175) and (148) is "only" that in equation (148) there occurred the vector potential of the total field.

§ 91. Theory of emission and absorption

We shall first of all apply the quantum theory of the interaction between radiation and matter to those problems which we discussed at the beginning of this chapter from the point of view of non-quantised radiation, that is, to absorption and emission.

Let there be at $t = 0$ a situation corresponding to a state (169) of the unperturbed system, that is, let at $t = 0$ the particle-system—or atom as we shall say for the sake of simplicity—be in the stationary state k_0 while there are present exactly n^0_λ right hand and $\bar{n}^0_\lambda$ left hand quanta corresponding to the wave vector $\boldsymbol{\sigma}_\lambda$. At that moment all coefficients γ are equal to zero except $\gamma_{k_0}(n^0_\lambda, \bar{n}^0_\lambda)$ which is equal to unity. Because of equations (171) the other γ will take on values different from zero and $|\gamma_{k_0}(n_\lambda^0, \bar{n}_\lambda^0)|^2$ will decrease. During a time interval within which $|\gamma_{k_0}(n_\lambda^0, \bar{n}_\lambda^0)|^2$ has changed but little from its initial value unity we get to a first approximation,

$$\left.\begin{aligned} &i\hbar\dot{\gamma}(n_\lambda, \bar{n}_\lambda) = G(k, n_\lambda, \bar{n}_\lambda; k_0, n_\lambda^0, \bar{n}_\lambda^0) \exp [i(E-E_0)t/\hbar],\\ &E_0 = E_{k_0} + \Sigma_\lambda h\nu_\lambda(n_\lambda^0 + \bar{n}_\lambda^0); \quad E = E_k + \Sigma_\lambda h\nu_\lambda(n_\lambda + \bar{n}_\lambda),\\ &\gamma_k(n_\lambda, \bar{n}_\lambda)\\ &= G(k, n_\lambda, \bar{n}_\lambda; k_0, n_\lambda^0, \bar{n}_\lambda^0)\{1 - \exp [i(E-E_0)t/\hbar]\}/(E-E_0). \end{aligned}\right\} \quad (176)$$

In this first approximation we need consider only the terms in G_{op} which are linear in the $\boldsymbol{\alpha}_\lambda, \ldots$ From the matrix representations (96) and (97) of the $\boldsymbol{\alpha}_\lambda, \boldsymbol{\beta}_\lambda, \boldsymbol{\alpha}_\lambda^*, \boldsymbol{\beta}_\lambda^*$ it follows immediately that only those matrix elements $G(k, n_\lambda, \bar{n}_\lambda; k_0, n_\lambda^0, \bar{n}_\lambda^0)$ of the perturbation energy will be different from zero for which the number of light quanta, n_λ $\bar{n}_\lambda$, of one kind—that is, of given λ and given sense of polarisation—decreases or increases by one. Physically this means a *quantum jump*: *After a short time there exists a probability that the particle system has emitted or absorbed one light quantum.* Equation (176) shows how this probability increases with time. From the occurrence of a resonance denominator $E-E_0$ it follows that this probability is appreciably different from zero only when the total energy of the system remains practically the same during the transition; this means that *practically only such transactions will occur where the atom jumps from one stationary state to another and where the energy lost or gained is exactly equal to the energy of the emitted or absorbed quantum.* We have thus obtained the *quantum mechanical interpretation of Bohr's frequency rule* (§ 18). If the initial situation were characterised by a large number of non-vanishing coefficients γ, all of them corresponding to the same k-value, k_0, (atom in a definite state, radiation field arbitrary), it is easily seen that the result remains the same.

To determine the probability for a transition we must calculate the matrix elements G of the part of G_{op} linear in the $\boldsymbol{\alpha}_\lambda$. There are four types of these according to whether n_λ or $\bar{n}_\lambda$ is changed by $+1$ or -1. For the time being we drop the spin terms of equation (174), that is, we use for $G_{\text{op}}^{(1)}$ the equation,

$$G_{\text{op}}^{(1)} = -c^{-1}(\Sigma_i e_i \mathbf{v}_i^0 \cdot \mathfrak{A}_i'^{\,\text{ext}}).$$

Using equations (167), (96), and (97) we get

$$\left.\begin{aligned} G(k, n_\lambda^0 - 1; k_0, n_\lambda^0) &= \frac{i}{c}\left(\frac{\hbar c}{\Omega\sigma_\lambda^3}\right)^{1/2} n_\lambda^{0\,1/2}\big(\sum_i e_i \mathbf{v}_i^0 . \mathfrak{c}_\lambda \exp[2\pi i(\boldsymbol{\sigma}_\lambda \cdot \boldsymbol{x}_i)]\big)_{kk_0}, \\ G(k, \bar{n}_\lambda^0 - 1; k_0, \bar{n}_\lambda^0) &= \frac{i}{c}\left(\frac{\hbar c}{\Omega\sigma_\lambda^3}\right)^{1/2} \bar{n}_\lambda^{0\,1/2}\big(\sum_i e_i \mathbf{v}_i^0 \cdot \mathfrak{c}_\lambda^* \exp[2\pi i(\boldsymbol{\sigma}_\lambda \cdot \boldsymbol{x}_i)]\big)_{kk_0}; \end{aligned}\right\} \quad (177)$$

$$\left.\begin{aligned} G(k, n_\lambda^0 + 1; k_0, n_\lambda^0) &= -\frac{i}{c}\left(\frac{\hbar c}{\Omega\sigma_\lambda^3}\right)^{1/2} (n_\lambda^0 + 1)^{1/2}\big(\sum_i e_i \mathbf{v}_i^0 \cdot \mathfrak{c}_\lambda^* \exp[-2\pi i(\boldsymbol{\sigma}_\lambda \cdot \boldsymbol{x}_i)]\big)_{kk_0}, \\ G(k, \bar{n}_\lambda^0 + 1; k_0, \bar{n}_\lambda^0) &= -\frac{i}{c}\left(\frac{\hbar c}{\Omega\sigma_\lambda^3}\right)^{1/2} (\bar{n}_\lambda^0 + 1)^{1/2}\big(\sum_i e_i \mathbf{v}_i^0 \cdot \mathfrak{c}_\lambda \exp[-2\pi i(\boldsymbol{\sigma}_\lambda \cdot \boldsymbol{x}_i)]\big)_{kk_0}; \end{aligned}\right\} \quad (178)$$

here the four expressions derive successively from the $\boldsymbol{\alpha}_\lambda$, the $\boldsymbol{\beta}_\lambda$, the $\boldsymbol{\alpha}_\lambda^*$, and the $\boldsymbol{\beta}_\lambda^*$ terms; also the left hand sides are meant to indicate that all light quantum numbers which are not written down are unchanged under the transition. The matrix elements on the right hand sides refer to the unperturbed atom and they can immediately be expressed by means of the *matrix-vector field of the electrical current density*. [see equations (8) and (9)] which was discussed extensively in § 82. If $\sigma_\lambda \mathfrak{a}_\lambda$ denotes either $\mathfrak{c}_\lambda$ or $\mathfrak{c}_\lambda^*$ we have

$$\begin{aligned} &(\Sigma_i e_i \mathbf{v}_i^0 \cdot \sigma_\lambda \mathfrak{a}_\lambda \exp[\pm 2\pi i(\boldsymbol{\sigma}_\lambda \cdot \boldsymbol{x}_i)])_{kk_0} \\ &\qquad = \sigma_\lambda(\mathfrak{a}_\lambda \cdot \int \exp[\pm 2\pi i(\boldsymbol{\sigma}_\lambda \cdot \boldsymbol{x})](\varrho \mathbf{v})_{kk_0} d\boldsymbol{x}); \quad |\mathfrak{a}_\lambda|^2 = 1. \end{aligned} \quad (179)$$

The further evaluation is also similar to the one in the case of the absorption of light. In particular we note the similarity of equations (176) and (18) even though the quantities γ have different meanings in those two equations.

In the case of *absorption* we have $E_k > E_{k_0}$. The probability that at time t the atom is in the state k and that at the same time a right hand light quantum with a wave vector $\boldsymbol{\sigma}_\lambda$ within a given element of solid angle $d\boldsymbol{\omega}$ is absorbed is given by the sum of the absolute squares of the coefficients $\gamma_k(n_\lambda^0 - 1, \bar{n}_\lambda^0)$. The summation is over all wave vectors $\boldsymbol{\sigma}_\lambda$ with directions within $d\boldsymbol{\omega}$ and with an absolute value satisfying approximately the relation $E_k - E_{k_0} \doteq h\nu_\lambda = hc\sigma_\lambda$. If Ω is sufficiently large, the different vectors $\boldsymbol{\sigma}_\lambda$ lie very close to each other. We assume that the numbers n_λ^0 are such that we can in a natural way define for each vector $\boldsymbol{\sigma}$ an average value, $\langle n_0(\boldsymbol{\sigma})\rangle$, of those quantum numbers n_λ^0 for which $\boldsymbol{\sigma}_\lambda$ lies close to $\boldsymbol{\sigma}$*. We shall also assume that $\langle n_0(\boldsymbol{\sigma})\rangle$ is a continuous function of $\boldsymbol{\sigma}$. The summation mentioned a moment ago can now be replaced by an integral over $\boldsymbol{\sigma}$-space and multiplication by Ω. The result is

* The quantity $\langle n_0(\boldsymbol{\sigma})\rangle$ may well be small compared to one.

$$\sum_{\lambda(d\boldsymbol{\omega})}|\gamma_k(n_\lambda^0-1,\bar{n}_\lambda^0)|^2=\sum_{\lambda(d\boldsymbol{\omega})}|G(k,n_\lambda^0-1;k_0,n_\lambda^0)|^2\left|\frac{1-\exp[i(E_k-E_{k_0}-h\nu_\lambda)t/\hbar]}{E_k-E_{k_0}-h\nu_\lambda}\right|^2$$

$$=\frac{<n_0>}{\hbar c\sigma}|(\mathfrak{a}\cdot\int\exp[2\pi i(\boldsymbol{\sigma}\cdot\boldsymbol{x})](\varrho\boldsymbol{v})_{kk_0}\,d\boldsymbol{x})|^2\sigma^2\,d\boldsymbol{\omega}\int\left|\frac{1-e^{i\Delta t}}{\Delta}\right|\frac{d\Delta}{2\pi c}$$

$$=(<n_0>\sigma/\hbar c^2)|(\mathfrak{a}\cdot\int\exp[2\pi i(\boldsymbol{\sigma}\cdot\boldsymbol{x})](\varrho\boldsymbol{v})_{kk_0}d\boldsymbol{x})|^2t\,d\boldsymbol{\omega}. \quad (180)$$

Following Einstein we put this number equal to [see eq. (20)],

$$\Sigma_{\lambda(d\boldsymbol{\omega})}|\gamma|^2=B_{k_0\to k}(\boldsymbol{\sigma},\mathfrak{a})I(\nu_{kk_0},\boldsymbol{\omega})t\,d\boldsymbol{\omega}/c, \quad (181)$$

where $I(\nu,\boldsymbol{\omega})d\nu d\boldsymbol{\omega}/c$ is the density in space of the right hand circularly polarised radiation energy at the position of the atom with a direction $\boldsymbol{\omega}$ of the radiation within $d\boldsymbol{\omega}$ and a frequency between ν and $\nu+d\nu$. If we remember that the total (right hand) radiation energy $\Sigma_\lambda n_\lambda^0 h\nu_\lambda=\Omega hc\int<n_0>\sigma^3 d\boldsymbol{\omega}\,d\sigma$ is uniformly distributed over Ω *, so that $I(\nu,\boldsymbol{\omega})=hc\sigma^3<n_0>$, *we find for B exactly the same value as before* [see eq. (21)],

$$B_{k_0\to k}(\boldsymbol{\sigma},\mathfrak{a})=(2\pi\hbar^2\nu_{kk_0}^2)^{-1}|(\mathfrak{a}\cdot\int\exp[2\pi i(\boldsymbol{\sigma}\cdot\boldsymbol{x})](\varrho\boldsymbol{v})_{kk_0}d\boldsymbol{x})|^2;$$
$$\nu_{kk_0}>0,\qquad\sigma=\nu_{kk_0}/c. \quad (182)$$

This equation was derived here for right hand light, that is, $\mathfrak{a}=\mathfrak{c}/\sigma$, and this means that we separate explicitly the left- and right hand part of the energy of the radiation field. We note now first of all that equation (182) would also hold if we were interested in the absorption of left hand light; in that case we put $\mathfrak{a}=\mathfrak{c}^*/\sigma$. However, it is then also possible to prove that *equation* (182) *is valid for any direction of polarisation.* The calculations are simplest if we use the transformed operators $\boldsymbol{\delta}_\lambda$, $\boldsymbol{\epsilon}_\lambda$, $\boldsymbol{\delta}_\lambda^*$, and $\boldsymbol{\epsilon}_\lambda^*$ (see § 87) instead of the $\boldsymbol{\alpha}_\lambda,\ldots$ to describe the radiation field, and if we choose the $\boldsymbol{\delta}_\lambda,\ldots$ in such a way that they correspond for each $\boldsymbol{\sigma}_\lambda$ to two (arbitrary) mutually perpendicular directions of polarisation $\mathfrak{a}$ and $\mathfrak{b}$. The general expression for $\mathfrak{F}_{\text{op}}$ becomes more complicated, but the expressions for the electrical field and for the vector potential $\mathfrak{A}'^{\text{ext}}_{\text{op}}$ occurring in the perturbation remain as simple as before. We need replace only the $\boldsymbol{\alpha}_\lambda,\boldsymbol{\alpha}_\lambda^*$ by $\boldsymbol{\delta}_\lambda,\boldsymbol{\delta}_\lambda^*$, the $\mathfrak{c}_\lambda$, $\mathfrak{c}_\lambda^*$ by $\sigma_\lambda\mathfrak{a}_\lambda$, $\sigma_\lambda\mathfrak{a}_\lambda^*$, the $\boldsymbol{\beta}_\lambda,\ldots$ in equation (167). The rest of the calculation proceeds as before, but the n_λ, $\bar{n}_\lambda$ now refer to the two new directions of polarisation.

The reason why for the calculations of this section the particular

* The expectation value of the corresponding terms in the energy density is independent of position.

advantage of the right hand and left hand quanta does not appear * is that the perturbation is described by $\mathfrak{A}'^{\text{ext}}_{\text{op}}$ which contains the $\boldsymbol{\alpha}$ and $\boldsymbol{\beta}$ always in the combination $\boldsymbol{\alpha}\mathfrak{c}+\boldsymbol{\beta}\mathfrak{c}^*$. This state of affairs continues when we now take into account the *influence of the spin.* Its influence is described by the second term in equation (174). We have written this equation immediately in such a form that the vector potential occurs explicitly (see § 82). As a result its kk_0 matrix elements can immediately be described by means of the matrix vector field of the spin current. It is easily seen that—in accordance with § 82—the expression for B remains valid, if the field is completed by the spin current: $(\varrho\mathbf{v})_{kk_0} \rightarrow (\varrho\mathbf{v})_{kk_0}+(\varrho\mathbf{v})^{\text{spin}}_{kk_0}$.

In two respects the present calculation of the absorption can be appreciably enlarged upon. To begin with we have made special assumptions about the initial state of the radiation field. If we generalise by letting in the initial state many coefficients γ in equation (170) be different from zero (general radiation field), we should expect that Einstein's description of the absorbing transitions remains valid, and that equation (182) continues to hold. The situation is, however, very complicated since now the radiation field may have different intensities at different points in space and since its intensity at the position of the atom may fluctuate very rapidly—partly unsystematically, partly systematically (atom is an intermittent radiation field). All of this is in principle described by the values of the γ's, but it may sometimes be very difficult to see it. We should, for instance, investigate in detail the question whether there exist systematic (or unsystematic) relations between the different initial values of the γ's. Even in the limiting case where the classical calculation holds this problem would be very complicated**, and we shall not consider it here. The related question of where in space and how the decrease of the radiation energy due to absorption occurs will not be discussed either—even though this discussion would not be very difficult in the case of our particular initial state.

* Their particular character did enter in the simple Lorentz invariant formulation (without potentials) of the field equation. It would also enter in the present discussion if we were interested in a discussion of the magnetic component $\mathfrak{H}$ of the radiation field, as $\mathfrak{H}$ contains the $\boldsymbol{\alpha}$ and $\boldsymbol{\beta}$ in the combination $\boldsymbol{\alpha}\mathfrak{c}-\boldsymbol{\beta}\mathfrak{c}^*$.

** In the classical calculations of § 82 the assumption of random phases of the different light components played an important part. It is interesting to note that in our particular assumptions about the initial state the question of random phases does not yet occur.

Secondly there is the question of *Einstein's radiation impulse.* If we assume that the absorbing particle system is a free atom, it is no longer correct to use discrete stationary states; the different possible values of the total momentum of the atom correspond to a continuous range of stationary states. Even if we were dealing with a system with discrete energy states, as far as the motion of its centre of gravity is concerned, k in equation (176) should be considered to be a continuous variable corresponding to the various values of the total momentum after the transition. We need now a more detailed analysis of the matrix elements of equation (177). This analysis shows that *transitions occur only in connexion with such changes in the momentum of the atom that the momentum* $h\boldsymbol{\sigma}$ *of the absorbed quantum is added (after the absorption) to the initial momentum of the atom.* This effect was predicted by Einstein on the basis of his conception of the corpuscular nature of radiation. In quantum mechanics it follows immediately from the *general law of the conservation of momentum,* the general validity of which in the quantum theory of radiation we could have proved explicitly by using the equations of § 90 (commutability of the energy and the operator of the total momentum). The transfer of momentum to matter during the absorption (and also during scattering; see § 93) is the correspondence analogue of the classical *radiation pressure.*

If $\boldsymbol{X}$ is the position of the centre of gravity of the system, the wave functions of two stationary states, which differ only in that the total momentum is zero in the one case and equal to $h\boldsymbol{\Sigma}$ in the other case, will differ only by a factor $\exp\,[2\pi i(\boldsymbol{\Sigma}\cdot\boldsymbol{X})]$ (de Broglie wave of the centre of gravity), $\Phi_2 = \exp\,[2\pi i(\boldsymbol{\Sigma}\cdot\boldsymbol{X})]\Phi_1$. The Φ_i are here functions of the relative coordinates of the particles.

The matrix elements in equations (177) and (178) are evaluated most conveniently by first integrating over the relative coordinates (with respect to the centre of gravity). The action of the $\boldsymbol{v}_i^0$ on the factors $\exp\,[2\pi i(\boldsymbol{\Sigma}\cdot\boldsymbol{X})]$ will often be negligibly small, but can be taken into account if necessary. The integral obtained in this way must finally be multiplied by a factor $\Omega^{-1}\int\exp\,[2\pi i(-\boldsymbol{\Sigma}_k+\boldsymbol{\Sigma}_{k_0}\cdot\boldsymbol{X})]\exp[2\pi i(\boldsymbol{\sigma}_\lambda\cdot\boldsymbol{X})]\,d\boldsymbol{X}$. The factor Ω^{-1} indicates that we assume that the atom is enclosed—as is the radiation—in a cube of volume Ω. Each Φ contains then a normalisation factor $\Omega^{-1/2}$ for the de Broglie wave of the centre of gravity and the $\boldsymbol{\Sigma}$ are restricted to the same discrete values as the $\boldsymbol{\sigma}$ (§ 86).

We see now that the integral is equal to zero except when $\boldsymbol{\Sigma}_{k_0}+\boldsymbol{\sigma}_\lambda = \boldsymbol{\Sigma}_k$, when the integral is equal to unity. This means that the total momentum is conserved during the absorption.

In practical applications the mass of the absorbing system will be so large that the change of momentum during the absorption leads to an extremely small change in velocity. Because of this equation (182) can still be applied. The calculations proceed as if the centre of gravity of the system is fixed in space [and the definition of the current density matrix $(\varrho \mathbf{v})_{kk_0}$ must be changed accordingly, compare the footnote at the discussion of eqq. (10) to (13)].

Equation (182) can be applied immediately to the *photoelectric effect*. Here also the final state of the atom after the absorption will correspond to a continuous set of parameters. We do not wish to enter into a detailed discussion.

In the case of *emission* we have $E_k < E_{k_0}$, and we must apply equations (178). By a calculation similar to equations (180) we can evaluate the probability that at t the atom is in a state k while at the same time a right hand light quantum with a wave vector within an element of solid angle $d\boldsymbol{\omega}$ has been emitted. The result is (compare eq. (180)]

$$\left.\begin{aligned}&\sum_{\lambda(d\boldsymbol{\omega})} |\gamma_k(n_\lambda^0+1)|^2\\ &=\frac{(\langle n_0\rangle+1)\sigma}{\hbar c^2}|(\mathfrak{a}^*\cdot\int \exp\left[-2\pi i(\boldsymbol{\sigma}\cdot\boldsymbol{x})\right](\varrho\mathbf{v})_{kk_0}\,d\boldsymbol{x})|^2 t\,d\boldsymbol{\omega}\\ &=[(\langle n_0\rangle+1)\sigma/\hbar c^2]|(\mathfrak{a}\cdot\int\exp[2\pi i(\boldsymbol{\sigma}\cdot\boldsymbol{x})](\varrho\mathbf{v})_{k_0k}d\boldsymbol{x}])|^2 t\,d\boldsymbol{\omega}.\end{aligned}\right\} \quad (184)$$

The term $\langle n_0\rangle$ in the factor $\langle n_0\rangle+1$ is proportional to the radiation density and corresponds to Einstein's *negative absorption* (*or induced emission*); the term 1 corresponds to the *spontaneous emission* which would be expected in the quantum theory of radiation. If we introduce Einstein's A's and B's by the equation

$$\begin{aligned}\Sigma_{\lambda(d\boldsymbol{\omega})}|\gamma_k(n_\lambda^0+1)|^2 = \{&B_{k_0\to k}(\boldsymbol{\sigma},\mathfrak{a})[I(\nu_{k_0k},\boldsymbol{\omega})/c]\\ &+A_{k_0\to k}(\boldsymbol{\sigma},\mathfrak{a})\}t\,d\boldsymbol{\omega},\end{aligned} \quad (185)$$

and use equation (184) we find for B

$$\left.\begin{aligned}B_{k_0\to k}(\boldsymbol{\sigma},\mathfrak{a}) &= (2\pi\hbar^2\nu_{k_0k})^{-1}|(\mathfrak{a}\,.\int\exp\left[2\pi i(\boldsymbol{\sigma}\cdot\boldsymbol{x})\right](\varrho\mathbf{v})_{k_0k}d\boldsymbol{x})|^2;\\ \sigma &= |\nu_{kk_0}|/c,\ \nu_{kk_0}<0,\end{aligned}\right\} \quad (186)$$

in agreement with equation (22), while we find for A

$$A_{k_0\to k}(\boldsymbol{\sigma},\mathfrak{a}) = (\nu_{k_0k}/\hbar c^3)|(\mathfrak{a}\,.\int\exp\left[2\pi i(\boldsymbol{\sigma}\cdot\boldsymbol{x})\right](\varrho\mathbf{v})_{k_0k}d\boldsymbol{x})|^2. \quad (187)$$

As in the case of real absorption these equations hold also, if $\mathfrak{a}$ refers to an arbitrary direction of polarisation. The influence of the spin can again be taken into account by adding the spin current.

The coefficient A of equation (187) is more specialised than the A_{k_0k} introduced by Einstein and discussed in § 84. The present A refers to those spontaneous transitions where a light quantum of given direction of polarisation is emitted within a given element of solid angle. It confirms exactly the assumption made in the correspondence theory of the emission of radiation that the emission corresponds to the classical emission of a *virtual source of radiation* $(\varrho \mathbf{v})_{k_0} = \Sigma_k 2 \operatorname{Re}\{(\varrho \mathbf{v})_{k_0 k} \exp(2\pi i \nu_{k_0 k} t)\} [\nu_{k_0 k} > 0]$ as far as intensity and polarisation are concerned (see § 84). By integrating over all directions and summing over the two kinds of polarisation we get immediately expression (56) for the total $A_{k_0 \to k}$ of the spontaneous transitions. There are, of course, a number of details where there are differences, corresponding to the quantum character of the radiation. These can be studied, for instance, in the case where there is initially no external radiation, by using the exact expressions for the coefficients γ. One can, for instance, look for a proof that the spontaneous radiation is really propagated with the velocity of light from the atom; this could be seen from an analysis of the corresponding classical radiation fields * (see § 87).

We must also take into account the conservation of momentum, if the atom is free. As in the analysis of the impulse of absorption we find that in equation (176) only those coefficients G are different from zero for which the total momentum is rigorously conserved. If therefore the emitted light quantum is found experimentally (for instance, photoelectric effect) to be active in a direction $\boldsymbol{\sigma}$ from the atom and if thus in that direction the possibility of finding a light quantum is realised, the atom must have undergone a *recoil* in the opposite direction **. This could be checked in principle by measuring the momentum of the atom before and after emission. It is possible to perform the measurements of the direction of the light quantum and of the momentum of the atom independently; however, if no other measurements make the check impossible, the results of these measurements should always be in agreement with the principle of the conservation of momentum ***.

Also the processes connected with an induced emission [eq. (186)]

* See also W. Heisenberg, *Z. Phys.*, **9**. 338, 1931.

** This effect, which follows immediately from the conception of the corpuscular nature of the radiation (needle-radiation) was discussed by Einstein in 1917. He showed the necessity of its occurrence together with that of the impulses during absorption if the Maxwell distribution is to be conserved.

*** See W. Heisenberg, *The physical principles of quantum theory,* University of Chicago Press, 1930. In principle we are dealing with a situation similar to the one where two free particles with initially well defined momenta have collided. This latter situation is a basic example of the quantum theoretical description of compound systems (§ 26).

involve a change in the momentum of the atom. During these processes the intensity of directed radiation of frequency ν_{k_0k} falling on the atom is increased (negative absorption) and the momentum of the atom in the direction opposite to the direction of the radiation is increased after the transition. The exact relations in the radiation field in the cases of induced and spontaneous emission are, as in the case of absorption, very complicated when the initial state of the radiation field is an arbitrary one.

We shall finally discuss briefly the question of the *natural* * *shape of spectral lines*. Let an atom be initially in an excited state k_0 and let there be no radiation. After some time there will be a positive probability (which is increasing with time) that there is present in space a light quantum of a frequency ν_λ which is approximately equal to ν_{k_0k}, where k is a state of lower energy to which the atom can make a radiative transition. We ask now what will be the probability distribution of the different ν_λ in the neighbourhood of ν_{k_0k} in the case where we have waited sufficiently long for the probability that the atom is still in the state k_0 to be practically equal to zero. We write $x = 2\pi(\nu_\lambda - \nu_{k_0k})$ and denote by $W(x)dx$ the relative probability that a light quantum is present with a frequency ν_λ between $\nu_{k_0k} + (x/2\pi)$ and $\nu_{k_0k} + [(x+dx)/2\pi]$. The function $W(x)$ describes the "shape" of a spectral line.

The solution of this problem requires the solution of the differential equation (171) with the boundary condition $\gamma_{k_0}(\ldots, 0, \ldots; \ldots, 0, \ldots) = 1$, all other $\gamma = 0$. Equation (176) gives us the solution for such small values of t that $|\gamma_{k_0}|$ is still practically equal to unity. We are now, however, interested in the solution for much larger values of t.

In the simple special case where k is the only state below k_0 which can be reached from k_0 by a radiative transition, and where from k there are no further transitions possible, the solution of the problem is contained in our discussion in § 55 of the time-proportional transition probabilities. According to equation (5.96) the probability of finding the atom in the state k_0 decreases exponentially, $|\gamma_{k_0}|^2 = \exp(-A_{k_0k}t)$, and according to equation (5.97) the function $W(x)$ is given by the equation

$$W(x) = \text{constant} \cdot [x^2 + \tfrac{1}{4}A^2_{k_0k}]^{-1}. \tag{188}$$

* The term "natural" indicates that experimentally observed spectral lines have a shape which is unfluenced by many other phenomena (Doppler effect, collision broadening, Stark effect, pressure broadening, self absorption, . . .)

In these equations A_{k_0k} is the Einstein coefficient of the total probability of a spontaneous transition. If k_0 is the first excited state of a linear harmonic oscillator of mass m, frequency ω, and charge e, and if k is the ground state, and if the extension in space of the states k_0 and k is small compared to the wave length c/ω, we find from equations (59) and (99) (put $eq_{1,0} = P_{k_0k}$),

$$A_{k_0\to k} = \frac{32\pi^3\omega^3}{3\hbar c^3}|P_{k_0k}|^2 = \frac{32\pi^3\omega^3}{3\hbar c^3}e^2\frac{\hbar}{4\pi\omega m} = \frac{8\pi^2e^2\omega^2}{3mc^3},$$

which is independent of $\hbar$. Substituting this into equation (188) we get exactly the natural line shape which in the classical theory of electrons is obtained for a harmonic oscillator damped by radiation.

Weiskopf and Wigner * have considered the more general case where several states k' can be reached from k_0 and where several states k'' can be reached from k. Instead of equation (188) they get

$$W = \text{constant}\cdot[x^2+\tfrac{1}{4}A^2]^{-1}, \quad A = \Sigma_{k'}A_{k_0k'}+\Sigma_{k''}A_{kk''}. \tag{189}$$

In the first sum $\Sigma_{k'}$ the term A_{k_0k} occurs; this term is the only one in the special case (188). We do not wish to discuss the details, but remark only that it is clear from equation (171) why the transition probabilities from k to lower states k'' will influence the line shape. A situation where there is present a light quantum of frequency approximately equal to ν_{k_0k} is not necessarily one where the atom is in the state k. After a sufficiently long time there will be only those states where the atom has reached the even lower states k'' and where one or more light quanta of other frequencies are present. The sums $\Sigma|\gamma|^2$ which occur in the evaluation of $W(x)$ are thus not only determined by the way in which $|\gamma_{k_0}|^2$ approaches zero exponentially $[|\gamma_{k_0}|^2 = \exp(-\Sigma_{k'}A_{k_0k'}t)$, the exponent given in the first part of the expression (189) for A; compare the discussion in § 55], but also on the time dependence of the various $|\gamma_k(1)|^2$, that is, on the life time of the state k [each $|\gamma_k(1)|^2$ would decrease as $\exp(-\Sigma_{k''}A_{kk''}t)$ if there were no transitions from the state k_0].

Equation (189) is valid only if x is of the order of magnitude of A. Because of its approximate nature (see § 89) the Dirac theory is unable to evaluate correctly the weak intensity of a spectral line far from its centre.

* V. Weisskopf and E. Wigner, *Z. Phys.*, **63**, 54, 1930.

§ 92. Scattering processes

In contrast to the calculations of § 91 the calculation of scattering of radiation by an atom requires second order perturbation theory. Through the interaction between the incoming radiation and the particles the situation of the total system is perturbed. Classically speaking a forced vibration is superimposed on the unperturbed movement of the particles in the atom. Because of this the atom will be a source of secondary radiation which shows certain properties of the original radiation. This secondary radiation, is however, again a consequence of the interaction between particles and radiation so that in toto we are dealing with a second order effect. We need not fear that the approximate character of the theory will cause us great difficulties. We know that mainly the fast periodic vibrations of the radiation field are incorrectly described (§ 89). However, classically the first part of the scattering process is the influence of the radiation on the atom, and this is correctly described. The particular character of the radiation theory entails, however, that certain spontaneous second order transitions will occur for which there are no immediate correspondence arguments.

The necessary calculations can be performed by means of equation (171). However, the expressions (177) and (178) for the G's are no longer sufficient; we must take into account the matrix elements of the second order terms in equation (174) which were neglected up to now.

We shall consider only the following special case. An atom in a stationary state is in a field of strictly monochromatic, strictly parallel radiation characterised by the wave vector $\boldsymbol{\sigma}_\lambda$ and the state of polarisation $\mathfrak{a}_\lambda$. It is true that our formulae refer to left and right hand light, but as we noted in the previous section they can also immediately be applied when we replace $\mathfrak{c}_\lambda(\mathfrak{c}_\lambda^*)$ in equations (177) and (178) by $\sigma_\lambda \mathfrak{a}_\lambda$ and $\sigma_\lambda \mathfrak{b}_\lambda^*$ ($\sigma_\lambda \mathfrak{b}_\lambda$ and $\sigma_\lambda \mathfrak{a}_\lambda^*$) where $\mathfrak{a}_\lambda$ is an arbitrary polarisation vector perpendicular to $\boldsymbol{\sigma}_\lambda$ $[|\mathfrak{a}_\lambda|^2 = |\mathfrak{b}_\lambda|^2 = 1$, $(\boldsymbol{\sigma}_\lambda \cdot \mathfrak{a}_\lambda) = (\boldsymbol{\sigma}_\lambda \cdot \mathfrak{b}_\lambda) = (\mathfrak{a}_\lambda^* \cdot \mathfrak{b}_\lambda) = 0]$. In the following we shall therefore talk about $\mathfrak{a}$ and $\mathfrak{b}$-quanta.

If the initial state of the atom is denoted by k_0 and the number of quanta present by n_λ^0 (this number must be of the order of magnitude of Ω if the radiation density at the position of the atom is to be appreciable), we see that the situation without perturbation is characterised by $\gamma_{k_0}(0, \ldots, n_\lambda^0, \ldots, 0, \ldots)$ being equal to unity and the other γ's being equal to zero. Let us first of all consider the influence of the first order perturbation terms.

Because of the properties of the G's of equations (177) and (178), in the first approximation only four kinds of $\gamma_k(n_\lambda, \bar{n}_\lambda)$ will become different from zero. Namely, those for which n_λ is increasing or decreasing by one, and those where either a new $\mathfrak{a}$-quantum of wave vector $\boldsymbol{\sigma}_{\lambda'}$ $(\neq \boldsymbol{\sigma}_\lambda)$ or an arbitrary $\mathfrak{b}$-quantum of wave vector $\boldsymbol{\sigma}_{\lambda''}$ is produced. From the differential equation (171) we can obtain the periodic solutions, and we find (the notation should be self explanatory),

$$\left.\begin{aligned}
i\hbar\,\dot{\gamma}_k(n_\lambda^0-1) &= i(\hbar/c\sigma_\lambda\Omega)^{1/2}\,n_\lambda^{0\,1/2}\,M^\lambda_{kk_0}\exp\,[2\pi i(-\nu_\lambda+\nu_{kk_0})t],\\
\gamma_k(n_\lambda^0-1) &= -i(n_\lambda^0/2\pi hc\sigma_\lambda\Omega)^{1/2}M^\lambda_{kk_0}\{\exp[2\pi i(-\nu_\lambda+\nu_{kk_0})t]-1\}/(-\nu_\lambda+\nu_{kk_0});\\
i\hbar\,\dot{\gamma}_k(n_\lambda^0+1) &= -i(\hbar[n_\lambda^0+1]/c\sigma_\lambda\Omega)^{1/2}N^\lambda_{kk_0}\exp[2\pi i(\nu_\lambda+\nu_{kk_0})t],\\
\gamma_k(n_\lambda^0+1) &= i([n_\lambda^0+1]/2\pi hc\sigma_\lambda\Omega)^{1/2}N^\lambda_{kk_0}\{\exp[2\pi i(\nu_\lambda+\nu_{kk_0})t]-1\}/(\nu_\lambda+\nu_{kk_0});\\
i\hbar\,\dot{\gamma}_k(n_\lambda^0, 1_{\lambda'}) &= -i(\hbar/c\sigma_{\lambda'}\Omega)^{-1/2}N^{\lambda'}_{kk_0}\exp[2\pi i(\nu_{\lambda'}+\nu_{kk_0})t],\\
\gamma_k(n_\lambda^0, 1_{\lambda'}) &= i(2\pi hc\sigma_{\lambda'}\Omega)^{1/2}N^{\lambda'}_{kk_0}\{\exp[2\pi i(\nu_{\lambda'}+\nu_{kk_0})t]-1\}/(\nu_{\lambda'}+\nu_{kk_0});\\
i\hbar\,\dot{\gamma}_k(n_\lambda^0, \bar{1}_{\lambda''}) &= -i(\hbar/c\sigma_{\lambda''}\Omega)^{1/2}O^{\lambda''}_{kk_0}\exp[2\pi i(\nu_{\lambda''}+\nu_{kk_0})t],\\
\gamma_k(n_\lambda^0, \bar{1}_{\lambda''}) &= i(2\pi hc\sigma_{\lambda''}\Omega)^{-1/2}O^{\lambda''}_{kk_0}\{\exp[2\pi i(\nu_{\lambda''}+\nu_{kk_0})\,t]-1\}/(\nu_{\lambda''}+\nu_{kk_0}).
\end{aligned}\right\}\quad(190)$$

The M, N, and O are the matrix elements

$$\left.\begin{aligned}
M^\lambda_{kk_0} &= \{\Sigma_i e_i(\mathbf{v}_i^0\cdot\mathfrak{a}_\lambda)\exp\,[2\pi i(\boldsymbol{\sigma}_\lambda\cdot\mathbf{x}_i)]\}_{kk_0},\\
N^\lambda_{kk_0} &= \{\Sigma_i e_i(\mathbf{v}_i^0\cdot\mathfrak{a}_\lambda^*)\exp\,[-2\pi i(\boldsymbol{\sigma}_\lambda\cdot\mathbf{x}_i)]\}_{kk_0},\\
O^\lambda_{kk_0} &= \{\Sigma_i e_i(\mathbf{v}_i^0\cdot\mathfrak{b}_\lambda^*)\exp\,[-2\pi i(\boldsymbol{\sigma}_\lambda\cdot\mathbf{x}_i)]\}_{kk_0}.
\end{aligned}\right\}\quad(191)$$

In equations (190) k refers to an arbitrary stationary state which may be identical with k_0. The expression $\nu_{kk_0} = (E_k-E_{k_0})/h$ can be negative, positive, or zero.

Equations (190) are substituted into equations (171), and we investigate whether in the second approximation there are γ's with such a time dependence that secular effects occur such as appeared in the case of absorption and emission in the first approximation. Such secular effects in the second approximation mean that certain final states k_1 of the total system which differ from the initial state both in the state of the light quanta field and in the state of the atom have a probability of being realised which increases with time. They are the necessary consequences of the occurrence of small resonance denominators. It is easily seen that these occur only when the final state has practically the same energy as the initial state. This should have been expected from the general validity of the principle of the conservation of energy. The simplest way to proceed is to look for possible final states and then to investigate which γ's of the first approximation [eqq. (**190**)] will contribute to the appearance of this final state. The calculation is, of course only complete when we have taken the second order terms in equation (**174**) into account.

According to equations (**190**) there is the possibility that even in first approximation small resonance denominators, that is, secular effects, will occur. We shall exclude this possibility as far as possible by assuming that the frequency ν_λ of the incoming radiation is not approximately equal to an absorption- or emission-frequency $|\nu_{kk_0}|$ of the atom in its initial state k_0. If spontaneous transitions to lower states are possible ($\nu_{kk_0} < 0$) we cannot prevent the occurrence of resonance denominators in the last two sets of equations (**190**). We may formally proceed as if those resonance denominators will not influence the final results. However, in fact they play an important part in a more detailed analysis of the final expressions.

Since in a second application of equation (**171**) the matrix elements G of equations (**177**) and (**178**) will occur again, the final state k_1 will differ from the intermediate states (**190**) by the addition or disappearance of one light quantum. There are thus finally the following possibilities. (We have written down in an easily understood notation those γ's which will play a role. It must be noted that for each wave vector $\boldsymbol{\sigma}$, except for the one of the incoming radiation, the polarisation vector $\mathfrak{a}$ can be chosen arbitrarily. It is often advisable to choose $\mathfrak{a}$ to be a continuous function of $\boldsymbol{\sigma}$)

I. *Final state k_1 identical with initial state k_0: Rayleigh scattering.*

One of the original ($\boldsymbol{\sigma}_\lambda$, $\mathfrak{a}_\lambda$)-quanta has disappeared; another light quantum of the same frequency but in a (practically always) different direction has appeared. There are two possibilities:

(a) the new quantum is an $\mathfrak{a}$-quantum: $\nu_{\lambda'} \doteq \nu_\lambda$; $\boldsymbol{\sigma}_{\lambda'} \neq \boldsymbol{\sigma}_\lambda$: $\gamma_{k'}(n_\lambda - \mathbf{1}, \mathbf{1}_{\lambda'})$;

(b) the new quantum is a $\mathfrak{b}$-quantum: $\nu_{\lambda''} \doteq \nu_\lambda$; $\boldsymbol{\sigma}_{\lambda''} \neq \boldsymbol{\sigma}_\lambda$ or $\boldsymbol{\sigma}_{\lambda''} = \boldsymbol{\sigma}_\lambda$: $\gamma_{k_1}(n_\lambda - \mathbf{1}, \overline{\mathbf{1}}_{\lambda''})$.

II. *Final state k_1 not identical with initial state k_0; $E_{k_1} > E_{k_0}$*, $\nu_{k_1k_0} > 0$.*

One of the original quanta has disappeared; another light quan-

* For the sake of simplicity we do not consider the case $E_{k_1} = E_{k_0}$ which can occur when E_{k_0} is degenerate.

tum of *smaller* frequency in an arbitrary direction has appeared: *Smekal-Raman effect*. There are again two possibilities:

(a) $\nu_{\lambda'} \doteq \nu_\lambda - \nu_{k_1 k_0}$; $\mathfrak{a}$-quantum, $\gamma_{k_1}(n_\lambda - 1, 1_{\lambda'})$;

(b) $\nu_{\lambda''} \doteq \nu_\lambda - \nu_{k_1 k_0}$; $\mathfrak{b}$-quantum, $\gamma_{k_1}(n_\lambda - 1, \overline{1}_{\lambda''})$.

III. *Final state k_1 not identical with initial state k_0*; $E_{k_1} < E_{k_0}$, $\nu_{k_1 k_0} < 0$.

We must differentiate between

IIIA. One of the original light quanta has disappeared; another light quantum of larger frequency in an arbitrary direction has appeared (antistokes Smekal-Raman effect). Two possiblities:

(a) $\nu_{\lambda'} \doteq \nu_\lambda - \nu_{k_1 k_0}$; $\mathfrak{a}$-quantum, $\gamma_{k_1}(n_\lambda - 1, 1_{\lambda'})$;

(b) $\nu_{\lambda''} \doteq \nu_\lambda - \nu_{k_1 k_0}$; $\mathfrak{b}$-quantum, $\gamma_{k_1}(n_\lambda - 1, \overline{1}_{\lambda''})$.

IIIB. The number of the original light quanta has increased by one, and another light quantum has appeared. This effect can occur only when $|\nu_{k_1 k_0}| > \nu_\lambda$. Two possibilities:

(a) $\nu_{\lambda'} \doteq -\nu_\lambda - \nu_{k_1 k_0}$; $\mathfrak{a}$-quantum, $\gamma_{k_1}(n_\lambda + 1, 1_{\lambda'})$;

(b) $\nu_{\lambda''} \doteq -\nu_\lambda - \nu_{k_1 k_0}$; $\mathfrak{b}$-quantum, $\gamma_{k_1}(n_\lambda + 1, \overline{1}_{\lambda''})$.

The effect I can be interpreted as elastic scattering of light by an atom. Smekal * was the first to consider the possibility of effects II and IIIA; Kramers and Heisenberg ** showed the necessity of their occurrence by a correspondence argument, and effect IIIB occurred as a necessary complement in their discussion. The effects II and IIIA were found experimentally in 1928 by Raman (scattering by liquids) and Mandelstamm and Landsberg (scattering by crystals).

If we drop the assumption that there is originally only one kind of light quanta, there are also secular processes possible where two light quanta are absorbed leaving the atom in a higher energy state ***.

We shall consider the three processes Ia, IIa, and IIIAa, involving $\gamma_{k_1}(n_\lambda - 1, 1_{\lambda'})$. If Δ is given by the equation

$$\Delta = 2\pi(\nu_{\lambda'} + \nu_{k_1 k_0} - \nu_\lambda), \tag{192}$$

we get by substituting expressions (190) into equation (171)

$$\begin{aligned} i\hbar\dot{\gamma}_{k_1}(n_\lambda - 1, 1_{\lambda'}) = \{\Sigma_k[G(k_1, n_\lambda^0 - 1, 1_{\lambda'}; k, n_\lambda^0 - 1)\gamma(n_\lambda^0 - 1) \\ + G(k_1, n_\lambda^0 - 1, 1_{\lambda'}; k, n_\lambda^0, 1_{\lambda'})\gamma(n_\lambda^0, 1_{\lambda'})] + F_{k_1 k_0}^{\lambda\lambda'}\} \exp(i\Delta t), \end{aligned} \tag{193}$$

* A. Smekal, *Naturwiss.*, **11**, 873, 1923.

** H. A. Kramers and W. Heisenberg, *Z. Phys.*, **31**, 681, 1925

*** M. G. Mayer, *Ann. Phys.*, **9**, 273, 1931.

The first two terms within braces can be explained as follows. The process in which we are interested may be described by either absorption of a λ-quantum followed by the emission of a λ'-quantum, or the emission of a λ'-quantum followed by the absorption of a λ-quantum. Note that this is not necessarily a description of the physical processes involved, but rather an interpretation of the mathematical expressions. The $\hat{\gamma}$ in equation (193) are the coefficients of the braces in the expressions (190) for the γ's. In using the $\hat{\gamma}$ we have effectively dropped the -1 terms in the γ's, since these will not give rise to resonance denominators. The summation is over all stationary states k of the atom. The extra term $F^{\lambda\lambda'}_{k_1, k_0}$ refers to the second order term $\Sigma_i (e_i^2/2m_i c^2)\mathfrak{A}_i'^{\text{ext}\,2}$ of the perturbation energy (174). Inspection of equation (167) shows that $\mathfrak{A}_i'^{\text{ext}\,2}$ is a sum of products each of two of the operators $\boldsymbol{\alpha}_\lambda$, $\boldsymbol{\beta}_\lambda$, $\boldsymbol{\alpha}^*_\lambda$, and $\boldsymbol{\beta}^*_\lambda$ which will contribute to the scattering processes considered. In equation (193) $F^{\lambda\lambda'}_{k_1, k_0}$ derives from the product $\boldsymbol{\alpha}_\lambda \boldsymbol{\alpha}^*_\lambda$.

If we use the abbreviated notation

$$\begin{aligned} R^{\lambda\lambda'}_{k_1, k_0} &= [\Sigma_i (e_i^2/m_i) \exp\{2\pi i(\boldsymbol{\sigma}_\lambda - \boldsymbol{\sigma}_{\lambda'} \cdot \boldsymbol{x}_i)\}]_{k_1 k_0}, \\ S^{\lambda\lambda'}_{k_1, k_0} &= [\Sigma_i (e_i^2/m_i) \exp\{-2\pi i(\boldsymbol{\sigma}_\lambda + \boldsymbol{\sigma}_{\lambda'} \cdot \boldsymbol{x}_i)\}]_{k_1 k_0}, \end{aligned} \tag{194}$$

and make use of equations (177) and (178) we get from equation (193)

$$\begin{aligned} i\hbar\dot{\gamma}(n_\lambda - 1, 1_{\lambda'}) = (\hbar^2 n_\lambda^0/\nu_{\lambda'}\nu_\lambda \Omega^2)^{1/2}[-\Sigma_k\{[N^{\lambda'}_{k_1, k}\; M^\lambda_{kk_0}/h(-\nu_\lambda + \nu_{kk_0})] \\ + [M^\lambda_{k_1 k} N^{\lambda'}_{kk_0}/h(\nu_{\lambda'} + \nu_{kk_0})]\} + R^{\lambda\lambda'}_{k_1 k_0}(\mathfrak{a}_\lambda \cdot \mathfrak{a}^*_{\lambda'})] \exp(i\Delta t). \end{aligned} \tag{195}$$

A similar equation will hold for the processes Ib, IIb, and IIIAb where a $\mathfrak{b}$-quantum is scattered, but this equation is of no particular interest because $\mathfrak{a}_\lambda$ is arbitrary.

For the processes IIIBa, corresponding to a double emission, we have a similar equation,

$$\begin{aligned} i\hbar\dot{\gamma}_{k_1}(n_\lambda + 1, 1_{\lambda'}) = (\hbar^2[n_\lambda^0 + 1]/\nu_{\lambda'}\nu_\lambda \Omega^2)^{1/2}[\Sigma_k\{[N^{\lambda'}_{k_1 k} N^\lambda_{kk_0}/h(\nu_\lambda + \nu_{kk_0})] \\ + [N^\lambda_{k_1 k} N^{\lambda'}_{kk_0}/h(\nu_{\lambda'} + \nu_{kk_0})]\} - S^{\lambda\lambda'}_{k_1 k_0}(\mathfrak{a}^*_\lambda \cdot \mathfrak{a}^*_{\lambda'})] \exp(i\Delta t). \end{aligned} \tag{196}$$

We must now sum the absolute squares of the $\gamma_{k_1}(n_\lambda - 1, 1_{\lambda'})$ over a small interval of wave vectors $\boldsymbol{\sigma}_{\lambda'}$ which span an element of solid angle $d\boldsymbol{\omega}$ and the absolute values of which are approximately equal to $|\nu_\lambda \mp \nu_{k_1 k_0}|/c$. We obtain then the intensity of the scattering of the type considered in that particular direction, or, more precisely, the probability that during a time interval t a light quantum of frequency $\nu_{\lambda'}$ is emitted within the element of solid angle $d\boldsymbol{\omega}$. From equation (180) and

$$i\dot{\gamma} = T \exp(i\Delta t) \tag{197}$$

we have

$$\Sigma_{(d\boldsymbol{\omega})}|\gamma|^2 = \Sigma_{(d\boldsymbol{\omega})}|T|^2|(1-e^{i\Delta t})/\Delta|^2 = |T|^2\, \Omega\sigma_{\lambda'}^2 d\boldsymbol{\omega} \int |(1-e^{i\Delta t})/\Delta|^2\, d\sigma_\lambda^{\cdot}$$
$$= (\Omega\sigma_{\lambda'}^2/c)|T|^2 t d\boldsymbol{\omega}. \tag{198}$$

In the case (195) the scattering intensity is exactly proportional to n_λ^0, that is, to the intensity of the incoming radiation which is given by the equation

$$I/h\nu_\lambda = cn_\lambda^0/\Omega. \tag{199}$$

The *cross section* $Q_{\lambda'}d\boldsymbol{\omega}$ for the scattering of frequency $\nu_{\lambda'} = \nu_\lambda - \nu_{kk_0}$ (processes I, II, and IIIa) is now obtained by dividing expression (198) by expression (199) and t, $Q_{\lambda'}d\boldsymbol{\omega} = (\Omega^2\nu_{\lambda'}^2/n_\lambda^0 c^4)|T|^2 d\boldsymbol{\omega}$. Using equations (197) and (195) we have finally

$$Q_{\lambda'}d\boldsymbol{\omega} = (\nu_{\lambda'}/\nu_\lambda c^4)|-\Sigma_k\{[N_{k_1 k}^{\lambda'} M_{kk_0}^\lambda / h(\nu_{kk_0}-\nu_\lambda)]$$
$$+[M_{k_1 k}^\lambda N_{kk_0}^{\lambda'}/h(\nu_{kk_0}+\nu_{\lambda'})]\}+R_{k_1 k_0}^{\lambda\lambda'}(\mathfrak{a}_\lambda \cdot \mathfrak{a}_{\lambda'}^*)|^2 d\boldsymbol{\omega}, \tag{200}$$
$$E_{k_1} \gtreqless E_{k_0}, \quad \nu_{\lambda'} = \nu_\lambda - \nu_{k_1 k_0}.$$

Instead of $\nu_{kk_0}+\nu_{\lambda'}$ we can clearly also write $\nu_{kk_1}+\nu_\lambda$.

In the case (196) the scattering is clearly proportional to n_λ^0+1; the 1 means spontaneous emission. Even if there were no radiation present, the atom could go from the state k_0 to the state k_1 emitting at the same time two light quanta of frequencies ν_λ and $\nu_{\lambda'}$ with $\nu_\lambda+\nu_{\lambda'} = \nu_{k_0 k_1}$. From a correspondence point of view there is no reason for such a twofold emission; up to a point this effect is a consequence of the fact that the basic canonical equations are only approximately valid. The question of the physical meaning and observability of these processes and their relation to the spontaneous single emissions leads to questions beyond the scope of the present theory.

The term n_λ in equation (196) corresponds to scattering processes accompanied by two-fold emission. The corresponding cross section is

$$Q_{\lambda'}d\boldsymbol{\omega} = (\nu_{\lambda'}/\nu_\lambda c^4)|-\Sigma_k\{[N_{k_1 k}^{\lambda'} N_{kk_0}^\lambda / h(\nu_{kk_0}+\nu_\lambda)]$$
$$+[N_{k_1 k}^\lambda N_{kk_0}^{\lambda'}/h(\nu_{kk_0}+\nu_{\lambda'})]\}+S_{k_1 k_0}^{\lambda\lambda'}(\mathfrak{a}_\lambda^* \cdot \mathfrak{a}_{\lambda'}^*)|^2 d\boldsymbol{\omega}, \tag{201}$$
$$E_{k_1} < E_{k_0}, \quad \nu_{\lambda'} = \nu_{k_0 k_1} - \nu_\lambda.$$

Again we may write $\nu_{kk_1}-\nu_\lambda$ for $\nu_{kk_0}+\nu_{\lambda'}$ in the denominator.

We have now obtained exact equations for the occurrence of light scattered by any atomic system. We note that the expressions M and N are given by equations (191) and R and S by equations (194). They depend on the wave vectors $\boldsymbol{\sigma}_\lambda$ and $\boldsymbol{\sigma}_{\lambda'}$ and the polarisation vectors $\mathfrak{a}_\lambda$ and $\mathfrak{a}_{\lambda'}$ of the incoming and scattered radiation.

In the practically important *limiting case where the dimensions a of the scattering system are small compared to the wave lengths of the incoming and the scattered light* the expressions for M, N, R, and S simplify. If we choose the origin of the system of coordinates somewhere inside the atom, all factors of the form $\exp[\pm 2\pi i(\boldsymbol{\sigma}\cdot\boldsymbol{x})]$ can be put equal to unity and we have [see eq. (24)] $(a\sigma_\lambda \ll 1$, $a\sigma_{\lambda'} \ll 1)$

$$M^\lambda_{kk_0} \doteq (\dot{\boldsymbol{P}}_{kk_0}\cdot\mathfrak{a}_\lambda) = 2\pi i\nu_{kk_0}(\boldsymbol{P}_{kk_0}\cdot\mathfrak{a}_\lambda),\ N^\lambda_{kk_0} \doteq (\dot{\boldsymbol{P}}_{kk_0}\cdot\mathfrak{a}^*_\lambda) = 2\pi i\nu_{kk_0}(\boldsymbol{P}_{kk_0}\cdot\mathfrak{a}^*_\lambda), \tag{202}$$

where $\boldsymbol{P}$ is the total electrical polarisation of the atom. The expressions for R and S are even simpler; they become the unit matrix, apart from a factor,

$$R^{\lambda\lambda'}_{k_1k_0} = S^{\lambda\lambda'}_{k_1k_0} = \Sigma_i(e_i^2/2m_i)\,\delta_{k_1k_0} \quad (a\sigma_\lambda \ll 1,\ a\sigma_{\lambda'} \ll 1). \tag{203}$$

We can now modify equation (200) slightly. From the commutation relations (3.77) for particles it follows that

$$\Sigma_k(P^{(x)}_{k_1k}\dot{P}^{(x)}_{kk_0} - \dot{P}^{(x)}_{k_1k}P^{(x)}_{kk_0}) = i\hbar\Sigma_i(e_i^2/m_i), \ldots;$$
$$\Sigma_k(P^{(x)}_{k_1k}\dot{P}^{(y)}_{kk_0} - \dot{P}^{(y)}_{k_1k}P^{(x)}_{kk_0}) = 0, \ldots,$$

and that

$$R^{\lambda\lambda'}_{k_1k_0}(\mathfrak{a}_\lambda\cdot\mathfrak{a}^*_{\lambda'})$$
$$= (2\pi/\hbar)\,\Sigma_k\{(\boldsymbol{P}_{k_1k}\cdot\mathfrak{a}_\lambda)(\boldsymbol{P}_{kk_0}\cdot\mathfrak{a}^*_{\lambda'})\nu_{kk_0} - (\boldsymbol{P}_{k_1k}\cdot\mathfrak{a}^*_{\lambda'})(\boldsymbol{P}_{kk_0}\cdot\mathfrak{a}_\lambda)\nu_{k_1k}\}. \tag{204}$$

In this equation $\mathfrak{a}_\lambda$ and $\mathfrak{a}^*_{\lambda'}$ can be interchanged. Using equations (202) and (204) we get from equation (200)

$$Q_{\lambda'}d\boldsymbol{\omega} = \frac{4\pi^2\nu^3_{\lambda'}\nu_\lambda}{\hbar^2c^4}\left|\sum_k\left[\frac{(\boldsymbol{P}_{k_1k}\cdot\mathfrak{a}^*_{\lambda'})(\boldsymbol{P}_{kk_0}\cdot\mathfrak{a}_\lambda)}{\nu_{kk_0}-\nu_\lambda} + \frac{(\boldsymbol{P}_{k_1k}\cdot\mathfrak{a}_\lambda)(\boldsymbol{P}_{kk_0}\cdot\mathfrak{a}^*_{\lambda'})}{\nu_{kk_1}+\nu_\lambda}\right]\right|^2 d\boldsymbol{\omega} \tag{205}$$

Equation (201) can be similarly rewritten.

For the occurrence of a given term in the Σ_k in equation (205) it is necessary that both $\boldsymbol{P}_{k_1k}$ and $\boldsymbol{P}_{kk_0}$ are different from zero, that is, that both the spontaneous (dipole) transition from k_0 to k and the one from k to k_1 are possible. This is usually stated as follows: *a given scattering process is possible only when there are intermediate states* *combining optically with both the initial and the final state.* It is immaterial whether the direct transition from the initial to the final state is possible; in many cases occurring in practice this transition is forbidden by selection rules *.

If we apply equation (205) to an harmonic oscillator, each state "combines" with only one other state with a quantum number either one larger or one smaller, since only in that case are the matrix elements of the polarisation different from zero. One can then show that Rayleigh scattering is the only possibility and that its intensity is exactly the same as in the classical theory of electrons.

Another limiting case of equations (200) and (201) obtains in the limit where the frequency ν_λ of the incoming radiation is very large compared to all frequencies $|\nu_{kk_0}|$ and $|\nu_{kk_1}|$ referring to intermediate states which are essential in evaluating the (convergent) sums Σ_k. The classical analogue to this limiting case is when the frequency of the incoming radiation is large compared to the "vibrational frequencies" of the particles of the scattering system; the theory of electrons shows that the particles scatter in first approximation as if they were free (Rayleigh scattering only). The same result holds also in quantum theory. This is dus to the fact that in the limit the sums Σ_k and the terms R, S are all very small except the diagonal terms $R^{\lambda\lambda'}_{k_0 k_0}$ which correspond to the scattering by free particles **.

§ 93. Conservation of momentum in scattering processes; Compton effect

If a free atom is scattering radiation we must take into account the conservation of momentum—as in the case of absorption (§ 91) and emission. Even if the stationary states with respect to the centre of gravity are discrete ones, the states k_0, k, k_1 will still belong to a continuous set corresponding to the arbitrariness of the total momentum. As in § 91 we introduce the assumption that in the initial and final state the total momentum has exactly defined values $h\mathbf{\Sigma}_{k_0}$ and $h\mathbf{\Sigma}_{k_1}$ corresponding to a de Broglie wave of the centre of gravity in the cube Ω—which means that we assume the position of the atom to be totally undetermined. If we also assume the "intermediate" state k to be characterised by a well defined momentum

* See, for instance, G. Placzek, *Handb. Radiologie*, **6**$_2$, 209, 1934.

** See H. A. Kramers, *Physica* (*Old series*), **5**, 369, 1926 where the connexion with the so-called *Thomas-Kuhn sum rule* (W. Thomas, *Naturwiss.*, **13**, 627, 1927; W. Kuhn, *Z. Phys.*, **33**, 408, 1925) is discussed.

$h\boldsymbol{\Sigma}_k$, it can be seen that the matrix elements M in equation (200) [see eq. (191)] are different from zero only provided the following conditions are satisfied *,

$$M^{\lambda}_{kk_0} \neq 0, \text{ if } \boldsymbol{\Sigma}_k = \boldsymbol{\sigma}_\lambda + \boldsymbol{\Sigma}_{k_0};\ M^{\lambda}_{k_1 k} \neq 0\ ,\ \text{if } \boldsymbol{\Sigma}_{k_1} = \boldsymbol{\sigma}_\lambda + \boldsymbol{\Sigma}_k,$$

that is, the momentum in the intermediate state must be equal to the momentum of the initial state plus the momentum of a light quantum of wave vector $\boldsymbol{\sigma}_\lambda$. Similarly the N of equations (200) and (201) are different from zero only if the following conditions are satisfied,

$$N^{\lambda}_{k_1 k} \neq 0, \text{ if } \boldsymbol{\Sigma}_{k_1} + \boldsymbol{\sigma}_\lambda = \boldsymbol{\Sigma}_k;\ N^{\lambda}_{kk_0} \neq 0, \text{ if } \boldsymbol{\Sigma}_k + \boldsymbol{\sigma}_\lambda = \boldsymbol{\Sigma}_{k_0}.$$

In the framework of the formal description of scattering processes mentioned in the last section where each scattering process was considered to consist of two elementary processes each involving *one* light quantum only, it is seen that in each elementary process momentum is strictly conserved (the same was not true of the energy). The matrix elements R and S in equations (200) and (201), which do not involve intermediate states, are also different from zero only if the difference of momentum in the initial and the final state is exactly compensated by the two light quanta involved in the scattering process.

It is important to note that this exact conservation of momentum involves a *correction of the scattering formulae* (200) and (201). This is due to the fact that in evaluating the sum (198) over all light quanta $h\nu_{\lambda'}$ which were scattered within the element of solid angle $d\boldsymbol{\omega}$ we assumed $\nu_{k_1 k_0}$, and thus also the energy of the final state, to be fixed. However, a change in $\boldsymbol{\sigma}_{\lambda'}$ involves the same change (but with opposite sign) in $\boldsymbol{\Sigma}_{k_1}$ and thus a change in E_{k_1}. In changing the sum $\Sigma_{(d\boldsymbol{\omega})}|(1-e^{i\Delta t})/\Delta|^2$ into an integral over Δ,

$$\Sigma_{(d\boldsymbol{\omega})}|(1-e^{i\Delta t})/\Delta|^2 = \Omega\sigma^2_{\lambda'}d\boldsymbol{\omega}\int|(1-e^{i\Delta t})/\Delta|^2 d\sigma_{\lambda'},$$

we can no longer replace $d\sigma_{\lambda'} = c^{-1}d\nu_{\lambda'}$ simply by $d\Delta/2\pi c$, but we must use the definition (192) of Δ to write $d\Delta = 2\pi(d\nu_{\lambda'} + d\nu_{k_1 k_0}) = 2\pi d\nu_{\lambda'}[1+(dE_{k_1}/hc\,d\sigma_{\lambda'})]$. Since according to the principles of wave mechanics (see § 4) $dE_{k_1} = h(\boldsymbol{v}_{k_1}\cdot d\boldsymbol{\Sigma}_{k_1})$, where $\boldsymbol{v}_{k_1}$ is the group

* If in evaluating $M^{\lambda}{}_{kk_0}$ of equation (191) the particle coordinates are expressed in terms of the relative coordinates with respect to the centre of gravity and the centre of gravity coordinates $\boldsymbol{X}$, we are after integrating over the relative coordinates left with an integral containing $\boldsymbol{X}$ in the form $\Omega^{-1}\exp[2\pi i(-\boldsymbol{\Sigma}_k + \boldsymbol{\sigma}_\lambda + \boldsymbol{\Sigma}_{k_0}\cdot\boldsymbol{X})]$ (see § 91).

velocity, that is, the classical velocity in the final state, and since the conservation of momentum entails that $d\boldsymbol{\Sigma}_{k_1} = -d\boldsymbol{\sigma}_{\lambda'}$, we see that equation (198) must be multiplied by a factor $\{1-[(\boldsymbol{v}_{k_1}\cdot\boldsymbol{\sigma}_{\lambda'})/c\sigma_{\lambda'}]\}^{-1}$ The scattering equations (200) and (201) must thus be corrected as follows,

$$Q_{\lambda'}^{\text{corr}} = Q_{\lambda'}^{\text{incorr}}\{1-[(\boldsymbol{v}_{k_1}\cdot\boldsymbol{\sigma}_{\lambda'})/c\sigma_{\lambda'}]\}. \tag{206}$$

If the system in the initial state is at rest, this correction is usually negligibly small.

If we apply equations (200) and (206) to a system consisting of one free electron, we get the *Compton effect*. If we use the non-relativistic approximation and neglect spin, the stationary states φ_k of the unperturbed system are completely determined by the momentum $h\boldsymbol{\Sigma}_k$ of the electron, $\varphi_k = \Omega^{-1/2}\exp[2\pi i(\boldsymbol{\Sigma}_k\cdot\boldsymbol{x})]$, and the velocity operator is given simply by $\boldsymbol{v} = \boldsymbol{p}/m = \hbar\boldsymbol{\nabla}/mi$. From equations (191) we get for the matrix elements $M_{kk_0}^{\lambda}$ and $N_{kk_0}^{\lambda'}$,

$$M_{kk_0}^{\lambda} = \frac{-e\hbar}{im\Omega}\int \exp\,[-2\pi i(\boldsymbol{\Sigma}_k\cdot\boldsymbol{x})](\boldsymbol{\nabla}\cdot\boldsymbol{\mathfrak{a}}_{\lambda})\exp\,[2\pi i(\boldsymbol{\sigma}_{\lambda}+\boldsymbol{\Sigma}_{k_0}\cdot\boldsymbol{x})]d\boldsymbol{x}$$

$$= -\frac{eh}{m}(\boldsymbol{\sigma}_{\lambda}+\boldsymbol{\Sigma}_{k_0}\cdot\boldsymbol{\mathfrak{a}}_{\lambda}),$$

or, $$M_{kk_0}^{\lambda} = -(eh/m)(\boldsymbol{\Sigma}_{k_0}\cdot\boldsymbol{\mathfrak{a}}_{\lambda}),\quad N_{kk_0}^{\lambda'} = -(eh/m)(\boldsymbol{\Sigma}_{k_0}\cdot\boldsymbol{\mathfrak{a}}_{\lambda}^{*}).$$

This situation is simplest when the electron is initially at rest ($\boldsymbol{\Sigma}_{k_0} = 0$), since in that case the scattering is determined by the term involving R. From equation (194) we have

$$R_{k_1k_0}^{\lambda\lambda'} = (e^2/m\Omega)\int\exp\,[2\pi i(-\boldsymbol{\Sigma}_{k_1}+\boldsymbol{\sigma}_{\lambda}-\boldsymbol{\sigma}_{\lambda'}+\boldsymbol{\Sigma}_{k_0}\cdot\boldsymbol{x})]d\boldsymbol{x} = e^2/m.$$

We find for the cross section after a simple change in the correction term (202),

$$Q_{\lambda'}d\boldsymbol{\omega} = (\nu_{\lambda'}/\nu_{\lambda})(e^4/m^2c^4)\{1-[(\boldsymbol{\Sigma}_{k_1}\cdot\boldsymbol{\sigma}_{\lambda'})/\sigma_{\mathrm{C}}\sigma_{\lambda'}]\}^{-1}|(\boldsymbol{\mathfrak{a}}_{\lambda}\cdot\boldsymbol{\mathfrak{a}}_{\lambda'}^{*})|^2d\boldsymbol{\omega}, \tag{207}$$

where σ_{C} is given by equation (209) below. The momentum vectors $h\boldsymbol{\sigma}_{\lambda}$, $h\boldsymbol{\sigma}_{\lambda'}$, and $h\boldsymbol{\Sigma}_{k_1}$ lie necessarily in one plane. If the light quantum

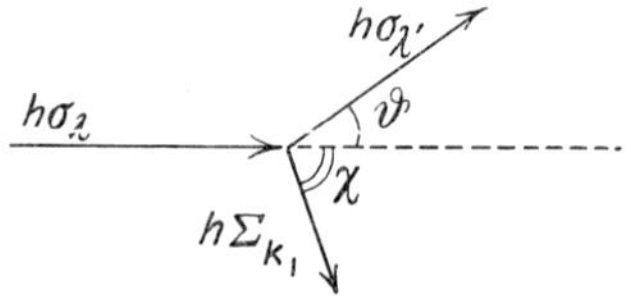

Fig. 14

is scattered over an angle ϑ and if the electron flies away at an angle χ (see fig. 14), we get from the conservation laws,

$$\left.\begin{array}{ll}\left.\begin{array}{l}\text{(a)}\\ \text{(b)}\end{array}\right\}\ \text{(momentum)} & \left\{\begin{array}{r}\sigma_{\lambda'}\sin\vartheta = \Sigma_{k_1}\sin\chi,\\ \sigma_\lambda-\sigma_{\lambda'}\cos\vartheta = \Sigma_{k_1}\cos\chi,\end{array}\right.\\ \text{(c)}\quad \text{(energy)} & c\sigma_\lambda = c\sigma_{\lambda'}+h\Sigma_{k_1}^2/2m,\ \text{or},\ 2\sigma_C(\sigma_\lambda-\sigma_{\lambda'})=\Sigma_{k_1}^2.\end{array}\right\} \tag{208}$$

Here we have introduced the Compton wave number σ_C, the reciprocal of the Compton wave length λ_C,

$$\sigma_C = \lambda_C^{-1} = mc/h, \qquad \lambda_C = 0.024\ \text{ÅU}. \tag{209}$$

Equation (208c) is valid only in the approximation in which the kinetic energy of the electron after the collision is small compared with the rest energy mc^2, that is, if $\Sigma_{k_1} \ll \sigma_C$ $(\sigma_\lambda-\sigma_{\lambda'} \ll \sigma_C)$. The sum of the squared equations (208a) and (208b) combined with equation (208c) gives us the *Compton-shift*,

$$\begin{aligned}2\sigma_C(\sigma_\lambda-\sigma_{\lambda'}) &= \sigma_\lambda^2-2\sigma_\lambda\sigma_{\lambda'}\cos\vartheta+\sigma_{\lambda'}^2 \doteqdot 2\sigma_\lambda^2(1-\cos\vartheta),\\ \sigma_{\lambda'}^{-1} &\doteqdot \sigma_\lambda^{-1}+\sigma_C^{-1}(1-\cos\vartheta);\quad \lambda'-\lambda \doteqdot \lambda_C(1-\cos\vartheta).\end{aligned} \tag{210}$$

This equation is clearly valid only, if the wave number of the incoming light itself is small compared to σ_C, $\sigma_\lambda \ll \sigma_C$. The angle ϑ can take on all values between 0 and π. While ϑ increases from 0 to π, χ decreases from $\frac{1}{2}\pi$ to 0, according to the approximate equation, $\tan\chi \doteqdot \sin\vartheta/(1-\cos\vartheta)$.

The state of polarisation of the scattered light is, according to equation (207), determined by the absolute value of $(\mathfrak{a}_\lambda \cdot \mathfrak{a}_\lambda^*)$ and is thus the same as that in the case of the scattering according to the classical theory of J. J. Thomson. However, the intensity of the scattered light is less by a factor $(\nu_{\lambda'}/\nu_\lambda)^2\{1-[(\mathbf{\Sigma}_{k_1}\cdot\boldsymbol{\sigma}_{\lambda'})/\sigma_C\,\sigma_{\lambda'}]\}^{-1}$ $\doteqdot (\nu_{\lambda'}/\nu_\lambda)^3 \doteqdot 1-3(\sigma_\lambda/\sigma_C)(1-\cos\vartheta)$. The value of $|(\mathfrak{a}_\lambda\cdot\mathfrak{a}_\lambda^*)|^2$ averaged over all possible polarisations of the λ-quantum and summed over two mutually perpendicular directions $\mathfrak{a}_{\lambda'}$ is equal to $\frac{1}{2}(1+\cos^2\vartheta)$ times the Thomson expression for the scattering of unpolarised light *. We get thus for the total cross section $Q(\sigma_\lambda)$ for absorption of unpolarised radiation of wave number σ_λ by an electron the equation

$$Q(\sigma_\lambda)=\frac{e^4}{m^2c^4}\oint\left[1-2\frac{\sigma_\lambda}{\sigma_C}(1-\cos\vartheta)\right]\tfrac{1}{2}(1+\cos^2\vartheta)d\boldsymbol{\omega}=\frac{8\pi e^4}{3m^2c^4}\left[1-2\frac{\sigma_\lambda}{\sigma_C}+..\right]. \tag{211}$$

* The simplest procedure is to resolve the light into linear vibrations parallel and perpendicular to the plane containing the momenta; the first kind of vibrations gives the contribution $\frac{1}{2}\cos^2\vartheta$ and the second kind the contribution $\frac{1}{2}$.

The classical cross section of the Thomson theory appears as the first approximation. The quantum mechanical cross section is, however, smaller than the classical one for large values of σ_λ. Equation (211) is only approximate.

Klein and Nishina * used the *strictly relativistic Dirac theory* to obtain exact equations for the scattering by free electrons correct for arbitrary wave numbers of the incoming light. They used the semi-classical method of treating scattering processes which we shall discuss in § 94; however, Waller ** has shown that the same final result is derived by using the quantum theory of the interaction between radiation and matter. One can again use equation (200) [and the correction (206)], but we must use the simple expression (175) for the interaction operator instead of expression (173). As a consequence the term with R (scattering by free electrons in the non-relativistic approximation) no longer occurs in equation (200). The role of these terms which in the preceding calculations were exclusively responsible for the scattering by an electron initially at rest, is taken over by those terms in equation (200) corresponding to intermediate states k of negative energy ***; however, also the terms corresponding to intermediate states of positive energy are different from zero. Because of the simple form of the velocity operator in the Dirac theory the matrix elements M and N in equations (191) which are necessary for the evaluation of the cross section, are extremely simple, especially when we represent the possible stationary states using equations (6.176) and when we use always circularly polarised light (and as before assume the electron to be initially at rest). For a given state of polarisation of the scattered light and a given direction of the spin after the scattering, the expression which is to be squared in equation (200) will contain four terms since there are only two possible values of the momentum and two possible signs of the energy for the intermediate state. By summing over the two possibilities of polarisation and over the two spin directions after the scattering we get finally for the cross section for scattering over an angle ϑ the Klein-Nishina equation,

* O. Klein and Y. Nishina, *Z. Phys.*, **52**, 853, 869, 1929.

** I. Waller, *Z. Phys.*, **61**, 837, 1930.

*** If an electron in a field of force is described by the Dirac theory (for instance, the hydrogen atom, see § 66) and if one calculates the scattering, the role of the terms with R is in the same way taken over by the terms involving intermediate states of negative energy.

$$Q_{\lambda'} d\boldsymbol{\omega} = \frac{e^4}{m^2c^4}\frac{\sigma_{\lambda'}^2}{\sigma_\lambda^2}\left\{\frac{1}{2}\left(\frac{\sigma_\lambda}{\sigma_{\lambda'}}+\frac{\sigma_{\lambda'}}{\sigma_\lambda}\right)-\frac{|(\mathfrak{a}_\lambda \cdot \boldsymbol{\sigma}_{\lambda'})|^2}{\sigma_{\lambda'}^2}\right\}d\boldsymbol{\omega}. \tag{212}$$

The relation (210) between σ_λ and $\sigma_{\lambda'}$ is now rigorously valid *,

$$\sigma_{\lambda'}^{-1} = \sigma_\lambda^{-1}+\sigma_{\mathrm{C}}^{-1}(1-\cos\vartheta), \quad \lambda'-\lambda = \lambda_{\mathrm{C}}\cos\vartheta.$$

If necessary we can use this relation to eliminate $\sigma_{\lambda'}$ from equation (212).

Equation (212) refers to an arbitrary polarisation vector $\mathfrak{a}$ of the incoming radiation. Averaging over two mutually perpendicular directions of polarisation we get $\sigma_{\lambda'}^{-2}\overline{|(\mathfrak{a}_\lambda \cdot \boldsymbol{\sigma}_{\lambda'})|^2} = \frac{1}{2}\sin^2\vartheta$. We shall not give here the details leading to equation (212).

Heitler** gives in detail Dirac's elegant method of determining $Q_{\lambda'}$. In this method one does not evaluate explicitly the separate terms in equation (200) but the result of the necessary summations over the different final states and averaging over the initial states is obtained immediately by considering the traces of certain operators acting on the spin variables. (The trace of an operator F_{op} is the sum of its eigenvalues and can be written in the form Trace $F_{\mathrm{op}} = \Sigma_\mu F_\mu = \Sigma_m F_{mm}$ [see eq. (4.50)] where F_{mm} can refer to any representation).

The explicit calculation by means of the method indicated here is hardly less simple and has the advantage of giving information about a number of details. For instance, we obtain in the case of right hand incoming light and different final states from equation (200) the following expressions

Scattered light	Direction of spin after scattering	A
right hand	parallel to $\boldsymbol{\Sigma}_{k_1}$	$C\sin^2\frac{1}{2}\vartheta(b\sin\frac{1}{2}\chi - a\cos\frac{1}{2}\chi)$
left hand	parallel to $\boldsymbol{\Sigma}_{k_1}$	$C\sin\frac{1}{2}(\vartheta+\chi)\cos\frac{1}{2}\vartheta\cdot b$ $- S\cos\frac{1}{2}\chi\cos\frac{1}{2}\vartheta(a\cos\frac{1}{2}\vartheta + b\sin\frac{1}{2}\vartheta)$
right hand	antiparallel to $\boldsymbol{\Sigma}_{k_1}$	$S\cos\frac{1}{2}(\vartheta+\chi)\cos\frac{1}{2}\vartheta\cdot b$ $+ C\sin\frac{1}{2}\chi\cos\frac{1}{2}\vartheta(a\cos\frac{1}{2}\vartheta + b\sin\frac{1}{2}\vartheta)$
left hand	antiparallel to $\boldsymbol{\Sigma}_{k_1}$	$S\sin^2\frac{1}{2}\vartheta(b\cos\frac{1}{2}\chi + a\sin\frac{1}{2}\chi)$

where $C = \cos\frac{1}{2}\psi$, $S = \sin\frac{1}{2}\psi$, $(\sin\psi)^{-1} = 1+\sigma_{\mathrm{C}}^{-1}(\sigma_\lambda-\sigma_{\lambda'})$, and $Q_{\lambda'} = (\sigma_{\lambda'}/\sigma_\lambda)^2 (e^4/m^2c^4)(\sin\psi)^{-1}2^{1/2}|A|^2$. The spinor $\begin{pmatrix}a\\b\end{pmatrix}$ describes the spin state of the electron before the scattering, the incoming light is moving in the direction of the z-axis and the x-axis is in the plane of the momenta, which means that the four Dirac components (see § 65) are $2^{-1/2}(a, b, a, b)$. Equation (212) is obtained by summing the coefficients of $|a|^2$ and $|b|^2$ and dividing by 2 ($\mathfrak{a}_\lambda^2 = 0$).

Equation (212) can also be applied to the scattering of hard X- or γ-rays by atoms, provided the electrons in the atoms may be considered to be loosely bound. In that way it was possible to

* In the relativistic case equation (208c) must be replaced by $(\sigma_{\mathrm{C}} - \sigma_\lambda - \sigma_{\lambda'})^2 = \sigma_{\mathrm{C}}^2 + \Sigma_{k_1}^2$, while equations (208a) and (208b) are unchanged.

** W. Heitler, *Quantum theory of radiation*, Oxford University Press, 1954.

verify beautifully, the validity of this formula both as regards the angular distribution of the scattered radiation and as regards the total absorption *.

§ 94. Semi-classical theory of scattering processes

Equations (200) can also be obtained by a method ** similar to the semi-classical theory of the spontaneous transitions (§ 84). At the beginning of § 83 we drew attention to the fact that scattering can be treated purely classically as follows. First of ali we determine the perturbation of the motion of the particles in a system by the external radiation field, and then we calculate the influence of this perturbation on the emission of radiation. The semi-classical method consists in first of all evaluating the influence of a (non-quantised) external radiation field on the stationary states of an atomic system ***. The wave functions in the perturbed states may be denoted by ψ_k; they will in zeroth approximation be equal to exp $(-iE_k t/\hbar)$ times the energy eigenfunctions φ_k of the unperturbed system,

$$\psi_k = \varphi_k \exp(-iE_k t/\hbar) + \psi_k^{(1)}, \tag{214}$$

and $\psi_k^{(1)}$ will be proportional to the amplitude of the incoming radiation. We can now construct the following time dependent matrix of the current operator $\frac{1}{2}\Sigma_i e_i(F_i \boldsymbol{v}_i + \boldsymbol{v}_i F_i)$ defined by equation (10),

$$\begin{aligned}\{\varrho\boldsymbol{v}(t)\}_{kl} &= \tfrac{1}{2}\Sigma_i[(\psi_k^* e_i \boldsymbol{v}_i \psi_l + \psi_l e_i \boldsymbol{v}_i^* \psi_k^*)\,\Pi_{j(\neq i)}\, d\boldsymbol{x}_j]_{\boldsymbol{x}_i=\boldsymbol{x}}\\ &= (\varrho\boldsymbol{v})_{kl} \exp(2\pi i\,\nu_{kl} t) + \{\varrho\boldsymbol{v}(t)\}_{kl}^{(1)},\end{aligned} \tag{215}$$

and *we consider the real current field*

$$[\varrho\boldsymbol{v}]_{kl} = 2\,\mathrm{Re}\{\varrho\boldsymbol{v}(t)\}_{kl} = \{\varrho\boldsymbol{v}(t)\}_{kl} + \{\varrho\boldsymbol{v}(t)\}_{lk}, \quad k \neq l; \tag{216}$$

$$[\varrho\boldsymbol{v}]_{kl} = \{\varrho\boldsymbol{v}(t)\}_{kk}, \quad k = l, \tag{217}$$

as a source of classical radiation. We now state that the separate components of this radiation can, indeed, be observed; they correspond to a transition from the state k to the state l, or vice versa.

Consideration of § 84 [in particular of eq. (53)] shows that this

* For details, see W. Heitler, *Quantum theory of radiation*, Oxford University Press, 1954.

** O. Klein, *Z. Phys.*, **40**, 407, 1927; W. Gordon, *Z. Phys.*, **41**, 117, 1927.

*** For the sake of simplicity we do not discuss in this section the details connected with the conservation of momentum. These details played, of course, an important role in the original considerations by Klein and Nishina of the Compton effect (see Klein, loc. cit., Gordon, loc. cit.).

statement is identical with the prescription for the calculation of spontaneous emission in the case where there is no external radiation. In that case the transitions are always from a higher to a lower energy state of the atom. The incoming radiation produces now an additional *source of scattered radiation,*

$$[\varrho \mathbf{v}]^{\text{scatt}}_{kl} = \{\varrho \mathbf{v}(t)\}^{(1)}_{kl} + \{\varrho \mathbf{v}(t)\}^{(1)}_{lk}, \text{ if } k \neq l; \; = \{\varrho \mathbf{v}(t)\}^{(1)}_{kk}, \text{ if } k = l.$$

Whether a component corresponds to a transition $k \to l$ or to a transition $l \to k$ must be determined by considering the energy balance.

We shall indicate how we can obtain in this way the scattering equations (200) and (201). Let the incoming radiation be a plane monochromatic wave of wave vector $\boldsymbol{\sigma}_\lambda$, frequency $\nu_\lambda = c\sigma_\lambda$, and complex amplitude $q\,\mathfrak{a}_\lambda$ [see eqq. (2) and (14)]. We need consider only the first order terms in the perturbation energy (3),

$$H^{(1)}_{\text{op}} = -(2c)^{-1}\{q(\mathfrak{a}_\lambda \cdot \Sigma_i e_i \mathbf{v}_i \exp[2\pi i(\boldsymbol{\sigma}_\lambda \cdot \mathbf{x}_i)]) \exp(-2\pi i \nu_\lambda t)$$
$$+ q^*(\mathfrak{a}^*_\lambda \cdot \Sigma_i e_i \mathbf{v}_i \exp[-2\pi i(\boldsymbol{\sigma}_\lambda \cdot \mathbf{x}_i)]) \exp(2\pi i \nu_\lambda t)\}. \qquad (218)$$

We shall write ψ in equation (214) in the form

$$\psi = \Sigma_m \gamma_m \varphi_m \exp(-iE_m t/\hbar), \; i\hbar\dot{\gamma}_m = \Sigma_l H^{(1)}_{ml} \exp(2\pi i \nu_{ml} t)\gamma_l,$$

and we shall assume that in zeroth approximation $\gamma_l = \delta_{lk}$. We then get

$$\psi_k = \varphi_k \exp(-iE_k t/\hbar) + (2c)^{-1}\Sigma_m\{qM^\lambda_{mk}\varphi_m[\exp(-2\pi i\nu_\lambda t)/h(\nu_{mk}-\nu_\lambda)]$$
$$+ q^* N^\lambda_{mk}\varphi_m[\exp(2\pi i \nu_\lambda t)/h(\nu_{mk}+\nu_\lambda)]\}\exp(-iE_k t/\hbar), \qquad (219)$$

where M and N are again defined by equations (191). If we substitute this expression into equation (215) we must take into account that not only the ψ-operators, but also the $\mathbf{v}_i$-operators contain a perturbation term deriving from the vector potential of the incoming radiation,

$$\mathbf{v}_i = \mathbf{v}^0_i - (e_i/2m_i c)\{q\,\mathfrak{a}_\lambda \exp[2\pi i\{(\boldsymbol{\sigma}_\lambda \cdot \mathbf{x}_i) - \nu_\lambda t\}]$$
$$+ q^*\,\mathfrak{a}^*_\lambda \exp[-2\pi i\{(\boldsymbol{\sigma}_\lambda \cdot \mathbf{x}_i) - \nu_\lambda t\}]\}.$$

We get for the matrix $\{\varrho \mathbf{v}(t)\}^{(1)}_{kl}$ which determines the source of scattered radiation

$$\{\varrho \mathbf{v}(t)\}^{(1)}_{kl} = (q/2c)\exp[2\pi i(\nu_{kl}-\nu_\lambda)t][\Sigma_m\{[(\varrho\mathbf{v})_{km}M^\lambda_{ml}/h(\nu_{ml}-\nu_\lambda)]$$
$$+ [M^\lambda_{km}(\varrho\mathbf{v})_{ml}/h(\nu_{mk}+\nu_\lambda)]\} - \mathfrak{a}_\lambda \zeta_{kl}\exp[2\pi i(\boldsymbol{\sigma}_\lambda \cdot \mathbf{x})]]$$
$$+ (q^*/2c)\exp[2\pi i(\nu_{kl}+\nu_\lambda)t][\Sigma_m\{[(\varrho\mathbf{v})_{km}N^\lambda_{ml}/h(\nu_{ml}+\nu_\lambda)]$$
$$+ N^\lambda_{km}(\varrho\mathbf{v})_{ml}/h(\nu_{mk}-\nu_\lambda)]\} - \mathfrak{a}^*_\lambda \zeta_{kl}\exp[-2\pi i(\boldsymbol{\sigma}_\lambda \cdot \mathbf{x})]], \qquad (220)$$

where the real scalar field

$$\zeta_{kl}(\boldsymbol{x}) = (d\boldsymbol{x})^{-1}\int \varphi_k^* \Sigma_i F_i(e_i^2/m_i)\,\varphi_l, \tag{221}$$

is $(d\boldsymbol{x})^{-1}$ times the matrix element of the sum of all e^2/m values of those particles which are within $d\boldsymbol{x}$.

These equations show the occurrence of scattered radiation with frequencies $|\nu_{kl}-\nu_\lambda|$ and $|\nu_{kl}+\nu_\lambda|$ exactly as the calculations involving quantised radiation required. Let us first consider the case where the states k and l are identical. Now $\nu_{kl}=0$, and we have Rayleigh scattering during which the atom remains in its initial state (case I of § 92). If, however, $\nu_{kl}>0$, the terms involving q^* correspond to antistokes' Raman radiation (case IIIA, transition $k\to l$), the terms involving q correspond either ($\nu_\lambda>\nu_{kl}$) to a normal Raman effect (case II, transition $l\to k$), or ($\nu_\lambda<\nu_{kl}$) to one of these scattering processes where two light quanta are produced (case IIIB, transition $k\to l$).

If we evaluate, using equations (40), the classical intensity and polarisation of this scattered radiation, the expressions in equation (220) must be multiplied by a retardation factor of the type $\mathfrak{a}_\lambda^* \exp[-2\pi i(\boldsymbol{\sigma}_\lambda\cdot\boldsymbol{x})]$ and must be integrated over the whole of space. We obtain expressions completely identical with equations (200) and (201). In particular the terms involving R and S in those equations which derived from the second order perturbation correspond to the terms involving ζ_{kl} in equation (220), which derived from the change in the meaning of the velocity operators.

We can show the correspondence of the calculations of this section with the purely classical theory of the scattering of light by bound particles in a manner similar to the discussion of the semi-classical theory of spontaneous radiation (see § 85). However, we shall not do this here *.

§ 95. Coherent scattering; dispersion

If the initial and final states are the same in equations (200) and (201), the scattered light has the same frequency as the incoming

* See H. A. Kramers and W. Heisenberg, *Z. Phys.*, **31**, 681, 1925. In this paper they derive a formula equivalent to equation (220) in the limiting case where the atomic system is small compared to the wavelength of the light [all factors $\exp\{\pm 2\pi i(\boldsymbol{\sigma}\cdot\boldsymbol{x})\}$ are put equal to unity]. The paper was based on the calculation of scattering in the classical theory, and the exact expression for the "scattering" moment of the atom was guessed by using a correspondence argument to interpret the derivatives occurring in the classical theory as differences.

radiation and we are dealing with Rayleigh radiation. Another name for this radiation is *coherently scattered radiation*, as we are here dealing with the quantum theoretical analogue of that kind of classical scattered radiation which played such an important role in Lorentz' theory of electrons for the *atomistic interpretation of optical phenomena*. As is well-known, classical theory shows that the radiation scattered by an elastically bound particle will everywhere show definite phase relations with the incoming radiation. The resultant interference causes, for instance, the gradual loss of intensity and phase shift which is suffered by radiation in its direction of propagation when it passes through a medium containing many of such oscillators. The same scattering by separate oscillators will produce reflexion and diffraction at a boundary.

We should expect that the quantum theory of these macroscopic phenomena should give the same results as the classical theory, at least as far as the essential features are concerned. We must note, however, immediately that the method of the previous sections for evaluating scattering can hardly be very suitable for investigating these phenomena. The classical treatment is particularly simple and elegant in the case of scattering of a monochromatic component of radiation by a stationary radiation field, that is, a radiation field which is strictly periodic in time. Such a stationary state never entered our quantum theoretical treatment. We assumed that originally there was no scattered radiation; after a while there is an (increasing) probability of finding a scattered quantum, and this probability was calculated. The occurrence of several scattered quanta after a longer period was not explicitly considered. If we wish to obtain in quantum theory a stationary description of scattering, dispersion, . . ., the decomposition of the radiation in terms of plane waves—as we have used up to now—is not very suitable *. We must rather consider a decomposition of the radiation field in such a way that the different components from the start are adapted, by using the classical results, to the scattering processes to be investigated. We must then consider the coefficients of these components as quantum mechanical operators, that is, as q-numbers.

It would lead us too far to discuss this theory in great detail.

* There are in the literature instructive examples of how one can obtain the essential features of the classical coherence—and propagation—phenomena in a non-stationary, strictly quantum theoretical calculation. See, for instance, G. Wentzel, *Handb. Phys.*, **24₁**, 758, 1933; W. Pauli, *Handb. Phys.*, **24₁**, 210, 1933.

However, we shall discuss the case of the *Laue-Bragg reflexion* of X-rays, and see how the equations developed in the previous sections show the correctness of the classical considerations of coherence in particular cases. We consider the application of equation (220) to the calculation of Rayleigh scattering of X-rays by a small crystal. We can put $k = l$ and we assume for the sake of simplicity that the influence of the sums Σ_m on the scattering is negligible compared to the influence of the terms involving ζ (we can nearly always assume the scattering to be due to free electrons),

$$[\varrho \mathbf{v}]_{kk} = \{\varrho \mathbf{v}(t)\}_{kk} = -\mathrm{Re}\,(q/c)\exp(-2\pi i\nu_\lambda t)\,\mathfrak{a}_\lambda \zeta_{kk} \exp[2\pi i(\boldsymbol{\sigma}_\lambda \cdot \boldsymbol{x})]. \tag{222}$$

According to equation (221) ζ_{kk} is a spatial scalar field which is practically equal to e^2/m times the average electron density (the nuclei produce practically no scattering). Such a source of secondary radiation occurs, however, also in the classical theory of X-ray reflexion. In particular the factor $\exp[2\pi i(\boldsymbol{\sigma}_\lambda \cdot \boldsymbol{x})]$ sees to it that each part of the crystal contributes with the correct phase.

To describe *dispersion phenomena in diffracting media* one uses generally the following semi-classical method. According to equations (217), (220), and (191) the source of Rayleigh scattering by an atomic system in a stationary state k is given by the equation

$$\left.\begin{aligned} [\varrho \mathbf{v}]_{kk} &= \mathrm{Re}\,(q/c)\exp(-2\pi i\nu_\lambda t)\,\boldsymbol{W}, \\ \boldsymbol{W} &= \Sigma_m\{[(\varrho \mathbf{v})_{km}(\mathfrak{a}_\lambda \cdot \textstyle\int \exp[2\pi i(\boldsymbol{\sigma}_\lambda \cdot \boldsymbol{x})](\varrho \mathbf{v})_{mk}\,d\boldsymbol{x})/h(\nu_{mk}-\nu_\lambda)] \\ &\quad + [(\varrho \mathbf{v})_{mk}(\mathfrak{a}_\lambda \cdot \textstyle\int \exp[2\pi i(\boldsymbol{\sigma}_\lambda \cdot \boldsymbol{x})](\varrho \mathbf{v})_{km}\,d\boldsymbol{x})/h(\nu_{mk}+\nu_\lambda)] \\ &\quad - \mathfrak{a}_\lambda \zeta_{kk}\exp[2\pi i(\boldsymbol{\sigma}_\lambda \cdot \boldsymbol{x})]\}, \end{aligned}\right\} \tag{223}$$

where the vector potential and the electrical field of the incoming radiation are described by the equations

$$\mathfrak{A} = \mathrm{Re}\,q\,\mathfrak{a}_\lambda \exp\{2\pi i[(\boldsymbol{\sigma}_\lambda \cdot \boldsymbol{x}) - \nu_\lambda t]\},$$
$$\mathfrak{E} = \mathrm{Re}\,(2\pi i\nu_\lambda q/c)\mathfrak{a}_\lambda \exp\{2\pi i[(\boldsymbol{\sigma}_\lambda \cdot \boldsymbol{x}) - \nu_\lambda t]\}.$$

The scattering atomic system may represent the whole medium. We now use the line of thought of the *semi-phenomenological method of the classical theory of electrons*. If a plane monochromatic wave transverses the medium, we can obtain the Maxwell equations describing this phenomenon macroscopically by *averaging* the equations of the theory of electrons,

$$\begin{gathered}\operatorname{curl} \mathfrak{H} = c^{-1}(\partial \mathfrak{E}/\partial t) + 4\pi \varrho \mathbf{v}/c, \quad \operatorname{curl} \mathfrak{E} = -c^{-1}\partial \mathfrak{H}/\partial t, \\ \overline{\mathfrak{E}} = \operatorname{Re} \mathfrak{E}_0 \exp\{2\pi i[(\boldsymbol{\sigma}\cdot\boldsymbol{x}) - \nu t]\},\ \overline{\mathfrak{H}} = \operatorname{Re} \mathfrak{H}_0 \exp\{2\pi i[(\boldsymbol{\sigma}\cdot\boldsymbol{x}) - \nu t]\}.\end{gathered} \tag{225}$$

First of all we average $\varrho \mathbf{v}$ over a plane of constant light phase * and call $\overline{\varrho \mathbf{v}}$ that part which is periodic in time with the same frequency ν as the radiation and which is proportional to the amplitude of $\overline{\mathfrak{E}}$

$$\overline{\varrho \mathbf{v}} = \frac{\partial}{\partial t} \operatorname{Re} (p_{\text{lin}} \mathfrak{E}_0 e^{2\pi i[(\boldsymbol{\sigma}.\boldsymbol{x}) - \nu t]}) = -2\pi \nu \operatorname{Re} i (p_{\text{lin}} \mathfrak{E}_0 e^{2\pi i[(\boldsymbol{\sigma}.\boldsymbol{x}) - \nu t]}). \tag{226}$$

The expression $p_{\text{lin}} \mathfrak{E}_0$ defines a linear vector function of $\mathfrak{E}_0$, $(p_{\text{lin}} \mathfrak{E}_0)_k = \Sigma_l p_{kl} \mathfrak{E}_{0l}$, where k and l denote x, y, and z. The tensor components p_{kl} still depend on $(\boldsymbol{\sigma} \cdot \boldsymbol{x})$ that is, on the position of the planes of constant phase. If they are averaged over all the positions of these planes in the (assumedly homogeneous) medium we call them *polarisation coefficients* and denote them by $\bar{p}_{kl}$; $\bar{p}_{\text{lin}} \mathfrak{E}_0$ is the complex amplitude of the polarisation. From the equations

$$[\boldsymbol{\sigma} \wedge \mathfrak{H}_0] = -(\nu/c)(1 + 4\pi \bar{p}_{\text{lin}}) \mathfrak{E}_0, \quad [\boldsymbol{\sigma} \wedge \mathfrak{E}_0] = (\nu/c) \mathfrak{H}_0, \tag{227}$$

we can derive the details of the propagation of light in the medium**.

The $\bar{p}_{kl}$ depend not only on the frequency, but also on the wave vector and polarisation vector of the $\mathfrak{E}_0$. $\mathfrak{H}_0$-wave. If $p_{\text{lin}} \mathfrak{E}_0$ reduces to the product of $\mathfrak{E}_0$ with a constant p, the *constant of polarisation*, $(1 + 4\pi \bar{p})^{1/2}$ is the (complex) index of refraction of the (isotropic) medium,

$$\bar{p}_{xx} = \bar{p}_{yy} = \bar{p}_{zz} = \bar{p}, \qquad \bar{p}_{xy} = \bar{p}_{yz} = \bar{p}_{zx} = 0, \qquad n^2 = 1 + 4\pi \bar{p}.$$

For media with an index of refraction practically equal to unity, we may assume when evaluating $\bar{p}_{\text{lin}}$ that the external radiation field operating at a point in the medium has just the amplitude $\mathfrak{E}_0$. In "dense" media this will in general not be the case (Lorenz-Lorentz formula, double breaking); for the sake of simplicity we shall neglect these complications in applying equation (223).

* This prescription must particularly be followed for problems where the linear dimensions of the molecules in the disperging medium are comparable to the wave length (natural rotation of the plane of polarisation), and one should not start to average over small elements of space. In the case of crystals "molecules" must be replaced by "regions within which the electrons show an appreciable dynamic interaction".

** We see that only the electrical polarisation of the medium enters into the discussion; a magnetisation of frequency ν does not enter. The fact that nevertheless one obtains correct solutions for certain dispersion problems (see L. Rosenfeld, *Z. Phys.*, **52**, 161, 1928) is fortuitous.

Comparing equations (223) and (224) with equation (226) we see that we have now

$$\mathfrak{E}_0 = (2\pi i \nu_\lambda q/c)\,\mathfrak{a}_\lambda, \quad \bar{p}_{\text{lin}}\,\mathfrak{a}_\lambda = (2\pi\nu_\lambda)^{-2}\,\overline{\boldsymbol{W}\exp\left[-2\pi i(\boldsymbol{\sigma}_\lambda \cdot \boldsymbol{x})\right]}. \quad (228)$$

The factor $\exp\left[-2\pi i(\boldsymbol{\sigma}_\lambda \cdot \boldsymbol{x})\right]$ takes into account the fact that in equation (225) we should first of all average over a plane of constant phase. The bar in equation (228) can thus now be interpreted to indicate simply averaging over the homogeneous medium.

The simplest application of equation (228) is to the case of scattering by free electrons so that only the term involving ζ_{kk} in equation (223) must be considered. We have $\bar{p}_{\text{lin}}\,\mathfrak{a}_\lambda = -(2\pi\nu_\lambda)^{-2}\,\mathfrak{a}_\lambda\bar{\zeta}_{kk}$ and we see that $\bar{p}_{\text{lin}}$ reduces to a constant of polarisation $\bar{p}$. Since according to equation (221) $\bar{\zeta}_{kk}$ is equal to e^2/m times the number, N, of free electrons per unit volume, we have for the index of refraction $n = (1+4\pi\bar{p})^{1/2} = [1-(Ne^2/\pi m\nu^2)]^{1/2}$ in accordance with the classical dispersion formula for free electrons applicable to X-rays.

We do not wish to discuss the complications which occur, if equations (228) are applied to a crystal (even apart from the complications arising from the fact that we are dealing with a dense medium). If we are dealing with identical separate molecules or atoms which are randomly distributed (dispersion in gases or vapours), we must calculate first of all the average value (228) for a given orientation of the molecule and then average *either* over all orientations [or rather over all different stationary states of the same energy but corresponding to different orientations (quantisation of the directions in space)] *or* over all possible directions of the incoming light. After that we must multiply by the number of molecules per unit volume.

In most applications with visible light the dimensions of the molecules can be neglected compared to the wave length and we find for the polarisation coefficient the simple expression * (see § 82),

$$\bar{p}_{\text{lin}}\,\mathfrak{a}_\lambda = \frac{N}{(2\pi\nu_\lambda)^2}\left[\sum_m\left\{\frac{\dot{\boldsymbol{P}}_{km}(\mathfrak{a}_\lambda\cdot\dot{\boldsymbol{P}}_{mk})}{h(\nu_{mk}-\nu_\lambda)} + \frac{\dot{\boldsymbol{P}}_{mk}(\mathfrak{a}_\lambda\cdot\dot{\boldsymbol{P}}_{km})}{h(\nu_{mk}+\nu_\lambda)}\right\} - \mathfrak{a}_\lambda\sum_i\frac{e_i^2}{m_i}\right], \quad (229)$$

where N is the number of particles per unit volume. We get thus for the constant of polarisation

* Equation (229) can also be applied to magnetic rotation in crystals.

$$\bar{p} = N < \frac{1}{(2\pi\nu_\lambda)^2}\left[\sum_m \left\{\frac{|(\mathfrak{a}_\lambda \cdot \dot{\boldsymbol{P}}_{mk})|^2}{h(\nu_{mk}-\nu_\lambda)} + \frac{|(\mathfrak{a}_\lambda \cdot \dot{\boldsymbol{P}}_{km})|^2}{h(\nu_{mk}+\nu_\lambda)}\right\} - \sum_i \frac{e_i^2}{m_i}\right] >_{\text{Av}}$$
$$= N < \Sigma_m \{[|(\mathfrak{a}_\lambda \cdot \boldsymbol{P}_{mk})|^2/h(\nu_{mk}-\nu_\lambda)] + [|(\mathfrak{a}_\lambda \cdot \boldsymbol{P}_{km})|^2/h(\nu_{mk}+\nu_\lambda)]\} >_{\text{Av}}.$$

The last equation is obtained by a calculation similar to the one used in deriving equation (205). If we take the average over all possible directions of polarisation of the incoming radiation, we get from the fact that $\overline{|(\mathfrak{a}_\lambda \cdot \boldsymbol{P}_{mk})|^2} = \frac{1}{3}|\boldsymbol{P}_{km}|^2$ the equation

$$\bar{p} = \frac{Ne^2}{4\pi^2 m_{\text{el}}} \sum_m \frac{f_{km}}{\nu_{mk}^2 - \nu_\lambda^2}, \quad f_{km} = \frac{8\pi^2 m_{\text{el}}}{3e^2 h} \nu_{mk} |\boldsymbol{P}_{km}|^2 = \frac{m_{\text{el}}}{\pi e^2} B_{k\to m} h\nu_{mk}, \tag{230}$$

where $-e$ and m_{el} are the charge and mass of the electron.

The correctness of the semi-classical method of determining radiation phenomena in media has not yet been proved generally. Wentzel * has proved it for a model where the medium consists of harmonic oscillators. The difficulties encountered for a more general model are connected with the fact that the semi-classical method must certainly be modified as soon as the wave length of the incoming radiation falls in a region of absorption of the medium. There occurs a kind of resonance and the medium is optically excited under the influence of the radiation. At the end of this section we shall briefly discuss *resonance scattering*, but we shall not discuss the particular problem of its role for macroscopic absorption and dispersion phenomena in media.

If k denotes the ground state of the molecule (or atom) all ν_{mk} are positive (frequencies of absorption lines). The number f_{km} are called the (*oscillator*) *strengths* of the transitions and they depend on the Einstein B's in the way indicated in equation (230). They were introduced to illustrate the correspondence with the classical theory of electrons. According to this theory we find for a medium where each molecule contains one electrically and isotropically bound electron with eigenfrequency ν,

$$\bar{p} = (Ne^2/4\pi^2 m_{\text{el}})(\nu^2 - \nu_\lambda^2)^{-1}. \tag{231}$$

The number f_{km} can thus be considered to be the "effective number" per molecule of elastically and isotropically bound electrons with eigenfrequency ν_{km}. If k in equation (230) is not the ground state, there may be one or several negative absorption frequencies ν_{mk} corresponding to spontaneous transitions $k \to m$. There can now also occur in equation (230) dispersion terms with negative f-values which have no classical counterpart. If equation (230) is

* G. Wentzel, *Helv. Phys. Acta*, **6**, 89, 1933.

applied to a *one-dimensional* harmonic oscillator of charge $-e$, mass m_{el}, and eigenfrequency ν, and if we denote the corresponding quantum number by k, we have $f_{k,k+1} = \frac{1}{3}(k+1)$, $f_{k,k-1} = -\frac{1}{3}k$, all other f equal to zero, and $\nu_{k+1,k} = \nu_{k,k-1} = \nu$. For each state k we get $\bar{p} = (Ne^2/4\pi^2 m_{el})[\frac{1}{3}/(\nu^2-\nu_\lambda^2)]$ as in the classical theory.

The formulae of the scattering- and of the dispersion theory lose their validity when a resonance denominator [$\nu_{kk_0}-\nu_\lambda$ or $\nu_{kk_0}+\nu_{\lambda'}$ in eq. (200), $\nu_{kk_0}+\nu_\lambda$ or $\nu_{kk_0}+\nu_{\lambda'}$ in eq. (201)] is very small or even equal to zero. This means *either* that the frequency ν_λ of the *incoming* radiation is equal to an absorption- [eq. (200)] or an emission- [eq. (201)] frequency, *or* that the frequency $\nu_{\lambda'}$ of the *scattered* radiation is equal to an emission frequency of the scattering system in the state k_0. It can easily be understood from physical arguments that in those cases something peculiar will occur. In the first case the incoming radiation produces real absorption- or emission jumps, and in the second case the system would anyhow emit radiation of frequency $\nu_{\lambda'}$. In general it will be true that equations (200) and (201) will no longer be valid as soon as a resonance denominator becomes of the same order of magnitude as the natural line width. To discuss what will happen exactly in such cases, we must return to equation (171) and the situation becomes very complicated.

In the case of coherent radiation we meet with the problem of resonance scattering or *resonance fluorescence*. If an atomic system is bombarded by light containing the frequency of an absorption line, light of this frequency will be scattered particularly strongly, even if not infinitely strongly as would follow formally from equations (200) or (229). Both as regards intensity and polarisation, and as regards the influences of external fields resonance scattering shows various peculiarities which have been the subject of many experimental investigations. The discussion of resonance fluorescence through elastically bound particles is rather simple in the classical theory of electrons. The term in the equations of motion of the particles describing the radiation damping plays a deciding role. The approximate character of the quantum theoretical treatment of radiation entails that the effects of radiation damping are expressed in a manner completely different from that of the classical theory. It has turned out that many experimental details of resonance fluorescence can only be explained satisfactorily on the basis of a quantum theoretical treatment. We refer to the literature for details *.

Another important point which we shall not discuss is the role of the *conservation of momentum in coherent scattering*, especially with reference to macroscopic radiation phenomena. If this is taken into account rigorously by means of quantum theory, the apparent inconsistencies occurring when we apply too naively the concept of radiation particles will disappear.

* See G. Wentzel, *Handb. Phys.*, **24$_1$**, 755, 1933; W. Heitler, *Quantum theory of radiation,* Oxford University Press, 1954.

INDEX